Advances in Geographic Information Science

The series aims to: present current and emerging innovations in GIScience; describe new and robust GIScience methods for use in transdisciplinary problem solving and decision making contexts, illustrate GIScience case studies for use in teaching and training situations, analyze GIScience tools at the forefront of scientific research, and examine the future of GIScience in an expanding knowledge-based economy. The scope of the series is broad and will encompass work from all subject disciplines that develop and use an explicitly spatial perspective for analysis and problem-solving.

Advances in Geographic Information Science

More information about this series at http://www.springer.com/series/7712

Pravat Kumar Shit • Hamid Reza Pourghasemi
Gouri Sankar Bhunia • Pulakesh Das
Adimalla Narsimha
Editors

Geospatial Technology for Environmental Hazards

Modeling and Management in Asian Countries

Springer

Editors
Pravat Kumar Shit
PG Department of Geography
Raja N. L. Khan Women's College (Autonomous)
Midnapore, West Bengal, India

Hamid Reza Pourghasemi
Department of Natural Resources and Environmental Engineering
Shiraz Univ, College of Agriculture
Shiraz, Iran

Gouri Sankar Bhunia
RANDSTAD India Pvt Ltd
New Delhi, Delhi, India

Pulakesh Das
World Resources Institute India
New Delhi, Delhi, India

Adimalla Narsimha
School of Environmental Science and Engineering
Chang'an University
Xi'an, Shaanxi, China

ISSN 1867-2434 ISSN 1867-2442 (electronic)
Advances in Geographic Information Science
ISBN 978-3-030-75196-8 ISBN 978-3-030-75197-5 (eBook)
https://doi.org/10.1007/978-3-030-75197-5

This Springer imprint is published by the registered company Springer Nature Switzerland AG
The registered company address is: Gewerbestrasse 11, 6330 Cham, Switzerland

Dedicated to beloved teachers and parents

Preface

Environmental hazards such as landslide, floods and flash floods, forest or wildland fire, and tropical cyclones are the most devastating natural disasters causing massive damages to natural and man-made features. Natural hazards are a major threat to human life (injury or death of human and animal life), properties (agricultural area, yield production, building, and homes), and infrastructures (bridges, roads, railways, urban infrastructures). The damage that can occur due to such disasters leads to huge economic loss and brings pathogens into urban environments that cause microbial development and diseases. The natural and social hazards are discontinuing the development of human society and sustainability.

The book demonstrates the geospatial technology approach to data mining techniques, data analysis, modeling, risk assessment, visualization, and management strategies in different aspects of natural and social hazards. This book has considered 25 chapters associated with risk assessment, mapping, and management strategies of environmental hazards. It covers major topics such as: Landslide Susceptibility, Arsenic Contaminated Groundwater, Earthquake Risk Management, Open Cast Mining, Soil loss, Flood Susceptibility, Forest Fire Risk, Malaria prevalence, Flood inundation, Socio-Economic Vulnerability, River Bank Erosion, and Socio-Economic Vulnerability. The content of this book will be of interest to researchers, professionals, and policymakers, whose work involves environmental hazards and related solutions.

We are thankful to all the authors who have meticulously completed their documents on a short announcement and paid in building this a very edifying and beneficial publication. We do believe that this will be an opportune book for the Geographers, Ecologists, Environmental Scientist, Hydrologist, Geospatial Scientist, Remote Sensing and GIS experts, Agriculture Scientist, and other fields of environmental hazards and management including the research scholars, environmentalists, and policymakers.

Midnapore, West Bengal, India — Pravat Kumar Shit
Shiraz, Iran — Hamid Reza Pourghasemi
New Delhi, Delhi, India — Gouri Sankar Bhunia
New Delhi, Delhi, India — Pulakesh Das
Xi'an, Shaanxi, China — Adimalla Narsimha

Acknowledgements

We are obliged to a chunk of experts for their kind support and valuable time to evaluate the chapters succumbed for insertion in this book. We are very much thankful to our respected teacher's Prof. Malay Mukhopadhya, Prof. Sunando Bandyopadhyay, Prof. L.N. Satpati, Prof. Ashis Kumar Paul, Prof. Ramkrishna Maiti, Prof. Soumendu Chatterjee, Prof. Nilanjana Das Chatterjee, Prof. Dilip Kr. Pal, Prof. Sanat Kumar Guchait, Prof. N.C. Jana, Dr. Jatishankar Bandopadhay, and Dr. Ratan Kumar Samanta for sharing their experiences, useful suggestions, continuous encouragement, and immense support throughout the work.

We would like to thank the anonymous reviewers, acted as independent referees. Their inputs were consistently constructive and have substantially improved the quality of the final product. Dr. Pravat Kumar Shit would like to thank Dr. Jayasree Laha, Principal, Raja N.L Khan Women's College (Autonomous), Midnapore, for her administrative support to carry on this project. We also acknowledge the Department of Geography, Raja N.L. Khan Women's College (Autonomous), for providing the logistic support and infrastructure facilities.

We would also like to thank Zachary Romano and Aaron Schiller, whose love, encouragement, and support kept us motivated up to the final shape of the book. Finally, the book has been several years in the making, and we, therefore, want to thank family and friends for their continuous support.

This work would not have been possible without constant inspiration from my students, lessons from my teachers, enthusiasm from my colleagues and collaborators, and support from my family.

Disclaimer

The authors of individual chapters are solely responsible for ideas, views, data, figures, and geographical boundaries presented in the respective chapters in this book, and these have not been endorsed, in any form, by the publisher, the editor, and the authors of forewords, preambles, or other chapters.

Contents

About the Editors

Pravat Kumar Shit is an Assistant Professor at the PG Department of Geography, Raja N. L. Khan Women's College (Autonomous), West Bengal, India. He received his M.Sc. and Ph.D. degrees in Geography from Vidyasagar University and PG Diploma in Remote Sensing and GIS from Sambalpur University. His research interests include applied geomorphology, soil erosion, groundwater, forest resources, wetland ecosystem, environmental contaminants and pollution, and natural resources mapping and modeling. He has published 9 books (7 books in Springer) and more than 60 papers in peer-reviewed journals. He is currently the editor of the GIScience and Geo-environmental Modelling (GGM) Book Series, Springer-Nature.

Hamid Reza Pourghasemi is an Associate Professor of Watershed Management Engineering in the College of Agriculture, Shiraz University, Iran. He has a B.Sc. in Watershed Management Engineering from the University of Gorgan (2004), Iran; an M.Sc. in Watershed Management Engineering from Tarbiat Modares University (2008), Iran; and a Ph.D. in Watershed Management Engineering from the same University (Feb 2014). His main research interests are GIS-based spatial modeling using machine learning/data mining techniques in different fields such as landslide, flood, gully erosion, forest fire, land subsidence, species distribution modeling, and groundwater/hydrology. Also, Hamid Reza works on multi-criteria decision-making methods in natural resources and environment. He has published more than 170 peer-reviewed papers in high-quality journals and 4 books in Springer and Elsevier as Editor. Also, he is an active reviewer in more of 75 international journals.

Gouri Sankar Bhunia received his Ph.D. from the University of Calcutta, India, in 2015. His Ph.D. dissertation work focused on disease transmission modeling using geospatial technology. His research interests include health geography, environmental modeling, risk assessment, data mining, urban planning, and information retrieval using geospatial technology. He is an Associate Editor and on the editorial boards of three international journal in Health GIS and Geosciences. Currently, he is involved various Smart City Planning programme in India. He is also working as a visiting faculty in a private university of West Bengal. He has worked as a 'Resource Scientist' in Bihar Remote Sensing Application Centre, Patna (Bihar, India). He is the recipient of the Senior Research Fellow (SRF) from Rajendra Memorial Research Institute of Medical Sciences (ICMR, India) and has contributed to multiple research programmes kala-azar disease transmission modeling, development of customized GIS software for kala-azar 'risk' and 'non-risk' area, and entomological study. He has published more than 60 papers in reputed peer-reviewed national and international journals and 3 books in Springer.

Pulakesh Das is currently working in World Resources Institute India (WRII), New Delhi, India. Previously, he was teaching as an Assistant Professor in the Department of Remote Sensing and GIS, Vidyasagar University, Midnapore, West Bengal, India. He has received his Ph.D. degree from the Indian Institution of Technology (IIT) Kharagpur, India, in July 2019. He completed his M.Sc. (2012) in Remote Sensing and GIS and B.Sc. (2010) in Physics from the Vidyasagar University, Midnapore, West Bengal, India. His primary research area includes Land Use Forest Cover (LUFC) modeling, Hydrological Modeling, Forest Cover Dynamics and Climate Change, Digital Image Processing, Microwave Remote Sensing for Soil Moisture and Forest Biomass

Estimation, Plant Biophysical Characterization, etc. He has published more than 15 research articles in reputed peer-reviewed journals and 6 book chapters.

Adimalla Narsimha received a Ph.D. from the University College of Science, Osmania University, Hyderabad, India, in 2012. His Ph.D. dissertation work focused on geochemistry of groundwater in Basara area, Adilabad district, Andhra Pradesh. His research interests include groundwater contaminations, environmental modeling, risk assessment, natural resources mapping and modeling, data mining and information retrieval using geospatial technology. Now, he is working as a postdoctoral researcher in the School of Environmental Science and Engineering, Chang'an University, China. Dr. Adimalla has published more than 60 articles in various journals in Scopus Indexed Journals.

Chapter 1
Geospatial Technology for Multi-hazard Risk Assessment

Gouri Sankar Bhunia and Pravat Kumar Shit

Abstract Many parts of the globe face numerous natural disasters, including terrible earthquakes, terrible landslides, epidemics, drought and/or flooding. In recent decades, disasters have impacted the world becoming more. A higher occurrence of intense hydro-meteorological activities, most likely due to climate change, and the rise of susceptible populations, may be the key reasons for this progression. Risk eradication approach, with an accent on risk evaluation, risk mapping and threat assessment, which both have a significant spatial aspect, should be achieved further in order to reduce disaster menace. Integration of remote sensing products and Geographic Information Systems (GIS) has converted an automated, well-developed and effective disaster risk management technique today. The present chapter highlighted a critical and detailed overview of recent multi-hazard risk analysis performed using remotely sensed data and geospatial techniques, as it permits participants to be intricate in numerous phases of prototypical development. This chapter also represented the methodology of machine learning and crowd sourcing, particularly for multi-hazard modelling, as a very valuable tool for risk management and disaster vindication.

Keywords Risk mapping · GIS · Machine learning · Crowd sourcing · Disaster management

G. S. Bhunia
Department of Geography, Seacom Skills University, Birbhum, West Bengal, India

P. K. Shit (✉)
PG Department of Geography, Raja N.L. Khan Womens College (Autonomous), Midnapore, West Bengal, India

P. K. Shit et al. (eds.), *Geospatial Technology for Environmental Hazards*, Advances in Geographic Information Science, https://doi.org/10.1007/978-3-030-75197-5_1

1.1 Introduction

'Multi-hazard' is a concept used by the United Nations as a collective initiative to encourage risk prevention and emergency management in the form of bearable growth and Agenda 21 (Pourghasemi et al., 2019), which would lead to the sustainable development of the earth (Keesstra et al., 2018). A multi-hazard strategy describes multiple threats with varying probabilities and intensities (Eshrati et al., 2015). In terms of rescue process operations during a disaster, chaos, misguidance, misallocation of workers and misunderstanding are creating immense costs, from the rescue time to the casualty toll. The lack of viable contingency planning or new evacuation plans restricts emergency management initiatives. Hazards can be described as possible risks to individuals, infrastructure and the environment resulting from 'the intersections of human systems, natural processes and technology systems' (Cutter 2003). They are 'risks that can or may not escalate to a disaster or catastrophe and have been called after a disaster/emergency that could be preceded' (Haddow et al. 2008). Such hazardous causes may emerge due to geological, meteorological, oceanographic, hydrological or biological systems on Earth, or technological intervention. Conversely, as pre-disaster prevention and readiness steps often form a critical part of the current emergency response strategy, so the scope can possibly be extended to include the word 'planning and response'. Furthermore, while natural and human threats cause disasters, they do not need to be covert single instances. Many impoverished people, particularly in the development of nations, frequently experience 'repeated shocks to their families and their livelihood. . .which can entrench any opportunities to stockpile resources and savings', which are making similar and often concurrent (Wisner et al., 2004). These continuing humanitarian crises may be exacerbated by natural and human hazards but also are aggravated and amplified by human behaviours and social forces.

Risk may be described as 'the combination of potentially dangerous (hazard, identified by possibility, severity and spatial size) and vulnerable elements (peoples, facilities, etc.) of a potentially harmful phénomene within the conceptual framework of the study' (Glatron and Beck 2008). Therefore, risk arises only if all hazards and vulnerabilities are active. Vulnerability can be defined to be either the degree of susceptibility to a bio-physical (geographical/physical) hazard or the social capacity to predict and overcome an environmental problem (social vulnerability) (Cutter 2006). Appraisals of risk and susceptibility form an essential part of the continuum of hazard and disaster management (HADM) of the pre-disaster process. A number of variables, including a rapidly rising world population and highly uncertain environment and economic factors, ensure that so many fragile societies are anticipated to be at risk of natural and technical hazards in the future (IFRCRCS, 2003).

Strategies for multi-hazard risk management can be primarily divided into two methods: (a) evaluating specific threats in a certain terrestrial setting autonomously (Carpignano et al., 2009; Schmidt et al., 2011) and (b) appraising probable communications and/or cascade paraphernalia among the different conceivable hazardous measures (Garcia-Aristizabal et al., 2015; Zhang & Zhang, 2017). As a consequence,

Fig. 1.1 Framework of multi-hazard risk analysis and management using geospatial tool

in natural disaster prevention and planning, applied multi-hazard evaluation has been under-emphasized. More significantly, most scientific investigations have focused exclusively on national hazards forecasting and monitoring, disregarding the human effects and exposure aspect (Fig. 1.1). In geographical literature and threats, risk (R) has been understood as the amalgamation of hazard exposure (H) and societal susceptibility (V). This association can be articulated in a 'pseudo-equation' of $R = H \times V$ (Wisner et al., 2004). There is, however, no chance if there is no association of a risky population and vulnerable population in a specific area.

An increasing number of Earth observation capabilities in the low-earth near to polar orbit have been developed in recent decades, improving our capacity to detect solid earth hazards. Almost 40 million Landsat scenes have been accessed via the US Geological Survey portal since late 2008, when Landsat Earth observation images were released publicly to all subscribers free of charge, and the rate of downloads is still growing. However, the majority of satellite observations today have restricted range and compatibility, since they are dominated by the different priorities of national space programmes. Both active radar (synthetic aperture radar-SAR) and passive optical imaging systems (multi-spectral) have undergone this continued expansion of satellite capability. The Copernicus programme of the European Space Agency (ESA) and daily Sentinel-1/2 acquisitions have been ideal for measuring such data collection more regular. The user base (abetted by the growth and facility of open software processing toolboxes) has been suggestively increased by the free and open data policies. A geographical methodology in risk management promotes disaster risk reduction (DRR) by vital data on risk source locations, future impact areas and the regional spread of (vulnerable) populations and hazardous infrastructure (Greiving et al. 2006). Spatial frameworks can also help to identify suitable areas of facilities for disaster prevention and help in relocation, response, distribution of resources and exculpatory policymaking. Consequently, the use of satellite data, GIS and social, demographic and economic data open to the public has the ability to promote modelling activities in terms of improved spatial precision, computational power, scientific rigour of quantitative techniques and profit-making information exchange (Bishop et al., 2012; Hoque et al., 2018). For example, at the International Conference on Satellite EO and Geohazards, EO implementations for threat detection, quantification and tracking for protection, resiliency, emergency management, post-emergency and recovery activities and prevention measures were an integral part of the discussions on community priorities. Moreover, Copernicus EMS provides vast volumes of DRR data and is also a very useful resource for end-users in this area. Searching at international satellite EO and DRR projects, a host of programmes, including the Geohazards Supersites and Natural Laboratories, the GEODARMA project, the GEO EO4SDG and the CEOS Working Group Disasters, is sponsored by both the Group on Earth Observation (GEO) and the Committee on Earth Observing Satellites (CEOS). In the past 60 years, an integrated and diverse area of infectious disease modelling have been developed, and our understanding of human and specific disease transmission processes, involving risk factors, pathogens and spatio-temporal disease distribution patterns has been progressed (Riley, 2007). In order to target limited preventive measures, monitoring and control services, recognizing geographical dynamics of anthropological menace of exposure to vector-borne disease agents is important (e.g. geographic vaccination targeting; administering drugs or information campaigns; using vector proliferation mitigation sentinel sites; and defining locations for pesticides to be used more efficiently). These modelling approaches vary from biostatistical strategies to large biophysical, ordinary differential equation (ODE)-dependent agent models to ecological niche models (Corley et al., 2014).

Moreover, space observations are specifically identified as significant contributions to the policy of disaster risk mitigation. Sendai Framework has goals for interventions, such as 'reinsurance understanding', 'sustainability reduction investing', and 'preparedness enhancement of disasters', for example, both of which can be assisted by strengthened earth observations (Fig. 1.2).

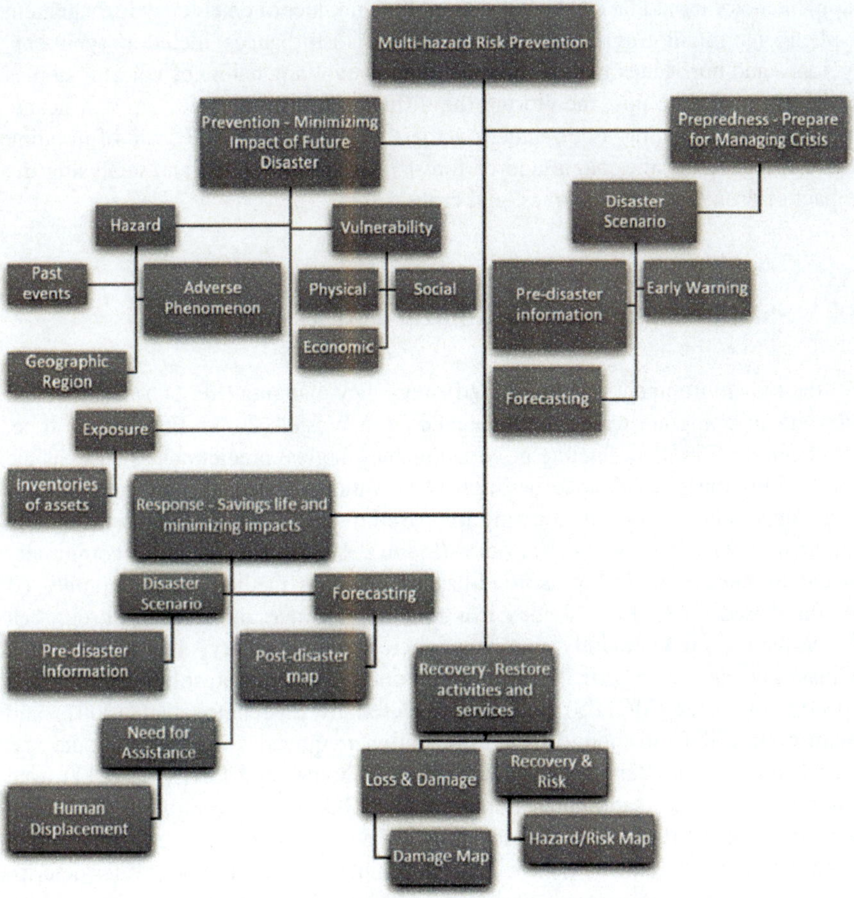

Fig. 1.2 User decision framework for multi-hazard risk assessment using geospatial technology

In certain cases, however, neither space-based nor in situ and airborne measurements explicitly help disaster strategic planning: instead, an indirect layer of research is recognized, which in turn tells consumers about risk management (Salichon et al., 2007). This review of the literature seeks to offer a detailed description of the technological innovations developed to minimise and avoid risk caused by natural hazards. Three academic databases which are widely used in literature reviews were chosen by the study: Scopus, Google Scholar and Web of Science. As search criteria, the key terms flood, drought, landslide, water conservation, catastrophe, sustainability, climate change and adaptation were used in accordance with other comprehensive research papers in the area. The aim of this chapter is to offer a critical and detailed overview of recent multi-hazard risk analysis performed using satellite data and geospatial methods. It provides a description of the inventory/detection of natural hazards, mapping and monitoring, vulnerability and risk mitigation on a variety of scales using geospatial data. Extents from these satellites offers useful supplementary input that can be considered for a number of emergency management applications: monitoring the intensity of tropical disturbances, including typhoons, cyclones and hurricanes worldwide; tracking the evolving nature of volcanic domes in the event of flare-ups; monitoring the diffusion of ash released during volcanic eruptions; investigating, often under overcast weather, the spatial scale of flooding regions; measuring the magnitude of forest fires and oil spills; and analysing the impact of droughts on soils, trees, and crops.

1.2 Satellite Data and Multi-hazard Assessment

For the EO environment, satellite-based emergency mapping (SEM) or fast risk and recovery mapping are widely implemented (Van Westen, 2013). For the first time, data from the TIROS-1 satellite provided meteorological predictions as early as the 1960s. This massive advance in Earth Observation opened up new horizons for catastrophe menace control, allowing improved tracking, awareness and eventually anticipation of meteorological hazards (Manna, 1985). The scientific community would be pleased to follow a flood-hazard mapping method focused purely on satellite-based observation as they it is reliable, accurate and relevant throughout the universe, predominantly in areas where ground surveys are not possible (Giustarini et al., 2015). Multi-temporal Moderate-Resolution Imaging Spectroradiometer (MODIS) imagery was used by Sakamoto et al. (2007) and Islam et al. (2010) to monitor flood inundation frequencies in the Cambodia and the Mekong Delta (Bangladesh), respectively. Thomas and Leveson (2011) used Landsat imagery to map annual floods in Australia, while the mixture of thermal (ASTER) and SAR (ENVISAT) data was efficiently used after the 2011 Tohoku (Japan) Tsunami to elucidate the overall flood intensity and to track water deteriorating in the subsequent weeks. Due to its synoptic view and rate of measurements, specifically in high mountain regions, remote sensing data and image processing methods can be used for in-depth risk mapping and monitoring. For the development of landslide susceptibility maps and landslide hazard index, Golovko et al. (2017) used several satellite data (e.g. LANDSAT, SPOT, ASTER, IRS-1C, LISS-III, and RapidEye) and automatic identification systems. For landslide detection and mapping, Lu et al. (2011) used Quickbird remotely sensed data. To establish multi-temporal landslide susceptibility maps, Guzzetti et al. (2012) used aerial imagery, high-resolution DEM (LiDAR) and satellite images (e.g. Landsat-7, IRS, IKONOS-2, Quickbird-2, WorldView-2, and GeoEye-1/2). Along with current landslide catalogue maps and the SAR and interferometric synthetic aperture radar (InSAR) image play an important role for up-dating the landslide inventory through incorporation of auxiliary data. The creation of digital elevation models (DEMs), for instance, those created from Indian remote sensing satellite (IRS) P5 images and TerraSAR-X/TanDEM-X images by InSAR-X, is among the most useful applications, such as the assessment of erosion, landslide and topographic multi-temporal differences (Du et al., 2017). For landslide change detection analysis, Hölbling et al.

(2015) used remotely sensed SPOT-5 data, while Kang et al. (2017) used ALOS/PALSAR imagery and InSAR techniques for landslide detection. The National Oceanographic and Atmospheric Administration (NOAA) satellite data perceived by the Advanced Very High-Resolution Radiometer (AVHRR) was considered to gather the prevalence and track the periodic outbreak of cholera in order to establish the early warning system in Bangladesh (Lobitz et al., 2000).

Since, topography is among the most crucial components in most hazard assessment, the development of a DEM and geomorphometric evaluation is crucially important. Current topographic maps, topographic levelling, Electronic Distance Measurement (EDM), differential Global Positioning Systems (DGPS) measurements, digital photogrammetry, Interferometric Synthetic Aperture Radar (InSAR) and Light Detection and Ranging (LiDAR) can be used to obtain elevation data. GTOPO30 (Hastings & Dunbar, 1998) and Shuttle Radar Topographic Mission (SRTM) (Farr & Kobrick, 2000) are the key origins of global DEMs used in hazards assessment and risk analysis. Global optically derived optical topography is also applicable at 30 m with Panchromatic Remote-Sensing Instrument for Stereo Mapping (PRISM) Advanced World-3D Advanced (AW3D) and higher-resolution PRISM datasets accessible for the trade. The topography is based on Advanced Spaceborne Thermal Emission and Reflection Radiometer (ASTER).

1.3 Spatial Modelling and Multi-hazard Risk Assessment

Over the last few decades, experimental multi-hazard experiments have been more popular (Pourghasemi & Kerle, 2016), but in the twenty-first century, they are still a problem for scientists and researchers (UN, 2002). To prepare hazard maps using geospatial technology, many approaches and models have been used (Adab et al., 2013, Teodoro and Duarte 2013; Hembram et al. 2011). Multi-faceted computational approaches, like multiple adaptive regression splines, have recently been studied for risk evaluation (Pourghasemi et al., 2018), logistic regression (Arabameri et al., 2018) and generalized additive models (Ravindra et al., 2019). In New Zealand, Schmidt et al. (2011) developed a multi-risk modelling methodology, providing an adaptable version of software that enables academicians to 'plug in' real progressions of interest. Pourghasemi et al. (2019) conducted a multi-hazard risk evaluation focused on artificial intelligence techniques in Fars Province, Iran. There has also been discussion of artificial intelligence and machine learning. In addition, the analytical hierarchy process (Youssef, 2015) unified with multi-criteria decision analysis (MCDA) and spatial decision support system (Ghorbanzadeh et al., 2018) has been used in evaluations of multi-hazard risk analysis. Soft-computing models, such as artificial neural networks (Yilmaz & Ercanoglu, 2019), fuzzy logic (Vakhshoori & Zare, 2016), decision trees (Wang et al., 2016), support vector machine (SVM) (TienBui et al., 2016), random forest (RF) (Youssef et al., 2016), deep learning method (Xiao et al., 2018), adaptive neuro-fuzzy inference system (ANFS) (Chen et al., 2019), kernel logistic regression (TienBui et al., 2016) and ensembles of ANFIS (RazaviTermeh et al., 2018), have been used for multi-hazard analysis and modelling. The ILWIS GIS framework assessment module allows and directs users to carry out spatial multi-criteria assessments for multi-hazard analysis. The input is a series of maps that depict the criteria's spatial representation. They are congregated, uniform and weighted in a 'criteria tree'. The product is one or more 'composite index map(s)' implying the realization of the model implemented. Several research organizations around the world have many comprehensive spatial tools, such as HAZUS, a GIS-based natural disaster analysis method built to determine flood risk; HEC-FDA, a computer programme to support crop engineers by susceptibility study of flood risk mitigation strategy. The AHP is one of the most common and effective hazard modelling techniques, such as flood forecasting, visualization and analysis of complex issues (Chen et al., 2011). In addition to AHP, the other suitable models for hazard analysis are MCDA (Samanta et al., 2016), weights of evidence (WofE) (Rahmati et al., 2016a, b), logistic regression (LR) (Tehrany et al., 2015), adaptive neuro-fuzzy interface method (Sezer et al., 2011), artificial neural networks (ANN) (Tiwari & Chatterjee, 2010) and FR model (Rahmati et al., 2016a, b). Gómez-Limón et al. (2003) proposed an innovative approach for weighting the parameters used for aversion to agricultural risk in decision-making. On a single scale, it uses numerical ratings based on multiple choices. In this approach, by utility value functions, all divergent parameters are converted into a single 0–1 scale to take the ultimate judgement. Models that show evidence of the disease's epidemiological (population-level) features were listed together as 'Epidemiological'. Variables directly addressing the agent or pathology have been classified under 'Etiology'. 'Geospatial modelling' included models that used satellite data such as the 'normalised difference vegetation index' (NDVI), to quantity of live green vegetation in a target area. 'Quantifiable' methods, such as tillage processes, were also recognized in some models, along with 'Temporal' and 'Agricultural' factors. The associations between mechanistic disease transmission factors can be used to explain the geographical spread of disease risk (Krefis et al., 2011), early warning systems, or to create mechanistic vector population and disease transmission models. In designing logical regression models based on Multi-satellite Precipitation Analysis (TMPA) and documenting a clear correlation in those areas between tropical rainfall and the MVEV outreach, Schuster et al. (2011) use remote sensing (RS) data of the Project of Tropical Rainfall Measurement Mission (TRMM). Landsat data are used to assess and compare the green leaf area index (LAI) of rice fields to the Aedes mosquito density build-up. In 2006, Glass et al. created a logistic regression model based on 1992–2005 Landsat Thematic Mapper (TM) imagery to estimate the menace of Hanta virus pulmonary syndrome (HPS) in 2006 and recorded that augmented rainfall in northern New Mexico and southern Colorado raises the risk of HPS following the previous drought years. The spreads of VBDs, viruses, reservoirs and vectors can be modelled by using mathematical and statistical models. Numerical models use a single reckoning or series of equations that feign or describe a system and/or predict that system's potential behaviour, while statistical models use simulation approaches that include assembling, evaluating and/or interpreting datasets (e.g. regressions) (Sadeghieh et al., 2020). New mapping apps, such as Google Earth™ and MS Virtual Earth™, offer simple and easy-to-use capabilities to produce not only spatial data overlaid on pre-existing satellite imagery or map simulations but also complex space-time sequence representations that can be played as animated films.

1.4 Integration Between Data Science and Geoinformatics: A Practical Guide to Manage Multi-hazard

The immediate need for hazard control and protection is illustrated by significant and irreparable disruption to agriculture, transport, bridges and many other facets of urban infrastructure (Mojaddadi et al., 2017). Moreover, it has been made apparent by the COVID-19 pandemic that biological and environmental hazards converge and raise the complexity of the overall impacts of disasters on communities and economies. However, in order to grab the interactions of natural and biological threats or control the aspects of connectiveness and rippling impacts on social, economic, and environmental environments, emergency response and risk analytics have been sluggish. The deficiency of viable catastrophe planning or new evacuation plans restricts alternative management initiatives. Alternatively, administrators need to be wary of all flaws and restrictions. Inadequate utilization and deployment of money is a big complication in emergency operations. To stop making the incorrect decisions that could cost the lives of individuals, emergency management must provide a clear view of the emergency circumstances reliant upon reliable data. Around the same time, dwelling on irrelevant data and deserting the flow of information due to too much meaningless data causes unsuccessful effects, for example, misidentifying the actual first responder.

Around 1.8 ZB of data was generated in late 2011, according to IDC (2014). Globally, around 1.2 ZB (1021) is generated as electronic data by various sources each year (Hilbert & López, 2011). The data is projected to hit 40 ZB by 2020 (Sagiroglu, 2013). Artificial Intelligence (AI) uses computational, mathematical approaches by software scripts and strategies to simplify decision-making that data centres use without clear instructions to the algorithm to execute a given human task effectively. Present shortcomings should be resolved in the risk assessment of roles and silos and in datasets, studies, simulation and control. Decisions historically reached largely 'from the gut' or by benchmarking would become data-driven and systemic. The increasing use of data science and machine learning (ML) are increasingly becoming one of the world's greatest challenges for computer-driven enterprises, data scientists and legal staff. Today there is enormous interest in corporate data processing, primarily due to 'the dawn of the Big Data era' (Ekeu-wei & Blackburn, 2018).

Geospatial Artificial Intelligence (GeoAI) incorporates space science (e.g. geographic information systems or GIS), AI, data processing and high-performance computing techniques to derive useful knowledge from big spatial data. GeoAI reflects a multi-hazard intelligence-based area that combines location to extract actionable knowledge that can be used to enhance risk management. The

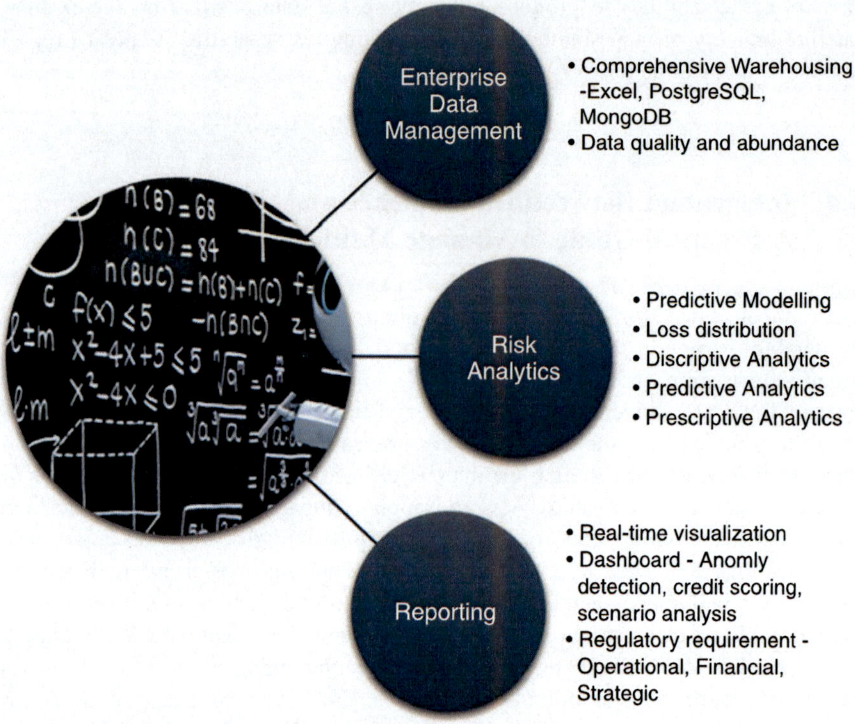

Fig. 1.3 Role of GeoAI in risk management

application of novel outlets of huge volume of spatial data, for instance, social media, automated health reports, satellite data, and personal sensors, and the advancement of public health science (particularly in the framework of 'smart and safe cities') are a prominent idea through GeoAI applications at population and discrete level, providing new possibilities to address traditional questions more comprehensively. Figure 1.3 shows probable application of GeoAI in multi-hazard risk management. It is imperative to consider the timeline and context of enterprise data management in order to make good design choices for the emerging technology of today and tomorrow and to prevent making past mistakes. Sensitive master data is now included in the continuing rapid increase of data volume and use across various silos, both on-site and in the cloud, and in a number of data formats. Most importantly, IoT sensors and systems are new and effective forms of geo-tagged big data generation implemented in urban societies (Kamel Boulos & Al-Shorbaji, 2014).

To process, interpret and make sense of such huge amount of spatial big data in real time, it is therefore important and indispensable to implement robust GeoAI technologies. Mojaddadi et al. (2017) proposed an ensemble approach (the frequency ratio (FR) methodology combined with a radial basis function and support vector machine (SVM)) that exhibited utility in GIS-based flood modelling to produce flood likelihood indices for the catchment of the Damansara River in Malaysia. An increasing number of researches have demonstrated the possibility of using machine learning (ML)-based algorithms with spatial datasets and satellite images to create regional-scale landslide susceptibility models, such as decision trees (DT) (Tsai et al., 2013), entropy- and evolution-based algorithms (Kavzoglu et al., 2015), fuzzy-theory (Zhu et al., 2014), neural-fuzzy systems (Xu et al., 2015; Bui et al., 2012), random forest algorithms (Lai & Tsai, 2019), and the advancement of computational resources, geospatial data and technologies. For instance, to predict everyday particulate matter <2.5 μm in diameter (PM2.5) in the USA, a neural network was used to employ numerous predictors, including satellite-based optical aerosol depth (AOD) from the Moderate Resolution Imaging Spectroradiometer (MODIS) (Di et al., 2016). In another research to resolve the lack of building maps in less developed nations for development targets associated to emergency relief and poverty reduction, WorldView-2 satellite data and voluntary geographic information (VGI) were implemented to deep learning (convolutions neural networks or CNNs) to mechanize map creation for buildings in Nigeria (Yuan et al., 2018). Spence et al. (2016) also examined recent developments in social media recruitment, data analysis and public desires and preferences measurement. In China, a geographically weighted gradient boosting machine (GW-GBM) algorithm was employed to model $PM_{2.5}$ acquaintances, allowing for spatial non-stationarity using spatial smoothing kernels in relations between predictors and $PM_{2.5}$ (Zhan et al., 2017). GeoAI has been employed in epidemiology to identify and examine the geographical spread of viruses and to explore the consequence of location-based influences on the outcome of diseases. For example, machine learning (K-means clustering) was employed to evaluate spatio-temporal gestational age trends at distribution for 145 million births in over 3000 US counties from 1971 to 2008 using the National Centre for Health Statistics Natality Files to encourage the generation of hypotheses relevant to the aetiology of preterm births (Byrne et al., 1992). Researchers aimed to better examine the principle of the prevalence of HIV based on computer-based algorithm (support vector regression) in the Ivory Coast of Africa to derive mobility and communication data from rectified cell phone data (Brdar et al., 2016). Deep learning in genetics has been extended to fields of research for example functional genomics (e.g. envisaging the arrangement specificity of DNA- and RNA-binding proteins) (Zou et al., 2019).

Figure 1.4 demonstrates a risk assessment system that helps data science and regulatory teams to build faster, more precise and more compliant ML models. For example, data scientists could be better placed to explain key desirable effects, whereas legal workers could describe particular undesired results that could give rise to legal liability. Defence lines relate to the functions and responsibilities of data scientists and those engaged in the ML development, rollout and auditing phase. The management of this data infrastructure, from the data pipeline to the model, is one of the most important and most neglected facets of ML governance. Understanding the outputs of a model is essential to monitor its health and any potential threats, both during preparation and while in deployment.

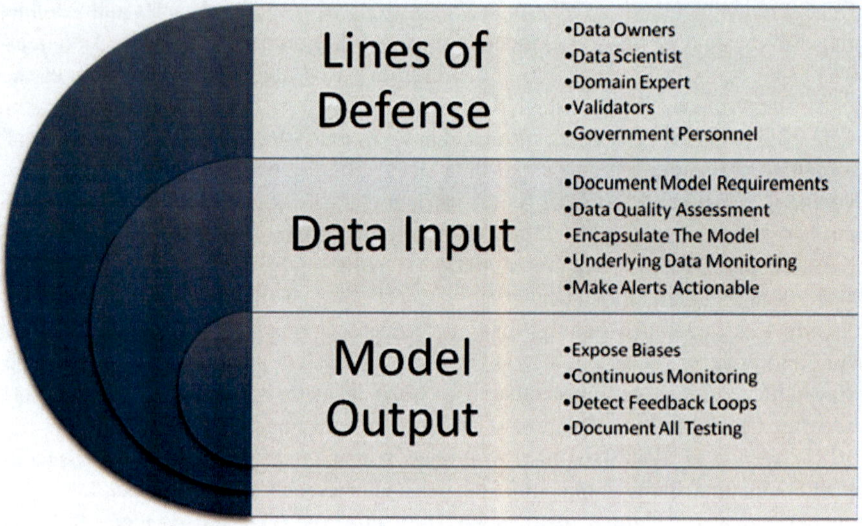

Fig. 1.4 Role of Artificial Intelligence in risk assessment strategy

1.5 Integration with Crowdsourcing and Geoinformatics on Multi-hazard Risk Assessment

As social networks advance, academic projects often concentrate on the use of social media for emergency relief. The main explanation is that social networks can provide not only rich data but also almost real-time information. Social networks build worlds where comments, photos and videos are exchanged within seconds, with 1.79 billion monthly active Facebook users and 500 million daily tweets (Sarvari et al., 2019). Panagiotopoulos et al. (2016) focused on using social media (Twitter) to convey threats to the public in order to help raise visibility or discourage public response from increasing. Two outlooks on risk and emergency message and the Social Amplification of Risk Framework (SARF) (Kasperson et al., 1988) and the Crisis and Emergency Risk Communication Model (CERC) are merged in the theoretical portion of this analysis (Reynolds & Seeger, 2005). Further investigations and innovative technologies have been carried out to reliably identify disaster information, such as machine learning, big data analysis and image processing. However, a detailed view of threatened fields is needed for data fusion. Fry and Binner (2016) investigated compartmental analysis and straightforward evacuation simulation. They model the actions of individuals and the influence of social media with maximal counter-strategies. They developed a Bayesian algorithm for maximal evacuation.

Unlike crowdsourced social media information discussed in the earlier segment, the word 'crowdsensing' is employed here to pronounce methods that rely on dedicated software systems to capture precise and organised information, as well as to leverage citizens' interpretive and analytical skills and local awareness (Gebremedhin et al., 2020). Several 'crowdsensing' schemes were developed, including devoted mobile disaster control and earth observation apps (Ferster & Coops, 2013). One example of the Ushahidi framework technology is the Flood Citizen Observatory prototype deployed in Brazil to permit people to report on the local status of river levels, flooded areas and the effects of flooding (Horita et al., 2015). Some programs, like Did You Feel It?, are explicitly constructed for disaster situations. U.S. Geological Survey (USGS) is used DYFI report to measure earthquake shaking intensity. The meteorological Phenomenon Identification Near the Ground (mPING), which tracks meteorological measurements and permits operators to display observations, was set up by the US National Oceanic and Atmospheric Administration (NOAA).NOAA uses mPING data to increase its dual-polarization radar and improve winter weather models, while ground-based meteorological measurements are necessary to verify that the radar has correctly calculated the amount of precipitation (Hultquist & Cervone, 2020). A particular category of knowledge and sharing portal is another crowdsourced geoinformation: the collaborative version of geographic features to conform with internet-based digital maps. This category includes the well-established Wikimapia and OpenStreetMap (OSM) platforms, in addition to the 'crowdsourcing' portion of the widespreadGoogleMaps framework, known as GoogleMapMaker (de Albuquerque et al., 2016). Such imagery is a very useful source of knowledge to be used by mappers, and it also helps volunteers from all around the world to participate, not just those who are specifically in the impacted regions.

1.6 Conclusion

Crowdsourced geographic information (CGI) has tremendous capacity not just to deal with the impacts of earthquakes but also to take proactive steps to boost metropolitan areas' exposure to natural hazards and extreme events. When vast volumes of data remain to be apprehended and gathered, data protection issues remain paramount. A range of approaches for GeoAI implementations are currently being used to assist risk management phases, such as risk recognition, risk estimation and risk assessment. In order to create machine learning models capable of supplying inputs to conventional risk management strategies, historical and real-time data are also used. Ethical mechanisms are therefore important to adequately warn research participants about risks and to protect individual privacy. In addition, in future research, the use of CGI in extenuation and planning phases should be stressed. For example, this could be achieved by exploiting initial instances of using CGI from concerted maps to sustenance catastrophe risk management practises, such as defining essential infrastructures to facilitate emergency planning. For the future work, we should follow the existing decision support system with suitable deep learning algorithms and IoT architectures. Future studies should build on existing GeoAI technologies, including location-based modelling geographies that have not previously been collected at a high spatio-temporal tenacity, or recently evolving spatial volume of data source engineering, to open novel study opportunities and accelerate our knowledge of multi-hazard risk.

References

Adab, H., DeviKanniah, K., & Solaimani, K. (2013). Modeling forest fire risk in the northeast of Iran using remote sensing and GIS techniques. *Natural Hazards, 65*, 1723–1743.

Arabameri, A., Pradhan, B., Rezaei, K., Yamani, M., Pourghasemi, H. R., & Lombardo, L. (2018). Spatial modelling of gully erosion using Evidential Belief Function, Logistic Regression and new ensemble EBF-LR algorithm. *Land Degradation & Development, 29*, 4035–4049.

Bishop, M. P., James, L. A., Shroder, J. F., Jr., & Walsh, S. J. (2012). Geospatial technologies and digital geomorphological mapping: Concepts, issues and research. *Geomorphology, 137*(1), 5–26.

Brdar, S., Gavric, K., Culibrk, D., & Crnojevic, V. (2016). Unveiling spatial epidemiology of HIV with mobile phone data. *Scientific Reports, 6*, 19342.

Bui, D. T., Pradhan, B., Lofman, O., Revhaug, I., & Dick, O. B. (2012). Landslide susceptibility mapping at HoaBinh province (Vietnam) using an adaptive neuro-fuzzy inference system and GIS. *Computational Geosciences, 45*, 199–211.

Byrne, D. E., Sykes, L. R., & Davis, D. M. (1992). Great thrust earthquakes and aseismic slip along the plate boundary of the Makran Subduction Zone. *Journal of Geophysical Research: Solid Earth, 97*, 449–478. https://doi.org/10.1029/91JB02165

Carpignano, A., Golia, E., Di Mauro, C., Bouchon, S., & Nordvik, J. P. (2009). A methodological approach for the definition of multi-risk maps at regional level: First application. *Journal of Risk Research, 12*(3–4), 513–534.

Chen, Y. R., Yeh, C. H., & Yu, B. (2011). Integrated application of the analytic hierarchy process and the geographic information system for flood risk assessment and flood plain management in Taiwan. *Natural Hazards, 59*(3), 1261–1276.

Chen, W., Panahi, M., Tsangaratos, P., Shahabi, H., Llia, L., Panahi, S., Li, S. J., Jaafari, A., & Ahmad, B. B. (2019). Applying population-based evolutionary algorithms and a neuro-fuzzy system for modelling landslide susceptibility. *Catena, 172*, 212–231.

Corley, C. D., Pullum, L. L., Hartley, D. M., Benedum, C., Noonan, C., Rabinowitz, P. M., et al. (2014). Disease prediction models and operational readiness. *PLoS One, 9*(3), e91989. https://doi.org/10.1371/journal.pone.0091989

Cutter, S. L. (2003). The vulnerability of science and the science of vulnerability. *Annals of the Association of American Geographers, 93*(1), 1–12.

de Albuquerque, J. P., Eckle, M., Herfort, B., & Zipf, A. (2016). Crowdsourcing geographic information for disaster management and improving urban resilience: An overview of recent developments and lessons learned. In C. Capineri, M. Haklay, H. Huang, V. Antoniou, J. Kettunen, F. Ostermann, & R. Purves (Eds.), *European handbook of crowdsourced geographic information* (pp. 309–321). Ubiquity Press. https://doi.org/10.5334/bax.w

Di, Q., Kloog, I., Koutrakis, P., Lyapustin, A., Wang, Y., & Schwartz, J. (2016). Assessing PM2.5 exposures with high spatiotemporal resolution across the continental United States. *Environmental Science & Technology, 50*(9), 4712–4721.

Du, Y., Xu, Q., Zhang, L., Feng, G., Li, Z., Chen, R., et al. (2017). Recent landslide movement in Tsaoling, Taiwan tracked by TerraSAR-X/TanDEM-X DEM time series. *Remote Sensing, 9*, 353.

Ekeu-wei, I. T., & Blackburn, G. A. (2018). Applications of open-access remotely sensed data for flood modelling and mapping in developing regions. *Hydrology, 5*, 39. https://doi.org/10.3390/hydrology5030039

Eshrati, L., Mahmoudzadeh, A., & Taghvaei, M. (2015). Multi hazards risk assessment, a new methodology. *International Journal of Health System and Disaster Management, 3*(2), 79–88. https://doi.org/10.4103/2347-9019.151315

Farr, T. G., & Kobrick, M. (2000). Shuttle Radar Topography Mission produces a wealth of data. *Eos Transactions American Geophysical Union, 81*, 583–585.

Ferster, C. J., & Coops, N. C. (2013). A review of earth observation using mobile personal communication devices. *Computers & Geosciences, 51*, 339–349. http://www.sciencedirect.com/science/article/pii/S0098300412003184

Fry, J., & Binner, J. M. (2016). Elementary modelling and behavioural analysis for emergency evacuations using social media. *European Journal of Operational Research, 249*(3), 1014–1023. https://doi.org/10.1016/j.ejor.2015.05.049

Garcia-Aristizabal, A., Gasparini, P., & Uhinga, G. (2015). Multi-risk assessment as a tool for decision-making. In *Urban vulnerability and climate change in Africa*. Springer.

Gebremedhin, E. T., Basco-Carrera, L., Jonoski, A., Iliffe, M., & Winsemius, H. (2020). Crowdsourcing and interactive modelling for urban flood management. *Journal of Flood Risk Managemen, 13*, e12602. https://doi.org/10.1111/jfr3.12602

Ghorbanzadeh, O., Feizizadeh, B., & Blaschke, T. (2018). Multi-criteria risk evaluation by integrating an analytical network process approach into GIS-based sensitivity and uncertainty analyses. *Geomatics, Natural Hazards and Risk, 9*(1), 127–151.

Giustarini, L., Chini, M., Hostache, R., Pappenberger, F., & Matgen, P. (2015). Flood Hazard Mapping Combining Hydrodynamic Modeling and Multi Annual Remote Sensing data. *Remote Sensing, 7*(10), 14200–14226.

Glatron, S., & Beck, E. (2008). Evaluation of socio-spatial vulnerability of citydwellers and analysis of risk perception:Industrial and seismic risks in Mulhouse. *Natural Hazards and Earth System Sciences, 8*(5). https://doi.org/10.5194/nhess-8-1029-2008

Golovko, D., Roessner, S., Behling, R., Wetzel, H.-U., & Kleinschmit, B. (2017). Evaluation of remote-sensing-based landslide inventories for hazard assessment in southern Kyrgyzstan. *Remote Sensing, 9*, 943. https://doi.org/10.3390/rs9090943

Gómez-Limón, J. A., Arriaza, M., & Riesgo, L. (2003). An MCDM analysis of agricultural risk aversion. *European Journal of Operational Research, 151*, 569–585.

Greiving, S., Fleischhauer, M., & Lückenkötter, J. (2006). A Methodology for an integrated risk assessment of spatially relevant hazards. *Journal of Environmental Planning and Management, 49*(1), 1–19. https://doi.org/10.1080/09640560500372800

Guzzetti, F., Mondini, A. C., Cardinali, M., Fiorucci, F., Santangelo, M., & Chang, K.-T. (2012). Landslide inventory maps: New tools for an old problem. *Earth-Science Reviews, 112*, 42–66. https://doi.org/10.1016/j.earscirev.2012.02.001

Haddow, G. D., Bullock, J. A., & Coppola, D. P. (2008). *Introduction to emergency management, 3rd edn*. Burlington, MA: Elsevier.

Hastings, D. A., & Dunbar, P. K. (1998). Development and assessment of the global land 1 km base elevation digital elevation model (GLOBE). *International Archives of Photogrammetry and Remote Sensing, 32*(4), 218–221.

Hembram, T. K., Paul, G. C., & Saha, S. (2011). Comparative analysis between morphometry and geo-environmental factor based soil erosion risk assessment using weight of evidence model: a study on jainti river basin, eastern India. *Environmental Processes, 6*(1). https://doi.org/10.1007/s40710-019-00388-5

Hilbert, M., & López, P. (2011). The world's technological capacity to store, communicate, and compute information. *Science, 332*(6025), 60–65.

Hölbling, D., Friedl, B., & Eisank, C. (2015). An object-based approach for semi-automated landslide change detection and attribution of changes to landslide classes in northern Taiwan. *Earth Science Informatics, 8*, 327–335. https://doi.org/10.1007/s12145-015-0217-3

Hoque, M. A. A., Phinn, S., Roelfsema, C., & Childs, I. (2018). Assessing tropical cyclone risks using geospatial techniques. *Applied Geography, 98*, 22–33.

Horita, F. E. A., Albuquerque, J. P., Degrossi, L. C., Mendiondo, E. M., & Ueyama, J. (2015). Development of a spatial decision support system for flood risk management in Brazil that combines volunteered geographic information with wireless sensor networks. *Computers & Geosciences, 80*, 84–94. https://doi.org/10.1016/j.cageo.2015.04.001

Hultquist, C., & Cervone, G. (2020). Integration of crowdsourced images, USGS networks, remote sensing, and a model to assess flood depth during hurricane florence. *Remote Sensing, 2020*(12), 834. https://doi.org/10.3390/rs12050834

IDC. 2014. Analyze the future. http://www.idc.com/

IFRCRCS. (2003). Disaster types. International Federation of Red Cross and Red Crescent Societies (IFRCRCS). Website: http://www.ifrc.org/what/disasters/types/

Islam, A. S., Bala, S. K., & Haque, M. A. (2010). Flood inundation map of Bangladesh using MODIS time-series images. *Journal of Flood Risk Management, 3*, 210–222.

Kamel Boulos, M. N., & Al-Shorbaji, N. M. (2014). On the Internet of Things, smart cities and the WHO healthy cities. *International Journal of Health Geographics, 13*, 10.

Kang, Y., Zhao, C., Zhang, Q., Lu, Z., & Li, B. (2017). Application of InSAR techniques to an analysis of the Guanling landslide. *Remote Sensing, 9*(10), 1046. https://doi.org/10.3390/rs9101046

Kasperson, R. E., Renn, O., Slovic, P., Brown, H. S., Emel, J., Goble, R., Kasperson, J. X., & Ratick, S. (1988). The social amlplification of risk: A conceptual framework. *Risk Analysis, 8* (2), 177–187. https://doi.org/10.1111/j.1539-6924.1988.tb01168.x

Kavzoglu, T., Sahin, E. K., & Colkesen, I. (2015). Selecting optimal conditioning factors in shallow translational landslide susceptibility mapping using genetic algorithm. *Engineering Geology, 192*, 101–112. https://doi.org/10.1016/j.enggeo.2015.04.004

Keesstra, S., Mol, G., de Leeuw, J., Okx, J., de Cleen, M., & Visser, S. (2018). Soil-related sustainable development goals: Four concepts to make land degradation neutrality and restoration work. *Land, 7*(4), 133.

Krefis, A. C., Schwarz, N. G., Nkrumah, B., Acquah, S., Loag, W., et al. (2011). Spatial analysis of land cover determinants of malaria incidence in the Ashanti Region, Ghana. *PLoS One, 6*, e17905.

Lai, J. S., & Tsai, F. (2019). Improving GIS-based landslide susceptibility assessments with multi-temporal remote sensing and machine learning. *Sensors (Basel)., 19*(17), 3717. https://doi.org/10.3390/s19173717

Lobitz, B., Beck, L., Huq, A., Wood, B., Fuchs, G., et al. (2000). Climate and infectious disease: Use of remote sensing for detection of Vibrio cholerae by indirect measurement. *Proceedings of the National Academy of Sciences of the United States of America, 97*, 1438–1443.

Lu, P., Stumpf, A., Kerle, N., & Casagli, N. (2011). Object-oriented change detection for landslide rapid mapping. *IEEE Geoscience and Remote Sensing Letters, 8*(4), 701–705. https://doi.org/10.1109/LGRS.2010.2101045

Manna, A. J. (1985). 25 years of TIROS satellites. *Bulletin of the American Meteorological Society, 66*(4), 421–423.

Mojaddadi, H., Pradhan, B., Nampak, H., Ahmad, N., & Ghazali, A. H. B. (2017). Ensemble machine-learning-based geospatial approach for flood risk assessment using multi-sensor remote-sensing data and GIS. *Geomatics, Natural Hazards and Risk, 8*(2), 1080–1102. https://doi.org/10.1080/19475705.2017.1294113

Panagiotopoulos, P., Barnett, J., Bigdeli, A. Z., & Sams, S. (2016). Social media in emergency management: Twitter as a tool for communicating risks to the public. *Technological Forecasting and Social Change, 111*, 86–96. https://doi.org/10.1016/j.techfore.2016.06.010

Pourghasemi, H. R., & Kerle, N. (2016). Random forests and evidential belief function-based landslide susceptibility assessment in western Mazandaran Province, Iran. *Environmental Earth Sciences, 75*(3), 1–17.

Pourghasemi, H. R., Gayen, A., Park, S., Lee, C. W., & Lee, S. (2018). Assessment of landslide prone areas and its zonation using logistic regression, Logit Boost, and Naïve Bayes machine learning algorithms. *Sustainability, 10*(10), 3697.

Pourghasemi, H. R., Gayen, A., Panahi, M., Rezaie, F., & Blaschke, T. (2019). Multi-hazard probability assessment and mapping in Iran. *Science of the Total Environment, 692*, 556–571.

Rahmati, O., Pourghasemi, H. R., & Zeinivand, H. (2016a). Flood susceptibility mapping using frequency ratio and weights-of-evidence models in the Golastan Province, Iran. *Geocarto International*. https://doi.org/10.1080/10106049.2015.1041559

Rahmati, O., Zeinivand, H., & Besharat, M. (2016b). Flood hazard zoning in Yasooj region, Iran, using GIS and multi-criteria decision analysis. *Geomatics, Natural Hazards and Risk, 7*(3), 1000–1017. https://doi.org/10.1080/19475705.2015.1045043

Ravindra, K., Rattan, P., Mor, S., & Aggarwal, A. N. (2019). Generalized additive models: Building evidence of air pollution, climate change and human health. *Environment International, 132*, 104987.

RazaviTermeh, S. V., Kornejady, A., Pourghasemi, H. R., & Keesstra, S. (2018). Flood susceptibility mapping using novel ensembles of adaptive neuro fuzzy inference system and metaheuristic algorithms. *Science of the Total Environment, 615*, 438–451.

Reynolds, B., & Seeger, M. W. (2005). Crisis and emergency risk communication as an integrative model. *Journal of Health Communication, 10*(1), 43–55. https://doi.org/10.1080/10810730590904571

Riley, S. (2007). Large-scale spatial-transmission models of infectious disease. *Science, 316*, 1298–1301.

Sadeghieh, T., Waddell, L. A., Ng, V., Hall, A., & Sargeant, J. (2020). A scoping review of importation and predictive models related to vector-borne diseases, pathogens, reservoirs, or vectors (1999–2016). *PLoS One, 15*(1), e0227678. https://doi.org/10.1371/journal.pone.0227678

Sagiroglu, S.D.. 2013. Big data: A review. In *Proceedings of the International Conference on Collaboration Technologies and Systems (CTS'13)*, pp. 42–47, San Diego, CA.

Sakamoto, T., van Nguyen, N., Kotera, A., Ohno, H., Ishitsuka, N., & Yokozawa, M. (2007). Detecting temporal changes in the extent of annual flooding within the Cambodia and the Vietnamese Mekong Delta from MODIS time-series imagery. *Remote Sensing of Environment, 109*, 295–313.

Salichon, J., Le Cozannet, G., Modaressi, H., Hosford, S., Missotten, R., McManus, K., Marsh, S., Paganini, M., Ishida, C., Plag, H. P., Labrecque, J., Dobson, C., Quick, J., Giardini, D., Takara, K., Fukuoka, H., Casagli, N., & Marzocchi., W. (2007). 2nd IGOS Geohazards Theme report, BRGM.

Samanta, S., Koloa, C., Pal, D. K., & Palsamanta, B. (2016). Flood risk analysis in lower part of Markham River based on multi-criteria decision approach (MCDA). *Hydrology, 3*(3), 29.

Sarvari, P. A., Nozari, M., & Khadraoui, D. (2019). The potential of data analyticsin disaster management. In F. Calisir et al. (Eds.), *Industrial engineering in the big data era, lecture notes in management and industrial engineering*. Springer. https://doi.org/10.1007/978-3-030-03317-0_28

Schmidt, J., Matcham, I., Reese, S., King, A., Bell, R., Henderson, R., Smart, G., Cousins, J., Smith, W., & Heron, D. (2011). Quantitative multi-risk analysis for natural hazards: A framework for multi-risk modelling. *Natural Hazards, 58*(3), 1169–1192.

Schuster, G., Ebert, E. E., Stevenson, M. A., Corner, R. J., & Johansen, C. A. (2011). Application of satellite precipitation data to analyse and model arbovirus activity in the tropics. *International Journal of Health Geographics, 10*, 8. https://doi.org/10.1186/1476-072X-10-8

Sezer, E. A., Pradhan, B., & Gokceoglu, C. (2011). Manifestation of an adaptive neuro-fuzzy model on landslide susceptibility mapping: Klang Valley Malaysia. *Expert Systems with Applications, 38*(7), 8208–8219.

Spence, P. R., Lachlan, K. A., & Rainear, A. M. (2016). Social media and crisis research: Data collection and directions. *Computers in Human Behavior, 54*, 667–672. https://doi.org/10.1016/j.chb.2015.08.045

Tehrany, M. S., Pradhan, B., & Jebur, M. N. (2015). Flood susceptibility analysis and its verification using a novel ensemble support vector machine and frequency ratio method. *Stochastic Environmental Research and Risk Assessment, 29*, 1149–1165. https://doi.org/10.1007/s00477-015-1021-9

Teodoro, A. C., & Duarte, L. (2013). Forest fire risk maps: A GIS open source application – a case study in Norwest of Portugal. *International Journal of Geographical Information Science, 27*(4), 699–720.

TienBui, D., Tuan, T. A., Klempe, H., Pradhan, B., & Revhaug, I. (2016). Spatial prediction models for shallow landslide hazards: A comparative assessment of the efficacy of support vector machines, artificial neural networks, kernel logistic regression, and logistic model tree. *Landslides, 13*, 361–378.

Tiwari, M. K., & Chatterjee, C. (2010). Uncertainty assessment and ensemble flood forecasting using bootstrap based artificial neural networks (BANNs). *Journal of Hydrology, 382*(1), 20–33.

Thomas, J., & Leveson, N. G. (2011). Performing hazard analysis on complex, software- and human intensive systems, in Proc. of the 29th ISSC Conference about System Safety.

Tsai, F., Lai, J.-S., Chen, W. W., & Lin, T.-H. (2013). Analysis of topographic and vegetative factors with data mining for landslide verification. *Ecological Engineering, 61*, 669–677. https://doi.org/10.1016/j.ecoleng.2013.07.070

UN. 2002. Johannesburg Plan of Implementation of the World Summit on Sustainable Development Technical report, United Nations.

Vakhshoori, V., & Zare, M. (2016). Landslide susceptibility mapping by comparing weight of evidence, fuzzy logic, and frequency ratio methods. *Geomatics, Natural Hazards and Risk, 7*(5), 1731–1752.

Van Westen, C. J. (2013). 3.10 remote sensing and GIS for natural hazards assessment and disaster risk management. In *Treatise on geomorphology* (pp. 259–298). Elsevier.

Wang, L.-J., Guo, M., Sawada, K., Lin, J., & Zhang, J. (2016). A comparative study of landslide susceptibility maps using logistic regression, frequency ratio, decision tree, weights of evidence and artificial neural network. *Geosciences Journal, 20*, 117–136.

Wisner, B., Blaikie, P., Cannon, T., & Davis, I. (2004). *At risk: Natural hazards, people's vulnerability and disasters*. Routledge.

Xiao, L., Zhang, Y., & Peng, G. (2018). Landslide susceptibility assessment using integrated deep learning algorithm along the China-Nepal Highway. *Sensors, 18*(12), 4436.

Xu, K., Guo, Q., Li, Z., Xiao, J., Qin, Y., Chen, D., & Kong, C. (2015). Landslide susceptibility evaluation based on BPNN and GIS: A case of Guojiaba in the Three Gorges Reservoir Area. *International Journal of Geographical Information Science, 29*, 1111–1124. https://doi.org/10.1080/13658816.2014.992436

Yilmaz, I., and M. Ercanoglu. 2019. Landslide inventory, sampling and effect of sampling strategies on landslide susceptibility/hazard modelling at a glance. In *Natural hazards GIS-based spatial modeling using data mining techniques*, pp. 205–224.

Youssef, A. M. (2015). Landslide susceptibility delineation in the Ar-Rayth area, Jizan, Kingdom of Saudi Arabia, using analytical hierarchy process, frequency ratio, and logistic regression models. *Environment and Earth Science, 73*, 8499–8518.

Youssef, A. M., Pourghasemi, H. R., Pourtaghi, Z. S., & Al-Katheeri, M. M. (2016). Landslide susceptibility mapping using random forest, boosted regression tree, classification and regression tree, and general linear models and comparison of their performance at Wadi Tayyah Basin, Asir Region, Saudi Arabia. *Landslides, 13*, 839–856.

Yuan, J., Roy Chowdhury, P. K., McKee, J., Yang, H. L., Weaver, J., & Bhaduri, B. (2018). Exploiting deep learning and volunteered geographic information for mapping buildings in Kano, Nigeria. *Scientific Data, 5*, 180217.

Zhan, Y., Luo, Y., Deng, X., Chen, H., Grieneisen, M. L., Shen, X., Zhu, L., & Zhang, M. (2017). Spatiotemporal prediction of continuous daily PM2.5 concentrations across China using a spatially explicit machine learning algorithm. *Atmospheric Environment, 155*, 129–139.

Zhang, L., & Zhang, S. (2017). Approaches to multi-hazard landslide risk assessment. In *Geotechnical safety and reliability*.

Zhu, A.-X., Wang, R., Qiao, J., Qin, C.-Z., Chen, Y., Liu, J., Du, F., Lin, Y., & Zhu, T. (2014). An expert knowledge-based approach to landslide susceptibility mapping using GIS and fuzzy logic. *Geomorphology, 214*, 128–138. https://doi.org/10.1016/j.geomorph.2014.02.003

Zou, J., Huss, M., Abid, A., Mohammadi, P., Torkamani, A., & Telenti, A. (2019). A primer on deep learning in genomics. *Nature Genetics, 51*(1), 12–18.

Chapter 2
GIS-Based Landslide Susceptibility Mapping in Eastern Boundary Zone of Northeast India in Compliance with Indo-Burmese Subduction Tectonics

Arnab Sengupta and Sankar Kumar Nath

Abstract The Eastern Boundary Zone of Northeast India, comprising the Indian States of Manipur, Mizoram and Nagaland, suffers immensely under the impact of frequent devastating landslides that results in widespread damage and casualty. A rough estimate of the decadal intensity of landslides from an inventory spanning over half a century calls for systematic assessment of landslide hazard and risk in the region for its effective mitigation and management. Landslide Susceptibility Zonation is the most fundamental step in that direction wherein spatial distribution of Landslide Susceptibility Index (LSI) is established through integrating nineteen causative factors, viz. surface geology, landform, lineament density, elevation, distance to lineament, slope angle, aspect, drainage density, distance to drainage, terrain ruggedness index, plan and profile curvature, normalized difference vegetation index, landuse/landcover, distance to road, road density, rainfall, earthquake epicentre proximity and peak ground acceleration rationally on GIS platform in 1:50,000 scale by following a multivariate statistics-based Logistic Regression (LR) procedure. This classifies the terrain into None, Low, Moderate, High, Very High and Severe susceptible zones on a raster map display, which is inevitably validated through statistical accuracy test by drawing a comparison with the 30% landslide inventory test dataset which exhibited 73% accuracy level. This landslide susceptibility map will invariably help the urban planners and the decision-makers in effective landslide risk mitigation and spatial design.

Keywords Landslide Susceptibility Zones · Indo-Barmic Subduction · Northeast India · Eastern Boundary Zone

---

A. Sengupta · S. K. Nath (✉)
Department of Geology and Geophysics, Indian Institute of Technology Kharagpur, Kharagpur, West Bengal, India
e-mail: nath@gg.iitkgp.ac.in

P. K. Shit et al. (eds.), *Geospatial Technology for Environmental Hazards*, Advances in Geographic Information Science, https://doi.org/10.1007/978-3-030-75197-5_2

2.1 Introduction

Landslide is a process associated with the downward movements of soil, rock, debris, artificial fill and/or a mixture of all of these. Landslide occurs due to both physical and man-made activities. Physical activities consist of an impending earthquake, volcanic eruptions, tectonic activities, torrential rainfall, storm, etc., while man-made activities include unscientific construction, unmannered tourism and so on. Rainfall-induced landslides are very common around the globe. Current estimates of landslide impacts suggest that they cause thousands of fatalities annually (Froude and Petley (2018); Petley (2012) and economic loss worth billions of US dollars (Dilley et al., 2005; CRED, n.d.) as shown in Fig. 2.1(a). Global warming, climate change and rising temperature are expected to trigger landslides, especially in mountainous regions with snow and ice and cover. In India, a total of 0.42 million km^2 or 12.6% of the landmass is imperil due to landslides of which an

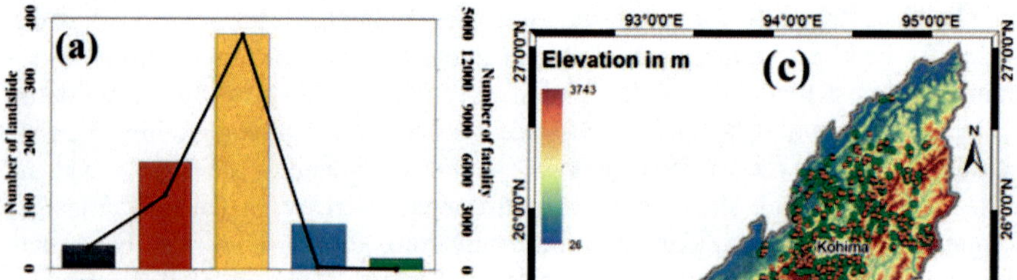

Fig. 2.1 Location map of the study region. (**a**) Continent-wise landslide inventory data (bar graph) and line graph are showing the number of fatalities (source: CRED, n.d.). (**b**) Landslide inventory map of the Indian subcontinent. (**c**) Landslide inventory (training and testing) data in the Eastern Boundary Zone of Northeast India comprising of the Indian States of Manipur, Mizoram and Nagaland

approximate 0.18 million km^2 falls is in the Northeast Himalayan province of India; an approximate 0.14 million km^2 lays in the Northwest Himalaya; 0.09 million km^2 is in the Western Ghats and Konkan hills; and an approximate 0.01 million km^2 is in the Eastern Ghats causing an estimated damage cost of around 2–5 billion US dollars and 25% of annual deaths as depicted in Fig. 2.1(b).

The Eastern Boundary Zone in Northeast India comprising the Indian States of Manipur, Mizoram and Nagaland is a landslide-prone region as depicted in Fig. 2.1 (c). The study region covers around 60,025.56 km^2 comprising of 12 small towns. It is surrounded by the Tertiary hills of Mizoram, Manipur and Nagaland with a maximum altitude of 3743 m with steep to moderate slopes. Geologically, the region is classified into pre-Cambrian to Quaternary era. Tertiary rocks of the Disang and Baraingroup that consists of shale and sandstone are most predominant in the territory, which on weathering becomes platy and splintery, proving the most ideal state for landsliding. The terrain is also seismogenic being one of the most active regions of the world and according to BIS (2002); it falls under Seismic Zone V with frequent moderate to large magnitude earthquakes visiting the terrain causing extensive damage to both life and property.

Landslide susceptibility mapping in the Eastern Boundary Block of Northeast India, using various algorithms, has already been attempted by several researchers, viz. Laldintluanga et al. (2016); Pathak (2016); Balamurugan et al. (2016); Lallianthanga and Lalbiakmawia (2013a, 2013b, 2014); Lallianthanga et al. (2013); Barman and Srinivasa Rao (2019); Lallianthanga and Laltanpuia (2014); Balamurugan and Ramesh (2016); Pachuau (2019); Khatsu and Van Westen (2005); Roy et al. (2019); Sema et al. (2017); and Singh et al. (2011). A comprehensive literature review unfolds that earlier works have been performed at the site-specific scale on a slope-slope basis and not in the regional scale. In this study, we, however, considered the entire Eastern Boundary Zone of the Northeast India comprising the Indian States of Manipur, Mizoram and Nagaland as a unit tectonic block for an overall understanding of the probability of landslide occurrence in the terrain using an ensemble of Remote Sensing-GIS for chocking out a pre-disaster landslide risk mitigation strategy to be put in place and to perform precursory damage estimation for insurance coverage purposes.

2.2 Data and Methodology

Spatial and Non-spatial Data

In order to achieve the GIS-based landslide susceptibility zonation, there is a requirement of spatial and non-spatial data as illustrated in Table 2.1.

Table 2.1 Spatial and non-spatial data used in the present study

Data	Causative layers	Source
Geology map	Surface geology map	Geological Survey of India
Lineament map	Lineament density and distance to lineament	Dasgupta et al. (2000), National Mission on Geomorphological and Lineament Mapping, http://bhuvan.nrsc.gov.in/gis/thematic/index.php
Road Network	Road density and distance to road	Open Street Map and Google Earth
Earthquake catalogue	Epicentre proximity	Nath et al. (2017), USGS, IMD and ISC
Rainfall data	Rainfall map	India Meteorological Department and Tropical Rainfall Measuring Mission
ALOS PALSAR DEM (30 m)	Slope angle, slope aspect, landform, elevation, Drainage density, distance to drainage, plan curvature, Terrain ruggedness index	Japan Aerospace Exploration Agency
Landsat 8 (30 m)	Normalized difference vegetation index	United States Geological Survey
GlobCover land cover map	Land use/land cover	European Space Agency
Seismic shaking	Surface consistent peak ground acceleration	Nath and Thingbaijam (2012)

Multivariate Statistics: Logistic Regression (LR)

Among a wide range of statistical methods proposed for the assessment of landslide susceptibility distribution, Logistic Regression (LR) has proven to provide one of the most reliable classification technique (Reichenbach et al., 2018; Guzzetti et al., 2006; Mancini et al., 2010; Hadmoko et al., 2017; Bai et al., 2011; Mathew et al., 2009; Lee & Pradhan, 2007; Nandi & Shakoor, 2010). LR simulates the probability of a certain class or event. It uses a logistic function to model a binary dependent variable even though many complex extensions exist. The aim of LR model is to establish a relation between the existing and the absent landslides. The advantage of this method is that the dependent variable can have only two values, i.e. occurring or non-occurring, and those predicted values can be interpreted as the probability since they are inhibited to lay in the interval between 0 and 1 Dai and Lee (2002). In the Logistic Regression analysis, there are some dependent variables that correlate with an independent variable. The predicted value ranges from 0 to 1, and it can be defined as the landslide susceptibility index. The road map of the algorithm worked out in the present computation has been depicted in Fig. 2.2. The index can be defined by the following formulations:

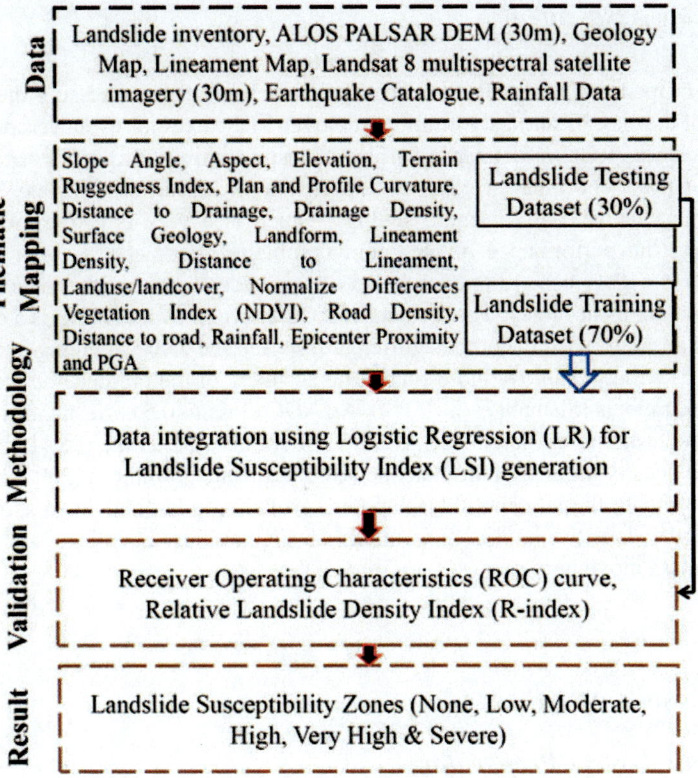

Fig. 2.2 Road Map of the algorithm worked out in the LR protocol employed here

$$Z = \beta_0 + \beta_1 X_1 + \beta_2 X_2 + \cdots + \beta_n X_n \quad (2.1)$$

Here Z denotes the linear logistic regression model. β_0 is a constant. β_1, β_2 are the corresponding coefficients of each of the respective contributing factors that indicate their contribution to landslide susceptibility. X_1, X_2 ,… X_n are the independent variables.

The probability index is calculated to predict the landslide hazard zone index; the possibility of occurrence and its intensity as

$$P = 1/(1 + \ell^{-z}) \quad (2.2)$$

Here P is the probability of landslide hazard index and Z takes any value from $-\infty$ to $+\infty$.

Accuracy Assessment

Receiver Operating Characteristic (ROC) is widely used for assessing the performance of the classification algorithms, employed in the extreme event computations. In geoscience, ROC is defined as a plot of test sensitivity or True Positive Rate (TPR) as the y-coordinate versus its 1-specificity or False Positive Rate (FPR) as x-coordinate at various threshold settings, which is a very effective method for evaluating the performance of dichotomy problems (Park et al., 2004; Fawcett, 2006). It is widely used in the validation of landslide susceptibility maps and also for estimating its accuracy. The area under ROC curve known as AUC is a common metric that can be used to compare different tests and the values, ranging from 0.5 to 1 which is widely employed to estimate the accuracy of the presence or absence of predictive models (Shahabi et al., 2014). An AUC close to 0.5 corresponds to a poor diagnostic test, and the larger the AUC, the more accurate is the test. The relative landslide density index (R-index) defined by Baeza and Corominas (2001) has been used to validate the susceptibility mapping results. R-index is defined as the ratio between the density of mass movements of a given susceptibility class and the overall mass movement density.

2.3 Results and Discussion

Thematic Layers Preparation

Landslide Inventory

Landslide inventory map of the terrain is prepared through multispectral satellite image interpretation, Google Earth imageries, published literature and reports from various government agencies, viz. Geological Survey of India (GSI); Nagaland State Disaster Management Authority; Disaster Management and Rehabilitation Department, Manipur; Remote Sensing Application Centre; and Bhuvan Portal developed by Indian Space Research Organization (n.d.). In the present study, multi-temporal satellite data and Google Earth imageries have been extensively used for the demarcation of a landslide accessible or inaccessible region in the hilly terrain. An extensive field survey has also been conducted to enlarge the inventory database as well as to validate the existing landslide inventory database. Through standard image analysis and field survey, a total of 4206 landslides have been identified. The landslide inventory database is randomly divided into subsets of 70% for training and 30% for testing as depicted in Fig. 2.1(c).

Morphometric Causative Factors

The topography is an essential factor for landslide susceptibility mapping that limits the density and spatial extent of landslides. The crucial morphological factors for the substantial causes of landslides are slope angle, elevation, aspect, plan and profile curvature, distance to drainage, drainage density and terrain ruggedness index (TRI). Slope failure is a very significant issue for landslide occurrence and is associated with slope movement due to gravitational forces (Catani et al., 2005). A slope angle is defined as an angle between the surface of the earth and a horizontal datum (Huang et al., 2017). At local scales, it affects the concentration of moisture and the level of pore pressure and is often used to resolve detailed patterns of instability. At larger scales, it controls regional hydraulic continuity and is considered an essential factor for GIS-based landslide susceptibility mapping (Guzzetti et al., 1999; Dai & Lee, 2002; Ohlmacher & Davis, 2003). The monotony of landslide occurrences can be defined by a morphometric slope based on the Topographic Gradient which is generated from the ALOS PALSAR digital elevation model (DEM). It is observed from the landslide inventory and slope angle database that the steeper the slope, the probability of mass failure increases in the region. In the present study, the slope angle varies from 0° to 76.6342°, as depicted in Fig. 2.3(a).

Elevation is another morphometric causative factor for landslide susceptibility mapping as several geological and geomorphological processes control it (Pourghasemi et al., 2012; Pradhan & Kim, 2014; Youssef et al., 2015). Landslides usually occur at intermediate elevation since slopes tend to be covered by a layer of thin colluvium that is prone to landslides (Dai & Lee, 2002). The altitude in the region is seen to vary from 26 to 3743 m, as shown in Fig. 2.3(b).

On the other hand, aspect is also pondered to have an augmented role in the mass movement, and it identifies the steepest downslope across a surface. The constraints associated with the Slope Aspect, such as the degree of saturation, discontinuities, drying winds and exposure to sunlight, may regulate the manifestation of a landslide. The slope aspect map is also obtained from ALOS PALSAR DEM and classified into nine standard directions, viz. flat, north, northeast, east, southeast, south, southwest, west and northwest, respectively, as depicted in Fig. 2.3(c).

The plan and profile curvature is defined as the rate of change of slope gradient or aspect, usually in a particular direction (Dikau, 1988; Wilson & Gallant, 2000; Nefeslioglu et al., 2008). The curvature value is evaluated by calculating the reciprocal value of the radius of curvature. Curvature is described as a contour formed by intersecting a horizontal plane with the surface. The impact of curvature on the slope erosion process is the convergence or divergence of water during downhill flow. Curvature has been used for landslide susceptibility mapping by Ayalew et al. (2004); Dikau (1988); Wilson and Gallant (2000); Nefeslioglu et al. (2008); Chen et al. (2017) and Ding et al. (2017). This parameter constitutes one of the causative factors in the present investigation as a vital factor, controlling landslide occurrences. The plan and profile curvatures have been prepared by using high-resolution DEM data, as depicted in Fig. 2.3(d and e).

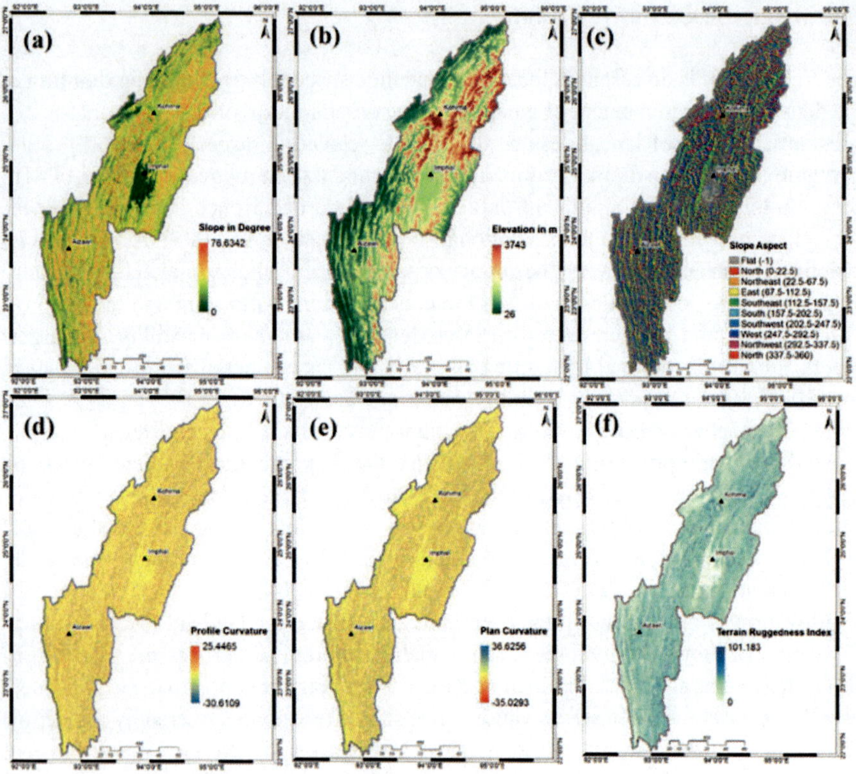

Fig. 2.3 GIS raster maps exhibit the morphometric causative factors of the terrain, viz. (**a**) Slope angle, (**b**) Elevation, (**c**) Slope Aspect, (**d**) Profile Curvature, (**e**) Plan Curvature and (**f**) Terrain Ruggedness Index (TRI)

The terrain ruggedness index is defined by the altitude variation between the adjacent cells of a digital elevation model (Alkhasawneh et al., 2013). The process determines the difference in altitude values between a centre cell and the surrounding cells. Then it squares each of the eight elevation difference values to make them all positive and then averages the squares. The terrain ruggedness index is then derived by taking the square root of this average, as shown in Fig. 2.3(f).

Drainage network is another causative factor for landslide occurrence and has been renowned as a topographic characteristic of fundamental importance. As the density of stream linkage reveals the geological, topographical, soil and the vegetation control, drainage network is chosen to simultaneously contemplate the undercutting of a hydrographic system for the role of inappropriate drainage (Shahabi et al., 2014). The proximity of the steep slope to the drainage network is an additional essential element controlling the slope stability because the streams adversely erode the material of the lower portion and make the proliferation of water level (Shahabi et al., 2014). The total length of the stream in a given section throughout its area provides drainage density, which has been calculated from the

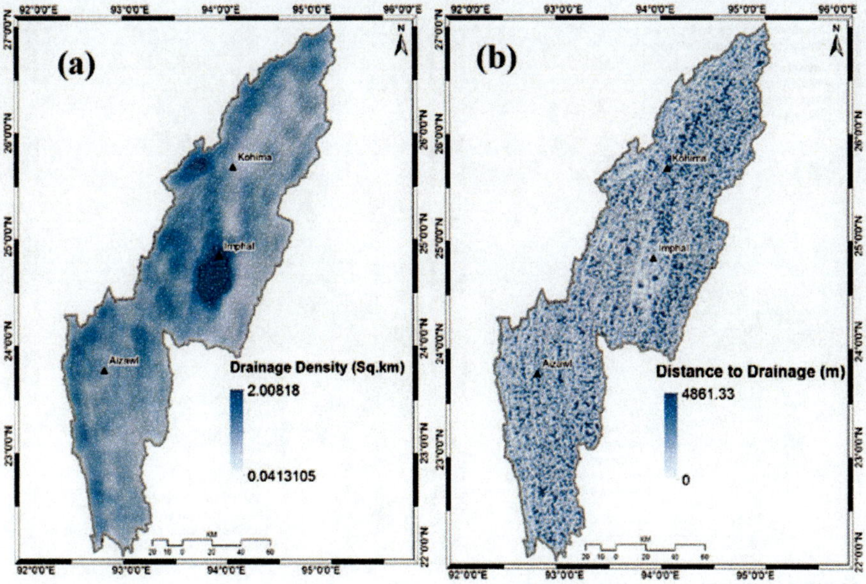

Fig. 2.4 GIS maps exhibit the morphometric causative factors of the terrain, viz. (**a**) Drainage Density and (**b**) Distance to Drainage

drainage network and is seen to vary from 0.0413105 to 2.00818 km^2 as depicted in Fig. 2.4(a). The drainage proximity varies between 0 and 4861.33 m as depicted in Fig. 2.4(b).

Geological Causative Factors

The surface geological attribution is considered as an independent variable in the present study. Various geological formations have different compositions and structures, which contribute to the strength and permeability of rocks and soils. The stronger rocks give more resistance to the driving forces as compared to the weaker rocks and hence are less prone to landslides. The major geological formation of the terrain belongs to Belt of Schuppen designated as (1) accretionary prism; (2) ophiolite/melange, (3) accetionary complex; (4) alluvial fill along foredeep; (5) alluvial fill along superposed basin; (6) crystalline complex overprinted by Himalayan fold-thrust movement; (7) and cover rocks of frontal belt affected by fold-thrust movement during the terminal phase of Himalayan geology (8) as shown in Fig. 2.5(a).

Landform defines the spatial topological interactions of landforms which involve segregating the terrain into intangible spatial objects such as chronology, composition and features. The numerous geomorphological features of the landscape have been derived from ALOS PALSAR DEM wherein various types of landforms, viz. plains, valleys, open slopes, upper slopes mesas, mountaintops high ridges, upland

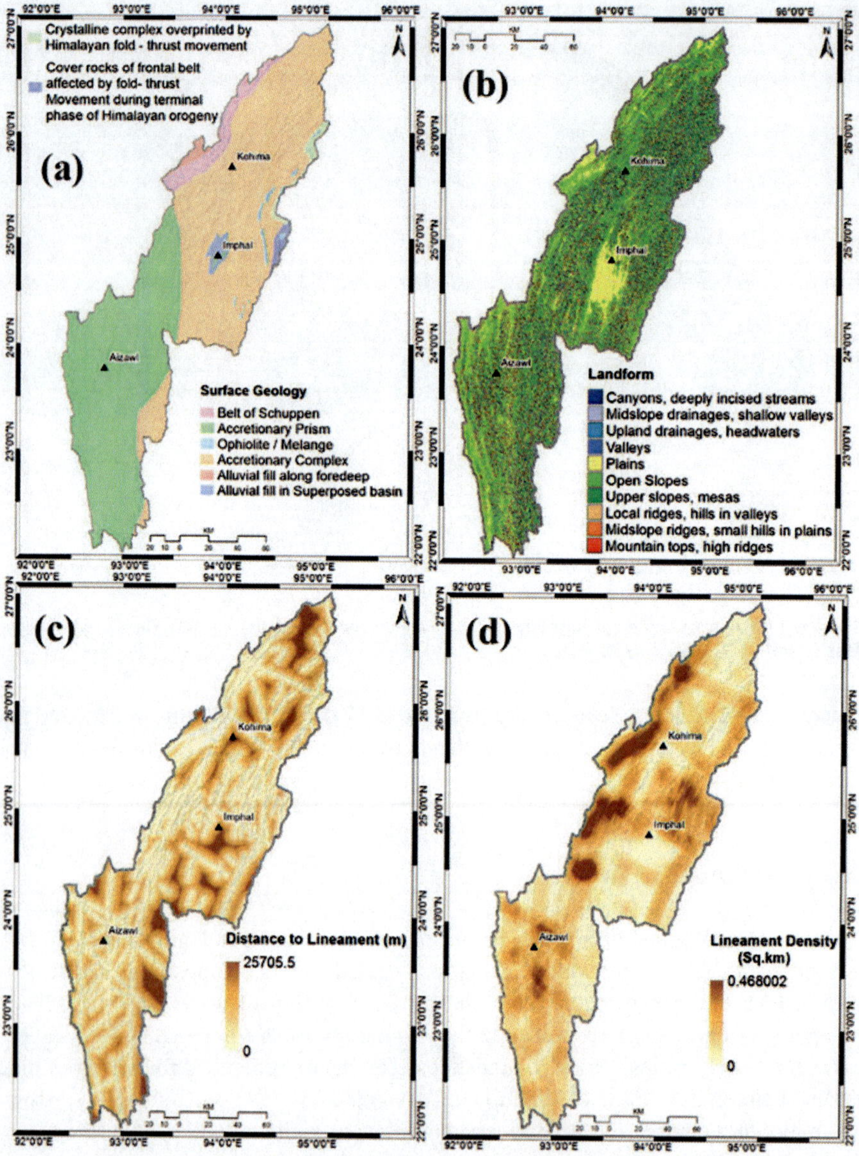

Fig. 2.5 GIS raster maps exhibit the geological causative factors of the terrain, viz. (**a**) Surface Geology, (**b**) Landform, (**c**) Distance to Lineament and (**d**) Lineament Density

drainages headwaters, midslope ridges small hills in plains, local ridges hills in valleys and canyons deeply incised streams and midslope drainages shallow valleys respectively following Jenness (2006) as depicted in Fig. 2.5(b).

The lineament is also an important contributing factor for the occurrence of a landslide as the degree of intense deformation, fracturing, fissures and weathering play crucial roles in causing slope failure (Bui et al., 2012). The disposition of structural discontinuities about slope inclination and direction has a great influence on the stability of slopes, which includes the extent of parallelism between the steepness and direction in the dip of discontinuity of the slope. Distance to lineament varies between 0 and 25,705.5 m, as shown in Fig. 2.5(c). The intensity of rock fracturing can be epitomized by lineament density, which is inevitable for the development of muffled passages over an area; therefore, the lineaments in the study region are mapped and analysed by lineament density and is seen to vary from 0 to 0.468 km^2 as depicted in Fig. 2.5(d).

Environmental Causative Factors

Normalized Differences Vegetation Index (NDVI) is an important causative factor for the movement of rainfall-induced landslides. Changes in vegetation cover often result in modified landslide behaviour (Van Beek, 2002; Wilkinson et al., 2002; Glade, 2003; Peduzzi, 2010). It is also a virtuous gage for the probability of mass movement. Vegetation roots penetrate the soil and increase their shear strength. The combination of recorded electromagnetic reflectance in near-infrared and red wavelength is highly correlated with the photosynthetic activity and the density of vegetation cover (Peduzzi, 2010). The NDVI map is prepared from Landsat 8 image through band ratio technique, i.e. near-infrared - Red/near-infrared + red, in which the index value ranges from −0.255 to 0.731. It is observed that the value of NDVI is comparatively higher in landslide-prone areas where there is a dense vegetation cover, such as the areas with heavy rainfall and the soil with the wet condition (Vakhshoori & Zare, 2016; Sonawane & Bhagat, 2017) as depicted in Fig. 2.6(a).

Landuse/landcover (LULC) plays a crucial role in the stability of the topographic gradient. Forest controls continuous water flow and regular infiltration; on the other hand, cropland and agricultural land affect slope stability owing to saturation of covered soil (Devkota et al., 2013; Regmi et al., 2014). In the present study, LULC map has been modified from GlobCover (2009) land cover map and classified into eight major LULC classes such as irrigated croplands, rainfed croplands, croplands/vegetation, vegetation/croplands, closed to open broadleaved evergreen or semi-deciduous forest, closed broadleaved deciduous forest, open broadleaved deciduous forest, closed needleleaved evergreen forest, closed to open mixed broadleaved and needleleaved forest, mosaic forest-shrubland/grassland, mosaic grassland/forest-shrubland, closed to open shrubland, closed to open grassland, artificial areas and water bodies as depicted in Fig. 2.6(b).

The road segmentation is a significant spot of anthropogenic instability and numerous road construction activities such as quarrying of soil, striking of additional load, vertical segmentation of slopes, dam construction and vegetation removal may lead to some tensional cracks due to an increase in stress on the back of the slope

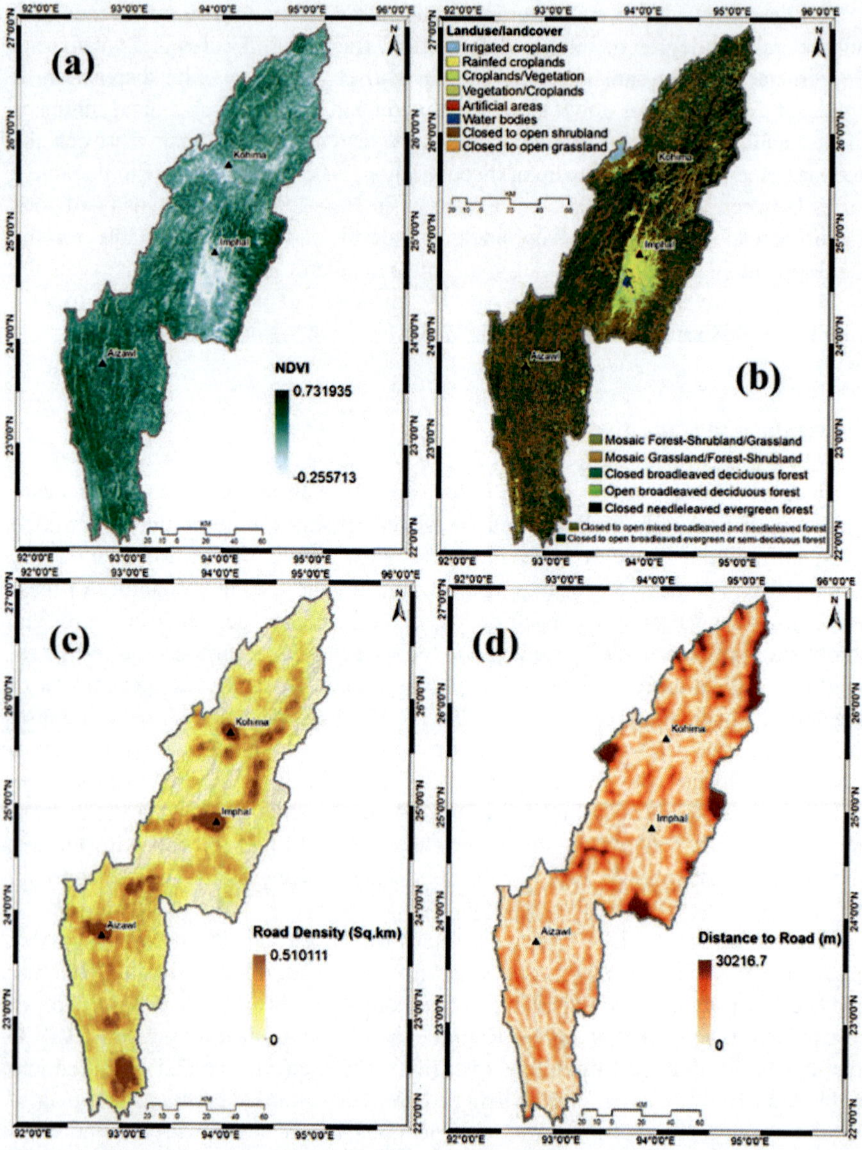

Fig. 2.6 GIS maps exhibit environmental causative factors of the terrain, viz. (**a**) Normalized Differences Vegetation Index (NDVI), (**b**) Landuse/landcover (LULC), (**c**) Road Density and (**d**) Distance to Road

which frequently serves as a cradle for the occurrence of landslides (Saadatkhah et al., 2014). In order to determine the effect of the road on the stability of slopes, various buffer zones are created on the path of the road from which the road density varies from 0 to 0.510 km^2 as shown in Fig. 2.6(c) and distance to road map has been prepared on GIS platform and classified between 0 to 30216.7 m as depicted in Fig. 2.6(d). It is also observed that there is a significant correlation between the extent of landslides and the distance to the regional road system.

Triggering Causative Factors

In the Eastern Boundary Block of Northeast India, both the earthquake and rainfall are the major triggering factors responsible for landslide. The entire terrain falls under the active Himalayan seismogenic zone. The terrain has been affected by more than 800 μ to large-magnitude earthquakes (Nath et al., 2017). It is observed that the epicentre proximity plays a major role in the occurrence of a co-seismic landslide in this region. An epicenter proximity map has been generated using the Euclidean distance tool in the GIS platform, as shown in Fig. 2.7(a) that exhibits a variation from 0 to 34,096.1 m.

Surface consistent peak ground acceleration (PGA) is another important factor responsible for triggering co-seismic landslides in the region. The intensity of ground-shaking calculated from the maximum acceleration representing the seismic hazard level in the region is a severe factor in the co-seismic landslide. The PGA with a 10% probability of exceedance in 50 years with a return period of 475 years has been adopted from Nath and Thingbaijam (2012), which shows a variation of PGA 0.558–0.944 g as depicted in Fig. 2.7(b). In general, there is greater vibration near the epicentre, where many of the co-seismic landslides generally occur.

Rainfall is another triggering factor for the occurrence of a landslide because it controls the water content in the soil. The amount of precipitation and the number of landslides is directly proportional to the altitude of the terrain (Sabatakakis et al., 2013). The average annual rainfall distribution map has been prepared using inverse distance weighted (IDW) interpolation technique by considering the rainfall data of

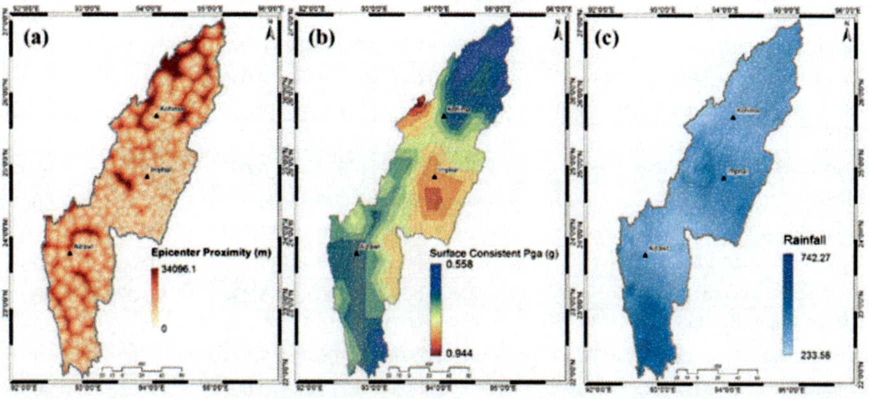

Fig. 2.7 GIS raster maps exhibit the triggering causative factors of the terrain: (**a**) Epicentre Proximity, (**b**) Surface Consistent Peak Ground Acceleration (PGA) with 10% probability of exceedance in 50 years with a return period of 475 years and (**c**) Average Annual Rainfall

the last 60 years (1950–2010), with a variation up to 233.58–742.27 mm/year as shown in Fig. 2.7(c). As per the data, it is observed that in the southern part of the region, the intensity of rainfall is high, while in the northern part, there is a low intensity of the precipitation.

Landslide Susceptibility Zonation Mapping by Using Multivariate Regression

In the present study, processing of data and factors has been carried out on the GIS Platform, while the statistical analysis by Logistic Regression has been performed using the Statistical Package for Social Sciences (SPSS). In the first step, 19 causative factors have been used for the independent variables dataset. All the causative factors have been exported to comma-separated values (CSV) format and imported into the statistical platform to achieve the regression coefficients, as shown in Table 2.2. The landslide inventory training dataset which is represented as a

Table 2.2 Causative factors and their coefficients derived through multivariate statistics-based logistic regression

Causative factors	β	S.E.	Wald	df	Sig.	Exp (β)
Landuse/landcover (LULC)	0.211	0.228	0.853	1	0.356	1.234
Surface geology	4.420	0.443	99.708	1	0.000	83.076
Rainfall (mm/year)	−0.004	0.000	105.028	1	0.000	0.996
PGA (g)	−6.029	0.518	135.398	1	0.000	0.002
Slope angle (degree)	0.052	0.007	58.140	1	0.000	1.053
Distance to drainage (m)	0.000	0.000	5.530	1	0.019	1.000
Terrain ruggedness index (TRI)	−0.006	0.012	0.246	1	0.620	0.994
Road density (km^2)	1.445	0.460	9.864	1	0.002	4.240
Profile curvature	0.029	0.031	0.872	1	0.350	1.029
Plan curvature	−0.031	0.031	0.943	1	0.332	0.970
Normalized difference vegetation Index (NDVI)	−3.007	0.432	48.456	1	0.000	0.049
Lineament density (km^2)	1.601	0.623	6.602	1	0.010	4.959
Elevation (m)	0.000	0.000	6.070	1	0.014	1.000
Epicentre proximity (m)	0.000	0.000	2.464	1	0.117	1.000
Distance to road (m)	0.000	0.000	137.183	1	0.000	1.000
Distance to lineament (m)	0.000	0.000	3.238	1	0.072	1.000
Drainage density (km^2)	−0.902	0.326	7.644	1	0.006	0.406
Aspect	−0.001	0.000	8.321	1	0.004	0.999
Landform	−0.005	0.115	0.002	1	0.964	0.995
Constant	2.309	0.650	12.613	1	0.000	10.061

S.E. standard error, *Wald* Wald chi-square values, *df* degree of freedom, *Sig.* significance, *Exp(β)* exponentiated coefficient

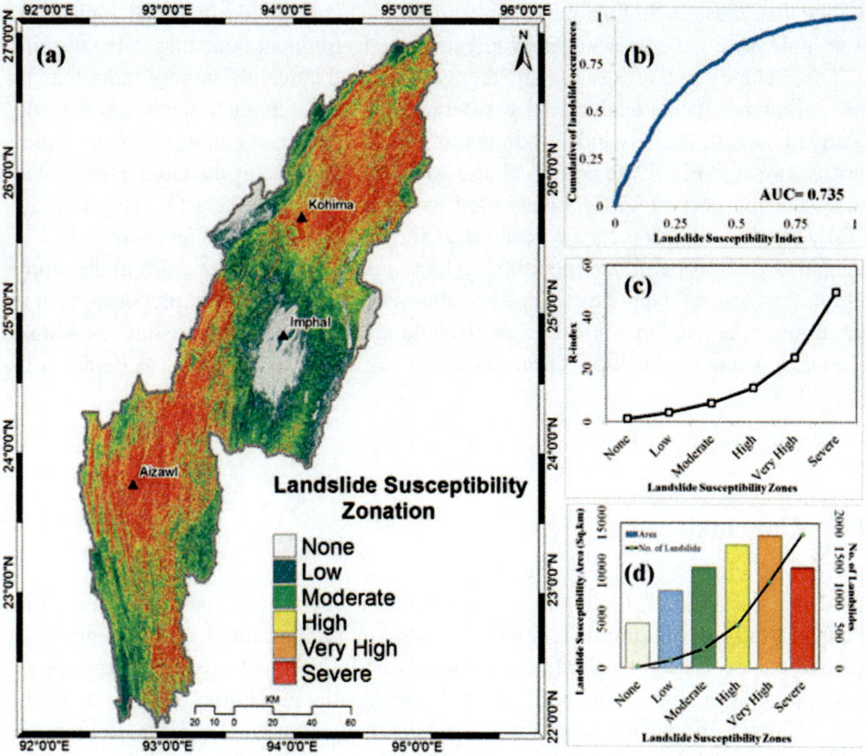

Fig. 2.8 (**a**) Landslide susceptibility map of the Eastern Boundary Block comprising of the States of Nagaland, Manipur and Mizoram of Northeast India classified into six susceptible zones; (**b**) ROC curve for the susceptibility map; graphs show the cumulative landslide occurrences versus landslide susceptibility index with AUC = 0.735, (**c**) relative landslide density index (R-index) of the LR-derived landslide map and (**d**) graph exhibiting the number of landslides (line graph) and landslide susceptibility zones area (bar graph)

dependent variable has been extracted after rasterizing polygons and then coding the cells falling in the landslide areas. In the LR analysis, cells could get attributes providing information on the presence or absence of the landslide phenomena within 30 × 30 m pixels. After integrating the coefficients, using both Eqs. (2.1) and (2.2), the proneness to landslide has been spatially distributed in the region. The logistic regression method is used to develop a landslide susceptibility zonation map for indicating locations where the probabilities of landslide occurrence varies from 0.0 to 1.0. Numbers closer to 1 indicate the probability of landslide occurrences. The susceptibility index map has been classified as none susceptible covering around 4622 km^2, low susceptibility covering around 8000 km^2, moderate susceptibility covering around 10,422 km^2, high susceptibility covering around 12,708 km^2, very high susceptibility covering around 13,702 km^2 and severe susceptibility covering around 10,415 km^2 as shown in Fig. 2.8(a). The urban, semi-urban centres and other lifeline facilities, including major transportation facilities like national highways, state highways, have been severely affected due to frequent landslide in the terrain.

The accuracy statistics of logistic regression-based landslide susceptibility map in the terrain has been evaluated by receiver operating characteristics (ROC) and relative landslide density index (r-index) in the area versus the number of landslides plot as shown in Fig. 2.8(c and d). The accuracy of the model is developed in the area under the roc curve (AUC) values vary from 0.5 to 1.0 in the ROC statistics. Its ability to reliability predicts the occurrence and non-occurrence of an event, which is defined by the probabilistic model of AUC to apply the ROC statistics in the study region for testing landslide inventory dataset which has been prepared to use randomly selected landslide events from landslide and non-landslide locations. The AUC value of the ROC curve for LR is estimated to be 0.735, as depicted in Fig. 2.8(b).

2.4 Conclusion

Landslide susceptibility mapping is considered the most important step onward in landslide hazard mitigation and management in the terrain. Logistic regression method has been used to describe the spatial distribution of landslide susceptibility zonation on a medium scale of 1:50,000 with a spatial resolution of 30 m $\times$ 30 m. The accuracy assessment of multivariate statistics-based Logistic Regression (LR) technique is established by using receiver operating characteristics, relative landslide density index and landslide area versus number of landslide graph and by matching the susceptibility map with testing inventory dataset. Logistic regression method indicates that land use/land cover, surface geology, slope angle, road density and lineament density establish the most important conditioning factors in causing mass movement in the terrain. The susceptibility zonation map provides information that led to a major improvement in the understanding of the causes for densely scattering of landslides in the terrain. The landslide susceptibility zonation map prepare in the present study can help as a reference for city planners, architects and geotechnical engineers in land use planning and slope management.

References

Alkhasawneh, M. S., Ngah, U. K., Tay, L. T., Isa, M., Ashidi, N., & Al-batah, M. S. (2013). Determination of important topographic factors for landslide mapping analysis using MLP network. *The Scientific World Journal, 2013*, 1–13. https://doi.org/10.1155/2013/415023

Ayalew, L., Yamagishi, H., & Ugawa, N. (2004). Landslide susceptibility mapping using GIS-based weighted linear combination, the case in Tsugawa area of Agano River, Niigata Prefecture, Japan. *Landslides, 1*(1), 73–81. https://doi.org/10.1007/s10346-003-0006-9

Baeza, C., & Corominas, J. (2001). Assessment of shallow landslide susceptibility by means of multivariate statistical techniques. *Earth Surface Processes and Landforms: The Journal of the British Geomorphological Research Group, 26*(12), 1251–1263. https://doi.org/10.1002/esp.263

Bai, S., Lü, G., Wang, J., Zhou, P., & Ding, L. (2011). GIS-based rare events logistic regression for landslide-susceptibility mapping of Lianyungang, China. *Environmental Earth Sciences, 62*(1), 139–149.

Balamurugan, G., Ramesh, V., & Touthang, M. (2016). Landslide susceptibility zonation mapping using frequency ratio and fuzzy gamma operator models in part of NH-39, Manipur, India. *Natural Hazards, 84*(1), 465–488.

Barman, B. K., & Srinivasa Rao, K. (2019). Landslide hazard susceptibility mapping of upper Tuirial watershed, Mizoram using Remote Sensing and GIS techniques. *International Journal of Research and Analytical Reviews, 6*(1), 1624–1630.

Bhuvan Portal developed by Indian Space Research Organization. https://bhuvan-app1.nrsc.gov.in/disaster/disaster.php?id=landslide_monitor

Bui, D. T., Pradhan, B., Lofman, O., Revhaug, I., & Dick, O. B. (2012). Spatial prediction of landslide hazards in HoaBinh province (Vietnam): A comparative assessment of the efficacy of evidential belief functions and fuzzy logic models. *Catena, 96*, 28–40. https://doi.org/10.1016/j.catena.2012.04.001

Catani, F., Casagli, N., Ermini, L., Righini, G., & Menduni, G. (2005). Landslide hazard and risk mapping at catchment scale in the Arno River basin. *Landslides, 2*(4), 329–342. https://doi.org/10.1007/s10346-005-0021-0

Chen, W., Xie, X., Wang, J., Pradhan, B., Hong, H., Bui, D. T., & Ma, J. (2017). A comparative study of logistic model tree, random forest, and classification and regression tree models for spatial prediction of landslide susceptibility. *Catena, 151*, 147–160. https://doi.org/10.1016/j.catena.2016.11.032

CRED; EMDAT: http://www.emdat.be

Dai, F. C., & Lee, C. F. (2002). Landslide characteristics and slope instability modeling using GIS, Lantau Island, Hong Kong. *Geomorphology, 42*(3–4), 213–228. https://doi.org/10.1016/S0169-555X(01)00087-3

Dasgupta, S., Sural, B., Harendranath, L., Mazumadar, K., Sanyal, S., Roy, A., Das, L. K., Misra, P. S., & Gupta, H. (2000). *Seismotectonic atlas of India and its environs*. Geological Survey of India, Calcutta, India.

Devkota, K. C., Regmi, A. D., Pourghasemi, H. R., Yoshida, K., Pradhan, B., Ryu, I. C., & Althuwaynee, O. F. (2013). Landslide susceptibility mapping using certainty factor, index of entropy and logistic regression models in GIS and their comparison at Mugling–Narayanghat road section in Nepal Himalaya. *Natural Hazards, 65*(1), 135–165. https://doi.org/10.1007/s11069-012-0347-6

Dikau, R. (1988). Case studies in the development of derived geomorphic maps. *Geologisches Jahrbuch A, 104*, 329–338.

Dilley, M., Chen, R. S., Deichmann, U., Lerner-Lam, A. L., & Arnold, M. et al. (2005). *Natural disaster hotspots – a global risk analysis*. Report of the International Bank for Reconstruction and Development/The World Bank and Columbia University: 132.

Ding, Q., Chen, W., & Hong, H. (2017). Application of frequency ratio, weights of evidence and evidential belief function models in landslide susceptibility mapping. *Geocarto International, 32*(6), 619–639.

Fawcett, T. (2006). An introduction to ROC analysis. *Pattern Recognition Letters, 27*(8), 861–874.

Froude, M. J., & Petley, D. N. (2018). Global fatal landslide occurrence from 2004 to 2016. *Natural Hazards and Earth System Sciences, 18*(8), 2161–2181.

Glade, T. (2003). Landslide occurrence as a response to land use change: A review of evidence from New Zealand. *Catena, 51*(3–4), 297–314. https://doi.org/10.1016/S0341-8162(02)00170-4

GlobCover. (2009). *Global land cover map*. http://due.esrin.esa.int/page_globcover.php

Guzzetti, F., Carrara, A., Cardinali, M., & Reichenbach, P. (1999). Landslide hazard evaluation: A review of current techniques and their application in a multi-scale study, Central Italy. *Geomorphology, 31*(1), 181–216. https://doi.org/10.1016/S0169-555X(99)00078-1

Guzzetti, F., Reichenbach, P., Ardizzone, F., Cardinali, M., & Galli, M. (2006). Estimating the quality of landslide susceptibility models. *Geomorphology, 81*(1–2), 166–184.

Hadmoko, D. S., Lavigne, F., & Samodra, G. (2017). Application of a semiquantitative and GIS-based statistical model to landslide susceptibility zonation in Kayangan Catchment, Java, Indonesia. *Natural Hazards, 87*(1), 437–468.

Huang, M. H., Fielding, E. J., Liang, C., Milillo, P., Bekaert, D., Dreger, D., & Salzer, J. (2017). Coseismic deformation and triggered landslides of the 2016 Mw 6.2 Amatrice earthquake in Italy. *Geophysical Research Letters, 44*(3), 1266–1274. https://doi.org/10.1002/2016GL071687

Jenness, J. (2006). Topographic Position Index (tpi_jen.avx) extension for ArcView 3.x, v. 1.3 a. Jenness Enterprises.

Khatsu, P., & Van Westen, C. J. (2005, November). Urban multi-hazard risk analysis using GIS and remote sensing: A case study from Kohima Town, Nagaland, India. In *Proceedings of the 26th Asian Conference on Remote Sensing*, pp. 7–11.

Laldintluanga, E. H., Lalbiakmawia, F., & Lalbiaknungi, E. R. (2016). International Journal of Engineering Sciences& Research Technology Landslide Hazard Zonation along State Highway between Aizawl City and Aibawk Town, Mizoram, India using Geospatial Techniques.

Lallianthanga, R. K., & Lalbiakmawia, F. (2013a). Landslide Hazard zonation of Aizawl district, Mizoram, India using remote sensing and GIS techniques. *International Journal of Remote Sensing & Geoscience, 2*(4), 14–22.

Lallianthanga, R. K., & Lalbiakmawia, F. (2013b). Micro-level landslide hazard zonation of Saitual Town, Mizoram, India Using Remote Sensing and GIS Techniques. *International Journal of Engineering Sciences & Research Technology, 2*(9), 2531–2546.

Lallianthanga, R. K., & Lalbiakmawia, F. (2014). Landslide susceptibility zonation of Kolasib District, Mizoram, India using remote sensing and GIS techniques. *International Journal of Engineering Sciences & Research Technology, 3*(3), 1402–1410.

Lallianthanga, R. K., & Laltanpuia, Z. D. (2014). Landslide hazard zonation mapping of Hnahthial Town, Mizoram, India. *Using Remote Sensing & GIS.*

Lallianthanga, R. K., Lalbiakmawia, F., & Lalramchuana, F. (2013). Landslide hazard zonation of Mamit town, Mizoram, India using remote sensing and GIS techniques. *International Journal of Geology, Earth & Environmental Sciences, 3*(1), 184–194.

Lee, S., & Pradhan, B. (2007). Landslide hazard mapping at Selangor, Malaysia using frequency ratio and logistic regression models. *Landslides, 4*(1), 33–41.

Mancini, F., Ceppi, C., & Ritrovato, G. (2010).GIS and statistical analysis for landslide susceptibility mapping in the Daunia area (Italy).

Mathew, J., Jha, V. K., & Rawat, G. S. (2009). Landslide susceptibility zonation mapping and its validation in part of Garhwal Lesser Himalaya, India, using binary logistic regression analysis and receiver operating characteristic curve method. *Landslides, 6*(1), 17–26.

Nandi, A., & Shakoor, A. (2010). A GIS-based landslide susceptibility evaluation using bivariate and multivariate statistical analyses. *Engineering Geology, 110*(1–2), 11–20.

Nath, S. K., & Thingbaijam, K. K. S. (2012). Probabilistic seismic hazard assessment of India. *Seismological Research Letters, 83*(1), 135–149.

Nath, S. K., Mandal, S., Adhikari, M. D., & Maiti, S. K. (2017). A unified earthquake catalogue for South Asia covering the period 1900–2014. *Natural Hazards, 85*(3), 1787–1810. https://doi.org/10.1007/s11069-016-2665-6

Nefeslioglu, H. A., Duman, T. Y., & Durmaz, S. (2008). Landslide susceptibility mapping for a part of tectonic Kelkit Valley (Eastern Black Sea region of Turkey). *Geomorphology, 94*(3–4), 401–418. https://doi.org/10.1016/j.geomorph.2006.10.036

Ohlmacher, G. C., & Davis, J. C. (2003). Using multiple logistic regression and GIS technology to predict landslide hazard in northeast Kansas, USA. *Engineering Geology, 69*(3–4), 331–343. https://doi.org/10.1016/S0013-7952(03)00069-3

Pachuau, L. (2019). Zonation of landslide susceptibility and risk assessment in Serchhip town, Mizoram. *Journal of the Indian Society of Remote Sensing, 47*(9), 1587–1597.

Park, S. H., Goo, J. M., & Jo, C. H. (2004). Receiver operating characteristic (ROC) curve: Practical review for radiologists. *Korean Journal of Radiology, 5*(1), 11–18. https://doi.org/10.3348/kjr.2004.5.1.11

Pathak, D. (2016). Knowledge based landslide susceptibility mapping in the Himalayas. *Geoenvironmental Disasters, 3*(1), 8.

Peduzzi, P. (2010). Landslides and vegetation cover in the 2005 North Pakistan earthquake: A GIS and statistical quantitative approach. *Natural Hazards and Earth System Sciences, 10*, 623–640. https://doi.org/10.5194/nhess-10-623-2010

Petley, D. (2012). Global patterns of loss of life from landslides. *Geology, 40*(10), 927–930.

Pourghasemi, H. R., Pradhan, B., & Gokceoglu, C. (2012). Application of fuzzy logic and analytical hierarchy process (AHP) to landslide susceptibility mapping at Haraz Watershed, Iran. *Natural Hazards, 63*(2), 965–996. https://doi.org/10.1007/s11069-012-0217-2

Pradhan, A. M. S., & Kim, Y. T. (2014). Relative effect method of landslide susceptibility zonation in weathered granite soil: A case study in Deokjeok-ri Creek, South Korea. *Natural Hazards, 72*(2), 1189–1217. https://doi.org/10.1007/s11069-014-1065-z

Regmi, A. D., Devkota, K. C., Yoshida, K., Pradhan, B., Pourghasemi, H. R., Kumamoto, T., & Akgun, A. (2014). Application of frequency ratio, statistical index, and weights-of-evidence models and their comparison in landslide susceptibility mapping in Central Nepal Himalaya. *Arabian Journal of Geosciences, 7*(2), 725–742. https://doi.org/10.1007/s12517-012-0807-z

Reichenbach, P., Rossi, M., Malamud, B. D., Mihir, M., & Guzzetti, F. (2018). A review of statistically-based landslide susceptibility models. *Earth-Science Reviews, 180*, 60–91.

Roy, J., Saha, S., Arabameri, A., Blaschke, T., & Bui, D. T. (2019). A novel ensemble approach for landslide susceptibility mapping (LSM) in Darjeeling and Kalimpong districts, West Bengal, India. *Remote Sensing, 11*(23), 2866.

Saadatkhah, N., Kassim, A., & Lee, L. M. (2014). Qualitative and quantitative landslide susceptibility assessments in Hulu Kelang area, Malaysia. *EJGE, 19*(47), 545–563.

Sabatakakis, N., Koukis, G., Vassiliades, E., & Lainas, S. (2013). Landslide susceptibility zonation in Greece. *Natural Hazards, 65*(1), 523–543. https://doi.org/10.1007/s11069-012-0381-4

Sema, H. V., Guru, B., & Veerappan, R. (2017). Fuzzy gamma operator model for preparing landslide susceptibility zonation mapping in parts of Kohima Town, Nagaland, India. *Modeling Earth Systems and Environment, 3*(2), 499–514.

Shahabi, H., Khezri, S., Ahmad, B. B., & Hashim, M. (2014). Landslide susceptibility mapping at central Zab Basin, Iran: A comparison between analytical hierarchy process, frequency ratio and logistic regression models. *Catena, 115*, 55–70. https://doi.org/10.1016/j.catena.2013.11.014

Singh, C. D., Behera, K. K., & Rocky, W. S. (2011). Landslide susceptibility along NH-39 between Karong and Mao, Senapati district, Manipur. *Journal of the Geological Society of India, 78*(6), 559–570.

Sonawane, K., & Bhagat, V. (2017). Improved change detection of forests using Landsat™ and ETM data. *Remote Sensing of Land 1*(1), 18–40. https://doi.org/10.21523/gcj1.17010102

Vakhshoori, V., & Zare, M. (2016). Landslide susceptibility mapping by comparing weight of evidence, fuzzy logic, and frequency ratio methods. *Geomatics Natural Hazards and Risk, 7*(5), 1731–1752. https://doi.org/10.1080/19475705.2016.1144655

Van Beek, L. P. H. (2002). *Assessment of the influence of changes in land use and climate on landslide activity in a Mediterranean environment*. Doctoral dissertation.

Wilkinson, P. L., Anderson, M. G., & Lloyd, D. M. (2002). An integrated hydrological model for rain-induced landslide prediction. *Earth Surface Processes and Landforms: The Journal of the British Geomorphological Research Group, 27*(12), 1285–1297. https://doi.org/10.1002/esp.409

Wilson, J. P., & Gallant, J. C. (2000). *Terrain analysis: Principles and applications*. Wiley.

Youssef, A. M., Pradhan, B., Jebur, M. N., & El-Harbi, H. M. (2015). Landslide susceptibility mapping using ensemble bivariate and multivariate statistical models in Fayfa area, Saudi Arabia. *Environmental Earth Sciences, 73*(7), 3745–3761.

Chapter 3
Social Vulnerability of Arsenic Contaminated Groundwater in the Context of Ganga-Brahmaputra-Meghna Basin: A Critical Review

Satabdi Biswas, Satiprasad Sahoo, and Anupam Debsarkar

Abstract The most alarming part of inorganic arsenic contamination is its silent killing ability which has an adverse impact on human society. Anthropogenic activities trigger threat from bio-physical to social vulnerability. The Ganga-Meghna-Brahmaputra (GMB) basin has been the worst sufferer for the last four decades. This review paper tries to focus on the impacts and consequences of arsenic calamity, assessment of the risk through Geographical Information System (GIS) and a feasible way-out involving rain water harvesting (RWH) with special reference to India. Arsenic poisoning creates a huge burden for rural people. Identification of various dimensions of arsenic coverage has been a difficult task which made GIS an important tool for the assessment of social vulnerability. However, the rural Indian mass is yet to become fully aware of the severity of the arsenic-related risk. They are still consuming the poison through drinking water for the last four decades without even knowing the treatment protocols. RWH is one of the easy way-outs to combat the situation of the arsenic risk, especially for the poor socio-economic rural households. Thus, to prevent further damages, awareness creation, proper medical care with due endeavours from national and international levels are required.

Keywords Arsenic contamination · Risk assessment · GIS · RWH

S. Biswas
Department of Geography, Mrinalini Datta Mahavidyapith, Kolkata, West Bengal, India

S. Sahoo (✉)
Department of Geography , Jadavpur University, Kolkata, West Bengal, India

A. Debsarkar
Department of Civil Engineering, Jadavpur University, Kolkata, West Bengal, India

P. K. Shit et al. (eds.), *Geospatial Technology for Environmental Hazards*, Advances in Geographic Information Science, https://doi.org/10.1007/978-3-030-75197-5_3

3.1 Introduction

The most significant and alarming aspect of inorganic arsenic (As) toxicity is its silent but for the extensive impact on society. As per WHO (2001), 140 million people in 50 countries have been consuming arsenic-contaminated water above the WHO safe limit of 0.01 ppm. As, the 'king of poison', is a highly toxic element and is naturally found in air, weathering of rocks, soil and groundwater of shallow aquifers (Aurora, 2005; Brinkel et al., 2009). More than 2.5 billion global populations rely on groundwater for drinking purposes. Nearly 108 countries with more than 230 million people have been suffering from As disaster (Shaji et al., 2020). The arsenic-contaminated aquifers are generally found in parts of younger orogenic belts and deltaic plains of the world, such as the western USA, central Mexico, Argentina, the Pannonian Basin, Inner Mongolia, the Indus Valley, the Ganges-Brahmaputra delta and the Mekong River and Red River deltas (Podgorski & Berg, 2020; Ghosh et al., 2020). There is a wide difference between developed and developing countries in terms of the impact of arsenic toxicity. For instance, developed countries like the USA have the same problem in alluvial areas, but the impact of it is not the same compared to the developing countries (Acharyya, 2002). In developing countries like India, Pakistan, Nepal and Bangladesh, unrestrained irrigation with shallow tube-well has been responsible for lowering the water table with arsenic contamination (Alcamo et al., 2000; CGWB, 2013). Under such a complex bio-physical arsenic-contaminated situation, the most vulnerable section, i.e. poor people of Southeast Asia are exposed to high-level risk (Hoque et al., 2019). However, the Ganga-Meghna-Brahmaputra (GBM) basin, i.e. in Bangladesh and India, are the worst cases of recent times (Levien, 2011; Chakraborty et al., 2020). For instance, more than 70 million people in Bangladesh are exposed to the toxicity of arsenic through drinking water (Ghosh et al., 2020). Simultaneously, in India, 20 states (West Bengal, Jharkhand, Bihar, Uttar Pradesh, Assam, Gujarat, Haryana, Madhya Pradesh, Punjab, Arunachal Pradesh, Karnataka, Tamil Nadu, Himachal Pradesh, Telangana, Andhra Pradesh, Orissa, Nagaland, Tripura, Manipur, Chhattisgarh) and 4 union territories (Delhi, Daman and Diu, Puducherry and Jammu and Kashmir) have the largest mass poisoning happened due to arsenic contamination in shallow groundwater (Sharma et al., 2014; CGWB, 2013; Shaji et al., 2020). West Bengal, Jharkhand, Chhattisgarh, Bihar, Uttar Pradesh, Assam, Arunachal Pradesh and Manipur are the major states in India having As contamination at a higher level (>10 μg/L) (Puri et al., 2014). As per BIS (2012), half of the Indian people have been affected by excessive iron and As contamination. However, West Bengal is the worst affected state, which has recently turned into an issue of global concern as a high concentration of arsenic patches are found in 79 blocks of eight districts (CGWB, 2013). It has been estimated that nearly 16.26 million people out of 91.28 million are at high risk in West Bengal (Chatterjee et al., 2009). However, to date, 85% of Indian rural domestic water requirements are fulfilled by groundwater (Suhag, 2016). Arsenic is not a new thing; rather hydro-geochemical evolution revealed that it occurred in the entire Ganga Basin with a spatial variation since the historical past. Unfortunately, the severity increased in such a way that it gradually became a global concern in the last four decades.

Most of the recent studies on As contamination were based on the assessment of groundwater quality or character of the sediment. For example, a study by Shaji et al. (2020) established a correlation between aquifer types with arsenic intoxication based on geological analysis of peninsular India. The result revealed that 90% and 10% of As contamination was found in the unconsolidated alluvial terrain and hard rock terrain, respectively. In the hard rock aquifer states (e.g. Karnataka and Chhattisgarh), As contamination happened due to sulphide mineralization and acid volcanic association. Singh et al. (2020) tried to assess the anthropogenic effect of arsenic and its probable vulnerability based on 171 seasonal groundwater samples collected for 2015–2016 in Darbhanga district, Bihar in the Ganga Flood plain. The result showed that agrochemicals, viz. calcium nitrate, calcium phosphate, As-bearing compounds and bleaching powder, applied over the surface get diluted and mobilized into groundwater by potential monsoon recharge. The pre-monsoon drafting should be regulated to restrict the high As concentration in the groundwater. Another hydrochemical study on STW water in Bangladesh conducted by Edmunds et al. (2015) established that the excessive tapping of groundwater for potable and irrigation purposes is a matter of serious concern and is yet to be understood well by the stakeholders. Richards et al. (2020) collected 273 samples of pre- and post-monsoonal groundwater from 5 to 180 m depth in38 districts of Bihar, India, to assess the harmful effect of geogenic contamination of As, uranium (U), and other elements on human health. The result showed that As, iron (Fe) and manganese (Mn) were positively correlated with each other, and As was inversely correlated with the depth of the aquifer. Saha and Sahu (2016) studied the similarities and differences between the Middle Ganga Plain (MGP) and the Bengal Basin based on the hydrogeological and geochemical assessment of shallow aquifers (8.0 m below the ground). The As contamination was noticed along the River Ganga and other Himalayan tributaries and sub-tributaries, i.e. the Ghaghra, the Gandak, the Kosi and the Mahananda. The rainwater carried organic carbon in the form of clay plugs, increased microbial processes, spread the anoxic front and released As in groundwater through infiltration and percolation by natural recharge in monsoon. They found that the newer alluvium areas of MBP having Pleistocene brownish yellow sediment had low concentration of As in groundwater after assessing the transmissivity of the aquifer. They recommended cement sealing for the wells of middle clay layers to stop downward leakage of As from the top aquifer. Patel et al. (2019) studied sediment samples of monsoon and post-monsoon seasons of the Subarnsiri-Dikrong-Ranganadi River system, Upper Brahmaputra floodplain, India, to assess the effect of leaching of As and fluoride (F^-) by annual flooding events. The results showed that the highest As and Fe were found in the raw sediments of the Ranganadi river that fell sharply in the post-monsoon season. They concluded that the total level of As is not only the prime determiner of groundwater contamination but the local anthropogenic influence also disturbed the fluvial environment. Almost similar observations were made by Das et al. (2018), who assessed the hydrochemical quality of groundwater and sediment samples in the Brahmaputra floodplains (BFP), India. The result found the strongest relationship between As and Fe in the upper BFP followed by the lower and middle BFP. They observed a definite trend of gradual increase in As and Fe due to easier access to shallow aquifer. This is responsible for the increase in non-cancer health risks among the 835 children in the BFP. Chen et al. (2020) also establish a geochemical analysis based on 23 soil samples in Suzhou University, China. After the spatial analysis of the samples, the results found that the Cu and As were slightly contaminated due to the use of chemical fertilizer in agricultural activities. Janardhana Raju (2012) assessed 68 borehole sediment and groundwater samples of the Ganga in Varanasi, India. However, borehole samples of the eastern side of Ganga showed high As and Fe concentration in newer alluvium sediments of the Holocene period. Thus, without knowing the quality of the shallow tube wells (40–70 m) water, the rural people had been using arsenic-contaminated water for drinking and irrigation purposes in the affected nine villages. Though no arsenical skin lesions were noticed in their survey, they concluded that continuous consumption of arsenic contaminated water without necessary precaution would increase the cases of arsenic victims in these villages in the near future.

Majorities of the recent approaches involved an assessment of groundwater quality or lithology-based analysis for assessment of the severity of As. Most of the researchers recommended structural management or government initiatives, alternative water sources, i.e. RWH for restricting As position but yet to emphasize generating awareness of potable water quality, the role of the governance, sharing the threat of As and its probable impact over the poor rural households. In this perspective, a critical review has been presented in this paper on arsenic-related social hazards to clarify a few important key aspects. The aspects of the paper are organized as (1) impact and consequence of arsenic as a bio-physical social hazard, (2) application of GIS techniques for arsenic assessment and (3) significance of water governance with RWH as a mitigation measure for arsenic. This paper also helps to address one of the most widely asked questions, i.e. whether RWH could be a feasible solution to arsenic contamination, especially for Asian countries.

3.2 Materials and Methods

Generally, mixed methods were adopted for the representation of the arsenic risk in the GIS environment. The ultimate goal was to detect the vulnerability of its effects on the population. In some cases, researchers combined quantitative data (spatial and attribute data) with qualitative data (questionnaire survey) for designing the problem formulation, data manipulation, analysis and interpretation (Hassan et al., 2003; Bhatia et al., 2014; Singh & Vedwan, 2015). The quantitative data (primary or published tube-well water data) were combined with another set of quantitative medical data (data about arsenicosis-affected persons). Then other sets of local or regional geostatistics data were transferred into the GIS platform to generate arsenic hazard-related thematic maps for spatial analysis (Hassan et al., 2003). The geostatistical method helped to detect the accuracy of the present status of the calamity and also predicted future needs. This further helped to take certain measures by the local authorities to manage the present groundwater scenario. Numerous researches used field variables (hydro-geochemical analysis of laboratory tested tube-well water samples, sediments samples, soil samples, lithology characters data) and secondary data (climate data, published groundwater data) to develop models by assigning weightage and ratings to correlate and assess the arsenic vulnerability by GIS overlay analysis (Hassan et al., 2003; Puri et al., 2014; Ghosh et al., 2020; Podgorski & Berg, 2020). Qualitative analysis was done based on hydrochemical properties of groundwater, and the results obtained were used to develop statistical model, viz. kriging, Thiessen polygon, DRASTIC Model, Random Forest model, Logistic Regression model and Hydrostratigraphic model for prediction of the sensitivity of the coverage (Puri et al., 2014; Mehrotra et al., 2016; Ghosh et al., 2020). Statistical modeling was carried out based on satellite data along with climate and arsenic data to highlight the As risk (Podgorski & Berg, 2020; Chakraborty et al., 2020).

3.3 Results and Discussion

Impact and Consequences of Arsenic as Bio-physical Social Hazard

More than 90% of arsenic pollution is geogenic (Ghosh et al., 2020). In the GMB plain, the alluvial soil has high agricultural potentiality. The uncontrolled increase in population and associated demands, i.e. food grains, irrigation, industrial and drinking purpose implied heavy drafting which worsens the condition. Easy drafting of groundwater, subsidized electricity, less expense of boring of tube wells (TWs), availability of loans from a bank and dependency of 'Boro paddy' (winter rice) along with other crops are responsible for unethical major share drafting of groundwater common in West Bengal and Bangladesh (Banerjee & Jatav, 2017; Hoque et al., 2019). Being agrarian-based countries, most of the poor farmers are forced to cultivate throughout the year. Hence, intensive use of groundwater is required for cultivation. Wealthy farmers are drafting groundwater 24 h a day (UNDP, 2006). However, the farmers are engaged in a race of drilling deeper and deeper with bore wells and fall into money owing traps (Banerjee & Jatav, 2017). In this way, the atmospheric oxygen enters into groundwater while drafting the same. As a result, the groundwater level declines continuously with increasing As contamination (Acharyya, 2002). Further, the application of chemical fertilizers and insecticides throughout the year especially the increase of winter rice cultivation is causing qualitative and quantitative degradation of water. Glendenning (2009) mentioned that in India, water extraction by shallow tube-wells benefited farmers in the short term, but this practice makes their land barren in the long run. The unplanned urbanization influenced the significant changes in land use/land cover (LULC). These LULC changes decreased the natural phenomena, i.e. water bodies, vegetation and wetland largely with a consequent increase of impervious area. On the other side, it is responsible for less infiltration capacity of water (Patra et al., 2018). Thus faster urbanization deteriorates groundwater recharge (DDWS, 2011). However, the groundwater has been still treated as an individual property. Thus, over-exploitation with contamination has taken place in several areas (DDWS, 2011). About 85% of the supply of drinking water in India is based on groundwater (DDWS, 2011). The unethical easy electrified pumps in shallow tube-well (STW) tap the groundwater for drinking purposes (Puri et al., 2014). A huge number of private shallow hand pump tube-wells were installed by individual households on their premises recently, the main source of rural drinking water with an excess of As and iron. Rural households tap this contaminated drinking water, and millions of people have been suffering from qualitative water stress varying from health to social (Chatterjee et al., 2009; Bhatia et al., 2014; Hoque et al., 2019). Since 2004, villages of West Bengal had experienced the huge growth of STW with the coming up of the National Policy of As Mitigation (NPAM) (Banerjee & Jatav, 2017). This further increased the gravity of the arsenic problem (CGWB, 2007). However, various Government Departments monitor observatory wells and do acknowledge these issues. In this monitoring process, they often identify and put a red cross on those contaminated tube-wells. Despite knowing the fact, local people are compelled to consume this contaminated water as they do not have any other alternative. Thus the situation was turned into a major socio-ecological risk (Hoque et al., 2019; Biswas et al., 2020). For instance, in coastal Bangladesh, the privately funded tube wells increased four times, compared to 78% in 2018, whereas the population grew by 4% during the past decade (Hoque et al., 2019). The groundwater is safe in terms of waterborne diseases, but at the same time, it gradually lifted the As to the top layer of the surface (Hassan et al., 2003). These hand-pumps covered millions of rural people at risk of arsenic contamination, which was one of the key health problems of the twenty-first century (Bhatia et al., 2014). Two major additional risks were there. One of which was the dietary habit of dependence on rice, both in India and Bangladesh. Gilbert-Diamond et al. (2011) indicated that rice consumption had been one of the main reasons for harmful arsenic exposures to the human body, based on a study of 229 pregnant women. Women in this sample survey had exposures to arsenic via their home tap water concentration ranging from $\leq$0.07 μg/L to nearly 100 μg/L and rice-based food habit (Gilbert-Diamond et al., 2011). They documented a positive relationship between rice consumption and urinary arsenic excretion. Secondly, the mushrooming of private bottled mineral water industries resulted in significant investment and exposure in shallow tube wells in countries like India and Bangladesh. The market entrepreneurship of bottled water industries packaged water over the last few decades without maintaining the WHO standard. Rural innocent people trust private bottled water without any concern and consume the same blindly. This is the recent hazards for an increase in arsenic vulnerability (Dave, 2016). This made a tremendous impact on groundwater particularly in peri-urban villages (Banerjee & Jatav, 2017). Thus, the quality of drinking water did matter a lot for several serious public health problems. The arsenic from underground shallow aquifer silently enters into an ecosystem and responsible for the increase of various diseases (Dave, 2016; Sharma et al., 2014; Bhattacharya et al., 2019; Sinha & Prasad, 2020). The high level of As exposure for a prolonged time has been associated with serious public health hazards, e.g. skin disorders; cardiovascular diseases; respiratory problems; complications of gastrointestinal tract; liver, kidney and bladder disorders reproductive failure neurotoxicity; and even cancer. Thus arsenic contaminated groundwater is grabbing our society slowly. WHO (2018) also noted that 1 in every 100 additional cancer deaths could be caused when people are exposed to contaminated drinking water. However, the rural households are the worst sufferers for the geogenicarsenic poisoning due to lack of proper diet (Mukherjee et al., 2009; Bartram et al., 2015). Arsenicosis cases are worsened by malnutrition, poor socio-economic status, illiteracy, food habit and prolong consumption of arsenic-contaminated water (PCI, 2007). Having no other option, poor people became silent victim of this hardship, and they were forced to enter into a vicious circle,where people further dip into acute poverty generation after generation. However, As contamination in groundwater was first reported in Chandigarh, India (Datta, 2015), and the second case was reported in West Bengal (Garat et al., 1984). Arsenic contamination of drinking water has also several indirect effects apart from the clinical symptom such as economic and social impacts, i.e. human productivity loss, treatment cost, human capital loss and many more (Bhattacharya et al., 2019). Another study by Brinkel et al. (2009) stated that arsenicosis patients face dual problems. Firstly, they face serious social impact such as marriage-related problems, problems of unemployment, social instability, social discrimination and rejection by community and sometimes from own families. Secondly, the patients suffer from mental retardation and disabilities like physical, cognitive, psychological and speech impairments. An arsenic-affected person is still being treated as a social stigma. It had a cascading effect that involved the entire family of the victims (Bhattacharya et al., 2019). Unaffected people were generally scared about arsenic victims. They usually avoided and isolate arsenic patients from the society (Hassan et al., 2003). Thus the mental health condition of arsenicosis sufferers resulted in deep depression where consequences might end up with social loss (Ghosh et al., 2020) (Fig. 3.1).

Application of GIS Techniques for Arsenic Assessment

In absence of any immediate mitigation action or awareness campaign, the people of the study area will be affected by mass poisoning and exposure to fatal diseases (Hassan et al., 2003). Thus, the assessment of arsenic vulnerability is necessary for understanding the risk. The extension of risks may be multidimensional such as economic loss, health loss, loss of opportunities or decline in the socio-economic status of their livelihood (Singh & Vedwan, 2015). Assessment of these risks along with the toxicity of human is a complicated task, not possible to measure directly. Thus various proxy data were used to capture the magnitude of this harm (Hassan et al., 2003). The general trend is to develop various thematic layers of maps of local aquifers to know the severity (DDWS, 2011). The groundwater is available, but it is often the case that it is contaminated by As and Fe pollutants. To make mass awareness about this social risk, need to depend on low-cost and timeless technology. Normally, the spatial study of arsenic-related groundwater needs thousands of water samples with time-series data, and that is also a time-consuming and expensive process. Due to the lack of adequate testing facilities, it is nearly impossible to

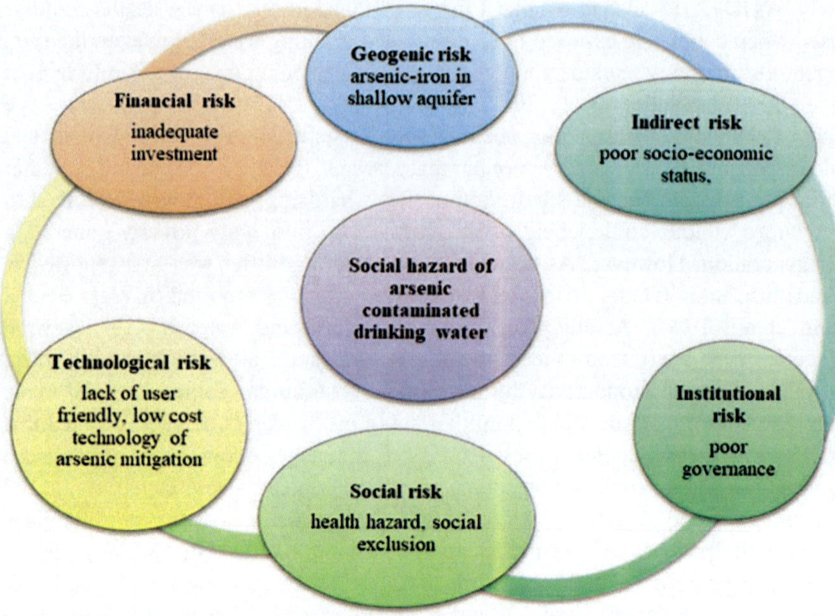

Fig. 3.1 Risks associated with arsenic-contaminated groundwater

collect huge data from vast rural regions. Here comes the importance of GIS for spatial mapping (Ghosh et al., 2020). GIS can be done by the limited sample points, thus easy to develop an As concentration zoning map, which ultimately helps to identify the risk zones exposed to As contamination and then quantify the magnitude of contamination (Ghosh et al., 2020). Thus, GIS is used to identify arsenic risk, an extension of exposures, and spatial zoning of risk, and it further helps to assess the vulnerability, find out the weakness of adaptive capacities and, overall, helps to resolve the issues. The GIS as a tool helps to clarify the situations and plan accordingly to the benefit of the common people (Mehrotra et al., 2016). Thus, like the absence of the As detection sensors, GIS-based techniques can be used for assessment at the block or ward level (Puri et al., 2014). Such GIS techniques help mark out the magnitude of vulnerability based on various proxies, i.e. environmental, bio-physical, natural, lithologies, aquifer characters, hydrology, water samples and other samples (rice, urine, water, soil, diseases even anthropological factors, etc.) (Chakraborty et al., 2020). Ultimately the GIS could easily enable to estimate the visual representation of the population at risk over any specified area. GIS is also helpful to give an alert of groundwater by producing the groundwater vulnerability mapping. Thus it helps to frame out planning by the government or authorities to take decisions to manage safe drinking water supply (Puri et al., 2014). Prediction-based vulnerability maps could be developed in a faster way by GIS tool which is invaluable for the planning of the highly arsenic severity areas (Singh & Vedwan, 2015). The recent trend of using GIS mapping in arsenic contamination usually adopted a regional scale-based assessment and highlighted the spatial heterogeneity. Charlet et al. (2007) found a high level of As presence in wells and that caused the widespread poisoning through drinking water in Chakdaha, West Bengal. They developed a spatial distribution map of arsenic with a depth of the aquifer. Results showed vertical transfer of (As) arsenic happened from shallow to deep wells (150 m) during the dry season. However, those deep wells were marked as high quality of drinking water free from arsenic previously. That was a real threat to the local population. Different methods have been used to evaluate the arsenic-related groundwater mishap. Such the random forest machine-learning model was analysed based on geospatial environmental parameters including 50,000 global data points of groundwater arsenic concentration and household groundwater usages data. The arsenic prediction model estimated that 94–220 million people were exposed to high concentrations. Among them, the majority (94%) of the people were residents in Asian countries (Podgorski & Berg, 2020), whereas Puri et al. (2014) used the DRASTIC model to access the groundwater vulnerability in Bardhaman district, West Bengal, India. The results showed that the study area was severely affected by As. A mixed approach was taken by Chakraborti et al. (2018). They include people's perceptional data on the risk of As, i.e. the opinion of the presence and functionality of government, interpersonal trust, and trust in institutional working along with As water data. Based on the above-mentioned data, they developed a GIS thematic map to capture the underlining adaptive capacity of the exposed communities. Another block-level mixed study was done by Chakraborty et al. (2020), who adopted a 'hybrid multi-modeling approach' based on both hydro-stratigraphic parameters (aquifer characters, geology, geomorphology) and anthropogenic parameters in 25 districts of the transboundary area of the Ganges River delta shared between India and Bangladesh. A high-resolution regional-scale hydro-stratigraphic model of the aquifer system was developed with the help of 2883 geo-referenced borehole lithologies. The result showed that 19 districts were fallen under the category of more than 25% of high As-hazard zones, while 7 districts (28%) were exposed to more than 75% extent of severity. Total 30.3 million people of the Ganges River delta were exposed to a high level of As-concentration (>10 μg/L) through drinking water. Another study by Bhatia et al. (2014) was conducted on 21children having age group of 5–10 years in a marginalized village community of Khaptolain Bihar, India. They assessed the geo-chemical analysis with health impacts using GIS over lay thematic maps. A contour map for arsenic was developed on the basis of drinking water samples from 20 private shallow (15–35 m) hand pump. The result showed that 57% and 25% of the tapping aquifers were responsible for more than 200 and 397 ppb arsenic concentrations, respectively. The children of the study village were under high risk of getting cancer with continued exposure. A study by Ghosh et al. (2020) was based on empirical methodology with interpolation approach ('Thiessen polygon and Kriging'). They also developed blockwise arsenic contamination map based on seasonal well data from the period 2006–2008 through GIS platform in North 24 Parganas, West Bengal, India. The result revealed that the unaffected blocks of 2006 gradually became significantly affected in the year of 2008. As testing was performed by field test kit (FTK) from 522 villages in Bahraich, India. They developed GIS-based arsenic-contaminated zoning map based on the tested drinking water samples. The zoning map showed 45.71% of high probability of arsenic concentration in Kaisarganj and Jarwal blocks and few newer villages and Gram Panchayats. It also identified to the coverage of the problem (Mehrotra et al., 2016). Another mixed approached was adopted by Singh and Vedwan (2015) in Bihar, India. They used biophysical, socio-economic and demographic factors for identifying the community's arsenic coverage by 'composite vulnerability index' and statistical analysis by PCA. They generated unique set of visual maps like social, bio-physical and environmental vulnerability map based on more than 30,000 published tested results of As concentrations in drinking water of Bihar, India. This helped to mark community vulnerability profiles for drawing of arsenic-contaminated groundwater. This study revealed that nine million population was found to be at risk in five districts of the state including Vaishali, Samastipur, Darbhanga, Purnia and Katihar. The highest As-affected population (63%) was found in Khagaria district, covering a total number of three blocks. The results implied that demographically and socio-economically, poor people were highly vulnerable as poor health would be more sensitive to arsenicosis-related health problems. The literacy rate was found to be a very important component to reduce total vulnerability. The expensive As mitigation plan would network in any of these districts. The literacy rate, female literacy rate, rural population, population growth, population below the poverty line, scheduled caste population, infant mortality rate, incidence of flood and drought, concentration of arsenic, fluoride and nitrate, lithology and the lithology-related geological formation were found to be most important variables by PCA for composite vulnerability of the communities. A geostatistical approach was used by Hassan et al. (2003) for detection of arsenic magnitude in a covering area of 17.26 km^2 in Bangladesh and West Bengal, India. A mixed data of spatial and questionnaire survey was used for development of arsenic spatial distribution maps through Kriging method. They collected data related to the arsenicosis patients, their water-consuming habits and period of exposure to the arsenic-contaminated drinking water. The result showed that about 95.50% (358) of tube wells were contaminated with <0.003 to 0.600 mg/L arsenic concentration out of 375 tube wells after analysis by spatial interpolation method. The west and northeast of the study area were more contaminated than southwest part.

The magnitude of arsenic assessment becomes more visual with GIS whatever the initially adopted methods were showing Table 3.1.

Significance of Water Governance with RWH as a Mitigation Measure for Arsenic

After detecting the arsenic contamination in the 1980s, the West Bengal Government had taken three types of mitigation measures, i.e.:

Table 3.1 Arsenic-contaminated groundwater-related studies using GIS

Author	Study area and sample size	Methods	Remarks
Hassan et al. (2003)	West Bengal (India) and Bangladesh, 11,000 inhabitant	Quantitative mixed approach, i.e. special information, 375 TW water data, households questionnaire survey (HHS), health-related data	GIS-based data processing for identifying the magnitude of As problem regions such as arsenic isoline map, three-dimensional arsenic concentration map, special As magnitude map Found 200 TWs (53.33%) highly affected (<0.003 mg/L), overall, 95.50% TWs were contaminated by As
Puri et al. (2014)	Bardhaman district, West Bengal, India, sample size not mentioned	Qualitative method i.e., DRASTIC model based on giving weightage to hydrogeological parameters (depth of the aquifer, recharge, aquifer media, soil, topography, etc.)	Developed thematic maps, i.e. groundwater vulnerability assessment, results revealed study area severely affected by high As concentration
Bhatia et al. (2014)	Bihar, India, 916 population	A mixed approach, 20 TW water samples, questionnaire survey done with the mothers of affected children	GIS overlay map focused on calculating cancer risk, hazard index Found 1.6 ha of area (57%) under extremely high (<200 ppb) As toxicity, predicted 5–10 years of children would be under highly vulnerable of getting affected with cancer
Singh and Vedwan (2015)	15 districts of Bihar, India. Nine million population	The quantitative approach developed a composite vulnerability index based on biophysical, socio-economic, demographic and perception-based information, also used PCA	Overlay maps for quantifying the arsenic vulnerability maps (i.e. As risk zoning, As a vulnerable population, environmental vulnerability, socio-economic demographic map, health, geological, composite vulnerability maps), the first component of PCA was the adaptive capacity of HHS Found 4.4 million of the population in 5 districts with <1000 μg/L As concentration

(continued)

Table 3.1 (continued)

Author	Study area and sample size	Methods	Remarks
Mehrotra et al. (2016)	Uttar Pradesh, India. Sample size not mentioned	The qualitative approach 30,216 hand pumps water samples were tested by field kits applying a blanket approach	Village-level thematic mapping by GIS arsenic affected villages, As zoning map Found 52.06% and 10.86% samples had 10–40 μg/L and >50 μg/L As contamination respectively in two blocks
Hoque et al. (2019)	Coastal Bangladesh, population 58,000.	Mixed method qualitative, i.e. TW water sample, depth of TW, log data and quantitative, i.e. HHS interview data, used PCA	Various thematic mapping for risk assessment applied, i.e. (aquifer quality, water supply infrastructure, sources of drinking water) Found water risk increased as of salinity, flooding and uncontrolled growth of private shallow TWs
Ghosh et al. (2020)	North 24 Parganas, India, sample size not mentioned	The empirical methodology used was based on water samples of TWs, Thiessen polygon and Kriging; a future trend was assessed by statistical analysis	Developed spatial distribution of As concentration map with the help of Thiessen polygon and Kriging, As zoning map of different seasons, predicted future trend through a regression model Found previously unaffected block significantly affected within 2 years. The regression model predicted after 10 years another 2 or 3 blocks were affected if the same trend would be followed
Podgorski and Berg (2020)	Global scale	The mixed approach used, random forest machine-learning model based on geospatial environmental parameters (including 50,000 groundwater samples, arsenic data, depth data) and country based domestic groundwater users HHS data	Developed a global prediction map of arsenic exceeding 10 μg/L having less than 100 m depth Found 220 million global people were affected, among them 94% were Asian Found high probability of affected zones were in central, south and Southeast Asia including Indus and GMB plains

(continued)

Table 3.1 (continued)

Author	Study area and sample size	Methods	Remarks
Chakraborty et al. (2020)	Shared India and Bangladesh with 110 million people	Used 'hybrid multi-modeling' approach based on hydro-stratigraphic, geomorphology, anthropology, bio-geo-chemical factors, statistical methods and artificial intelligence, Random Forest, Boosted regression tree and Logistic Regression	The transboundary model was developed for the prediction of As hazard map, aquifer connectivity map, aquifer permeability map, silt, and clay thickness map. Found 30.3 million people were exposed to As, predicted probability of As by population (in high hazard zone 76% districts) predicted in high As hazard zones

1. As a short-term measure, the installation of numerous hand pump tube wells and ring wells were made into deeper aquifers.
2. Arsenic treatment units were attached with existing tube wells. Introduction of arsenic removal plants in existing groundwater-based piped water supply schemes were made for the medium-term measures. The large diameter deeper aquifer tube wells were fixed for existing or new groundwater based piped water supply.
3. There was12 mega piped-surface-water-supply schemes (PWSS) from the Ganga and another 338-groundwater based piped water supply schemes (PWSS) for arsenic affected areas were still running.

Modified Sujapur-Sadipur model, Gobordanga Model, community-based arsenic removal plant for multi-village water supply schemes in North 24 Parganas, West Bengal were also functional (DDWS, 2011; Rana, 2013; Bhattacharya et al., 2019). Total 28394.56 km^2 alluvial zone in the Ganga-Brahmaputra-Meghna plain of Indian part was recommended for artificial recharge (CGWB, 2013). People did not understand the 'safe water' issues; thus they did not argue for having safe water as their right (Dave, 2016). Another important reason was that mass awareness regarding arsenic-related health effects was very low (Sinha & Prasad, 2020). As an example in the technological park at Baruipur, West Bengal, most of the arsenic removal plants were found abandoned because of a poor sense of belonging, willingness, awareness, etc. (PCI, 2007). However, the public standpoint on deep tube well schemes did not yield the desired health impacts as households continued drinking contaminated groundwater by private hand pumps or wells situated at their premises (PHED, 2018). In recent past, several studies were dealt with groundwater contamination issues with various arsenic corrective technologies mainly removal of arsenic from groundwater using filters, exploration deeper or alternative aquifers, treatment of the aquifer itself, installation of nano-filter, dilution method by artificial recharge to groundwater and conjunctive use of RWH and groundwater, etc. (CPCB,

2008; Singh et al., 2014; Abhinav et al., 2017; Zakhar et al., 2018; Shaji et al., 2020). However, the quality of the water remained a matter of concern. Sarkar et al. (2010) pointed out that in 1997, Bengal Engineering Science and Technology and Lehigh University in the USA introduced community-level arsenic removal units and fortunately that decreased arsenic contamination in affected villages of south Bengal. The high cost of maintenance and installation were the main reasons for failure after the detection of the calamity of near about 32 years. To date, the government is yet to provide a simple low-cost technology to encounter arsenic, while presently available and widely used arsenic mitigation filter system further damages soil, surface water and the local ecosystem due to unplanned open disposal (Dey et al., 2014). The above-mentioned reasons further indicated a weak policy implementation system of arsenic contamination, and it certainly did not trickle down to the marginalized poor rural people (Bhowmick et al., 2018). The water-related governance issues yet to be addressed adequately (MWR, 2012). Thus it was necessary to reform strict water policy and simplistic user-friendly technology, involving mass in the arsenic mitigation plan, giving the incentive to encourage the community to manage their local aquifers were some effective measures (Ghosh et al., 2020). Effective regulations were required at national and international levels to prevent future arsenic-based health hazards (Sinha & Prasad, 2020). Even with the mapping of local aquifer, water quality information was not shared with the communities which was another big issue (Dave, 2016). It was necessary to change the present habit from tapping groundwater to switching to new sources like RWH to avoid deadly diseases and arrest the declining of ground water levels in over-exploited areas (Dey et al., 2014). RWH also helped to dilute the aquifers (DDWS, 2011). The needs of safe drinking water should be the first priority for any water supply scheme. The main goals of the government were to ensure water security to reduce arsenic related diseases. Thus, the Government of West Bengal fixed a target to supply surface water of 70 liters per capita per day (lpcd) in rural areas through Vision 2020 for giving the priority on the arsenic contaminated areas (PHED, 2018). Conjunctive use of RWH and safe groundwater was recommended to provide safe drinking water. However, the main focus was to move away from high cost arsenic treatment plan towards RWH (DDWS, 2011). The RWH was not so popular even in India and Bangladesh, and a wide gap existed between legislation of the rule/policies and its implementation. Bhattacharya et al. (2019) stated that the government needed to immediately take few measures under different programs of Government of India, viz. MSDP/BaDP, Panchayat Raj, etc., as a part of arsenic mitigation plan. For instance, the Government of West Bengal formed Task Force (2005–2006) as long-term measures. However, the Kolkata and Haldia industrial areas already depleted the piezometric surface. Under such circumstances, arsenic pollution had put a huge burden on rural households. It happened in many ways. First, the unpopular Government programme, lack of proper institutional efforts, ignorance of socio-economic and cultural background of affected communities, lack of integrated approach in water planning and lack of funds were some loopholes. These all posed fundamental questions on the failure of governmental policy about RWH (Banerjee & Jatav, 2017; Patra et al., 2018; PHED, 2018; Ghosh et al., 2020). But the supply of safe water entirely depended on governance, political will, investment, international cooperation awareness and acceptance (Asare, 2004; Fakult, 2013; Wutich & Brewis, 2014).

The concept of RS and GIS had been an effective recent tool for selecting sites and planning suitable artificial recharging structures to get the best result (Sharma et al., 2014; Jha et al., 2014; Mahmood & Hossain, 2017). For instance, Jha et al. (2014) identified 83 sites for artificial recharge by farm pond and percolation tanks after the development of land use/land cover (LULC) map. They used the IRS-P6 LISS-IV image and DEM to assess the potential recharge sites for domestic, livestock, irrigation and groundwater recharge purposes. Verma (2016) also identified artificial recharge sites based on GIS and GPS mapping in the different watershed areas of Chhattisgarh, India. The artificial recharge structures, i.e. percolation tank and check dams, were recommended, and these ultimately helped to reduce the demand for the main water supply (groundwater) and also helped to save water, energy and money. Gomez and Teixeira (2017) found 40% of drinking water demand could be saved by RWH. Mahmood and Hossain (2017) developed model-based GIS maps to determine the feasibility of domestic rainwater harvesting (DRH) in the South Asian region. They recommended DRH for Bangladesh, Sri Lanka, from the Himalayan range to North-Eastern, Central, Eastern and coastal parts of Southern India. It could satisfy yearly 7.5 lpcd for drinking and cooking purposes. A study by Mukherjee et al. (2015) investigated the in-situ groundwater storage for Punjab, Haryana, Uttar Pradesh, Bihar and West Bengal, India, after using RS and GIS from the year 2005–2013. They developed potential groundwater recharge zones using GIS-based hydro-geological databases (trends of precipitation, usable groundwater, groundwater storage with analysis of satellite imagery) with a mathematical model for artificial recharge, whereas CPWD (2002) set up an SPG project at Dwarka, New Delhi, India, having a total area of 47.5 ha mainly to augment groundwater in urban areas. Another scheme for the artificial recharge of groundwater was set up at Faridabad in Hariyana to restrict the decline of groundwater along with awareness generation among the common people for proper management of RWH. A study was performed on the LULC changes with multicriteria analysis (MCA) to identify a potential zone for the construction of water reservoirs for RWH with recharge (Kar et al., 2020). However, UN-Habitat (2015) mentioned that in India, RWH was a part of state policy. In Chennai, Delhi and Bangalore, RWH was made mandatory. In the state of West Bengal, RWH was made mandatory for the construction of new buildings in urban areas under West Bengal Municipal (Building) Rules (GWB, 2007) but not in rural areas. Meanwhile, CGWB (2011) started that the identification of artificial recharge areas with suitable structures in different states including West Bengal in its VIII plan (1992–1997) to handle the contaminated groundwater situation. Further, the central groundwater board started the experiment of artificial recharge of the aquifer in 1970. From the eighth plan, rooftop RWH was introduced in West Bengal and during the tenth plan; a demonstration was implemented through NGOs in 100 rural schools. Various techniques of RWH such as injection well with rooftop RWH was proposed for potential recharge of the confined aquifer areas which indirectly enhanced the quality of groundwater (Sekar & Randhir, 2007; CGWB, 2013). For instance, installation of an arsenic removal plant (ARP) with RWH at the Sujapur, West Bengal, found that the concentration of arsenic and iron was reduced from 0.2 mg/L to 0.03 mg/L and 1.7 mg/L to 0.25 mg/L, respectively (Studer & Liniger, 2013). The central groundwater board (CGWB, 2007) of India already started the RWH with artificial recharge schemes into permeable strata of shallow depth. On average yearly, 5500 to 34.50 lakh cubic meter of runoff was successfully recharged in selected areas of Arunachal Pradesh, Assam, Bihar, Chandigarh, Gujrat, Hariyana, Jharkhand and Uttar Pradesh by RWH. Similarly, in Himachal Pradesh, Karnataka, Punjab, Tamilnadu, Madhya Pradesh, Orissa and Maharashtra, a substantial amount of water was recharged through a combination of percolation tanks, watershed structures along with recharge wells and rooftop rainwater harvesting (RRH). RWH with artificial recharge by a combination of farm pond, Nala bunds, and sub-surface dykes were capable to rise by 0.15 m of water table successfully in the districts of Purulia, Bankura and Birbhum, in the western part of West Bengal, though Rajasthan had a prestigious historical background to practice RWH. Consequently, RWH was predominant in India, Jordan and other parts of Asia, Italy and South Africa since the late 1900s (Debusk & Hunt, 2012). Providing arsenic-free safe drinking water to huge rural masses had been a major challenge to the government, planners and executors (Ghosh et al., 2020). Md Rana (2013) mentioned that the modified Sujapur-Sadipur model, Gobordanga model and community-based arsenic removal plant for multi-village water supply schemes for North 24 Parganas of West Bengal Government were very much time taking and high-cost projects. CGWB (2013) suggested that RRH could supply domestic water requirements, not for water-scarce areas but water excess areas. The RWH had been user-friendly, low cost and an alternative technology for arsenic mitigation. The CGWB estimated up to 70% of groundwater recharge would be possible with 100 m^2 roof in the regions having 780 mm of average monsoon rainfall. Another 55–275m^3 harvested water could be managed to meet the demand of a five-member family for 100 to 500 days. Except for Darjeeling, other districts did not implement RWH successfully in West Bengal (PHED, 2018). However, before the use of harvested water in domestic sector solar technology, rapid sand filters, Filtration Absorption Disinfection (FAD) purification system for turbidity, COD, DOC, *E. coli* and total coliforms were the best options for maintaining the microbiological free harvested water supply (Helmreich & Horn, 2009; Naddeo et al., 2013). RWH was a local solution for proving safe drinking water by a pond and supplying it with piped water after purification. Ponds had been a good source of drinking water provided proper planning and motivation of local people (Adham et al., 2018). One of the main objectives of the National Water Mission of India (Government of India, 2013) was to publish comprehensive water quality-quantity data in the public domain. Other objectives were to publish the area-wise impact assessment of vulnerability and promote the concern of the state's water conservation, augmentation, and preservation policies among citizens. It also highlighted traditional water conservation systems (i.e. RWH), mandatory water audits, incentivizing by giving awards for water conservation, efficient use and efficient irrigation practice (MWR, 2012).

Again Ministry of Drinking Water and Sanitation of India (MDWS, 2013) decided to cover at least 55% of rural households under piped water schemes by 2017. Among them, 35% would have water connections within households thereby decreasing public tap water use by less than 20% and hand pump used by less than 45% to mitigate arsenic calamity. Thus, PHED (2013) suggested RWH and prepared a document namely 'Master Plan for Artificial Recharge to Ground Water-2013' to provide information about area-specific artificial recharge techniques. This plan would construct a 1.11 crore artificial recharge structure including rooftop RWH in urban and rural areas. It was estimated that 85.565 MCM of surplus runoff would be harnessed to augment groundwater. Water security was determined by the complex interactions among water resources, governance systems, infrastructure development and user needs (Hoque et al., 2019). Bhowmick et al. (2018) described that there had been still a lack of well-planned effort for the mitigation of arsenic risk. Thus, good governance should introduce transparent information about water resources (quality and quantity), careful water management with RWH, include law enforcement to prevent social isolation, maintain equity and social justice for affected people. Existing Acts might have to be modified accordingly to get a mass response from the affected community to build good institutional coordination for the underprivileged people (Brinkel et al., 2009; DDWS, 2011; Hoque et al., 2019). By sensitizing the local, arranging rehabilitation programs to generate employment opportunities and providing accurate health information and supportive counseling process, it could be possible to overcome the bio-physical socio crisis of arsenic.

3.4 Recommendation for Sustainable Groundwater Management

It is indeed a great challenge ahead of the policymakers and engineers to ensure the supply of arsenic safe water understanding the magnitude of the arsenic poisoning. Many alternatives are safer, but none of them is suitable or affordable compared to shallow tube wells. At the same time, the traditional water sources, i.e. large-diameter dug wells, ponds and lakes, are also getting polluted. Considering the magnitude and extent of the problem, following recommendations are suggested.

- To make people aware of the calamity, the extent of arsenic poisoning is required to be estimated through vulnerability mapping based on GIS technology. This may include the present scenario of the quality as well as quantity wise status of the groundwater. Sharing of these thematic maps and models with the affected community would be the simplest but effective form of regular public awareness campaign.
- The land use/land cover (LULC) maps should be published on the local level to restrict any drastic change of the same and augment the natural recharge.

- The launch of the proper treatment protocols involving locals and ensuring the availability of adequate medical personnel are essential to support the mental and physical health of the affected population. Alternate job opportunities are to be ensured to restrict any excessive drafting of groundwater.
- RRH should be made mandatory even for small rooftops in rural areas with low-cost water treatment technology (slow sand filter). Poor people should be given incentives to install the RRH. Social acceptance of the scheme is essential to make it a success.
- As a long-term resolution, a global strategy is to be formulated to eradicate the hazard of arsenic.

3.5 Conclusions

The outcome of the review emphasizes the careful analysis of arsenic risk identification, causes and consequences. There is still a gap in the awareness of the impact of biophysical aspects of arsenic contamination as affected people have not been taking RWH seriously. It is not even popular among educated people. The project of real investment for RWH depends on the acceptance of common people. RRH may be of much help as an alternative solution to the arsenic calamity in rural areas. To overcome the arsenic-iron risk, the responsible authorities should arrange a water safety plan including an awareness programme. Also, periodical updating of mapping with qualitative data on local aquifers and providing proper know-how of the RWH scheme to the people remains a great challenge. The present study, in that sense, would help the policymakers and concerned authorities to delineate a proper guideline for remediation of the arsenic-related problem.

References

Abhinav, Navin, S., Shankar, P., Kumar, R., Ali, M., Verma, S., et al. (2017). Arsenic contamination of groundwater and human blood in Vaishali District of Bihar, India: Health Hazards. *International Journal of Advanced Research, 5*(8), 2092–2100. https://doi.org/10.21474/ijar01/5276

Acharyya, S. K. (2002). Arsenic contamination in groundwater affecting major parts of southern West Bengal and parts of western Chhattisgarh: Source and mobilization process. *Current Science, 82*(6), 740–744.

Adham, A., Sayl, K. N., Abed, R., Abdeladhim, M. A., Wesseling, J. G., Riksen, M., et al. (2018). A GIS-based approach for identifying potential sites for harvesting rainwater in the Western Desert of Iraq. *International Soil and Water Conservation Research, 6*(4), 297–304. https://doi.org/10.1016/j.iswcr.2018.07.003

Alcamo, J., T. Henrichs, and T. Rösch. 2000. World's Water in 2025. In Kassel World Water Series 2. Retrieved from http://www.usf.uni-kassel.de/usf/archiv/dokumente/kwws/kwws.2.pdf

Asare, Y. B. O. (2004). *Household Water Security and Water Demand in the Volta Basin of Ghana*. PhD Thesis, pp. 1–19. Retrieved July 17, 2016, from http://www.zef.de/fileadmin/template/Glowa/Downloads/thesis_osei.pdf

Aurora, P. E. (2005). Focus arsenic: A global poison. *Environmental Health Perspectives, 113*, 1–16. https://doi.org/10.1289/ehp.113-a378

Banerjee, P., & Jatav, M. (2017). *Thematic paper on urbanisation and ground water use: Socio-ecomonic system mapping*. South Asia Consortium for Interdisciplinary Water Resources Studies (SaciWATERs). Retrieved from http://saciwaters.org/shiftinggrounds/pdfs/Thematic%20report%20on%20urbanisation%20and%20ground%20water%20use.pdf

Bartram, J., Baum, R., Coclanis, P. A., Gute, D. M., Kay, D., Mcfadyen, S., et al. (2015). Bradley classification of disease transmission routes for water-related hazards. In *Routledge handbook of water and health*. Routledge. https://doi.org/10.4324/9781315693606.ch03

Bhatia, S., Balamurugan, G., & Baranwal, A. (2014). High arsenic contamination in drinking water hand-pumps in Khap Tola, West Champaran, Bihar, India. *Frontiers in Environmental Science, 2*, 1–8. https://doi.org/10.3389/fenvs.2014.00049

Bhattacharya, A., Sarani, C., Kumar Roy, A., Karthik, D., Singh, A., Kumari, S., et al. (2019). An analysis of Arsenic Contamination in the Groundwater of India, Bangladesh and Nepal with a Special Focus on the Stabilisation of Arsenic-Laden Sludge from Arsenic Filters. *Electronic Journal of Geotechnical Engineering, 24*(Bund.01), 1–34.

Bhowmick, S., Pramanik, S., Singh, P., Mondal, P., Chatterjee, D., & Nriagu, J. (2018). Arsenic in groundwater of West Bengal, India: A review of human health risks and assessment of possible intervention options. *Science of the Total Environment, 612*, 148–169. https://doi.org/10.1016/j.scitotenv.2017.08.216

BIS. (2012). *Indian Standards Drinking Water Specifications IS 10500:2012*. Bureau of Indian Standard, Indian Standards Drinking Water Specifications, 2(May), 11. Retrieved from http://cgwb.gov.in/Documents/WQ-standards.pdf

Biswas, S., Debsarkar, A., & Pal, M. (2020). Water insufficiency, health hazards and rainwater harvesting in North 24 Parganas, West Bengal, India: Results of a socio-economic survey. *Arthaniti: Journal of Economic Theory and Practice*, 097674792096339. https://doi.org/10.1177/0976747920963399

Brinkel, J., Khan, M. H., & Kraemer, A. (2009). A systematic review of arsenic exposure and its social and mental health effects with special reference to Bangladesh. *International Journal of Environmental Research and Public Health, 6*(5), 1609–1619. https://doi.org/10.3390/ijerph6051609

CGWB. (2007). *Manual on artificial recharge of groundwater* (pp. 1–14). Ministry of Water Resource.

CGWB. (2011). *Profile of North 24 Parganas District, West Bengal* (pp. 1–16). Ministry of Water Resource.

CGWB. (2013). *Master plan for artificial recharge to ground water in India*. Retrieved from http://cgwb.gov.in/documents/MasterPlan-2013.pdf

Chakraborti, D., Singh, S. K., Rahman, M. M., Dutta, R. N., Mukherjee, S. C., Pati, S., & Kar, P. B. (2018). Groundwater arsenic contamination in the ganga river basin: A future health danger. *International Journal of Environmental Research and Public Health, 15*(2), 1–19. https://doi.org/10.3390/ijerph15020180

Chakraborty, M., Sarkar, S., Mukherjee, A., Shamsudduha, M., Ahmed, K. M., Bhattacharya, A., & Mitra, A. (2020). Modeling regional-scale groundwater arsenic hazard in the transboundary Ganges River Delta, India and Bangladesh: Infusing physically-based model with machine learning. *Science of the Total Environment, 748*, 141107. https://doi.org/10.1016/j.scitotenv.2020.141107

Charlet, L., Chakraborty, S., Appelo, C. A. J., Roman-Ross, G., Nath, B., Ansari, A. A., et al. (2007). Chemodynamics of an arsenic "hotspot" in a West Bengal aquifer: A field and reactive transport modeling study. *Applied Geochemistry, 22*(7), 1273–1292. https://doi.org/10.1016/j.apgeochem.2006.12.022

Chatterjee, D., Nath, B., Jana, J., Goswami, A., & Chakraborty, S. P. M. (2009). Geochemistry and speciation of solid and aqueous phase arsenic in the Bengal delta plain aquifers. *Bhu-Jal News, 24*(2–3), 59–70.

Chen, S., Wu, C., Hong, S., & Chen, Q. (2020). Assessment, distribution and regional geochemical baseline of heavy metals in soils of densely populated area: A case study. *International Journal of Environmental Research and Public Health, 17*(7). https://doi.org/10.3390/ijerph17072269

CPCB. (2008). Status of groundwater quality in India. In *Groundwater quality series*. Retrieved from https://cpcb.nic.in/openpdffile.php?id=UHVibGljYXRpb25GaWxlLzc3NF8xNTQ0NDI3NjQ3X0dXUVMtMS5wZGY=

CPWD. (2002). *Rainwater harvesting and conservation*. Retrieved from http://cpheeo.gov.in/upload/uploadfiles/files/Rainwater%20Harvesting%20Manual-CPWD.pdf

Das, N., Das, A., Sarma, K. P., & Kumar, M. (2018). Provenance, prevalence and health perspective of co-occurrences of arsenic, fluoride and uranium in the aquifers of the Brahmaputra River floodplain. *Chemosphere, 194*(March), 755–772. https://doi.org/10.1016/j.chemosphere.2017.12.021

Datta, N. (2015). Evaluating impacts of watershed development program on agricultural productivity, income, and livelihood in Bhalki Watershed of Bardhaman District, West Bengal. *World Development, 66*, 443–456.

Dave, S. N. (2016). *Rural drinking water situation: Challenges and opportunities in West Bengal*. (March 2013). Retrieved from www.mdws.gov.in/sites/default/files/annualre-

DDWS. (2011). Strategic Plan—2011-2022 "Ensuring Drinking Water Security In Rural India."

Debusk, K., & Hunt, W. F. (2012). *Rainwater harvesting: a comprehensive review of literature*. Retrieved from http://www.lib.ncsu.edu/resolver/1840.4/8170.

Dey, T. K., Banerjee, P., Bakshi, M., Kar, A., & Ghosh, S. (2014). Groundwater Arsenic Contamination in West Bengal: Current Scenario, Effects and Probable Ways of Mitigation. *International Letters of Natural Sciences, 13*, 45–58. https://doi.org/10.18052/www.scipress.com/ilns.13.45

Edmunds, W. M., Ahmed, K. M., & Whitehead, P. G. (2015). A review of arsenic and its impacts in groundwater of the Ganges-Brahmaputra-Meghna delta, Bangladesh. *Environmental Sciences: Processes and Impacts, 17*(6), 1032–1046. https://doi.org/10.1039/c4em00673a

Fakult, D. (2013, September). Water harvesting for integrated water resources management and sustainable development in Khartoum State.

Garat, R., Chakraborty, A. K., Dey, S. B., & Saha, K. C. (1984). Chronic arsenic poisoning from tube-well water. *Journal of the Indian Medical Association, 82*(1), 34–35.

Ghosh, M., Pal, D. K., & Santra, S. C. (2020). Spatial mapping and modeling of arsenic contamination of groundwater and risk assessment through geospatial interpolation technique. *Environment, Development and Sustainability, 22*(4), 2861–2880. https://doi.org/10.1007/s10668-019-00322-7

Gilbert-Diamond, D., Cottingham, K. L., Gruber, J. F., Punshon, T., Sayarath, V., Gandolfi, A. J., et al. (2011). Rice consumption contributes to arsenic exposure in US women. *Proceedings of the National Academy of Sciences of the United States of America, 108*(51), 20656–20660. https://doi.org/10.1073/pnas.1109127108

Glendenning, C. (2009). Evaluating the impacts of Rainwater Harvesting (RWH). In *A case study Catchment: India*.

Gomez, Y., & Teixeira, L. (2017). Residential rainwater harvesting: Effects of incentive policies and water consumption over economic feasibility. *Resources, Conservation and Recycling, 127* (July), 56–67. https://doi.org/10.1016/j.resconrec.2017.08.015

Government of India. (2013). National rural drinking water programme: Movements towards ensuring people's drinking water security in rural India. In *Ministry for drinking water & sanitation*. Retrieved from http://indiawater.gov.in/imisreports/nrdwpmain.aspx

Government of West Bengal (GWB). (2007). The West Bengal Municipal (Building) Rules, 2007 (Vol. 1993). Retrieved June 25, 2019, from http://wbconsumers.gov.in/writereaddata/ACT%

20&%20RULES/Relevant%20Act%20&%20Rules/The%20WB%20Municipal%20_Building_%20Rules,%202007.pdf

Hassan, M. M., Atkins, P. J., & Dunn, C. E. (2003). The spatial pattern of risk from arsenic poisoning: A Bangladesh case study. *Journal of Environmental Science and Health Part A, 38* (1), 1–24. https://doi.org/10.1081/ESE-120016590

Helmreich, B., & Horn, H. (2009). Opportunities in rainwater harvesting. *Desalination, 248*(1–3), 118–124. https://doi.org/10.1016/j.desal.2008.05.046

Hoque, S. F., Hope, R., Arif, S. T., Akhter, T., Naz, M., & Salehin, M. (2019). A social-ecological analysis of drinking water risks in coastal Bangladesh. *Science of the Total Environment, 679*, 23–34. https://doi.org/10.1016/j.scitotenv.2019.04.359

Janardhana Raju, N. (2012). Arsenic exposure through groundwater in the middle Ganga plain in the Varanasi environs, India: A future threat. *Journal of the Geological Society of India, 79*(3), 302–314. https://doi.org/10.1007/s12594-012-0044-9

Jha, M. K., Chowdary, V. M., Kulkarni, Y., & Mal, B. C. (2014). Rainwater harvesting planning using geospatial techniques and multicriteria decision analysis. *Resources, Conservation and Recycling, 83*, 96–111. https://doi.org/10.1016/j.resconrec.2013.12.003

Kar, S., Sen, E., & Mukherjee, S. (2020). A geospatial technique-based site suitability analysis for construction of water reservoirs in Arsha and Balarampur Blocks, Purulia. *World Water Policy, 6*, 52–88. https://doi.org/10.1002/wwp2.12021

Levien, M. (2011). Special economic zones and accumulation by dispossession in India. *Journal of Agrarian Change, 11*(4), 454–483. https://doi.org/10.1111/j.1471-0366.2011.00329.x

Mahmood, A., & Hossain, F. (2017). Feasibility of managed domestic rainwater harvesting in South Asian rural areas using remote sensing. *Resources, Conservation & Recycling, 125*, 157–168.

MDWS. (2013). Manual for preparation of detailed project report for rural piped water supply schemes.

Mehrotra, A., Mishra, A., Tripathi, R. M., & Shukla, N. (2016). Mapping of arsenic contamination severity in Bahraich district of Ghagra basin, Uttar Pradesh, India. *Geomatics, Natural Hazards and Risk, 7*(1), 101–112. https://doi.org/10.1080/19475705.2013.871354

Mukherjee, S. C., Pati, S., Dutta, R. N., & Saha, K. C. (2009). Groundwater arsenic contamination situation in West-Bengal, India: A nineteen year study. *Bhujal News Quarterly Journal*, 1–40.

Mukherjee, A., Saha, D., Harvey, C. F., Taylor, R. G., Ahmed, K. M., & Bhanja, S. N. (2015). Groundwater systems of the Indian Sub-Continent. *Journal of Hydrology: Regional Studies, 4*, 1–14. https://doi.org/10.1016/j.ejrh.2015.03.005

MWR. (2012). National Water Policies 2012. *Government of India*. https://doi.org/10.1061/(ASCE)1052-3928(1988)114:4(408)

Naddeo, V., Scannapieco, D., & Belgiorno, V. (2013). Enhanced drinking water supply through harvested rainwater treatment. *Journal of Hydrology, 498*, 287–291. https://doi.org/10.1016/j.jhydrol.2013.06.012

Patel, A. K., Das, N., Goswami, R., & Kumar, M. (2019). Arsenic mobility and potential co-leaching of fluoride from the sediments of three tributaries of the Upper Brahmaputra floodplain, Lakhimpur, Assam, India. *Journal of Geochemical Exploration, 203*, 45–58. https://doi.org/10.1016/j.gexplo.2019.04.004

Patra, S., Sahoo, S., Mishra, P., Mahapatra, S. C., Islam, K., Jashimuddin, M., et al. (2018). A geospatial technique-based site suitability analysis for construction of water reservoirs in Arsha and Balarampur Blocks, Purulia. *World Water Policy, 7*(2), 37–47. https://doi.org/10.35940/ijeat.a1011.1291s319

PCI. (2007). Report of the Task Force on Formulating Action Plan for Removal of Arsenic Contamination in West Bengal. In *Government of India Planning Commission, New Delhi* (Vol. 001). Retrieved from http://planningcommission.gov.in/aboutus/committee/wrkgrp11/tf11_arsenics.pdf

PHED. (2013). Government of West Bengal Schedule and Guidelines for Official Use Only (Not for Sale).

PHED. (2018). West Bengal Drinking Water Sector Improvement Project. Retrieved from https://www.adb.org/sites/default/files/linked-documents/49107-006-ea.pdf

Podgorski, J., & Berg, M. (2020). Global threat of arsenic in groundwater. *Science, 368*(6493), 845–850. https://doi.org/10.1126/science.aba1510

Puri, S., Kumar, P., Rana, S., & Kr, B. (2014). GIS-Based Geospatial Mapping of Arsenic Polluted Underground Water in Purbasthali Block in Bardhaman, West Bengal. … and Computing (ICC- …, pp. 544–552. Retrieved from http://www.researchgate.net/publication/268291464_GIS-Based_Geospatial_Mapping_of_Arsenic_Polluted_Underground_Water_in_Purbasthali_Block_in_Bardhaman_West_Bengal

Rana, J. (2013). Arsenic contamination in West Bengal with reference to Malda District MD. *Juel Rana*. (2277).

Richards, L. A., Kumar, A., Shankar, P., Gaurav, A., Ghosh, A., & Polya, D. A. (2020). Distribution and geochemical controls of arsenic and uranium in groundwater-derived drinking water in Bihar, India. *International Journal of Environmental Research and Public Health, 17*(7). https://doi.org/10.3390/ijerph17072500

Saha, D., & Sahu, S. (2016). A decade of investigations on groundwater arsenic contamination in Middle Ganga Plain, India. *Environmental Geochemistry and Health, 38*(2), 315–337. https://doi.org/10.1007/s10653-015-9730-z

Sarkar, S., Greenleaf, J. E., Gupta, A., Ghosh, D., Blaney, L. M., Bandyopadhyay, P., et al. (2010). Evolution of community-based arsenic removal systems in remote villages in West Bengal, India: Assessment of decade-long operation. *Water Research, 44*(19), 5813–5822. https://doi.org/10.1016/j.watres.2010.07.072

Sekar, I., & Randhir, T. O. (2007). Spatial assessment of conjunctive water harvesting potential in watershed systems. *Journal of Hydrology, 334*, 39–52.

Shaji, E., Santosh, M., Sarath, K. V., Prakash, P., Deepchand, V., & Divya, B. V. (2020). Arsenic contamination of groundwater: A global synopsis with focus on the Indian Peninsula. *Geoscience Frontiers, 12*(3), 101079. https://doi.org/10.1016/j.gsf.2020.08.015

Sharma, A. K., Tjell, J. C., Sloth, J. J., & Holm, P. E. (2014). Review of arsenic contamination, exposure through water and food and low cost mitigation options for rural areas. *Applied Geochemistry, 41*, 11–33. https://doi.org/10.1016/j.apgeochem.2013.11.012

Singh, S. K., & Vedwan, N. (2015). Mapping composite vulnerability to groundwater arsenic contamination: an analytical framework and a case study in India. *Natural Hazards, 75*(2), 1883–1908. https://doi.org/10.1007/s11069-014-1402-2

Singh, S. K., Gosh, A. K., Kumar, A., Kislay, K., Kumar, C., Tiwari, R. R., Parwez, R., Kumar, N., & Imam, M. D. (2014). Groundwater arsenic contamination and associated health risks in Bihar, India. *International Journal of Environmental Research, 8*(1), 49–60. https://doi.org/10.22059/ijer.2014.693

Singh, A., Patel, A. K., Deka, J. P., & Kumar, M. (2020). Natural recharge transcends anthropogenic forcing that influences arsenic vulnerability of the quaternary alluviums of the Mid-Gangetic Plain. *NPJ Clean Water, 3*(1), 1–12. https://doi.org/10.1038/s41545-020-0075-5

Sinha, D., & Prasad, P. (2020). Health effects inflicted by chronic low-level arsenic contamination in groundwater: A global public health challenge. *Journal of Applied Toxicology, 40*(1), 87–131. https://doi.org/10.1002/jat.3823

Studer, M., & Liniger, H. (2013). Water harvesting. In *IFAD*. https://doi.org/10.4135/9781446247501.n4126

Suhag, R. (2016). Overview of ground water in India. In Water resources.

UNDP. (2006). *Human development report 2006: Summary*. https://doi.org/10.1016/S1352-0237(02)00387-8.

UN-Habitat. (2015). Rainwater Harvesting and utilisation. In *UN-Habitat*. Retrieved from http://www.unep.or.jp/ietc/publications/urban/urbanenv-2/index.asp

Verma, P. (2016). Study of ground water recharge planning for semi-critical blocks of Rajnandgaon and Kawardha districts of Chhattisgarh plains.

WHO. (2001). Environmental Health Criteria 224 Arsenic And Arsenic Compunds.

WHO. (2018). *Arsenic*. WHO. Retrieved January 12, 2020, from https://www.who.int/news-room/fact-sheets/detail/arsenic

Wutich, A., & Brewis, A. (2014). Food, water, and scarcity. *Current Anthropology, 55*(4), 444–468. https://doi.org/10.1086/677311

Zakhar, R., Derco, J., & Čacho, F. (2018). An overview of main arsenic removal technologies. *Acta Chimica Slovaca, 11*(2), 107–113. https://doi.org/10.2478/acs-2018-0016

Chapter 4
Impact Assessment of Open Cast Mining Activity in ADDA Region, Paschim Barddhaman on Land Surface Temperature

Debduti Dey, Chalantika Laha Salui, and Biplab Biswas

Abstract Open cast mining affects much on the physical environment adjoining to it. This study has been carried out in Asansol Durgapur Development Authority (ADDA) in West Bengal, analyzing such impact of open cast mines on the environmental issues especially the increasing land surface temperature along some elongated tracts by either excavation or dumping. Monitoring of the spatial extension of open cast mines (1999–2019) and analyzing its relationship, the spatial distribution of the land surface temperature in the ADDA region is the core objective of this work. Land surface temperature was examined by using Landsat 4–5 TM and Landsat 8 OLI and TIRS temporal satellite data by single channel algorithm. The outcome of surface temperature is validated with the field information. Supervised image classification technique with maximum likelihood method which was used to show the changes in land use/land cover along with the temporal expansion of opencast mining areas over time. The temporal changes in the distribution of mining sites undergone a detailed correlation study with the spatial variability of these two environmental quality parameters which came out with a strong correlation among them. The result indicates a significant positive correlation between the open cast mines and the spatial distribution of LST ($R^2 = 0.9578$), respectively. This considers the opencast mining activity as a major contributor to the environmental quality status of the adjacent area.

Keyword Open cast mining · Land surface temperature · Spatial variability · Correlation

---

D. Dey · B. Biswas
Department of Geospatial Science, The University of Burdwan, Barddhaman, West Bengal, India

C. L. Salui (✉)
Department of Mining Engineering, Indian Institute of Engineering Science and Technology (IIEST), Shibpur, Howrah, West Bengal, India

P. K. Shit et al. (eds.), *Geospatial Technology for Environmental Hazards*, Advances in Geographic Information Science, https://doi.org/10.1007/978-3-030-75197-5_4

4.1 Introduction

The area under opencast mining has been increasing at a high rate since the last few decades in Asansol Durgapur Development Authority (ADDA), a major mining belt of eastern India, mainly dominated by companies like Eastern coalfields limited (ECL), Bharat Coking Coal Limited (BCCL), and the Indian Iron & Steel Company (IISCO) in both the forms: Opencast as well as underground mining. It is mainly due to changing land use from agro-forestry to mining and its ancillaries such as warehouses, subsidence-prone areas, dumping areas, ash ponds, etc. As a consequence of prolonged mining, the area mostly turned into non-usable with infertile eroded soil.

This study mainly aimed at monitoring the continuous expansion of open cast mines and its correlation with the changing spatial distribution of surface temperature in this area. The analysis was executed on temporal geospatial secondary datasets of 1999, 2011, and 2019. For monitoring the expansion of opencast areas, land use/land cover spatial data is generated from the temporal satellite images of these three phases with proper field checks in 2019. The change detection highlights a prominent cluster wise conversion of vegetative to non-vegetative surface cover, which can be considered as a background behind the increase of surface temperature (Choudhury et al., 2018). The burning of coal and digging up the surface releases increase the land surface temperature as well as air temperature in the surrounding area. Several works were executed on such environmental issues related to mining activities. Choudhury et al. (2018) and Dutta et al. (2018) worked on the land-use dynamics as well as the trend of LST in mining regions. Such environmental depletion leads to the depletion of biodiversity of such areas. These effects are prominent in the eastern coal belt of India (Guha et al., 2012; Gangopadhyay et al., 2005; Kamila & Pal, 2015; Manna & Maiti, 2014). Remote sensing data can provide us with the output for a large coverage area at a time with equally distributed averaging cells; researchers like Karfa and Tah (2019) and Kamila and Pal (2015) used this technology extensively for land-use change monitoring and associated LST change analysis in their study. Mainly thermal band data were used for deriving LST pixels (Choudhury et al., 2018; Fawzi & Jatmiko, 2015).

After all the previous discussions, it is clear that the rising land surface temperature and air temperature caused by changing land use is a serious health issue for living bodies as well as a threat to the ecosystem. While going toward sustainability, it is important to nullify or decrease the level of threat, for which taking a proper plan is an immediate necessity. Now, to configure a proper plan, it is a primary need to understand the scenario of the problem area first; is there any threat or not, if yes, then how much.

As the objective of this research concentrates on the coal mining scenario, the coalfields of Paschim Bardhaman area of West Bengal is selected. This area has an age-long history of coal reserve and is under continuous mining of coal since 1774. In recent time, it covers a total of 6182.37 ha area and is extending in a very fast rate. With this extension, it is very likely to increase the LST.

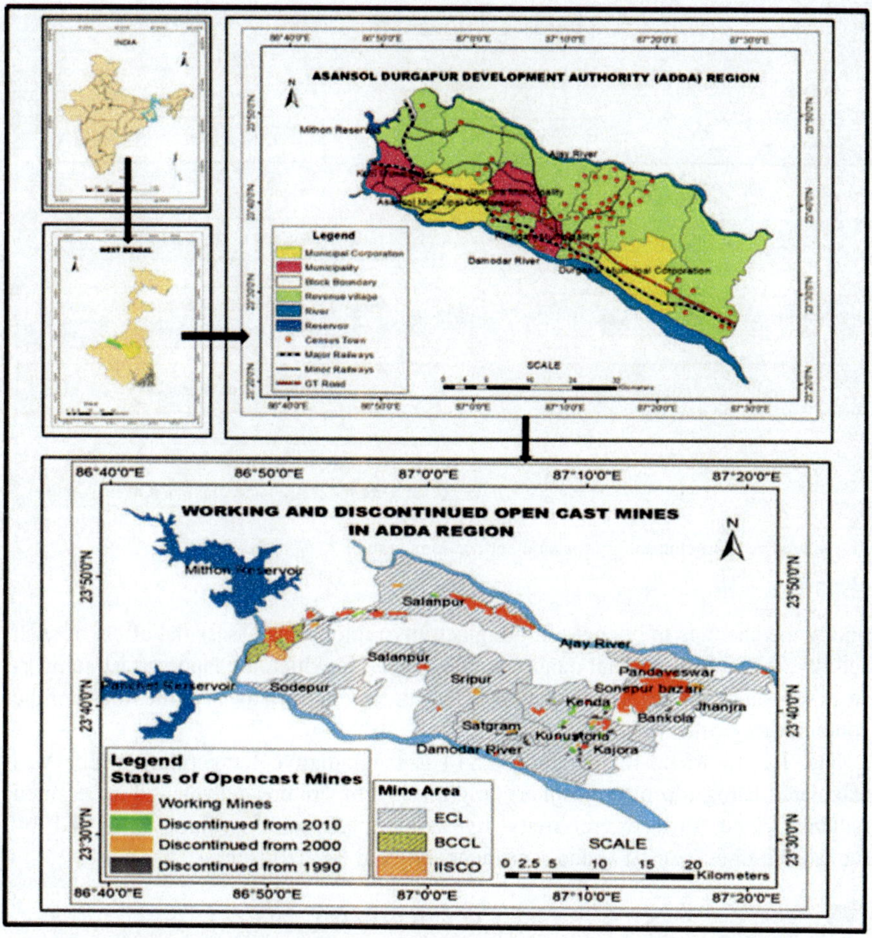

Fig. 4.1 Study area (ADDA region). Source: Choudhury et.al (2018), Eastern Coalfield Limited (ECL) Providing data and author map composition

Accounting those observations, coalfield areas of Paschim Bardhaman, along with the ADDA region, are selected as the study area of this research (Fig. 4.1).

4.2 Materials and Methods

Temporal satellite images (Landsat TM 4–5 for January 1999 and February 2011 and Landsat 8–OLI for January 2019) were used to generate the temporal land use/land cover thematic layers. Supervised classification method was applied with maximum likelihood algorithm for these three temporal images and field validation of the classification was made on 2019 image with an accuracy of 82%. These temporal land use/land cover datasets facilitate in monitoring the changes in land use and

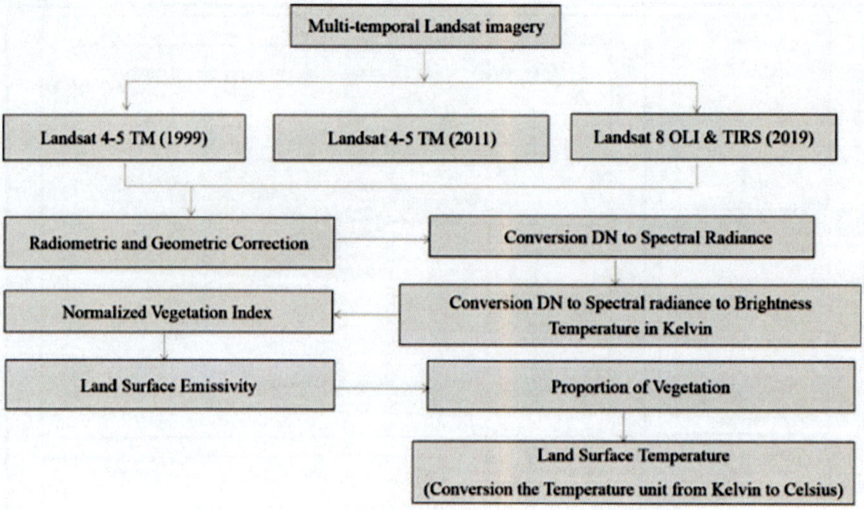

Fig. 4.2 General methodology for land surface temperature

finally the changes in opencast mining activity regions. To assess the effect of such activity, few environmental quality were assessed in which the land surface temperature was one, which was calculated for the study area from the satellite image of the contemporary time scale.

The Land surface temperature (LST) is the radiative temperature which was calculated using top of atmosphere brightness temperature, wavelength of emitted radiance, land surface emissivity by single channel algorithms proposed by Jeevalakshmi et al. (2017) and Anaandababu et al. (2018) (Fig. 4.2).

$$\text{LST} = (\text{BT}/1) + W * (\text{BT}/14380) * \ln\ (E)$$

where:

BT = top of atmosphere brightness temperature (°C)
W = wavelength of emitted radiance [Landsat 4–5 TM =10.895, Landsat 8 OLI & TIRS = 11.45]
E = land surface emissivity

Here, land surface emissivity (E) in formula-1 was calculated via the following formula for each temporal satellite images.

$$\text{Emissivity} = 0.004 * \text{PV} + 0.986$$

where PV = proportion of vegetation (proportion of vegetation or fractional vegetation cover) in formula-2 which was obtained from NDVI values for vegetation and soil according to with value various between 0.00 and 1.00.

$$PV = [(NDVI - NDVI\ min)/(NDVI\ max - NDVI\ min)]^2$$

where the Normalized Differential Vegetation Index (NDVI) in formula-3 is a standardized vegetation index which is calculated using near infrared (Band 5) and red (Band 4) bands.

$$NDVI = (NIR - RED)/(NIR + RED)$$

On the other hand, BT in formula-1 can be derived when spectral radiance data were converted to top of atmosphere brightness temperature using the thermal constant values in the Metadata file.

Conversion of Kelvin to Celsius

As the result is in Kelvin, the radiant temperature is revised by adding the absolute zero (−273.15 °C) to get the result in Celsius.

$$BT = K2/\ln\ (k1/L\lambda + 1) - 273.15$$

where:

BT = top of atmosphere brightness temperature (°C)
$L\lambda$ = TOA spectral radiance (Watts/(m^2 * sr * μm))

Using the radiance rescaling factor, thermal infrared digital numbers can be converted to TOA spectral radiance.

For the validation of the retrieval methods, the land surface temperatures were also collected in the ground by using an infrared thermometer (LASER—GM320) over seven stations, namely, Mohanpur, Sonepur Bazari, Khottadih, Kalipahari, Gourandi, Gourandi-Begunia, and North Searsole.

4.3 Results and Discussion

Land use/land cover data were generated for 1999, 2011, and 2019 with classes like build-up area, vegetation, mining area, agricultural land, barren land, river, reservoir/lakes/ponds, and river sand (Fig. 4.3). It showed a sharp increase in the mining area (open cast) from 1630.53 to 6182.37 ha (0.93 to 3.51%) during 1999–2019 (Fig. 4.4). This was associated with a significant areal increase of built-up area from 34,293.87 to 52,962.75 ha (19.48 to 30.08%), a steep fall of vegetation coverage from 23,129.01 ha to 15,309.45 ha (13.14% to 8.7%), as well as agricultural land from 100,666.44 ha to 84799.8 ha (57.15% to 48.15%). The field visit was

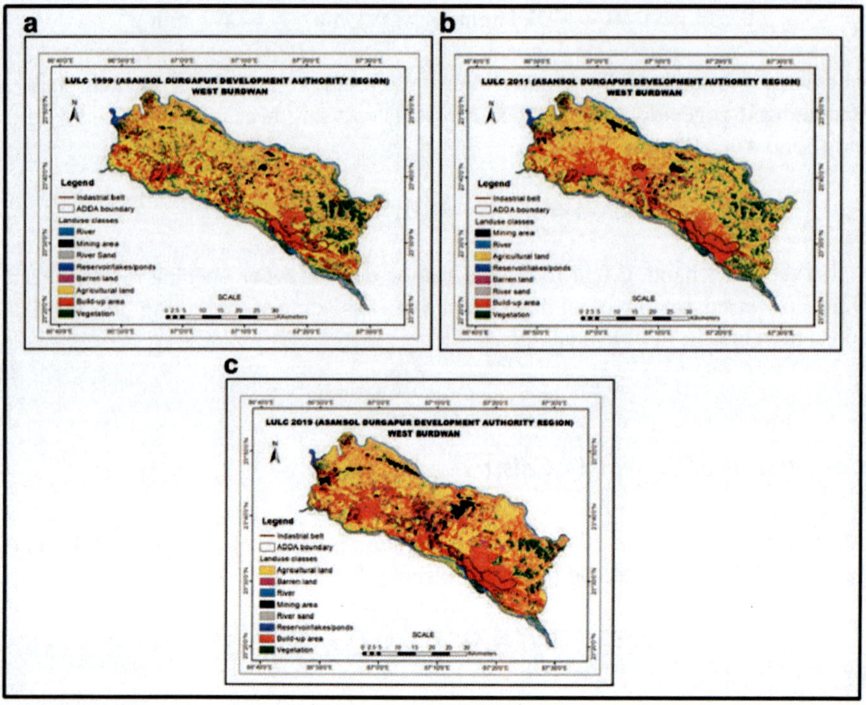

Fig. 4.3 Landsat TM 4 – 5 for January 1999 and February 2011 and Landsat 8 - OLI for January 2019 from USGS (https://earthexplorer.usgs.gov/)

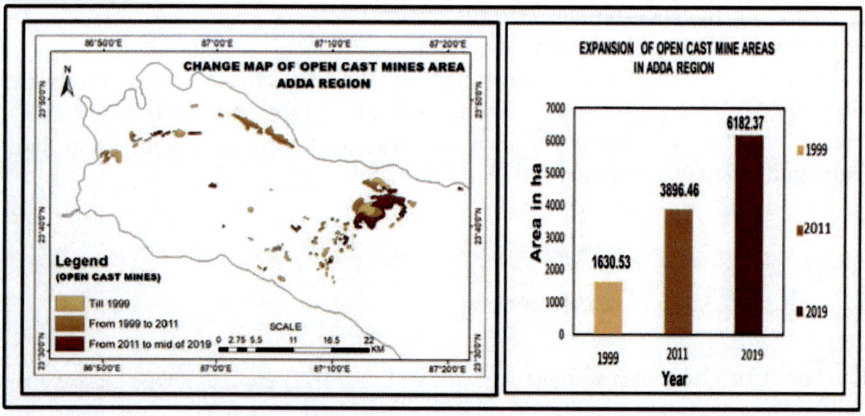

Fig. 4.4 Expansion of open cast mining activity in ADDA region

conducted for land use categories for 2019 with a detailed visit to open cast mining areas. Open cast mining areas on the classified data were also verified with ECL data for an accurate assessment.

Land Surface Temperature (LST)

As per the change analysis of LST from 1999 to 2019 (Fig. 4.5), it was ranging from 14 to 28 °C in January 1999, with the highest limit distributed over Sonepur Bazari, Mohanpur, and Khottadih mine areas. The temperature of these areas increased to a peak temperature of 32° during February 2011. It further increased to 38° as highest in January 2019.

On the other hand, Narayankuri, New Kenda, and Banshra open cast mines and part of BCCL, IISCO Coalfields were showing the medium LST from 23 to 26 °C with a surrounding LST of 21 °C in 1999. Active mining belts were showing a rise in temperature (up to 34°) in February 2011. It was showing temperature hot spots on Gourandi, Gourandi-Begunia, Jambad, Bonjemehari, and Sonepur Bazari OC with LST values ranging from 32 to 34 °C with surrounding average LST value of 30 °C. This distribution of land surface temperature over this area was showing an increasing trend even in January 2019 as per LST data retrieved from satellite images. The high-temperature pixels were distributed in some non-mining areas also. Those are basically due to the presence of industries and settlements. Hence, the temporal change of land surface temperature is shown to be closely associated with the rapid land use/land cover change (Ziaul & Pal, 2016) (Fig. 4.6).

By using an infrared thermometer (LASER—GM320) land surface temperature was measured in the field to validate the process of deriving LST values from remote sensing data. Figure 4.9 is showing the variation of LST measured and calculated

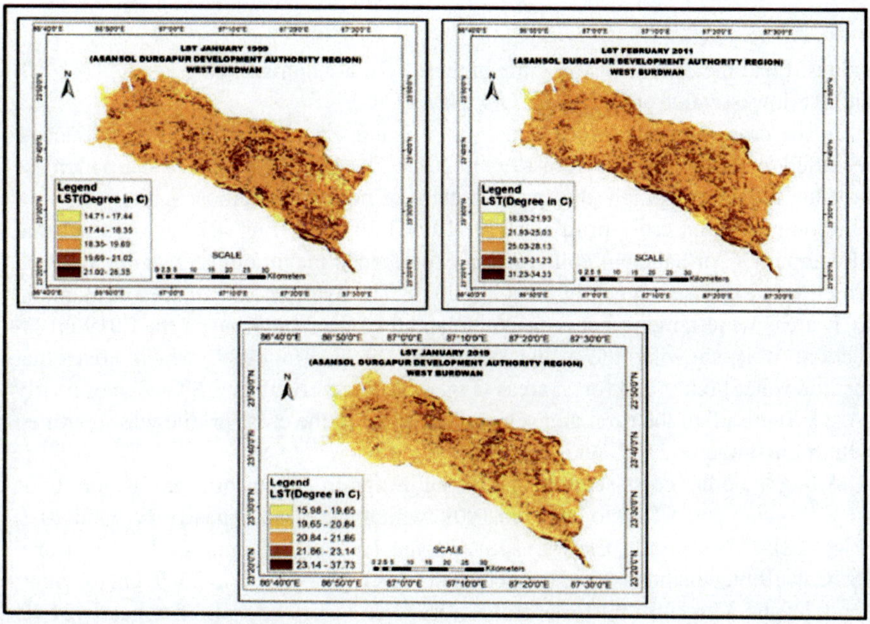

Fig. 4.5 Land surface temperature

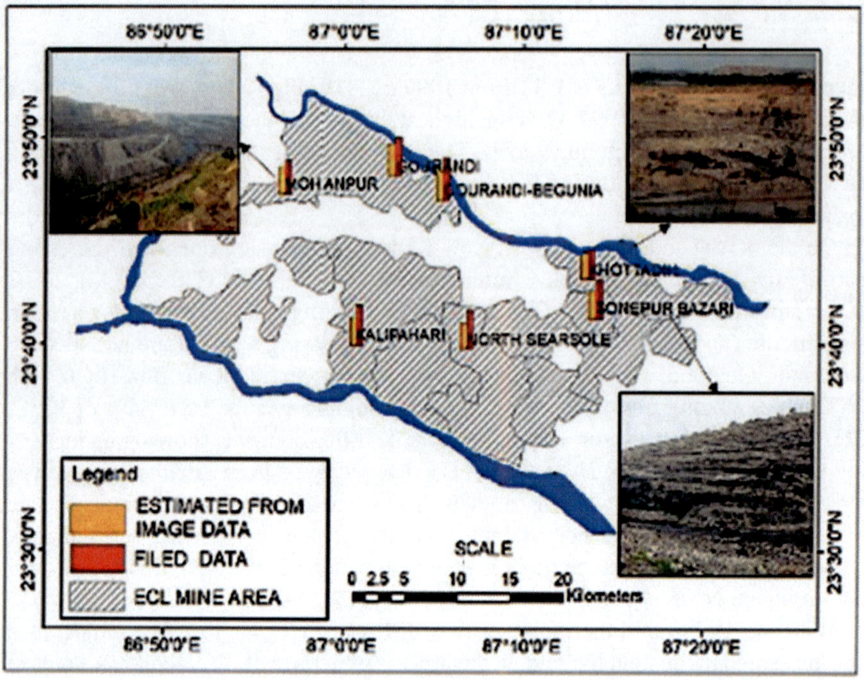

Fig. 4.6 Bar graph comparison between Field LST data and Estimated from image data of ADDA region, February 2019. Source: By Author

values. Field measurements give the highest LST at Gourandi-Begunia OC (43 °C) and the lowest value at Khottadih OC with 32 °C.

In the cross-sectional profiles in Figs. 4.7 and 4.8, the increasing trend can be recorded for a time interval from 1999 to 2019. Though it represents a sharp increase in value for the full graph, the spatial pattern of hotspots is almost static.

Along with the cross profile AB in 1999 LST data (Fig. 4.7), it is found that Mohanpur OC of Salanpur colliery area experienced maximum LST of about 28 °C which was increased to nearly 30 °C in 2019. The average LST of Kenda, Bankola ECL areas were ranging between 26 °C and 24 °C in 1999 but in the 2019 cross-section, it is showing the value of nearly 29 °C. In 1999, which areas that include water bodies and forest areas (Fig. 4.3), there giving an LST value of nearly 14 °C. The LST of the rural and semi-urban areas in the cross profile was identified within the range of 21 °C and 17 °C.

Along the other cross-section CD, Sripur, Satgram, Kunustoria, and Kajora ECL were showing an LST up to 24 °C in 1999, which increased to nearly 29 °C in 2019 (Fig. 4.8). Cross profile EF was showing the LST of Salanpur–ECL area nearly 18 °C in 1999, which is seen to be increased to as high as 35 °C in 2019. On the other hand, Sripur, Kenda, Kunustoria, and Kajora were experiencing LST of nearly 21 °C in January in 1999 which was ranging between 28 to 35 °C in 2019.

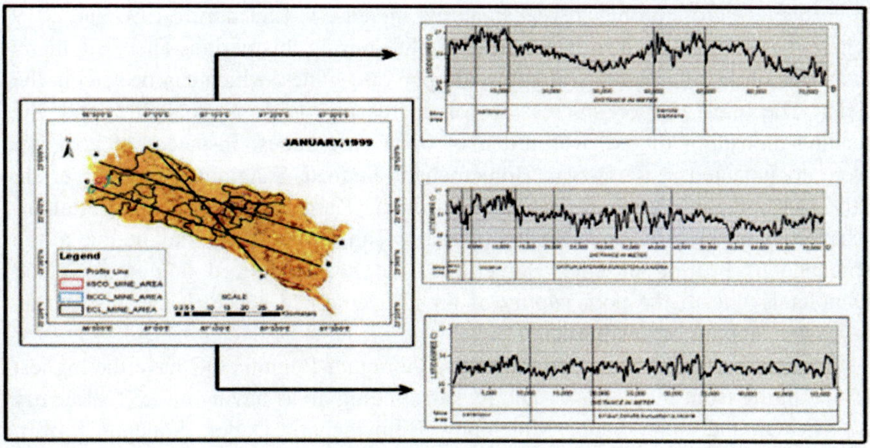

Fig. 4.7 LST profiles (A-B, C-D, and E-F) in January 1999

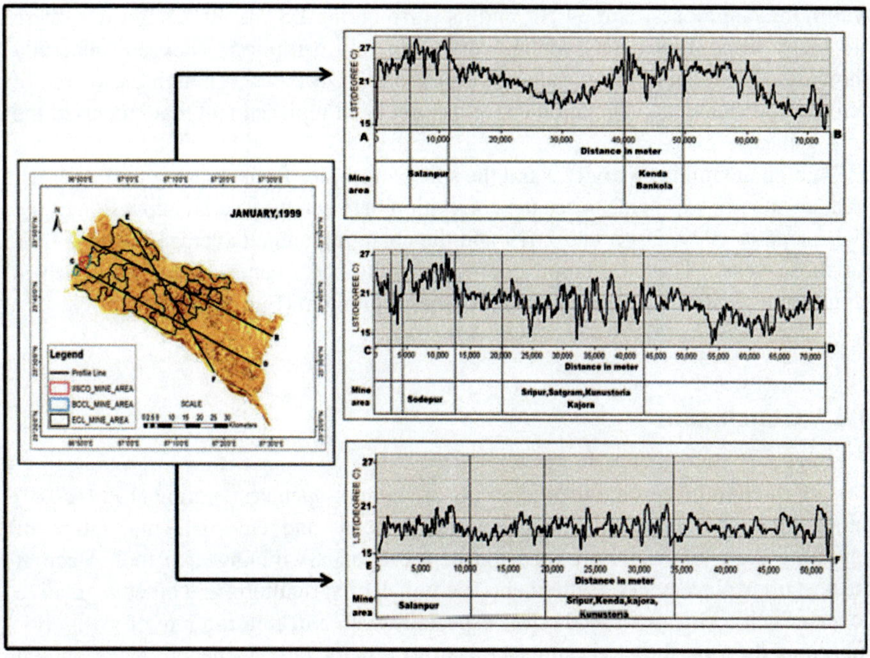

Fig. 4.8 LST profiles (A-B, C-D, and E-F) in January 2019

Hence, the cross profiles reveal an abrupt increase in LST during 1999 and 2019 where the mining area also increased to 2.58% during this period. There are many instances of the unauthorized digging of open cast mines, which can be seen in the field. The most important reason for increasing LST is coal fire and the loading-unloading of the overburden of open cast mines. Instances of coal fire were documented at Ramnagar, Bonjemehari, Jambad, Sangramgarh (Guha et al., 2012), Khottadih, Sonepur Bazari, Churuliya, Kamarkhola, Belbadh, Sitalpur, Khayerbandh, Jamuria, and Dhadka (Karfa & Tah, 2019). They may be due to the spontaneous heating of coal. Sometimes, the heat produced is not dissipated completely due to the poor supply of air in a confined space. It can also be an excessive amount of overburden.

As recorded in 2019, the Gourandi and Gourandi-Begunia OC have the highest LST ranging from 39 to 43 °C, and its surrounding areas having an LST of nearly 37 °C. On the other hand, Mohanpur, Bonjemehari, Dabor, Sonepur Bazari, Khottadih Madhabpur, Jambad, New Kenda, North Searsole OC, and parts of BCCL, IISCO Coalfields have the medium LST ranging from 34 to 38 °C, and its surrounding areas experienced LST of 36 °C. The LST of Madhabpur, Dalurbandh, Itapara, Madhabpur, Banshra, Narayankuri, Nimcha-Amkola, Bhanora west OC was within the range of 31 and 34 °C, and its surrounding LST is 30 °C. Besides, there are many discontinued OC where work is in the disruption. They are generally showing a moderately lower value of LST from 26 to 29 °C. Though these are not active mine areas, the vegetation cover has not been replaced and that increased the LST value.

Though the mining activity is not the sole contributor for increasing LST over the area, an attempt was made to see the correlation between the overall expansion of the OCM area in 1999, 2011, and 2019 and the changing annual average value of LST for those years. The correlation coefficient value (R^2) came as 0.9578 which is representing a strong positive correlation between them (Fig. 4.9).

4.4 Conclusion

This research indicates that the increasing area under open cast mining (1999–2019) has a significant impact on the economic, social, and environmental fabric of adjoining areas of ADDA. Expansion of the opencast mining area took place at the cost of displacement of settlements, demolish of agriculture, and other vegetative covers. Such a change in land cover especially with a decreasing rate of greeneries subsequently took the environment toward a quality fall. A change in the annual seasonal temperature also took a hike. To confirm this, this study got a correlation statistics between the temporal expansion of open cast mining areas and annual mean surface temperature change of the corresponding area. The analysis came with a very strong positive correlation between these factors. Mining area expansion has direct control over the NDVI change and resulting change in the mean surface temperature of this Asansol-Durgapur Development Region. This study also explored the

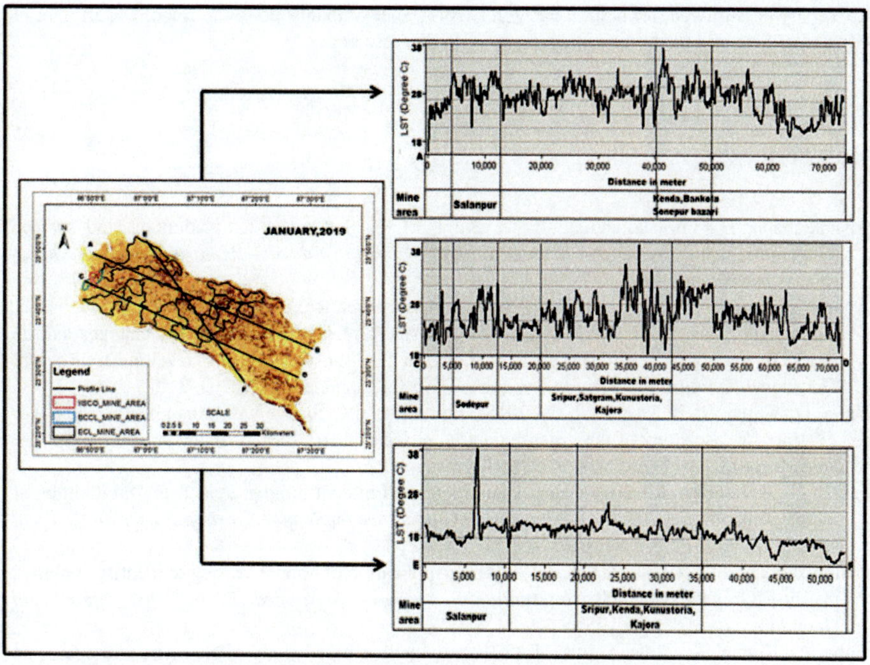

Fig. 4.9 Correlation between areal changes of mines (ha in %) and LST

correlation analysis result between LST and areal change (%) of coal mine areas ($R^2 = 0.95$) showed a positive relationship and a negative relationship between LST and NDVI ($R^2 = 0.89$). As a result the spatial relationship linear correlation between areal changes (%) in Mining, LST, and NDVI where R^2 value near about 1.

Sometimes, the exploration is complete in one region, the deactivated mine remains the same without the proper land reclamation and rebuilding of vegetative cover. Mining operations can't be the sole responsible factor behind the environmental quality deterioration, but rather industrial development along with urban expansion also contribute much. Therefore, these mining activities should be followed by proper environmental protection measures. Reclamation planning is one of the major environmental protection measures to reduce air temperature. After the completion of exploration, abandoned coal mines should be filled up with sand or soil, or water. Regaining of the soil fertility should be planned along with afforestation programs. Advanced mining technology and effective strategies to reduce air temperature around coal mining areas, as well as public awareness and education, should be encouraged.

Acknowledgments The authors are thankful to Eastern Coalfield Limited (ECL) for providing data access and their kind cooperation. The researcher is also thankful to the Director, IIEST, Shibpur, and Head of the Department of Mining Engineering for their full support to conduct this research. The authors are express indwelling gratitude to Department of Geospatial Science, The

University of Burdwan, for help to carry this work. Special thanks go to the Head, Department of Geospatial Sciences for his encouragement to do this research.

References

Anaandababu, D., Purushothaman, B., & Babu, M. S. S. (2018). Estimation of land surface temperature using LANDSAT 8. *International Journal of Advance Research, Ideas and Innovations in Technology, 4*, 177–186.

Anonymous. (n.d.). Retrieved Febuary, 2019, from https://earthexplorer.usgs.gov/

Choudhury, D., Das, K., & Das, A. (2018). Assessment of land use land cover changes and its impact on variations of land surface temperature in the Asansol-Durgapur Development Region. *The Egyptian Journal of Remote Sensing and Space Sciences, 22*(2), 203–218.

Dutta, D., Gupta, S., & Kishtawal, C. (2018). Linking LULC change with urban heat islands over 25 years: A case study of the urban-industrial city Durgapur, Eastern India. *Journal of Spatial Science*, 1–18.

Fawzi, N., & Jatmiko, R. (2015). Heat island detection in coal mining areas using multitemporal remote sensing. In *ACRS 2015—36th Asian Conference on Remote Sensing: Fostering Resilient Growth in Asia, Proceedings*.

Gangopadhyay, P., Dutt, K., & Saha, K. (2005). Application of remote sensing to identify coal fires in the Raniganj Coalbelt. *International Journal of Applied Earth Observation And Geoinformation., 8*, 188–195.

Guha, A., Kumar, K., & Kamaraju, M. (2012). A satellite-based study of coal fires and open-cast mining activity in Raniganj coalfield, West Bengal. *Current Science, 103*(9), 1096–1104.

Jeevalakshmi, D., Narayana Reddy, S., & Manikiam, B. (2017). Land surface temperature retrieval from LANDSAT data using emissivity estimation. *International Journal of Applied Engineering Research, 12*, 9679–9687.

Kamila, A., & Pal, S. (2015). Monitoring of land surface temperature and analyzing environmental prediction on Asansol and Durgapur sub-division, Burdwan district, West Bengal using Landsat imagery. *International Journal of Remote Sensing & Geoscience, 4*(1), 33–36.

Karfa, C., & Tah, S. (2019). COAL mine fire and temporal change in landuse landcover pattern of Raniganj Coal Field Areas in West Bengal using Rs and Gis technique. *International Journal of Research in Advent Technology, 7*(1), 18–31.

Manna, A., & Maiti, R. (2014). Opencast coal mining induced defaced topography of Raniganj Coalfield in India—Remote sensing and GIS-based analysis. *Journal of the Indian Society of Remote Sensing, 42*(4), 755–764.

Ziaul, S. k., & Pal, S. (2016). Image-Based surface temperature extraction and trend detection in an urban area of West Bengal, India. *Journal of Environmental Geography, 9*(3–4), 13–25.

Chapter 5
Effect of Land Use/Land Cover Change on Soil Loss in the Tropical River Catchment of Northeast India

Jatan Debnath and Nibedita Das(Pan)

Abstract Soil erosion becomes a common problem in tropical river basins like the Muhuri River basin where intensity of rainfall as well as degradation of forest, unscientific agricultural activities and excessive jhoom cultivation are more pronounced. It accelerates the rate of sediment deposition in the river which disturbs natural condition of the channel. Therefore, the present research aims to assess the spatio-temporal change of soil erosion status owing to land use/land cover change (LULC) in the Muhuri River basin. This study used land sat imageries of 1986 (TM) and 2019 (OLI) along with supervised classification technique to estimate the LULC change. Moreover, the model of Universal Soil Loss Equation (USLE) was applied to assess the soil erosion. The whole work was processed and computed under the geo-processing tool of ArcGIS 10.1. The result revealed that during the period 1986–2019, the area under dense forest, open forest, degraded forest, jhoom cultivation and water body were reduced by 92.05, 68.63, 5.05, 25.29 and 25.64%, respectively, while the extent of the agricultural land, rubber plantation, settlement and barren land increased by 14.47, 1600.39, 95.36 and 197.33%, respectively. As a result, the range of average annual soil loss of the Muhuri River was increased from 0 to 101.06 t ha^{-1} $year^{-1}$ (during 1986) to 0 to 110.08 t ha^{-1} $year^{-1}$ (during 2019). Obviously, the mean rate of soil erosion is associated with the LULC change of the study area, and therefore, in case of degraded forest, agricultural land, rubber plantation and jhoom cultivation, the mean soil loss was increased by 2.37, 1.22, 1.50 and 1.32 t ha^{-1} $year^{-1}$ correspondingly within the study period (1986–2019). Therefore, the findings of the research illustrate that declined natural forest cover along with increased settled area, barren land, agricultural land and existence of shifting cultivation increases the soil erosion potential in the river basin over the study period. Thus, there is a need of sustainable maintenance of watershed resources to control the sediment influx into the river.

J. Debnath (✉) · N. Das(Pan)
Department of Geography and Disaster Management, Tripura University, Agartala, Tripura, India

P. K. Shit et al. (eds.), *Geospatial Technology for Environmental Hazards*, Advances in Geographic Information Science, https://doi.org/10.1007/978-3-030-75197-5_5

Keywords LULC · Universal Soil Loss Equation (USLE) · Spatio-temporal change · Muhuri River basin

5.1 Introduction

Soil erosion is a natural phenomenon ensuing from detachment of the top soil due to some natural agents (wind and water) and anthropogenic activities (overgrazing, deforestation, forest fire, etc.). According to Angima et al. (2003), about 85% of land and degradation occurs owing to soil erosion, and Pandey et al. (2009) opined that LULC change is the main component for this. Since the last century, soil erosion expanded via anthropogenic activities becomes the most important environmental hassle all over the globe (Sharma et al., 2011). In India, almost 113.3 million hectare of land is affected by soil erosion and almost 5334 million tonnes of soil is being removed yearly owing to different reasons (Narayan & Babu, 1983), and almost 7.5% areas (0.20 million km^2) are effected by intense bank erosion (Ahmad et al., 2020). According to Bhattacharyya et al. (2016), inappropriate agricultural practices are mostly responsible for the soil erosion in India, which has reduced the soil fertility and created an effect on the food and livelihood security of the farmers. Generally, soil erosion is regulated by the nature of topography, soil properties, forest cover, land use, etc. Moreover, a group of aspects like slope steepness, heavy rainfall after long dry period, improper LULC pattern (e.g., degraded forest) and ecological disaster (e.g., shifting cultivation) along with some inherent characteristics of soil profile, such as thin upper layer and less organic content, make it more prone to erosion (Ganasri & Ramesh, 2016).

Nowadays, declining forest cover, growing settled area and inappropriate agricultural practice have accelerated soil erosion in the river catchments significantly. Intense soil erosion amplifies sedimentation in the channel bed and reservoir as well as affects floodplain morphology and its ecological functioning (Asselman & Middelkoop, 1995; Verstraeten & Poesen, 2002; Richards et al., 2002). Therefore, watershed management becomes an important task to the planners. To evaluate the annual soil loss and sediment yield of a river basin or a watershed and to recognize the erosion prospective zone at regional, as well as global scale, various models were used (Pandey et al., 2007; Dabral et al., 2008) among which USLE became widely accepted throughout the world (Wischmeier & Smith, 1978). The RS and GIS techniques are extensively used for its cost effectiveness and better accuracy to soil erosion estimation and its spatial distribution in broader areas (Shit et al., 2015; Phinzi et al., 2020). The USLE model associated with GIS provides a fruitful and better accuracy result than the traditional methods (Roy, 2019) and also helps for mapping the priority-wise erosion prone areas (Girmay et al., 2020; López-García et al., 2020; Fiener et al., 2020; Delgado, 2020). On the other hand, use of temporal satellite images to study the relation among the change of LULC and soil erosion becomes widespread in modern research (Jordan et al., 2005). Apparently, LULC change and its effects on the soil erosion potential in addition with sediment transport rate were studied by many researchers at temporal scale (Sharma et al., 2011; Esa et al., 2018; Huang et al., 2020).

The present research pursuits to estimate the amount of soil loss due to LULC change using USLE model with the useful resource of the GIS medhod. Through this study the contribution of each LULC element on soil erosion was justified. Moreover, the spatio-temporal analysis of soil erosion over the whole basin was supportive to identify the most affected part of the basin and useful for the engineers and the planners for formulating important action plans.

5.2 Regional Setting

The Muhuri River basin consists of three Rural Development Blocks and is located in the southern part of Tripura (Fig. 5.1). It lies among 23°10′ N to 23°25′ N latitude and 91°26′ E to 91°46′ E longitude occupied about 701.72 km^2 surface area. The River Muhuri originates from the Baramura-Deotamura hill range and enters Bangladesh after flowing for 53.3 km within Tripura. The basin experiences moist humid climate with medium to high rainfall (1500–2000 mm). Here the average summer temperature varies from 25° to 29 °C, whereas the winter temperature from 19 to 24 °C.

The western part of the basin is bounded by Bangladesh. Its total population is estimated to be more than 1,80,000 among which the immigrated Bengalese dominate over the indigenous community. The area is most densely populated with a density of 256 persons/km^2. Here majority of the tribal population live in the hilly part and there is predominance of Bengalese in the plain part.

5.3 Materials and Methods

Identification of LULC Change Map

The study of LULC change detection and its evaluation was accomplished by adopting a series of steps and processes which includes collection of satellite images, pre-processing, supervised classification using maximum likelihood algorithm and post classification comparison. In an effort to identify the changes in LULC of the study area, the researcher used the TM (Thematic Mapper) satellite image of 1986 as the base year and OLI (Operational Land Imager) image of 2019 as the current year. All the Landsat images were downloaded from the United States Geological Survey (USGS) website. The collected raw imageries were processed in ArcGIS 10.1. The image pre-processing was performed for contrast enhancement of the satellite images so that the pixel values can be distributed uniformly to avoid the radiometric distortion, and thus reliability of the pixels' brightness value increases. On the other hand, supervised classification technique with maximum likelihood algorithm was

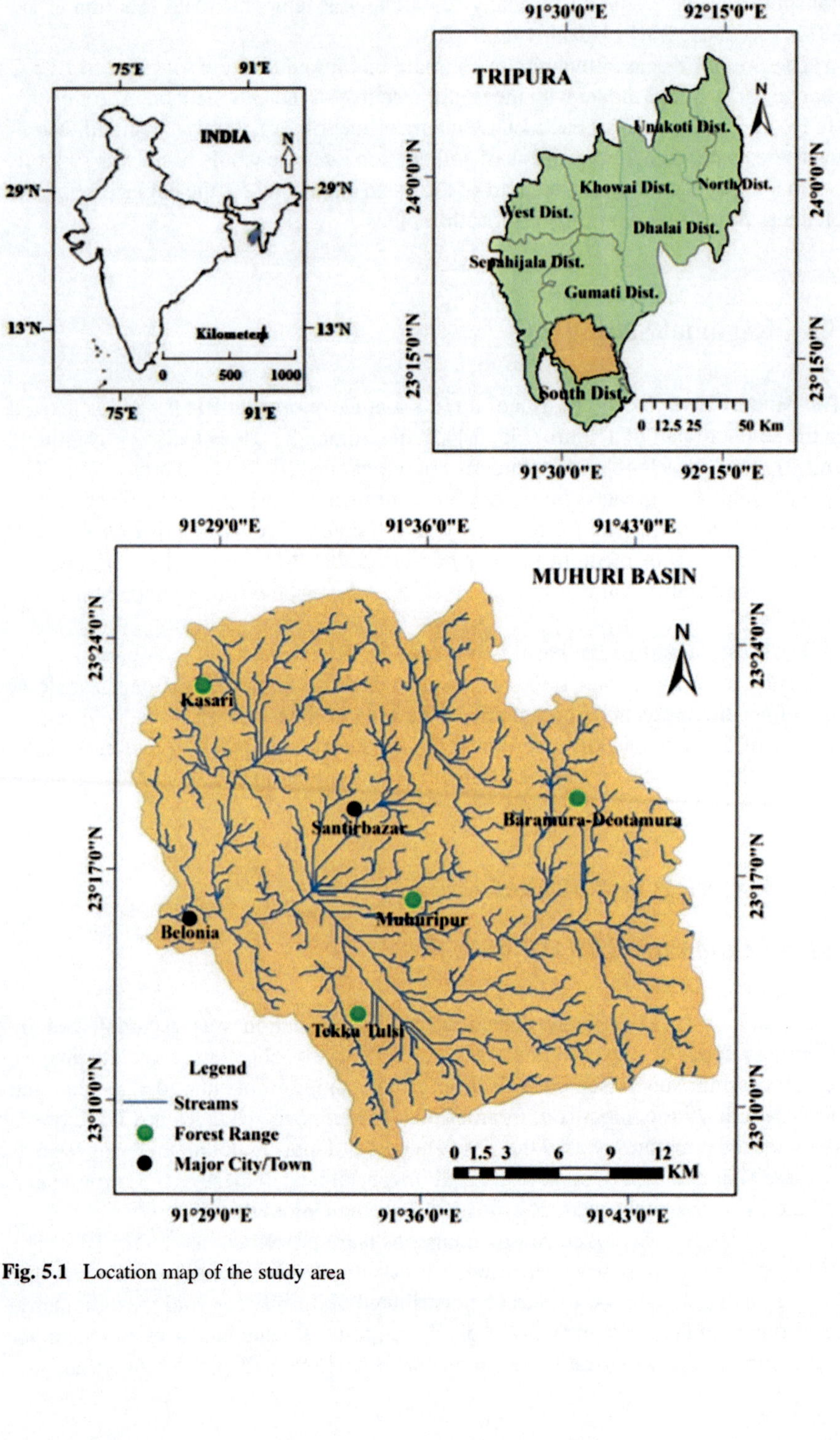

Fig. 5.1 Location map of the study area

used to prepare the LULC map from the collected imageries (Debnath et al., 2017a, b), and accuracy of the classified maps was assessed after field verification. Post classification comparison was completed to identify the changes of the LULC classes.

Confusion matrix was used for classification accuracy where producer's accuracy, user's accuracy, overall accuracy (Eq. 5.1) and Kappa accuracy (Eq. 5.2) were calculated using this matrix table (Congalton, 1991). These were calculated using the following formula:

$$\text{Overall accuracy} = \frac{\sum_{i=1}^{r} x_{ii}}{N} \times 100 \tag{5.1}$$

$$\text{Kappa accuracy} = \frac{N\sum_{i=1}^{r} x_{ii} - \sum_{i=1}^{r} (x_{i+} * x_{+i})}{N^2 - \sum_{i=1}^{r} (x_{i+} * x_{+i})} \tag{5.2}$$

where r is regarded as the number of rows in the matrix, x_{ii} expresses the total number of correctly classified pixels in row i and column i, x_{i+} and x_{+i} are the marginal totals of row i and column i, respectively, and N is the total number of pixels in the matrix table.

Estimation of Soil Loss Using USLE

USLE is an experimental Equation (Eq. 5.3) which predicts and estimates average annual soil loss from a particular area and its spatial distribution (Ahmed et al., 2017). The result of USLE equation depends on the five major factors which are expressed by the following equation:

$$A = (R \times K \times LS \times C \times P) \tag{5.3}$$

where A represents the mean annual soil loss in ton ha^{-1} year^{-1}, R is regarded as rainfall-runoff erosivity factor, K expresses the soil-erodibility factor, LS is the slope length and gradient factor, C is the crop-management factor and P is the support practice factor. All the factors were integrated in GIS environment to calculate the soil loss from the study area.

For estimation of soil loss of the Muhuri River basin for the years 1986 and 2019, the Landsat TM 1986 and Landsat OLI 2019 were used to prepare the LULC maps. These maps were utilized as crop management factor (C) and support practice factor (P) maps. Rainfall data of Belonia, Bokafa, Sabroom, Sonamura and Amarpur rain gauge stations were collected for 10 years' period (for 1986, data of 1976–1985 and for 2019, data of 2009–2018) to prepare the rainfall erosivity factor map (R). Whereas, so as to estimate the soil erodibility factor (K) map, the soil map of the

NBSS and LUP (1996) of 1:250,000 scale was used as the base map. Aster DEM of 30 m resolution was used to produce the slope length and gradient factor (LS) map.

Development of Database

Rainfall-Runoff Erosivity Factor (R)

Rainfall-runoff erosivity factor is one of the essential factors that influence soil erosion significantly. The R factor usually depicts the erosive power of rainfall at a particular place on account of the amount and intensity of rain (Uddin et al., 2016). It is computed (Eq. 5.4) with the help of the storm kinetic energy and the maximum 30 min rainfall intensity and facilitates to identify the effect of rain drop over a piece of land. As in the present study area such kind of meteorological data is unavailable, therefore, in this case the Rainfall erosivity factor was calculated using the formula

$$Ra = 79 + 0.363 \times P \tag{5.4}$$

where Ra is the annual R factor and P is Rainfall in mm. R factor data of these five rain gauge stations were processed using the ArcGIS software. In order to prepare the map Interpolation method was applied using the IDW tool.

Soil Erodibility Factor (K)

The soil erodibility factor refers to the quantitative analysis of the intrinsic erodibility strength of a specific sort of soil. It is normally measured on susceptibility to detachment of soil particles from land surface and flowing through runoff. The K factor ranges between 0 and 1, where 0 (zero) means minimal prone to erosion, while 1 signifies highly prone to erosion through water. The soil properties that influence K factor are soil texture like sand, silt and clay, organic matter, soil structure, and its permeability. Here the soil erodibility factor was evaluated with the help of soil erodibility nomograph which is depends on Geo-pedological Map of the NBSS and LUP, 1996 Govt. of India (Wischmeier & Smith, 1978).

Slope Length and Gradient Factor (LS)

Slope length (L) and slope gradient (S) factor is also known as topographic erosivity factor. Generally, with the steepening of slope, soil erosion also augmented due to increasing velocity of the surface runoff towards downhill direction (Pradeep et al., 2015).

The Aster DEM for the study area was used to compute the topographic erosivity factor using the ArcGIS and Arc Hydro extension tool. The algorithm (Eq. 5.5) used by Moore and Burch (1986) was applied to derive the soil erosion weightage for topographic factor as given below:

$$LS = \left[\left(\text{Flow accumulation} \times \frac{\text{Cell size}}{22.13} \right) 0.4 \times \left(\frac{\sin \text{slope}}{0.0896} \right) 1.3 \right] \tag{5.5}$$

where flow accumulation denotes the accumulated upslope contributing area for a specified cell, cell size = size of grid cell (30 m for the present research work) and sin slope = sin value of slope angle in degree.

Crop Management Factor (C)

This factor depends on the nature of LULC of an area. The C factor is especially used to reflect the erosion rate as the effect of cropping and other management. Moreover, this factor is highly related with canopy cover of a particular place. In general, soil erosion reduces in consequence with the growth of canopy cover (Shit et al., 2013). The land cover interrupts rainfall, permits penetration and decreases the kinetic energy of rainfall over the land surface. For the present study, Landsat TM, 1986, and Landsat OLI, 2019, were used to generate the C factor maps.

Support Practice Factor (P)

This factor usually reflects the ratio of soil loss from a particular practice to the corresponding loss with straight row ploughing up and down slope (Dabral et al., 2008). The higher value of P factor indicates the application of effective soil conservation measures (Prasannakumar et al., 2011). During field verification, it was observed that the soil conservation techniques were adopted in the agricultural fields only. As a result, the researcher considered agricultural and non-agricultural land separately and assigned the P value of 0.28 and 1.0 respectively (Fig. 5.5). These values were used in other literatures also by Pandey et al. (2009) and Ahmed et al. (2017).

5.4 Result and Discussion

LULC Change (1986–2019)

The present observation shows that in the year 1986 about 5.90, 28.59 and 46.87% areas of the basin were occupied by dense forest, open forest and degraded forest, respectively, whereas, 9.99, 1.46, 1.71, 3.41, 1.94 and 0.11% areas were under agriculture, rubber plantation, jhoom cultivation, settlement, water body and barren

Table 5.1 Areas under different LULC categories in the Muhuri River basin

LULC Classes	1986 Area (km^2)	1986 Area (%)	2019 Area (km^2)	2019 Area (%)	1986–2019 Changed area (km^2)	1986–2019 Changed area (%)
Dense forest	41.4	5.90	3.29	0.47	−38.11	−92.05
Open forest	200.63	28.59	62.93	8.97	−137.7	−68.63
Degraded forest	328.92	46.87	312.3	44.50	−16.62	−5.05
Agricultural land	70.14	10	80.29	11.44	10.15	14.47
Rubber plantation	***10.28***	***1.46***	***174.8***	***24.91***	***164.52***	***1600.39***
Jhoom cultivation	12.02	1.71	8.98	1.28	−3.04	−25.29
Settlement	23.93	3.41	46.75	6.66	22.82	95.36
Water body	13.65	1.95	10.15	1.45	−3.5	−25.64
Barren land	0.75	0.11	2.23	0.32	1.48	197.33
Total	701.72		701.72			

land, respectively. Again, in the year 2019 about 0.47, 8.97 and 44.50% areas were under dense forest, open forest and degraded forest, respectively. On the other hand, 11.44, 24.91, 1.28, 6.66, 1.45 and 0.32% areas were occupied by agriculture, rubber plantation, jhoom cultivation, settlement, water body and barren land, respectively (Table 5.1).

The temporal data sets of LULC of the study area indicate some considerable changes between the base year and the current year which were inflated by human, as well as natural factors. During the 1986–2019 periods of 33 years, the areas under dense forest, open forest and degraded forest were decreased by 92.05, 68.63 and 5.05%, respectively, whereas areas under rubber plantation and agriculture were increased significantly by 1600.39 and 14.47%, respectively. Moreover, area under shifting cultivation was decreased by 25.29% and settled area increased appreciably by 95.36% with reduction in water body by 25.64% (Table 5.1). During this period, the basin experienced a significant increase in barren land by 197.33%.

Rainfall Erosivity Factor (R)

In order to compare rainfall erosivity (R) of the years 1986 and 2019, the mean rainfall of the periods 1976–1985 and 2009–2018, respectively, were used. The R factor values of the year 1986 ranged between 843.69 and 987.28 MJ mm ha^{-1} h^{-1} $year^{-1}$, whereas during the year 2019, it became 786.55 and 892.15 MJ mm ha^{-1} h^{-1} $year^{-1}$ within the Muhuri basin (Table 5.2).

Table 5.2 Average annual rainfall and R value of the selected stations

	Average annual (1976–1985)		Average annual (2009–2018)	
Station	Rainfall (mm)	R-factor	Rainfall (mm)	R-factor
Matabari	2407.44	952.9	1949.18	795.2
Amarpur	2390.75	946.84	2048.22	786.55
Belonia	2476.62	978.01	2110.03	822.5
Sabroom	2502.15	987.28	2240.09	844.94
Bokafa	N.A.	N.A.	2129.18	892.15
Sonamura	2106.58	843.69	1972.98	851.89

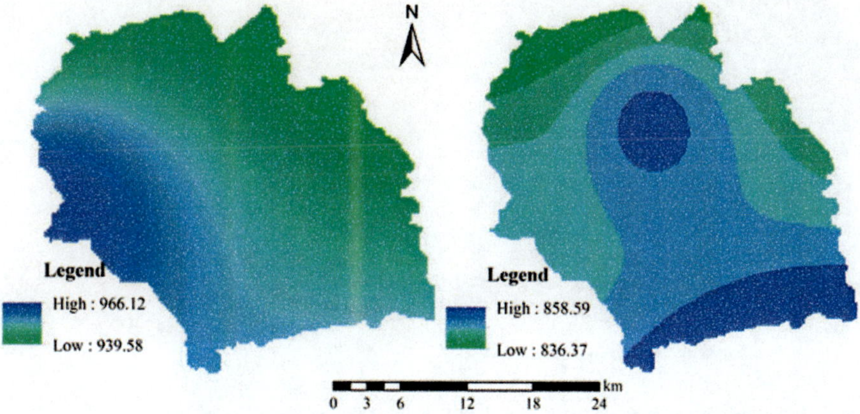

Fig. 5.2 Spatial distribution of R factor in the Muhuri River Basin for the years 1986 and 2019

The spatial distribution of R factor of 1986 indicates the concentration of highest R value in the western part and lowest value in the extreme northern part of the basin, whereas in the year 2019, it was observed in the central and the extreme northwestern portions of the basin, respectively (Fig. 5.2).

Soil Erodibility Factor (K)

Soil type varies according to the topographic and lithologic characters of the concerned area. The K value of each soil type of the study area was derived from the work of Ghosh et al. (2013), which was calculated using the nomograph (Wischmeier & Smith, 1978) of USLE (Table 5.3).

Higher the amount of *K* value more erosion prone is the area and vice versa. The highest *K* value (0.36) was observed in the central part of the basin in the soils of inter-hill valley, whereas the lowest value (0.16) was observed in the central and the

Table 5.3 K value of the study area according to the geo-pedological characteristics

Soil type	*K* factor
Soils of low relief structural hills and ridges	0.24
Soils of undulating plains with low mounds and narrow valleys	0.16
Soils of inter hill valleys	0.36
Soils of flood plains	0.34

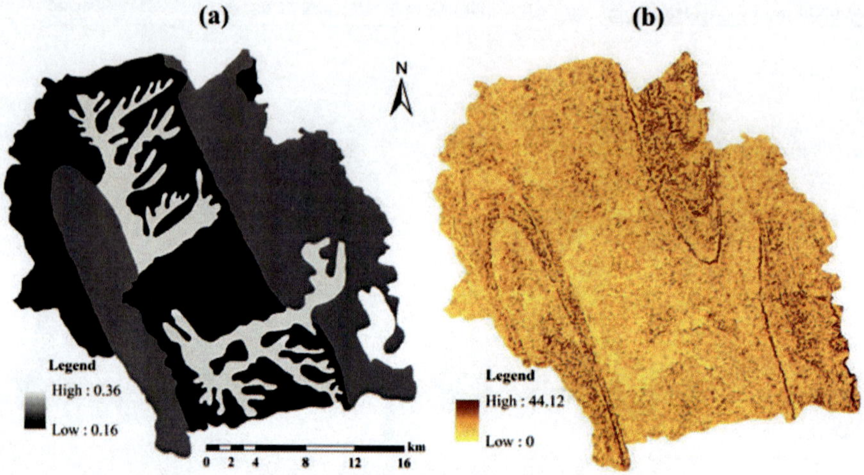

Fig. 5.3 Spatial distribution of (**a**) K and (**b**) LS in the Muhuri River basin

northwestern parts of the basin in the soils of undulating plains with low mounds and narrow valleys (Fig. 5.3a).

Slope Length and Gradient Factor (LS)

The LS value of the present study ranged between 0 to 44.02. Higher value was found in the upper and the lower catchments of the basin due to the presence of the Baramura-Deotamura hill range and the TekkaTulsi hill, respectively, whereas the minimum value was observed in the central part of the basin over the flood plains and inter-hill valleys (Fig. 5.3b).

Crop Management Factor (C)

The C values of the study area ranged between 0 to 1 (Table 5.4). The lower value was observed in the eastern part of the basin due to the presence of forest cover

Table 5.4 C value of different LULC classes of the basin

LULC class	C value	Source
Dense forest	0.004	Dabral et al. (2008)
Open forest	0.008	Dabral et al. (2008)
Degraded forest	0.008	Dabral et al. (2008)
Agricultural land	0.28	Dabral et al. (2008)
Rubber plantation	0.20	Sujaul et al. (2012)
Shifting cultivation	0.33	Dabral et al. (2008)
Settlement	1.00	Dabral et al. (2008)
Water body	0.00	Ganasri and Ramesh (2016)
Barren land	0.18	Pandey et al. (2007)

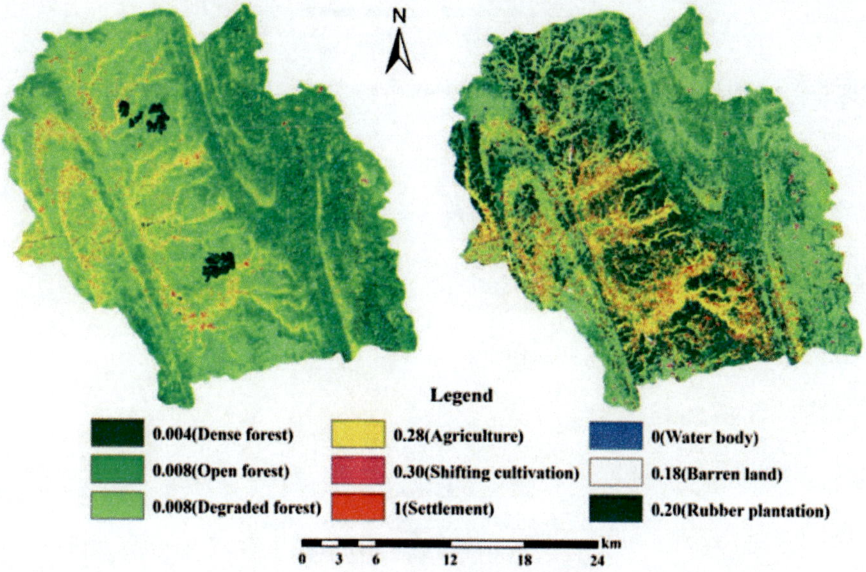

Fig. 5.4 Spatial distribution of C in the Muhuri River basin for the years 1986 and 2019

whereas the central and the western parts demonstrated higher values for the presence of agricultural land, settlement and rubber plantation. However, the occurrence of higher C value in the year 2019 in comparison to the year 1986 was due to the active interference of rubber plantation (Fig. 5.4).

Support Practice Factor (P)

Contour cultivation, strip cropping and terrace system are the most essential conservation practices revealed in the USLE. During field verification, only bundings in the agricultural lands was observed which was ineffective. Therefore, the *P* value was set to 0.28 for paddy fields and 1 for rest of the areas (Fig. 5.5).

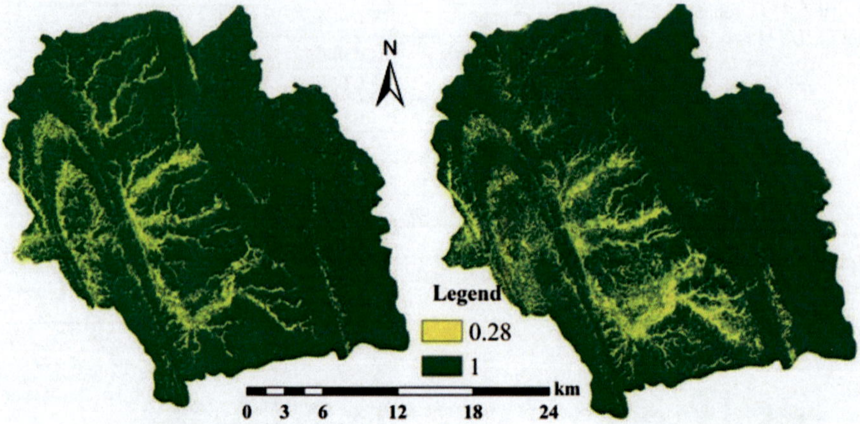

Fig. 5.5 Spatial distribution of P factor in the study area

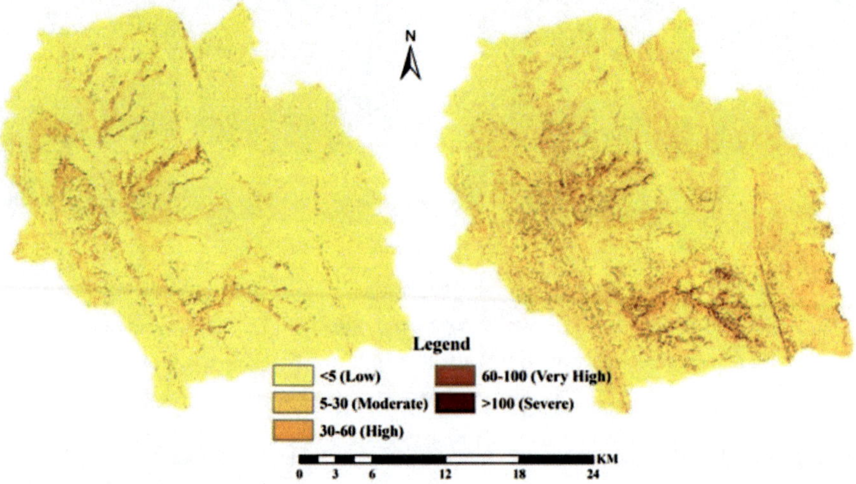

Fig. 5.6 Spatial distribution of annual soil loss of the study area for 1986 and 2019

Spatial and Temporal Features of Soil Erosion in the Muhuri River Basin During 1986 and 2019

Present study shows that during the year 1986 and 2019, the average annual soil loss from the basin was between 0 to 101.06 t ha^{-1} $year^{-1}$ and 0 to 110.08 t ha^{-1} $year^{-1}$, respectively (Fig. 5.6). The amount of annual soil loss has been categorised into five classes, such as low, medium, high, very high and severe. During 1986 and 2019, about 82.58 and 51.58% areas were under low soil erosion class, respectively (Table 5.5), where high vegetal cover in comparison to the open forests was found.

Table 5.5 Category-wise areas under soil erosion during 1986 and 2019

Average annual soil loss (t ha^{-1} year^{-1})	Erosion class	1986		2019		Change (1986–2019)	
		Area (ha.)	Area (%)	Area (ha.)	Area (%)	Area (ha.)	Area (%)
<5	Low	57,713.06	82.58	36,044.69	51.58	−21,668.37	−37.54
5–30	Moderate	7532.75	10.78	26,498.11	37.92	18,965.36	251.77
30–60	High	3071.7	4.39	4636.08	6.63	1564.38	50.93
60–100	Very high	1239.89	1.77	1903.94	2.72	525.79	42.41
>100	Severe	323.89	0.46	799.17	1.14	475.28	145.74

In these years, about 10.78 and 37.92% of the basin experienced moderate soil loss, 4.39 and 6.63% areas high soil loss, 1.77 and 2.72% very high and 0.46 and 1.14% areas severe soil loss (>100 t ha^{-1} year^{-1}), respectively.

During the periods 1986–2019, the areas under low soil loss were decreased to 37.54%. On contrary, the areas under moderate, high, very high and severe soil loss were significantly increased to 251.77, 50.93, 42.41 and 145.74%, respectively (Table 5.5). The central portion of the basin was mostly occupied by agricultural lands and settlements but very recently rubber plantation was flourished vigorously in this part (Fig. 5.6).

Such factors boosted the status of soil erosion in that part of the basin. As we know that the hilly parts of the basin are mainly affected by the traditional jhooming practised by the indigenous people and that's why the degraded forests still remain in high altitudes which ultimately become responsible for increasing soil erosion in that part.

Analysis of Relation Between Mean Soil Loss and LULC

Within the study period of 1986 to 2019, about 92.05, 68.63 and 5.05% areas under dense forest, open forest and degraded forest were decreased, respectively, whereas, agricultural land, area under rubber plantation, settled area and area under barren land were increased to 14.47, 1600.39, 95.36 and 197.33%, respectively.

The authors observed a relation between the LULC change and the rate of mean soil loss in the Muhuri River basin during the periods 1986–2019. With modification of the land cover, the erosion rate was changed significantly. The rate of mean soil loss in the areas under degraded forest, agriculture, rubber plantation, jhoom cultivation and settlement were 0.25, 11.02, 0.71, 3.55 and 12.85 t ha^{-1} year^{-1}, respectively, during the year 1986 and became 2.62, 12.24, 2.21, 4.87 and 13.25 t ha^{-1} year^{-1}, respectively, during 2019. On contrary, during 1986, the rate of mean soil loss from the areas under dense forest and open forest were 0.01 and 0.02 t ha^{-1} year^{-1}, respectively, but with decrease in areal coverage the soil loss became absent during 2019. In case of degraded forest, agricultural land, rubber

Table 5.6 LULC-wise mean annual soil loss of the study area

	1986		2019		1986–2019	Mean soil loss ($t\ ha^{-1}\ year^{-1}$)		
LULC	Area (ha)	Area (%)	Area (ha)	Area (%)	Change (%)	1986	2019	Change
Dense forest	4140	5.90	329	0.47	−92.05	0.01	0	−0.01
Open forest	20,063	28.59	6293	8.97	−68.63	0.02	0	−0.02
Degraded forest	32,892	46.87	3123	44.50	−5.05	0.25	2.62	2.37
Agricultural land	7014	9.99	8029	11.44	14.47	11.02	12.24	1.22
Rubber plantation	1028	1.46	1748	24.91	1600.39	0.71	2.21	1.50
Shifting cultivation	1202	1.71	898	1.28	−25.29	3.55	4.87	1.32
Settlement	2393	3.41	4675	6.66	95.36	12.85	13.25	0.40
Water body	1365	1.94	1015	1.45	−25.64	0	0	0
Barren land	75	0.11	223	0.32	197.33	1.71	2.39	0.68
Total	70,172	100	70,172	100				

plantation, jhoom cultivation, settlement and barren land the mean soil erosion was increased by 2.37, 1.22, 1.5, 1.32, 0.4 and 0.68 $t\ ha^{-1}\ year^{-1}$, respectively, whereas in case of dense forest and open forest, it was decreased considerably (Table 5.6).

Within the study period, all categories of soil erosion illustrated decreasing trend in case of the dense and open forests as areal coverage declined significantly. Again, in case of the degraded forest, the percentage of all erosion classes, except moderate and high class, were declined (Table 5.7). Although agricultural lands are of gentle slope but the absence of proper conservation practice augmented moderate, high and very high classes of erosion (Table 5.7). Rubber plantation was newly introduced in the study area and mainly found on moderate and steep slopes. Since it's a profit benefit income source, therefore, the area under rubber plantation increased significantly within the basin. During the year 1986, the area under rubber plantation was only 1.46%, whereas it became 24.91% in the year 2019. As a result, all categories of soil erosion except severe class were increased. However, despite decline in shifting cultivation by 25.29% areas under erosion classes were increased.

The settled areas which have considerable uncovered surface were revealed as most vulnerable to erosion. The areas under settlement were increased by 95.36% within the study period with increasing population pressure. Consequently, the areas under moderate, high, very high and severe erosion were augmented by 199.24, 215.20, 93.05 and 54.84%, respectively (Table 5.7). On the other hand, the barren lands were free from high, very high and severe erosion in both the years, while the susceptibility to low and moderate erosion were increased by 242.86 and 100%, respectively, with increasing the barren land from 0.11 to 0.32%. A clear relation between LULC change and amount of mean soil loss was evident in the study area. Growing anthropogenic activities had already altered the natural forest cover into

Table 5.7 Actual and changed area under different LULC classes with erosion status

Soil erosion class	Area (km^2)																	
	Dense forest		Open forest		Degraded forest		Agricultural land		Rubber plantation		Shifting cultivation		Settlement		Water body		Barren land	
	1986	2019	1986	2019	1986	2019	1986	2019	1986	2019	1986	2019	1986	2019	1986	2019	1986	2019
Low	40.65	3.27	194.43	63.33	339.93	229.82	15.91	13.12	7.56	139.8	2.4	2.53	8.76	5.57	0.0	0.0	0.56	1.92
Moderate	0.46	0.0	10.61	0.0	1.95	76.44	17.37	23.88	0.23	29.23	1.05	2.67	3.92	11.73	0.0	0.0	0.13	0.26
High	0.05	0.0	0.93	0.0	2.15	7.46	15.53	24.73	0.03	2.49	0.32	1.07	4.87	15.35	0.0	0.0	0.01	0.01
Very high	0.01	0.0	0.03	0.0	1.18	0.1	7.63	12.58	0.01	0.03	0.05	0.06	3.02	5.83	0.0	0.0	0.0	0.0
Severe	0.0	0.0	0.01	0.0	0.57	0.0	3.22	2.41	0.0	0.0	0.02	0.04	1.55	2.4	0.0	0.0	0.0	0.0
Soil erosion class	**Changed area**																	
	km^2	%	**km^2**	%	**km^2**	%	**km^2**	%	**km^2**	%	**km^2**	%	**km^2**	%	**km^2**	%	**km^2**	%
Low	−37.38	−91.96	−131.1	−67.43	−110.11	−32.39	−2.79	−17.54	132.24	1749.21	0.13	5.42	−3.19	−36.42	0.0	0.0	1.36	242.86
Moderate	−0.46	−100	−10.61	−100	74.49	3820	6.51	37.48	29	12608.7	1.62	154.29	7.81	199.24	0.0	0.0	0.13	100
High	−0.05	−100	−0.93	−100	5.31	246.98	9.2	59.24	2.46	8200	0.75	234.36	10.48	215.20	0.0	0.0	0.0	0.0
Very high	−0.01	−100	−0.03	−100	−1.08	−91.53	4.95	64.88	0.02	200	0.01	20	2.81	93.05	0.0	0.0	0.0	0.0
Severe	0.0	0.0	−0.01	−100	−0.57	−100	−0.81	−25.16	0.0	0.0	0.02	100	0.85	54.84	0.0	0.0	0.0	0.0

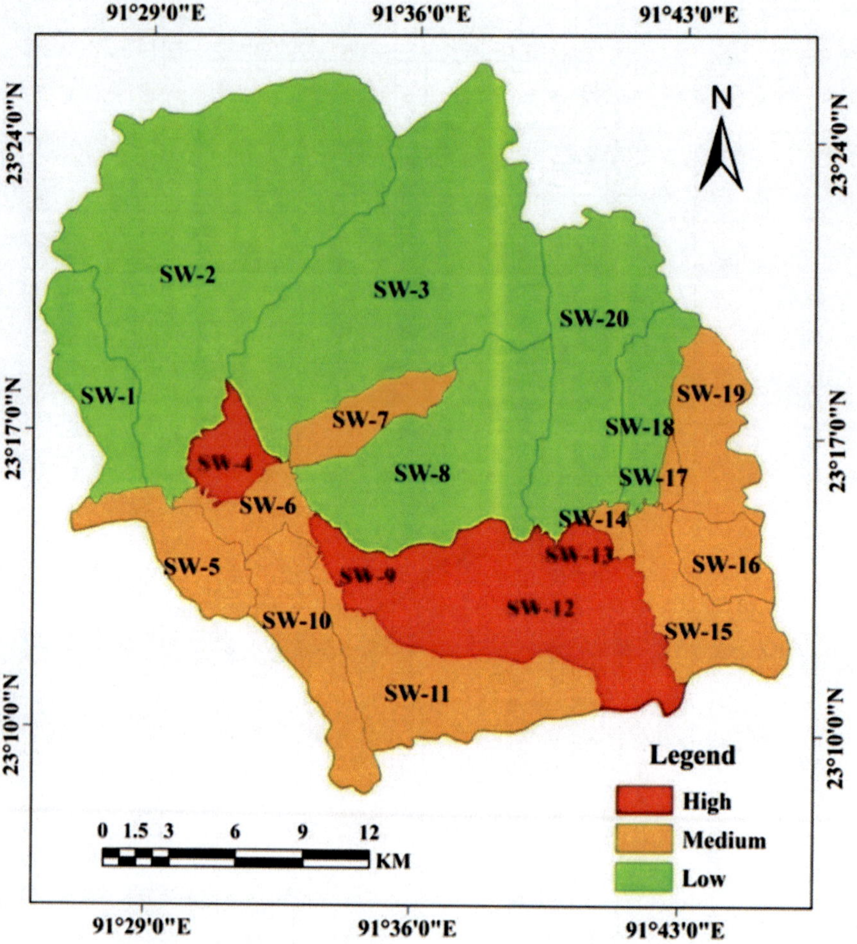

Fig. 5.7 Priority-wise sub-watersheds (SW) for control of soil erosion

degraded forest, as a result, areas under dense and open forests were decreased in the hilly areas and degraded forests still maintained higher position in the study area in general and in the hilly part in particular (Debnath et al., 2017a, b). Therefore, the combination of steep slope and degraded forest had accelerated the moderate, high and very high rate of soil loss in the hilly areas, although shifting cultivation had also contributed to moderate to severe erosion in the hilly part.

Rubber plantation was recognised as a substitute monoculture practice in the forest areas, mainly in the tilla lands with moderate to steep slope. Since the owners collect latex regularly and there is a chance of forest fire in plantation during dry season from the flushing leaves (Fig. 5.7a), the owners always try to keep the floor clean by clearing the secondary growth and removing the flushing leaves from the floor (Fig. 5.7b). Thus, the plantation floors always remain barren which promote soil erosion, especially during the monsoon season. According to Wu et al. (2001), conversion of tropical forests to rubber monoculture increases the rate of soil erosion significantly. The central and the western parts of the basin were occupied by rubber plantation and gradually extended towards the hilly part leading to widespread moderate to very high erosion. In addition, moderate to steep slopes in the plantation areas favour runoff when rainfall intensity becomes high. Therefore, the barren ground surfaces of the plantations experience soil loss. The increased demand for agricultural land owing to the growing population in the study area had consequent upon the clearing of forest lands.

During field verification, it was observed that the farmers became conscious about soil fertility and its importance in high yield. As a result, the area under severe erosion class was slightly decreased, but moderate, high and very high erosion class still persist in the central part of the basin. Prolonged dry period followed by high intensity rainfall contributes high risk of soil erosion in the settled areas. The area under current fallow increases during dry season on account of shortage of irrigation water and contributes to high erosion.

The assessment of soil erosion using satellite images and GIS technique becomes the most important means for evaluation and monitoring of the past and present scenario of a macro- as well as a micro-geographical area in repetitive timescale coverage. Obviously, the mean rate of soil erosion is related with the nature of LULC.

The major changes identified in the river basin were due to the introduction of monoculture activity and decrease of natural forest cover. The significant increase in soil erosion had accelerated sedimentation process in downstream of the Muhuri River.

Prioritization of the Sub-watersheds (SW) for the Management

Identification of the priority areas of the entire Muhuri River basin is crucial for taking decision on soil and water conservation with the forest management programme (Fig. 5.7). For this purpose, the researcher has considered the actual amount of erosion, change of the erosion amount within the period 1986 to 2019 due to LULC change and LULC-wise estimated erosion. In the present study, 18 SW of the Muhuri River basin have been prioritized into three classes: low, medium and high based on mean soil loss from them measured through USLE model (Table 5.8).

Low priority level: The sub-watersheds 1, 2, 3, 8, 18 and 20 come under low priority level class where mean soil loss ranges between 2.43 to 4.40 t ha^{-1} $year^{-1}$. Hence, it's indicating that these watersheds do not need instantaneous measurement to reduce the soil erosion.

Medium priority level: The sub-watersheds with the mean soil erosion range of 4.40–6.50 t ha^{-1} $year^{-1}$come under the medium priority level. The sub-watersheds 5, 6, 7, 10, 11, 14, 15, 16, 17 and 19 have come under this category. There is need of

Table 5.8 Priority ranking of SW on the basis of mean soil loss

	Soil loss (t ha^{-1} year^{-1})		
Code of sub-watershed	Mean	SD	Category
SW-1	3.08	3.61	Low
SW-2	3.35	4.33	Low
SW-3	4.02	5.24	Low
SW-4	7.07	7.58	High
SW-5	4.88	5.3	Medium
SW-6	5	5.42	Medium
SW-7	5.48	7.05	Medium
SW-8	3.73	4.32	Low
SW-9	6.90	7.20	High
SW-10	4.98	5.74	Medium
SW-11	4.41	5.7	Medium
SW-12	7.4	8.22	High
SW-13	7.7	9.87	High
SW-14	4.97	6.32	Medium
SW-15	5.11	5.5	Medium
SW-16	5.72	5.34	Medium
SW-17	4.17	2.61	Medium
SW-18	3.87	3.34	Low
SW-19	4.45	3.99	Medium
SW-20	2.43	2.96	Low

proper attention in order to check these sub-watersheds from conversion to the successive priority level.

High priority level: Four sub-watersheds i.e., SW-4, 9, 12 and 13 have given under high priority level due to high mean soil loss of more than 6.50 t ha^{-1} year^{-1}. Hence, immediate action plan should be needed in these watersheds to minimize the soil erosion.

The study of the LULC map shows that the SW having agriculture and settlement come under high, as well as medium category. All other sub-watersheds except SW-15, 16, 17 and 19 come under the high and medium priority level, mainly the zone of rubber plantation, which have spontaneous soil erosion, whereas those remaining sub-watersheds, located in the hilly areas and mostly under shifting cultivation, have ultimately achieved the medium priority level in the soil and water conservation map.

Strategies of Soil Conservation

In order to reduce the intensity of the soil erosion in the Muhuri River basin, few alternative strategies has been suggested in the present study. Since the study area is situated in the hilly part of NorthEast India; therefore shifting cultivation is mostly dominated in the upper catchment. This unscientific way of cultivation recognised as a vital factor for the soil erosion. Thus, terrace cultivation is recognised as a most effective conservation technique in the hilly part, which will reduce the surface runoff along with velocity of the rain water and reduced the soil erosion. According to Guo et al. (2019) and Dai et al. (2018), contour ridge tillage and cross ridge tillage are the most important techniques to reduce the soil erosion in the hilly region compare to the down slope tillage. The experimental study of Sharma et al. (2017) and Dai et al. (2018) suggested that intercropping and mulching practices can be a another techniques to reduce the soil erosion in the farm land of the hilly region.

Moreover, construction of check dams in the streams of the hilly parts can be effective measures to reduce the supply of eroded materials to the main channel which will reduce the problem of flood hazard in the lower parts of the basin. Identification of the suitable areas of the afforestation using the modern geospatial techniques like GIS and remote sensing and implementation of the proper afforestation programme can be a important measures in this watershed. This technique will increase the forest areas and reduce the soil erosion of the study area.

5.5 Conclusion

The role of human activities behind LULC change during last 33 years which led to change the soil erosion potential, i.e., either positive or negative have been revealed in the present study. Very limited parts of the watershed have protective land cover while most of the parts, mainly hilly areas, were affected by deforestation; tilla and plain areas are experienced by infrastructure and related land degradation which inherently enlarged the risk of the soil erosion. Obviously, the mean rate of soil erosion is associated with the LULC change of the study area, and therefore, in case of degraded forest, agricultural land, rubber plantation and jhoom cultivation, the mean soil loss was increased by 2.37, 1.22, 1.50 and 1.32 t ha^{-1} $year^{-1}$ correspondingly within the study period (1986–2019). The main reasons for such increase of soil erosion potential in the river basin over the study period were decreased natural forest cover, increased settled area, barren land, agricultural land and existence of shifting cultivation. However, this enhanced erosion accelerated sedimentation problem in downstream of the Muhuri River.

The study illustrated that the conversion of forest land into rubber plantation became a common phenomenon in the basin which eliminated the effective impediment to soil erosion. Hence, there is a need of management through proper programme implementation. Recently in order to develop jhum areas, a number of integrated approaches of mixed land use system become most suitable in the hilly part. The approaches include promotion of modern agriculture, intercropping diversification with local preference, jhum fallow management, tree farming, floriculture, agroforestry, watershed management, etc. (Jamir et al., 2004; Verma et al., 2017). Growth of vegetal cover on the barren surfaces and degraded forests is necessary to reduce the soil erosion potential of the river basin. Moreover, there is scope of further study within this study area where priority of sub-watershed and afforestation programme related researches could be carried out in order to facilitate the management plan.

References

Ahmad, N. S. B. N., Mustafa, F. B., Muhammad Yusoff, S. Y., & Didams, G. (2020). A systematic review of soil erosion control practices on the agricultural land in Asia. *International Soil and Water Conservation Research, 8*, 103–115.

Ahmed, I., Das(Pan), N., Debnath, J., & Bhowmik, M. (2017). An assessment to prioritise the critical erosion-prone sub-watersheds for soil conservation in the Gumti basin of Tripura, North-East India. *Environmental Monitoring and Assessment, 189*(600), 1–15.

Angima, S., Stott, D., Oneill, M., Ong, C., & Weesies, G. (2003). Soil erosion prediction using RUSLE for Central Kenyan highland conditions. *Agriculture, Ecosystems and Environment, 97* (1), 295–308.

Asselman, N. E. M., & Middelkoop, H. (1995). Floodplain sedimentation: Quantities, patterns, and processes. *Earth Surface Processes and Landforms, 20*, 481–499.

Bhattacharyya, R., Ghosh, B. N., Dogra, P., Mishra, P. K., Santra, P., Kumar, S., et al. (2016). Soil Conservation Issues in India. *Sustainability, 8*, 565. https://doi.org/10.3390/su8060565

Congalton, R. G. (1991). A review of assessing the accuracy of classifications of remotely sensed data. *Remote Sensing of Environment, 37*, 35–46.

Dabral, P. P., Baithuri, N., & Pandey, A. (2008). Soil erosion assessment in a hilly catchment of north-eastern India using USLE, GIS and remote sensing. *Water Resources Management, 22*, 1783–1798.

Dai, C., Liu, Y., Wang, T., Li, Z., & Zhou, Y. (2018). Exploring optimal measures to reduce soil erosion and nutrient losses in southern China. *Agricultural Water Management, 210*, 41–48. https://doi.org/10.1016/j.agwat.2018.07.032

Debnath, J., Das(Pan), N., Ahmed, I., & Bhowmik, M. (2017a). Channel migration and its impact on land use/land cover using RS and GIS: A study on Khowai River of Tripura, North-East India. *The Egyptian Journal of Remote Sensing and Space Sciences, 20*, 197–210.

Debnath, J., Das(Pan), N., Ahmed, I., & Bhowmik, M. (2017b). Chronological change of land use/land cover of the Muhuri River basin from 1972 to 2016, Tripura, North-East India. *Indian Journal of Science and Technology, 10*(22), 1–18.

Delgado, J. (2020). Estimation of soil loss using the USLE model and GIS tools in the Chill6n river basin, Lima, Peru. *South Sustainability, 1*(1), e007. https://doi.org/10.21142/SS-0101-2020-007

Esa, E., Assen, M., & Legass, A. (2018). Implications of land use/cover dynamics on soil erosion potential of agricultural watershed, North-western highlands of Ethiopia. *Environmental Systems Research, 7*, 21.

Fiener, P., Dostál, T., Krása, J., Schmaltz, E., Strauss, P., & Wilken, F. (2020). Operational USLE-based modelling of soil erosion in Czech Republic, Austria, and Bavaria—Differences in model adaptation, parametrization, and data availability. *Applied Sciences, 10*, 36–47. https://doi.org/10.3390/app10103647

Ganasri, B. P., & Ramesh, H. (2016). Assessment of soil erosion by RUSLE model using remote sensing and GIS—A case study of Nethravathi Basin. *Geoscience Frontiers, 7*, 953–961.

Ghosh, K., De, S. K., Bandyopadhyay, S., & Saha, S. (2013). Assessment of soil loss of the Dhalai River basin, Tripura, India using USLE. *International Journal of Geosciences, 4*, 11–23.

Girmay, G., Moges, A., & Muluneh, A. (2020). Estimation of soil loss rate using the USLE model for Agewmariayam Watershed, northern Ethiopia. *Agriculture & Food Security, 9*, 9. https://doi.org/10.1186/s40066-020-00262-w

Guo, S. F., Zhai, L. M., Liu, J., Liu, H. B., Chen, A. Q., Wang, H. Y., et al. (2019). Crossridge tillage decreases nitrogen and phosphorus losses from sloping farmlands in southern hilly regions of China. *Soil and Tillage Research, 191*, 48–56. https://doi.org/10.1016/j.still.2019.03.015

Huang, C., Yang, Q., Cao, X., & Li, Y. (2020). Assessment of the soil erosion response to land use and slope in the loess plateau—A case study of Jiuyuangou. *Water, 12*, 529. https://doi.org/10.3390/w12020529

Jamir, A., Tiwari, B. K., Choudhury, D., Yim, S. K., Singh, L. J., Roy, S., Nakaro, V., Darlong, V. T., & Wotsa, Q. (2004). *Farmers innovations in different shifting cultivation systems in the Eastern Himalayas: North East India*. Report Prepared for ICIMOD.

Jordan, G., van Rompaey, A., Szilassi, P., Csillag, G., Mannaerts, C., & Woldai, T. (2005). Historical land use changes and their impact on sediment fluxes in the Balaton basin (Hungary). *Agriculture Ecosystems and Environment, 108*, 119–133.

López-García, E. M., Torres-Trejo, E., López-Reyes, L., Flores-Domínguez, Á. D., Peña-Moreno, R. D., & López-Olguín, J. F. (2020). Estimation of soil erosion using USLE and GIS in the locality of Tzicatlacoyan, Puebla, México. *Soil and Water Research, 15*(1), 9–17.

Moore, I. D., & Burch, G. J. (1986). Physical basis of the length slope factor in the universal soil loss equation. *Soil Science Society of America, 50*(5), 1294–1298.

Narayan, D. V. V., & Babu, R. (1983). Estimation of soil erosion in India. *Journal of Irrigation and Drainage Engineering, 109*(4), 419–431.

Pandey, A., Chowdary, V. M., & Mal, B. C. (2007). Identification of critical erosion prone areas in the small agricultural watershed using USLE, GIS and remote sensing. *Water Resources Management, 21*, 729–746.

Pandey, A., Chowdary, V. M., & Mal, B. C. (2009). Sediment yield modelling of an agricultural watershed using MUSLE, remote sensing and GIS. *Paddy and Water Environment, 7*, 105–113. https://doi.org/10.1007/s10333-009-0149-y

Phinzi, K., Ngetar, N. S., & Ebhuoma, O. (2020). Soil erosion risk assessment in the Umzintlava catchment (T32E), Eastern Cape, South Africa, using RUSLE and random forest algorithm. *South African Geographical Journal, 103*(2), 139–162. https://doi.org/10.1080/03736245.2020.1716838

Pradeep, G. S., Ninu Krishnan, M. V., & Vijith, H. (2015). Identification of critical soil erosion prone areas and annual average soil loss in an upland agricultural watershed of Western Ghats, using analytical hierarchy process (AHP) and RUSLE techniques. *Arabian Journal of Geosciences, 8*, 3697–3711.

Prasannakumar, V., Vijith, H., Geetha, N., & Shiny, R. (2011). Regional scale erosion assessment of a sub-tropical highland segment in the Western Ghats of Kerala, South India. *Water Resources Management, 25*(14), 3715–3727.

Richards, K., Brasington, J., & Hughes, F. (2002). Geomorphic dynamics of floodplains: Ecological implications and a potential modelling strategy. *Freshwater Biology, 47*, 559–579.

Roy, P. (2019). Application of USLE in a GIS environment to estimatesoil erosion in the Irga watershed, Jharkhand, India. *Physical Geography, 40*(4), 361–383.

Sharma, A., Tiwari, K. N., & Bhadoria, P. B. S. (2011). Effect of land use land cover change on soil erosion potential in an agricultural watershed. *Environmental Monitoring and Assessment, 173*, 789–801.

Sharma, N. K., Singh, R. J., Mandal, D., Kumar, A., Alam, N. M., & Keesstra, S. (2017). Increasing farmer's income and reducing soil erosion using intercropping in rainfed maize-wheat rotation of Himalaya, India. *Agriculture, Ecosystems & Environment, 247*, 43–53. https://doi.org/10.1016/j.agee.2017.06.026

Shit, P. K., Bhunia, G. S., & Maiti, R. (2013). Assessing the performance of check dams to control rill-gully erosion: Small catchment scale study. *International Journal of Current Research, 5*(4), 899–906.

Shit, P. K., Nandi, A. S., & Bhunia, G. S. (2015). Soil erosion risk mapping using RUSLE model on Jhargram sub-division at West Bengal in India. *Modeling Earth Systems and Environment, 1*, 28.

Sujaul, I. M., Muhammad Barzani, G., Ismail, B. S., Sahibin, A. R., & MohdEkhwan, T. (2012). Estimation of the rate of soil erosion in the TasikChini Catchment, Malaysia using the RUSLE Model integrated with the GIS. *Australian Journal of Basic and Applied Sciences, 6*(12), 286–296.

Uddin, K., Murthy, M. S. R., Wahid, S. M., & Matin, M. A. (2016). Estimation of soil erosion dynamics in the Koshi Basin using GIS and remote sensing to assess priority areas for conservation. *PLoS One, 11*(3), 1–19.

Verma, P. K., Kumar, V., Chandra, A., & Thounaojam, B. (2017). Alternatives of shifting cultivation in North-eastern region of India. *Report and Opinion, 9*(12), 1–8.

Verstraeten, G., & Poesen, J. (2002). Regional scale variability in sediment and nutrient delivery from small agricultural watersheds. *Journal of Environmental Quality, 31*, 870–879.

Wischmeier, W. H., & Smith, D. D. 1978. *Predicting rainfall erosion loss: A guide to conservation planning*. Agricultural Handbook, No. 537, US, Department of Agriculture, Agricultural Research Service, Washington.

Wu, Z. L., Liu, H. M., & Liu, L. Y. (2001). Rubber cultivation and sustainable development in Xishuangbanna, China. *International Journal of Sustainable Development and World Ecology, 8*, 337–345.

Chapter 6
Application of Geoinformatics and AHP Technique to Delineate Flood Susceptibility Zone: A Case Study of Silabati River Basin, West Bengal, India

Debasis Ghosh, Monali Banerjee, Manas Karmakar, and Dayamoy Mandal

Abstract Among all the natural disasters, floods are the most common phenomena that cause huge obliteration to the human lives and socio-economic and cultural infrastructures. Silabati, a monsoon influenced river of West Bengal is well known for frequent flooding events in its lower basin areas. In the present study, an attempt has been made to delineate flood susceptible areas of Silabati river basin using AHP (Analytical Hierarchy Process) technique and geospatial technology. A total number of 11 physiographic, climatic, and anthropogenic factors (elevation, slope, flow accumulation, distance from river, drainage density, geomorphology, lithology, surface runoff, topographic wetness index, land use land cover, and curvature) are taken into consideration to prepare the flood susceptibility map of the study area. The map is categorized into five distinct flood susceptible zones, such as very high, high, moderate, low, and very low susceptible zones, and these zones cover 14.04%, 20.67%, 21.76%, 20.69%, and 22.84% of the total basin area, respectively. Keshpur, Ghatal, Chandrakona-I, Chandrakona-II, and Daspur-I Community Development (C.D.) blocks of West Medinipur district located in lower Silabati basin fall under very high and high flood susceptibility zones. The performance and efficiency of AHP are validated using Area Under Curve (AUC) method, which ensured significant accuracy (76.41%) of the study. A large number of people residing on lower Silabati basin along with several socio-economic and cultural structures get severely affected many times during floods. Therefore, this study may facilitate the formulation and implementation of flood management strategies in the vulnerable areas of Silabati river basin.

Keywords Geospatial technology · AHP · Causative factors · Flood susceptibility · Silabati river

---

D. Ghosh (✉) · M. Banerjee · M. Karmakar · D. Mandal
Department of Geography, University of Calcutta, Kolkata, West Bengal, India

P. K. Shit et al. (eds.), *Geospatial Technology for Environmental Hazards*, Advances in Geographic Information Science, https://doi.org/10.1007/978-3-030-75197-5_6

6.1 Introduction

Flood is the most common among all catastrophic environmental hazards; it causes massive damages to the natural and human resources. In respect to the human society, flood is the most devastating natural hazard because of its irreversible damage capacity to infrastructure, life loss, loss of agricultural, social disruption, etc. (Taylor et al., 2011; Dawod et al., 2011; Swain et al., 2020). The low-lying regions of the world are mostly facing the terrible form of floods due to water logging for incessant rainfall, exhaust of water beyond the carrying capacity of the stream, settlement in the front of swirling water, etc. (Ghosh & Kar, 2018; Natarajan et al., 2021). Globally, on an average, 170 million people are adversely affected by flood every year (Kowalzig, 2008; Das, 2019). In between 1998 and 2017, more than two billion people are affected due to the devastation caused by flood events across the world (World Health Organization, 2018), and 656 million USD economic loss with more than 1.4 lakh human life loss have occurred during this time span (Wallemacq & Rowena, 2018). A total number of 4731 people have lost their lives across the world only in floods occurred in 2016, where average annual death (2006–2015) in flood is 5709 people (Guha-Sapir et al., 2016; Chowdhuri et al., 2020). According to world disaster report (2015), Asia recorded highest number of affected and dead people due to flood around the world in between 2005 and 2014 (World Disasters Report 2015). Most of the developing countries of Asia bear the devastating results of flood, such as China, India, Pakistan, Bangladesh, Sri Lanka, Indonesia, etc. (Wallemacq & Rowena, 2018). In India, every year since 1953 to 2016, 7.19 million hectare area and 31.88 million people on an average are affected by flood. During this time period due to flood India has suffered a loss of 347,581.201 crore rupees and the increasing trend of loss with years is going on (Flood Damage Statistics, 2018). In 2017, a total number of 868 people lost their lives by flood during mid of august (Hindustan Times, 2017a). In the next year (August 2018) a massive flood hit Kerala, where 1.24 million people dislocated and 474 people died (Sphere India, 2018). This terrible scenario of adverse effects of flood does not change very much for West Bengal. The flood damage statistics of 2018 revealed that around 244.23 million people of West Bengal are affected by floods occurs in between 1953 and 2016, and the state has suffered with a loss of 43,954.792 crore rupees (Flood Damage Statistics, 2018).

Floods in India are mainly occurred because of natural (heavy monsoonal rainfall, cyclone, cloud blast, etc.) and manmade (large dam, siltation of river, sand mining, unsystematic management of drainage, etc.) factors (Şen, 2018; Chowdhuri et al., 2020; Sarkar & Mondal, 2020). In India, rural areas are facing huge agricultural loss, occasional loss of livestock and human life, etc. and urban areas are suffering with losses of domestic assets and infrastructures due to inundation of flood plain of the river, costal area, etc. (Şen, 2018). This amount of losses and damages from floods are increasing every year in relation with increasing trends of flood frequency (Flood Damage Statistics, 2018). In the previous two decades, the frequency of floods has increased by 40% worldwide (Hirabayashi et al., 2013; Khosravi et al., 2019).

Therefore, impeding an increase in loss and damages by flood are essential for the sake of human society and environment (Jongman, 2018; Genovese & Thaler, 2020). The analysis of flood events is very significant to reduce the potential damages of floods (Huang et al., 2008; Samanta et al., 2018). Pre-flood measure, flood forecasting, post-flood measures are the impost steps of flood analysis (Wanders et al., 2014; Chowdhuri et al., 2020). The first step of flood mitigation is the identification of flood vulnerable zone, which helps to forecasting the flood events (Sanyal & Lu, 2003). Therefore, flood susceptibility mapping is the important part of flood hazard mitigation system, and it is very useful for early warning system, mitigation of future floods events, reducing causalities, etc. (Tehrany et al., 2015; Ali et al., 2019; Sarkar & Mondal, 2020). The susceptibility of flood is determined by several geographical and flood influencing factors, such as geomorphology, geology, rainfall, slope, elevation, land use land cover, drainage density, flow accumulation, etc. (Kazakis et al., 2015; Das, 2018; Hong et al., 2018; Das, 2019). All these factors are analyzed to assess the flood susceptibility zones. Various researchers from different field have adopted different kinds of mathematical and statistical techniques to analyze these factors, such as Analytical Hierarchy Process (AHP) (Das, 2018; Ali et al., 2019; Yousuf Gazi et al., 2019: Jabbar et al., 2019), Frequency Ratio (FR) (Khosravi et al., 2016; Lee et al., 2018; Ali et al., 2019; Liuzzo et al., 2019; Rahman et al., 2019; Sarkar & Mondal, 2020), Logistic Regression (LR) (Pradhan, 2010; Tehrany & Jones, 2017; Lee et al., 2018; Liuzzo et al., 2019; Rahman et al., 2019), Evidential Belief Function (EBF) (Shafapour Tehrany et al., 2019b; Chowdhuri et al., 2020), Artificial Neural Network (ANN) (Jahangir et al., 2019; Rahman et al., 2019), Analytic Network Process (ANP) (Dano et al., 2019), Shannon's Entropy (SE) (Haghizadeh et al., 2017; Liuzzo et al., 2019), Statistical Index (SI) (Shafapour Tehrany et al., 2019a), Weight of Evidence (WoE) (Khosravi et al., 2016; Rahmati et al., 2016), and Support Vector Machine (SVM) (Tehrany et al., 2015, 2019b), etc. All the models have some advantages and limitations; thus, selection of best model for flood susceptibility is a tough work. The most favorable and preferred multi-criteria decision-making technique is the AHP techniques for flood susceptibility mapping, and a wide number of researchers have already adopted AHP with adequate accuracy (Lawal et al., 2012; Matori et al., 2014; Elkhrachy, 2015; Ghosh & Kar, 2018; Yousuf Gazi et al., 2019; Hammami et al., 2019; Hoque et al., 2019; Jabbar et al., 2019; Phrakonkham et al., 2019; Rahman et al., 2019; Vojtek & Vojteková, 2019; Mishra & Sinha, 2020; Subbarayan & Sivaranjani, 2020; Chakraborty & Mukhopadhyay, 2019). In this technique, the impact of flood provoking factors is assessed by the application of pair-wise comparison matrix and ranked all the factors to delineate priorities zones (Vojtek & Vojteková, 2019; Swain et al., 2020). Enumerating, weight computation, interpolating and integration methods are incorporated in this technique (Chen et al., 2011). In the recent time, the integration of different factor to delineate the susceptible zones of flood is done through the application of Geographical Information System (GIS) and Remote Sensing (RS).

In the recent past, various researchers from different disciplines successfully integrate multi-criteria decision-making tools and GIS for the delineation of flood vulnerable zones, such as Das (2019) worked on flood susceptibility mapping and hydro-geomorphic response to flood. Integrated application of AHP and GIS is used to fulfill the objective; Saha and Agrawal (2020) adopted AHP and GIS technique to assess the flood risk of Prayagraj district, India. In their study, flow accumulation factor is considered as the high influencing factor to food; Sarkar and Mondal (2020) conducted a study on Kulik river basin to delineate the flood vulnerability zone. The FR model has been successfully used to perform the quantification of determination potentiality of causative factors of flood and integrated using GIS platform; Swain et al. (2020) have tried to find out the flood susceptibility zone of a flood vulnerable zone of Bihar using AHP and GIS technique. In their study, a total number of 21 flood causative factors are considered; other studies are Ghosh and Kar (2018); Chowdhuri et al. (2020); Deepak et al. (2020); Natarajan et al. (2021); Malik and Pal (2021); Arya and Singh (2021); etc. So, widely accepted integration approach of AHP and GIS has been used in the present study to find the susceptible zone of flood in Silabati river basin.

In perspective of flood, Silabati river is one of the most vulnerable rivers in south Bengal (Government of West Bengal, 2019). Almost every year overflow of Silabati river causes flood, specifically in the areas of Banka, Khirpai, and Ghatal of lower Silabati river basin (IWAI, 2016; Das et al., 2020). Due to inundation of adjacent areas of Silabati, thousands of people have dislocated and lots of damages have occurred (Hindustan Times, 2017b). In between 1978 and 2016, a total number of 10 high magnitude floods hit the flood plain of Silabati basin (Dandapat & Panda, 2018). During the inundation period, the loss of resources in lower reach of Silabati is maximum (Das et al., 2020). A number of flood hazards have occurred in the recent past years (2011, 2013, 2014, 2015, 2016, and 2017) in Silabati River caused huge damages (NRSC, 2018). The narrow channel of lower Silabati River indulges the overflow of water and causing of embankment breach (Das et al., 2020). This event causing high inundation of area and people suffered with huge agricultural losses, which terrifies the people of several villages (News18, 2019). Additionally, every year some causeways are submerged under the water of Silabati River (Anandabazar Patrika Online, 2020). In this context, flood susceptibility zone delineation and level of vulnerability assessment are very much essential for flood management in Silabati river basin. Considering the flood prone characteristics and adverse effect of flood in Silabati River, the present study was conducted. Thus, the present study tried to find out the flood susceptibility zones of Silabati river basin using AHP and geoinformatics. It will help to provide better forecasting about inundation intensity to the floodplain dwellers and make a positive step in flood management of Silabati river basin.

6.2 Geographical Accounts of Silabati River Basin

The Silabati River is also known as "Silai" which is originated from plateau fringe region of Chota Nagpur plateau and meets with Dwarakeswar River near Ghatal of Paschim Medinipur district, West Bengal. Every year flooding characteristics cannot permit to imagine about the source of Silabati river; it is originated from a pond near Napara village (23°14′09.94″ N and 86°38′42.14″ E) of Puncha community development (C.D.) block of Purulia district. After origin, the river run about 185 km southeasterly and several tributaries, such as Joypanda, Kubai, Tamal, Betal, Parang, Champa, Tarajoli, Mulajor, Purandar and Amoor, etc. enters into the main river channel during it course. The entire Silabati basin covers an area of 4195 km^2 and located Purulia, Bankura and West Medinipur districts of West Bengal, India (Fig. 6.1). The geographical extension of the river basin is 22°23′N to 23°15′N and 86°38′E to 87°46′E. Geomorphologically, the basin is located in between Bengal plateau and stable-self part of Bengal basin. Topography of Silabati river

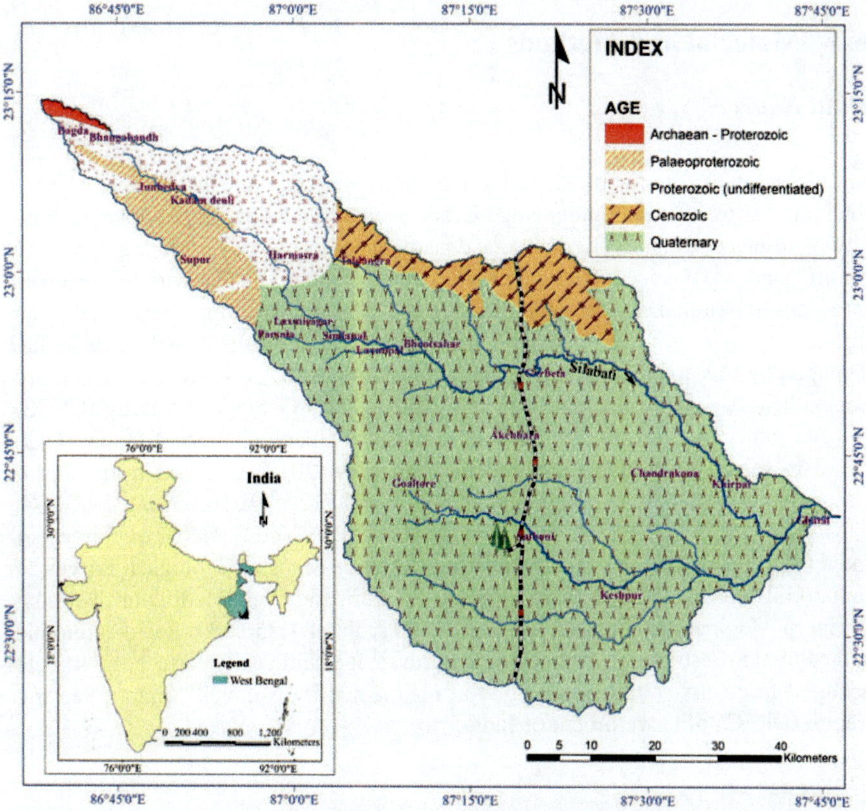

Fig. 6.1 Location map of Silabati river basin. Different symbols used within the basin indicate age of various geological formations. (Source: GSI and Google Earth Image)

basin gradually varies from upper reach (undulating surface of Chota Nagpur plateau) to lower reach (alluvium plain of Ganga delta) (Geological Survey of India, 2001). The isolated hillocks (average 150 m height) are found in the upper part, while isolated highlands are observed at the lower part. The high concentration of lateritic upland of Pleistocene and river flow along this upland formed the famous geomorphosite of West Bengal, i.e., bad land of Gangani (Mandal & Chakrabarty, 2021). The study area experiences sub-humid tropical climatic conditions, and receives annual 100–150 cm rainfall annually (Mahala, 2020). Most of the rainfall is occurred in Monsoon period. Almost in every year, lower reach of the basin encounters flood owing to physiographic characteristics and excessive rainfall over a short time span. It is quite evident from the structures of river bank protection near Talda, Khamardanga, Mathurakata, Gangra, Jhumka and Gugia villages (Bankura district) that the flood occurring nature of the river is very much severe. The important towns within the catchment area are Taldangra and Simlapal of Bankura district; Garbeta, Salboni, and Chandrakona of Paschim Medinipur district.

6.3 Material and Methods

Data Sources

In order to assess the flood risk and vulnerability mapping of Silabati river basin, a total number of 11 flood influencing factors, such as elevation, slope, flow accumulation, distance from river, drainage density, geomorphology, lithology, surface runoff, land use land cover, topographic wetness index, and curvature are taken into consideration. Data set of these selected causative factors is obtained from different types of sources, like Digital Elevation Model (DEM) of Shuttle Radar Topography Mission (SRTM) with spatial resolution of 1 Arc sec (acquired on 23rd September, 2014) downloaded from the United States Geological Survey (USGS) website (https://earthexplorer.usgs.gov); Landsat 8 Operational Land Image (OLI) of 30 m spatial resolution (acquired on 16th May, 2018) obtained from https://earthexplorer.usgs.gov; Topographical Maps (73I/12, 73I/16, 73J/13, 73M/4, 73M/8, 73M/12, 73N/1, 73N/2, 73N/5, 73N/6, 73N/9 and 73N/10) of Survey of India (SOI) on a scale of 1:50,000; District Resource Maps of Geological Survey of India (GSI) on a 1:250,000 scale (Purulia, Bankura and West Medinipur districts); geomorphological map derived from GSI on a scale of 1:250,000; Socio-economic and cultural infrastructures from Google Earth Image; and yearly surface runoff data acquired from https://bhuvan.nrsc.gov.in maintained by National Remote Sensing Centre (NRSC) of Government of India.

Preparation of Geospatial Layers

The flood susceptibility map of Silabati river basin is prepared using the mentioned factors (Fig. 6.2). All these factors are analyzed in geospatial platform, where geospatial layers of all these factors are prepared. The basin map of the study area is delineated from SRTM-DEM using the hydrology tools in ArcGIS software. Before the operation, SRTM-DEM has been projected with Universal Transverse Mercator (UTM), World Geodetic Survey 1984 (WGS-84) 45 North Zone. After the delineation, basin map was rectified with topographical maps. The elevation, slope, flow accumulation, topographic wetness index, curvature and drainage layer of Silabati river basin are also extracted from SRTM-DEM using spatial analyst tools in ArcGIS software. The drainage density map is prepared by using Inverse Distance Weighted (IDW) technique in ArcGIS. In the same way, buffer tools are used to prepare distance from river layer. The land use land cover layer has been prepared from Landsat 8 (OLI) satellite imagery using maximum likelihood algorithm of supervise classification in Erdas Imagine software. Geomorphology and lithology layer of the basin is prepared from geomorphological and geological map of GSI, respectively. The surface runoff layer of Silabati river basin is prepared from runoff data of NRSC and using the interpolation method (IDW) in ArcGIS software. The geomorphological and geological map has been geo-rectified with the UTM, WGS-84, 45N zone and prepared the required maps in ArcGIS 10.3.1 platform. Finally, all thematic layers are classified into five distinct categories following the Jenks natural break method.

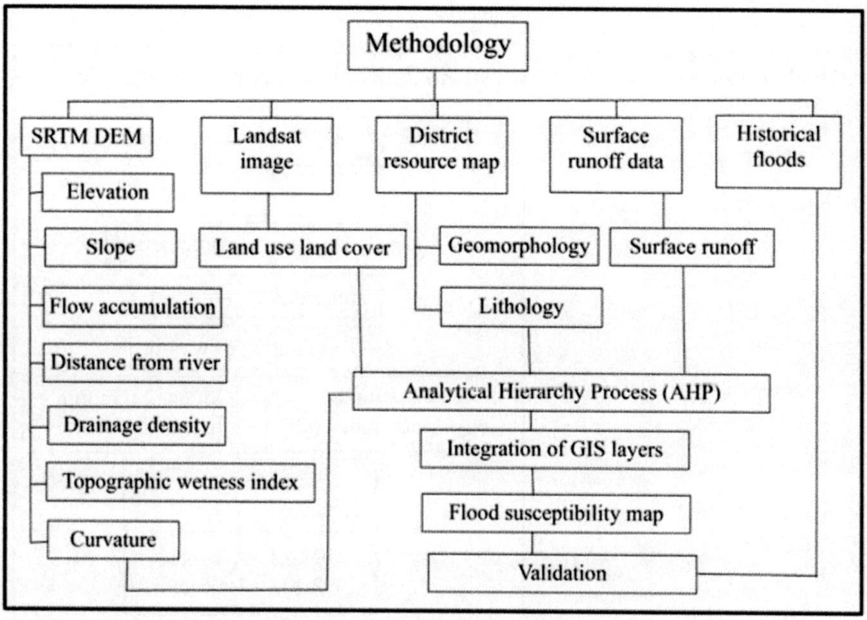

Fig. 6.2 Flowchart of the methodology adopted in the study

Analytical Hierarchy Process (AHP)

All the 11 flood causative factors are interrelated. In the present study, AHP technique is used to quantify the influencing capacity of each factor in flood occurrence through the assigning of weight to each factor (Fig. 6.3). The AHP technique is introduced by Saaty in 1980 (Saaty, 1980). It is a semi-quantitative multi-criteria decision-making approach, in which decisions are made through pair-wise comparison between different factors without inconsistency (Das, 2018; Chowdhuri et al., 2020; Ghosh et al., 2020a; Karmakar et al., 2021). In the field of environmental science several problems have severe complexity and it can be reduced by the application of AHP technique (Kannan, 2010; Çelik, 2019; Arya & Singh, 2021). Another advantage of the technique is the quantification of influencing factor by assigning weight for factor based on the relative importance of factor (Maity & Mandal, 2019; Waris et al., 2019; Ghosh et al., 2020a). In this method, weight assignment is done on the basis of expert judgment of decision makers, and assessment of consistency of judgment value by consistency index (CI) is the strongest part of the method (Das, 2019; Murmu et al., 2019; Karmakar et al., 2021). In this work, the AHP is consisted of two major parts, such as evaluation of causative factors and assign weight and integration of all these factors to compute flood susceptibility index. The evaluation of factors has been done in four steps. In the beginning, weight of each factor is assigned from the literature review, field visits, and expert judgments. Using the nine-point scale of Saaty weight of each factor is assigned based on the relative importance of a factor on the occurrence of

Fig. 6.3 Flowchart of analytical hierarchy process

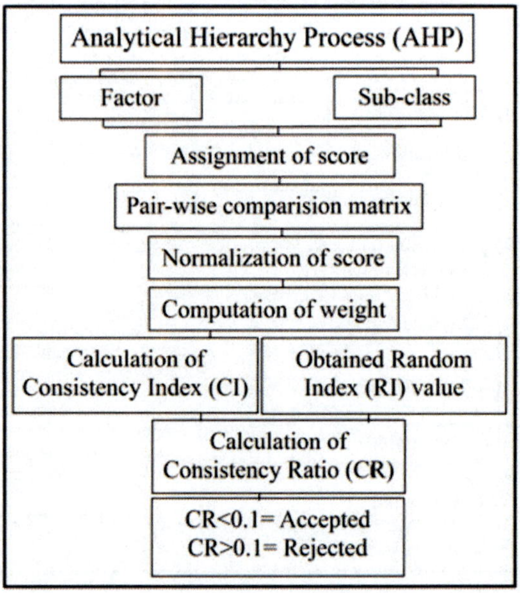

flood (Table 6.1). In the next step, comparison of all factors with each other is done by a matrix which is known as pair-wise comparison matrix (Eq. 6.1).

$$W = \begin{vmatrix} v_{11} & v_{12} & \cdots & v_{1n} \\ v_{21} & v & \cdots & v_{2n} \\ \cdots & \cdots & \cdots & \cdots \\ v_{n1} & v_{n2} & \cdots & v_{nn} \end{vmatrix} \tag{6.1}$$

Here, W is the pair-wise comparison matrix, v is the assign weight of ith factor. In this matrix, assignment of weight depends on the importance of the factor compare to other factor. Thus, the greater influence factor has been assigned to absolute number (Table 6.1) as weight per requirement and less importance factor assigned reciprocal of the absolute number in relation of the greater influence factor. The pair-wise comparison matrix is shown in Table 6.2. After that, estimated Eigen value of each factor is calculated from the matrix table (Table 6.2) by multiplying of all assigned weight of all factors in a row and then the Nth root of the product result is computed by using the Eq. (6.2).

$$EE = \sqrt[N]{V_a \times V_b \times V_c \times \ldots V_N} \tag{6.2}$$

Here, EE represent the estimated Eigen value of ith factor; V_a, V_b V_c V_n is the values of the row value of ith factor and N is the total number of factors. The principle Eigen value of each factor is computed using the Eq. (6.3), where Ev denotes the principle Eigen value.

$$Ev = \frac{\sqrt[N]{V_a \times V_b \times V_c \times \ldots V_N}}{\sum_{i=1}^{n} EE} \tag{6.3}$$

After the principle Eigen value ($\lambda_{\max}$) has been calculated, the consistency index (CI) is calculated using the Eq. (6.4). The CI value is divided by the RI value (Eq. 6.5) to compute the consistency ratio (CR) and verify the judgment coherence. The RI value is varying with number of parameter taken into consideration. In the consistency assessment, CR value must be less than 0.1 for acceptance of the judgment value of factor; if not, the pair-wise comparison matrix has to revise with new judgment values and this process will continue until the CR value comes down to less than 0.1 (Saaty, 1990).

$$\mathrm{CI} = \frac{\lambda \max - n}{n - 1} \tag{6.4}$$

$$\mathrm{CR} = \frac{\mathrm{CI}}{\mathrm{RI}} \tag{6.5}$$

Table 6.1 Nine-point scale of Saaty and Random consistency index

Scale of importance	1	2	3	4	5	6		7	8	9	
Definition	Equal	Weak	Moderate	Moderate plus	Strong	Strong plus		Very strong	Very very strong	Extreme	
Number of factors	1	2	3	4	5	6	7	8	9	10	11
RCI	0	0	0.58	0.90	1.12	1.24	1.32	1.41	1.45	1.49	1.51

Table 6.2 Pair-wise comparison matrix of causative factors

Factors	Curvature	Land use land cover	Topographic wetness index	Surface runoff	Lithology	Geomorphology	Drainage density	Distance from river	Flow accumulation	Slope	Elevation
Curvature	1	1/2	1/3	1/4	1/5	1/5	1/6	1/6	1/7	1/8	1/9
Land use land cover	2	1	1/2	1/3	1/4	1/4	1/5	1/5	1/6	1/7	1/8
Topographic wetness index	3	2	1	1/2	1/3	1/4	1/5	1/6	1/7	1/8	1/8
Surface runoff	4	3	2	1	1/1	1/2	1/3	¼	1/5	1/6	1/6
Lithology	5	4	3	1	1	1/1	1/2	1/3	1/4	1/5	1/6
Geomorphology	5	4	4	2	1	1	1/2	1/3	1/4	1/5	1/6
Drainage density	6	5	5	3	2	2	1	1/2	1/3	1/4	1/5
Distance from river	6	5	6	4	3	3	2	1	1/2	1/3	1/4
Flow accumulation	7	6	7	5	4	4	3	2	1	1/2	1/3
Slope	8	7	8	6	5	5	4	3	2	1	1/3
Elevation	9	8	8	6	6	6	5	4	3	3	1

The CR value of the present study is 0.054, which revealed that a consistency is present in the judgment values. Thereafter, weight is assigned to all sub-classes of each factor and the principle Eigen value is computed by considering same procedure of weight assigning.

Flood Susceptibility Index (FSI)

The flood susceptibility index (FSI) is computed using all the selected flood causative factors. The weighted linear combination method is adopted to calculate the FSI. In this method, weight of each factor is multiplied with the rank of sub-classes. The rank is assigned accordingly to sub-classes based on the weightage value of each sub-class of a factor (Table 6.3). Now, the Eq. (6.6) is used to calculate the FSI, where Ev_i is represented the principle Eigen value of ith factor and r_j is the rank of jth sub-classes of respective ith factor.

$$\text{FSI} = \sum_{i=1}^{n} \left(Ev_i \times r_j\right) \tag{6.6}$$

Finally, FSI value is classified into five distinct classes, very low, low, moderate, high, and very high. Based on this method, all the geospatial layers are integrated in ArcGIS software using the map algebra tool.

Accuracy Assessment

Accuracy assessment or validation of model is an integral part of any decision-making model. Without accuracy assessment, any simulation model is incomplete (Chung & Fabbri, 2003; Ghosh et al., 2020b; Sarkar & Mondal, 2020). There are several techniques used by different researchers across the world to validate the simulation models. However, the application of ground level data for accuracy assessment is the most accurate and convenient technique to validate such a model. In case of flood susceptibility assessment, use of historical data of inundated areas due to flood is also evolved as a significant data set for validation of the flood susceptibility map (Pradhan, 2010; Lee et al., 2018; Ali et al., 2019; Das, 2019). In the present study, a total number of 197 known flood sites are demarcated during the flooding season of 2017 using handheld GPS (Garmin eTrex-20) and inundation area identified from the flood map of 2017 provided by NRSC (collected from https://bhuvan.nrsc.gov.in). In this study, the Area Under Curve (AUC) is considered to evaluate the performance and efficiency of the AHP model. To construct the AUC, resultant flood susceptibility map is classified into 100 classes, and cumulative percentage of flood occurrence in different susceptibility classes is also computed. Based on calculation, the AUC is plotted to validate the model (Fig. 6.17).

Table 6.3 Allocation of weight to the selected parameters and their sub-classes

Factors	Weight	Sub-classes	Consistency ratio	Weight	Rank
Curvature	0.0137	0.33–3.37	0.0542	0.0329	1
		0.11–0.33		0.0636	2
		(−0.11)–0.11		0.1296	3
		(−0.33)–(−0.11)		0.2638	4
		(−2.37)–(−0.33)		0.5100	5
Land use land cover	0.0184	Vegetation	0.0098	0.0951	1
		Agricultural land		0.1599	2
		Build-up area		0.2777	3
		Water body		0.4671	4
Topographic wetness index	0.0222	4.19–7.73	0.0542	0.0329	1
		7.73–9.66		0.0636	2
		9.66–11.94		0.1296	3
		11.94–15.39		0.2638	4
		15.39–25.67		0.5100	5
Surface runoff (mm/day)	0.0372	93.70–109.26	0.0155	0.0615	1
		109.26–118.98		0.0975	2
		118.98–127.49		0.1602	3
		127.49–137.70		0.2634	4
		137.70–155.70		0.4174	5
Lithology	0.0476	Chota Nagpur gneissic complex	0.0066	0.1089	1
		Singhbhum group of rock		0.1089	1
		Unclassified metamorphic rock		0.1089	1
		Newer alluvium		0.2008	2
		Older alluvium		0.4724	3
Geomorphology	0.0521	Denudational hill and valley	0.0418	0.0271	1
		Pediment–pediplain complex		0.0510	2
		Older alluvial plain		0.1114	3
		Older flood plain		0.1114	3
		Active flood plain		0.2420	4
		Water body		0.4571	5
Drainage density (km/km^2)	0.0764	0.12–0.23	0.0358	0.0377	1
		0.23–0.29		0.0643	2
		0.29–0.34		0.1155	3
		0.34–0.40		0.2667	4
		0.40–0.50		0.5157	5
Distance from river (km)	0.1059	>2	0.0299	0.0419	1
		1.5–2		0.0730	2
		1–1.5		0.1317	3

(continued)

Table 6.3 (continued)

Factors	Weight	Sub-classes	Consistency ratio	Weight	Rank
		0.5–1		0.2398	4
		<0.5		0.5135	5
Flow accumulation	0.1492	0–168,106	0.0058	0.0427	1
		168,106–616,390		0.0750	2
		616,390–1,456,922		0.1429	3
		1,456,922–2,484,239		0.2697	4
		2,484,239–4,744,336		0.4696	5
Slope (degree)	0.1983	4.83–28.67	0.0111	0.0434	1
		3.15–4.83		0.0756	2
		1.91–3.15		0.1427	3
		0.67–1.91		0.2694	4
		0–0.67		0.4690	5
Elevation (m)	0.2784	137–227	0.0111	0.0434	1
		96–137		0.0756	2
		66–96		0.1427	3
		37–66		0.2694	4
		6–37		0.4690	5

6.4 Results and Discussion

Elevation

In general, low laying areas are more prone to flood than elevated areas (Nakajima & Umeyama, 2015). Thus, elevation is the prime factor for determining of flood susceptible areas (Das, 2018, 2019; Janizadeh et al., 2019). Flat low land area tends to have higher vulnerability to inundation than the area located at higher elevation (Das, 2018; Ali et al., 2019; Khosravi et al., 2016; Liuzzo et al., 2019; Rahman et al., 2019; Sarkar & Mondal, 2020; Vojtek & Vojteková, 2019). The prepared elevation map of Silabati river basin shows that the elevation varies from 6 to 227 m (Fig. 6.4). The highest elevation is found in the north-western part of the basin, and continuously decreases towards the south-east direction. The category of elevation of 37–66 m dominates the river basin with a share of 29.10% area followed by 6–37 m (27.38%) and 66–96 m (25.49%).

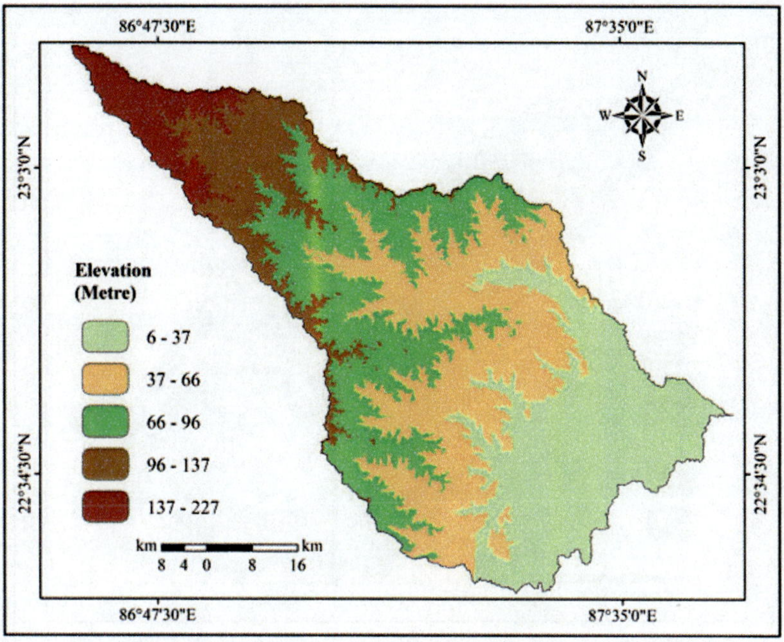

Fig. 6.4 Elevation map of Silabati river basin. (Source: SRTM DEM)

Slope

Slope refers to the amount of inclination of surface in respect to horizontal plane. This factor plays a crucial role in controlling surface runoff, infiltration process, and sub-surface flow (Ali et al., 2019; Das, 2018; Hammami et al., 2019). An area with gentle slope makes itself more vulnerable during flood as flat terrains are more susceptible to water stagnation compared to the area under steep slope (Hammami et al., 2019; Periyasamy et al., 2018). It is observed from the slope map that the slope ranges from 0° to 28.67° in the study area (Fig. 6.5). An area of 47.63% of the river basin experiences slope in between 0.67° and 1.91° followed by 1.91° to 3.14° (24.17%), and only an area of 1.19% belongs to the slope category of 4.83–28.63°.

Flow Accumulation

Flow accumulation is the concentration of flow in a pixel draining out from neighboring pixels (Das, 2018, 2019; Vojtek & Vojteková, 2019). Generally, flow accumulation is observed to be higher and lower in lower reach and upper reach, respectively, in a basin, since stream order and flow accumulation are positively

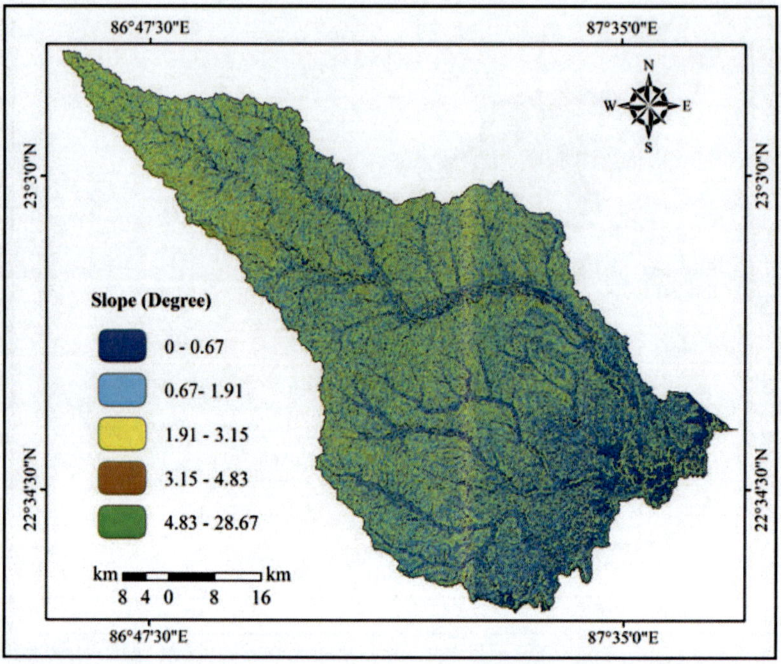

Fig. 6.5 Distribution of slope in Silabati river basin. (Source: SRTM DEM)

correlated (Das, 2018). An increase in flow accumulation enhances the propensity of flood risk (Vojtek & Vojteková, 2019). In this study, highest flow accumulation is seen in the lower reach of the basin due to flat terrain and very gentle slope (Fig. 6.6).

Distance from River

Distance of an area from an active channel is very significant in the field of flood risk mapping (Das, 2018). Areas near the active channels are more vulnerable to flood (Ali et al., 2019; Yousuf Gazi et al., 2019). Based on the drainage network map, a buffer analysis of active channels is carried out with an interval of 0.5 km using proximity analysis tool in ArcGIS software (Das, 2018). A total amount of 27.40% of river basin area lies within 1 km from the active river channel followed by 23.64% area (within 1–2 km), while 48.96% basin area lies greater than 2 km buffer (Fig. 6.7).

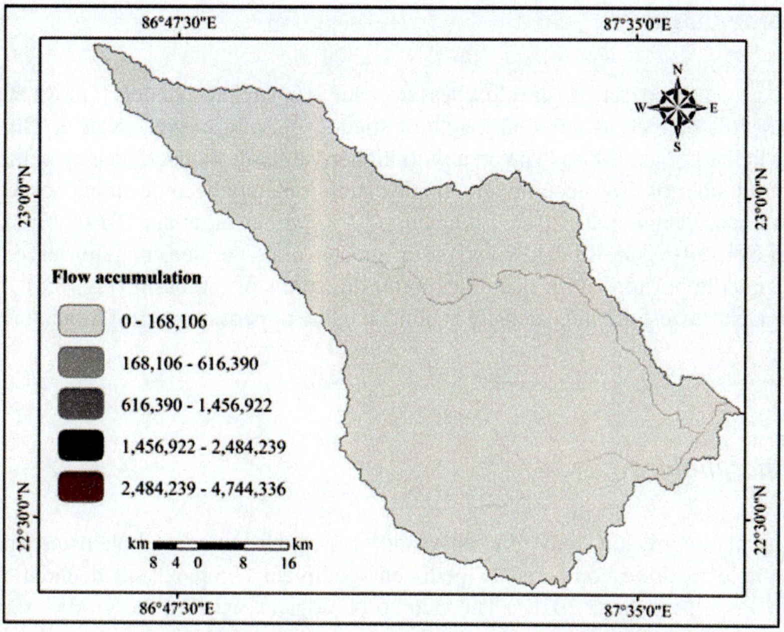

Fig. 6.6 Flow accumulation map of Silabati river basin. (Source: SRTM DEM)

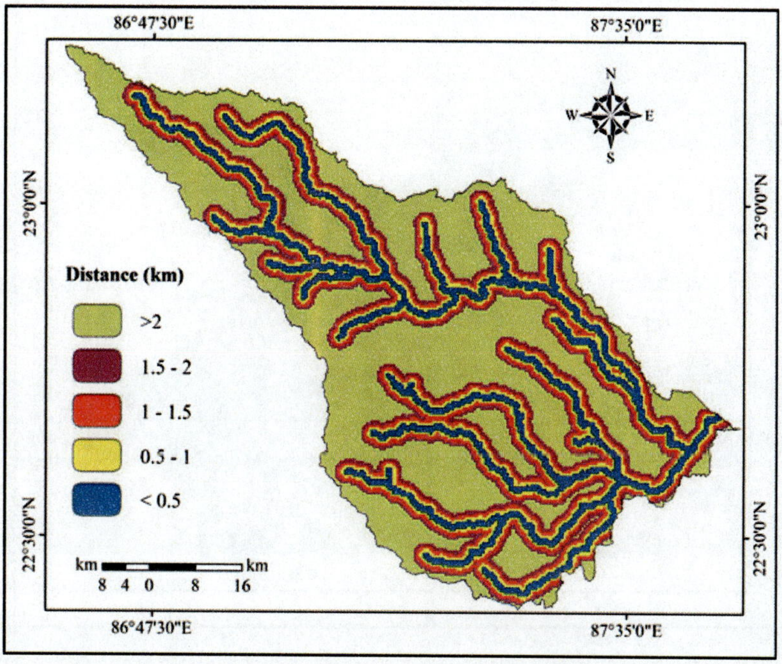

Fig. 6.7 Distance of basin area from active river channel in Silabati river basin

Drainage Density

Drainage density describes the closeness of spacing of stream channels (Ghosh et al., 2020a). It is defined as the total length of streams per unit of area (Kale & Gupta, 2001; Pallard et al., 2009). The area with higher value of drainage density exhibits greater number of flow accumulation paths, and ultimately becomes more susceptible to flood (Chapi et al., 2017; Elkhrachy, 2015; Purnawali et al., 2017; Vojtek & Vojteková, 2019). In Silabati river basin, greater drainage density is observed in lower middle portion and it decreases to all directions of the basin (Fig. 6.8). The major areas lowest drainage density is noticed in the uppermost area of Silabati river basin.

Geomorphology

Geomorphology is the study of earth's landforms. Low-lying flood plains are more susceptible to flood compared to pediment–pediplain complex and denudational hills (Das, 2018; Şen, 2018). The geomorphological map of the study area is extracted in ArcGIS environment using the geomorphological map published by GSI. The map reveals that an area of 82.11% of the total basin is formed by

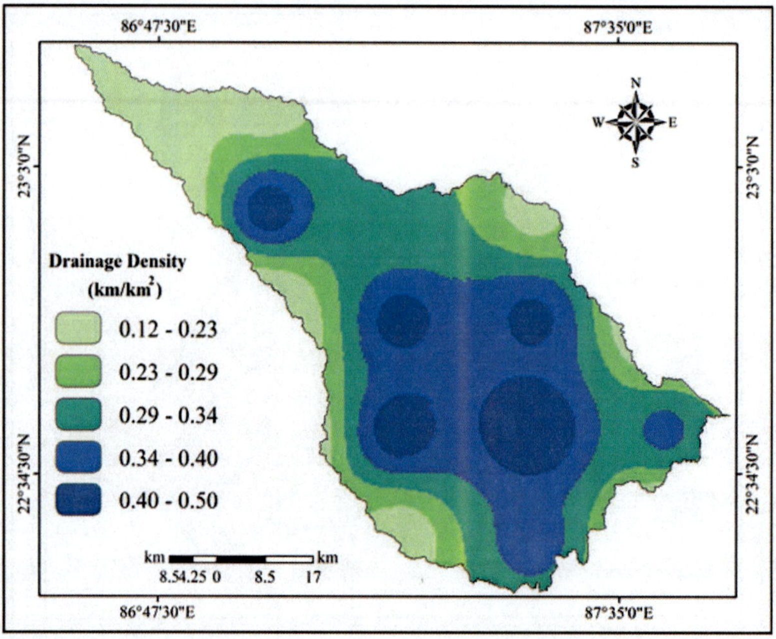

Fig. 6.8 Distribution of drainage density in Silabati river basin. (Source: SRTM DEM and Topographical map)

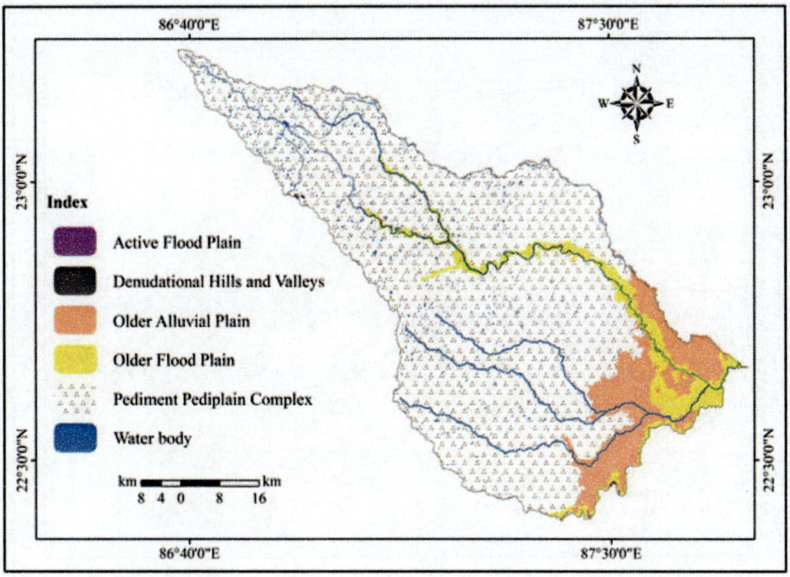

Fig. 6.9 Distribution of different geomorphological units in Silabati river basin. (Source: GSI)

pediment–pediplain complex, while 6.12% and 0.14% of areas are occupied by older flood plain and active flood plain, respectively (Fig. 6.9). Therefore, it is clear that only the lower basin area and areas along the main river channel are characterized with plain lands.

Lithology

Lithological map is used in flood susceptibility assessment due to diverse sensitivity of lithological units (Tien Bui et al., 2019). An area of hard rock lithology, many a time, is characterized by low drainage density and stream frequency (Kale & Gupta, 2001); hence the area faces less probability of being flooded (Tien Bui et al., 2019). A lithological map of the area is obtained from the District Resource Maps of GSI and processed in ArcGIS software. North-western part of the basin is covered with hard rock lithology, which is comprised of Chota Nagpur gneissic complex, Singbhum groups of rocks and metamorphosed basic rocks. Newer alluvium is found along the Silabati river and south-eastern part of the basin, while the older alluvium occupies around 70% of the total study area (Fig. 6.10).

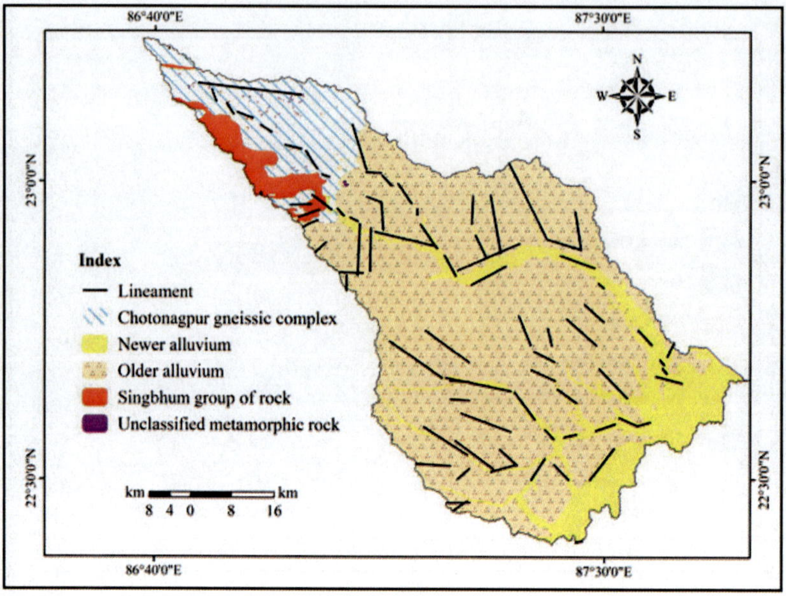

Fig. 6.10 Lithological map of Silabati river basin. (Source: GSI and https://bhuvan.nrsc.gov.in)

Surface Runoff

During precipitation, the water drains over the land surface following the local gradient and gravity. This process of water flow is called surface runoff (Uzor-Totty & Lawal, 2019). The interaction between precipitation and surface runoff depends on time and space. Surface runoff is influenced by both climatic factors (precipitation type, rainfall amount, intensity, duration, distribution, soil moisture resulting from earlier precipitation, direction of storm movement, evaporation, relative humidity, and seasons) and physical factors (vegetation, soil, elevation, slope, drainage area, basin shape, drainage network) (Uzor-Totty & Lawal, 2019). Moreover, anthropogenic activities, such as urbanization and other constructional activities, and its resultant impervious surfaces reduce the infiltration rate, increase surface runoff, and shorten runoff time into streams. Finally, all these increase the chance of higher magnitude and frequency of flood in nearby streams (Uzor-Totty & Lawal, 2019). To prepare a surface runoff map, maximum surface runoff of 4 years (2016–2019) is obtained from NRSC and interpolated using the Inverse Distance Weighted (IDW) method in ArcGIS software. At the end, the resultant map has been classified into five zones based on Jenks natural break method. The higher amount of surface runoff is recorded in the lower middle and lower reach of the basin, mainly in interfluves zones (Fig. 6.11).

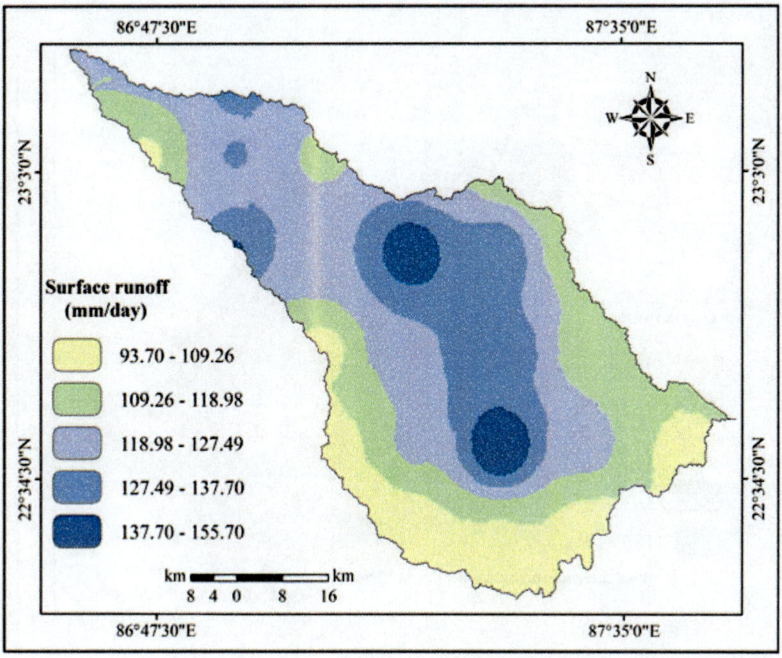

Fig. 6.11 Surface runoff map of Silabati river basin. (Source: https://bhuvan.nrsc.gov.in)

Topographic Wetness Index

Topographic wetness index is commonly used to evaluate the influence of topography in accumulation of flow or generation of runoff at any point of the basin area (Ali et al., 2019; Das, 2018, 2019; Moore et al., 1991; Sarkar & Mondal, 2020; Sørensen et al., 2006). It is expressed as TWI = ln ($a/$ tan B); where "TWI" refers to topographic wetness index, "a" is the specific basin area, and "B" is the local slope (Das, 2018, 2019; Sørensen et al., 2006; Tien Bui et al., 2019). The area with higher TWI value indicates high potentiality of flood event (Das, 2018; Tien Bui et al., 2019). It is evident that the higher TWI value is found in active flood plain region of Silabati river basin because of lower elevation of this area (Fig. 6.12).

Land Use Land Cover

Land use land cover has a significant role in determining surface runoff, which is directly related to flood event in catchment area (Phrakonkham et al., 2019). An area covered with vegetation reduces the intensity of surface runoff, and enhances the proliferation of infiltration process; whereas build-up area strongly impedes water percolation into the ground and hastens the surface flow (Das, 2018; Hammami et al., 2019; Roslee et al., 2018; Samanta et al., 2018; Sarkar & Mondal, 2020). To

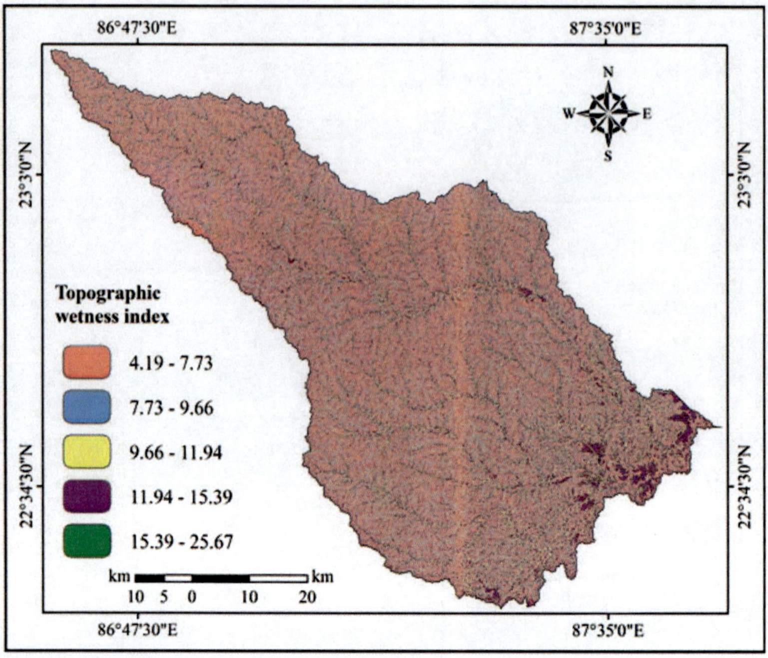

Fig. 6.12 Topographic wetness index map of Silabati river basin. (Source: SRTM DEM)

perceive the nature of land use in the study area, a supervised classification is done adopting maximum likelihood method in Erdas Imagine software (Das, 2019; Ghosh et al., 2020a; Ghosh et al., 2020b; Sarkar & Mondal, 2020). The land use pattern of Silabati river basin is dominated by agricultural land (76.79%), while forest cover is the second highest land cover occupying 20% of the total basin area. Here, Sal (Shorea Robusta), Shishu (Dalbergia), Palas (Butea Monosperma), Arjuna (Terminalia Arjuna), Sonajhuri (Acacia Auriculiformis) and Eucalyptus (Eucalyptus Globules), etc., plants are commonly found. Only 2.11% of the basin is covered by build-up area, while the water body is recorded in a tiny amount, i.e., 0.42% (Fig. 6.13).

Curvature

Curvature determines surface flow and infiltration process, and ultimately influences the incidence of flood (Das, 2018, 2019). It can be three types: (1) concave (positive curvature), (2) flat (zero curvature), and (3) convex (negative curvature). There are different opinions about the role of curvature in controlling the surface flow. Young and Mutchler (1969) advocated that a convex slope can produce much more runoff than the concave slope (Chapi et al., 2017; Young & Mutchler, 1969). Hudson and Kesel (2000) demonstrated that the area with the curvature value in between 1.0 and

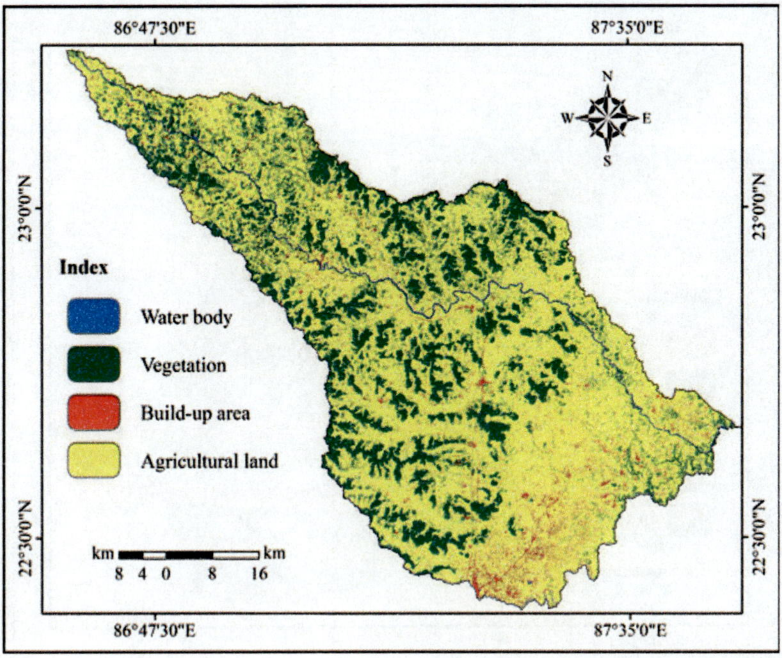

Fig. 6.13 Distribution of different types of land use land cover in Silabati river basin. (Source: Landsat 8 OLI Satellite Image)

2.0 is more vulnerable to flood (Das, 2019; Hudson & Kasel, 2000). The curvature category of −0.11 to 0.11 possesses an area of 45.72% followed by 0.11 to 0.33 (23.93%) and −0.33 to −0.11 (17.34%). It is the −2.37 to −0.33 category that shares only 2.89% of the basin area (Fig. 6.14).

Assessment of Flood Susceptibility

The final susceptibility map is generated based on integration of factor weight and sub-class rank using raster calculator in ArcGIS software. The derived map is categorized into five distinctive classes by Jenks natural break method. The classes are very high, high, moderate, low, and very low susceptibility of flood, and these classes cover 14.04%, 20.67%, 21.76%, 20.69%, and 22.84% of the total basin area, respectively (Fig. 6.15). The lower segment of the river basin is observed to be in vulnerable condition, since it belongs to the very high to high flood susceptible category. It is because the area has low elevation (6–30 m) and low slope (<2^0). This portion is distinguished by flat alluvial plain and well developed flood plain. The north-western part of the basin comes under very low to low flood susceptibility zone as the elevation is observed to be high (>100 m), and slope is seen more than 2^0.

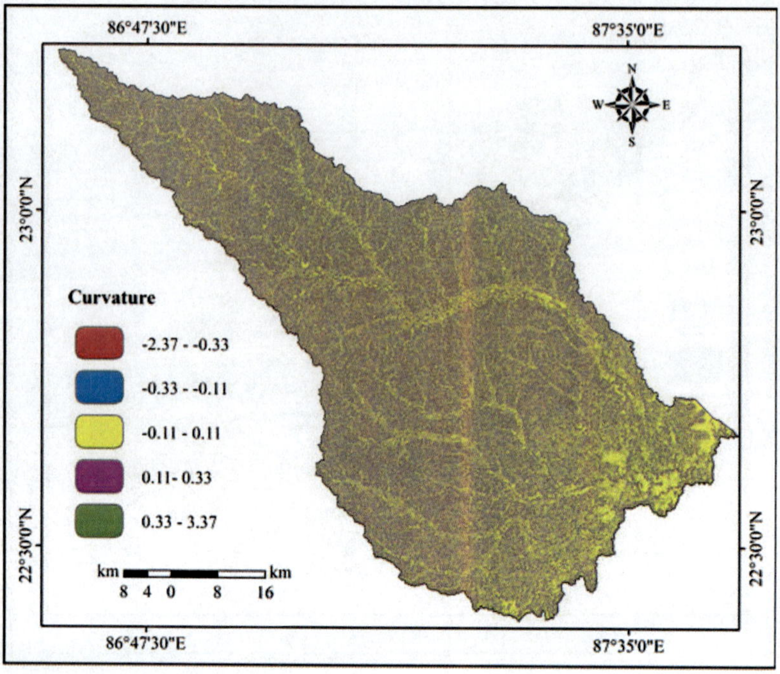

Fig. 6.14 Curvature map of Silabati river basin. (Source: SRTM DEM)

In this section, the river Silabati, flowing through the undulating topography of Chota Nagpur plateau, is characterized by narrow restricted valley, lower width-depth ratio, lacking of flood plain development, rough bed configuration and fewer number of bars; whereas the middle reach of the basin with moderate elevation shows moderate vulnerability to flood. It is observed that flood susceptibility varies from higher in south-east to lower in north-west direction (Fig. 6.15). Thus, the susceptibility of flood decreases as the elevation increases towards north-west direction.

The degree of flood vulnerability is greatly influenced by social and economic factors of an area (Działek et al., 2019). Very high and high flood susceptibility zones of the study area are mainly located at Keshpur, Ghatal, Chandrakona-I, Chandrakona-II and Daspur-I Community Development (C.D.) blocks of West Medinipur district (Fig. 6.15). Moreover, Garbeta-I, Garbeta-II, Garbeta-III, Salbani, and Medinipur C.D. blocks of West Medinipur district are also affected by flood. The population density and flood vulnerability are positively correlated to each other (Hoque et al., 2019; Sarkar & Mondal, 2020). These C.D. blocks have a population density ranging from 340 persons/km^2 to 1212 persons/km^2. High population density is found in Daspur-I (1212 persons/km^2), Ghatal (1016 persons/km^2), Chandrakona-II (819 persons/km^2), Chandrakona-I (702 persons/km^2), and Keshpur (702 persons/km^2). The Decadal (2001–2011) population growth rate varies from 13.20 to 21.37% in these blocks (Census of India, 2011). Therefore, a large number

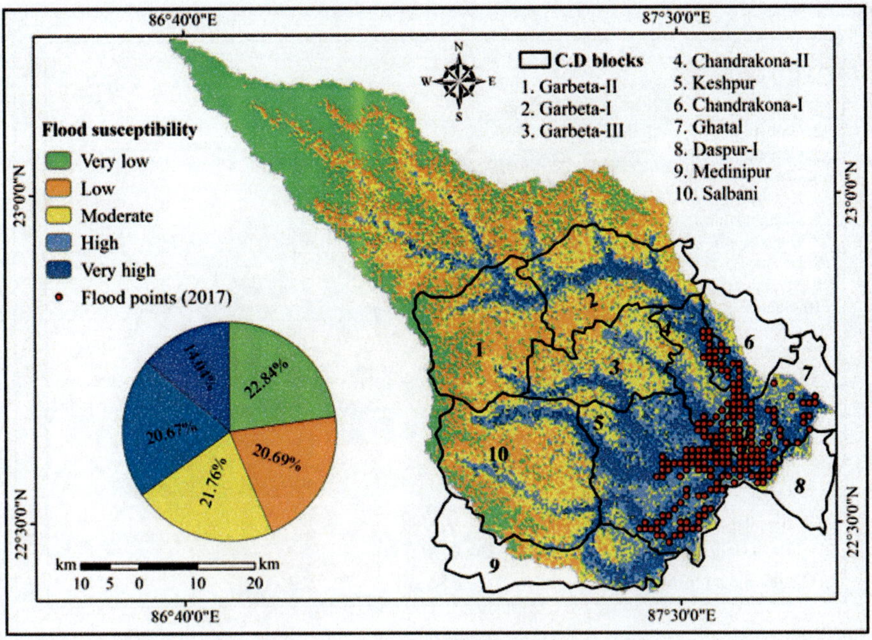

Fig. 6.15 Flood susceptibility map of Silabati river basin. The red dots in the map indicate flood points verified during 2017 flood

of people can be affected significantly by the floods. The areas with large number of households are more at-risk during occurrence of flood (Sarkar & Mondal, 2020). The household frequency ranges in between 71 and 261 households/km^2 in the studied C.D. blocks (Census of India, 2011). It is reported that the higher number of household frequency is recorded in Daspur-I (262 households/km^2), Ghatal (215 households/km^2), Chandrakona-II (175 households/km^2), Chandrakona-I (157 households/km^2), and Keshpur (142 households/km^2) (Census of India, 2011). Another important aspect needs to be mentioned here that, the flood vulnerability of an area gets momentum owing to its nature of house type (Hoque et al., 2019; Sarkar & Mondal, 2020). It is noticed that the percentage of permanent, semi-permanent, and temporary houses in these C.D. blocks varies from 6.39% to 37.95%, 40.76% to 78.83%, and 8.29% to 35.38%, respectively. The large number of temporary houses is located at Medinipur (35.38%) and Keshpur (18.26%) C.D. blocks; there are significant numbers of semi-permanent houses in Chandrakona-I (78.83%), Chandrakona-II (77.92%), Keshpur (75.17%), Garbeta-II (72.05%), and Garbeta-III (71.46%) (Census of India, 2011). During flood event, children and females become more vulnerable because of their incapacitation to move at once during the emergency evacuation situations (Hoque et al., 2019). The percentage of females and children among the studied C.D. blocks ranges from 48.44% to 49.53% and 11.08% to 13.52%, respectively (Census of India, 2011).

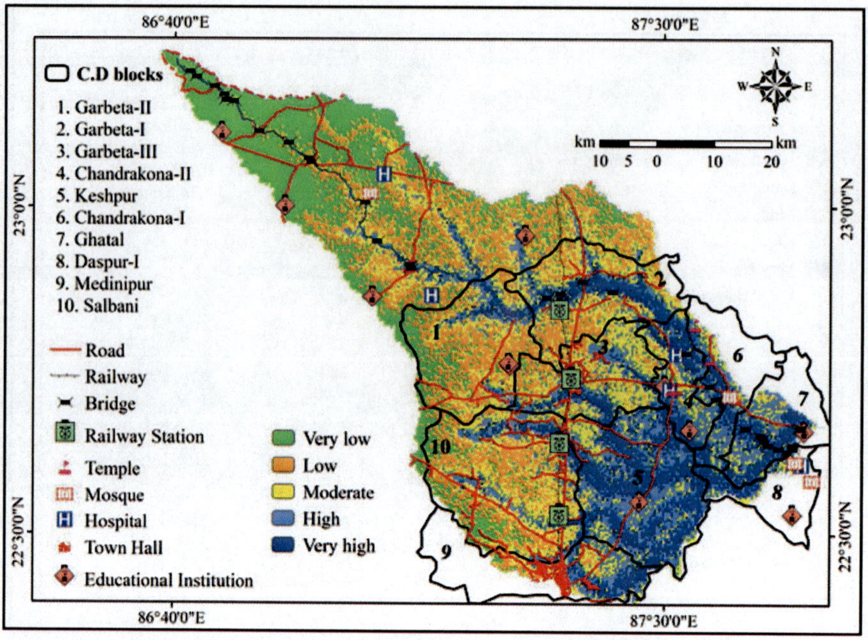

Fig. 6.16 Distribution of different socio-economic and cultural structures present within Silabati river basin overlie flood susceptibility map. (Source: Google Earth Image)

There is a negative relation between literacy rate and flood vulnerability (Hoque et al., 2019), and the average literacy rate of these C.D. blocks is calculated to be 67.28% (Census of India, 2011). The entire study area is agriculturally dependent, and flood incident brings the crop production to an end pushing thousands of men engaged in cultivation works in distress conditions (Hoque et al., 2019). It is to be noted that the agriculturally dependent populations of these C.D. blocks vary in between 59.90 and 83.67%, where the higher number of agriculturally dependent population is found in the C.D. blocks of Garbeta-II (83.67%), Chandrakona-II (83.55%), Keshpur (79.88%), Chandrakona-I (76.80%), Garbeta-I (76.37%), and Salbani (75.31%) (Census of India, 2011). There are several permanent and temporary bridges over Silabati river, and roads stretch for a considerable length within very high to moderate flood susceptibility zones (Figs. 6.16 and 6.17). Hence, it is clear that a large number of people along with many socio-economic infrastructures can significantly be affected by the flood events of Silabati river basin.

The geospatial modeling provided an effective way of flood management in the study area. Various spatial components of flood are identified and the extent of potential flooded area is quantified as well. It will facilitate the implementation of evacuation strategy, rehabilitation plan, and damage assessment during critical flood situation. It may also be effective in the development of policy, guidelines, and recommendation of land use planning.

Fig. 6.17 (**a**) Flood water spilling over the road near Bhelaidiha, Bankura on 25th July, 2017; (**b**) Flood situation near Simlapal, Bankura on 25th July, 2017 that submerged adjacent road; (**c** and **d**) Bank protection structures near Jhumka village, Bankura constructed by Irrigation & Waterways Department, Kangsabati Circle, Bankura to reduce vulnerability of river bank failure during high discharge events or floods

Validation of the Flood Susceptibility Map

In order to assess the validity of the flood susceptibility map, the AUC is computed, and the output value, i.e., 76.41% depicts that the accuracy level of the flood map prepared adopting AHP technique is well acceptable (Fig. 6.18). The value of AUC less than 50% is considered as inappropriate for flood vulnerability mapping, while the calculated value of AUC exceeding 75% is well accepted for the predicted model (Akgun et al., 2008; Egan, 1975; Ozdemir & Altural, 2013; Pedzisai, 2010; Saha, 2017; Sarkar & Mondal, 2020).

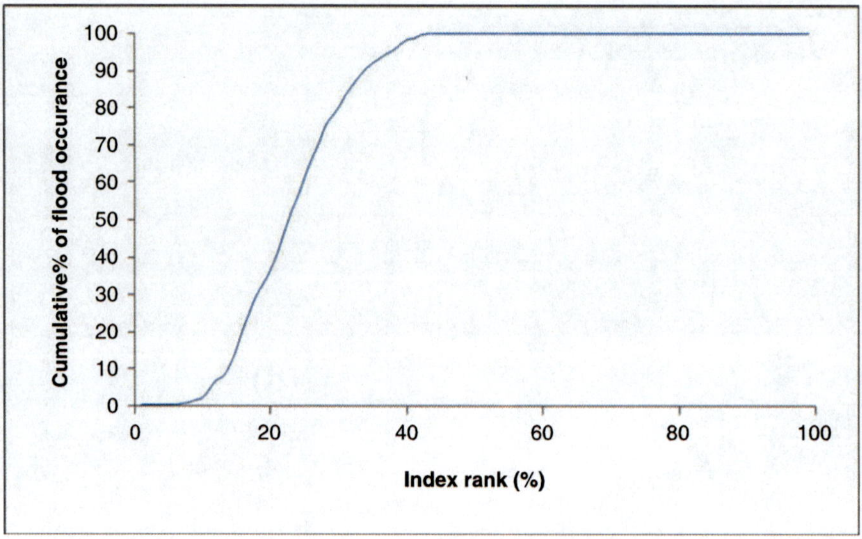

Fig. 6.18 Area under curve for validation of flood susceptibility map of Silabati river basin

6.5 Conclusion

To analyze and perceive the flood vulnerability of the Silabati river, three factors, such as elevation, slope, and flow accumulation turn out to be most dominating parameters out of adopted 11 factors considered in the present study. It is found that an area of 14.04 and 20.67% of the total basin comes under very high and high flood susceptibility zones, respectively, and most of this area lies in West Medinipur district. Amongst all C.D. blocks of the West Medinipur district, only ten are more or less affected by the flood events, but five C.D. blocks, namely: Keshpur, Ghatal, Chandrakona-I, Chandrakona-II, and Daspur-I are severely influenced by flood due to flat topography, very low elevation, and higher amount of flow accumulation compared to other C.D. blocks. As a result, a large number of populations have been witnessing and suffering from floods almost in every year for decades. Their misery is the consequences of losing shelters, crops, and cattle. To minimize the consequences of floods in Silabati river basin, implementation of different management strategies in higher flood susceptible areas is very much required. This research work may help the policy makers and implementing authorities to gather basic information related to the flood, including its vastness and areas under risk. It is suggested that C. D block level assessment of flood susceptibility can give accurate flood potentiality at local scale. Moreover, a flood simulation study can be helpful to quantify depth, velocity, and duration of flood and to access the surface water and groundwater interaction during flood.

Acknowledgments The authors would like to thank all national and international organizations for providing secondary data and maps. We are also thankful to Mr. Susanta Mandi, State Aided College Teacher, Department of Geography, Arsha College, Purulia, India, and our chauffeur, Sk. Firoz for their support during the field work.

References

Akgun, A., Dag, S., & Bulut, F. (2008). Landslide susceptibility mapping for a landslide-prone area (Findikli NE of Turkey) by likelihood-frequency ratio and weighted linear combination models. *Environmental Geology, 54*(6), 1127–1143. https://doi.org/10.1007/s00254-007-0882-8

Ali, S. A., Khatun, R., Ahmad, A., & Ahmad, S. N. (2019). Application of GIS-based analytic hierarchy process and frequency ratio model to flood vulnerable mapping and risk area estimation at Sundarban region, India. *Modeling Earth Systems and Environment, 5*, 1083–1102. https://doi.org/10.1007/s40808-019-00593-z

Anandabazar Patrika Online. (2020). *Submerged the causeway of Shilabati, the water entering the city*. Anandabazar Patrika Online.

Arya, A. K., & Singh, A. P. (2021). Multi criteria analysis for flood hazard mapping using GIS techniques: A case study of Ghaghara River basin in Uttar Pradesh, India. *Arabian Journal of Geosciences, 14*, 656. https://doi.org/10.1007/s12517-021-06971-1

Çelik, R. (2019). Evaluation of groundwater potential by GIS-based multicriteria decision making as a spatial prediction tool: Case study in the Tigris river batman-Hasankeyf sub-basin, Turkey. *Water, 11*, 2630. https://doi.org/10.3390/w11122630

Census of India. (2011). *C.D. Block Wise Primary Census Abstract Data (PCA)—West Bengal*. Directorate of Census, Government of India, New Delhi.

Chakraborty, S., & Mukhopadhyay, S. (2019). Assessing flood risk using analytical hierarchy process (AHP) and geographical information system (GIS): Application in Coochbehar district of West Bengal, India. *Natural Hazards, 99*, 247–274. https://doi.org/10.1007/s11069-019-03737-7

Chapi, K., Singh, V. P., Shirzadi, A., Shahabi, H., Bui, D. T., Pham, B. T., & Khosravi, K. (2017). A novel hybrid artificial intelligence approach for flood susceptibility assessment. *Environmental Modelling & Software, 95*, 229–245. https://doi.org/10.1016/j.envsoft.2017.06.012

Chen, Y.-R., Yeh, C.-H., & Yu, B. (2011). Integrated application of the analytic hierarchy process and the geographic information system for flood risk assessment and flood plain management in Taiwan. *Natural Hazards, 59*, 1261–1276. https://doi.org/10.1007/s11069-011-9831-7

Chowdhuri, I., Pal, S. C., & Chakrabortty, R. (2020). Flood susceptibility mapping by ensemble evidential belief function and binomial logistic regression model on river basin of eastern India. *Advances in Space Research, 65*, 1466–1489. https://doi.org/10.1016/j.asr.2019.12.003

Chung, C.-J. F., & Fabbri, A. G. (2003). Validation of spatial prediction models for landslide hazard mapping. *Natural Hazards, 30*, 451–472. https://doi.org/10.1023/B:NHAZ.0000007172.62651.2b

Dandapat, K., & Panda, G. K. (2018). A geographic information system-based approach of flood hazards modelling, Paschim Medinipur district, West Bengal, India. *Jàmbá, 10*(1), 518. https://doi.org/10.4102/jamba.v10i1.518

Dano, U., Balogun, A.-L., Matori, A.-N., Wan Yusouf, K., Abubakar, I., Said Mohamed, M., Aina, Y., & Pradhan, B. (2019). Flood susceptibility mapping using GIS-based analytic network process: A case study of Perlis, Malaysia. *Water, 11*, 615. https://doi.org/10.3390/w11030615

Das, S. (2018). Geographic information system and AHP-based flood hazard zonation of Vaitarna basin, Maharashtra, India. *Arabian Journal of Geosciences, 11*, 576. https://doi.org/10.1007/s12517-018-3933-4

Das, S. (2019). Geospatial mapping of flood susceptibility and hydro-geomorphic response to the floods in Ulhas basin, India. *Remote Sensing Applications: Society and Environment, 14*, 60–74. https://doi.org/10.1016/j.rsase.2019.02.006

Das, U., Bajpai, R., & Chakrabortty, D. (2020). River regulation and associated geo-environmental problems: A case study of lower reaches of Shilabati river basin, West Bengal, India. *International Journal of Ecology and Environmental Sciences, 2*, 233–240.

Dawod, G. M., Mirza, M. N., & Al-Ghamdi, K. A. (2011). Gis-based spatial mapping of flash flood hazard in Makkah city, Saudi Arabia. *JGIS, 03*, 225–231. https://doi.org/10.4236/jgis.2011.33019

Deepak, S., Rajan, G., & Jairaj, P. G. (2020). Geospatial approach for assessment of vulnerability to flood in local self governments. *Geoenviron Disasters*, (7), 35.

Działek, J., Biernacki, W., Konieczny, R., Fiedeń, Ł., Franczak, P., Grzeszna, K., & Listwan-Franczak, K. (2019). Social vulnerability as a factor in flood preparedness. In *Understanding Flood Preparedness: Flood Memories, Social Vulnerability and Risk Communication in Southern Poland* (pp. 61–90). Springer International Publishing. https://doi.org/10.1007/978-3-030-04594-4_4

Egan, J. P. (1975). *Signal detection theory and ROC-analysis*. Academic.

Elkhrachy, I. (2015). Flash flood hazard mapping using satellite images and GIS tools: A case study of Najran city, kingdom of Saudi Arabia (KSA). *The Egyptian Journal of Remote Sensing and Space Science, 18*, 261–278. https://doi.org/10.1016/j.ejrs.2015.06.007

Flood Damage Statistics. (2018). State wise flood damage statistics and for the Country for the period 1953 to 2016.

Genovese, E., & Thaler, T. (2020). The benefits of flood mitigation strategies: Effectiveness of integrated protection measures. *AIMS Geosciences, 6*, 459–472.

Geological Survey of India (GSI). (2001). District Resource Map of Purulia, Bankura and Paschim Medinipur.

Ghosh, A., & Kar, S. K. (2018). Application of analytical hierarchy process (AHP) for flood risk assessment: A case study in Malda district of West Bengal, India. *Natural Hazards, 94*, 349–368. https://doi.org/10.1007/s11069-018-3392-y

Ghosh, D., Mandal, M., Karmakar, M., Banerjee, M., & Mandal, D. (2020a). Application of geospatial technology for delineating groundwater potential zones in the Gandheswari watershed, West Bengal. *Sustainable Water Resources Management, 6*, 14. https://doi.org/10.1007/s40899-020-00372-0

Ghosh, D., Mandal, M., Banerjee, M., & Karmakar, M. (2020b). Impact of hydro-geological environment on availability of groundwater using analytical hierarchy process (AHP) and geospatial techniques: A study from the upper Kangsabati river basin. *Groundwater for Sustainable Development, 11*, 100419. https://doi.org/10.1016/j.gsd.2020.100419

Government of West Bengal. (2019). *Annual flood report 2019*. Irrigation and Waterways Department.

Guha-Sapir, D., Hoyois, P., Wallemacq, P., & Below, R. (2016). *Annual disaster statistical review 2016: The numbers and trends*. CRED.

Haghizadeh, A., Siahkamari, S., Haghiabi, A. H., & Rahmati, O. (2017). Forecasting flood-prone areas using Shannon's entropy model. *Journal of Earth System Science, 126*, 39. https://doi.org/10.1007/s12040-017-0819-x

Hammami, S., Zouhri, L., Souissi, D., Souei, A., Zghibi, A., Marzougui, A., & Dlala, M. (2019). Application of the GIS based multi-criteria decision analysis and analytical hierarchy process (AHP) in the flood susceptibility mapping (Tunisia). *Arabian Journal of Geosciences, 12*, 653. https://doi.org/10.1007/s12517-019-4754-9

Hindustan Times. (2017a). *868 people killed in floods in 11 states: Govt*. Hindustan Times.

Hindustan Times. (2017b). *Bengal flood: IAF chopper winches 9 marooned villagers in dramatic rescue operation*. Hindustan Times.

Hirabayashi, Y., Mahendran, R., Koirala, S., Konoshima, L., Yamazaki, D., Watanabe, S., Kim, H., & Kanae, S. (2013). Global flood risk under climate change. *Nature Clim Change, 3*, 816–821. https://doi.org/10.1038/nclimate1911

Hong, H., Panahi, M., Shirzadi, A., Ma, T., Liu, J., Zhu, A.-X., Chen, W., Kougias, I., & Kazakis, N. (2018). Flood susceptibility assessment in Hengfeng area coupling adaptive neuro-fuzzy inference system with genetic algorithm and differential evolution. *Science of the Total Environment, 621*, 1124–1141. https://doi.org/10.1016/j.scitotenv.2017.10.114

Hoque, M., Tasfia, S., Ahmed, N., & Pradhan, B. (2019). Assessing spatial flood vulnerability at Kalapara upazila in Bangladesh using an analytic hierarchy process. *Sensors, 19*, 1302. https://doi.org/10.3390/s19061302

Huang, X., Tan, H., Zhou, J., Yang, T., Benjamin, A., Wen, S. W., Li, S., Liu, A., Li, X., Fen, S., & Li, X. (2008). Flood hazard in Hunan province of China: An economic loss analysis. *Natural Hazards, 47*, 65–73. https://doi.org/10.1007/s11069-007-9197-z

Hudson, P. F., & Kesel, R. H. (2000). Channel migration and meander-bend curvature in the lower Mississippi River prior to major human modification. *Geology, 28*(6), 531. https://doi.org/10.1130/0091-7613(2000)28<531:CMAMCI>2.0.CO;2

Inland Waterway Authority of India (IWAI). (2016). *Consultancy Services for preparation of Two Stage Detailed Project Report (DPR) of National Waterway 92, Cluster—I: Silabati River*. IWAI.

Jabbar, F. K., Grote, K., & Tucker, R. E. (2019). A novel approach for assessing watershed susceptibility using weighted overlay and analytical hierarchy process (AHP) methodology: A case study in Eagle Creek Watershed, USA. *Environmental Science and Pollution Research, 26*, 31981–31997. https://doi.org/10.1007/s11356-019-06355-9

Jahangir, M. H., Mousavi Reineh, S. M., & Abolghasemi, M. (2019). Spatial predication of flood zonation mapping in Kan River Basin, Iran, using artificial neural network algorithm. *Weather and Climate Extremes, 25*, 100215. https://doi.org/10.1016/j.wace.2019.100215

Janizadeh, S., Avand, M., Jaafari, A., Phong, T. V., Bayat, M., Ahmadisharaf, E., Prakash, I., Pham, B. T., & Lee, S. (2019). Prediction success of machine learning methods for flash flood susceptibility mapping in the tafresh watershed Iran. *Sustainability, 11*(19), 5426. https://doi.org/10.3390/su11195426

Jongman, B. (2018). Effective adaptation to rising flood risk. *Nature Communications, 9*, 1986. https://doi.org/10.1038/s41467-018-04396-1

Kale, V. S., & Gupta, A. (2001). *Introduction to geomorphology*. Orient Longman.

Kannan, V. (2010). Benchmarking the service quality of ocean container carriers using AHP. *Benchmarking, 17*, 637–656. https://doi.org/10.1108/14635771011076416

Karmakar, M., Banerjee, M., Mandal, M., & Ghosh, D. (2021). Application of AHP for groundwater potential zones mapping in plateau fringe terrain: Study from western province of West Bengal. In P. K. Shit, G. S. Bhunia, P. P. Adhikary, & C. J. Dash (Eds.), *Groundwater and society* (pp. 189–219). Springer International Publishing.

Kazakis, N., Kougias, I., & Patsialis, T. (2015). Assessment of flood hazard areas at a regional scale using an index-based approach and Analytical Hierarchy Process: Application in Rhodope–Evros region, Greece. *Science of the Total Environment, 538*, 555–563. https://doi.org/10.1016/j.scitotenv.2015.08.055

Khosravi, K., Nohani, E., Maroufinia, E., & Pourghasemi, H. R. (2016). A GIS-based flood susceptibility assessment and its mapping in Iran: A comparison between frequency ratio and weights-of-evidence bivariate statistical models with multi-criteria decision-making technique. *Natural Hazards, 83*, 947–987. https://doi.org/10.1007/s11069-016-2357-2

Khosravi, K., Shahabi, H., Pham, B. T., Adamowski, J., Shirzadi, A., Pradhan, B., Dou, J., Ly, H.-B., Gróf, G., Ho, H. L., Hong, H., Chapi, K., & Prakash, I. (2019). A comparative assessment of flood susceptibility modeling using multi-criteria decision-making analysis and machine learning methods. *Journal of Hydrology, 573*, 311–323. https://doi.org/10.1016/j.jhydrol.2019.03.073

Kowalzig, J. (2008). Climate, poverty, and justice: What the Poznań UN climate conference needs to deliver for a fair and effective global deal. *Oxfam Policy Practice Climate Change Resilience, 4*, 117–148.

Lawal, D. U., Matori, A.-N., Hashim, A. M., Yusof, K. W., & Chandio, I. A. (2012). *Detecting flood susceptible areas using GIS-based analytic hierarchy process*. IACSIT Press.

Lee, S., Lee, S., Lee, M.-J., & Jung, H.-S. (2018). Spatial assessment of urban flood susceptibility using data mining and geographic information system (GIS) tools. *Sustainability, 10*, 648. https://doi.org/10.3390/su10030648

Liuzzo, L., Sammartano, V., & Freni, G. (2019). Comparison between different distributed methods for flood susceptibility mapping. *Water Resources Management, 33*, 3155–3173. https://doi.org/10.1007/s11269-019-02293-w

Mahala, A. (2020). Land degradation processes of Silabati river basin, West Bengal, India: A physical perspective. In P. K. Shit, H. R. Pourghasemi, & G. S. Bhunia (Eds.), *Gully erosion studies from India and surrounding regions* (pp. 265–278). Springer International Publishing.

Maity, D. K., & Mandal, S. (2019). Identification of groundwater potential zones of the Kumari river basin, India: An RS & GIS based semi-quantitative approach. *Environment, Development and Sustainability, 21*, 1013–1034. https://doi.org/10.1007/s10668-017-0072-0

Malik, S., & Pal, S. C. (2021). Potential flood frequency analysis and susceptibility mapping using CMIP5 of MIROC5 and HEC-RAS model: A case study of lower Dwarkeswar River, Eastern India. *SN Applied Sciences, 3*, 31. https://doi.org/10.1007/s42452-020-04104-z

Mandal, R., & Chakrabarty, P. (2021). Badlands of gangani in West Bengal, India: An assessment on account of geotourism development. *International Journal of Geoheritage and Parks, 9*, 147–156. https://doi.org/10.1016/j.ijgeop.2021.02.001

Matori, A. N., Lawal, D. U., Yusof, K. W., Hashim, M. A., & Balogun, A.-L. (2014). Spatial analytic hierarchy process model for flood forecasting: An integrated approach. *IOP Conference Series: Earth and Environmental Science, 20*, 012029. https://doi.org/10.1088/1755-1315/20/1/012029

Mishra, K., & Sinha, R. (2020). Flood risk assessment in the Kosi megafan using multi-criteria decision analysis: A hydro-geomorphic approach. *Geomorphology, 350*, 106861. https://doi.org/10.1016/j.geomorph.2019.106861

Moore, I. D., Grayson, R. B., & Ladson, A. R. (1991). Digital terrain modelling: A review of hydrological geomorphological and biological applications. *Hydrological Processes, 5*(1), 3–30. https://doi.org/10.1002/hyp.3360050103

Murmu, P., Kumar, M., Lal, D., Sonker, I., & Singh, S. K. (2019). Delineation of groundwater potential zones using geospatial techniques and analytical hierarchy process in Dumka district, Jharkhand, India. *Groundwater for Sustainable Development, 9*, 100239. https://doi.org/10.1016/j.gsd.2019.100239

Nakajima, T., & Umeyama, M. (2015). A new concept for the safety of low-lying land areas from natural disasters. *Journal of Ocean Engineering and Marine Energy, 1*(1), 19–29. https://doi.org/10.1007/s40722-014-0002-2

Natarajan, L., Usha, T., Gowrappan, M., Palpanabhan Kasthuri, B., Moorthy, P., & Chokkalingam, L. (2021). Flood susceptibility analysis in chennai corporation using frequency ratio model. *Journal of the Indian Society of Remote Sensing, 49*, 1533–1543. https://doi.org/10.1007/s12524-021-01331-8

National Remote Sensing Centre (NRSC). (2018). Bhuvan | isro's geoportal | gateway to Indian earth observation | disaster services. In: *Bhuvan*. Retrieved May 22, 2021, from https://bhuvan-app1.nrsc.gov.in/disaster/disaster.php?id=flood_hz

News18. (2019). *Several villages are in fear of flooding due to the collapse of Shilabati dam*. News18.

Ozdemir, A., & Altural, T. (2013). A comparative study of frequency ratio weights of evidence and logistic regression methods for landslide susceptibility mapping: Sultan Mountains SW Turkey. *Journal of Asian Earth Sciences, 64*, 180–197. https://doi.org/10.1016/j.jseaes.2012.12.014

Pallard, B., Castellarin, A., & Montanari, A. (2009). A look at the links between drainage density and flood statistics. *Hydrology and Earth System Sciences, 13*(7), 1019–1029. https://doi.org/10.5194/hess-13-1019-2009

Pedzisai, E. (2010). *Rainfall-runoff modelling for flash floods in Cuongthinh catchment; Yen Bai Province: Vietnam*. International Institute for Geo-information Science and Earth Observation.

Periyasamy, A. P., Ramamoorthy, S. K., Rwawiire, S., & Zhao, Y. (2018). Sustainable wastewater treatment methods for textile industry. In S. S. Muthu (Ed.), *Sustainable Innovations in Apparel Production* (pp. 21–87). Springer, Singapore. https://doi.org/10.1007/978-981-10-8591-8_2

Phrakonkham, S., Kazama, S., Komori, D., & Sopha, S. (2019). Distributed hydrological model for assessing flood hazards in Laos. *JWARP, 11*, 937–958. https://doi.org/10.4236/jwarp.2019.118056

Pradhan, B. (2010). Flood susceptible mapping and risk area delineation using logistic regression, GIS and remote sensing. *Journal of Spatial Hydrology, 9*.

Rahman, M., Ningsheng, C., Islam, M. M., Dewan, A., Iqbal, J., Washakh, R. M. A., & Shufeng, T. (2019). Flood susceptibility assessment in Bangladesh using machine learning and multi-criteria decision analysis. *Earth Systems and Environment, 3*, 585–601. https://doi.org/10.1007/s41748-019-00123-y

Rahmati, O., Pourghasemi, H. R., & Zeinivand, H. (2016). Flood susceptibility mapping using frequency ratio and weights-of-evidence models in the Golastan Province, Iran. *Geocarto International, 31*, 42–70. https://doi.org/10.1080/10106049.2015.1041559

Roslee, R., Tongkul, F., Mariappan, S., & Simon, N. (2018). Flood hazard analysis (FHAN) using multi-criteria evaluation (MCE) in Penampang area, Sabah, Malaysia. *ASM Science Journal, 11* (3), 104–122.

Saaty, T. L. (1980). *The analytic hierarchy process: Planning, priority setting, resource allocation*. McGraw-Hill International Book Co.

Saaty, T. L. (1990). How to make a decision: The analytic hierarchy process. *European Journal of Operational Research, 48*, 9–26. https://doi.org/10.1016/0377-2217(90)90057-I

Saha, S. (2017). Groundwater potential mapping using analytical hierarchical process: a study on Md. Bazar Block of Birbhum District West Bengal. *Spatial Information Research, 25*(4), 615–626. https://doi.org/10.1007/s41324-017-0127-1

Saha, A. K., & Agrawal, S. (2020). Mapping and assessment of flood risk in Prayagraj district, India: A GIS and remote sensing study. *Nanotechnology for Environmental Engineering, 5*, 11. https://doi.org/10.1007/s41204-020-00073-1

Samanta, S., Pal, D. K., & Palsamanta, B. (2018). Flood susceptibility analysis through remote sensing, GIS and frequency ratio model. *Applied Water Science, 8*, 66. https://doi.org/10.1007/s13201-018-0710-1

Sanyal, J., & Lu, X. X. (2003). Application of remote sensing in flood management with special reference to monsoon Asia: A review. *Natural Hazards, 33*, 283–301. https://doi.org/10.1023/B:NHAZ.0000037035.65105.95

Sarkar, D., & Mondal, P. (2020). Flood vulnerability mapping using frequency ratio (Fr) model: A case study on Kulik river basin, Indo-Bangladesh Barind region. *Applied Water Science, 10*, 17. https://doi.org/10.1007/s13201-019-1102-x

Şen, Z. (2018). *Flood modeling, prediction and mitigation* (1st ed.). Springer International Publishing: Imprint: Springer.

Shafapour Tehrany, M., Kumar, L., Neamah Jebur, M., & Shabani, F. (2019a). Evaluating the application of the statistical index method in flood susceptibility mapping and its comparison with frequency ratio and logistic regression methods. *Geomatics, Natural Hazards and Risk, 10*, 79–101. https://doi.org/10.1080/19475705.2018.1506509

Shafapour Tehrany, M., Kumar, L., & Shabani, F. (2019b). A novel GIS-based ensemble technique for flood susceptibility mapping using evidential belief function and support vector machine: Brisbane, Australia. *Peer Journal, 7*, e7653. https://doi.org/10.7717/peerj.7653

Sørensen, R., Zinko, U., & Seibert, J. (2006). On the calculation of the topographic wetness index: evaluation of different methods based on field observations. *Hydrology and Earth System Sciences, 10*(1), 101–112. https://doi.org/10.5194/hess-10-101-2006

Sphere India. (2018). *Kerala floods joint detailed needs assessment report: In the aftermath of the floods in Kerala.* Sphere India Inter Agency Coordination Standard Operating Procedures (SOP).

Subbarayan, S., & Sivaranjani, S. (2020). Modelling of flood susceptibility based on GIS and analytical hierarchy process—A case study of Adayar river basin, Tamil Nadu, India. In I. Pal, J. von Meding, S. Shrestha, I. Ahmed, & T. Gajendran (Eds.), *An interdisciplinary approach for disaster resilience and sustainability* (pp. 91–110). Springer Singapore.

Swain, K. C., Singha, C., & Nayak, L. (2020). Flood susceptibility mapping through the GIS-AHP technique using the cloud. *IJGI, 9*, 720. https://doi.org/10.3390/ijgi9120720

Taylor, J., Lai, K. M., Davies, M., Clifton, D., Ridley, I., & Biddulph, P. (2011). Flood management: Prediction of microbial contamination in large-scale floods in urban environments. *Environment International, 37*, 1019–1029. https://doi.org/10.1016/j.envint.2011.03.015

Tehrany, M. S., & Jones, S. (2017). Evaluating the variations in the flood susceptibility maps accuracies due to the alterations in the type and extent of the flood inventory. *International Archives of the Photogrammetry, Remote Sensing and Spatial Information Sciences, XLII-4* (W5), 209–214. https://doi.org/10.5194/isprs-archives-XLII-4-W5-209-2017

Tehrany, M. S., Pradhan, B., Mansor, S., & Ahmad, N. (2015). Flood susceptibility assessment using GIS-based support vector machine model with different kernel types. *Catena, 125*, 91–101. https://doi.org/10.1016/j.catena.2014.10.017

Tien Bui, D., Hoang, N.-D., Pham, T.-D., Ngo, P.-T. T., Hoa, P. V., Minh, N. Q., Tran, X.-T., & Samui, P. (2019). A new intelligence approach based on GIS-based Multivariate Adaptive Regression Splines and metaheuristic optimization for predicting flash flood susceptible areas at high-frequency tropical typhoon area. *Journal of Hydrology, 575*, 314–326. https://doi.org/10.1016/j.jhydrol.2019.05.046

Uzor-Totty, A. E., & Lawal, O. (2019). Surface Runoff Dynamics across Imo River Basin: Implications for Flood Management. *Watershed Modelling & Flood Risk.* CDRMDS Regional Conference on Disaster Risk Reduction & Centre's 10th Year Anniversary: From Disaster Risk Vulnerability to Sustainability Actions

Vojtek, M., & Vojteková, J. (2019). Flood susceptibility mapping on a national scale in slovakia using the analytical hierarchy process. *Water, 11*, 364. https://doi.org/10.3390/w11020364

Wallemacq, P., & Rowena, H. (2018). *Economic losses, poverty and disasters 1998–2017.* Centre for Research on the Epidemiology of Disasters (CRED) and UNISDR.

Wanders, N., Karssenberg, D., de Roo, A., de Jong, S. M., & Bierkens, M. F. P. (2014). The suitability of remotely sensed soil moisture for improving operational flood forecasting. *Hydrology and Earth System Sciences, 18*, 2343–2357. https://doi.org/10.5194/hess-18-2343-2014

Waris, M., Panigrahi, S., Mengal, A., Soomro, M. I., Mirjat, N. H., Ullah, M., Azlan, Z. S., & Khan, A. (2019). An application of analytic hierarchy process (Ahp) for sustainable procurement of construction equipment: Multicriteria-based decision framework for Malaysia. *Mathematical Problems in Engineering, 2019*, 1–20. https://doi.org/10.1155/2019/6391431

World Disasters Report. (2015). *World Disasters Report Focus on local actors, the key to humanitarian effectiveness.* The International Federation of Red Cross and Red Crescent Societies.

World Health Organization (WHO). (2018). *Floods.* World Health Organization. Retrieved May 22, 2021, from https://www.who.int/westernpacific/health-topics/floods

Young, R. A., & Mutchler, C. K. (1969). Effect of slope shape on erosion and runoff. *Transactions of the American Society of Agricultural Engineers, 12*, 231–233.

Yousuf Gazi, M., Ashraful Islam, M., & Hossain, S. (2019). Flood-hazard mapping in a regional scale—Way forward to the future hazard atlas in Bangladesh. *Malaysian Journal Geosciences, 3*, 01–11. https://doi.org/10.26480/mjg.01.2019.01.11

Chapter 7
Soil Loss Estimation Using Models and Field Database in Lateritic Badlands, Eastern India: Evaluation and Validation

Sandipan Ghosh

Abstract The key purpose of this chapter is to determine the suitability and applicability of index-based erosion models for the precise estimation and prediction of annual soil erosion rates under the monsoon-dominated geo-climatic and land use systems. The study unit is represented as the lateritic badlands of Dwarka–Brahmani River Basin (Eastern India). The present study finds a variable range of annual erosion rates (8.12–24.01 kg m^{-2} $year^{-1}$ as measured data) at hillslope scale of watershed (i.e. basins of permanent gullies) using popular models of Revised Universal Soil Loss Equation (RUSLE) and Revised Morgan-Morgan-Finney (RMMF), sedimentation pits and field measured data (2016–2017). The important part of this experimental design and quantitative analysis, used to assess the effectiveness of models, is to compare the forecast given by model to field measured data. The regression analysis of experimental results show that there is a positive correlation and increment between measured and predicted erosion data in RUSLE modelling ($Y_c = 5.90 + 0.659\ X$, $R^2 = 0.521$), but an inverse relation and negative increment are observed in RMMF modelling ($Y_c = 16.27 + 0.162\ X$, $R^2 = 0.212$). The indices of model evaluation and testing statistics have confirmed the reliable performance (best fit to observed erosion rate) of RUSLE over RMMF. The potential erosion map of area depicts annual erosion rate beyond the tolerance limit (1.0 kg m^{-2} $year^{-1}$). It is estimated that the mean soil depth of 0.95 cm $year^{-1}$ is permanently lost from the surface of lateritic catchments, and the water erosion will require typically 176 years to erode the mean soil thickness of 1500 mm.

Keywords Soil erosion · Land degradation · Gully erosion · Laterite · Soil loss tolerance · RUSLE · RMMF model

S. Ghosh (✉)
Department of Geography, Chandrapur College, Purba Barddhaman, 713145, West Bengal, India

P. K. Shit et al. (eds.), *Geospatial Technology for Environmental Hazards*, Advances in Geographic Information Science, https://doi.org/10.1007/978-3-030-75197-5_7

7.1 Introduction

Soil erosion is regarded as one of pervasive geomorphic hazards in Anthropocene and taking immediate steps and management actions to preserve our soil resources should need no explanation (Bell, 2002; Lal, 2012; Poesen, 2018). Soil erosion is now referred to as the most important factor of land degradation and globally; about 30% of land areas are estimated to be degraded in this condition of environmental change, affecting almost 3.2 billion people (mainly Brazil, China, Ethiopia, Indian and Spain) (Wen & Deng, 2020). Erosion visibly degrades landscapes through exposure of sub-soil, presence of rills and gullies or the occurrence of dust storms. Land degradation can be defined as a negative trend in land potentiality, caused by direct or indirect human-induced processes including anthropogenic climate change, slope modification, deforestation and land use changes, expressed as long-term reduction or loss of at least one of the follows: biological productivity, ecological integrity or value humans and soil productivity (Nkonya et al., 2016). In the Anthropocene, soil losses by human activities (i.e. anthropogenic soil erosion) have also become very significant: e.g. tillage erosion, soil erosion by land leveling, soil quarrying, crop harvesting, explosion cratering and trench digging (Bocco, 1991; Poesen, 2019). During the past 60 years, many studies and researches have documented variable magnitude of soil erosion problems in different parts of the world (especially in India) (Table 7.1 and Fig. 7.1), expressed as billions of tons of eroded soil or billions dollars of erosion and sedimentation damage each year (Narayana & Babu, 1983; Bocco (1991); Kothyari, 1996; Lal, 1990; Singh et al., 1992; Wasson, 2003; Vente & Poesen, 2005; Pimentel, 2006; Reddy & Galab, 2006; Thakkar & Bhattacharyya, 2006; Kumar & Pani, 2013; Pimentel & Burgess, 2013; Sharda et al., 2013; Sharda & Dogra, 2013; Aulakh & Sidhu, (2015); Borrelli et al., 2017; Froechlich, 2018; Sharma, 2018; Poesen, 2018; Pennock, 2019).

Soil erosion is defined as the net long-term balance of all processes that detach soil particles and move it from its original location through sheet flow, rill and gully channels (Eekhout & Vente, (2019); Pennock, 2019). In the Indo-Gangetic Plain, the world's larger alluvial plain and other agricultural regions of India, soil erosion by water is the most serious cause of land degradation (Marzolff & Pani, 2019). It affects 64% of the estimated area of 147 m ha of degraded wasteland in the country (Marzolff & Pani, 2019). It estimated an annual average potential soil erosion amounting 35 Pg year^{-1} for 2001 and in 2012, an overall increase of 2.5% in soil erosion (Borrelli et al., 2017). RUSLE-based modelling approach predicts global potential soil erosion rates of 43 Pg year^{-1}, and due to climate change and land use transformation, average soil erosion can be increased from 30 to 66% in between 2015 and 2070 (Borrelli et al., 2020). About 5–7 million ha (12.4–17.3 million acre) of arable land in the world is degraded annually through various erosion processes, and out of 2 billion ha (4.9 billion acre) of degraded area in world, water erosion alone, being a global phenomenon, contributes about 55% (Fig. 7.1) (Sharda et al., 2010). Among the soil groups of India, red–lateritic soils (mostly alfisol, inceptisols and utlisols) and black soils (vertisols and vertic subgroups) acutely suffer due to

Table 7.1 Key information and findings on soil erosion issues in India

Sl. no.	Important facts and research outcomes	Source
1	Annual soil erosion is taking place at the rate of 16.35 t ha^{-1} $year^{-1}$	Narayana and Babu (1983)
2	Indo-Gangetic Plains of Punjab, Haryana, Uttar Pradesh, Bihar and West Bengal are affected by erosion rate of 5 to 10 t ha^{-1} $year^{-1}$	Singh et al. (1992)
3	About 20% of India's existing reservoirs will have lost 50% of their previous storage capacity due to soil loss and siltation	Kothyari (1996)
4	3.975 million ha of wastelands are severely affected by gullies and ravines	Yadav and Bhushan (2002)
5	Due to siltation, India is losing about 1.3 billion m^3 of storage capacity each year and to create this storage capacity India will require Rs. 1448 crores	Thakkar and Bhattacharyya (2006)
6	Paddy is the most affected among all crops in terms of both productions 4.3 million tonne and monetary loss of Rs. 24.4 billion	Sharda et al. (2010)
7	The Lower Gangetic Plain and eastern part of Chota Nagpur Plateau has soil loss tolerance level of 2.5–12.5 t ha^{-1} $year^{-1}$	Bhattacharyya et al. (2007); Mondal and Sharda (2011)
8	India suffers an annual loss of 13.3 million tonne in production of cereals, oilseeds and pulses due to water erosion	Sharda et al. (2013)
9	About 69.5% area of India has soil loss tolerance limit of <10 t ha^{-1} $year^{-1}$	Sharda and Dogra (2013)
10	About 5.4 million tone of fertilizer worth US $ 245 million is washed away by water erosion	Gulati and Rai (2014)
11	Erosion escalates the siltation rate of reservoirs in India–Maithon (1.076 mm $year^{-1}$), Panchet (0.631 mm $year^{-1}$), Tilaiya (2.792 mm $year^{-1}$), Tenughat (0.716 mm $year^{-1}$), Durgapur barrage (0.042v), Kangsabati (0.752 mm $year^{-1}$) and Massanjore (0.557 mm $year^{-1}$)	Central Water Commission (2015)
12	The soil pool loses 110 Mt Carbon into the atmosphere due to soil erosion. It is projected that 1% increase in rainfall intensity may increase the rainfall Erosivity by 2–6%. Annual loss due to soil degradation ranges from Rs. 89–232 billion	Bawa (2017)
13	1 mm loss of soil from one hectare land, an additional 1642 MJ of energy is expended, which is equivalent to about 91 kg of petrol	Sharda et al. (2019)
14	It is estimated from satellite images that 9593.06 km^2 land of India (17.09 km^2 land of West Bengal) is affected by gullies and ravines	National Remote Sensing Agency, NRSC (2019)
15	In a river basin of semi-arid region, soil erosion risk was assessed using RUSLE and frequency ratio probability algorithm to prioritize erosion susceptible areas	Gayen et al. (2020)

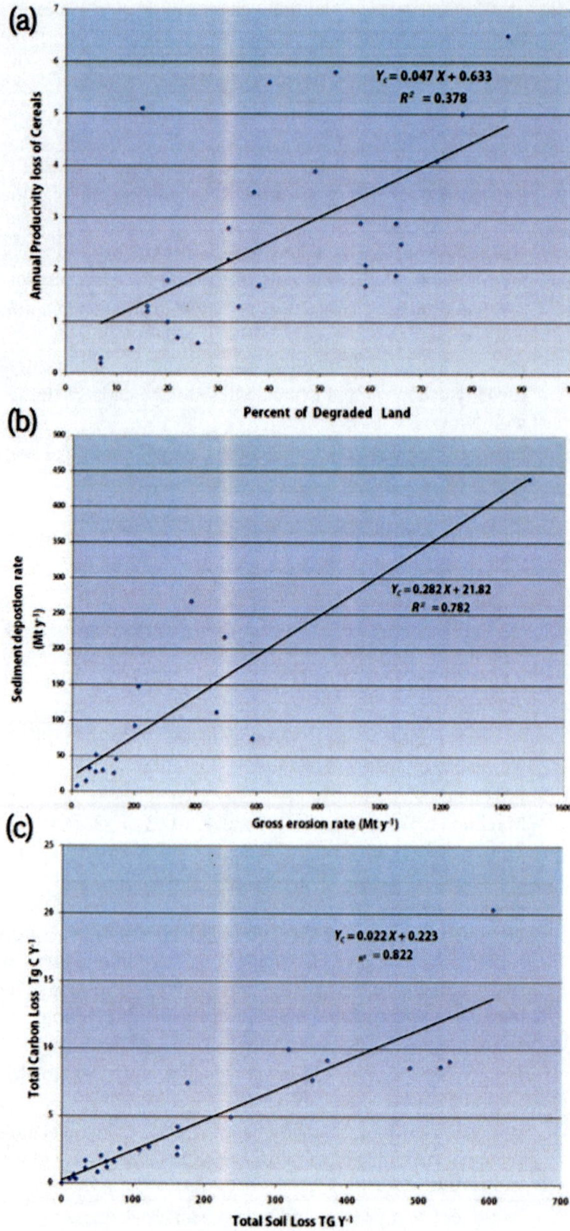

Fig. 7.1 Important research findings of soil erosion hazard in India—(**a**) with increasing percentage of state-wise degraded land the production loss of cereals is also rising, (**b**) there are a positive correlation and increasing trend in between gross erosion rates of river basins and sediment deposition rates of reservoirs and (**c**) total carbon loss from land is escalating with increasing soil loss in India

water erosion (Table 7.2). It is estimated that 120.72 million ha area is affected by various forms of land degradation and desertification in India with water erosion being chief contributor (68.4%) (Sharda et al., 2013). About 69.5% area of India has soil loss tolerance limit of <10 t ha^{-1} year^{-1}, while about 13.3% area has a soil loss tolerance limit of only up to 2.5 t ha^{-1} year^{-1} (Sharda & Dogra, 2013). In India

Table 7.2 Expected and average values of loss of soil productivity due to water erosion in different soils of India (Sharda et al. (2010)

Erosion class	Soil loss (t ha^{-1} $year^{-1}$)	Loss in productivity (%)		
		Alluvial soils	Black soils	Red soils
Very slight	<5	0.0	2.5	5.0
Slight	5–10	2.5	7.5	17.5
Moderate	10–20	7.5	17.5	37.5
Strong	20–40	17.5	37.5	60.0
Severe	>40	37.5	60.0	–

major rainfed crops suffer an annual production loss of 13.4 Mt due to water erosion which amounts to a loss of Rs. 305.32 billion in monetary terms (Ghosh et al., 2020).

It is now understood that soil erosion is a pertinent issue where the adage 'think globally, act locally' is clearly applicable (Toy et al., 2013). The essential purpose of quantitative assessment is that erosion control targeted toward the areas with the highest rates can markedly reduce erosion averages. Before taking any erosion protection measures, the estimation of annual erosion rate at plot to basin scale is the fundamental step towards achieving soil conservation and sustainable development (Toy et al., 2013). Models can serve a needful purpose of soil conservation which acts to make broad-scale erosion surveys in order to realize the existing problem over an erosion-prone lateritic region of tropical monsoon climate and to track changes in erosion over time (Nearing, 2013). Modelling and prediction of soil erosion by water has long legacy and preliminary popular studies published in various international journals probably seven decades ago using North American data sets (Bennett, 1939). The largest number of publications with the application of Revised Universal Soil Loss Equation (RUSLE) model has been found in the USA (274 papers), China (218 papers), Brazil (88 papers), India (67 papers), Spain (66) papers), etc. Up to 2017, 1556 research papers have been published at various spatial scales (1977–2017) (Alewell et al., 2019; Yanshuang et al., 2020). Many mathematical models categorized as empirical or index-based, conceptual, physically based or process-oriented are variable to estimate soil erosion at different spatial and temporal scales (Wischmeier & Smith, 1978; Renard et al., 1997; Morgan et al., 1998; Flanagan et al., 2001; Morgan, 2001; Merriti et al., 2003; Avwunudiogba & Hudson, 2014; James et al., 2017; Morgan & Duzant, 2008; Alewell et al., 2019; Pennock, 2019; Gayen et al., 2020; Yanshuang et al., 2020).

The data availability on land degradation, soil erosion rates and permissible soil loss limits is either qualitative or insufficient for proficient planning of conservation and management of erosion intensity at watershed or regional scale (Sharda et al., 2013). The criterion for judging whether the soil has potential risk of erosion or not is essentially required for adopting appropriate erosion control measures on grazing land, arable land, barren land and other land use systems (Sharda et al., 2013). Realistic assessment of erosion risk or soil loss rate thus constitutes the first step for understanding the ground reality of erosion and raising awareness among governmental and other stakeholders in a given region to adopt appropriate strategies for sustainable and efficient use of natural resources for the current and future generations (Sharda et al., 2013). Erosion protection measures should start from microscale to get long-term soil productivity and long-term sustainable agriculture in the developing counties, like India, where erosion protection technologies are limited by economic and other cultural conditions. In addition, it is necessary to state that the laterite terrain of West Bengal (known as *Rarh* Plain, i.e. the land of red soil) is severely dissected by the dense network of rills and gullies (Ghosh & Guchhait, 2017), developing badland topography, and there are very few databases of accurate annual erosion rates and empirical model applications. The lateritic *Rarh* region and plateau fringe (districts of Purulia, Bankura, Paschim Barddhaman and Paschim Medinipur) show lower T value ranging from 2.5 to 5.0 Mg ha^{-1} year^{-1} (Mondal & Sharda, 2011; Lenka et al., 2014). In West Bengal as a whole, about 88% of the area is identified as T value zone of 12.5 Mg ha^{-1} year^{-1} (Mondal & Sharda, 2011; Lenka et al., 2014). In this regard, this study can give few insights on the aspect of soil erosion modelling using minimal data inputs and measured plots at basin scale to estimate annual erosion rate in the lateritic badlands. Two major objectives of the study are set forth as follows:

(1) To estimate annual soil erosion rate using models and field experimental database
(2) To evaluate suitability and effectiveness of model in the study area

7.2 Geographical Setting of Study Area

The geomorphic unit of study is recognized as the badlands (interfluves) in between Brahmani (north) and Dwarka (south) rivers (encompassed by 24° 20′ N to 23° 40′ N, and 87° 26′ E to 88° 21′ E) (Fig. 7.2). This geomorphic region is recognized as plateau proper and plateau fringe of Chota Nagpur, prevailing the patches of laterite exposures and basaltic hills, and it is categorized as the northern part of the *Rarh* Plain (Biswas, 1987). Geologically, the interfluve is associated with the contiguous unit between Rajmahal Basalt Traps (RBT) (Early Cretaceous origin) and the Bengal Basin which exhibits shallow Quaternary alluvium deposits. The palaeogenesis of the deep weathering profiles under intense tropical wet–dry palaeoclimate on the basaltic surface formed hard ferruginous crust, i.e. *Ferricrete* (Palaeogene–Early Pleistocene) (Ghosh et al., 2020).

The sample study area of laterite interfluve (about 176 km^2, encompassed by 24°08′N to 24°14′ N and 87°38′ E to 87°44′ E) covers Shikaripara block (Dumka, Jharkhand) and Rampurhat I and Nalhati I blocks (Birbhum, West Bengal) (Fig. 7.3). Field study reveals successive occurrences of fresh quartz-normative tholeiite Rajmahal basalt, weathered coarse saprolite, kaolinite pallid zone, mottle zone and pisolitic ferricrete in the litho-sections (Ghosh & Guchhait, 2015). Each laterite section reflects both primary in situ-type palaeogenesis of high-level plateau laterites (Chorley et al., 1984) and secondary ex situ evolution of piedmont slope laterites which are prone of to water erosion, forming patches of badlands in the *Rarh* Plain.

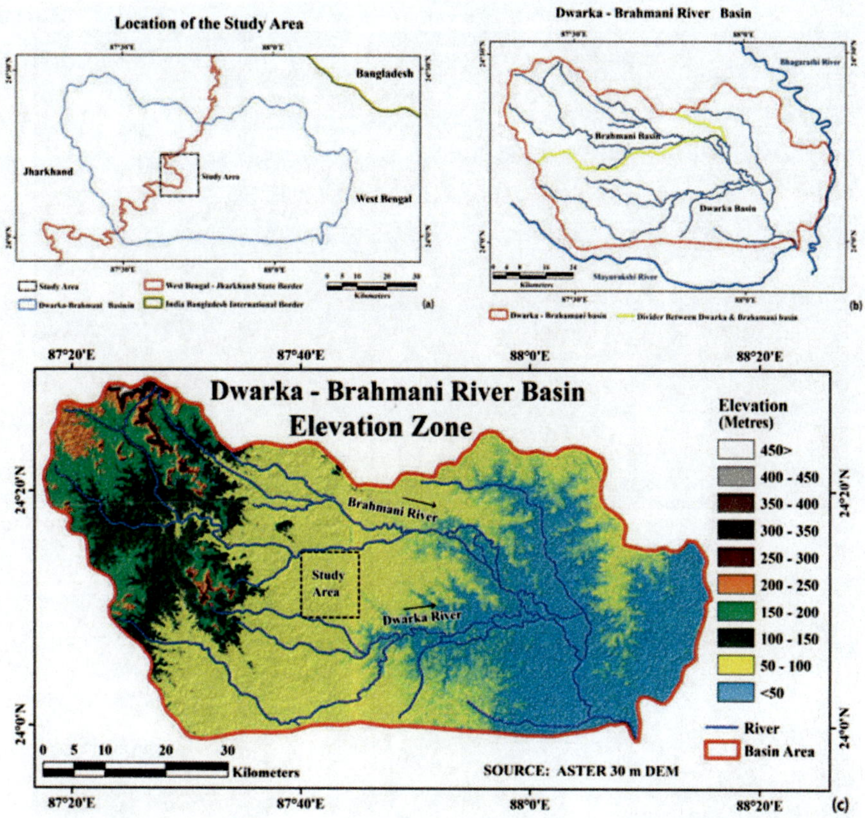

Fig. 7.2 Spatial extent and elevation zones of Dwarka-Brahmani Interfluve (Ghosh & Guchhait, 2020)

The climate of this region has been identified as sub-humid and sub-tropical monsoon type, receiving mean annual rainfall of 1300–1437 mm. The amount of rainfall is decreasing from western to eastern part. On the basis of 2010–2016 rainfall data, the mean annual rainfall of Paikor, Md. Bazar, Rampurhat and Mallarpur is 720.0 mm, 1176.0 mm, 1293.5 mm and 1372.8 mm respectively. The peak monsoon and cyclonic rainfall intensity of 21.51 mm h^{-1} (minimum) to 25.51 mm h^{-1} (maximum) are the most powerful climate factors to develop this lateritic badlands (Table 7.3) (Ghosh & Bhattacharya, 2012). The region has experienced intense thunderstorms during hot summer and prolonged rainfall during the tropical depression and cyclone.

In and around the study area, the soil series of Bhatina, Raspur and Jhinjharpur (Sarkar et al., 2007) has been developed in the present geo-climatic setting. Generally, thin solum is loamy skeletal and hypothermic in nature developing on the barren lateritic wastelands with sparse bushy vegetation and grass. The dark reddish to brown-coloured sandy clay loam of 0–16 cm (A horizon, maximum grass root

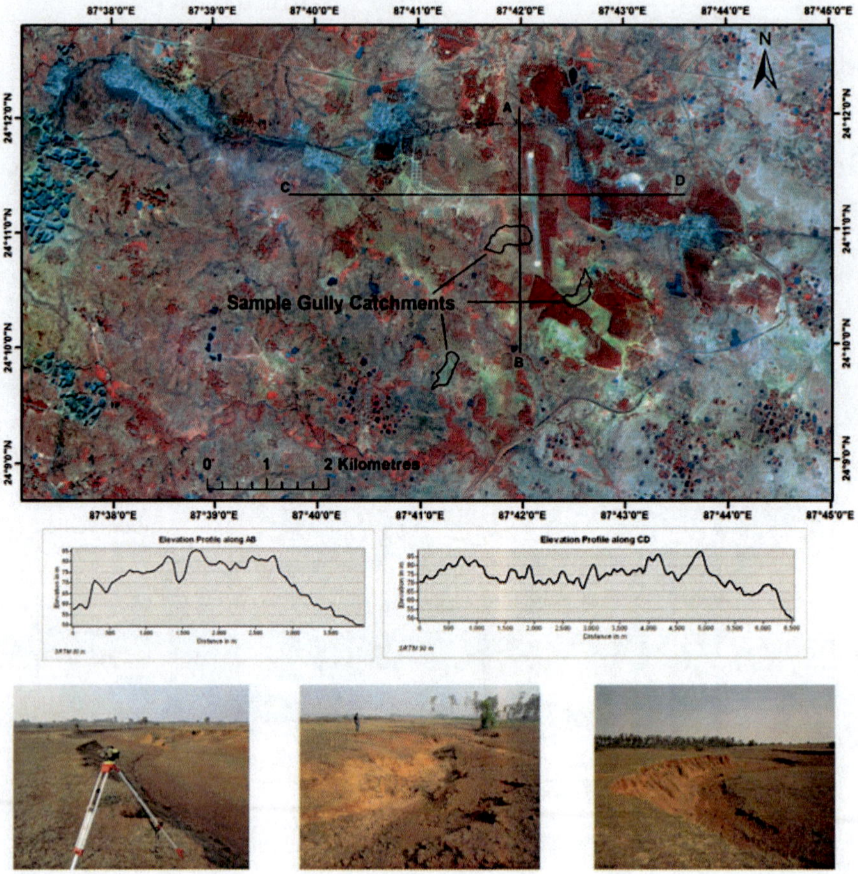

Fig. 7.3 Standard FCC IRS LISS IV image (Dec, 2015) of study area showing location of sample gully catchments, SRTM DEM elevation profiles and field photographs

Table 7.3 Climate–erosion relationship in the study area

Climatic phenomena	Effects on landforms and soil loss
1. Seasonal variation of temperature (about 15°–18 °C) and ground moisture	Encourage various processes of weathering, like block disintegration, formation of cracks and joints
2. High temperature range (max. 45 °C and min. 9 °C)	Lowering soil moisture and ground water table, loosening of soil particles, drying up of surface soils, reduction in soil cohesiveness
3. Season rainfall (from mid-June to October)	Weathered products and loose particles are removed from slope, favour lateritization
4. Short phase of heavy downpour within monsoon months	Development of badland topography, maximum erosion, tunnel erosion, mass wasting of valley sides and head cut migration of rill and gully

Source: Ghosh and Bhattacharya (2012)

Table 7.4 Estimated SCS-CN values of AMC II condition in the sample gully catchment 1, 2 and 3 on the basis of existing land use/land cover

Gully catchment 1						
HSG group	LULC	CN II	Area (m^2)	Product of CN II × area	CN II weighted	S (mm) II
C	Natural Vegetation	73	27,700	2,022,100	85.88	41.72
B	Grassland	86	41,150	3,538,900		
B	Bare surface	91	40,400	3,680,040		
Gully catchment 2						
HSG group	**LULC**	**CN II**	**Area (m^2)**	**Product of CN II × area**	**CN II weighted**	**S (mm) AMC II**
C	Natural Vegetation	73	25,900	1,890,700	85.52	42.97
B	Grassland	86	36,275	3,119,650		
B	Bare surface	91	56,150	5,109,650		
Gully catchment 3						
HSG group	**LULC**	**CN II**	**Area (m^2)**	**Product of CN II × area**	**CN II weighted**	**S (mm) AMC II**
C	Natural Vegetation	73	64,500	4,708,500	84.9644	44.92
B	Grassland	86	28,600	2,459,600		
B	Bare surface	91	122,950	11,188,450		

Note: *HSG* Hydrologic Soil Group, *LULC* land use/land cover, *CN* curve number, *S* maximum surface storage, *AMC* antecedent moisture condition
Source: Ghosh and Guchhait (2020)

zone) is developed over the fragmented secondary laterites. The loose secondary laterite (16–34 cm) is developed as cementation (low cohesion and weak structure) of derived materials over mottle and kaolinte horizon, and it is much prone to overland flow erosion, tunnel erosion and bank failure (Ghosh & Guchhait, 2020). The natural vegetation of the study area belongs to the tropical moist and dry deciduous type with few evergreen types. The observed natural vegetation species are Babul (*Acacia nilotica*), Bel (*Aegle marmelos*), Behara (*Terminalia bellirica*), Sal (*Shorea robusta*), Mahua (*Madhuca indica*), Khair (*Acacia catechu*), Khajur (*Phoenix sylvestris*), Jamun (*Syzygium cumini*), etc.

In this context, the land use classification and SCS-CN (Soil Conservation Service–Curve Number) data of three sample gully catchments are derived. In gully catchment 1 (basin area of 109,250 m^2), the principal land use/land cover is identified as natural vegetation (25.35%), grassland (37.67%) and bare laterite land (36.98%) (Table 7.4) (Ghosh & Guchhait, 2020). In gully catchment 2 (basin area of 118,325 m^2), the areal coverage of natural vegetation, grassland and bare laterite soil are 21.88%, 30.65% and 41.47%, respectively. In gully catchment 3 (basin area of 216,050 m^2), the total areal coverage of natural vegetation, grassland and bare laterite land are 29.85%, 13.23% and 56.94%, respectively (Ghosh & Guchhait, 2020). Applying the SCS-CN method (Chow et al., 1998; Mishra & Singh, 2003; Mishra et al., 2006; Bhunya et al., 2014; Gajbhiye et al., 2014; Srivastava & Imtiyaz, 2016; Singh, 2016) in three sample watersheds of gullies, it is found that on the basis of rainfall range of 42–137.2 mm, the sample watersheds can yield runoff of 40.02–118.0 mm in excess moisture condition of monsoon (Ghosh & Guchhait, 2020).

7.3 Methodology

The goal of United Nations Sustainable Development have new challenges and policy developments which provide opportunities for researchers and scholars to respond with more accurate assessments of erosion rates and solutions of erosion vulnerability, targeting negative trend of land degradation (Panagos & Katsoyiannis, 2019). To understand the hydro-geomorphic processes of soil erosion and to apply quantitative erosion models, the study demands an inter-disciplinary outlook, applying the methods of hydrology, geomorphology and statistics. The total methodology is combination of various sequential steps, viz. development of experimental design, data collection, model description, application and evaluation, soil loss tolerance, statistical analysis and thematic mapping (Fig. 7.4).

Experimental Design and Erosion Measurement at Hillslope Scale

The selection of erosion measurement sites to justify the application of erosion model poses a problem of sampling. Since it is not only possible to take measurements at each specific point in the landscape, it is important that the sample area should be representative of the catchment as high erosion prone zone (where maximum erosion is observed). From the field survey, it is observed that except permanent channels, the gully head slope (average slope 7°34′) is the key pathway of sediment transport to the main gully. In this lateritic terrain, the high erosion risk catchment of gully is firstly selected, and it has well-defined basin area (about 109,250–216,050 m^2) and dense network of gullies (7.57–8.33 km km^{-2}) (Fig. 7.5). Firstly, 18 gully heads of 3 basins (selected randomly within 17 basins at study area, based on high drainage density of greater than 7.5 km km^{-2}) were identified, and then 18 gully head slope elements (considering 2 m width of slope strip to incorporate soil–land use parameters) were selected, denoting S1–S18, respectively. The steepness of hillslope was measured using Leica Sprinter 150 m digital levelling instruments (accuracy – ± 0.7 mm of the 250 m distance) and other parameters of models were estimated in the recurrent field survey (2016–2018) and

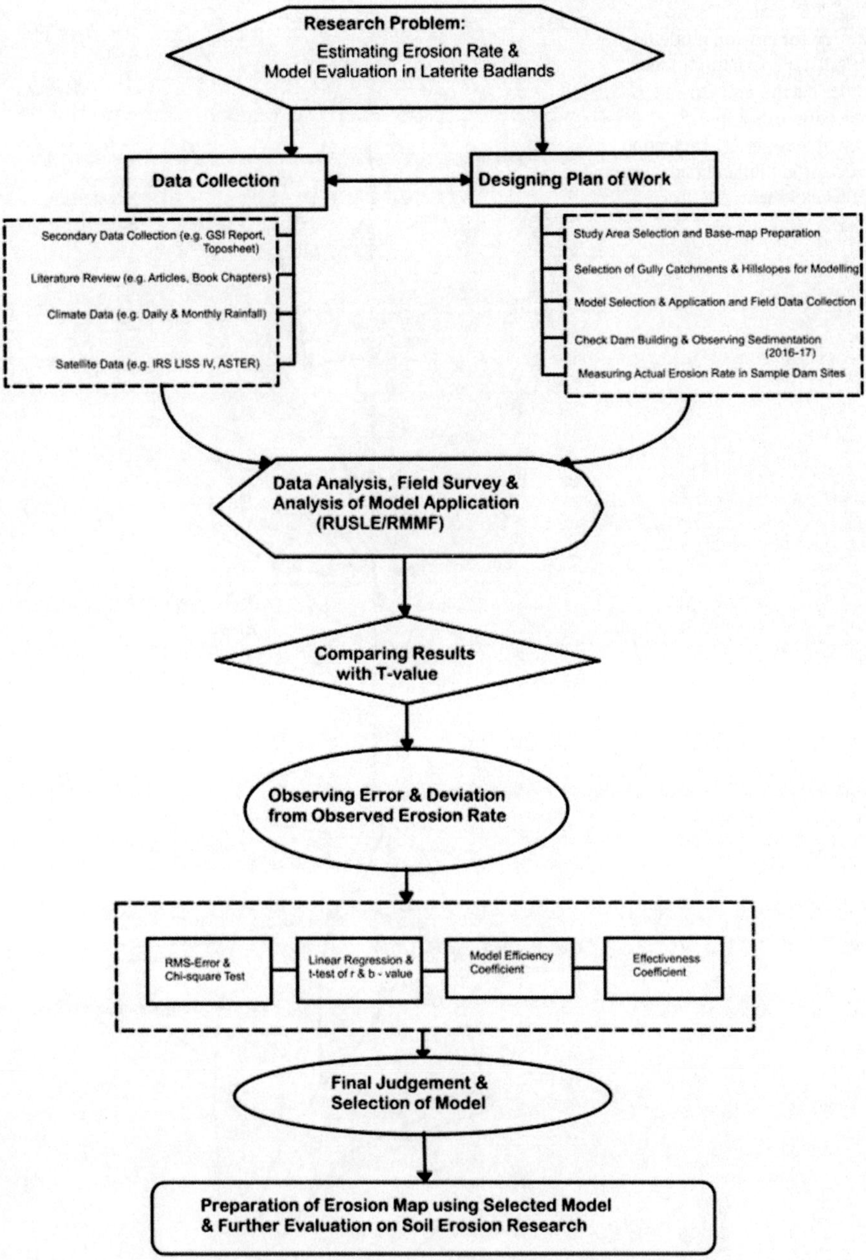

Fig. 7.4 Methodological flowchart of erosion model used in soil erosion research

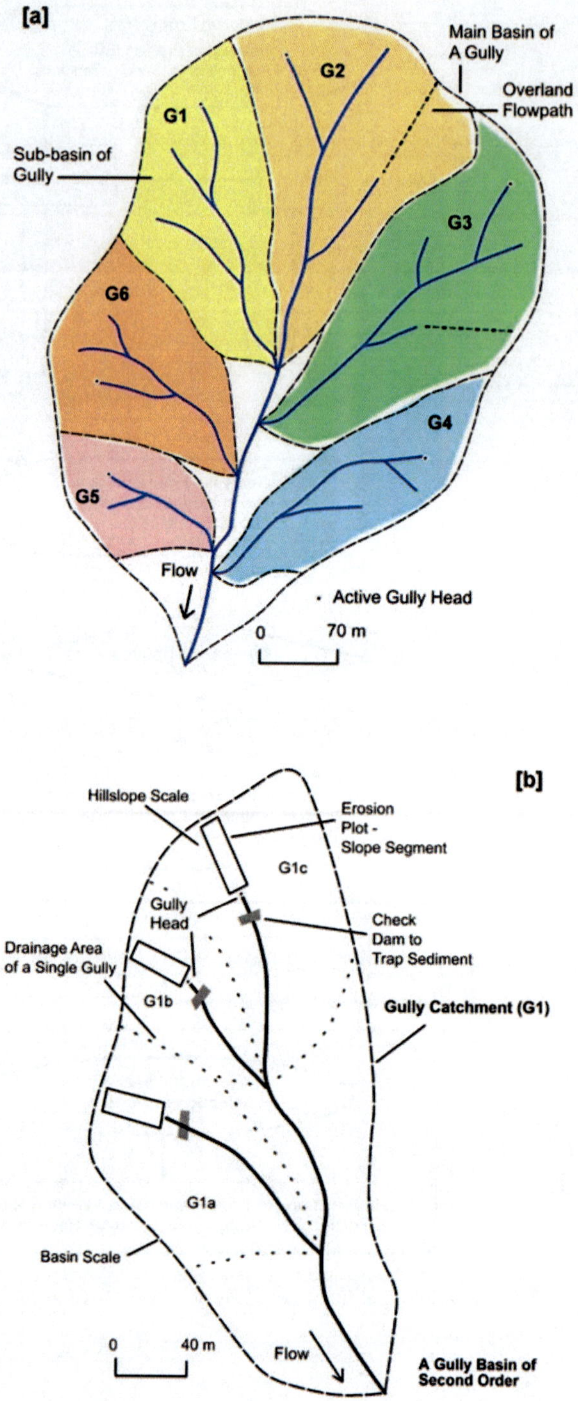

Fig. 7.5 Experimental design for erosion modelling at hillslope scale in a lower order catchment: (**a**) selection of sample catchments and (**b**) selection of erosion plot and dam sites in a catchment

Fig. 7.6 (**a**) Sample dam locations at gully headcuts in catchment 1, (**b**) development of temporary dam to trap sediments, (**c**) final structure of small dams below gully headcuts and (**d**) measuring the morphological parameters above dam

guided values of erosion models (Renard et al., 2011; Renard et al., 1997; Morgan, 2001; Morgan & Duzant, 2008). The total slope length is the overland flow part between the gully head and water divide. The steepness of slope elements varies from 3°45′to 11°06′, whereas slope length varies from 22.1 m to 106.8 m. Maintaining a certain distance (1.5 to 2 m) from active gully head, 18 check dams (used as sedimentation pits) were developed (denoting Dam 1 to Dam 18) at the base (i.e. gully floor) of representative slope elements to trap eroded sediments coming from upslope in a year (2016–2017) (Fig. 7.6).

Following the shape of gully channel, it was decided to built *V*-shape design using sand, cement and laterite boulders of irregular shape, and the gap between gully head base and dam was used as sedimentation pits to collect eroded materials. The dams were developed in January 2016 with the help of local manpower and resources. The constructed dams had a height range of 40–55 cm and width range of

92–190 cm. Mostly during monsoon period (June to October) of 2016, the eroded material of these slope elements or upstream drainage areas of gullies were trapped behind the dams. It is very needed to mention that occasional high sediment flux was observed during few extreme thunderstorms (5 times in 2016), having very high rainfall intensity of greater than 25.51 mm h^{-1} (April to June). Then after one year of observation, the sedimentation was measured in January 2017, and the mass volume was measured as multiplying the area of sedimentation behind dam and mean depth of sedimentation at 18 dam sites. The bulk density of eroded materials was calculated at laboratory (mean bulk density of materials is 1.717 gm cm^{-3}), and the mass weight of sedimentation materials was measured by multiplying the volume of mass (unit in kg) by bulk density. The observed rate of erosion (unit as kg m^{-2} $year^{-1}$) was measured by dividing the mass weight by strip area of slope element or erosion plot for one year (2016–2017).

It was calculated that in 18 dam sites, the estimated weight of trapped sediments (i.e. mostly ferruginous nodules and coarse sands) varies to a great extent due to activeness of water erosion, slope angle and overland flow length, ranging from 566 to 3581 kg. The observed annual erosion rate (O) of three sample catchments was finally measured as (a) 10.50–24.27 kg m^{-2} $year^{-1}$ (gully catchment 1), (b) 8.12–20.82 kg m^{-2} $year^{-1}$ (gully catchment 2) and (c) 11.87–20.82 kg m^{-2} $year^{-1}$ (gully catchment 3), respectively (Table 7.5). The average observed rate is near about 16.27 kg m^{-2} $year^{-1}$ which is much greater than the soil loss tolerance T-value of this region (i.e. 1.0 kg m^{-2} $year^{-1}$). Field survey and laboratory analysis suggest that erosion occurs on two types of soil texture—(1) sandy loam and (2) sandy clay loam (Table 7.6). Therefore, it can be said that the lateritic badlands of study area have high erosion risk (rendering organic rich top-soil development and increasing Fe-crusting, badlands area and degradation of biomass) and the region needs immediate protective measures to check erosion and land degradation at basin scale. After getting the measured erosion data, the analysis was carrying forward to fulfill the key purpose of study which was to compare the predicted data of erosion models (RUSLE and RMMF) with the observed data at field scale.

Secondary Data Collection

The key sources of main secondary data are regional soil report, geology report and other physical environmental report published by NBSS and LUP (National Bureau of Soil Service and Land Use Planning), Census of India, district gazetteer, official websites of IMD (Indian Meterological Department) Pune and Kolkata, Irrigation and Waterways Dept. of Govt. of West Bengal (IWD), Geological Survey of India (GSI), related e-books and e-journals. The topographical sheets of Survey of India (72 P/12/NE, R.F. 1:25,000 and 72 P/12, R.F. 1:50,000), District Resource Map of Geological Survey of India, District Planning Map of NATMO (National Atlas Thematic Mapping Organization) and Block map of Census of India are most important sources of spatial information (Ghosh & Guchhait, 2020). Landsat TM

Table 7.5 A brief summary of dam's parameters, sedimentation and observed rate of water erosion at 18 dam sites of three catchments (year of observation 2017–2018)

Gully catchments	Check dam	Width (cm)	Height (cm)	Mean sedimentation depth (m)	Measured mass of sediment (kg)	Rate of erosion kg m^{-2} $year^{-1}$ [mean = 16.27 kg m^{-2} $year^{-1}$]
Catchment 1	1	102	42	0.13	0566	14.10
	2	110	50	0.20	614	19.94
	3	87	41	0.25	2204	24.27
	4	114	48	0.15	1881	14.45
	5	98	35	0.11	1068	10.50
	6	94	30	0.23	3583	24.01
Catchment 2	7	98	37	0.22	0713	8.12
	8	110	41	0.27	3350	15.81
	9	108	46	0.15	2798	15.23
	10	122	48	0.11	1741	10.15
	11	107	45	0.19	1458	20.82
	12	125	51	0.23	2428	14.71
Catchment 3	13	98	32	0.19	2258	15.05
	14	105	45	0.23	1774	16.12
	15	102	48	0.27	2579	19.24
	16	112	54	0.14	2137	11.87
	17	94	40	0.17	2598	20.62
	18	90	38	0.20	1977	17.97

Table 7.6 Textural data of sample soils in the study area

Sample site	Location	Sand %	Silt %	Clay %	Organic matter %	Soil texture
1	24°11′06″N, 87°42′40″E	65.3	24.6	10.1	0.61	Sandy loam
2	24°10′57″N, 87°42′49″E	64.0	22.4	13.6	0.68	Sandy loam
3	24°11′23″N, 87°42′40″E	52.6	28.3	19.1	0.21	Sandy clay loam
4	24°11′51″N, 87°42′41″E	70.2	19.1	10.7	0.57	Sandy loam

and ETM+ (30 m resolution) images are downloaded from the website of Global Land Cove Facility (GLCF) and SRTM (Shuttle Radar Topography Mission, 90 m resolution), and ASTER (*Advanced Spaceborne Thermal Emission and Reflection* Radiometer, 30 m resolution) elevation data are downloaded from the websites of GLCF and Consortium for Spatial Information (CGIAR-CSI) (Ghosh & Guchhait, 2020). The spatial information is stored in Geographic Information System (GIS) and the thematic maps are prepared using GIS software (ArcGIS 9.2 and Erdas Image 9.1) (Ghosh & Guchhait, 2020).

In this case we have gathered the daily, monthly and annual rainfall data from three IWD (Irrigation and Waterways Department, Government of Wes Bengal) rain-gauge stations at Nalhati (24°17′25″N, 87°49′44″E), Rampurhat (24°10′13″N, 87°46′50″E) and Mollarpur (24°04′35″N, 87°42′36″E) which are situated at eastern part of study area, having areal distance of 18–25 km. The calculated mean annual rainfall for this region is 1510 mm in 2016 (maximum intensity of erosive rain is 25.21 mm h^{-1}), and the per day rainfall amount is 17.48 mm, considering total rainfall and rainy days in a year.

The base map is geo-referenced in UTM (Universal Transverse Mercator) projection with WGS-84 (World Geodetic Survey, 1984) datum. In the GIS framework, we have plotted the existing drainage of study area (derived from toposheet) on the ASTER (Advanced Spaceborne Thermal Emission and Reflection Radiometer) elevation map to depict the regional dissection of water divides (Ghosh & Guchhait, 2016). The locations of laterite exposures are mapped on the basis of field expeditions, toposheets, survey points of Garmin Montana 650 GPS receiver (with horizontal accuracy of ±3 m) and Google Earth Pro. Leica Geosystem Sprinter 150 m was used to measure the angle of slope facets (Ghosh & Guchhait, 2016). Alongside few cases (due to physical obstacles) from ASTER DEM (Digital Elevation Model), the slope length and angle (usually from gully headcut to water divide) is measured to judge the length of surface flow (responsible for gully erosion) (Ghosh & Guchhait, 2016).

Potential and Problem of Erosion Models

Models are of necessary simplifications of reality (Morgan, 2005). Researchers seek models that describe how the system functions in order to enlighten understanding of the system and how it responds to change (Morgan, 2005). It is not possible to take measurements at every point in the landscape, and it also takes time to build up a sufficient database and long-term measurements (Morgan, 2005). In order to overcome these deficiencies, models can be used to predict erosion under a wide range of conditions. Erosion models can be used as predictive tools for assessing soil loss for conservation planning, project planning and soil erosion inventories and for regulation, and it can be used a tools for understanding erosion processes and their interactions and for setting research priorities (Nearing et al., 1994). In selecting an erosion model, a rational decision must be made as to whether the model is to be used for on-site concerns (degradation of thinning of the soil profile) or off-site concerns (sediment yield or siltation of reservoirs) (Nearing, 2013).

The mathematical equations used in erosion models have five components: (1) independent variables, (2) dependent variables, (3) parameters, (4) mathematical operators and (5) a computation sequence and logic that link the equations within the model (Toy et al., 2013). The three major types of erosion models based on model structure are the regression-derived, index-based and process-based models (Table 7.7). One was to derive an erosion model that uses statistical regression procedures to fit an equation to a data set. The equation form and independent variables (factors) in the equation are selected to give the best fit to the experimental data as measured by a statistical goodness of fit (Toy et al., 2013). Every erosion model must represent how the four factors of climate, soil, topography and land use

Table 7.7 A short description of erosion models

Model type	Form	Derivation method	Strengths
Regression-derived	A single or a few equations having a for that best fits the data	Derived by fitting an equation(s) to an empirical database representing field conditions	Generally simple and easy to use; input values can be simple and easy to obtain
Index-based	Using indices, usually in a multiplicative form, to represent how climate, soil, topography and land use affect erosion	Values for indices determined from large empirical database representing field conditions	Simple and easy to use; input values can be simple and easy to obtain; very powerful in relation to simplicity and input values
Process-based	Represents individual erosion processes using simple steady-state equations	Equations derived from theory and empirical databases for erosion processes, validated against database representative of field conditions	Can be simple; represents main fundamental erosion processes; improved performance

affect soil loss and related variables (Toy et al., 2013). Toy et al. (2013) have suggested a simple form of erosion as follows (Eq. 7.1):

$$SL = CF.SF.TF.LUF \tag{7.1}$$

where SL = average annual soil loss, CF = climate factor, SF = soil factor, TF = topographic factor and LUF = land use factor.

Equation (7.1) is an index-based erosion model. Each variable in the equation is an index that represents the effect of that variable based on the value assigned to the index. In process-based or dynamic models, erosion occurs as a series of discrete events with different erosion amounts for each event because of differences in storms and land use conditions at each event (Toy et al., 2013). These models can track temporal variables by computing values at regular points through time between storm events. The physically based erosion models and regression models have until now not always provided very satisfying results for prediction of soil erosion and sediment yield (Poesen, 2018).

Problems of Using Models and Its Solution

It is found that lumped parameter models (i.e. empirical models) linked to GIS are practicable for conservation planning than sophisticated distributed parameter models. Lumped Parameter Models (LPMs) use averaging techniques to lump the influence of non-uniform spatial processes of a given area, such as a basin-averaged precipitation for run off computation (Torri & Borselli, (2012); Avwunudiogba & Hudson, 2014). The RUSLE is an empirical equation for predicting long-term average soil erosion from agricultural field under specific cropping and management practice. There are few hindrances or problems to implement distributed parameter or process-based models (like WEPP, EURSOEM etc.) in the study area. Three key problems are stated as follows (Boardman & Favis-Mortlock, 1998; Morgan and Nearing, (2011)):

(1) Does the amount of money and time devoted to collection of the data justify their application for simple watershed planning in humid tropical environments?
(2) Do communities in these region possess the institutional framework, personnel and financial commitment to undertake the long-term research necessary for implementations of process-based models?
(3) LPMs are more attractive in the immediate future because of the ease with which data requirements can be met and the greater suitability of these models for the socio-economic context of this region.

The models can be implemented in situations with limited data and parameter inputs and are particularly useful as a first step in identifying sources of sediment and nutrient generation (Merriti et al., 2003). Empirical models or index-based models are based primarily on the analysis of observations and seek to characterize response from the data. The feature of this class of models is their high level of spatial and temporal aggregation and their incorporation of a small number of casual variables (Merriti et al., 2003). In this study, an index-based model (Revised Universal Soil Equation, RUSLE) and a combined index-based and process-based model (Revised Morgan Morgan Finney model, RMMF) are applied to get predicated erosion rates, and then two models are compared to evaluate the suitability and effectiveness of each model in the erosion prone region of laterite terrain. The total workflow of erosion model selection, processing, application and analysis are completed in seven steps—(1) user requirements, (2) model selection, (3) developing core database, (4) expanding database, (5) model verification, (6) validating the model and (7) sensitivity analysis (Boardman & Favis-Mortlock, 1998; Nearing, 2013; Morgan, 2005; Morgan and Nearing, 2011; Toy et al., 2013).

Revised Universal Soil Loss Equation (RUSLE)

One of the main reasons why RUSLE type modelling is so widely used throughout the world is certainly its high degree of flexibility and data accessibility, a parsimonious parameterization, extensive scientific literature and comparability of results allowing to adapt the model to nearly every wind of condition and region of the world (Alewell et al., 2019). The precise description of RUSLE is found in the writing of Renard et al. (1997), predicting soil erosion by water for conservation planning in the geo-climatic condition of the USA. Chandramohan et al. (2015) have applied RUSLE, Unit Sediment Graph (USG) and Water Erosion Predication Project (WEPP) on small watersheds of Pamba River Basin (Kerala, India) to observe rainfall–runoff–sediment yield relationship, and they have found good applicability of RUSLE than other models. Similarly, Smith (1999), Sovrin (2003), Babu et al. (2004), Martin-Fernandez and Martinez-Nunez (2011), Jain and Das (2012), Sinha et al., (2012), Sinha and Joshi (2012), Bayramov et al. (2013), Kinnell (2014), Karydas et al. (2014), Devatha et al. (2015), Mondal et al. (2017) and Benavidez et al. (2018) have successfully applied RUSLE to assess erosion rate in different environmental settings, and they have found the suitability and effectiveness of RUSLE in comparison to other models, e.g. Soil Loss Estimation Model for Southern Africa (SLEMSA), Morgan Morgan Finney Model (MMF), Water Erosion Prediction Project (WEPP) and European Soil Erosion Model (EUROSEM). The applied version of RUSLE (Eq. 7.2) is mentioned as follows (Renard et al., 1997) (Table 7.8 and Fig. 7.7):

$$A = R\,K\,L\,S\,C\,P \tag{7.2}$$

where

- A is the computed soil loss per unit area (tons per acre per year); it can transformed into SI unit

Table 7.8 Operating parameters and functions of the RUSLE model

Description	Operating functions	Parameter definitions	Source
Rainfall Erosivity Index (R)	$R = (R_1 + R_2)/2$ $R_1 = P$ $(0.119 + 0.0873 \log_{10} I_m) . \log_{10} I_{30}$ $R_2 = 79 + 0.363\ P$	P is the mean annual rain-fall, I_m is the average rain-fall intensity (i.e. 25.21 mm h^{-1}), I_{30} is the maximum 30 min rain-fall intensity (i.e. 75 mm h^{-1}, recommended by Wischmeier & Smith, 1978)	Renard et al. (1997); Sarkar et al. (2005); Jha and Paudel (2010); Ganasri and Ramesh (2016); Benavidez et al. (2018)
Soil Erod-ibility Index (K)	$K = 1.2917$ $[2.1 \times 10^{-4} (12 - OM) M^{1.14} + 3.25 (s - 2) + 2.5 (p - 3)]/100$ $M = \%$ silt (100 − % clay)	OM is the percentage of organic matter in soil, M is the particle size parameter, s is the soil structure code and p is permeability code recommended by (Wischmeier &and Smith, 1978)	Sarkar et al. (2005); Bayramov et al. (2013)
Slope-Length Index (LS)	$LS = (L/22.13)^{0.5} . (0.065 + 0.045\ \theta + 0.0065\ \theta^2)$	L is the slope length (m) and θ is slope steepness in percent	Sarkar et al. (2005); Rahaman et al. (2015)

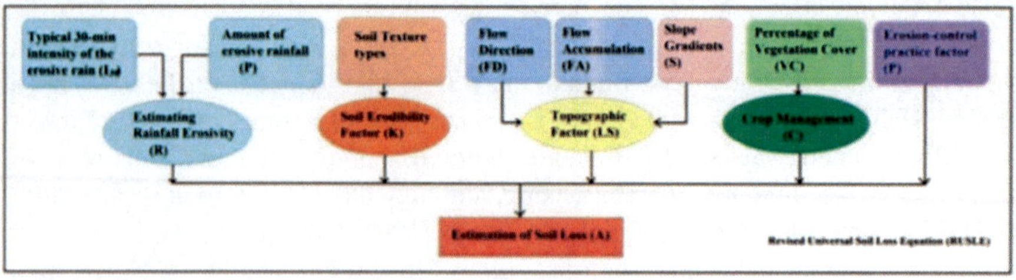

Fig. 7.7 Flowchart of data input and methods for RUSLE-based soil erosion modelling

- R, the rainfall and runoff factor, is the number of rainfall erosion index units, i.e. EI_{30}
- K, the soil erodibility factor, is the soil loss rate per erosion index unit for a specified soil as measured on a unit plot, which defined as a 72.6 ft length of uniform 9% slope continuously in clean-tilled fallow
- L, the slope-length factor, is the ratio of soil loss from the field slope length to that from a 72.6 ft length under identical conditions
- S, the slope-steepness factor, is the ratio of soil loss from the field slope gradient to that from a 9% slope under otherwise identical conditions
- C, the cover and management factor, is the ratio of soil loss from an area with specified cover and management to that from an identical area in tilled continuous fallow

- P, the support practice factor, is the ratio of soil loss with a support practice like contouring, strip cropping or terracing to that with straight-row farming up and down the slope

Revised Morgan–Morgan–Finney (RMMF) Model

Another popular model is the revised Morgan–Morgan–Finney (RMMF) model which was documented in the article of Morgan (2001), and its modifications were done by Morgan and Duzant (2008) to enable the effects of vegetation cover to be expressed through plant parameters. This model is also effectively applied in a variety of geo-climatic conditions (Sovrin, 2003; Mondal et al., 2011; Bayramov et al., 2013; Avwunudiogba & Hudson, 2014; Tesfahunegn et al., 2014; Efthimiou, 2019), and many workers (Jetten et al., 1994; Vente & Poesen, 2005; James et al., 2017; Choi et al., 2017) have given the results of model evaluation and additional modifications for the development and further applicability of RMMF model. The model validation was carried out by comparing predicted and observed values of annual runoff and erosion for 67 sites in 12 countries (Morgan et al., 1984). The model comprises a water phase and a sediment phase. Rainfall energy and runoff volume are estimated from annual rainfall amount in the water phase (Morgan et al., 1984; Morgan, 1986). In the sediment phase, erosion is taken to result from the detachment of soil particles by rainsplash and their transport by runoff (Morgan et al., 1984; Morgan, 1986). The revised version of the model is depicted as follows (Morgan, 2005) (Table 7.9 and Fig. 7.8).

Sensitivity Analysis

Two types of sensitivity indices are used in this study (Eq. 7.3): (a) Absolute Sensitivity (AS) and (b) Average Linear Sensitivity (ALS) (Nearing et al., 1989). The absolute sensitivity describes the rate of change in output with respect to a change in the value of input. The relative sensitivity describes the normalization of input and output in relation to their mean values, to produce an average linear sensitivity index. Now, ALS is widely popular in erosion prediction technology, and it can be described as follows (Morgan, 2005):

$$\mathrm{ALS} = [(O_2 - O_1)/O_\mathrm{m}]/[(I_2 - I_1)/I_\mathrm{m}] \tag{7.3}$$

where O_1 and O_2 are values of model output obtained with values of I_1 and I_2 of input and O_m and I_m represent the respective average values of the two input and output values. If ALS is greater than 1.0, then the input parameter is highly sensitive to change in output. Alongside the estimated error between (Eq. 7.4) measured and

Table 7.9 Operating parameters and functions of the RMMF model

Description	Operating functions	Parameter definitions	Source
Effective Rainfall (ER, mm)	$ER = R_a (1 - A_c)$	R_a = mean annual rainfall (mm); A_c = proportion of rainfall reaching soil surface considering canopy cover (0 to 1) in the basin	Morgan (2005); Morgan and Duzant (2008); Efthimiou (2019)
Leaf Drainage (LD, mm) and Direct throughfall (DT, mm)	$LD = (ER - CC)$, $DT = (ER - LD)$	CC = proportion of canopy cover (0 to 1)	
Kinetic energy of LD (KE_{LD}, J m^{-2})	$KE_{LD} = LD\ [(15.8 - P_H^{0.5}) - 5.87]$	P_H = plant height (m)	
Kinetic energy of DT (KE_{DT}, J m^{-2})	$KE_{DT} = DT\ (11.9 + 8.7 \log I)$	I = erosive rainfall intensity (mm h^{-1})	
Total kinetic energy (KE_T, J m^{-2})	$KE_T = KE_{LD} + KE_{DT}$	–	
Soil moisture storage capacity (R_C)	$R_C = 1000.MS.BD.EHD.(E_t/E_o)$	MS = soil moisture content at field capacity (% w/w); BD = bulk density of soil (Mg m^{-3}); effective hydrological EHD = effective hydrological depth (m); E_t/E_o = the ratio of actual to potential evapotranspiration	
Annual Runoff (I_r, mm)	$Q_r = ER.\exp(-R_c/R_o)$	I_o = mean daily rainfall (mm)	
Annual soil particle detachment by raindrop impact (F, kg m^{-2})	$F = K.\ KE\ .10^{-3}$	K = soil erodibility (g J^{-1})	
Annual soil particle detachment by runoff (H, kg m^{-2})	$H = ZQ^{1.5} \sin S\ (1\text{-}GC)\ 10^{-3}$ $Z = 1/0.5\ COH$	S = slope steepness; GC = proportion of ground cover (0–1); Z = resistance of soil; COH = soil cohesion (kPa)	
Total detachment (J, kg m^{-2})	$J = H + Z$	–	
Annual transport capacity of runoff (G, kg m^{-2})	$G = C\,Q^2 \sin S.10^{-3}$	I = the product of the C and P factors of RUSLE	

predicted values, it can be calculated by root mean square relative error (RMS-error) using the following equation (Morgan, 2005).

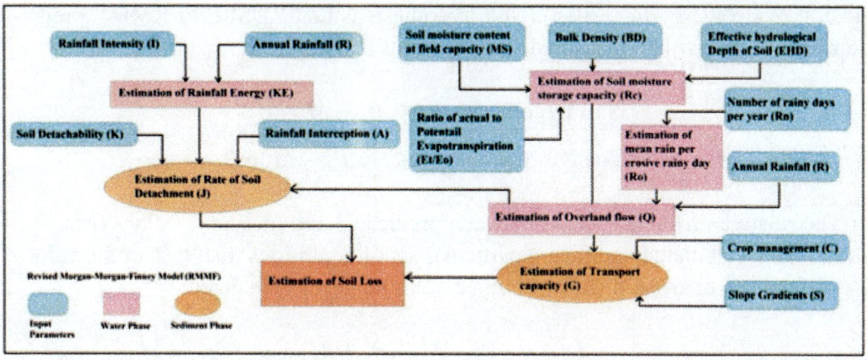

Fig. 7.8 Flowchart of data input and methods for RMMF-based soil erosion modelling

$$\text{RMS} - \text{error} = \sqrt{\left[{}_{i=1}\Sigma \left(X_{\text{obs}} - X_{\text{pred}}\right)^2 100/m\right]} \tag{7.4}$$

where *m* is the number of observations.

Model Validation

Validation is the process of ensuring that the model serves its intended purpose as described in the user requirements (Morgan, 2005; Toy et al., 2013), although an important part of validation is to determine how well the model fits measured data. Erosion models typically fit measured average annual soil loss with an uncertainty of about ±25% for moderate erosion rates of about 6–60 metric tons per hectare per year (Toy et al., 2013). The model efficiency coefficient (MEC), firstly proposed by Nash and Sutcliffe (1970), is now increasingly used an alternative to the correlation coefficient to express the performance of model (Morgan, 2011). Generally, a MEC value (Eq. 7.5) of greater than 0.5 is considered that the model performs satisfactorily in the region, and one should not expect values to exceed 0.7 (Quinton & Morgan, 1998; Morgan, 2005, 2011).

$$\text{MEC} = 1 - \Sigma\left(X_{\text{obs}} - X_{\text{pred}}\right)^2 / \Sigma\left(X_{\text{obs}} - X'_{\text{obs}}\right)^2 \tag{7.5}$$

where X_{obs} is the observed value, X_{pred} is the value predicted by the model and X'_{obs} is the mean of a set of observed values.

One of the important methods used to evaluate the effectiveness of soil erosion model is to compare the predictions given by the model to measured data from soil loss collected on plots taken under natural rainfall conditions (Nearing, 2013). A model 'effectiveness coefficient' was defined by Nearing (2013) for studies undertaken on large numbers of prediction versus measured data comparisons. This method provides a quantitative criterion for taking into account natural variability and uncertainty in measured erosion plot data when those data are used to evaluate erosion models (Nearing, 2013). Null hypothesis is that RUSLE or RMMF prediction (P_s) is equal to the measured value (M) for that case.

$$\text{Null hypothesis—}H_0 : P_s - M = 0$$

$$\text{Alternative hypothesis—}H_1 : P_s - M \neq 0$$

The relative difference (R_{diff}) between predicated and measured values (Eqs. 7.6 and 7.7) are calculated and then a particular set of conditions that 95% of the values for differences in erosion (fall within a certain range) is calculated.

$$R_{diff} = (P_s - M)/(P_s + M) \tag{7.6}$$

Relative difference values (Y-axis) are plotted against measured values (X-axis) to get a trend in the scatters.

$$R_{diff} = m \log_{10}(M) + b \tag{7.7}$$

The method of evaluation of a single data point may be extended to the larger data set and, from the analysis a model effectiveness coefficient (E_c), may be calculated. It is defined E_c as the fraction of simulation model predictions for which a model is effective in predicting the measured erosion, using the acceptance criteria. Using the 95% occurrence intervals from the replicated erosion data, it would result in a value, $E_{c(a=0.05)}$. The value of $E_{c(a=0.05)}$ signifies that the percentage of the difference between measured and predicted soil loss fell within the expected range of difference for two measured data points within the same population (Nearing, 2013). The procedure was as follows:

(1) List the measured and predicted data pairs.
(2) Calculate the relative difference between measured and predicted soil loss (R_{diff}).
(3) Compute the 95% occurrence interval as given by equation for each data point.
(4) Determine the number of predictions for which the R_{diff} value fell within the interval.
(5) Calculate $E_{c(a=0.05)}$ as the fraction of 'acceptable' predictions for the data set.

Statistical Analysis

For the statistical judgement and significant interrelationship of observed and predicted values, Chi-square test, linear regression, Pearson's product moment correlation, t-test of correlation and regression slope are applied (Table 7.10). χ^2 goodness of fit is used to determine where there is a statistically significant difference between expected frequencies and observed frequencies in sample population. The p-value of χ^2 test is also used to help us on the decision of rejection or acceptance of null hypothesis. The simple linear regression ($Y_c = a + bX$) and scatter plot gives an actual picture and trend of X-Y relationship which reflects the resemblance or

Table 7.10 Statistical parameter used in the study

Statistical parameter	Description and operating functions	Statements of null (H_0)/alternate (H_1) hypothesis
Chi-square test (χ^2 goodness of fit)	$\chi^2 = \Sigma (O_i - E_i)/E_i$, $n - 1$ degree of freedom O_i = observed erosion rate, E_i = predicted erosion rate of RUSLE or RMMF	H_0 ($O_i - E_i = 0$)—no difference between observed and predicted erosion rate H_1 ($O_i - E_i \neq 0$)—significant difference between observed and predicted erosion rate
Linear regression	$Yc = a + b X$ Y_c = predicted erosion rate of RUSLE or RMMF X = observed erosion rate a = intercept b = slope R^2 = coefficient of determination	–
Pearson's product moment correlation (r)	$r = \text{Cov}(X, Y)/\sigma_X \sigma_Y$ Cov (X, Y) = Covariance of X and Y σ_X = standard deviation of X σ_Y = standard deviation of Y Range = $+1 < r < -1$	–
t-test of b-value	$t_b = b/SE_b$ SE_b = standard error of b $SE_b = \sigma_X/\sigma_Y \sqrt{} (1 - r^n)/(n - 2)$ n = number of sample degree of freedom ($n - 2$)	H_0—regression slope (based on observed and predicted erosion rate) is significant, having close resemblance of X–Y relationship H_1—slope is insignificant
t-test of r-value	$t_r = r \sqrt{} (n - 2)/(1 - r)$ degree of freedom ($n - 2$)	H_0—there is a zero correlation H_1—there is a significant correlation, i.e. no zero
Confidence interval	$C_i = X_m \pm (Z. \sigma_X/\sqrt{n})$ X_m = mean of observed erosion rates Z = the Z value (1.96) for desired confidence level α ($\alpha_{0.05}$–95% confidence level) (obtained from normal curve)	–

association between observed and predicted erosion rates. The b-value (slope of trend line) reflects the amplitude of trend line to understand the interdependence of predicted values on the observed values. In addition correlation coefficient value (r) also confirms the degree of resemblance in the X-Y relationship. Then the estimated values of b and r are tested using t-test statistics at 0.05 significance level.

Soil Loss Tolerance

The term 'soil loss tolerance' (T value) denotes the maximum level of soil erosion that will permit a high level of crop productivity to be sustained economically and indefinitely (Wischmeier & Smith, 1978). The soil loss tolerance value (i.e. T value) has been defined as indication of how much erosion should be tolerated (Osman, 2014). For example, shallow soils over hard rock terrain have small T values. The concept of T value mainly described the maximum acceptable soil loss allowing a high level of productivity to be maintained for a long period, based on consideration of soil fertility and productivity (Li et al., 2009). A value for the rate of erosion alone is, however, of limited use without a corresponding value for an 'acceptable' or 'tolerable' rate (T-value) of erosion. Rates of tolerable soil loss calculated using soil production rates range from 0.2 to 2.2 t ha^{-1} $year^{-1}$ and tolerable rates based on maintenance of crop production range from approximately 1 to 11 t ha^{-1} $year^{-1}$ (Pennock, 2019). The low T value reflects likelihood of rill and gully formation and loss of plant nutrients by erosion. Here, the T-value is compared with the results of experiment to understand the erosion risk.

7.4 Results

Analysis of RUSLE Results

The input parameters of RUSLE are mean annual rainfall (P); average rainfall intensity (I_m); soil erodibility (K) based on soil organic matter content and percentage of sand, silt and clay particles; crop cover and management factor (C); and protective erosion control factor (P). Based on average rainfall data of three rain-gauge stations (collected from Irrigation and Waterways Department, Government of West Bengal), the mean P is estimated as 1510 mm in 2016–2017, and I_m is calculated as 25.52 mm h^{-1} for this climatic region. The analysis has assigned the Rainfall erosivity factor (R) of RUSLE modelling, i.e. 654 for this region. The mean K-factor of laterite terrain is estimated by soil texture and organic matter content of sample soils and the average K values of the catchments varies from 0.23 to 0.28 (Table 7.11). In general, coarse granular soil structure ($b = 7$) and moderate soil permeability ($c = 3$) are observed on the ferruginous soils. The length of slope elements or erosion plots varies from 22.1 to 106.8 m (length in between gully head and water divide), having 55–75% of bare lateritic stony surface with development of rills. The steepness of hillslope varies from 3° 45′ to 11° 06′, having average slope of 7° 14′ 30″ in the sample sites. It is observed that the land use/land cover of the catchments do not change too much throughout the year, and the region has minimum human disturbance. The C-factor is estimated as weighted value in respect of land use condition in three gully catchments, and it varies in each slope elements—(1) 0.61–0.91 (gully catchment 1), (2) 0.65–0.83 (gully catchment 2) and (3) 0.68–0.82 (gully catchment 3). The most important phenomenon is that the study area is not protected under any erosive control measures, except few patches of Acacia plantation. Therefore, in each slope element, the P-factor is regarded as 0.1 for RUSLE modelling.

Based on the above estimation of inputs, multiplied R, K, LS, C and P factors are taken to get potential or predicated values of annual soil erosion rate (A_P). A_P of three gully catchments are estimated as (1) 13.22–20.87 kg m^{-2} $year^{-1}$ (gully

Table 7.11 Input parameters sand predicted erosion rate of RUSLE model

Gully catchments	Check Dam	Slope length (m)	Slope degree	Gradient/steepness in percent	LS	K	R	P	C	A_P kg m^{-2} year^{-1} (mean = 16.633 kg m^{-2} year^{-1})
Catchment 1	1	22.1	10° 09′	11.27	1.33	0.28	654	0.1	0.61–0.91	15.89
	2	25.4	11° 06′	12.33	1.34					16.74
	3	45.4	8° 30′	9.44	1.53					18.7
	4	65	6° 11′	6.85	1.16					13.22
	5	74.5	5° 50′	6.63	1.19					13.07
	6	50.8	8° 05′	8.89	1.5					20.87
Catchment 2	7	44.2	4° 30′	5.2	0.67	0.23	654	0.1	0.65–0.83	7.86
	8	106.8	5° 30′	6.2	1.3					15.37
	9	84	7° 20′	8.15	1.68					19.71
	10	36.7	3° 45′	4.16	0.72					8.44
	11	35.2	8° 30′	9.4	1.33					15.72
	12	65	8° 45′	9.55	1.38					16.19
Catchment 3	13	75.2	5° 20′	6.1	1.07	0.28	654	0.1	0.68–0.82	16.06
	14	55.2	7° 30′	8.3	1.39					20.87
	15	62	7° 00′	7.7	1.33					19.97
	16	90.5	6° 40′	7.4	1.42					17.68
	17	58.1	8° 10′	9.07	1.63					24.47
	18	55	7° 30′	8.3	1.35					18.34

catchment 1), (2) 7.86–19.71 kg m^{-2} year^{-1} (gully catchment 2) and (3) 16.06–24.47 kg m^{-2} year^{-1} (gully catchment 3). It is obtained from database that A_P of hillslope yields maximum erosion value due to high LS-factor (>1.50). It is found that if the slope is recognized as short length and high steepness, it has high potential for erosion (at dam sites 1, 2, 6, 11 and 14). Based on 18 dam sites, the average A_P is 16.63 kg m^{-2} year^{-1} which is beyond the soil tolerance T-value limit (1.0 kg m^{-2} year^{-1}), showing high risk of erosion.

Analysis of RMMF Results

The climatic input parameters, i.e. mean annual rainfall 2016–2017 (*R*), number of rainy day (R_n) and mean rainfall (R_O), are based on the meteorological records of permanent stations. The effective rainfall (ER) (i.e. the remaining part of rainfall is stored and added in soil after leaf drainage and it has main role in water erosion) of the sample sites varies from 936 to 1057 mm, having leaf drainage (LD) of 88.34–290.16 mm and direct throughfall (DT) of 645.84–961.87 mm (Table 7.12). The parameters of topographic conditions (viz. slope angle, slope length and width) and soil–plant factors (i.e. soil surface roughness, canopy cover, ground cover, soil depth and plant height, etc.) are measured at field. Soil moisture content (MS), bulk density (BD), effective hydrological depth (EHD), soil erodibility index (K) and soil cohesion (COH) are measured by guide values of RMMF model (Morgan, 2001; Morgan & Duzant, 2008). The soil moisture at field capacity (R_c, % w/w) varies from

Table 7.12 Primary input parameters for RMMF model

Input data RMMF	ER	CC	LD	DT	R_c	GC	C
Catchment 1	936	0.13	121.68	814.32	9.96	0.12–0.44	0.41–0.78
		0.27	252.72	683.28	7.67		
		0.11	102.96	833.04	7.54		
		0.17	159.12	776.88	8.78		
		0.31	290.16	645.84	9.91		
		0.15	140.4	795.6	10.62		
Catchment 2	981	0.11	107.96	873.53	13.44	0.1–0.66	0.55–0.79
		0.09	88.34	893.16	12.67		
		0.14	137.41	844.09	14.72		
		0.15	147.22	834.27	11.64		
		0.09	88.34	893.16	15.9		
		0.17	166.85	814.64	14.48		
Catchment 3	1057	0.17	179.69	877.31	11.46	0.11–0.68	0.59–0.79
		0.11	116.27	940.73	10.72		
		0.15	158.55	898.45	10.25		
		0.12	126.84	930.16	12.67		
		0.09	95.13	961.87	11.55		
		0.19	200.83	856.17	10.72		

Table 7.13 Secondary input parameters for RMMF model

Input database RMMF	MS	E_t/E_o	EHD	BD	K	COH
Catchment 1	0.28	0.05–0.23	0.11–0.19	1.2	0.7	2
Catchment 2	0.28	0.25–0.32	0.08–0.12	1.2	0.7	2
Catchment 3	0.31	0.22–0.38	0.08–0.15	1.3	0.8	3

0.28 to 0.31 and other parameter is estimated as (1) actual to potential evapotranspiration (E_t/E_o, 0.05–0.38), bulk density (1.2–1.3 Mg m^{-3}), soil erobibility index (0.7–0.8), ground cover proportion (0.1–0.68), canopy cover proportion (0.41–0.79) and crop cover proportion (0.09–0.31) (Table 7.13).

Based on the above estimates of inputs, firstly we have estimated the potential detachment rate of soil particle by raindrop (F) which does not vary to great extent in the catchments, i.e. 14.73–17.24 kg m^{-2}. Then, potential detachment by runoff (H) is estimated in 18 sites, and it varies from 0.88 to 5.07 kg m^{-2}, and the runoff amount (Q) fluctuates from 721.63 to 973.08 mm. The addition of F and H gives the total water erosion rate of catchments (as sediment phase of RMMF model). So, the total detachment rate (J) varies from 16.67 to 21.28 kg m^{-2}. The J values is compared with the potential transport capacity by runoff (G) which are very high in this region and G varies from 30.29 to 84.21 kg m^{-2} in 18 dam sites. In this case, J value is much less than *G* value (i.e. transport capacity is much higher than the sediment supply rate), so the erosion process is transport limited (here J < G, J value recognizes annual soil erosion rate). Ultimately, the predicted value of annual soil erosion rate (S_P) varies from 16.67 to 21.28 kg m^{-2} $year^{-1}$ which exceeds the soil tolerance T-value limit of laterites (1.0 kg m^{-2} $year^{-1}$), showing high risk of erosion (Table 7.14).

Model Sensitivity Analysis

To measure the sensitivity of RUSLE, the maximum, minimum and average input parameters of rainfall (P), slope-length (LS), soil erodibility index (K) and crop cover (C) are used to estimate maximum, minimum and average output values of SE_P. On the slope element of gullies, all these input and output parameters are performed within the frame of RSULE (Table 7.15). The prime objective of sensitivity analysis is to measure the effects of variable input parameters on the output soil loss rate and to calculate the degree of sensitivity. Firstly, Average Linear Sensitivity (ALS) of P (R-factor) on the estimation of potential soil loss (SE_P) is 0.879. Then, ALS of LS-factor, K-factor and C-factor are 1.001, 0.999 and 1.001, respectively. So, it can be said that the LS-factor and C-factor (>1.0) are more sensitive in RSULE model to produce high deviation in SE_P values. K-factor is also highly sensitive, but the R-factor is not sensitive in RUSLE model. Therefore, to apply this model, we have to caution in measuring accurate slope angle, slope length, soil textural data and land use data.

Table 7.14 Summary of results and predicted erosion rates in RMMF model

Gully catchments	Check dam	Slope degree	Sin S	Runoff (mm)	Total KE Rainfall $J\ m^{-2}$	F $kg\ m^{-2}\ year^{-1}$	H $kg\ m^{-2}\ year^{-1}$	J $kg\ m^{-2}\ year^{-1}$	G $kg\ m^{-2}\ year^{-1}$	S_P $kg\ m^{-2}\ year^{-1}$ (mean = 18.63 $kg\ m^{-2}\ year^{-1}$)
Catchment 1	1	10° 09′	0.176	845.09	20,836	16.67	3.71	20.38	84.21	20.38
	2	11° 06′	0.192	965.74	18,950	15.16	5.07	20.23	78.79	20.23
	3	8° 30′	0.147	973.08	21,106	16.88	2.98	19.87	80.73	19.86
	4	6° 11′	0.107	905.25	20,298	17.24	1.63	18.87	35.95	18.87
	5	5° 50′	0.101	847.55	18,412	14.73	2.04	16.77	56.59	16.77
	6	8° 05′	0.14	813.21	20,567	16.45	2.56	19.01	60.18	19.01
Catchment 2	7	4° 30′	0.069	821.78	22,102	15.47	1.2	16.67	30.29	16.67
	8	5° 30′	0.088	854.99	22,383	15.66	1.98	17.64	45.67	17.64
	9	7° 20′	0.125	756.07	21,682	15.17	0.88	16.05	52.97	16.05
	10	3° 45′	0.057	905.77	21,542	15.08	0.93	16.01	36.94	16.01
	11	8° 30′	0.142	845.58	22,382	15.66	3.03	18.69	55.84	18.69
	12	8° 45′	0.139	829.96	21,262	14.88	2.82	17.7	55.53	17.7
Catchment 3	13	5° 20′	0.092	774.35	22,936	16.05	2.76	18.81	38.61	18.81
	14	7° 30′	0.13	808.23	23,844	16.69	4.59	21.28	60.29	21.28
	15	7° 00′	0.121	830.38	23,239	16.26	1.68	17.94	55.9	17.94
	16	6° 40′	0.116	721.63	23,692	16.58	2.28	18.86	47.72	18.86
	17	8° 10′	0.142	770.31	24,146	16.9	4.49	21.39	49.71	21.39
	18	7° 30′	0.133	808.48	22,633	15.84	4.44	20.28	53.89	20.28

Table 7.15 Average linear sensitivity analysis of RUSLE and RMMF

Input parameter	Maximum	Average	Minimum	ALS
RUSLE				
Rainfall (P)	1697	1523	1350	0.871
Slope-length factor (LS)	1.53	1.345	1.16	1.001
K-factor	0.31	0.25	0.19	0.999
C-factor	0.93	0.88	0.83	1.001
RMMF				
Rainfall (R)	1697	1523	1350	0.79
Slope steepness (S)	1106	832	558	1.001
K-factor	0.8	0.55	0.3	1.001
GC-factor	0.17	0.12	0.07	0.027

To measure the sensitivity of RMMF the maximum, minimum and average input parameters of rainfall (P), slope steepness (S), soil detachability index (K) and ground cover (GC) are used to estimate maximum, minimum and average output values of J. All the sensitivity analyses are done on the sample slope element of S1 (Table 7.15). The prime objective of sensitivity analysis is to measure the effects of sensible input parameters on the predicted values of soil loss, i.e. J, and to calculate the degree of sensitivity. It is found that the mentioned factors of S and CC are highly sensitive to erosion prediction accurately, because the both ALS value of K and CC is 1.001 which is greater than 1.0, i.e. highly sensitive index. The R-factor is moderately sensitive, as it values about 0.79, but GC factor is less sensitive. This analysis reflects that during the application and prediction of RMMF model, we should care about these input parameters.

Model Evaluation and Validation

In this part of model evaluation and validation, we have applied firstly absolute error, root mean square (RMS) error estimation, Chi-square test, model efficiency coefficient (MEC) and lastly scatter plot and linear regression ($Y_c = a + bx$), t-test of b value and product moment correlation (r) and coefficient of effectiveness (E_C) at 0.05 level of significance. The total statistical analysis is based on the measured erosion rate (O) and predicated erosion rate (A_P and S_P) with 18 sample size ($n = 18$).

Error Analysis

The absolute error between observed and predicted data is measured, showing positive anomaly (over estimation of erosion in response to observed rate) and negative anomaly (under estimation of erosion in response to observed rate) (Table 7.16). It is learned that 55.55% of predicated sample (i.e. ten dam sites)

Table 7.16 Summary of data error estimation, model validation and evaluation

O kg m^{-2} year^{-1} [m = 16.27]	A kg m^{-2} year^{-1} (m = 16.63)	S kg m^{-2} year^{-1} (m = 18.63)	(O-A)	RMS-error	MEC	E_c at 95% Con. Int.	(O-S)	RMS-error	MEC	E_c at 95% Con. Int.
14.1	15.89	20.38	−1.79	3.22	0.48	0.61	−6.28	4.45	0.22	0.38
19.94	16.74	20.23	3.2				−0.29			
24.27	18.7	19.86	5.57				4.41			
14.45	13.22	18.87	1.23				−3.42			
10.5	13.07	16.77	−2.57				−6.27			
24.01	20.87	19.01	3.14				5			
8.12	7.86	16.67	0.26				−8.55			
15.81	15.37	17.64	0.44				−1.83			
15.23	19.71	16.05	−4.48				−0.82			
10.15	8.44	16.01	1.71				−5.86			
20.82	15.72	18.69	5.1				2.13			
14.71	16.19	17.7	−1.48				−2.99			
15.05	16.06	18.81	−1.01				−3.76			
16.12	20.87	21.28	−4.75				−5.16			
19.24	19.97	17.94	−0.73				1.3			
11.87	17.68	18.86	−5.81				−6.99			
20.62	24.47	21.39	−3.85				−0.77			
17.97	18.34	20.28	−0.57				−2.31			

provide under estimation of erosion phenomena and 44.45% of data gives over estimation of erosion phenomena in RUSLE modelling. The value of absolute error varies from −5.81 to +5.57 from the observed data, and the estimated RMS-error is assigned as 3.22. In RMMF modelling, it is found that 77.17% of data sample (i.e. 14 dam sites) shows under estimation of erosion phenomena and 22.23% of data sample shows over estimation of erosion phenomena in respect to observed erosion rate. The absolute error value varies form −8.55 to +5.0 from the observed data, and the estimated RMS-error is assigned as 4.45. Therefore, it can be said that the RUSLE model gives less error than RMMF model.

Chi-Square (χ^2) Test Statistic

At 0.05 level of significance and 17 (n − 1) degree of freedom, the Chi-square (χ^2) test statistic sets forth the null hypothesis (H_0, $O - A_P$ or $S_P = 0$) which states that there is no difference between certain characteristics of a population, i.e. difference between predicted and observed value is zero and good correlation. The alternate hypothesis (H_1, $O - A_P$ or $S_P \neq 0$) reflects significant difference between predicted and observed value. The value of χ^2 statistic is assigned as 27.59 at 0.05 significance level with 17 degree of freedom (Table 7.17). The χ^2 statistic values of RUSLE and RMMF modelling are estimated, respectively, as 10.43 and 20.10 which are much lower than the tabulated χ^2 value at 0.05 level. Therefore, it is concluded that H_0 is accepted and H_1 is rejected. So, there is no significant difference between observed and predicted values in the study. Another statistic p-value of this test is used to

Table 7.17 Results of testing statistics

Statistical parameter	Tabulated testing statistical value	Calculated value		Remarks on hypothesis
		RUSLE	RMMF	
Chi-square (χ^2)	χ^2 statistic is assigned as 27.59 at 0.05 significance level with 17 degree of freedom	10.43	20.10	H_0 is accepted and H_1 is rejected (both models are accepted and predicted values resemblance with measured values)
t-test statistic of r value	t statistic is assigned as 2.120 at 0.05 significance level with 16 degree of freedom	5.44	3.37	H_0 is rejected and H_1 is accepted (both models are accepted and there is god correlation)
t-test statistic of b value	t statistic is assigned as 2.120 at 0.05 significance level with 16 degree of freedom	2.99	1.71	H_0 is rejected and H_1 is accepted (regression value of RUSLE give desired result than RMMF)
Confidence interval, $\alpha_{0.05}$	14.15 to 18.29 kg m^{-2} year^{-1} at 0.05 significance level	61% of sample fallen within interval	38% of sample fallen within interval	RUSLE model can provide more satisfactory results than RMMF

know the quantitative level of acceptance and large *p*-value means weakness of alternative hypothesis. The estimated *p*-value of RUSLE modelling is 0.8844 which reflects that H_0 is accepted, having 88.44% chance of getting desired results, but in case of RMMF modelling, there is only 26.90% chance (*p*-value –0.2690) of getting desired results at 0.05 significance level. Now, it can be said that according to χ^2 test statistic, RUSLE model gives significant good results than RMMF model in this analysis.

Model Efficiency Coefficient

Now, applying model efficiency coefficient (MEC) (Nash & Sutcliffe, 1970; Morgan & Duzant, 2008) into the relation between observed and predicted data, we have found two MEC values for two erosion models: (1) 0.48 (RUSLE) and (2) 0.22 (RMMF), respectively. The MEC >0.50–0.70 signifies good and satisfactory performance of model in reference to observed erosion results (Quinton, 1997; Morgan, 2011). The result shows that MEC value of RMMF is much lower than 0.50, but MEC value of RUSLE is much closer to 0.50. So, it can be decided that RUSLE model can be applied in this geo-climatic setting in place of RMMF model.

Linear Regression and *t*-Test Statistics

To get the trend of inter-relation between observed and predicted database, now the scatter plot and linear regression trend line ($Y_c = a + bX$) are prepared, taking observed data as *X* and predicted data of RUSLE and RMMF models as *Y*. It is finally estimated that the predicted values of RUSLE is statistically interrelated with the observed values ($A_P = 5.90 + 0.659\ O_E$), having good coefficient of determination (R^2) of 0.521 (i.e. inter-relation explained 52.10% in population) and notable slope (*b*) value of trend line, i.e. 0.659 (Fig. 7.9). The *b*-value of regression line (i.e. slope) reflects the quantitative judgment (indicating a change on response variable caused by a unit change of respective explanatory variable) of *Y* dependence on *X*. The *t*-test statistic of *b* value is 2.120 at 0.05 significance level with 16 ($n - 2$) degree of freedom (H_0: $b = 0$, *Y* does not depend on X; H_1 – *Y* depends on *X*). The estimated values of *t*-test statistic are 2.99 (RUSLE) and 1.71 (RMMF). This analysis reflects that test statistic of RUSLE *b*-value is greater than the tabulated *t*-value, and it means high dependence of predicted values on the observed values (i.e. RUSLE predicted values resemblance with observed erosion rates). In other case, another analysis reflects that test statistic of RMMF *b*-value is lower than the tabulated *t*-value, and it means low dependence of predicted values on the observed values (i.e. RMMF predicted values do not resemblance with the observed erosion rates). The Pearson's product moment correlation (*r*) of this analysis is estimated as 0.72 for RUSLE and 0.56 for MMF which reflect good

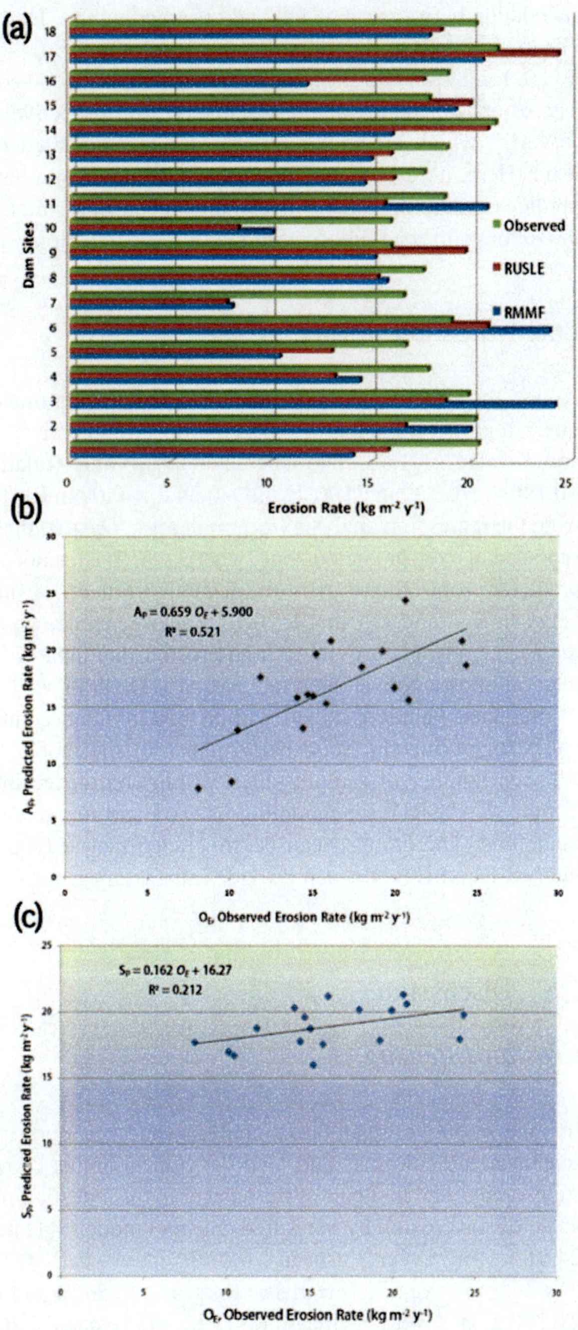

Fig. 7.9 Comparing the final results between observed and predicted erosion rates—(**a**) composite bar diagram showing dam site-wise result deviations in observed (O_E) and predicted erosion rates (RUSLE and RMMF), (**b**) predicted erosion rate (A_P) of RUSLE has close resemblance with O_E, having 52.1% of explanation in relationship and (**c**) predicted erosion rate (S_P) of RMMF has very weak resemblance with O_E, having only 21.2% of explanation in relationship

correlation between predicted and observed values. Here, again the *t*-test statistic of *r* value is 2.120 at 0.05 significance level with 16 ($n - 2$) degree of freedom (H_0: $b = 0$, *Y* does not correlate with *X*; H_1—there is a good correlation between *Y* and *X*). The estimated values of *t*-test statistics are, respectively, 5.44 (RUSLE) and 3.37 (RMMF) which are much greater than the tabulated *t*-value at 0.05 significance level. Here, it can be concluded that *r* value or correlation between observed and predicted value is statistically significant in this study, but the results of RUSLE modelling correlate highly with the observed erosion rates than RMMF modelling.

Effectiveness Coefficient

At last, the effectiveness coefficient (E_C) of erosion model is applied on the basis of linear regression database, 0.05 confidence interval of observed erosion rate (O_E) and Z-value of 1.96. The calculated R_{diff} value (relative difference) varies from +0.196 to −0.139 in RUSLE and + 0.34 to −0.09 in RMMF, respectively. It is found from the regression analysis ($R_{diff} = m \log_{10} O_E + b$) that 55.55% of RUSLE results is placed in over-predicted zone, whereas 77.7% results of RMMF is located in over-predicted zone (Fig. 7.10). It generally reflects, from the logarithmic relation between R_{diff} and O_E, that RMMF model generates an over-predicted result of the reality, i.e. always providing high erosion rate than observed rate. The confidence interval of observed erosion rate is 14.15–18.39 kg m^{-2} year^{-1}. If the large number of predicted values is fallen within this confidence interval, then E_C yields high value, signifying the good performance of the model. In general, E_C is the ratio between number of sample fallen within confidence interval and total number of sample. E_C of RUSLE modelling is 0.61 and the value is 0.38 in case of RMMF modelling. Therefore, it can be concluded that at 0.05 significance of confidence interval RUSLE model can provide satisfactory results in this region.

7.5 Discussion

Erosion Intensity

In spite of above quantitative analysis, one key question is always raised by soil scientists, agriculturists and land developers is that 'how serious is erosion in this study area'? The first part of the answer to this question involves establishing typical value of soil erosion by measured data and models: (1) field measured data −8.12 to 24.01 kg m^{-2} year^{-1} (mean 16.27 kg m^{-2} year^{-1}), (2) RUSLE data −7.86 to 24.47 kg m^{-2} year^{-1} (mean 16.68 kg m^{-2} year^{-1}) and (3) RMMF data −16.01 to 21.28 kg m^{-2} year^{-1} (mean 18.63 kg m^{-2} year^{-1}). It is needed to compare the research results with the T-value to understand the critical level of erosion which can be reduced to an acceptable limit using crop management and land management techniques. Erosion is natural geological process, and it is impossible to stop; instead

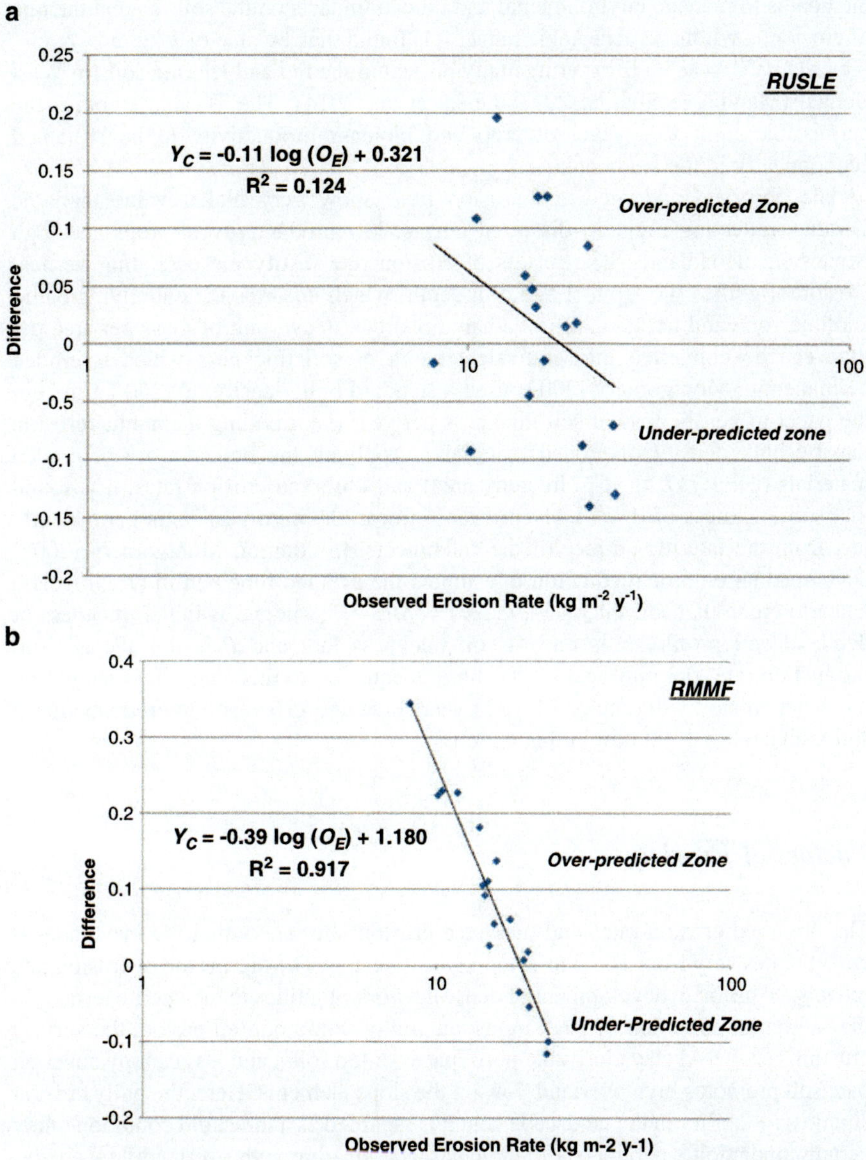

Fig. 7.10 Logarithmic relationship between relative differences of results (R_{diff}) and observed erosion rate: (**a**) high variations of R_{diff} (RUSLE) in relation to logarithmic increase of O_E, and (**b**) low variations of R_{diff} (RMMF) in relation to logarithmic increase of O_{E} and high numbers of over-predicted results

the goal is to manage environmental and human impacts on the soils so that the rate of erosion is within an acceptable range. It is found that T-value of 1 kg m^{-2} $year^{-1}$ (i.e. 10 t ha^{-1} $year^{-1}$) is experimentally proved in the red and lateritic soils of West Bengal (Mandal & Sharda, 2013; Lenka et al., 2014). The T-value signifies the permissible limit to a level of crop and biomass productivity to be sustained economically in the study area.

The observed and predicted erosion rates show very higher value than the T-value, reflecting rugged—dissected terrain, low productivity of crops and high expansion of badlands. The results of erosion rate justify the escalating vertical erosion of gullies (i.e. up to 1.5–3.5 m depth) which unearths the underlying pallid kaolinte zone and occasionally weathered bedrock. Now, unit of mass per area per time can be converted into equivalent depth of soil thickness which is eroded permanently. Montgomery (2007) used a standard bulk density of 1200 kg m^{-3} in the paper to get the loss of soil thickness per year (i.e. dividing the annual erosion rate by bulk density of eroded materials). Applying the bulk density of eroded materials (i.e. 1.717 kg m^{-3} in study area) and observed erosion rates, it is found that soil thickness of 0.47–1.41 cm $year^{-1}$ (mean 0.95 cm $year^{-1}$) is permanently lost from the lateritic surface of the catchments. In addition Montgomery (2007) developed an empirical equation to estimate the average time period (T_c, in years) taken to erode that soil thickness, viz. $T_c = S/E - P$ (where S is initial thickness or depth of soil profile, E is rate of soil thickness loss and P is the average soil production rate, 0.2 mm $year^{-1}$). Using this equation to this study, it is learnt that the water erosion will require 127–223 years (average 176 years) to erode the mean soil thickness of 1500 mm in this region.

Factors of Erosion

The observed erosion rates and predicted erosion rates show that the mean annual erosion rates vary from 16.27 to 18.63 kg m^{-2} $year^{-1}$. The high value of erosion rate reflects the ultimate development of dense network of gullies in the laterite terrain. In the saturation condition of peak monsoon and cyclonic rainfall period, the surface crusting (i.e. Fe-Al clay closes the pore spaces of top soils) and less canopy cover on bare soil promotes high overland flow on the slope elements. Here, the gully erosion signifies instability in the landscape, and it is regarded as a threshold condition under certain topographic parameters in the landscape, relating with overland flow erosivity and surface resistance of laterite terrain (Ghosh & Guchhait, 2020). High runoff, due to intense rainfall, is the primary trigger, but the local conditions such as slope morphometry (i.e. high concavity at the base of slope), land use (i.e. high proportion of bare soil cover and low proportion of canopy cover) and soil characteristics (high erodibility and surface crusting) control the triggering of gully erosion (Ghosh & Guchhait, 2020). It is found that 52.51% of gullies are affected by overland flow erosion (slope steepness, S, 1.2–5.2° and drainage area, A, 2129.1–10513.9 m^2), while 27.96% belongs to landslide erosion (S 5.2–9.5° and A 457.1–5702.5 m^2)

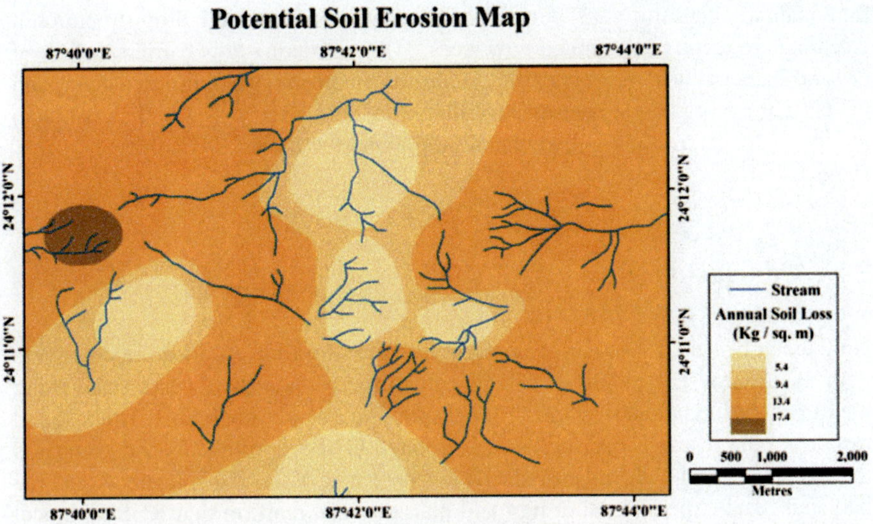

Fig. 7.11 RUSLE erosion map depicting (rill and inter-rill erosion) spatial coverage of different zones of annual soil erosion rate (kg m^{-2} $year^{-1}$), considering 17 catchments of gullies

(Ghosh & Guchhait, 2020). This experimental study and model evaluation is suggested that instead of using RMMF model, the RUSLE model can be applied for this lateritic region to estimate potential rate of annual soil loss. Therefore, based on the 17 sub-catchments of gullies (a part of study area) and RUSLE modelling (considering 118 gully head slope), an erosion map is developed to depict the potential annual rate of soil loss due to rill and inter-rill erosion in the lateritic region. The erosion map (Fig. 7.11) shows that the western and eastern part is very much susceptible to soil erosion (greater than 9.4 kg m^{-2} $year^{-1}$) due to high LS factor and bare soil cover, but the erosion rate (less than 9.4 kg m^{-2} $year^{-1}$) is much lower in the central part, because this part is covered with Acacia plantation, Sal forest, aerodrome pavement and relatively low LS factor. Also it is understood that the whole region is under very high erosion risk, because the erosion rate is beyond the acceptable T value limit (i.e. 1 kg m^{-2} $year^{-1}$).

The most vulnerable sites of water erosion (need to be protected) are identified as (Ghosh et al., 2020): (1) the region above the gully heads where the rills have tendency to converge; (2) high steepness (>5°) and long stretch of convex slope (>70 m); (3) the region having high bareness of slope and surface crusting promotes more runoff; and (4) bank failure due to mass wasting, pipe flow, flow convergence at heads and undercutting by channel flow. The fundamental problem to control soil erosion is centred on the on-site management of too much runoff water in short span of torrential rains or thunderstorms. To check channel erosion, the prime focus of erosion control strategies should be placed on five aspects: (1) reducing discharge rate through good growth of vegetation at catchment and water retention basins; (2) minimizing channel grade through construction of check dams and rock chutes for enhancing deposition; (3) controlling headcut erosion through drop structures at catchment to recue concentrated flow areas; (4) constructing flow barriers (as gravel or sand bags or loose rock piles) to control downstream sediment movement; and (5) promoting vegetative measures (plantation of trees and grasses on bare surface) to control splash, rill and gully erosion throughout the basin.

7.6 Conclusion

Now it can be said at last that the present research work has fulfilled the objectives with mentioning the region as a high potential erosion risk at basin scale using measured data and models (16.27–18.63 kg m^{-2} year^{-1}). Using limited database and resources, the research has successfully applied the erosion models and compared the results against field measured erosion data to get the real picture of laterite badlands. The experimental design and plan of work reaffirm that RUSLE model gives desire results in comparison to RMMF model with very high model efficiency coefficient (0.48) and effectiveness coefficient (0.61). The predicted values of RUSLE (A_P) follow the field measured data (O_E), with a positive correlation ($r = 0.72$) and trend line ($Y_c = 5.9 + 0.659\ O_E$) which is not very resemblance in case of RMMF model. It is learned from the analysis that the logarithmic relation, between R_{diff} (relative difference) and observed erosion rate (O_E), reflects more over-prediction of erosion results (i.e. yielding high predicted erosion rate than O_E) in case of RMMF modelling than RUSLE. The measured data of erosion rate confirms the vulnerability and high erosion risk of the region against T-value (1.0 kg m^{-2} year^{-1}) of laterite soils. It is found that the mean soil thickness of 0.95 cm per year is permanently lost from the surface of gully catchments. Applying RUSLE model in the whole study area, it is estimated that the region is dissected by annual erosion rates of 5.25–18.12 kg m^{-2} year^{-1}.

The most challenging task is to apply rightly an erosion model understanding the geo-environmental conditions which can be measured accurately through input variable functions of that model. The more inputs in field-based data collection, innovative techniques, flexibility of model application and better understanding of hydro-geomorphic processes will help to get good prediction in the soil erosion research. No model can give exact results in comparison to observed data at plot scale or basin scale, but high expertise and fine tuning of advanced model can provide sufficient inference on the erosion rates. It is understood that there is a need of further research to apply RUSLE or RMMF models in different parts of India for the applicability and validity of model. Validation of any erosion model can only be done or justified scientifically if the GIS-based model data will evaluate with the observed data taken at field plots. At last, the present study reveals that anyone cannot blindly exercise any erosion model and prepare any thematic map of soil erosion spatially in any particular region without evaluating the scale effect and error statistics of predicted values in comparison to field measured data.

References

Alewell, C., Borrelli, P., Meusburger, K., & Panagos, P. (2019). Using the USLE: Chances, challenges and implications of soil erosion modelling. *International Soil and Water Conservation Research, 7*, 203–225. https://doi.org/10.1016/j.iswcr.2019.05.004

Aulakh, M. S., & Sidhu, G. S. (2015). Soil degradation in India: Causes, major threats and management options. In *MARCO Symposium 2015—Next Challenges of Agro-Environmental Research in Monsoon Asia, National Institute for Agro-Environmental Sciences (NIAES)*, Japan, pp. 151–156.

Avwunudiogba, A., & Hudson, P. F. (2014). A review of soil erosion models with special reference to the needs of humid tropical mountainous environments. *Journal of Sustainable Development, 3*(4), 299–310. https://doi.org/10.14207/ejsd.2014.v3n4p299

Babu, R., Bhyani, B. L., & Kumar, N. (2004). Assessment of erodibility status and refined iso-erodent map of India. *India Journal of Soil Conservation, 32*, 171–172.

Bawa, A. K. (2017). Mitigating land degradation due to water erosion. *National Academy of Agricultural Science, New Delhi, Policy Paper, 88*, 1–19.

Bayramov, E., Buchroithner, M. F., & McGurty, E. (2013). Differences of MMF and USLE models for soil loss prediction along BTC and SCP pipelines. *Journal of Pipeline Systems Engineering and Practice, 4*(1), 81–96. https://doi.org/10.1061/(ASCE)PS.1949-1204.0000117

Bell, F. G. (2002). *Geological hazards: Their assessment, avoidance and mitigation*. Taylor & Francis.

Bennett, H. H. (1939). A permanent loss to New England soil erosion resulting from the hurricane. *Geographical Review, 29*(2), 196–204. https://doi.org/10.2307/209942

Bhattacharyya, T., Babu, R., Sarkar, D., Mandal, C., Dhyani, B. L., Nagar, A. P. (2007). Soil and crop productivity model in humid sub-tropical India. *Current Science 93*(10), 1397–1403.

Bhunya, P. K., Jain, S. K., Singh, P. K., & Mishra, S. K. (2014). A simple conceptual model of sediment yield. *Water Resource Management, 24*, 1697–1716. https://doi.org/10.1007/s11269-009-9520-4

Biswas, A. (1987). Laterities and Lateritoids of Bengal. In V. S. Datye, J. Diddee, S. R. Jog, & C. Patial (Eds.), *Exploration in the tropics* (pp. 157–167). K.R. Dikshit Felicitation Committee.

Boardman, J., & Favis-Mortlock, D. (1998). Modelling soil erosion by water: Some conclusions. In J. Boardman & D. Favis-Mortlock (Eds.), *Modelling soil erosion by water* (pp. 515–520). Springer. https://doi.org/10.1007/978-3-642-58913-3_39

Bocco, G. (1991). Gully erosion: Processes and models. *Progress in Physical Geography, 15*(4), 392–406. https://doi.org/10.1177/030913339101500403

Borrelli, P., Robinson, D. A., Fleischer, L. R., Lugato, E., Ballabio, C., Alewell, C., Meusburger, K., Modugnos, S., Schutt, B., Ferro, V., Bagarellow, V., Osst, K. V., Montanarella, L., & Panagos, P. (2017). An assessment of the global impact of 21st century land use change on soil erosion. *Nature Communications, 8*, 1–13. https://doi.org/10.1038/s41467-017-02142-7

Borrelli, P., Robinson, D. A., Panagos, P., Lugato, E., Yang, J. E., Alewell, C., Wuepper, D., Montanarella, L., & Ballabio, C. (2020). Land use and climate changes impacts on global soil erosion by water (2015–2070). *Proceedings of the National Academy of Sciences of the United States of America, 117*(36), 21994–22001. https://doi.org/10.1073/pnas.2001403117

Central Water Commission. (2015). *Compendium on silting of reservoirs in India*. CWC.

Chandramohan, T., Venkatesh, B., & Balchand, A. N. (2015). Evaluation of three soil erosion models for small watersheds. *Aquatic Procedia, 4*, 1227–1234. https://doi.org/10.1016/j.aqpro.2015.02.156

Choi, K., Huwe, B., & Reineking, B. 2017. *Commentary on 'Modified MMF (Morgan-Morgan-Finney) model for evaluating effects of crops and vegetation cover on soil erosion' by Morgan and Duzant (2008)*. Retrieved from https://arxiv.org/abs/1612.08899

Chorley, R. J., Schumm, S. A., & Sugden, D. (1984). *Geomorphology*. Methuen.

Chow, V. T., Maidment, D. R., & Mays, L. W. (1998). *Applied hydrology*. McGraw-Hill Education.

Devatha, C. P., Despande, V., & Renukaprasad, M. S. (2015). Estimation of soil loss using USLE model for Kulhan watershed, Chhattisgarh—A case study. *Aquatic Procedia, 4*, 1429–1436. https://doi.org/10.1016/j.aqpro.2015.02.185

Eekhout, J. P. C., & Vente, J. D. (2019). How soil erosion model conceptualization affects soil loss projections under climate change. *Progress in Physical Geography*. https://doi.org/10.1177/0309133319871937

Efthimiou, N. (2019). Development and testing of the Revised Morgan-Morgan-Finney (RMMF) soil erosion model under different pedological datasets. *Hydrological Sciences Journal, 64*(9), 1095–1116. https://doi.org/10.1080/02626667.2019.1623896

Flanagan, D. C., Ascough, J. C., II, Nearing, N. A., & Laflen, J. M. (2001). The Water Erosion Prediction (WEPP) model. In R. S. Harmon & W. W. Doe (Eds.), *Landscape erosion as evolution modeling*. Springer. https://doi.org/10.1007/978-1-4615-0575-4_7

Froechlich, D. C. (2018). Estimating reservoir sedimentation at large dams in India. *E3S Web of Conference, 40*, 1–8. https://doi.org/10.1051/e3sconf/20184003042

Gajbhiye, S., Mishra, S. K., & Pandey, A. (2014). Relationship between SCS-CN and sediment yield. *Applied Water Science, 4*, 363–370. https://doi.org/10.1007/s13201-013-0152-8

Ganasri, B. P., & Ramesh, H. (2016). Assessment of soil erosion by RUSLE model using remote sensing and GIS—A case study of Nethravathi Basin. *Geoscience Frontiers, 7*, 953–961. https://doi.org/10.1016/j.gsf.2015.10.007

Gayen, A., Saha, S., & Pourghasemi, H. R. (2020). Soil erosion assessment using RUSLE model and its validation by FR probability model. *Geocarto International, 35*(15), 1750–1768.

Ghosh, S., & Bhattacharya, K. (2012). Multivariate erosion risk assessment of lateritic badlands of Birbhum (West Bengal, India): A case study. *Journal of Earth System Sciences, 121*(6), 1441–1454. https://doi.org/10.1007/s12040-012-0243-1

Ghosh, S., & Guchhait, S. K. (2015). Characterization and evolution of primary and secondary laterites in northwestern Bengal Basin, West Bengal, India. *Journal of Palaeogeography, 4*(2), 203–230. https://doi.org/10.3724/SP.J.1261.2015.00074

Ghosh, S., & Guchhait, S. K. (2017). Estimation of geomorphic threshold in permanent gullies of lateritic terrain in Birbhum, West Bengal, India. *Current Science, 113*(3), 478–485. https://doi.org/10.18520/cs/113/i03/478-485

Ghosh, S., & Guchhait, S. K. (2020). Erosion threshold and SCS-CN based runoff and sediment yield modelling in the gullies of Dwarka–Brahmani Interfluve, West Bengal, India. In P. K. Shit, H. R. Pourghasemi, & G. S. Bhunia (Eds.), *Gully erosion studies from India and surrounding region* (pp. 45–68). Springer Nature. https://doi.org/10.1007/978-3-030-23243-6_4

Ghosh, S., & Guchhait, S. K. (2016). Geomorphic threshold estimation for gully erosion in the lateritic soil of Birbhum, West Bengal, India. *Soil Discussion*. https://doi.org/10.5194/soil-2016-48

Ghosh, S., Guchhait, S. K., Allahi, R. A., Bera, S., & Roy, S. (2020). Geomorphic character and dynamics of gully morphology, erosion and management in laterite terrain: Few observations from Dwarka-Brahmani Interfluve, eastern India. *Geology, Ecology and Landscapes, doi*. https://doi.org/10.1080/24749508.2020.1812148

Gulati, A., Rai, S. C. (2014). Cost estimation of soil erosion and nutrient loss from a watershed of the Chotanagpur Plateau, India. *Current Science 107*(4):670–674.

Jain, M. K., & Das, D. (2012). Estimation of sediment yield and areas of soil erosion and deposition for watershed prioritization using GIS and remote sensing. *Water Resource Management, 24*, 2091–2112. https://doi.org/10.1007/s11269-009-9540-0

James, C., Ascough, I. I., Dennis, C. F., Tatarko, J., Nearing, M. A., & Kipka, H. (2017). *Soil erosion modeling and conservation planning* (Agronomy Monograph 59). Wiley. https://doi.org/10.2134/agronmonogr59.2013.0011

Jetten, V., Roo, A., & Favis-Mortlock, D. (1994). Evaluation of field-scale and catchment-scale soil erosion models. *Catena, 37*, 521–541. https://doi.org/10.1016/S0341-8162(99)00037-5

Jha, M. K., & Paudel, R. C. (2010). Erosion predictions by empirical models in a mountainous watershed in Nepal. *Journal of Spatial Hydrology, 10*(1), 89–102.

Karydas, C. G., Panagos, P., & Gitas, I. (2014). Classification of water erosion models according to their geospatial characteristics. *International Journal of Digital Earth, 7*(3), 229–250. https://doi.org/10.1080/17538947.2012.671380

Kinnell, P. I. A. (2014). Applying the QREI30 index within the USLE modeling environment. *Hydrological Processes, 28*(3), 591–598. https://doi.org/10.1002/hyp.9591

Kothyari, U. C. (1996). Erosion and sedimentation problems in India. *International Association of Hydrological Sciences, 236*, 531–540.

Kumar, H., & Pani, P. (2013). Effects of soil erosion on agricultural productivity in semi-arid regions: The case of lower Chambal valley. *Journal of Rural Development, 32*(2), 165–184.

Lal, R. (1990). Soil erosion and land degradation: The global risk. In R. Lal & B. A. Stewart (Eds.), *Soil degradation volume II soil advances in soil science* (pp. 129–172). Springer-Verlag. https://doi.org/10.1007/978-1-4612-3322-0_4

Lal, R. (2012). Climate change and soil degradation mitigation by sustainable management of soils and other natural resources. *Agricultural Research, 1*(3), 199–212. https://doi.org/10.1007/s40003-012-0031-9

Lenka, N. K., Mondal, D., & Sudhishri, S. (2014). Permissible soil loss limits for different physiographic regions of West Bengal. *Current Science, 107*(4), 665–670.

Li, L., Du, S., Wu, L., & Liu, G. (2009). An overview of soil loss tolerance. *Catena, 78*, 93–99. https://doi.org/10.1016/j.catena.2009.03.007

Mandal, D., Sharda, V. N. (2013). Appraisal of soil erosion risk in the Eastern Himalayan region of India for soil conservation planning. *Land Degradation and Development 24*(5), 430–437.

Martin-Fernandez, L., & Martinez-Nunez, M. (2011). An empirical approach to estimate soil erosion risk in Spain. *Science of the Total Environment, 409*, 3114–3123. https://doi.org/10.1016/j.scitotenv.2011.05.010

Marzolff, I., & Pani, P. (2019). Dynamic and pattern of land leveling for agricultural reclamation of erosional badlands in Chambal Valley (Madhya Pradesh, India). *Earth Surface Processes and Landforms, 43*(2), 524–542. https://doi.org/10.1002/esp.4266

Merriti, W. S., Letcher, R. A., & Jakeman, A. J. (2003). A review of erosion and sediment transport models. *Environmental Modelling & Software, 18*, 761–799. https://doi.org/10.1016/S1364-8152(03)00078-1

Mishra, S. K., & Singh, V. P. (2003). *Soil Conservation Service Curve Number (SCS-CN) methodology*. Kluwer Academic Publishers.

Mishra, S. K., Tyagi, J. V., Singh, V. P., & Singh, R. (2006). SCS-CN-based modelling of sediment yield. *Journal of Hydrology, 324*, 301–322. https://doi.org/10.1016/j.jhydrol.2005.10.006

Mondal, A., Khare, D., & Kundu, S. (2011). A comparative study of soil erosion modelling by MMF, USLE and RUSLE. *Geocarto International, 33*(1), 89–103. https://doi.org/10.1080/10106049.2016.1232313

Mondal, A., Khare, D., Kundu, S., Mukherjee, S., Mukhopadhyay, A., & Mondal, S. (2017). Uncertainty of soil erosion modelling using open source high resolution and aggregated DEMs. *Geoscience Frontiers, 8*, 425–436. https://doi.org/10.1016/j.gsf.2016.03.004

Mondal, D., & Sharda, V. N. (2011). Assessment of permissible soil loss in India employing a quantitative bio-physical model. *Current Science, 100*(3), 383–390.

Morgan, R. P. C. (1986). *Soil Erosion and Conservation*. Longman Scientific and Technical.

Morgan, R. P. C. (2001). A simple approach to soil loss prediction: A revised Morgan-Morgan-Finney model. *Catena, 44*, 305–322. https://doi.org/10.1016/S0341-8162(00)00171-5

Morgan, R. P. C. (2005). *Soil erosion and conservation*. Blackwell Publishing.

Morgan, R. P. C. (2011). Model development: A user's perspective. In R. P. C. Morgan & M. A. Nearing (Eds.), *Handbook of erosion modelling* (pp. 9–32). Wiley.

Morgan, R. P. C., & Duzant, J. H. (2008). Modified MMF (revised Morgan-Morgan-Finney) model for evaluating effects of crops and vegetation cover on soil erosion. *Earth Surface Processes and Landforms, 32*, 90–106. https://doi.org/10.1002/esp.1530

Morgan, R. P. C., Morgan, D. D. V., & Finney, H. J. (1984). A predictive model for the assessment of soil erosion risk. *Journal of Agricultural Engineering Research, 30*, 245–253. https://doi.org/10.1016/S0021-8634(84)80025-6

Morgan, R. P. C., Quinton, J. N., Smith, R. E., Givers, G., Poesen, J., & Auerswald, K. (1998). The European Soil Erosion Model (EUROSEM): A process-based approach for predicting soil loss from fields and small catchments. *Earth Surface Processes and Landforms, 28*(3), 591–598. https://doi.org/10.1002/(SICI)1096-9837(199806)23:6<527::AID-ESP868>3.0.CO;2-5

Morgan, R. P. C., Nearing, (2011). Handbook of Erosion Modelling (ed). Wiley-Blackwell, New York.

Narayana, D. V. V., & Babu, R. (1983). Estimation of soil erosion in India. *Journal of Irrigation and Drainage Engineering, 109*(4), 419–434. https://doi.org/10.1061/(ASCE)0733-9437(1983)109:4(419)

Nash, J. E., & Sutcliffe, J. V. (1970). River flow forecasting through conceptual model. *Journal of Hydrology, 10*, 282–290. https://doi.org/10.1016/0022-1694(70)90255-6

Nearing, M. A. (2013). Soil erosion and conservation. In J. Wainwright & M. Mulligan (Eds.), *Environmental modelling: Finding simplicity in complexity* (pp. 365–378). John Wiley & Sons.

Nearing, M. A., Foster, G. R., Lane, L. J., & Chaves, H. M. L. (1989). A process-based soil erosion model for USDA-Water Erosion Prediction Project Technology. *Transactions of the American Society of Agricultural Engineers, 32*, 1587–1593.

Nearing, M. A., Lane, L. J., & Lopes, V. L. (1994). Modelling soil erosion. In R. Lal (Ed.), *Soil Erosion: Research and Methods* (pp. 127–156). Soil and Water Conservation Society.

Nkonya, E., Mirzabaev, A., & Braun, J. (2016). Economics of land degradation and improvement: An introduction and overview. In E. Nkonya, A. Mirzabaev, & J. Braun (Eds.), *Economics of land degradation and improvement: A global assessment for sustainable development* (pp. 1–14). Springer Open.

NRSC. (2019). Wasteland Atlas of India. National Remote Sensing Centre and Department of Land Resource, Hyderabad.

Osman, K. T. (2014). *Soil degradation, conservation and remediation*. Springer.

Panagos, P., & Katsoyiannis, A. (2019). Soil erosion modelling: The new challenges as the results of policy developments in Europe. *Environmental Research, 172*, 470–474.

Pennock, D. (2019). *Soil erosion: The greatest challenge for sustainable soil management*. FAO.

Pimentel, D. (2006). Soil erosion: A food and environmental threat. *Environment Development and Sustainability, 8*, 119–137. https://doi.org/10.1007/s10668-005-1262-8

Pimentel, D., & Burgess, M. (2013). Soil erosion threatens food production. *Agriculture, 3*, 443–463. https://doi.org/10.3390/agriculture3030443

Poesen, J. (2011). Challenges in gully erosion research. *Landform Analysis, 17*, 5–9.

Poesen, J. (2018). Soil erosion in the Anthropocene: Research needs. *Earth Surface Processes and Landforms, 43*(1), 64–84. https://doi.org/10.1002/esp.4250

Poesen, J. (2019). *Soil erosion in the Anthropocene: Do we still need more research?* (pp. 1–2). Proceedings of the Global Symposium on Soil Erosion, FAO.

Quinton, J. N .(1997). Reducing predictive uncertainty in model simulations: a comparison of two methods using the European Soil Erosion Model (EUROSEM). *Catena 30*(2-3), 101–117.

Quinton, J. N., & Morgan, R. P. C. (1998). EUROSEM: An evaluation with single event data from the C5 watershed, Oklahoma, USA. In J. Boardman & D. Favis-Mortlock (Eds.), *Modelling soil erosion by water* (pp. 65–74). Springer. https://doi.org/10.1007/978-3-642-58913-3_7

Rahaman, S. A., Aruchamy, S., Jegankumar, R., & Ajeez, S. A. (2015). Estimation of annual average soil loss, based on RUSLE model in Kallar watershed, Bhavari Basin, Tamil Nadu, India. *ISPRS Annals of the Photogrammetry, Remote Sensing and Spatial Information Sciences, 2*, 207–214. https://doi.org/10.5194/isprsannals-II-2-W2-207-2015

Reddy, V. R., & Galab, S. (2006). Looking beyond the debt trap. *Economic and Political Weekly, 41*(19), 1838–1841. https://doi.org/10.2307/4418195

Renard, K. G., Foster, G. A., McCool, D. K., & Yoder, D. C. (1997). *Predicting soil erosion by water: A guide to conservation planning with the revised universal soil loss equation*. USDA Agriculture Handbook No. 703.

Renard, K. G., Yoder, D. C., Lightle, D. T., & Dabney, S. M. (2011). Universal soil loss equation and revised universal soil loss equation. In R. P. C. Morgan & M. A. Nearing (Eds.), *Handbook of erosion modelling* (pp. 9–32). Wiley.

Sarkar, D., Nayak, D. C., Dutta, D., & Dhyani, B. L. (2005). *Soil erosion of West Bengal*. National Bureau of Soil Survey and Land use Planning, NBSS Publication 117, Nagpur.

Sarkar, D., Nayak, D. C., Dutta, D., & Gajbhiye, K. S. (2007). *Optimizing land use of Birbhum District (West Bengal) Soil Resource Assessment*. NBSS & LUP, NBSS Publ.130, Nagpur.

Sharda, V. N., & Dogra, P. (2013). Assessment of productivity of monetary losses due to water erosion in rainfed crops across different states of India for prioritization and conservation planning. *Agricultural Research, 21*(4), 382–392. https://doi.org/10.1007/s40003-013-0087-1

Sharda, V. N., Dogra, P., & Prakash, C. (2010). Assessment of production losses due to water erosion in rainfed areas of India. *Journal of Soil and Water Conservation, 65*(2), 79–91. https://doi.org/10.2489/jswc.65.2.79

Sharda, V. N., Mandal, D., & Dogra, P. (2019). Assessment of cost of soil erosion and energy saving value of soil conservation measuring in India. *Indian Journal of Soil Conservation, 47* (1), 1–6.

Sharda, V. N., Mandal, D., & Ojasvi, P. R. (2013). Identification of soil erosion risk areas for conservation planning in different states of India. *Journal of Environmental Biology, 34*, 219–226.

Sharma, R. K. (2018). Soil loss setbacks to Indian agriculture. *Acta Scientific Agriculture, 2*(6), 95–97.

Singh, G., Babu, R., Narain, P., Bhushan, L. S., & Abrol, I. P. (1992). Soil erosion rates in India. *Journal of Soil and Water Conservation, 47*(1), 97–99.

Singh, V. P. (2016). *Chow's handbook of applied hydrology*. Mc-Graw-Hill Education.

Sinha, D., & Joshi, V. U. (2012). Application of Universal Soil Loss Equation (USLE) to recently reclaimed badlands along the Adula and Mahalung Rivers, Pravara basin, Maharashtra. *Journal Geological Society of India, 80*, 341–350.

Sinha, R., Jain, V., Tandon, S. K., & Chakraborty, T. (2012). Large river systems of India. *Proceedings of the Indian National Science Academy, 78*(3), 1–17. https://doi.org/10.16943/ptinsa/2016/48482

Smith, H. J. (1999). Application of empirical soil loss models in southern Africa: A review. *South African Journal of Plant and Soil, 16*(3), 158–163. https://doi.org/10.1080/02571862.1999.10635003

Sovrin, J. (2003). A test of three soil erosion models incorporated into a geographical information system. *Hydrological Processes, 17*, 967–977. https://doi.org/10.1002/hyp.1174

Srivastava, R. K., & Imtiyaz, M. (2016). Testing of coupled SCS curve number model for estimating runoff and sediment yield for eleven watersheds. *Journal Geological Society of India, 88*, 627–636. https://doi.org/10.1007/s12594-016-0529-z

Tesfahunegn, G. B., Tamene, L., & Vlek, P. L. G. (2014). Soil erosion prediction using Morgan-Morgan-Finney model in a GIS environment in northern Ethiopia catchment. *Applied and Environmental Soil Science*. https://doi.org/10.1155/2014/468751

Thakkar, H., & Bhattacharyya, S. (2006). *Reservoir siltation in India: Latest studies*. SANDRP.

Torri, D., & Borselli, L. (2012). Water erosion. In P. M. Huang & M. S. Li (Eds.), *Handbook of soil sciences: Resource management and environmental impacts* (pp. 22.1–22.9). CRC Press.

Toy, T. J., Foster, G. R., & Renard, K. G. (2013). *Soil erosion: Processes, prediction, measurement and control*. Wiley.

Vente, J., & Poesen, J. (2005). Predicting soil erosion and sediment yield at the basin scale: Scale issues and semi-quantitative models. *Earth Science Reviews, 71*, 95–125. https://doi.org/10.1016/j.earscirev.2005.02.002

Wasson, R. J. (2003). A sediment budget for the Ganga–Brahmaputra catchment. *Current Science, 84*(8), 1041–1047.

Wen, X., & Deng, X. (2020). Current soil erosion assessment in the Loess Plateau of China: A mini-review. *Journal of Cleaner Production, 276*, 123091. https://doi.org/10.1016/j.clepro.2020.123091

Wischmeier, W. H., & Smith, D. D. (1978). *Predicting rainfall erosion losses—A guide to conservation planning*. USDA Agriculture handbook No. 537.

Yadav, R. C., & Bhushan, L. S. (2002). Conservation of gullies in susceptible riparian areas of alluvial soil region. *Land Degradation and Development, 13*, 201–219. https://doi.org/10.1002/ldr.493

Yanshuang, Z., Jing, L., Bibib, L., Jie, L., & Mingxin, L. (2020). Process of soil erosion study method. In *IOP Conf. Series: Earth and Environmental Science*, p. 526. https://doi.org/10.1088/1755-1315/526/1/012008

Chapter 8
Forest Fire Risk Zone Mapping in Tropical Forests of Saranda, Jharkhand, Using FAHP Technique

Sk Mujibar Rahaman, Masjuda Khatun, Sanjoy Garai, Pulakesh Das, and Sharad Tiwari

Abstract The forest fire has severe environmental and societal consequences causing millions of monetary losses every year in the form of loss of forest resources, animals, and man-made infrastructures globally. Mapping and monitoring of forest fire and its severity are essential to examine the loss of forest cover resources, environmental degradation, release of carbon, etc. The present study attempts to demarcate the forest fire-prone zones in Saranda forests, Jharkhand state, India, which houses Asia's largest Sal forest area (769 km^2). The Sentinel 2A multispectral satellite data and ALOS PALSAR digital elevation model (DEM) data were used to identify the forest-fire prone zones employing the fuzzy analytic hierarchy process (FAHP). The adopted method indicated a high modelling accuracy (overall 88% and kappa coefficient 84%). The study identified that about 77% area of the total forest area is under moderate to very high risk of a forest fire. The study suggests that the dense forest areas, which are characterized by high humidity and residing at higher altitudes, are less prone to a forest fire risk. Alternatively, the open and moderately dense forests at drier regimes are more prone to a forest fire. The developed maps are essential for forest cover management and preparedness to minimize the consequences of a forest fire. Various initiatives such as awareness programs, safeguarding forests from human interventions, formulation of forest fire task forces, and afforestation of native species in the open and disturbed forests in the moist areas are required to mitigate the forest fire risk in the Saranda forests.

Keywords Forest fire · FAHP · Sentinel 2A · ALOS PALSAR · Saranda forest

S. M. Rahaman · M. Khatun · S. Garai · S. Tiwari (✉)
Institute of Forest Productivity, Ranchi, Jharkhand, India
e-mail: tiwaris@icfre.org

P. Das
World Resource Institute, New Delhi, India
e-mail: pulakesh.das@wri.org

P. K. Shit et al. (eds.), *Geospatial Technology for Environmental Hazards*, Advances in Geographic Information Science, https://doi.org/10.1007/978-3-030-75197-5_8

8.1 Introduction

The forest fire is described as uncontrolled burning of vegetation in a forested landscape. It is caused by various anthropogenic and natural factors including drought and warm climate and in some instances due to the occurrence of lightning (Taylor et al., 2008; Westerling et al., 2006). The climate-driven factors are the major causes of the forest fire, where the paleoclimatic studies have indicated higher fires accumulation during the prolonged drier period (Clark, 1988). The dry condition and heat waves have a direct influence on plant growth and humidity (Morgan et al., 2008). The increased dryness or drought condition due to climate alteration has significantly increased the fire events in the past few decades. Besides, the anthropogenic disturbances have induced changes in the land use/land cover (LULC) pattern, which is one of the prime reasons for increased forest fire (Running, 2006). The human interventions in the form of landscape development, land clearing for various purposes including agriculture or shifting cultivation in hilly terrains, and negligence during the tourism and other human activities are also regarded as causes of a forest fire (Behera et al., 2018).

The forest fire significantly contributes to the modification of the ecosystem structure. The fire severity determines the loss of vegetation cover and thereby the biodiversity and ecosystem productivity (Pausas, 2004; Piñol et al., 1998). Every year, fauna and flora in vast stretches of landscape are removed due to wildfire. It also has societal impacts via deteriorating human health and damages infrastructure. The forest fire releases enormous tree carbon into the atmosphere and significantly alters the regional to the global carbon budget (Spracklen et al., 2007). In the past few decades, several devastating forest fire instances have been reported globally such as Australian bush fire (2002, 06, 12, 13, 19–20), Russian wildfire (2003, 15), Northwest territories fires (2014), British Columbia wildfire (2017), Siberian wildfire (2019), Amazon rainforest wildfire (2019), California wildfire (2020), etc. (Iemima, 2018; Luke & McArthur, 2020; Pierce & Meyer, 2008)). In India, forest fire incidents are also frequent, where the recently reported incidences are Uttarakhand forest fire (2016) and Bandipur forest fire (2019) were the most devastating (Milton, 2019; Upadhyay, 2020).

Satellite remote sensing data provides a variety of indicators to assess the forest fire-driven changes in forest cover, structural attributes, biochemical properties, etc. The multitemporal analysis (i.e., pre- and post-fire images) allows to examine the changes in forest and land cover, wherein the GIS analysis facilitates the identification of drivers. Visual image interpretation of the satellite imagery allows the manual identification of fire burnt areas. Moreover, the changes in vegetation indices (spectral enhancement), e.g., Normalized Difference Vegetation Index (NDVI; normalized difference between near-infrared and red band) and Enhanced Vegetation Index (EVI; modified NDVI with canopy background soil correction factor) during the pre- and post-fire event, allow automatic identification. Several indices have been developed for effective burned area mapping, such as Normalized Burn Ratio (NBR; normalized difference between NIR and SWIR band), Normalized

Difference Moisture Index (NDMI); the pre- and post-event NBR (dNBR) and its relativized NBR (RdNBR) (Chuvieco et al., 2002; Key & Benson, 2005; Miller & Thode, 2007; Kolden et al., 2015). The Moderate Resolution Imaging Spectroradiometer (MODIS) and Medium Resolution Imaging Spectrometer (MERIS) sensor generate the global burnt area products at frequent intervals (1–3 days intervals) at a coarser resolution of 500 m and 30 m, respectively (Giglio et al., 2003; Alonso-Canas & Chuvieco, 2015), whereas the Landsat satellite data are widely used for forest fire mapping at moderate resolution (30 m) for the past five decades at various scales (Escuin et al., 2008; Long et al., 2019).

The forest fire occurrence maps created with the satellite data allows analyzing the causes of fire events integrating several response variables and proxies in the Geographical Information System (GIS) environment. The geospatial approach is recognized as a reliable method in developing fire susceptibility map in India, where a large portion of forest land is being occupied and managed by the forest dwellers (Jain et al., 1996; Roy et al., 1991). The statistical analysis of the past events and drivers enables examining the relative influence of the causative factors and allows to create a fire risk zone map (Chuvieco et al., 2010; Núñez-Regueira et al., 2000). Previous studies have identified several factors as biologic, physiographic, and anthropogenic. The biologic factors have significant impacts on forest fire incidence, where the forest fires have differential interactions depending on the species diversity and forest type (Kodandapani et al., 2008). Land use/land cover (LULC), vegetation density, and moisture content of vegetation determine the burnable fuel for forest fire expansion and severity (Biranvand et al., 2011; Adab et al., 2013; Siachalou et al., 2009). The moisture condition of vegetation plays a very influential factor in the spread of forest fire, where high moisturized areas are less prone to burning and vice-versa (Siachalou et al., 2009). The indices on vegetation greenness and moisture content as NDVI and NDMI, respectively, are widely used as the satellite data derived proxies (Serrano et al., 2000). Thin layers of a canopy with a higher reflectance in a short-wave infrared (SWIR) band signifies high moisture content, whereas the higher reflectance in NIR band indicates lower moisture content (Siachalou et al., 2009).

The physiographic variables as elevation and slope are linked with the wind and direction that act as stimulating factors and regulates the fire spread (Gao et al., 2011; Jaiswal et al., 2002; Weise & Biging, 1997). The downward spread of forest fire is slow as compared to the spread of intensity toward a higher slope (Kushla & Ripple, 1997). The moisture content determined by the amount of incident solar energy in an area is correlated with the aspect. The sun-facing aspects create favorable condition for the higher rate of fire spread owing to higher sunlight, heat, low clamminess, low fuel vapors, and heavy winds (Anderson, 1982; Prasad et al., 2008). The anthropogenic influences encourage the fire spread rate due to closeness to settlements and roads (Avila-Flores et al., 2010). Fire risk is higher as it offers more chance for unpredicted human-made explosions, due to the more forceful human actions (Alencar et al., 2004).

Jaiswal et al. (2002) employed the LISS-III data derived from forest type, slope, settlement, and road network map to map the forest fire risk zone in Gorna

Subwatershed, Madhya Pradesh, India. They have observed the high and very high risk in around 30% area of the total forest area. Kumari and Pandey (2020) integrated several factors such as fuel type, bare soil index, slope, aspect, elevation, distance from road, and settlement using the AHP technique for forest fire risk analysis in Palamau Tiger Reserve Forest, Jharkhand. They have observed a good agreement comparing with the MODIS and SNPP-VIRRS product and identified the high and very high-risk zone in about 43% of the area. Sharma et al. (2012) applied the Crisp AHP (CAHP) and Fuzzy AHP (FAHP) techniques for forest fire risk mapping in the Taradevi forest range of the Shimla Forest Division, India. The resultant map showed high accuracy with the forest fire observed data points and identified about 6.89 and 9% area as the very high-risk zone for the CAHP and FAHP technique, respectively. Kayet et al. (2020) compared the Frequency Ratio (FR) model and AHP technique for the forest fire risk mapping in Melghat Tiger Reserve forest, India. The validation with the Forest Survey of India (FSI) fire occurrence point data indicated an overall accuracy of 81% and 79% for the FR and AHP technique, respectively. Adab et al. (2013) employed the MODIS data product to compare the accuracies observed for various methods as Hybrid Fire Index, Structural Fire Index, and Fire Risk Index for northeast Iran. The receiver operating characteristic (ROC) curve indicated high accuracy of 76.7% for the hybrid fire index. Ahmad and Goparaju (2017) analyzed the forest fire hotspot districts in Jharkhand state, India, from 2005 to 2016 and reported the Paschim Singhbhum district as the most forest fire affected district.

In the present study, we have attempted to assess the forest fire risk in the Saranda Forest Division of Jharkhand using Fuzzy Analytic Hierarchy Process (FAHP) model and GIS. The objective of the study was to study the factors responsible for a forest fire in the Saranda Forest Division and to create a knowledge base that would enable better planning and management strategies to combat the future forest fire risk.

8.2 Study Area

The present study was conducted in the Saranda Forest Division (SFD) of West Singhbhum district, Jharkhand, India (Fig. 8.1). Saranda means "Land of Seven hundred Hills." It is Asia's largest dense Sal (*Shorea robusta*) forest and lies between 22° 22′ 58.83″ N to 22° 0′ 36.62″ N and 84° 58′ 47.35″ E to 85° 25′ 38.22″ E geographic extent and spread over about 1003 km^2 of geographical area. Jharkhand has about 29% of its geographical area under forest cover. Both protected forests (61%) and reserved forests (19%) are common to forest fires (Mishra, 2013). Previous studies have identified significant changes in forest cover in the Saranda Forest Division from 1992 to 2014 and built-up and mining area expansion replacing the dense and open forests. About 9% forest reduction with equivalent mining and built-up area expansion was reported by Kayet and Pathak (2015). The analysis

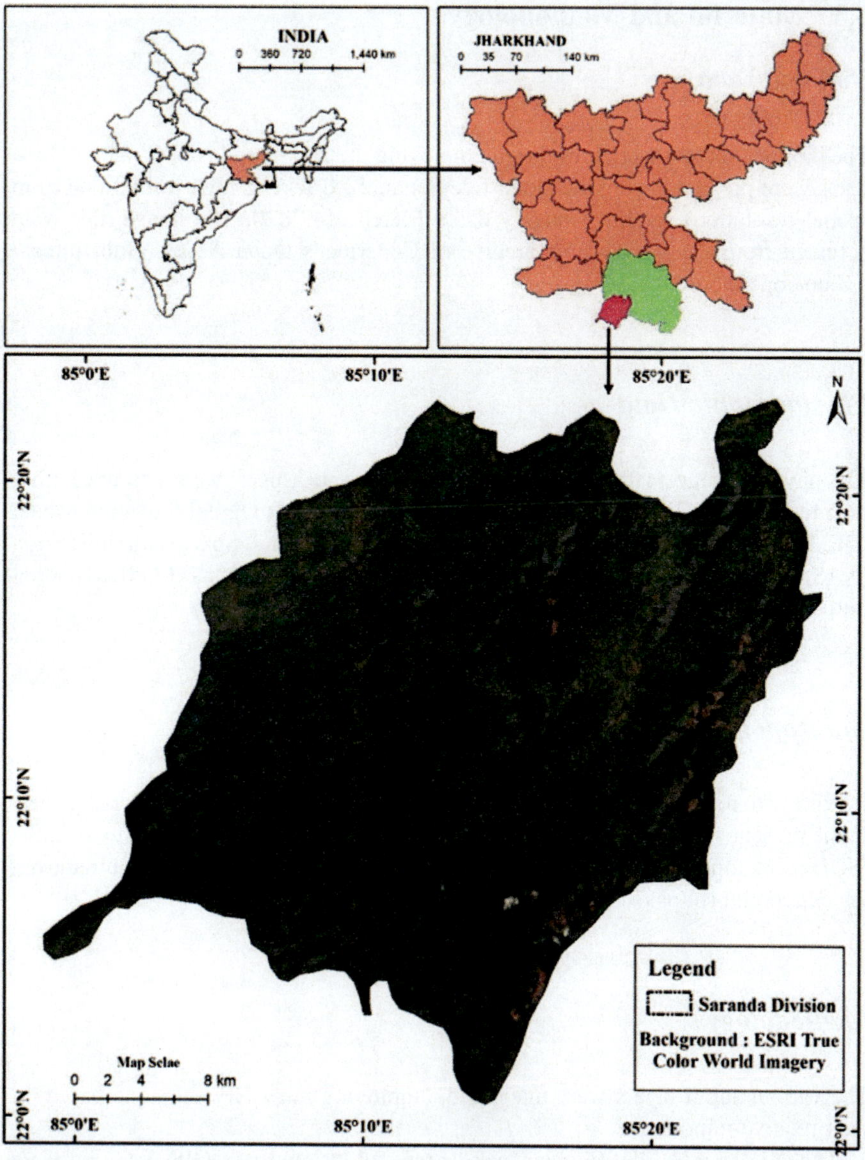

Fig. 8.1 Location map of the study area

carried out by the FSI (Forest Survey of India, 2019) estimated about 0.21%, 2.18%, and 9.16% area of Jharkhand falls under the extreme, very high, and highly fire-prone zones of the total forest cover in Jharkhand state.

8.3 Material and Methodology

Biologic Data

The biologic factors were introduced employing the NDVI and NDMI maps. These maps were prepared using the cloud-free Sentinel-2B level 2 high-resolution (10 m spatial resolution) satellite imagery dated March 28, 2020. The image tiles were accessed from the open-source archive of Copernicus Open Access Hub (https://scihub.copernicus.eu/).

Physiographic Data

The physiographic factors, i.e., elevation, slope, and aspect, were extracted from high resolution (12.5 mspatial resolution) ALOS PALSAR Digital Elevation Model (DEM) datasets acquired from National Aeronautics and Space Administration (NASA) Earth Observing Systems Data and Information Systems (EOSDIS) portal (https://search.asf.alaska.edu/#/?dataset=ALOS).

Anthropogenic Data

Proximity to road and settlement is sensitive factor for mapping forest fire risk zones. In the present study, road and settlement maps were prepared using Survey of India (SOI) 1:50k topographical map (73F3, 73F4, 73F7, 73F8, and 73B16) acquired from the data portal (https://soinakshe.uk.gov.in/).

Methodology

The various input layers were integrated, employing the AHP technique based on their relative importance.

The NDVI and NDMI maps were prepared using the QGIS 3.14 software employing the following formulae (Eqs. 8.1 and 8.2, respectively).

$$\mathrm{NDVI} = \frac{\mathrm{NIR} - \mathrm{RED}}{\mathrm{NIR} + \mathrm{RED}} \tag{8.1}$$

$$\mathrm{NDMI} = \frac{\mathrm{NIR} - \mathrm{SWIR}}{\mathrm{NIR} + \mathrm{SWIR}} \tag{8.2}$$

The NDVI layer was used to create the map vegetation and non-vegetation cover, where the NDVI threshold of 0.2 was considered for segregation (Al-doski, 2013). The leaf water content of the vegetation was assessed by the NDMI. Lower NDMI value indicates dry vegetation which is more flammable than fresh vegetation that has high NDMI value. Thus, alternative lower and higher weights were assigned to higher and lower moisture conditions areas.

The physiographic factors as elevation, slope, and aspect were derived from the ALOS DEM datasets using QGIS 3.14 software "raster terrain analysis" tool. Forest fire influences are higher in the lower altitude due to favorable climatic conditions; therefore, higher weightage was assigned to lower elevation (Jaiswal et al., 2002). The higher vertical slope increases the likelihood of increasing the spread of forest fires due to water loss and more efficient convection prewarming (Kushla & Ripple, 1997). Consequently, high and low slopes were assigned high and low weights, respectively. In the northern hemisphere, the south and the north-facing slope is exposed to highest and least sunlight, respectively, thus assigned contrasting higher and lower weightage (Kumari & Pandey, 2020).

Anthropogenic factors like distance from road and settlement were digitized from SOI topographical map on 1:50k scale. In the forest fire risk index, roads play both positive and negative roles. Roads offer more human access to forested areas enabling more human interference and increased chances of a spark of fire; alternatively, the roads create a barrier that prevents fire spreading. Areas close to the settlements and roads were assigned with higher weightage values. All the input variables were converted into raster format with uniform cell size and were projected for further processing.

Fuzzy Analytical Hierarchy Process (FAHP) Model

The FAHP is the combination of Analytic Hierarchy Process (AHP) and fuzzy logic. It follows a similar process to the AHP method. In FAHP, the arrays of the AHP are scaled into the fuzzy triangle scale to retrieve the importance of the input variables (Putra et al., 2018). Using a fuzzy triangle scale, fuzzified pairwise comparison matrix was prepared, as shown in Table 8.2.

Estimation of Geometric Mean

For each criterion, the geometric mean ($\widetilde{r}_i$) fuzzy comparison values are calculated as shown in the Eq. (8.3) (Buckley, 1985).

$$\widetilde{r}_i = \left(\prod\nolimits_{j=1}^{n} \widetilde{d}_{ij}\right)^{\frac{1}{n}}, \quad i = 1, 2, 3, \ldots, n \tag{8.3}$$

Estimate Weight ($\widetilde{w}_i$):

$$\widetilde{w}_i = \widetilde{r}_i \otimes \left(\widetilde{r}_1 \otimes \widetilde{r}_2 \otimes \cdots \otimes \widetilde{r}_n\right)^{-1} \tag{8.4}$$

Defuzzified (M_i):

$$M_i = \left(\frac{lw_i + mw_i + uw_i}{3}\right) \tag{8.5}$$

Normalized Fuzzy Weight(N_i):

$$N_i = \frac{M_i}{\sum_{i-1}^{n} M_i} \tag{8.6}$$

Forest Fire Risk Index

The input parameters as NDVI, NDMI, elevation, slope, aspect, distance to road, and distance to settlement were assigned different weights ranging from 2 to 10 based on their comparative importance to fire risk (Table 8.1). Lower and higher weightage value indicates the relative lower and higher hazard induced by the drivers. The mathematical equation used for calculating the forest fire risk index is shown below:

$$\text{FFRI} = (V_m * 38) + (V_i * 20) + (e * 3) + (S * 14) + (a * 4) + (R * 12) + (s * 10) \tag{8.7}$$

Where V_m is Normalised Difference Moisture Index, V_i is Normalised Difference Vegetation Index, e is elevation, S is slope, a is aspect, R is distance from road, s is distance to settlement, and 38, 20, 3, 14, 4, 12, 10 is the weight of the factor estimated by using the FAHP model (Tables 8.2, 8.3, and 8.4 and Eqs. 8.3–8.6).

For validation, the Suomi NPP and Visible Infrared Imaging Radiometer (SNPP-VIRS) forest fire dataset were used. The data were acquired from the Level-1 and Atmosphere Archive and Distribution System (LAADS) Distributed Active Archive

Table 8.1 Weights assigned to each variable and classes for forest fire risk modelling

Variables	Classes	Rating of risk
NDMI (−1 to +1)	<0, 0–0.16, 0.16–0.26, 0.26–0.36, >0.36	10, 8, 6, 4, 2
NDVI (−1 to +1)	<0.2, >0.2	4, 10
Elevation (m)	<350, 350–550, >550	10, 6, 2
Slope (°)	<5, 5–10, 10–15, 15–20, >20	2, 4, 6, 8, 10
Aspect(°)	North, East, West, South	4, 6, 8, 10
Proximity to Road(m)	<500, 500–1000, 1000–1500, 1500–2000, >2000	10, 8, 6, 4, 2
Proximity to Settlement(m)	<500, 500–1000, 1000–1500, 1500–2000, >2000	10, 8, 6, 4, 2

Table 8.2 Fuzzified pair-wise comparison matrix

Criteria	NDMI	NDVI	Elevation	Slope	Aspect	Proximity to road	Proximity to settlement
NDMI	**1, 1, 1**	4, 5, 6	5, 6, 7	5, 6, 7	6, 7, 8	5, 6, 7	5, 6, 7
NDVI	$\frac{1}{6}, \frac{1}{5}, \frac{1}{4}$	**1, 1, 1**	5, 6, 7	3, 4, 5	5, 6, 7	3, 4, 5	3, 4, 5
Elevation	$\frac{1}{7}, \frac{1}{6}, \frac{1}{5}$	$\frac{1}{7}, \frac{1}{6}, \frac{1}{5}$	**1, 1, 1**	1, 2, 3	$\frac{1}{3}, \frac{1}{2}, \frac{1}{1}$	$\frac{1}{7}, \frac{1}{6}, \frac{1}{5}$	$\frac{1}{7}, \frac{1}{6}, \frac{1}{5}$
Slope	$\frac{1}{7}, \frac{1}{6}, \frac{1}{5}$	$\frac{1}{5}, \frac{1}{4}, \frac{1}{3}$	$\frac{1}{3}, \frac{1}{2}, \frac{1}{1}$	**1, 1, 1**	$\frac{1}{3}, \frac{1}{2}, \frac{1}{1}$	$\frac{1}{6}, \frac{1}{5}, \frac{1}{4}$	$\frac{1}{6}, \frac{1}{5}, \frac{1}{4}$
Aspect	$\frac{1}{8}, \frac{1}{7}, \frac{1}{6}$	$\frac{1}{7}, \frac{1}{6}, \frac{1}{5}$	1, 2, 3	1, 2, 3	**1, 1, 1**	$\frac{1}{6}, \frac{1}{5}, \frac{1}{4}$	$\frac{1}{6}, \frac{1}{5}, \frac{1}{4}$
Proximity to road	$\frac{1}{7}, \frac{1}{6}, \frac{1}{5}$	$\frac{1}{5}, \frac{1}{4}, \frac{1}{3}$	5, 6, 7	4, 5, 6	4, 5, 6	**1, 1, 1**	1, 2, 3
Proximity to settlement	$\frac{1}{7}, \frac{1}{6}, \frac{1}{5}$	$\frac{1}{5}, \frac{1}{4}, \frac{1}{3}$	5, 6, 7	4, 5, 6	4, 5, 6	$\frac{1}{3}, \frac{1}{2}, \frac{1}{1}$	**1, 1, 1**

Table 8.3 Estimated geometric mean

Criteria	Geometric mean ($\widetilde{r}_i$)
NDMI	3.95, 4.63, 5.29
NDVI	1.96, 2.40, 2.85
Elevation	0.28, 0.36, 0.47
Slope	1.42, 1.67, 2.10
Aspect	0.34, 0.45, 0.57
Proximity to road	1.13, 1.43, 1.75
Proximity to settlement	0.96, 1.18, 1.50
Total (T)	10.04, 12.12, 14.52
Reverse (T^{-1})	0.10, 0.08, 0.07

Table 8.4 Estimated fuzzy weight

Criteria	Fuzzy weight ($\widetilde{w}_i$)	Defuzzified (M_i)	Normalized fuzzy weight (N_i)
NDMI	0.27, 0.38, 0.53	0.39	0.38
NDVI	0.14, 0.20, 0.28	0.21	0.20
Elevation	0.02, 0.03, 0.05	0.03	0.03
Slope	0.10, 0.14, 0.21	0.15	0.14
Aspect	0.02, 0.04, 0.06	0.04	0.04
Proximity to road	0.08, 0.12, 0.17	0.12	0.12
Proximity to settlement	0.07, 0.10, 0.15	0.10	0.10

Center (DAAC) website (https://ladsweb.modaps.eosdis.nasa.gov/). For validation, 222 points were created randomly, where the respective values were extracted from the forest fire risk map and SNPP-VIRS to compute the frequency and confusion matrix.

8.4 Result

The forest density map is shown in Fig. 8.2, which indicates the dominant occurrence of moderately dense forest (519 km^2) followed by open forest (281 km^2), grassland (83.65%), and dense forest (35.92%), whereas non-forest areas are estimated in 17.13 km^2 area. The factor maps are shown in Fig. 8.2. The NDMI map indicated higher moisture content for the grassland-dominated regions, as observed in the eastern and western part of the study area (Fig. 8.3a). The NDVI index map indicates the dominance of densely vegetated areas except for the eastern and northwestern part mostly occupied by grassland, settlement, and mining areas (Fig. 8.3b), which are also mostly in the lower altitudes (Fig. 8.3c). The corresponding slope map indicates lower slope in these regions and a higher slope in the rest of the study area (Fig. 8.3d). The aspect map indicates the face of the slope (Fig. 8.3e), which is categorized in four types as east, west, north, and south. The south-facing slope is given a higher value, and the north-facing slope is given a

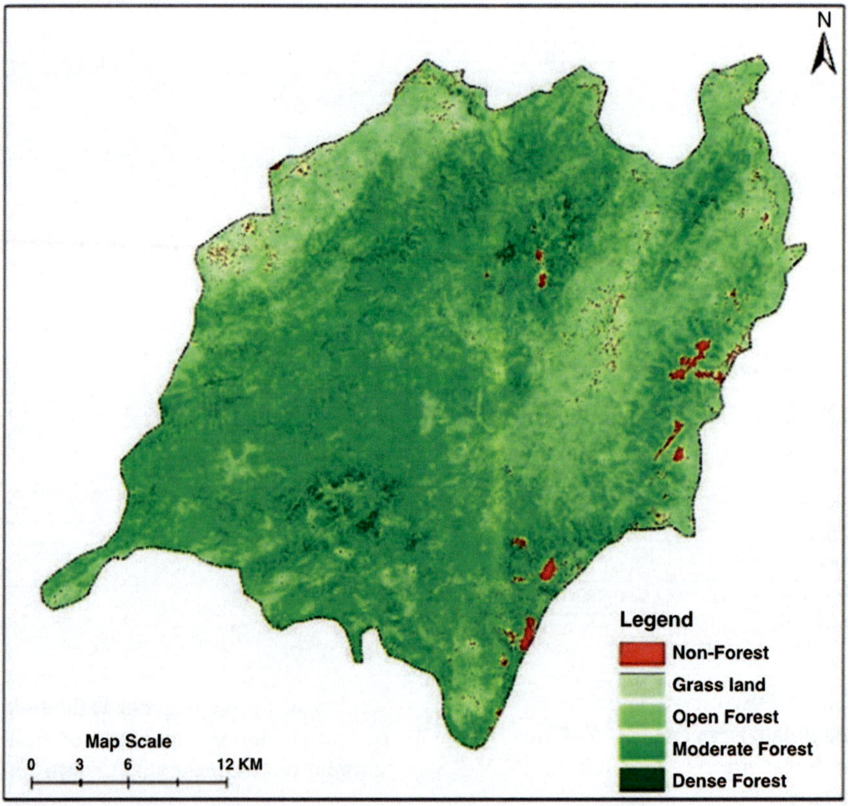

Fig. 8.2 Forest cover map

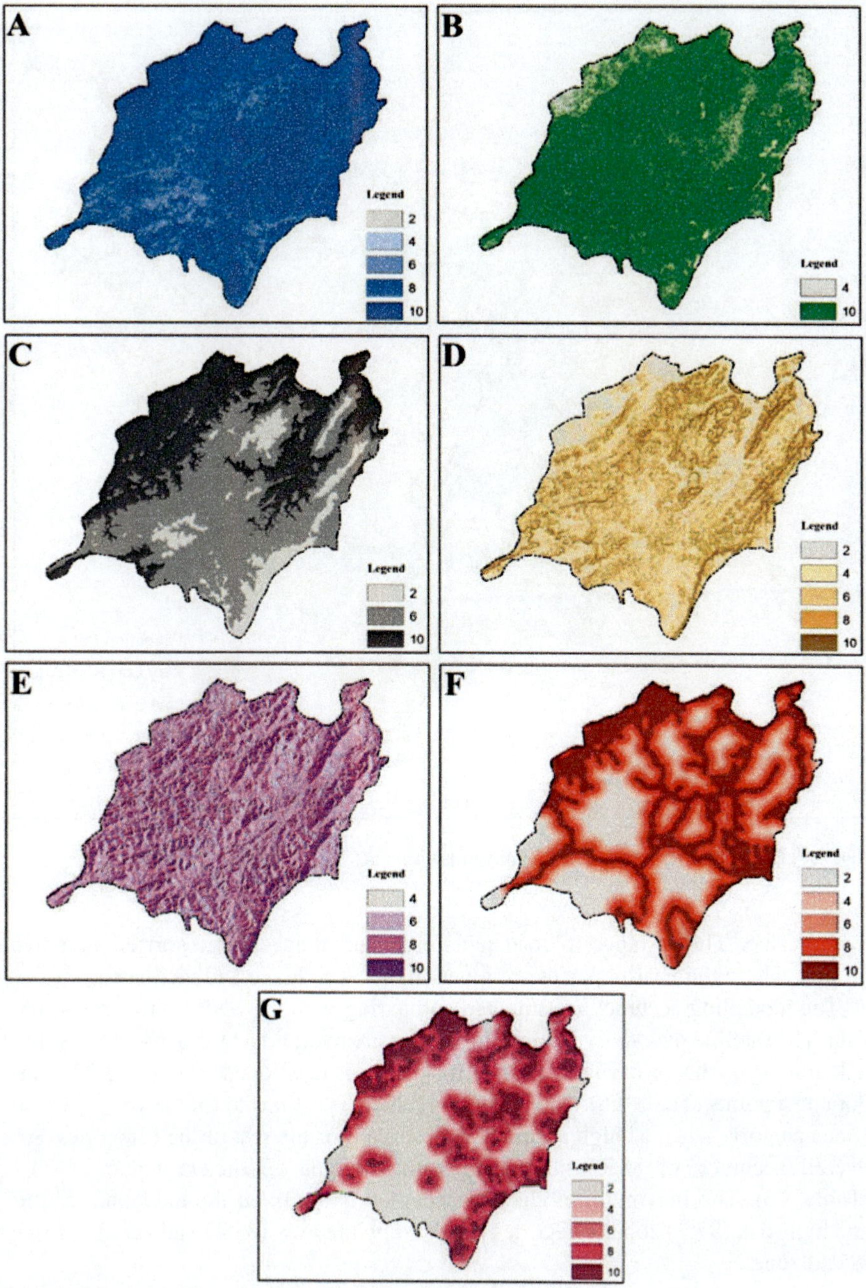

Fig. 8.3 Maps depicting: (**a**) Moisture Index, (**b**) Vegetation Index, (**c**) Elevation, (**d**) Slope, (**e**) Aspect, (**f**) Proximity to Road, and (**g**) Proximity to Settlement

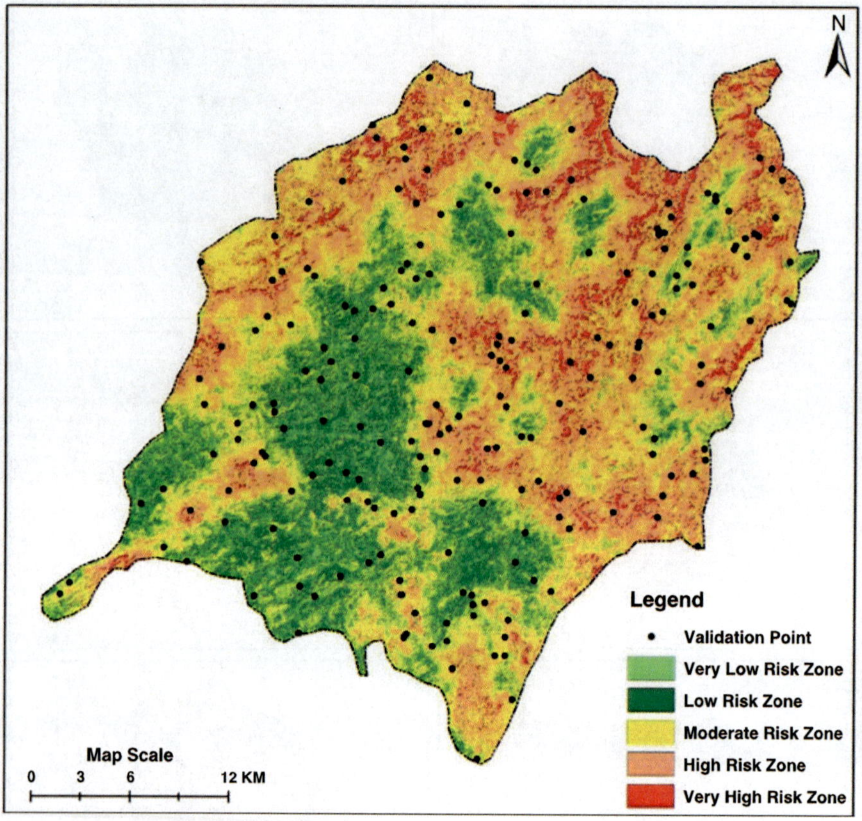

Fig. 8.4 Forest Fire Risk Index map with validated points

lower value. The distance to road and settlement map are categorized into five groups, where the nearby places are given higher weightage and vice-versa.

The modelling accuracy is estimated comparing with the SNPP-VIRS forest fire data. The satellite data-derived forest fire points are overlaid on the derived forest fire risk zone map shown in Fig. 8.4. An error matrix is developed employing 222 random data points (Table 8.5). The lowest agreement is estimated for the very low-risk zone category, whereas high accuracy is observed for the rest of the categories. An overall accuracy of 88% is obtained with a kappa coefficient value of 0.84 (Table 8.6). The maximum producer's accuracy is observed the moderate- (98%) and high-risk (97%) zones categories followed by the low- (88%) and very high-risk (82%) zone.

The entire study area is classified into five risk zones, i.e., very high risk, high risk, moderate risk, low risk and very low risk. The area falling under different risk zones is calculated and is depicted in Table 8.7. Dominant forest covers in the eastern, northern, and western regions are estimated under moderate to high fire risk.

Table 8.5 Error matrix between the observed and predicted category

		Observed					
	Fire risk	Very low	Low	Moderate	High	Very high	Total
Predicted	Very low	8	1	0	0	0	9
	Low	2	37	1	0	0	40
	Moderate	4	1	88	0	1	94
	High	5	2	0	51	2	60
	Very high	2	2	0	2	13	19
	Total	21	43	89	53	16	222

Table 8.6 Error and accuracy estimate

Fire risk	Commission error	User accuracy	Omission error	Producer accuracy	Overall accuracy	Kappa value
Very low	11	89	24	76	88	84
Low	7	93	12	88		
Moderate	7	93	2	98		
High	15	85	3	97		
Very high	31	69	18	82		

Table 8.7 Area of forest fire risk zone

Risk zone	Area (km^2)	Percentage (%)
Very low	31.98	3.19
Low	201.38	20.07
Moderate	406.99	40.58
High	297.24	29.64
Very high	65.41	6.52

Alternatively, the forest covers in the low to very low-risk zones are mostly observed in the central and southern parts of the study area. The very high risk is estimated only 6.52% of the study area, whereas high- and moderate-risk zones are estimated in 29.64% and 40.58% area. The low-risk zone is estimated in 20.07%, whereas 3.9% area is estimated under the very low risk. The outcome of the study suggested that higher risk are mostly found in low-lying plain areas characterized by very low humidity, high slopes, ascending south, close to roads, and settlements. In contrast, areas characterized by high humidity, lesser land slope, and undisturbed forests are comparatively less prone to the occurrence and spread of forest fire.

Very-high fire risk is observed for open and moderately dense forests due to suitable biological, anthropogenic, and topographical conditions like less soil and vegetation moisture condition, gentle slope condition, low elevation, proximity to roads and settlements, and south-facing slope. In contrast, dense forests due to less human interference and its distribution along high altitudes fall under the low-risk zone. The moderately and open forest occupied the majority of the study area followed grassland, dense forest, and non-forest areas. In the dense forest, the

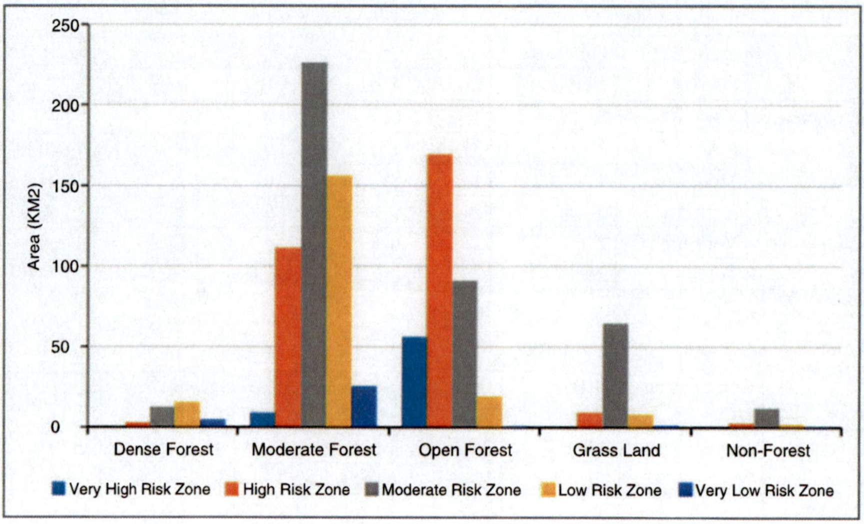

Fig. 8.5 Graph showing forest cover wise forest fire risk zone

majority of the area is estimated in lower-risk zone followed by moderate risk. Out of the total moderately dense forest of 519.6 km^2, the moderate risk zone is estimated in 226.38 km^{2}area followed by the low- (156.34 km^2) and high-risk zone (111.67 km^2) (Fig. 8.5). In the open forest, more than 60% of the area is estimated in the high-risk zone, whereas 32% area is estimated in the moderate risk-zone. The majority of areas under grassland are estimated in the moderate-risk zone (56.31 km^2) followed by nearly 5 km^2 and 4.5 km^2 under the high- and low-risk zones.

8.5 Discussion

Understanding the factors leading to an ecosystem vulnerable is important to assess ecosystem susceptibility (Chuvieco & Congalton, 1989). Almost 89% of the forest fires in tropical regions took place during the drier periods, e.g., March and April (Ahmad et al., 2018). West Singhbhum is the highest forest fire-prone district of Jharkhand state and experiences more than 30% of total forest fire of the state in about 17% of the state forest cover (Ahmad & Goparaju, 2017). The present study revealed that the open forests and moderate forest cover area form the majority forest cover, which is highly vulnerable to forest fire occurrence. Consequently, these parts are more dominated by human activities and experiencing an increase in surface temperature (Kayet et al., 2020). Human interventions could be one of the major drivers that ignite the forest fire (Dong et al., 2005). The study revealed that landscapes falling under the grassland category are prone to forest fire occurrence. The forest and grassland areas near the roads and settlements are more prone to fire detonation because accidental fires can be caused by forest dweller and the tourists (Jaiswal et al., 2002). In comparison, the dense forests are estimated less prone to forest fire; which could be due to less human intervention and being distributed at a higher altitude. At higher altitudes, fire intensity remains less acute due to higher precipitation (Chuvieco & Congalton, 1989). Further, being densely distributed, the area is characterized by compactness, comparatively low temperature, and high moisture content, leading to a reduced risk of fire. Another major factor that contributes significantly to the spread of forest fire is wind. The densely distributed compact undisturbed forests act as a barrier to wind flow and thus reduce the forest fire intensity or spread. In contrast, moderately dense and open forest areas due to suitable biological, anthropogenic, and topographical conditions like low moisture, gentle slope, low elevation, open spaces, and proximity to roads and settlements provide an ideal environment for the occurrence of a forest fire. The Fuzzy AHP technique is successfully applied for the present study as indicated by the comparison (overall accuracy 88%) with the satellite-based observations. In comparison to FAHP, CAHP, and Dong model, FAHP and CAHP performed better (Sharma et al., 2012) than Dong model (Eskandari & Miesel, 2017). The study revealed that about ~23% area of Saranda forest falls under low- to very low-risk zone class and the remaining area ~77% that forms the majority of the Saranda forest division falls under moderate to very high risk prone zone and is vulnerable to forest fire incidents.

8.6 Control Measures

As the majority of the forest fire incidents are mostly caused by human activities, the prevention or control of forest fire can be achieved through people's participation. Identification and prioritization of the fire prone areas and region specific causes are more important. Creating awareness among local inhibitors living in the proximity of forests is very important. Various studies have suggested that the open and disturbed forest that are more prone to forest fire occurrence. These spaces shall be filled through afforestation programs based on the selection of native tree species supporting a high moisture environment. Formation of trained task forces equipped to deal with any situation of forest fire needs to be conformed and well-deployed in fields. Forest departments need to be equipped with modern firefighting equipment. Creation of small water resources like ponds and similar structures can be developed in the proximity of high-risk areas. Regular field monitoring, trained staff, and awareness among locals are some of the initiatives that can help to combat the forest fire.

8.7 Conclusion

The present study attempted to study the forest fire proneness in Saranda forests. The adopted methodology provided a reliable output well verified with the satellite-based observations. The study suggested that about 769 km^2 area of the study area, which is close to approximately ~77% of the total geographical area of Saranda forest division is vulnerable to moderate to very high risk of forest fire incidence. The various input maps and final forest risk zone map could be highly useful for the forest and biodiversity managers. The study exhibited that areas falling under vegetation class, i.e., grassland, open and moderately dense forests, are more prone to forest fire occurrence and dense forest areas are comparatively less prone to forest fire risk. The Saranda forest is one of the major Sal forests of Asia and is a hub to diverse species of economic and medicinal importance. Thus, the area needs special attention to safeguard it from increasing human interventions. Additionally, future studies may include assessment of the invasive species spread in the Saranda forests, as the presence of the bushy type of invasive weeds not only alter the native resources but also act as fire stimulation agents. An integrated approach in the form of research, awareness, and prioritization of area needs to be carried out to strengthen the management strategies to mitigate the increased risk of a forest fire.

References

Adab, H., Kanniah, K. D., & Solaimani, K. (2013). Modeling forest fire risk in the northeast of Iran using remote sensing and GIS techniques. *Natural Hazards, 65*(3), 1723–1743. https://doi.org/10.1007/s11069-012-0450-8

Ahmad, F., & Goparaju, L. (2017). Geospatial Assessment of Forest Fires in Jharkhand (India). *Indian Journal of Science and Technology, 10*(21), 1–7. https://doi.org/10.17485/ijst/2017/v10i21/113215

Ahmad, F., Uddin, M. M., & Goparaju, L. (2018). An evaluation of vegetation health and the socioeconomic dimension of the vulnerability of Jharkhand state of India in climate change scenarios and their likely impact: A geospatial approach. *Environmental and Socio-Economic Studies, 6*(4), 39–47. https://doi.org/10.2478/environ-2018-0026

Al-doski, J. (2013). NDVI differencing and post-classification to detect vegetation changes in Halabja City, Iraq. *IOSR Journal of Applied Geology and Geophysics, 1*(2), 01–10. https://doi.org/10.9790/0990-0120110

Alencar, A. A. C., Solórzano, L. A., & Nepstad, D. C. (2004). Modeling forest understory fires in an eastern amazonian landscape. *Ecological Applications, 14*(4 Suppl), 139–149. https://doi.org/10.1890/01-6029

Alonso-Canas, I., & Chuvieco, E. (2015). Global burned area mapping from ENVISAT-MERIS and MODIS active fire data. *Remote Sensing of Environment, 163*, 140–152. https://doi.org/10.1016/j.rse.2015.03.011

Anderson, H. E. (1982). *Aids to determining fuel models for estimating fire behavior* [Grass, shrub, timber, and slash, photographic examples, danger ratings]. In USDA Forest Service general technical report INT—Intermountain Forest and Range Experiment Station (USA).

Avila-Flores, D., Pompa-Garcia, M., Antonio-Nemiga, X., Rodriguez-Trejo, D. A., Vargas-Perez, E., & Santillan-Perez, J. (2010). Driving factors for forest fire occurrence in Durango State of

Mexico: A geospatial perspective. *Chinese Geographical Science, 20*(6), 491–497. https://doi.org/10.1007/s11769-010-0437-x

Behera, M. D., Gupta, A. K., Barik, S. K., Das, P., & Panda, R. M. (2018). Use of satellite remote sensing as a monitoring tool for land and water resources development activities in an Indian tropical site. *Environmental Monitoring and Assessment, 190*(7), 401.

Biranvand, A., Babaei, K. S., & Kiadaliri, H. (2011). *Investigation the ecological factors affecting fire spread in forest ecosystems (case study: Kakareza-Lorestan)*. USDA Forest Service General Technical Report INT-Intermountain Forest and Range Experiment Station (USA).

Buckley, J. J. (1985). Fuzzy hierarchical analysis. *Fuzzy Sets and Systems, 17*(3), 233–247. https://doi.org/10.1016/0165-0114(85)90090-9

Chuvieco, E., Aguado, I., Yebra, M., Nieto, H., Salas, J., Martín, M. P., Vilar, L., Martínez, J., Martín, S., Ibarra, P., de la Riva, J., Baeza, J., Rodríguez, F., Molina, J. R., Herrera, M. A., & Zamora, R. (2010). Development of a framework for fire risk assessment using remote sensing and geographic information system technologies. *Ecological Modelling, 221*(1), 46–58. https://doi.org/10.1016/j.ecolmodel.2008.11.017

Chuvieco, E., Martin, M. P., & Palacios, A. (2002). Assessment of different spectral indices in the red-near-infrared spectral domain for burned land discrimination. *International Journal of Remote Sensing, 23*(23), 5103–5110.

Chuvieco, E., & Congalton, R. G. (1989). 0034-4257-2889-2990023-0.Pdf. *Remote Sensing of Environment, 159*(29), 147–159.

Clark, J. S. (1988). Effect of climate change on fire regimes in northwestern Minnesota. *Nature, 334* (6179), 233–235.

Dong, X., Li-min, D., Guo-fan, S., Lei, T., & Hui, W. (2005). Forest fire risk zone mapping from satellite images and GIS for Baihe Forestry Bureau, Jilin, China. *Journal of Forestry Research, 16*(3), 169–174. https://doi.org/10.1007/bf02856809

Escuin, S., Navarro, R., & Fernández, P. (2008). Fire severity assessment by using NBR (Normalized Burn Ratio) and NDVI (Normalized Difference Vegetation Index) derived from LANDSAT TM/ETM images. *International Journal of Remote Sensing, 29*(4), 1053–1073. https://doi.org/10.1080/01431160701281072

Eskandari, S., & Miesel, J. R. (2017). Comparison of the fuzzy AHP method, the spatial correlation method, and the Dong model to predict the fire high-risk areas in Hyrcanian forests of Iran. *Geomatics, Natural Hazards and Risk, 8*(2), 933–949.

Forest Survey of India. (2019). India State of Forest Report (ISFR) 2019 (Issue 11.12 Jharkhand 11.12). Retrieved from https://fsi.nic.in/isfr19/vol2/isfr-2019-vol-ii-jharkhand.pdf

Gao, X., Fei, X., & Xie, H. (2011). Forest fire risk zone evaluation based on high spatial resolution RS image in Liangyungang Huaguo Mountain Scenic Spot. In *ICSDM 2011—Proceedings 2011 IEEE International Conference on Spatial Data Mining and Geographical Knowledge Services*, pp. 593–596. https://doi.org/10.1109/ICSDM.2011.5969116.

Giglio, L., Descloitres, J., Justice, C. O., & Kaufman, Y. J. (2003). An enhanced contextual fire detection algorithm for MODIS. *Remote Sensing of Environment, 87*(2–3), 273–282.

Iemima, P. (2018). *Largest brush and forest fires in recorded history*. Worldatlas. Retrieved from https://www.worldatlas.com/articles/largest-brush-and-forest-fires-in-recorded-history.html

Jain, A., Ravan, S. A., Singh, R. K., Das, K. K., & Roy, P. S. (1996). Forest fire risk modelling using remote sensing and geographic information system. *Current Science*, 928–933.

Jaiswal, R. K., Mukherjee, S., Raju, K. D., & Saxena, R. (2002). Forest fire risk zone mapping from satellite imagery and GIS. *International Journal of Applied Earth Observation and Geoinformation, 4*(1), 1–10. https://doi.org/10.1016/S0303-2434(02)00006-5

Kayet, N., Chakrabarty, A., Pathak, K., Sahoo, S., Dutta, T., & Hatai, B. K. (2020). Comparative analysis of multi-criteria probabilistic FR and AHP models for forest fire risk (FFR) mapping in Melghat Tiger Reserve (MTR) forest. *Journal of Forestry Research, 31*(2), 565–579.

Kayet, N., & Pathak, K. (2015). Remote sensing and GIS based land use/land cover change detection mapping in Saranda Forest, Jharkhand, India. *International Journal of Earth Sciences, 3*(10), 1–6.

Key, C., & Benson, N. (2005). Landscape assessment: Ground measure of severity, the Composite Burn Index; and remote sensing of severity, the Normalized Burn Ratio. In *FIREMON: Fire effects monitoring and inventory system 2004*.

Kodandapani, N., Cochrane, M. A., & Sukumar, R. (2008). A comparative analysis of spatial, temporal, and ecological characteristics of forest fires in seasonally dry tropical ecosystems in the Western Ghats, India. *Forest Ecology and Management, 256*(4), 607–617. https://doi.org/10.1016/j.foreco.2008.05.006

Kolden, C. A., et al. (2015). Limitations and utilisation of Monitoring Trends in Burn Severity products for assessing wildfire severity in the USA. *International Journal of Wildland Fire, 24* (7), 1023–1028.

Kumari, B., & Pandey, A. C. (2020). Geo-informatics based multi-criteria decision analysis (MCDA) through analytic hierarchy process (AHP) for forest fire risk mapping in Palamau Tiger Reserve, Jharkhand state, India. *Journal of Earth System Science, 129*(1). https://doi.org/10.1007/s12040-020-01461-6

Kushla, J. D., & Ripple, W. J. (1997). The role of terrain in a fire mosaic of a temperate coniferous forest. *Forest Ecology and Management, 95*(2), 97–107. https://doi.org/10.1016/S0378-1127(97)82929-5

Long, T., Zhang, Z., He, G., Jiao, W., Tang, C., Wu, B., et al. (2019). 30 m resolution global annual burned area mapping based on Landsat Images and Google Earth Engine. *Remote Sensing, 11* (5), 489.

Luke, & McArthur (2020). Bushfire History, South Australian Country Fire Service. South Australian Country Fire Service. Retrieved from https://www.cfs.sa.gov.au/about-cfs/history-of-the-cfs/bushfire-history/

Miller, J. D., & Thode, A. E. (2007). Quantifying burn severity in a heterogeneous landscape with a relative version of the delta Normalized Burn Ratio (dNBR). *Remote Sensing of Environment, 109*(1), 66–80.

Milton, L. (2019). Karnataka: Bandipur fire doused, forest official confirms act of sabotage. *Times of India*. Retrieved from https://timesofindia.indiatimes.com/city/mysuru/karnataka-bandipur-fire-doused-forest-official-confirms-act-of-sabotage/articleshow/68145439.cms

Mishra, A. K. N. (2013). More forests burning in Jharkhand. *Times of India*. Retrieved from https://timesofindia.indiatimes.com/city/ranchi/More-forests-burning-in-Jharkhand/articleshow/19504839.cms?utm_source=contentofinterest&utm_medium=text&utm_campaign=cppst&pcode=461

Morgan, P., Heyerdahl, E. K., & Gibson, C. E. (2008). Multi-season climate synchronized forest fires throughout the 20th century, northern Rockies, USA. *Ecology, 89*(3), 717–728. https://doi.org/10.1890/06-2049.1

Núñez-Regueira, L., Proupín-Castiñeiras, J., & Rodríguez-Añón, J. A. (2000). Design of risk index maps as a tool to prevent forest fires in the hill-side zone of Galicia (NW Spain). *Bioresource Technology, 73*, 123–131.

Pausas, J. G. (2004). Peninsula (Mediterranean Basin). *Climatic Change, 63*, 337–350.

Pierce, J., & Meyer, G. (2008). Long-term fire history from alluvial fan sediments: The role of drought and climate variability, and implications for management of Rocky Mountain forests. *International Journal of Wildland Fire, 17*(1), 84–95.

Piñol, J., Terradas, J., & Lloret, F. (1998). Climate warming, wildfire hazard, and wildfire occurrence in coastal eastern Spain. *Climatic Change, 38*(3), 345–357. https://doi.org/10.1023/A:1005316632105

Prasad, V. K., Badarinath, K. V. S., & Eaturu, A. (2008). Biophysical and anthropogenic controls of forest fires in the Deccan plateau, India. *Journal of Environmental Management, 86*(1), 1–13. https://doi.org/10.1016/j.jenvman.2006.11.017

Putra, M. S. D., Andryana, S., & Fauziah, & Gunaryati, A. (2018). Fuzzy analytical hierarchy process method to determine the quality of gemstones. *Adv. Fuzzy Syst., 2018*. https://doi.org/10.1155/2018/9094380

Roy, P. S., Ranganath, B. K., Diwakar, P. G., Vohra, T. P. S., Bhan, S. K., Singh, I. J., & Pandian, V. C. (1991). Tropical forest typo mapping and monitoring using remote sensing. *International Journal of Remote Sensing, 12*(11), 2205–2225. https://doi.org/10.1080/01431169108955253

Running, S. W. (2006). Is global warming causing more, larger wildfires? *Science, 313*(5789), 927–928. https://doi.org/10.1126/science.1130370

Serrano, L., Ustin, S. L., Roberts, D. A., Gamon, J. A., & Peñuelas, J. (2000). Deriving water content of chaparral vegetation from AVIRIS data. *Remote Sensing of Environment, 74*(3), 570–581. https://doi.org/10.1016/S0034-4257(00)00147-4

Sharma, L. K., Kanga, S., Nathawat, M. S., Sinha, S., & Pandey, P. C. (2012). Fuzzy AHP for forest fire risk modeling. *Disaster Prevention and Management: An International Journal.*

Siachalou, S., Doxani, G., & Tsakiri-Strati, M. (2009). Integrating remote sensing processing and GIS to fire risk zone mapping: A case study for the Seih-Sou Forest of Thessaloniki. In *Proceedings of the 24th International Cartographic Conference (ICC), January 2016*, pp. 1–10. Retrieved from http://icaci.org/documents/ICC_proceedings/ICC2009/html/nonref/6_10.pdf

Spracklen, D. V., Logan, J. A., Mickley, L. J., Park, R. J., Yevich, R., Westerling, A. L., & Jaffe, D. A. (2007). Wildfires drive interannual variability of organic carbon aerosol in the western U.S. in summer. *Geophysical Research Letters, 34*(16), 2–5. https://doi.org/10.1029/2007GL030037

Taylor, A. H., Trouet, V., & Skinner, C. N. (2008). Climatic influences on fire regimes in montane forests of the southern Cascades, California, USA. *International Journal of Wildland Fire, 17*(1), 60–71. https://doi.org/10.1071/WF07033

Upadhyay, V. (2020). Second-worst in 16 yrs, fires destroy 2521 ha forest cover in Uttarakhand. *Times of India*. Retrieved from http://timesofindia.indiatimes.com/articleshow/69660263.cms?utm_source=contentofinterest&utm_medium=text&utm_campaign=cppst_prime

Weise, D. R., & Biging, G. S. (1997). A qualitative comparison of fire spread models incorporating wind and slope effects. *Forest Science, 43*(2), 170–180.

Westerling, A. L., Hidalgo, H. G., Cayan, D. R., & Swetnam, T. W. (2006). Warming and earlier spring increase Western U.S. forest wildfire activity. *Science, 313*(5789), 940–943. https://doi.org/10.1126/science.1128834

Chapter 9
Impact of Land Use/Land Cover Changes on Climate Change Parameters

Ratnakar Swain

Abstract Urbanization and industrialization are responsible for a variety of environmental issues like air pollution, increased runoff and subsequent flooding, increase in temperature, and deterioration of water quality. It is evident that for environment management and decision-making process especially in climate impact analysis, the study of land use/land cover (LULC) plays a vital role. The aim of this study is analyzing the climate change response due to land use/land cover (LULC) changes in the eastern India's Brahmani River Basin that experienced a fast increase in industrialization and deforestation in the recent decades. Herein, the Landsat satellite images were collected from the United States Geological Survey (USGS) from year 1975 to 2018 and processed in ERDAS Imagine software. The whole LULC mapping involves (1) geo-referencing, (2) mosaicking, (3) sub-setting on the basis of Area of Interest (AOI), (4) development of signature files, and (5) classification. The supervised classification method is followed herein to classify the study area with delineated classes such as water bodies, sand, barren/crop land, forest area, and built-up area. The study reveals that the major land cover in the study area is dense forest which decreases from 71.70% to 14.85% from year 1975 to 2018. The second major category of land is barren/crop land, which was increased by 30% due to development in agricultural technology, irrigation facilities. The third category of land cover is built-up area which increases by 32.73% from year 1975 to 2018 due to man-made activities. The sand comes under the fourth category which has slightly increased 6.57% to 7.05%. The least area covered by water bodies which is the fifth category of land cover was 0.33% in the year 1975 which increased to 1.65% in the year 1999 due to the construction of Rengali Dam and Samal Barrage but subsequently decreased to 0.87% in 2018. Hence, it is verified that the industrialization and development activities cause heavy deforestation in the catchment. It is also proved that the remote sensing satellite data can be efficiently used for the spatio-temporal changes in LULC in real time.

R. Swain (✉)
Department of Civil Engineering, National Institute of Technology, Rourkela, Odisha, India
e-mail: swainrk@nitrkl.ac.in

P. K. Shit et al. (eds.), *Geospatial Technology for Environmental Hazards*, Advances in Geographic Information Science, https://doi.org/10.1007/978-3-030-75197-5_9

Keywords LULC · ERDAS · Remote sensing · Brahmani River Basin · Landsat satellite

9.1 Introduction

Land use and land cover (LULC) analysis plays a crucial role in studying climate change parameters. As per the US National Research Council (NRC, 2005), LULC should be studied for the climate change analysis. Due to changes in vegetation, urbanization, and agriculture land, the greenhouse gases, CO_2 emission, CO_2 absorption, and other atmospheric changes occur (NRC, 2005). There is a huge impact of LULC changes on climatic conditions like temperature, precipitation, humidity, etc. in global and regional scale. NRC report also suggests that changes in landscape and human settlement have greater influence on ecological cycle and energy cycle which has link with climatic response. Different organization like International Geosphere-Biosphere Programme (IGBP) and the Global Energy and Water Cycle Experiment (GEWEX) also gives importance to LULC on climate change studies. In literature, it is proved that the deforestation and urbanization are mainly responsible for changes in daily peak rainfall and trends of rainfall (Kishtawal et al., 2010; Petchprayoon et al., 2010; Swain & Sahoo, 2015; Swain et al., 2018).

Petchprayoon et al. (2010) described changes in stream flow in the Yom River Basin due to change LULC particularly urbanization. Similarly, Sertel et al. (2010) verify the impact of land cover changes on climate parameters like temperature. Moreover, the changes in the island of Indonesia were studied by Tokairin et al. (2010) using meso-scale model and found that the human settlement and deforestation are prime reasons of the climate change. Also, Costa and Pires (2010) studied the tropical forest region of South America and concluded that the deforestation over the period of time changes the rainfall pattern and quantity. Lawrence and Chase (2010) studied the simulation of global climate with existing and maximum possible vegetation without any that anthropogenic LULC changes. In the southern region of the USA, agricultural especially the winter crop like wheat is the main responsible for the change in regional surface temperature as stated by Ge (2010). The effect of Three Gorges Dam is also being analyzed to study the changes in hydro-meteorological characteristics of the basin (Xiao et al., 2010).

From India's perspective, various studies have been conducted in different catchments of India to analyze the impact of LULC changes on climate change parameters. Hengade and Eldho (2016) studied both climate and LULC changes using VIC model in Ashti River Basin of India. Similarly, Khan et al. (2019) conducted a study on peninsular river basin in India to estimate future scenario of flow and sediment load under changes in climate and LULC. Moreover, the impact of LULC change on streamflow is assessed by Sinha et al. (2020a, 2020b). Sinha et al. (2020a, 2020b) used hydro-meteorological data of river basin of Western Ghat, India, and found there is a significant impact of both climate and LULC on streamflow in river. Impact of LULC and climate changes on river streamflow were also analyzed by Chanapathi and Thatikonda (2020) in Krishna River Basin, India. Chanapathi and Thatikonda (2020) and Garg et al. (2017) prove that the present and future scenario of LULC changes plays a major role in changing the river streamflow. With the streamflow characteristics, the sediment yield is directly affected by the change in climate and LULC in Netravati River Basin, Western Ghats, India, as shown by Sinha et al. (2020a, 2020b) and Sinha and Eldho (2018). Dynamics of hydrometeorological parameters such as evapotranspiration, surface runoff base flow and infiltration also is altered by the change in LULC in eastern India' river basin (Das et al., 2018). Furthermore, impact of LULC changes on streamflow response was studied in Tons River Basin, India, by Kumar et al. (2018), and in Nethravathi River Basin, India, by Babar and Ramesh (2015). A similar study is conducted in Tapi River Basin to analyze the streamflow as well as sediment yield. Nilawar and Waikar (2019) found serious impacts of climate change and LULC on streamflow and sediment concentration Purna River Basin, India.

Many studies have been conducted by several authors across the globe and concluded that the disturbance in the LULC highly impacts the regional climate condition (Takahashi et al., 2010; Xiao et al., 2010; Strengers et al., 2010; Lawrence & Chase, 2010; Swain & Sahoo, 2015). Herein, the Brahmani River Basin of Odisha state of Eastern India is chosen due to various anthropogenic activities over the last 4–5 decades. Especially after the construction of Rengali Dam in the Brahmani River, there are huge agricultural activities which disturb the LULC of the basin. Moreover, the vast industrial setup in the basin is also the prime reason of human settlement which is responsible for deforestation and vegetation change (Swain & Sahoo, 2017a, b). In this study, the LULC changes have been studied over the last four decades. Impact of the LULC changes is verified with the change in climate change parameters like extreme temperature, extreme daily rainfall pattern, and extreme relative humidity. In the last, the peak streamflow at Jenapur a downstream gauging station in the Brahmani River is analyzed.

9.2 Methodology

Multispectral LULC classification is widely used in planning and management of the forest area and urban and rural area over the long period especially after development of various satellites (Hord, 1982). Nowadays due to development of different remote sensing satellites of various spatio-temporal resolutions, LULC mapping becomes more informative and also cost effective. Remote sensing techniques play a vital role in predicting LULC changes accurately (Comber et al., 2005; Townshend, 1992).

Recently, satellite images are widely used for the large-scale studies like LULC, landscape mapping, hydrological studies, etc. Herein, the freely available Landsat satellite images are used for the LULC mapping due to its 30 m × 16-day spatio-temporal resolution. All the Landsat-6, Landsat-7 TM, and Landsat-8 ETM as per the availability over the time period and study area are used for making the LULC

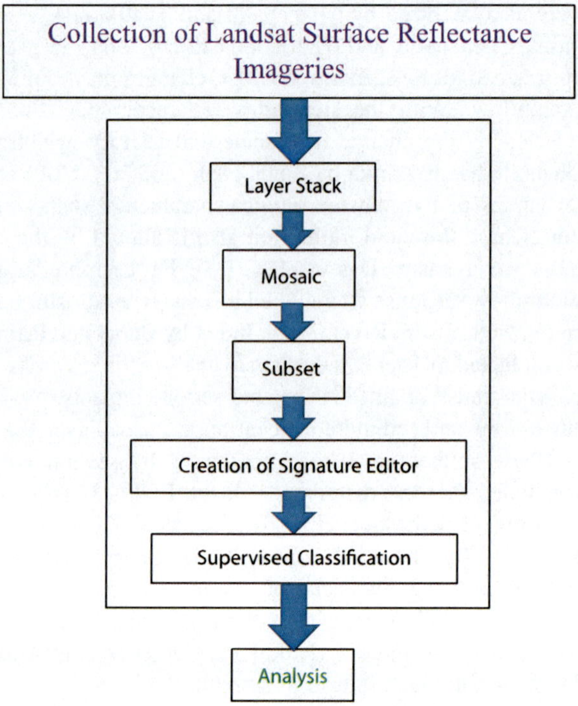

Fig. 9.1 Procedure of supervised LULC classification

mapping. ERDAS IMAGINE software is used for geo-referencing, mosaicking, and sub-setting of the images on the basis of area of interest (AOI). The LULC classification is carried out using supervised image classification algorithm. In this algorithm, firstly, pixel signature is assigned and then five types of landscape are provided based upon the digital number (DN) (Fig. 9.1). The delineated classes were crop land, sand, water bodies, built-up areas, and deciduous broad leaf forest which are identified in the study area. Maximum likelihood method of supervised classification is flowed here in which the classification is done based upon the training sets provided and field knowledge. The procedure to carryout supervised classification using ERDAS IMAGINE is described in Fig. 9.1. Herein, five types of signature are provided such as (1) barren/crop land, (2) sand, (3) water bodies, (4) built-up areas, and (5) dense forest.

After LULC mapping, the important climate change parameters such as extreme precipitation, maximum temperature, peak relative humidity, and maximum streamflow are taken into consideration for the analysis. Over the period of LULC study, all these hydro-meteorological data are plotted in time-series, and peak values are used in the analysis. The nonlinear regression-based trend analysis of the climate parameter is carried out to study the impact of LULC. The extreme values of climate parameters such as annual maximum temperature, maximum daily rainfall, and annual maximum relative humidity are taken into trend analysis. The daily average temperature and relative humidity over the study period are also considered to find any regional change in pattern. Since relative humidity and temperature have a vital role in changing precipitation, the daily peak over the years is added in the analysis. LULC also plays major role in change in infiltration and surface runoff. Therefore, the change in peak streamflow over the study period at the basin outlet is also considered in the analysis. It is evident that the industrial, domestic, and agricultural demand also affect streamflow characteristics along with LULC change. Therefore, it is difficult to relate streamflow with LULC change.

9.3 Study Area and Data Collection

Herein, the Brahmani River Basin, the second largest river of Odisha, is considered for the study due to vast industrialization and deforestation of its catchment. The Brahmani River is formed by the confluence of the Sankh and South Koel rivers, near the major industrial town of Rourkela at 22 15′N and 84 47′. After flowing through the districts such as Cuttack, Angul, Jajpur, Kendrapara, and Sundargarh, the Brahamani River forms a delta with the Baitarani River near Bay of Bengal. The total length of the rivers including its constituent streams is about 799 km of which 541 kilometers are in Orissa. Catchment area of the basin is about 39,033 km^2 (15,071 mi^2) in Odisha alone. The Brahmani River Basin is located in northern latitude of 20°28′ to 23°35′ and east longitude of 83°52′ to 87°03′. In the Brahmani River, there are many hydraulic structures constructed in the last 4–5 decades. Among the major structures are Rengali Dam and Samal Barrage located near Angul and Talcher town of the basin, respectively. The mean annual rainfall of the basin is around 1460 mm. 70–80% of rainfall occur only during southwest monsoon in the period of June–October. In the Bay of Bengal, low pressure and cyclonic precipitation is common throughout the year. The summer climate is hot, and gradually the peak temperature increases every year. The winter climate is medium cold, and temperature ranges from 10 to 15 °C. The peak temperature during summer varies from 38 to 50 °C (Figs. 9.2 and 9.3).

The Landsat satellite images of the study area were collected from USGS Earth Explore website (https://earthexplorer.usgs.gov/). The 33 years (1979–2012) daily stage and discharge data available at Jenapur gauging stations were collected from the Water Resources Information System (WRIS), India. The 0.50 lat-long gridded daily meteorological data for the study area for the same period were also collected from the Climate Forecast System Reanalysis (CSFR), India. The daily scale meteorological variables of maximum temperature, minimum temperature, rainfall, and relative humidity were downloaded for the study period from 1979 to 2014 (https://climatedataguide.ucar.edu/climate-data/climate-forecast-system-reanalysis-cfsr). To get the average LULC of the study area representing the study period, the available cloud-free Landsat images of November 19, 1975, and January 9, 2018, were downloaded from http://www.earthexplorer.usgs.gov. The various images collected are mentioned in Table 9.1.

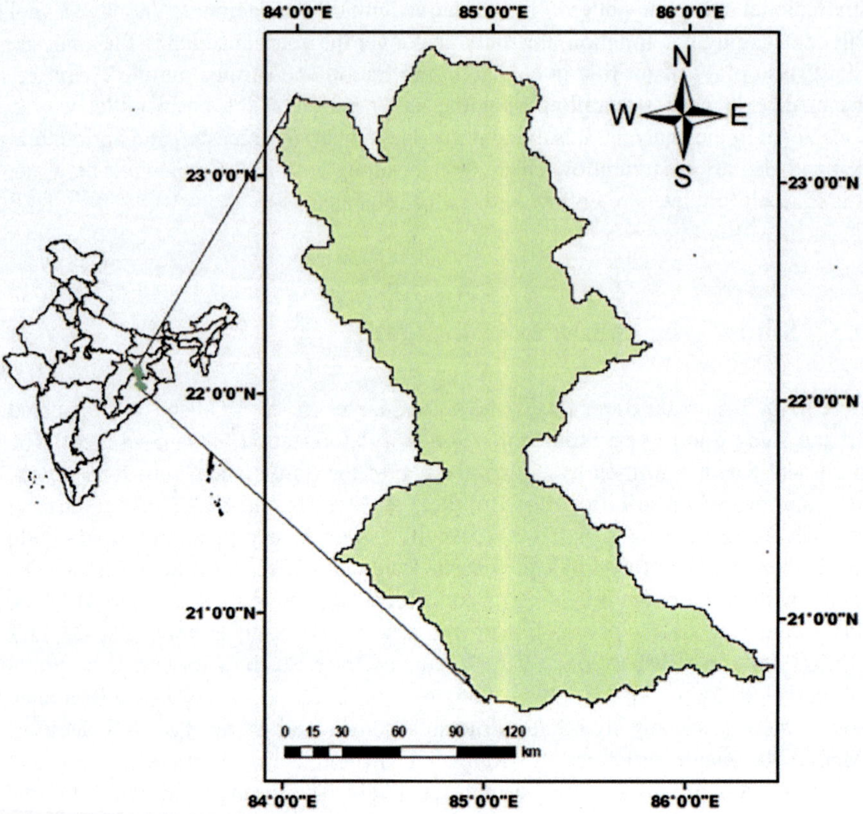

Fig. 9.2 Index map of Brahmani River Basin

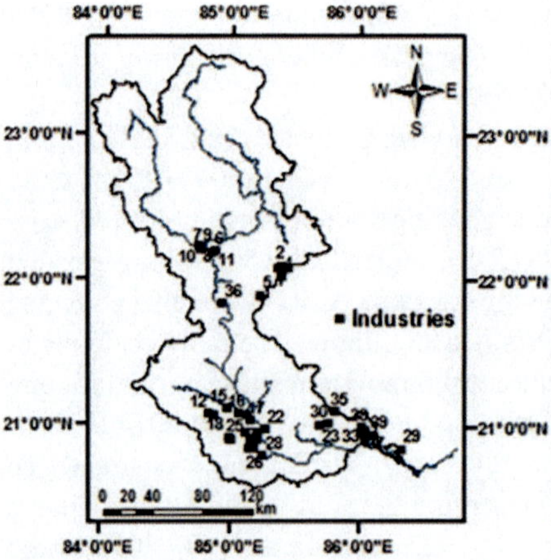

1- I K I W L	20- I.W.S.S Mel
2- Essel Mining and Industries Ltd	21- Bharatpur Coal Mines
3- JSPL Pellet Plant	22- Rana Sponge Iron Ltd.
4- TISCO Plant, Joda	23- Sukinda Chromite Mine
5- Essel Mining & Industries Ltd	24- Talcher Thermal Power Corporation
6- Slag Dumping Area	25- Jindal steel & Power ltd
7- Poly Refractories Pvt Ltd	26- NALCO
8- Britannia Industries Ltd	27- NALCO CPP
9- Utkal Metalics Ltd	28- Bhusan Steel
10- Ifgl Refractories Limited	29- Ferochrome plant
11- RSPL	30- KVK Nilachal Power Pvt. Ltd
12- Machhakuta Coal Mine	31- Kathpal Chromite Mine
13- MCL	32- Jindal steel
14- NTPC Kanhia	33- TATA steel
15- Jindal India Thermal Power Limited	34- Maithan ispat ltd
16- OPCL 25 MW Power Plant	35- IFCAL
17- Bindal Sponge Ltd	36- Daitari Ore Mines
18- Monnet Power Plant	37- B.R Iron Sponge Plant
19- Fertilizer Corporation of India Ltd.	

Fig. 9.3 Location of major industries over Brahmani River Basin (Swain & Sahoo, 2017b)

Table 9.1 Specification of Landsat surface reflectance imageries used

Data	Period	Date
Landsat-MSS	1970–1980	November 19, 1975
Landsat-TM	1980–1990	February 8, 1988
Landsat-ETM	1990–2000	November 29, 1999
Landsat-2005	2000–2005	November 5, 2005
Landsat 8	2015–2020	January 9, 2018

9.4 Result and Discussion

LULC Classification

The supervised classification of LULC of the Brahmani River Basin over the years 1975–2018 are shown in Fig. 9.4a–c. The maroon color represents the built-up area, dark green color represents the forest area, blue color shows the water bodies, yellow color shows the crop land, and white shows the sand. The areas under each classification were calculated in km^2 and expressed in percentage (%). The areas under water bodies, sand, barren/crop land, forest area, and built-up area were subsequently entered in the table along with their respective dates. Classified images depict spatial distributional pattern of land use/cover of the Bramhani Basin for the year 1975–2018.

These data reveal that in 1975, about 71.70% (29,517 km^2) of area were under forest, 27.41% (11,284 km^2) under crop/barren land, 00.00% (0 km^2) under built-up land and 0.33% (134.34 km^2) under water body, and 0.56% (230.30 km^2) under sand. During 1988, the area under these land categories was found about 14.85% (22,369 km^2) forest, 34.7% (14,273 km^2) crop/barren land, 7.13% (2933 km^2) sand, 2.69% (1105.29 km^2) built-up land, and 1.11% (457.63 km^2) water body, whereas in 1999, it was found that about 39.39% (16,205 km^2) belongs to forest, 54.79% (22,538 km^2) crop/barren land, 2.87% (1182 km^2) sand, 1.3% (532.96 km^2) built-up land, and 1.65% (679 km^2) water body. During 2005 the area of deep forest was 29.18% (10,836 km^2), the area of crop/barren land 62.06% (23,044 km^2), the area of sand 0.12% (45.89 km^2), area of built-up land 7.49% (8454 km^2), and 1.13% (562.49 km^2) under water body. During 2018, the area under these land categories was found about 54.38% (5514 km^2) under forest, 51.46% (19,107 km^2) under crop/barren land, 0.08% (31.40 km^2) sand, 32.73% (12,153 km^2) under built-up land, and 0.87% (322.46 km^2) under water body.

Analysis of Change in LULC

The LULC change from year 1975 to 2018 is shown in Figs. 9.5 and 9.6. Figure 9.1 shows that the area of dense forest was high in 1975, i.e., 71.70%; eventually it was decreasing linearly to only 14.85% at the current time period. So, there is a huge loss of vegetation, and ultimately it has a great impact on the climate and rainfall. Also, it

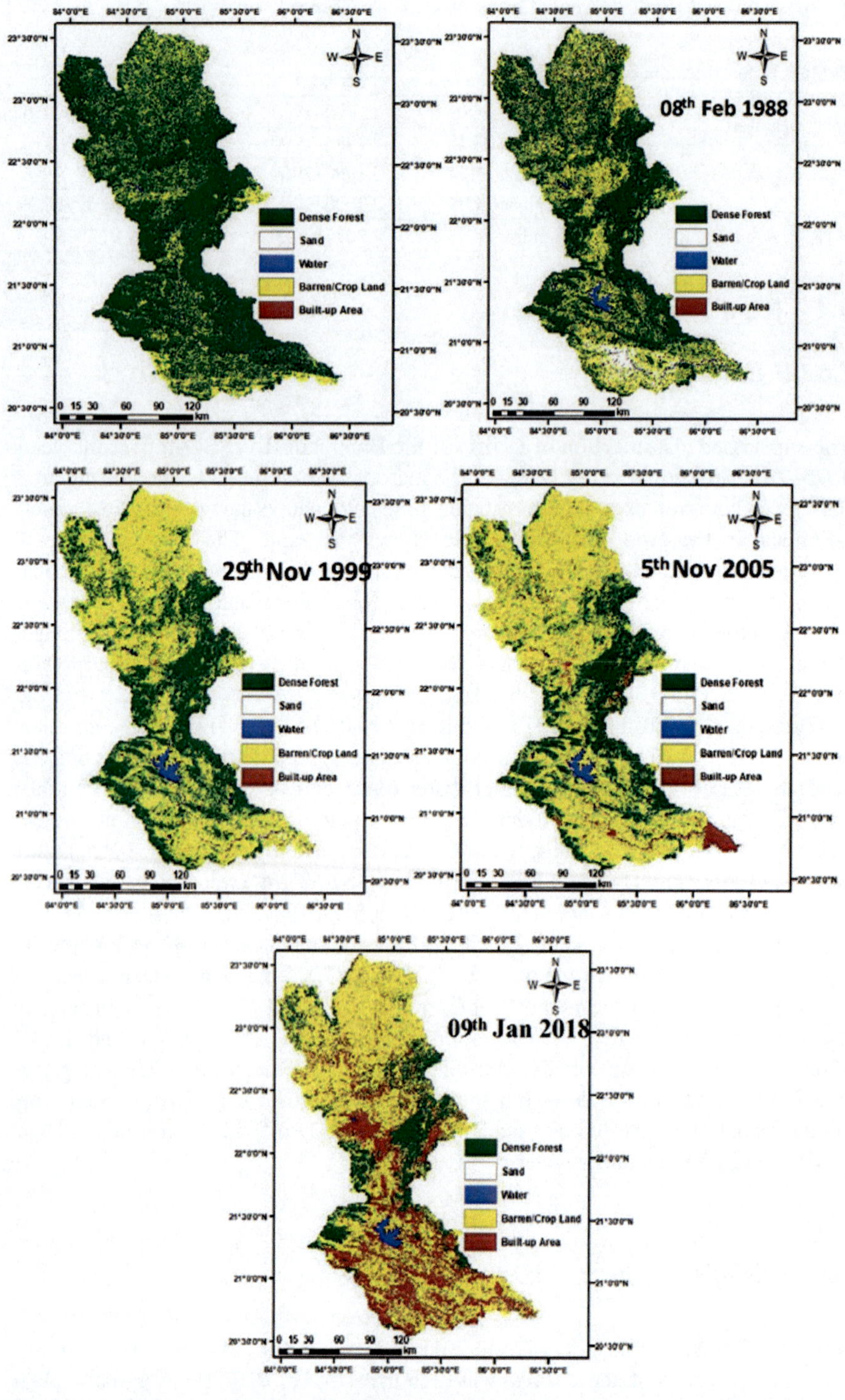

Fig. 9.4 (**a**) LULC classification of the Brahmani River Basin of dated November 19, 1975, and February 8, 1988. (**b**) LULC classification of the Brahmani River Basin of dated November 29, 1999, and November 05, 2005. (**c**) LULC classification of the Brahmani River Basin of dated January 9, 2018

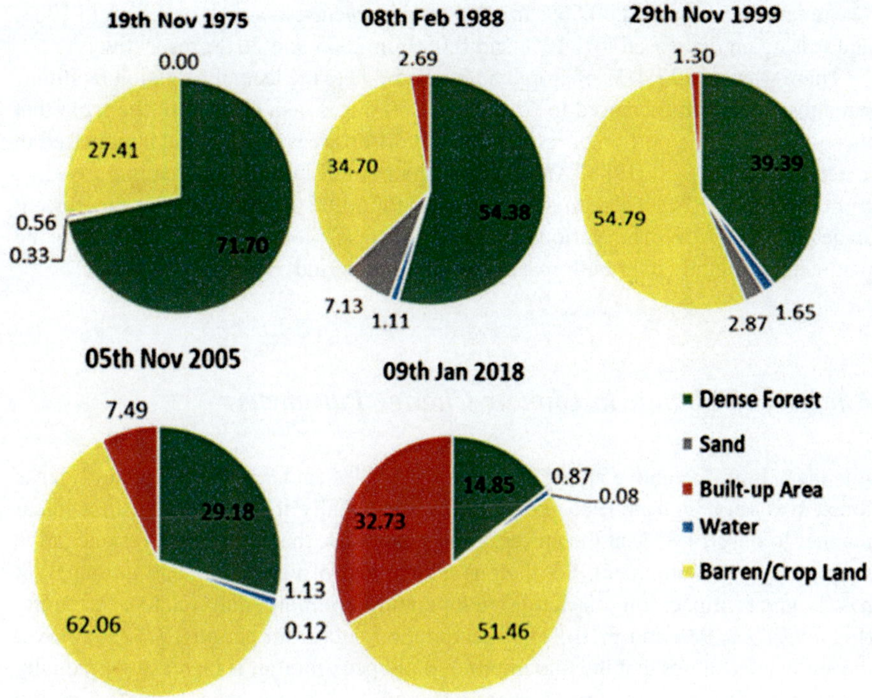

Fig. 9.5 Representation of LULC changes of Brahmani River Basin (%)

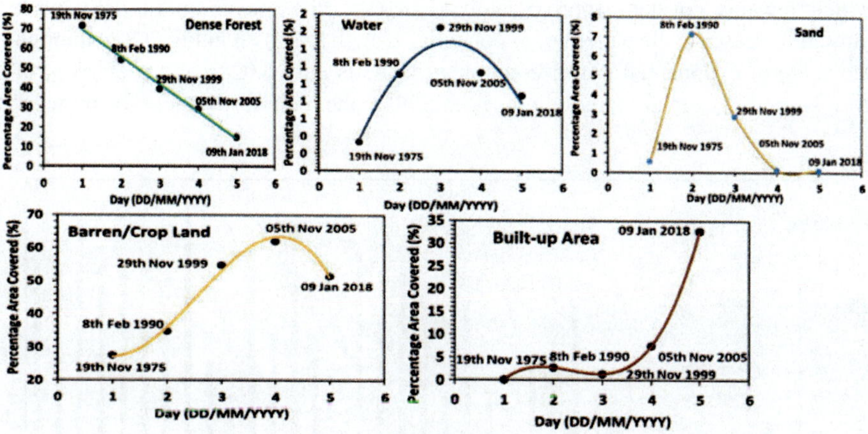

Fig. 9.6 Graphical representation of change in different land cover from 1975 to 2018

is clear from the table that in 1975, the water bodies are only 0.33% of the total area, but it is somehow increased to 1.65% in 1999 due to the construction of Rengali Dam and Samal Barrage. Then it is again decreased to 0.87% in current time period.

The area of sand was only 0.56% in 1975, and it reached to a pick of 7.13% in 1988, and it is again decreased to 1.13% and 0.08% in 2005 and 2018, respectively.

There was only 27.41% of crop land in 1975 due to the lack of irrigation facilities, but subsequently it increased to 62.06% in 2005. It is also clear from the table that there is a negligible portion of residential or built-up areas in 1975, but it increased in a steadily to 2.69% in 1988. And there is again a fall in built-up area in 1999, i.e., from 2.69 to 1.30%. Thereafter it increased to 7.49% in 2005, and at the current stage, it is 32.73%. The various graphs were also plotted for the change in the parameters of land cover with respect to the time period.

Analysis of Change in Climate Change Parameters

It is clear from the above analysis and from Figs 9.5 and 9.6 that the area of dense forest was very high in 1975, i.e., 71.70%; eventually it is decreasing in a linear manner to only 14.85% at the current time period. So, this indicates a very negative impact on our environment. So, there is a huge loss of vegetation, and ultimately it has a great impact on discharge, temperature, rainfall, and relative humidity (Figs. 9.7, 9.8, 9.9, and 9.10). The precipitation data from period 1979–2014 over the study area shows that the maximum of daily peak rainfall is increasing gradually (Fig. 9.7). The average first 5-year data and last 5-year data from the period 1979–2014 shows that there is 45% increase in peak rainfall. However, the average annual rainfall over the Brahmani River Basin is 1460 mm (Swain & Sahoo, 2015) which remains constant approximately. However, the maximum of daily peak rainfall increases in the pattern of exponential growth as shown in Fig. 9.7. Although the value of exponential power component value is only 0.0286 and coefficient of correlation (R^2), small increase in peak rainfall is the main responsible for frequent

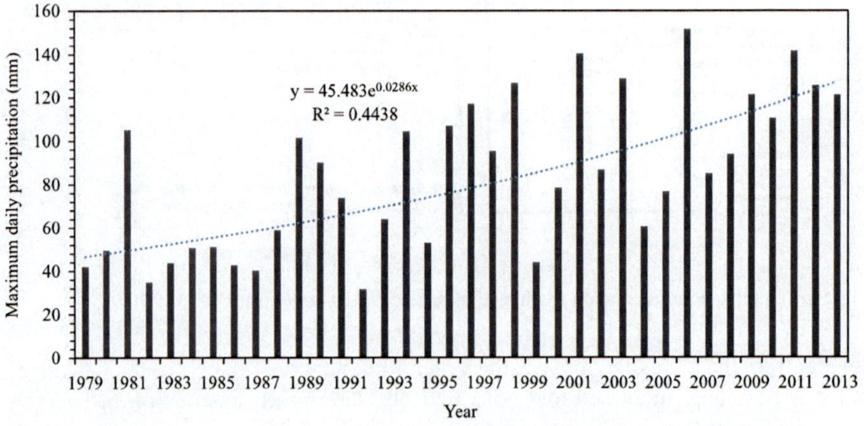

Fig. 9.7 Change in maximum daily maximum rainfall (mm) over the years 1979–2014

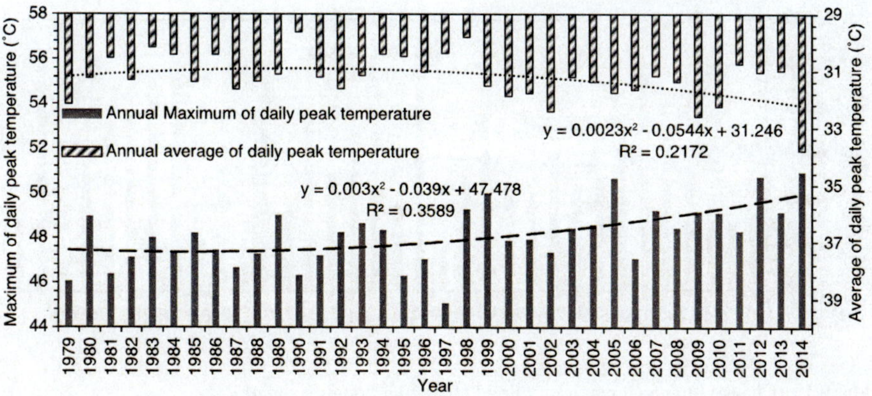

Fig. 9.8 Change in maximum and average of daily peak temperature (°C) over the years 1979–2014

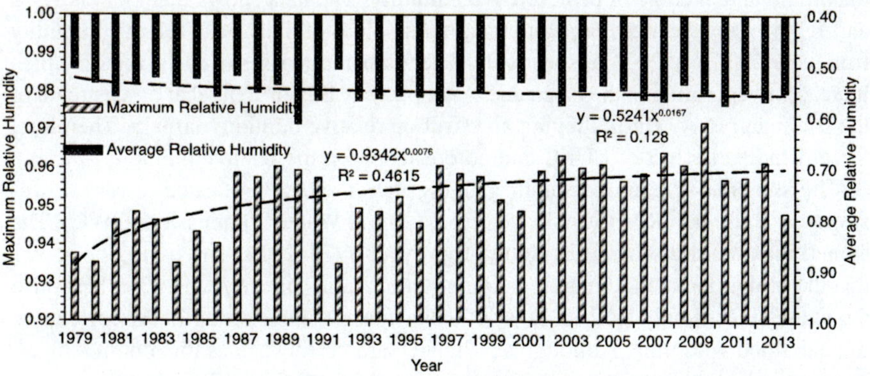

Fig. 9.9 Change in maximum and average of daily peak relative humidity (%) over the years 1979–2014

flood in the whole region. Rainfall water couldn't get sufficient percolation time for the infiltration and so surface runoff increases due high intensity of rainfall which causes sudden flood in the basin and downstream of the basin.

The maximum and average of daily peak temperature data is plotted as shown in Fig. 9.8. The peak temperature is one of the main parameters of climate change. Both the trend lines show that the maximum and average of peak temperature increase with nonlinear temporal pattern as shown in Fig. 9.8. Here also it is observed that the values of coefficients of the regression equation and R^2 are very small. However, the increase in peak temperature is a matter of concern although it is gradual. The data shows that there is 8.6% and 4.6% increase in maximum and average of daily peak temperature respectively from year 1979 to year 2014. This increase in temperature can be correlated with deforestation and hence with LULC in the local and global

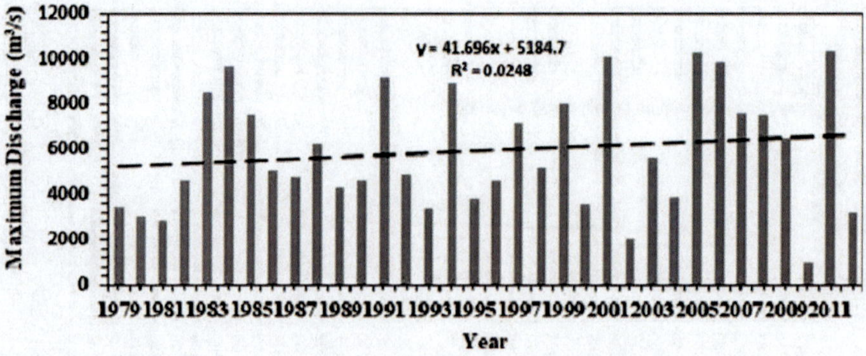

Fig. 9.10 Change in annual maximum flood at Jenapur gauging station

scale. Similarly, the maximum and average of daily peak relative humidity is plotted as shown in Fig. 9.9. Here also similar trend of temporal change is found in both maximum and average of peak relative humidity. The data shows that there is 9.9% and 1.14% increase in average and maximum values of daily peak relative humidity from year 21,978 to 2014, respectively. The temperature rises over the region which increases evaporation in the basin and water body of Bay of Bengal at downstream of the Brahmani River Basin, having an effect on relative humidity largely. Therefore, there is indirect effect of LULC and deforestation on the relative humidity.

The streamflow data of Jenapur gauging station which is located toward downstream of Brahmani River is collected from Central Water Commission (CWC). The annual maximum discharge is plotted from year 1978–2014. The trend line shows that the peak flood at the Jenapur gauging station varies linearly with time. Here also the value of R^2 is only 0.0248. Low R^2 value doesn't show strong linear correlation of peak flood with time. Although all climate parameters such as maximum temperature, relative humidity, and precipitation changes with the LULC change over the region in local as well as global scale, the streamflow data depends indirectly on change in LULC. It is evident that in a basin, the change in streamflow at any gauging station depends upon many factors such as LULC change, precipitation, soil type, water demand, etc. Due to increase in agricultural and industrial growth along with human settlement in the Brahmani River Basin, the demand of water also increased drastically. Therefore, the peak flood data is likely to be biased if only the effect of LULC change is considered. Therefore, peak flood data at Jenapur station is considered to be not following any trend line. Although LULC change have major impact on the peak flood, other important factor like water demand needs to be taken into consideration in the study.

9.5 Conclusion

The LULC study reveals that there are major changes in the LULC over the years 1975–2018 in the study area especially dense forest cover. The area under dense forest decreased by 57% due to an increase in urbanization and agriculture. The second major category of land in the study area is barren/crop land, which was increased by 30% due to development in agricultural technology, irrigation, transportation facilities, etc. The third major category of land in the study area is built-up area which is increasing by 32.73%. During the study period (i.e., 1975–2018), built-up area has been increased due to advancement of communication, transportation, and agricultural activities. The area under fourth category of land, i.e., the sand, has increased at the first stage by 6.57%, and in the later stage it is decreased by 7.05%. Hence, the present study proves that remote sensing technology can be efficiently used for the spatio-temporal changes in catchment cover no cost and in real time.

From this study, we clearly found the decrease in about 56.85% of forest land in just 36 years and its serious impact on climate, temperature, rainfall, relative humidity, and discharge by analyzing the previous 36 years (1979–2014) data. So, we have to stop the rapid deforestation and to increase afforestation to reduce the unwanted weather change and hence to reduce global warming and other calamities. The climate parameters maximum and average of daily peak temperature and relative humidity data from year 1978–2014 shows an increasing trend. Similarly, maximum of daily peak rainfall data over the study area shows increasing trend. In this study, it is proved that the change in LULC and deforestation have direct impact on these climate parameters. The streamflow data is also studied at Jenapur gauging station and the change in peak flood also is observed. However, the change in peak flood is also affected by change in agricultural, industrial, and municipal water demand. Therefore, due to all these different factors, the trend in peak flood could not be defined.

However, the LULC change is not only responsible for change in climate change parameters. The climate change parameters also affected by global climate change which is due to large-scale pollution and deforestation globally as well as locally. This study is carried out on a regional scale. Therefore, there is a need to analyze the impact of global LULC changes along with pollution on the climatic parameters which can be addressed in future study. Additionally, the changes in river streamflow are largely affected by various local demands such as municipal, industrial, and agricultural demand. Therefore, these demand factors should take into consideration along with LULC to detect the impact on streamflow. These research gaps can be addressed in a future study.

Acknowledgments The authors would like to thank the USGS research team for allowing us to download the surface reflectance products of Landsat from their website http://www.earthexplorer.usgs.gov.

References

Babar, S., & Ramesh, H. (2015). Streamflow response to land use-land cover change over the Nethravathi River Basin, India. *Journal of Hydrologic Engineering, 20*(10), 05015002.

Chanapathi, T., & Thatikonda, S. (2020). Investigating the impact of climate and land-use land cover changes on hydrological predictions over the Krishna river basin under present and future scenarios. *Science of the Total Environment, 721*, 137736.

Comber, A., Fisher, P., & Wadsworth, R. (2005). What is land cover? *Environment and Planning B: Planning and Design, 32*(2), 199–209.

Costa, M. H., & Pires, G. F. (2010). Effects of Amazon and Central Brazil deforestation scenarios on the duration of the dry season in the arc of deforestation. *International Journal of Climatology, 30*, 1970–1979.

Das, P., Behera, M. D., Patidar, N., Sahoo, B., Tripathi, P., Behera, P. R., et al. (2018). Impact of LULC change on the runoff, base flow and evapotranspiration dynamics in eastern Indian river basins during 1985–2005 using variable infiltration capacity approach. *Journal of Earth System Science, 127*(2), 1–19.

Garg, V., Aggarwal, S. P., Gupta, P. K., Nikam, B. R., Thakur, P. K., Srivastav, S. K., & Kumar, A. S. (2017). Assessment of land use land cover change impact on hydrological regime of a basin. *Environmental Earth Sciences, 76*(18), 1–17.

Ge, J. (2010). MODIS observed impacts of intensive agriculture on surface temperature in the southern Great Plains. *International Journal of Climatology, 30*, 1994–2003.

Hengade, N., & Eldho, T. I. (2016). Assessment of LULC and climate change on the hydrology of Ashti Catchment, India using VIC model. *Journal of Earth System Science, 125*(8), 1623–1634.

Hord, R. M. (1982). *Digital image processing of remotely sensed data*. Elsevier.

Khan, M., Sharma, A., & Goyal, M. K. (2019). Assessment of future water provisioning and sediment load under climate and LULC change scenarios in a peninsular river basin, India. *Hydrological Sciences Journal, 64*(4), 405–419.

Kishtawal, C. M., Niyogi, D., Tewari, M., Pielke, R. A., Sr., & Shepherd, J. M. (2010). Urbanization signature in the observed heavy rainfall climatology over India. *International Journal of Climatology, 30*, 1908–1916.

Kumar, N., Singh, S. K., Singh, V. G., & Dzwairo, B. (2018). Investigation of impacts of land use/land cover change on water availability of Tons River Basin, Madhya Pradesh, India. *Modeling Earth Systems and Environment, 4*(1), 295–310.

Lawrence, P. J., & Chase, T. N. (2010). Investigating the climate impacts of global land cover change in the Community Climate System Model (CCSM). *International Journal of Climatology, 30*, 2066–2087.

National Research Council. (2005). *Population, land use, and environment: Research directions*. National Academies Press.

Nilawar, A. P., & Waikar, M. L. (2019). Impacts of climate change on streamflow and sediment concentration under RCP 4.5 and 8.5: A case study in Purna river basin, India. *Science of the Total Environment, 650*, 2685–2696.

Petchprayoon, P., Blanken, P. D., & Ekkawatpanit, C. (2010). Hydrological impacts of land use/land cover change in a large river basin in central-northern Thailand. *International Journal of Climatology, 30*, 1917–1930.

Sertel, E., Robock, A., & Ormeci, C. (2010). Impacts of land cover data quality on regional climate simulations. *International Journal of Climatology, 30*, 1942–1953.

Sinha, R. K., & Eldho, T. I. (2018). Effects of historical and projected land use/cover change on runoff and sediment yield in the Netravati river basin, Western Ghats, India. *Environmental Earth Sciences, 77*(3), 1–19.

Sinha, R. K., Eldho, T. I., & Subimal, G. (2020a). Assessing the impacts of land use/land cover and climate change on surface runoff of a humid tropical river basin in Western Ghats, India. *International Journal of River Basin Management*, 1–12.

Sinha, R. K., Eldho, T. I., & Subimal, G. (2020b). Assessing the impacts of land cover and climate on runoff and sediment yield of a river basin. *Hydrological Sciences Journal, 65*(12), 2097–2115.

Strengers, B., Müller, C., Schaeffer, M., Haarsma, R., Severijns, C., Gerten, D., Schaphoff, S., van den Houdt, R., & Oostenrijk, R. (2010). Assessing 20th century climate–vegetation feedbacks of land-use change and natural vegetation dynamics in a fully coupled vegetation–climate model. *International Journal of Climatology, 30*, 2055–2065.

Swain, R., & Sahoo, B. (2015). Variable parameter McCarthy-Muskingum flow transport model for compound channels accounting for distributed non-uniform lateral flow. *Journal of Hydrology, 530*, 698–715. https://doi.org/10.1016/j.jhydrol.2015.10.030

Swain, R., & Sahoo, B. (2017a). Mapping of heavy metal pollution in river water at daily time-scale using spatio-temporal fusion of MODIS-Aqua and Landsat satellite imageries. *Journal of Environment Management, 192*, 1–14. https://doi.org/10.1016/j.jenvman.2017.01.034

Swain, R., & Sahoo, B. (2017b). Improving river water quality monitoring using satellite data products and a genetic algorithm processing approach. *Sustainability of Water Quality and Ecology, 9–10*, 88–114. https://doi.org/10.1016/j.swaqe.2017.09.001

Swain, R., Sahoo, B., & Perumal, M. (2018). An embedded VPMM-AD model for riverine transient flow and non-reactive contaminant transports. *Journal of Hydrology, 563*(2018), 711–725. https://doi.org/10.1016/j.jhydrol.2018.06.025

Takahashi, H. G., Yoshikane, T., Hara, M., Takata, K., & Yasunari, T. (2010). High-resolution modelling of the potential impact of land-surface conditions on regional climate over Indochina associated with the diurnal precipitation cycle. *International Journal of Climatology, 30*, 2004–2020.

Tokairin, T., Sofyan, A., & Kitada, T. (2010). Effect of land use changes on local meteorological conditions in Jakarta, Indonesia: Toward the evaluation of the thermal environment of mega cities in Asia. *International Journal of Climatology, 30*, 1931–1941.

Townshend, J. R. (1992). *Improved global data for land applications. A proposal for a new high-resolution data set*. Report of the Land Cover Working Group of IGBP-DIS. Global Change Report (Sweden).

Xiao, C., Yu, R., & Fu, Y. (2010). Precipitation characteristics in the Three Gorges Dam vicinity. *International Journal of Climatology, 30*, 2021–2024.

Chapter 10
Application of Geospatial Technique in Analysis of Malaria Prevalence in an Endemic Area of Ranchi, India

Manoj Prasad Gandhi, Fayma Mushtaq, Afaan Gulzar Mantoo, and Mili Ghosh Nee Lala

Abstract The fatality and endemic nature of malaria is one of the prime causes of mortality and morbidity in the Tribal states of India. Despite many advanced control measures, India contributes up to 1000 deaths per year. This present study was conducted to determine the impact of topographic factors, socio-economic status, and climatic conditions on malaria prevalence in Ranchi district of Jharkhand state. The Landsat-derived indices including NDVI and NDWI have been used along with the Primary Health Care center's data to spatially assess the malaria incidences. Malaria hotspots have been identified by using the criteria developed by the NVBDCP, Ministry of Health and Family Welfare, India, based on Slide Positive Rate (SPR), Annual Parasite Incidence (API), and Annual Blood Examination Rate (ABER). The Bundu and Sonahatu blocks have been identified as the vulnerable areas for malaria, where topographical variation and socio-economic factors played an important role to aggravate the malaria incidences along with climatic factors. The year 2005 was found to be worst affected observing 17 deaths. Elevation range of 200–400 m was found as favorable for malaria occurrences. Rainfall and average maximum temperature were found to be positively correlated with API in monsoon months, whereas relative humidity was found to be positively correlated with SPR. This work illustrates the efficacy of geomatics to assess the spatio-temporal relationship among malaria host, agent, and environmental factors.

Keywords Malaria · Socio-economic · NDVI · NDWI · Jharkhand · Landsat

M. P. Gandhi
Department of Education, +2 Gandhi Smarak H/S, Chand, Bihar, India

F. Mushtaq (✉) · A. G. Mantoo
Department of Ecology, Environment and Remote Sensing, Government of Jammu and Kashmir, Srinagar, India

M. G. N. Lala
Department of Remote Sensing, Birla Institute of Technology, Mesra, Ranchi, Jharkhand, India

P. K. Shit et al. (eds.), *Geospatial Technology for Environmental Hazards*, Advances in Geographic Information Science, https://doi.org/10.1007/978-3-030-75197-5_10

10.1 Introduction

In the twenty-first century, the mosquito-borne malaria is an endemic throughout the tropical and subtropical regions of the world. Malaria results in enormous socio-economic deprivations and is a major cause of the grim scenario of poverty in India as well as in many developing countries (Dhiman, 2009; Gallup & Sachs, 2001; Sarkar et al., 2019). In addition to extermination of people, malaria creates a substantial economic burden globally, especially for poor people living in remote areas of the world (Amegah et al., 2013; Enayati & Hemingway, 2010; Rai et al., 2014; Sachs & Malaney, 2002; Sharma, 2003; Worrall et al., 2005). In the north-eastern India people reside under deprived health environmental conditions, which enhances the possibility to get affected by the malarial transmission (Dev et al., 2003). Several researchers have mentioned that there is a high incidence of malaria disease in the northeastern states of India (Dhiman et al., 2010; Nath et al., 2013; Rabha et al., 2011; Yadav et al., 2012). The World Health Organization (referred as "WHO" hereafter) reveals that malaria caused 584 thousand deaths throughout the world in the year 2013 (WHO, 2014). WHO reports indicate that approximately 300 million cases of acute malaria cases have been recorded every year. Despite the high level of control measures, this disease of antiquity remains prevalent.

A vector-borne disease (VBD) such as malaria generates an unacceptable and terrible health burden on developing nations (Ceccato et al., 2005; Ratmanov et al., 2013; Mutheneni et al., 2014). In Southeast Asia, India has the highest malaria burden (61%), followed by Myanmar and Indonesia (World Malaria Report, Fact Sheet, 2012). However, malaria is not homogenous in its distribution and has a major regional burden variation. Malaria is the major health problem faced by the tribal states of India including Assam (Yadav et al., 2014), Orissa (Daash et al., 2009), Jharkhand (Saxena et al., 2014), Madhya Pradesh (Bharti et al., 2020), and north-western Rajasthan (Kochar et al., 1997). The tribal population of India contributes nearly about 30% of total malaria cases with more than 60% of *Plasmodium falciparum*, and 50% deaths owing to malaria (Srivastava et al., 2009). National Vector Borne Disease Control Programme (NVBDCP) India indicated that the low socio-economic development, difficult terrains, and less developed infrastructure are the major attributes responsible for the malaria eruption in the country. The malaria control measures and efforts initiated in India have been presented in Fig. 10.1.

In the 1940s, DDT was introduced to control malaria epidemic in India. Malaria pervasiveness remained low in India throughout the 1960s and then experienced a slight revival in the 1970s (Cutler et al., 2010). Malaria prevalence or its vector proliferation is primly attributed to anthropogenic activities, immigration behavior, and performance of control measures in an endemic area. In the state of Jharkhand, studies conducted by the National Institute of Malaria Research (NIMR), India, indicated that *An. fluviatilis* has been found dichlorodiphenyl trichloroethylene (DDT) resistant/tolerant, whereas *An. culicifacies* shows resistant nature for DDT and malathion and the secondary vector *An. annularis* was found resistant to DDT, susceptible to deltamethrin, and tolerant to malathion. Jharkhand is one of the states

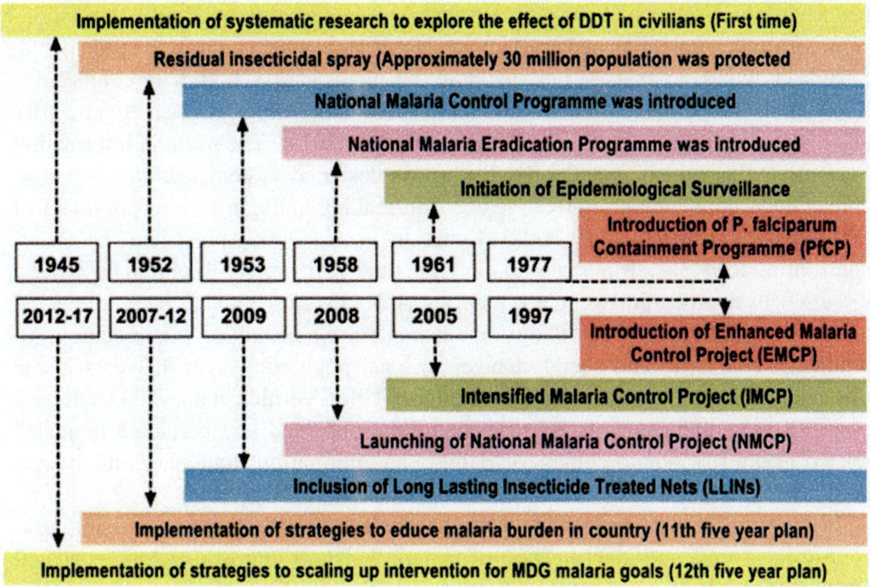

Fig. 10.1 Historical and a new future prospect related particularly to the malaria disease in India from the year 1945 to 2017

of India where regions with Annual Parasite Incidence (i.e., a malariomatric index to express malaria cases per thousand population) (API) >5 can be seen. *Anopheles fluviatilis*, *Anopheles culicifacies*, and *Anopheles annularis* are some of the malaria vectors which are more prevalent in the Ranchi district of Jharkhand, India (Saxena et al., 2012).

Multitudinous research work has been conducted in the tribal states of India (such as Haryana, Gujarat, Assam, Jharkhand, etc.) to identify the malarial hotspots (Bharti et al., 2020; Saxena et al., 2014; Srivastava et al., 1999; Srivastava et al., 2009; Yadav et al., 2014;). Sonitpur district in Assam (Nath et al., 2012, 2013), Gwalior city, Koraput district of Orissa (Daash et al., 2009), and Udalguri district in Assam (Yadav et al., 2012) are some of the studied regions of India. In order to explore malaria vector transmissions, the ecological and entomological studies have been performed very well (Ahmad et al., 2011; Smith et al., 2013); spatial modeling of vector-borne diseases (Machault et al., 2011) and effectiveness of the malaria control measures (Karema et al., 2012) are some of the important researches in the field of malaria preventions.

Globally immense progressions have been made in the direction of malaria eradication. Despite of the advance and fast intensification of malaria control measures, geographically half of the world remains at risk of contracting this disease (Gething et al., 2011). In this view the Geographical Information System (GIS) played an important role in better understanding of spatial distribution of malaria burden (Craig et al., 1999; Omumbo et al., 2004; Saxena et al., 2014). GIS provides better visualization of malaria prevalence (e.g., Kleinschmidt et al., 2000), relationship with other potential malaria factors (Hu et al., 1998; Hightower et al., 1998; Saxena et al., 2014), malaria risk mapping (Clements et al., 2006; Kirk et al., 2015; Snow et al., 1999), development of malaria surveillance index (Cohen et al., 2010), and malaria ecotoping (Kaewwaen & Bhumiratana, 2015). The Malaria Information Systems (MIS) which is a PC-based purpose-designed system was developed to collect the accurate data for effective decision-making in the malarious provinces of South Africa (Carrin et al., 2002). Louis Jr. et al. (2005) presented the issues confronting Roll Back Malaria with the influential Italian drive to eliminate malaria disease between the mid-twentieth centuries. A detailed research has been conducted to study the altering epidemiology about malaria by means of interpretations of political, ecological, and social changes in local population over 40 years in the Cascades region of southwestern Burkina Faso (Giles-Vernick et al., 2011). Oloukoi et al. (2014) conducted a study to examine the observed and predicted trends of related health risk due to climatic variability and mitigation strategies at the household level in Nigeria.

Presently, GIS has been widely used to assess the malaria transmission risk, control measures, and identification of malaria hotspots and to develop geospatial attributes for anti-malarial plants (Srivastava et al., 2004; Qayum et al., 2014, 2015). The surveillance of malaria vectors, outbreak investigations, and potential environmental factors and socio-economic variable can be mapped using a GIS frame. Global malaria risk distribution maps have been prepared using GIS that helps to combat with any calamity in any region of the world. It has been stated that malaria shows a positive correlation with proximity to wetland areas (Van der Hoek et al., 2003; Zhou et al., 2012). In the year 2004, Hay et al. (2004) presented a single global malaria risk distribution map using overlay operations in GIS.

The real burden of malaria in India is still not known, and this jeopardizes the desirable outcomes despite extensive planning and resource allocation for malaria eradication. The present work is an effort to identify the areas of malarial transmission risk in Ranchi based on defined risk categories related to the various climatic, topographic, and ecological indicators in a GIS environment. The study attempts to address suspected underreporting of malaria cases in the neglected region where the ecological malaria suitability is high, while case detection is low. The generated ecology-based geospatial malaria risk model should help the identification of hotspots and contribute to judicious planning and management of the malaria situation in all potentially endemic areas, which is of particular importance in the elimination phase. Knowledge of malaria and socio-economic upliftment are key factors in adopting the appropriate intervention strategies. Keeping in mind that there are many risk factors that influence vulnerability to malaria including proper knowledge about malaria transmission and prevention, demography, and socio-economic status of different population groups, the present study was undertaken to identify factors predisposing to malaria in highly endemic areas in Ranchi district. This work represents a GIS-based assessment of environmental and socio-economic drivers which facilitates to the malaria combating agencies for targeting the vector control measures in the Jharkhand state of India. Identification of malaria hotspot in this plateau region at the block and village level has been also attempted with the help of ancillary survey data. The findings will help to understand the patterns of malaria epidemiology in district Ranchi and would present an understanding how the malarial parasite is circulating constantly and which area requires more effective control measures to combat it.

10.2 Study Area

Ranchi, the capital of Jharkhand state, is situated in the southern part of Chotanagpur plateau, India, consisting an area of 7698 km^2 with a population of around 29,14,253. Ranchi district has 14 administrative blocks with an average literacy rate of 87.68%, and 13% of the total population contributes as below the age of 6 years. This district is the 46th largest city in India with geographical extent from 23°05' N and 84°52′ E to 23°24' N and 85°54′ E and 23°43' N and 84°58′ E to 23°23' N and 85°22′ E. The average elevation of the study area is 650 m and western region having a relatively higher elevation compared to eastern part. Ranchi has a rolling topography with approximately 159.14 ha of dense tropical forest. This area mainly lies in humid sub-tropical climate zone "Cfa," but unique topographical settings and dense forest provide a pleasant climate to this region. The "Subarnarekha" River and tributaries of "South Koel" are the life-lines for Ranchi region. In the summer temperature varies from 20 to 40 °C, whereas winter experiences a range of temperature from 1 to 25 °C. This region receives 80% of the total precipitation during monsoon, and the average annual rainfall of the district recorded as 56.34 inches. The state of Jharkhand is malaria endemic and attributed approximately 7% of the total cases of malaria disease in India. Apart from that, the district of Ranchi has been identified as a malaria vulnerable district in India. *Anopheles culicifacies*, *An. annularis*, and *An. fluviatilis* have been identified as the major vectors of malaria in Jharkhand state (Saxena et al., 2012). The "Bundu" PHC of the Ranchi district has been identified as the most malaria-affected area (Saxena et al., 2014). The lowlands of the district Ranchi have been found to be suitable for paddy cultivation, whereas the uplands provide conditions for orchards and cultivation of pulse, millet, and vegetables. The study area is shown in Fig. 10.2.

10.3 Methodological Framework

This section describes the collection of spatial and non-spatial data, details of study design, and image pre-processing steps undertaken to generate the spatio-temporal analysis of malaria prevalence in Ranchi district: first, the temporal design decided for identification of malaria prevalent regions; second, input satellite images have been described along with elevation data information; third, the strategy for ancillary data collection, discussions about criteria for finding malaria hotspot, and related

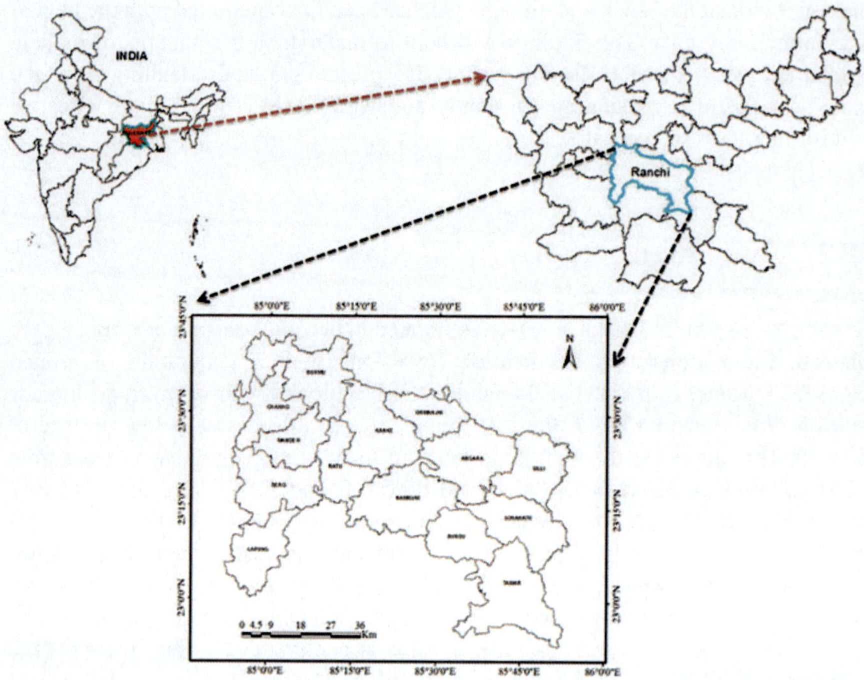

Fig. 10.2 Geographical location of the study area

calculations have been presented; and fourth, the thematic layer generation pertaining to the malaria hotspot surveillance in the Ranchi district. Lastly, vegetation condition assessment and Water Indexing have been performed to explore the environmental influences in malaria aggravation in the study area.

Study Design (Temporal) Strategy

The present study has been carried out for the year 2000–2010. Remote sensing and GIS have been used in the identification of malaria-affected regions in Ranchi district. The study has been designed as dividing the total study years into four different time periods (subgroups), i.e., T1-Study of Malaria Epidemic (SME) (2000–2002); T2-SME (2003–2005); T3-SME (2006–2008); and T4-SME (2008–2010).

Satellite Images, Elevation, and Ancillary Data Information

In GIS modeling framework a collection of well-defined secondary database is required to describe the relationship among various spatial and non-spatial attributes. In this view, malaria incidence data has been collected from the State Malaria Office, Ranchi, Jharkhand, for the given time periods. The Primary Health Centre (PHC) and Health Sub Centre (HSC) wise annual statistics regarding malaria disease data has been obtained for the year 2000–2010. In the present work Landsat TM image of November 2005 has been used. The satellite image which has been used in analysis was converted to the reflectance units. The Survey of India Toposheet of 1:50,000 and 1:250,000 scales has been considered for the base map generation. Elevation information has been extracted by using ASTER Digital elevation model data. Rainfall, temperature, and humidity records have been utilized to explore the relationship between occurrence of malaria incidences and environmental influence, especially pertaining to the climate convenience of malaria disease. The present work basically aimed to explore the influence of socio-demographic and environmental characteristics over the malaria incidences in Ranchi district. In this connection block wise demographical data (i.e., status of families below poverty line (BPL), district literacy rate, and population density) was taken into consideration.

Malaria Hotspots Identification Criteria

An ample number of malaria indicators have been defined by the World Health Organization. The main problem is to select the appropriate "Core Indicators" that can provide an accurate information about targeted population by a specific vector-borne disease. In the present study, criteria developed by the NVBDCP, Ministry of Health and Family Welfare, India, have been used to identify the highly malaria-prone areas in Ranchi region. The criteria are as follows:

1. Number of deaths owing to the malaria disease (based on clinical diagnosis or microscopic confirmation).
2. Annual Parasite Incidence (API) must be 5 or >5 during the last 3 years.
3. Annual Blood Examination Rate (ABER) must be 10% or higher during any of the years from the last 3 years.
4. Malaria case reports contribute to the proportion of *P. falciparum* as 30% or higher, provided the SPR = 3% or higher during any of the years from the last 3 years.

Calculation of Core Indicators

The prime objective of surveillance is to identify alteration in trend or distribution in malaria diseases. It is the basis of vector-borne disease control measure strategy. There are several indicators for malaria surveillance which have been proposed, i.e., Annual Falciparum Incidence (AFI), Annual Malaria Incidence (AMI), Malaria Death Rate (MDR), Crude Death Rate (CDR), and Under-Five Mortality. In spite of that some standard principal indicators have also been documented for malaria surveillance, i.e., Annual Blood Examination Rate (ABER), Annual Parasite Incidence (API), and Slide Positivity Rate (SPR). The Slide Falciparum Rate (SFR) and Plasmodium Falciparum (PF%) are two other decisive indicators for the malaria disease assessment. From the aforementioned indicators selectively ABER, API, SPR, and PF% presented in Fig. 10.3 have been calculated as follows:

Annual Blood Examination Rate (ABER)—ABER gives the total number of blood smears examined for Malaria Parasites per 100 populations in a year and expressed as % of population. The operational efficiency of the control measures can be optimized by using ABER. The status about accurate Annual Parasite Incidence

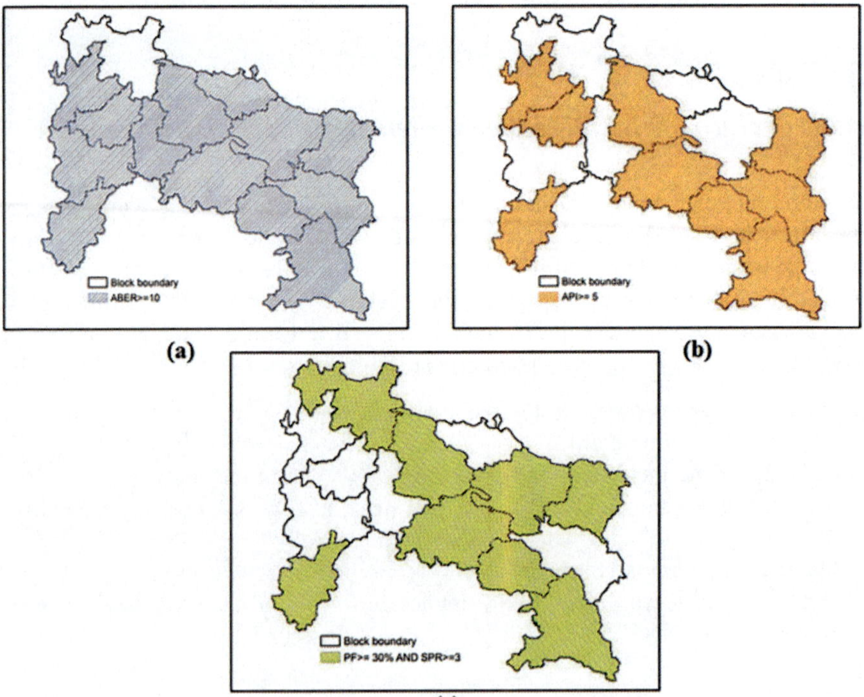

(a) (b) (c)

Fig. 10.3 Maps of core indicators for malaria surveillance in Ranchi during T4-SME generated using hotspot criteria: (**a**) ABER ≥ 10 in any of the last 3 years; (**b**) API ≥ 5 in any of the last 3 years; (**c**) PF % > 30 in any of last three years and SPR ≥ 3

(API) fundamentally depends upon the ABER. It has been calculated by Eq. (10.1) as given below

$$\text{ABER} = (\text{Number of Slides Examined/Population}) \times 100 \quad (10.1)$$

Annual Parasite Incidence (API)—API defines total number of blood smears positive for malaria parasite (MP) per 1000 population in a given year for a given area. It depends on the adequacy of case detection mechanism of ABER. API has been calculated by using Eq. (10.2) as given below:

$$\text{API} = (\text{Total Positive/Total Population}) \times 1000 \quad (10.2)$$

Slide Positive Rate (SPR)—SPR expresses the total number of blood slides found positive for malaria parasite per 100 number of blood slides examined. It is more reliable than API even for areas where ABER fluctuates by year to year and has been calculated by using Eq. (10.3)

$$\text{SPR} = \frac{\text{No. of Blood Smears found Positive for Malaria Parasite}}{\text{Total Blood Silde Examination}} \times 100 \quad (10.3)$$

Plasmodium Falciparum Percentage (PF%)—It is the total number of blood smears found positive for P. falciparum per 100 number of blood smears positive for malaria parasite. PF% facilitates the relative proportion of *P. falciparum* infection and trends in relation to the total case load. It has been calculated by Eq. (10.4)

$$\text{PF\%} = \frac{\text{Total No. of Blood Smears Positive for P.falciparum}}{\text{Total No. of Blood Smears Positive for Malaria Parasite}} \times 100 \quad (10.4)$$

Development of Thematic Layers

A GIS-based framework facilitates the overlay operations that can be very useful to interpret the relationship among the malaria surveillance indices. Thematic layers developed by using ArcGIS software provide the indispensable information about the malaria-affected region over the temporal as well as spatial scale. Based on certain pre-defined criteria (Sect. 3.3), a series of thematic maps have been generated.

Socio-Environmental Condition Assessment

This study aimed to explore the influence of environmental and socio-economic factors on malaria prevalence in the Ranchi region. In this view, it was necessary to assess the regional vegetation and water resources condition of the area under investigation. Normalized Difference Vegetation Index (NDVI) and Normalized Difference Water Index (NDWI) have been developed to assess the regional environment. Further, BPL data and census 2001 and 2011 demographic data were collected to analyze % BPL population, % literacy, and population density in the study area.

10.4 Results and Discussion

Statistical Analysis

During the total time-period considered, the PF% was found highest (62.11%) in the year 2003; interestingly no death has been reported in this year. In year 2005, 17 deaths were reported despite the low PF% (43%) as compared to PF% in 2003 (Table 10.1). An increasing trend in API (from 1.03 to 4.73) has been observed during 2003–2009, whereas the PF% values have been found to be decreased from 60.11% to 30.15% in the years of 2003–2006, respectively (Fig. 10.4a, b).

A block wise malaria incidence trend has also been analyzed from year 2000 to 2010. It was observed that the Bundu, Lapung, and Silli were the most malaria-affected administration blocks of Ranchi district in the years 2008, 2001, and 2009, respectively (Fig. 10.5a, b). It has also been found that most of the malaria cases occurred in the month of August and September due to continuous rainfall credited to the Monsoons. Owing to the regular precipitation the local waterbodies and water channels receive plenty of water that supports the mosquito breeding events. From the detailed examination of the spatial distribution of malaria surveillance

Table 10.1 Malaria case situation of district Ranchi during 2000–2010

Year	ABER	API	SPR	SFR	PF%
2000	2.51	2.38	9.50	4.89	51.49
2001	1.97	2.26	11.49	5.83	50.75
2002	1.58	1.92	12.20	6.60	54.10
2003	0.90	1.03	11.36	7.06	62.11
2004	2.31	1.69	7.31	3.01	41.20
2005	7.92	3.27	4.13	1.77	42.77
2006	2.96	2.42	8.17	2.46	30.15
2007	3.25	3.39	10.43	4.26	40.83
2008	5.93	4.66	7.85	3.95	50.33
2009	9.77	4.73	4.85	2.44	50.28
2010	9.72	2.76	2.84	1.16	40.88

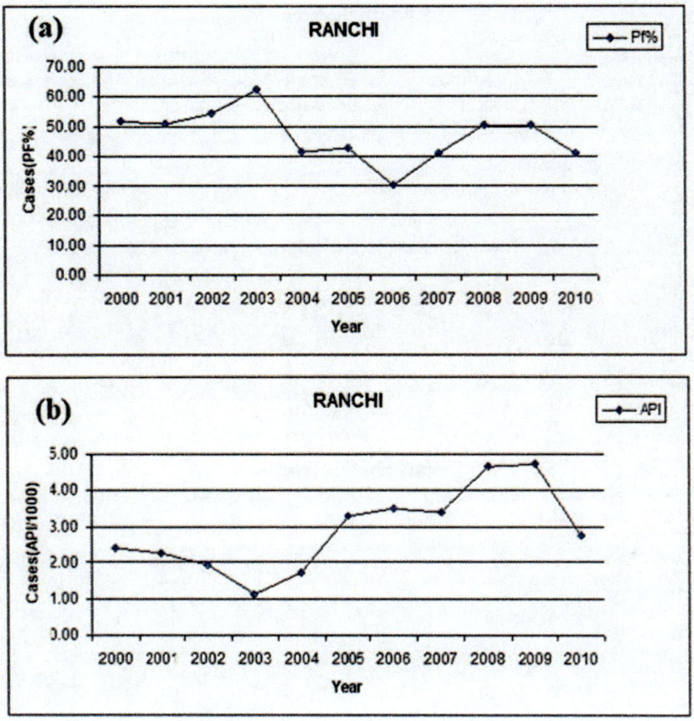

Fig. 10.4 (**a**) Yearly variation in *Plasmodium falciparum* (PF%) during the years from 2000 to 2010; (**b**) yearly variation in Annual Parasite Incidence (API) during the years from 2000 to 2010

parameters over the Ranchi district, the administrative block Bundu has been identified as the most malaria-affected region in the district.

Identification of Malaria Hotspot Region

As the identification of malaria hotspot regions was the prime objective for the present study, malaria-affected blocks of Ranchi district have been identified using thematic layers generated in GIS environment based on the criteria defined in Sect. 3.3. It has been observed that the Bundu block was the most frequently affected by malaria during the study periods (Fig. 10.6a–f). Most of the malaria hotspots have been observed in the southeast and northeastern region of the Ranchi district. At the level of HSCs from the year 2005 to 2007, Bundu and Sonahatu blocks have been identified as the most vulnerable area for malaria (Fig. 10.6f). A detailed investigation revealed that a total of 68 villages from Bundu and 40 villages from Sonahatu block have been under serious impacts of malaria incidences.

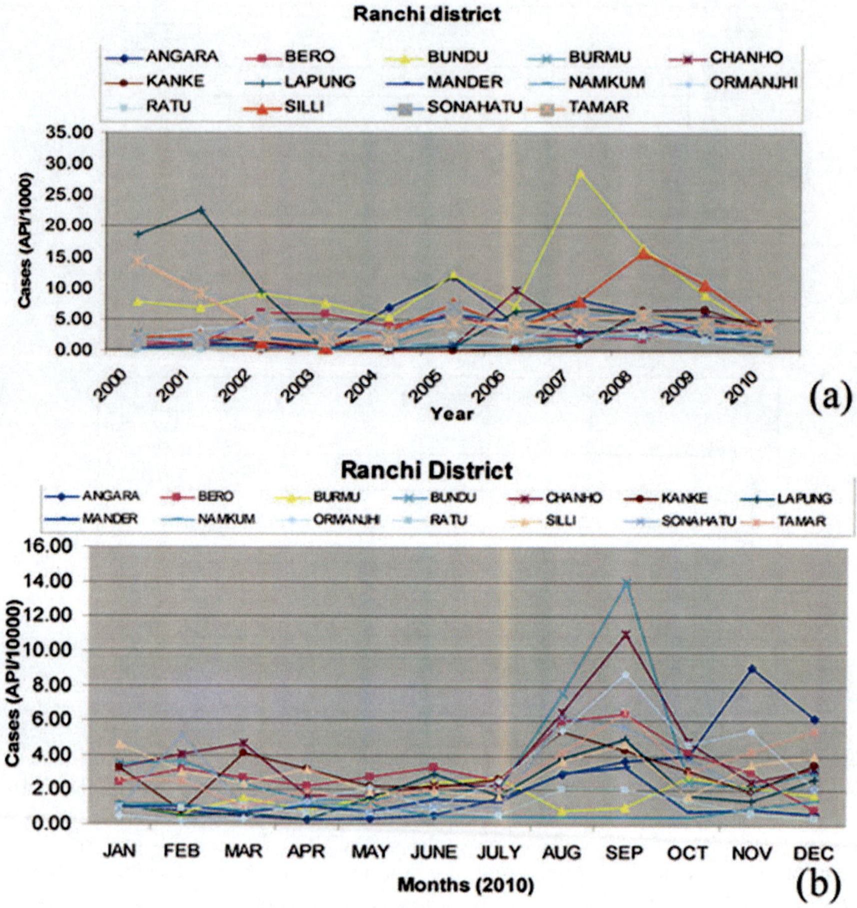

Fig. 10.5 (**a**) Block wise yearly variation in malaria incidences of Ranchi district during the years from 2000 to 2010; (**b**) block wise monthly variation in malaria incidences of Ranchi district during the year 2010

Analysis of Factors Responsible for Malaria Incidence

Topography and Malaria

Topography is one of the decisive parameters for occurrence of malaria incidences in plateau region. During summer season malaria vectors have shown a tendency of shifting their habitats from lower to higher elevations. Figure 10.7a, b depicts that areas having elevation up to 400 meters in Ranchi district are highly prone to malaria disease due to availability of favorable conditions for breeding of the malaria parasites, i.e., temperature, vegetation cover, and humidity. Lower elevated areas are closely related to the streams/swamps and forest that develop a positive impact for suitability of malaria vector and their growth as water is not washed away when it

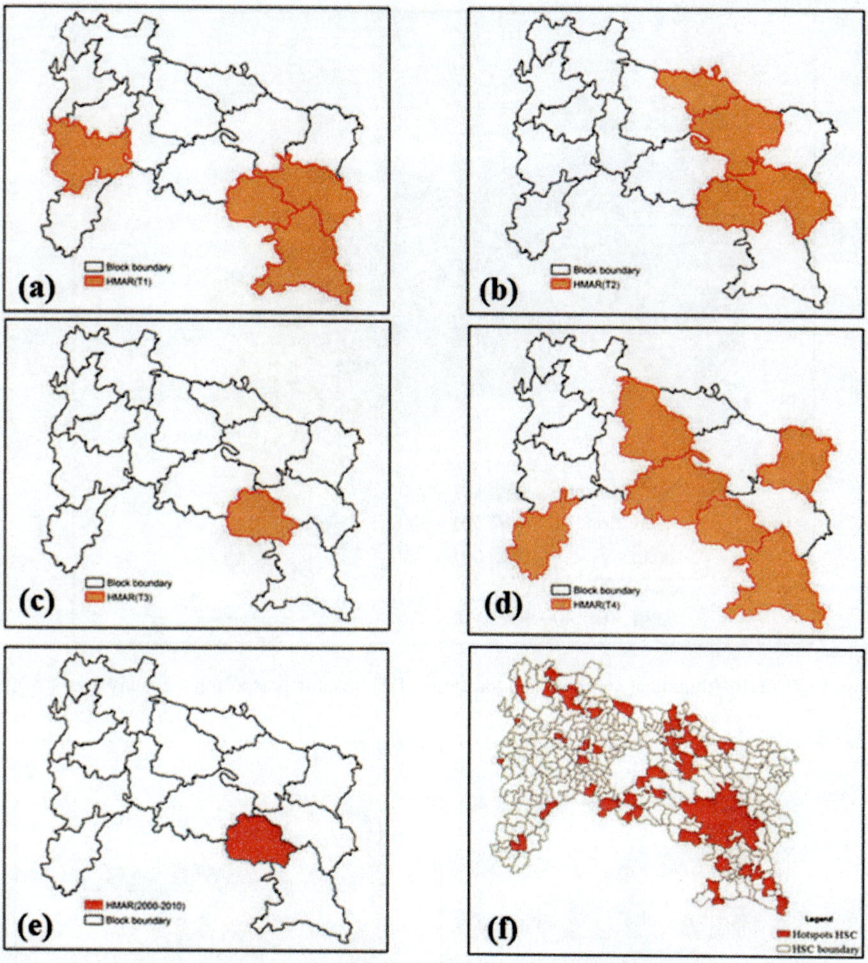

Fig. 10.6 (**a–e**) Highly affected regions by malarial disease in the district of Ranchi from 2000 to 2010; (**f**) highly affected regions by the malarial disease at the Health Sub-Centre level in Ranchi district during 2005–2007

rains as in the mountainous areas. At higher elevations temperature decreases which decreases the rate of parasite reproduction; therefore areas having higher elevation are less prone to malaria in the study area. The vulnerability of malaria incidences with respect to altitude variation at the HSC level has been presented in Fig. 10.7a. The Bundu block (300–600 m) has shown favorable habitat conditions for the malaria vector in this region.

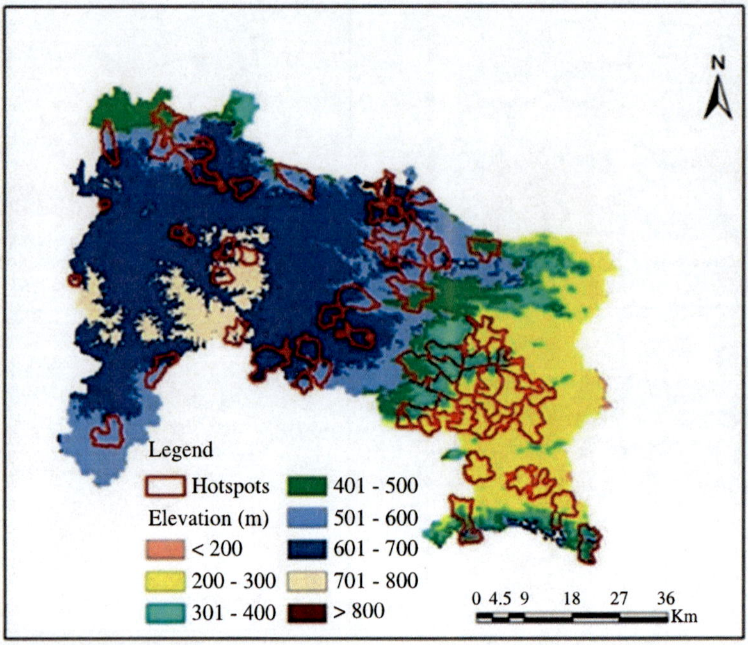

Fig. 10.7 (**a–b**) Altitudinal variation and locations of the malaria hotspots in the study area

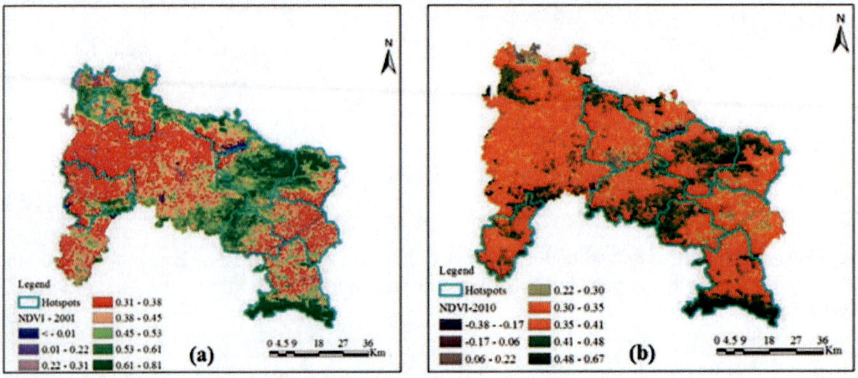

Fig. 10.8 (**a–b**) NDVI map and locations of the malaria hotspots in the Ranchi district

Vegetation and Malaria

The NDVI, as a measure of photosynthesis, can be used as a proxy for suitability of mosquito development. In the present study, areas with having NDVI values from 0.3 to 0.5 shows a high suitability or malaria prevalence in this area (Fig. 10.8a, b). The cropland habitats coupled with its humid conditions support rapid multiplication of malaria vectors. The Bundu block identified under NDVI ranges of 0.3–0.6 during

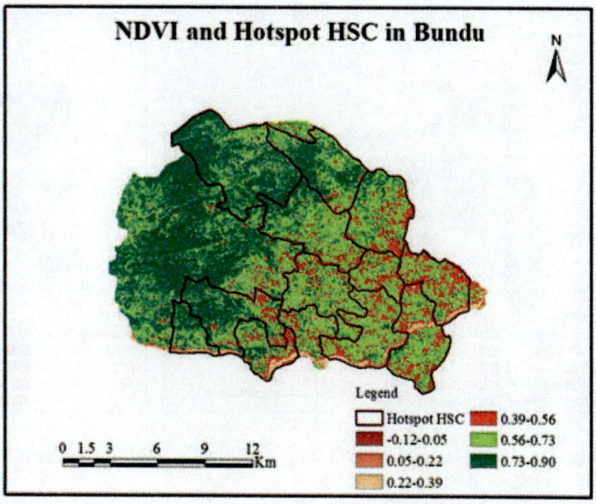

Fig. 10.9 Normalized difference vegetation map of Bundu block with HSC level hotspots

2000–2010 has been found favorable for malaria vector breeding as most of the areas of the block were under paddy cultivation and with occurrence of grasslands (Fig. 10.9). Form the vegetation map it has been observed that Bundu block is highly prone to malaria incidences.

Water Bodies and Malaria

The waterlogged area is one of the major factors for providing favorable conditions for breeding of the malaria parasite and may include riverbeds, oases, drainages, and other man-made water sites. It provides a perfectly suitable site for mosquitoes to lay their eggs. In the present work it has been observed that Ranchi district shows a frequent presence of water bodies. NDWI values of 0.1–0.8 have been observed during study period which acts as a favorable habitat for malaria vectors (Fig. 10.10a, b). The Bundu, Tamar, and Angara blocks have high NDWI values as compared to other blocks and are most vulnerable to malaria disease.

Meteorological Factors and Malaria

Different meteorological parameters such as rainfall, temperature, and humidity with API values have been determined for Ranchi district for the years of 2000–2010. The interaction between temperature and rainfall was found highly responsible for the seasonal characteristic of malaria transmission in this area. Among the several environmental factors, rainfall was found to be well correlated with malaria incidences (Fig. 10.11a). The relative humid condition was associated with increased vector longevity and greater frequency of feeding. The relationship between the aforementioned meteorological parameters and the occurrence of malaria incidences

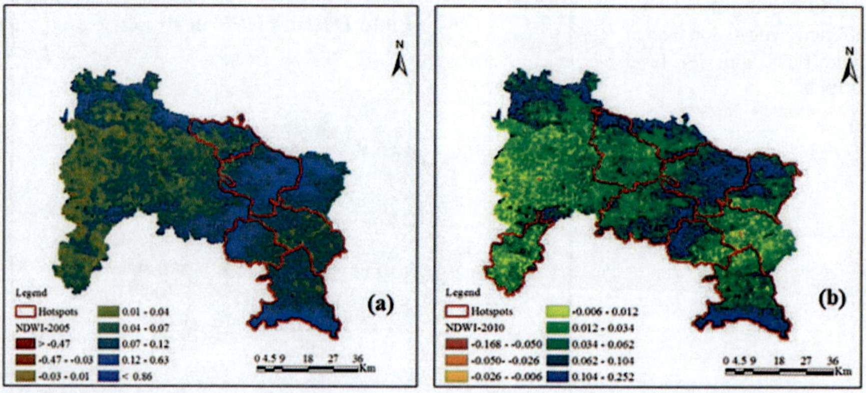

Fig. 10.10 (**a–b**) The NDWI map and locations of the malaria hotspots in the Ranchi district

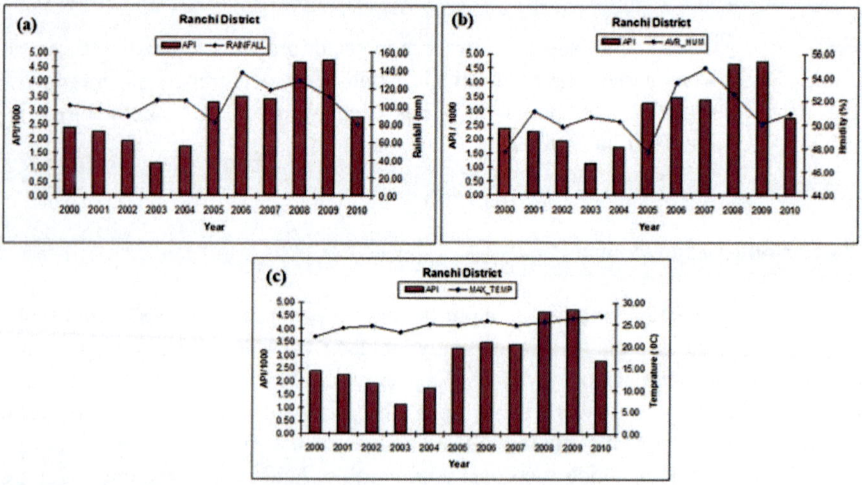

Fig. 10.11 (**a**) Relationship between the rainfall and API; (**b**) relationship between the relative humidity and API in Ranchi; (**c**) relationship between the temperature and API in Ranchi

has been shown in Fig. 10.11a–c. After detailed investigation it was found that there exists a strong correlation between API and rainfall for the year of 2005, 2008, 2009, and 2010. The API and humidity were also found highly correlated for the same time period.

Socio-Economic Factors and Malaria

The average API and different socio-economic data of different blocks of Ranchi district have been investigated. It has been observed that most of the administrative blocks show a positive relationship in between API and %BPL status data, whereas a

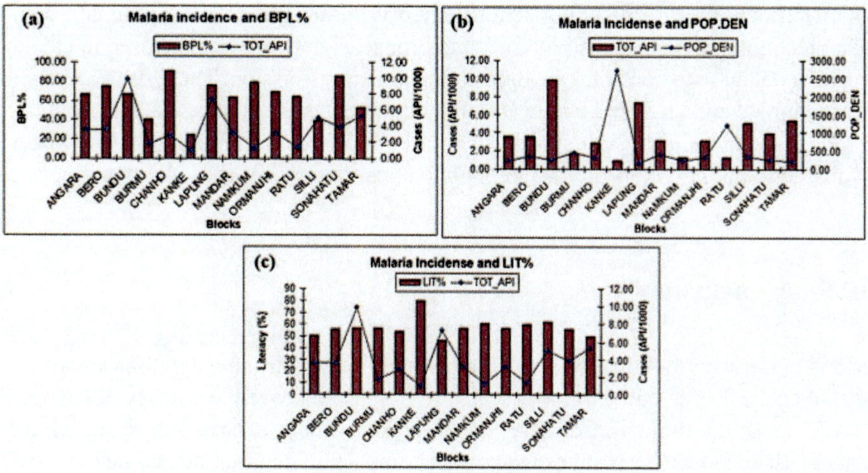

Fig. 10.12 (**a**) Relationship between the BPL% (Year-2010) and API; (**b**) relationship between the population density (Census-2011) and API; (**c**) relationship between percentage literacy (Census-2011) and API

poor relationship was found in case of blocks named Chanho, Namkum, and Ratu (Fig. 10.12a). In Bundu block, the %BPL during the year 2010 was >60%, and it was found that the total API was also very high (10 per 1000 population approx.). It is worth mentioning that in Kanke block, the value of API per 1000 population was very low with respect to the block population density along with low value of %BPL (Fig. 10.12a, b). Kanke block's population has not been much triggered by the malaria incidence owing to the high literacy percentage instead of its population density (Fig. 10.12c). It implies that education is one of the very important factors that can change the current malaria status in the country. The possible reasons behind the vulnerability and highly frequent malaria incidences in the Bundu block have been found as a combined impact of comparatively lower literacy rate and socio-economic condition of the block.

Role of GIS in the Management of Malaria

GIS has been used in many studies as a powerful tool for monitoring public health in various geographical locations (Clarke et al., 1996; Hay et al., 2009; Tanser & Le Sueur, 2002). Use of climate data, topographical data, medical reports, and questionaries can be processed using a GIS package to generate information for any epidemic. The information may include hotspot zonation and locating Public Health Centre/hospitals. GIS can be used in rapid estimation of the spatial distribution of malaria prevalence and incidence, medical facilities, and disseminating data to the concerned authorities. The community-based data generated using GIS is valuable as an operational tool for planning actions against the malaria risk. GIS coupled with other technologies such as the Internet and broadcast networks can be used to create a spatial database to constrain the limits of malaria transmission. These databases can be assembled and analyzed by a team of geographers, statisticians, epidemiologists, biologists, and public health specialists. Furthermore, the information can be made available to public via mediums like mobile apps, radios, TV sets, and newspapers.

10.5 Conclusion

GIS is a powerful tool that can assist in the control of environmental diseases where spatial analysis is essential to focus scarce resources, improve the efficacy of control, and decrease the burden of disease. In the present study malaria prevalence in the Ranchi district has been investigated. The impact of environmental factors and socio-economic conditions on malaria incidences has been investigated. Based on criteria defined by ICMR India, malaria hotspot mapping has been performed. The Bundu administrative block has been found as the most affected area by malaria parasite in this region during the last decade (2000–2010). The monsoon period and post-monsoon time have a great possibility to aggravate the malarial situation in this region. As malaria control measures studies rely mainly on ground-based surveys, the present work would facilitate as baseline research for the ongoing malaria eradication program in India. It could be helpful to achieve the malaria elimination target in India by the year 2022. The present work reveals that the environmental settings and socio-economic factors along with the availability of disease control measures can have a decisive impact on malaria eradication program. This study, although limited to a particular area, can serve as a model for a larger area to highlight malaria hotspots and to monitor the pace of control programs, which is not otherwise feasible. GIS-based models can be used, even at village or cluster level, to pin point the malaria hotspots, and information can be updated and retrieved easily. The methodology has the potential to be applied to other diseases and areas of health that will benefit from being able to display and analyze spatial relationships based on accurate and reliable data. Using the extrapolation technique for current malaria incidence as well as in the past, and the hotspot identification used in this study, malaria occurrence could be predicted in the future, and policy makers could be advised accordingly for effective and optimal distribution of governmental aid for malaria control. The conclusion is that modeling such as this can help determine spatio-temporal prediction and mapping of malaria incidence to aid in the design and administration of appropriate interventions. The study does not suggest any specific control measure, but emphasizes the need to concentrate malaria surveillance and monitoring programs in these malaria hotspots. More studies should be carried out, such as on bed net usage, the relationship between household presence of trees and malaria, and others. The map generated can be further linked with the vector density, vector breeding sites, physio-chemical factors, socio-economic status, and available health resources to provide excellent framework for disease management.

Acknowledgments The authors are grateful to Mr. J. D. Singh, State Malaria Office RCH Namkum, Ranchi, for providing the malaria disease data sets and Dr. A. Jangbahadur Singh, Mr. L. B. Singh, and Mrs. R. K. Vidyutma, District Malaria Office, Ranchi, for providing the HSC wise monthly malaria statistics of Ranchi district. We are also very thankful to the Department of Remote Sensing, Birla Institute of Technology, Mesra, Ranchi, for providing necessary facilities for this research work.

References

Ahmad, R., Ali, W., Nor, Z. M., Ismail, Z., Hadi, A., Ibrahim, M., & Lim, L. (2011). Mapping of mosquito breeding sites in malaria endemic areas in Pos lenjang, Kuala Lipis, Pahang, Malaysia. *Malaria Journal, 10*, 361.

Amegah, A. K., Damptey, O. K., Sarpong, G. A., Duah, E., Vervoorn, D. J., & Jaakkola, J. J. K. (2013). Malaria infection, poor nutrition and indoor air pollution mediate socioeconomic differences in adverse pregnancy outcomes in Cape Coast, Ghana. *PLoS One, 8*, e69181.

Bharti, P. K., Rajvanshi, H., Nisar, S., Jayswar, H., Saha, K. B., Shukla, M. M., & Lal, A. A. (2020). Demonstration of indigenous malaria elimination through track-test-treat-track (T4) strategy in a malaria elimination demonstration project in Mandla, Madhya Pradesh. *Malaria Journal, 19*(1), 1–12.

Carrin, M., Bronwyn, C., Colleen, F., & Brian, S. (2002). The use of a GIS-based malaria information system for malaria research and control in South Africa. *Health & Place, 8*, 227–236.

Ceccato, P., Connor, S. J., Jeanne, I., & Thomson, M. C. (2005). Application of geographical information systems and remote sensing technologies for assessing and monitoring malaria risk. *Parassitologia, 47*, 81–96.

Clarke, K. C., McLafferty, S. L., & Tempalski, B. J. (1996). On epidemiology and geographic information systems: A review and discussion of future directions. *Emerging Infectious Diseases, 2*(2), 85.

Clements, A. C., Lwambo, N. J., Blair, L., Nyandini, U., Kaatano, G., Kinung'hi, S., Webster, J. P., Fenwick, A., & Brooker, S. (2006). Bayesian spatial analysis and disease mapping: Tools to enhance planning and implementation of a schistosomiasis control programme in Tanzania. *Tropical Medicine & International Health, 11*, 490–503.

Cohen, A. A., Dhingra, N., Jotkar, R. M., Rodriguez, P. S., Sharma, V. P., & Jha, P. (2010). The summary index of malaria surveillance (SIMS): A stable index of malaria within India. *Population Health Metrics, 8*, 1–12.

Craig, M. H., Snow, R. W., & Le Sueur, D. (1999). A climate-based distribution model of malaria transmission in sub-Saharan Africa. *Parasitology Today, 15*, 105–111.

Cutler, D., Fung, W., Kremer, M., Singhal, M., & Vogl, T. (2010). Early-life malaria exposure and adult outcomes: Evidence from malaria eradication in India. *American Economic Journal: Applied Economics, 2*, 72–94.

Daash, A., Srivastava, A., Nagpal, B. N., Saxena, R., & Gupta, S. K (2009). Geographical information system (GIS) in decision support to control malaria-a case study of Koraput district in Orissa, India. *Journal of Vector Borne Diseases, 46*(1), 72–74.

Dev, V., Bhattacharyya, P. C., & Talukdar, R. (2003). Transmission of malaria and its control in northeastern region of India. *The Journal of the Association of Physicians of India, 51*, 1073–1076.

Dhiman, S., Gopalakrishnan, R., Goswami, D., Baruah, I., & Singh, L. (2010). Malaria epidemiology along indo-Bangladesh border in Tripura state, India. *The Southeast Asian Journal of Tropical Medicine and Public Health, 41*, 1279–1289.

Dhiman, S. K. (2009). Malaria control: Behavioural and social aspects. *DRDO Science Spectrum, 2009*, 183–186.

Enayati, A., & Hemingway, J. (2010). Malaria management: Past, present and future. *Annual Review of Entomology, 55*, 569–591.

Gallup, J. L., & Sachs, J. D. (2001). The economic burden of malaria. *The American Journal of Tropical Medicine and Hygiene, 64*, 85–96.

Gething, P. W., Patil, A. P., Smith, D. L., Guerra, C. A., Elyazar, I. R., Johnston, G. L., Tatem, A. J., & Hay, S. I. (2011). A new world malaria map: Plasmodium falciparum Endemicity in 2010. *Malaria Journal, 10*, 378.

Giles-Vernick, T., Traoré, A., & Sirima, S. B. (2011). Malaria, environmental change, and a historical epidemiology of child hood 'cold fevers': Popular interpretation s from southwestern Burkina Faso. *Health & Place, 17*, 836–842.

Hay, S. I., Guerra, C. A., Gething, P. W., Patil, A. P., Tatem, A. J., Noor, A. M., et al. (2009). A world malaria map: Plasmodium falciparum endemicity in 2007. *PLoS Medicine, 6*(3), e1000048.

Hay, S. I., Guerra, C. A., Tatem, A. J., Noor, A. M., & Snow, R. W. (2004). The global distribution and population at risk of malaria: Past, present, and future. *The Lancet Infectious Diseases., 4*, 327–336.

Hightower, A., Ombok, M., Otieno, R., Odhiambo, R., Oloo, A. J., Lal, A. A., Nahlen, B. L., & Hawley, W. A. (1998). A geographic information system applied to a malaria field study in western Kenya. *American Journal of Tropical Medicine and Hygiene, 58*, 266–272.

Hu, H. P. S., Salazar, N. P., Thimasarn, K., Wu, Y., Yang, H., & Zhu, D. (1998). Factors influencing malaria endemicity in Yunnan Province, PR China. *The Southeast Asian Journal of Tropical Medicine and Public Health, 29*, 191–200.

Kaewwaen, W., & Bhumiratana, A. (2015). Landscape ecology and epidemiology of malaria associated with rubber plantations in Thailand: Integrated approaches to malaria ecotoping. *Interdisciplinary Perspectives on Infectious Diseases*, 1–17.

Karema, C., Aregawi, M. W., Rukundo, A., Kabayiza, A., Mulindahabi, M., Fall, I. S., Gause, K., Williams, R. O., Lynch, M., Cibulskis, R., Fidele, N., Nyemazi, J.-P., Ngamije, D., Umulisa, I., Newman, R., & Binagwaho, A. (2012). Trends in malaria cases, hospital admissions and deaths following scale-up of anti-malarial interventions, 2000-2010, Rwanda. *Malaria Journal, 11*, 236.

Kirk, K. E., Haq, M. Z., Alam, M. S., & Haque, U. (2015). Geospatial technology: A tool to aid in the elimination of malaria in Bangladesh. *ISPRS International Journal of Geo-Information, 4*, 47–58.

Kleinschmidt, I., Bagayoko, M., Clarke, G. P., Craig, M., & Le Sueur, D. (2000). A spatial statistical approach to malaria mapping. *International Journal of Epidemiology, 29*, 355–361.

Kochar, D., Kumawat, B. L., Karan, S., Kochar, S. K., & Agarwal, R. P. (1997). Severe and complicated malaria in Bikaner (Rajasthan), western India. *The Southeast Asian Journal of Tropical Medicine and Public Health, 28*(2), 259–267.

Louis, F. A., Jr., Gilberto, C., & Mario, C. (2005). Lessons learned from malaria: Italy's past and sub-Sahara's future. *Health & Place, 11*, 67–73.

Machault, V., Vignolles, C., Borchi, F., Vounatsou, P., Pages, F., Briolant, S., Lacaux, J. P., & Rogier, C. (2011). The use of remotely sensed environmental data in the study of malaria. *Geospatial Health, 5*, 151–168.

Mutheneni, S. R., Upadhyayula, S. M., Kadiri, M. R., & Nishing, K. (2014). Malaria prevalence in Arunachal Pradesh—A northeastern state of India. *The American Journal of Tropical Medicine and Hygiene, 91*(6), 1088–1093.

Nath, M. J., Bora, A., Talukdar, P. K., Das, N. G., Dhiman, S., Baruah, I., & Singh, L. (2012). A longitudinal study of malaria associated with deforestation in Sonitpur district of Assam, India. *Geocarto International, 27*, 79–88.

Nath, M. J., Bora, A. K., Yadav, K., Talukdar, P. K., Dhiman, S., Baruah, I., & Singh, L. (2013). Prioritizing areas for malaria control using geographical information system in an endemic district of Assam, India. *Public Health, 127*, 572–580.

Oloukoi, G., Bob, U., & Jaggernath, J. (2014). Perception and trends of associated health risks with seasonal climate variation in Oke-Ogun region, Nigeria. *Health & Place, 25*, 47–55.

Omumbo, J. A., Hay, S. I., Guerra, C. A., & Snow, R. W. (2004). The relationship between the Plasmodium falciparum parasite ratio in childhood and climate estimates of malaria transmission in Kenya. *Malaria Journal, 3*, 1–8.

Qayum, A., Arya, R., Kumar, P., & Lynn, A. M. (2015). Socio-economic, epidemiological and geographic features based on GIS-integrated mapping to identify malarial hotspots. *Malaria Journal, 192*, 1–19.

Qayum, A., Lynn, A., & Arya, R. (2014). Traditional knowledge system based GIS mapping of antimalarial plants: Spatial distribution analysis. *Journal of Geographic Information System, 6*, 478–491.

Rabha, B., Goswami, D., Dhiman, S., Das, N. G., Talukdar, P. K., Nath, M. J., Baruah, I., Bhola, R. K., & Singh, L. (2011). A cross sectional investigation of malaria epidemiology among seven tea estates in Assam, India. *Journal of Parasitic Diseases, 36*, 1–6.

Rai, P. K., Nathawat, M. S., & Rai, S. (2014). Using the information value method in a geographic information system and remote sensing for malaria mapping: A case study from India. *Journal of Innovation in Health Informatics, 21*(1), 43–52.

Ratmanov, P., Mediannikov, O., & Raoult, D. (2013). Vector borne diseases in West Africa: Geographic distribution and geospatial characteristics. *Transactions of the Royal Society of Tropical Medicine and Hygiene, 107*, 273–284.

Sachs, J., & Malaney, P. (2002). The economic and social burden of malaria. *Nature, 415*, 680–685.

Sarkar, S., Singh, P., Lingala, M. A. L., Verma, P., & Dhiman, R. C. (2019). Malaria risk map for India based on climate, ecology and geographical modelling. *Geospatial Health, 14*(2).

Saxena, R., Das, M. K., Nagpal, B. N., Srivastava, A., Gupta, S. K., Kumar, A., et al. (2014). Identification of risk factors for malaria control by focused interventions in Ranchi district, Jharkhand, India. *Journal of Vector Borne Diseases, 51*(4), 276–281.

Saxena, R., Nagpal, B. N., Das, M. K., Srivastava, A., Gupta, S. K., Kumar, A., Jeyaseelan, A. T., & Baraik, V. K. (2012). A spatial statistical approach to analyze malaria situation at micro level for priority control in Ranchi district, Jharkhand. *The Indian Journal of Medical Research, 136*, 776–782.

Sharma, V. P. (2003). Malaria and poverty in India. *Current Science, 84*, 513–515.

Smith, M., Macklin, M. G., & Thomas, C. J. (2013). Hydrological and geomorphological controls of malaria transmission. *Earth-Science Reviews, 116*, 109–127.

Snow, R. W., Craig, M. H., Deichmann, U., & Le Sueur, D. (1999). A preliminary continental risk map for malaria motality among African children. *Parasitology Today, 15*, 99–104.

Srivastava, A., Nagpal, B. N., Joshi, P. L., Paliwal, J. C., & Dash, A. P. (2009). Identification of malaria hot spots for focused intervention in tribal state of India: A GIS based approach. *International Journal of Health Geographics, 8*, 30.

Srivastava, A., Nagpal, B. N., Saxena, R., & Sharma, V. P. (1999). Geographical information system as a tool to study malaria receptivity in Nadiad taluka, Kheda district, Gujarat, India. *The Southeast Asian Journal of Tropical Medicine and Public Health, 30*, 650–656.

Srivastava, A., Nagpal, B. N., Saxena, R., Wadhwa, T. C., Mohan, S., Siroha, G. P., et al. (2004). Malaria epidemicity of Mewat region, district Gurgaon, Haryana, India, a GIS based study. *Current Science, 86*, 1297–1303.

Tanser, F. C., & Le Sueur, D. (2002). The application of geographical information systems to important public health problems in Africa. *International Journal of Health Geographics, 1*(1), 1–9.

Van der Hoek, W., Konradsen, F., Amerasinghe, P. H., Perera, D., Piyaratne, M. K., & Amerasinghe, F. P. (2003). Towards a risk map of malaria for Sri Lanka: The importance of house location relative to vector breeding sites. *International Journal of Epidemiology, 32*, 280–285.

WHO. (2012). *World Health Organization: World Malaria Report 2012 Fact Sheet.*

WHO. (2014). *World Health Organization: World Malaria Report 2014 Switzerland.*

Worrall, E., Basu, S., Hanson, K., 2005. Is malaria a disease of poverty? A review of the literature. Tropical Medicine & International Health 10, 1047–1059.

Yadav, K., Dhiman, S., Rabha, B., Saikia, P. K., & Veer, V. (2014). Socio-economic determinants for malaria transmission risk in an endemic primary health Centre in Assam, India. *Infectious Diseases of Poverty, 3*(1), 1–8.

Yadav, K., Nath, M. J., Talukdar, P. K., Saikia, P. K., Baruah, I., & Singh, L. (2012). Malaria risk areas of Udalguri district of Assam, India: A GIS-based study. *International Journal of Geographical Information Science, 26*, 123–131.

Zhou, S. S., Zhang, S. S., Wang, J. J., Zheng, X., Huang, F., Li, W. D., et al. (2012). Spatial correlation between malaria cases and water-bodies in Anopheles sinensis dominated areas of Huang-Huai plain, China. *Parasites & Vectors, 5*, 106.

Chapter 11
Assessment on Social Vulnerability to Adapt the Hindrances of Natural Hazards in Purba Medinipur District, West Bengal, India

Sumita Gayen and Ismael Vallejo Villalta

Abstract Social vulnerability is the resistance power of a community that influences its ability to cope with the impact of natural hazards which depicts the difficulties of advancement in local livelihood. Vulnerability not only depends on the frequency and intensity of hazards but also on the socio-economic conditions of local people in a region. Strong socio-economic infrastructure of a region can protect people from destructions of natural hazards and make the region less vulnerable. The present chapter represents block wise assessment of social vulnerability in Purba Medinipur district, located in eastern part of India. To determine the resistance power of the district, block wise data have been utilized. Eighteen indicators were identified and principal component analysis technique has been employed to calculate the social vulnerability index. After principal component analysis, 18 indicators were reduced to six components. The result also shows that Moyna block is most vulnerable and Haldia, Sutahata, Contai-I blocks are socio-economically least vulnerable. This assessment helps to identify communities; those may need support in preparing for hazards or recovering after hazards.

Keywords Social vulnerability · Natural hazards · Indicators · Principal component analysis · Vulnerable zone

11.1 Introduction

Social vulnerability is a pre-existing condition of the population that influences its ability to prepare for, respond to, and recover from hazard events (Cutter et al., 2003). It is the complex set of some characteristics that include a person's initial

S. Gayen (✉) · I. V. Villalta
Departamento de Geografía Física y Análisis Geográfico Regional, Universidad de Sevilla, Sevilla, Spain
e-mail: sumgay@alum.us.es

P. K. Shit et al. (eds.), *Geospatial Technology for Environmental Hazards*, Advances in Geographic Information Science, https://doi.org/10.1007/978-3-030-75197-5_11

well-being, livelihood and resilience, self-protection, social-protection, and social and political networks and institutions (Cannon et al., 2003). Social vulnerability is a measure of the capability of some elements to withstand events of a certain physical character (Weichselgartner, 2001).

Social vulnerability assessment is a process of identifying, quantifying, and analyzing the vulnerability factors in a region. It has been documented that by reducing human vulnerability to natural hazards, the impact of hazards could be minimized. In present time, social vulnerability assessment is a common approach for both regional (Dwyer et al., 2004; Müller et al., 2011) and country levels (Cutter et al., 2003; Holand et al., 2011; Frigerio et al., 2018). Dwyer et al. (2004) selected 13 vulnerability indicators and 2 hazard indicators and measured the social vulnerability of Perth in Western Australia. Müller et al. (2011) analyzed urban vulnerability of flood for the capital city of Chile, Santiago de Chile. Cutter et al. (2003) constructed Social Vulnerability Index (SoVI) of the United States based on data of the year 1990. At first, 42 variables were used by Cutter et al. (2003) and then were reduced to 11 independent factors. Holand et al. (2011) classified the SoVI into two groups, Socio-economic Vulnerability Index (SeVI) and Built Environment Vulnerability Index (BEVI), to measure the social vulnerability in Norway. Frigerio et al. (2018) identified 16 indicators and used data of 1991, 2001, and 2011 census years and created social vulnerability index of Italy.

In India, social vulnerability assessment has been conducted both in regional and national levels. Table 11.1 represents some works of social vulnerability analysis conducted in India.

The study area of the present chapter is Purba Medinipur district of West Bengal state of India. In the last decade a lot of studies have been conducted in Sagar Island of West Bengal. Mondal (2013), Mukherjee et al. (2019), and Mondal et al. (2020) have selected their study areas for different vulnerability assessment in Sagar Island. But there is no such vulnerability assessment work performed at block level in Purba Medinipur district. The present work has been assessed through 18 indicators that describe the demographic, social, economic, and health condition of the district. The aim of this indicator-based assessment is to measure the block level hazard resistance power of Purba Medinipur district of West Bengal.

11.2 History of Natural Hazards in Purba Medinipur District and Relationship with Social Vulnerability

Purba Medinipur district is located (Fig. 11.1) in the southern part of West Bengal, India. It was formed on 1st January 2002 after the partition of Medinipur district into Purba Medinipur district and Paschim Medinipur district. It is surrounded by Rupnarayan and Hooghly rivers in the east, Paschim Medinipur district in the west, Howrah and Paschim Medinipur district by the north, and south by the Bay of Bengal. In the south-west corner, this district shares a common border with the

Table 11.1 Social vulnerability assessment in India

Authors	Title	Study area	Objective	Indicators used	Author's source of data/year
Maiti et al. (2015)	Assessment of social vulnerability to climate change in the eastern coast of India	4 states (West Bengal, Odisha, Andhra Pradesh, and Tamil Nadu) including 29 eastern coastal districts	Social vulnerability to climate change	Population density, decadal growth rate, rural literacy rate, rural population to total population, rural households availing banking service, households not having drinking water sources at their home premises, households having houses in dilapidated condition, etc.	Census of India / 2011
Ahmad et al. (2016)	Flood hazard zonation and vulnerability assessment of Greater Srinagar, J and K, India	Greater Srinagar, Jammu and Kashmir	Flood hazard zonation and vulnerability assessment	Housing density, population density, literate population, primary working population, female population, number of households and working population	Census of India/ 2001 and 2011
Maiti et al. (2017a)	An assessment of social vulnerability to climate change among the districts of Arunachal Pradesh, India	Arunachal Pradesh	Social vulnerability to climate change	Population density, decadal growth rate, rural literacy rate, rural population to total population, per capita income, rural households not having drinking water sources in their home premises etc.	Demographic data were taken from Census of India/ 2011
Vittal and Karmakar (2019)	A comprehensive social vulnerability analysis at a national scale	India	Social vulnerability analysis	Total population, female population, main agricultural and cultivator population, children population (population < 6 years), rural	Census of India/ 2011

(continued)

Table 11.1 (continued)

Authors	Title	Study area	Objective	Indicators used	Author's source of data/year
				population, number of households, illiterate population, illiterate female population, SC and ST population, SC and ST female population, marginal workers (including cultivators, agricultural laborers, household industry), non-working population, working population, female working population, literate population, and female literate population	
Kumar and Bhattacharjya (2020)	Study of integrated social vulnerability index $SoVI_{int}$ of hilly region of Uttarakhand, India	Hilly region of Uttarakhand state	Integrated social vulnerability index	Literacy rate, population density, elderly, children under 15, female population, poverty rate, agricultural workers, no. of tourist visited, unemployment, etc.	Previous research work, survey, etc.
Jaman et al. (2021)	Assessment of impact of cyclone hazard on social vulnerability of Bhadrak District of Odisha State during Phailin Cyclone in 2013 and Titli Cyclone in 2018 using multi-criteria analysis and geospatial techniques	Bhadrak district, Odisha state	Impact of cyclone hazard on social vulnerability	Population under 06, female population, SC population, St population, illiterate population, main-workers, non-workers	Census of India/ 2011

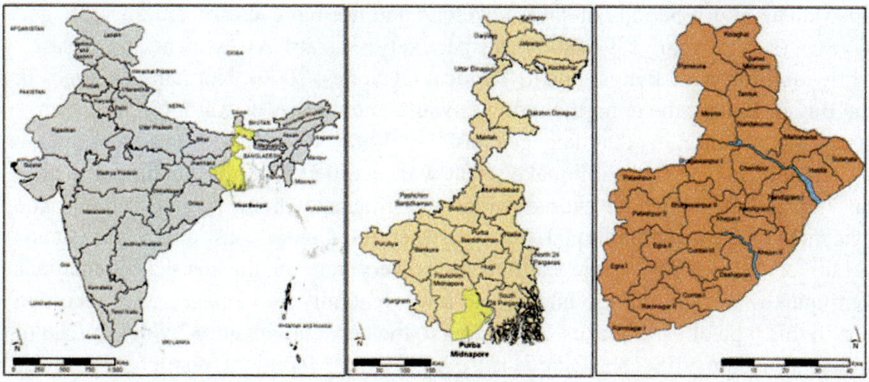

Fig. 11.1 Location map of Purba Medinipur district

state of Orissa. The district comprises of 25 blocks and 5 municipalities. The district headquarter is located in Tamluk.

Geographically Purba Medinipur district is situated between 21°36′35″N and 22°57′10″N and 86°33′50″E and 88°12′40″E longitude (Census of India, 2011). The district is a part of lower Ganga plain. Costal features like estuary, tidal flats, mud flats, salt pans, and sand dunes are observed in Purba Medinipur district. Haldi, Rasulpur, Keleghai, Kangsabati, and Rupnarayan are the main rivers of the district. All the rivers of Purba Medinipur district flow from west to east according to the slope of the region. The climate of Purba Medinipur district is of tropical monsoon type.

Natural hazards are natural processes or phenomena that occur in the biosphere and may constitute a damaging event (ISDR, 2004). Purba Medinipur district is a natural hazard prone region of West Bengal. The district is affected by several types of natural hazards mainly flood, tropical cyclone, earthquake, and drought. Flood is a very common event and is the main natural disaster in Purba Medinipur district. Almost all blocks of this district have suffered due to flood. Mainly floods occurred during south-west monsoon period from middle of June to September. Some parts of the district are affected by flood in almost every year. Kaliaghai and New Cossey rivers are mainly responsible for flood in the district. The district is a part of Ganga delta. Most of the rivers of the district are small in character. Flood of the district becomes voluminous because of the shape of the catchment area, its steep slope starting from a high level plateau area in the neighboring states and sloping sharply down to a flat terrain near the outfall.

Purba Medinipur district touches Bay of Bengal. All the coastal blocks, Ramnagar-I, Ramnagar-II, Contai-I, Deshapran, Khejuri-II, and Nandigram-I are cyclone-prone blocks in the district. The district has two cyclone seasons—pre-monsoon (April–May) and post-monsoon (October–November). The peak cyclone frequency is found in the months of May and November. Shallow water of Bay of Bengal, low flat coastal topography, and funnel shape of the coastline are the main causes of huge damage even in moderate intensity of cyclone (McBride,

1995). The return periods of severe cyclone and normal cyclones are 38 years and 4 years (survey year: 1891–2004), respectively (Hazard Assessment and Disaster Mitigation for West Bengal due to Tropical Cyclones, 2006). Normally cyclones in the Bay of Bengal move northward, westward, and north-westward. Cyclone brings strong winds and rains that damage coastal region.

The most devastated earthquake so far was recorded in Purba Medinipur district on 15 April 1964 which caused widespread damage in the Contai region. The epicenter of the 1964 earthquake was between the Contai coast and Sagar islands (Maiti et al., 2017b). Large earthquake is very rare in the district. Sometimes earthquakes are felt, but the intensity is low. Generally no damages were recorded due to this type of earthquakes. According to the new classification by the Bureau of Indian Standards [IS-1893 (Part-1): 2002], Purba Medinipur district falls under seismic zone III which is moderate intensity zone (Source: Ministry of Earth Science). The intensity on Modified Mercalli scale (MM) is VII for Zone III.

The experience of drought is also common in the district. Drought affects the population frequently and causes damage to the limited agriculture in the area. In 2010–2011, 598 mouzas of Tamluk, Contai, and Egra sub-divisions are affected by drought-like situation. All the mouzas have Kharif Paddy production below 50% due to this situation (Source: Office of the Deputy Director of Agriculture [Admn.], Purba Medinipur).

Summing up, the main natural hazards of Purba Medinipur district are flood and cyclone. After independence of India, there have been 46 major and medium floods and 23 cyclones which attacked the region (undivided Medinipur district, survey year: 1948–2015). In a hazard assessment, normally intensity, frequency, and probability of occurrence are evaluated. Relationship between destruction of hazard and socio-economic condition of people are closely related. Socio-economic condition plays a key role in hazard and social vulnerability analysis. For example, one region may be very highly vulnerable for flood but the density of population is very low, and another region is highly vulnerable for flood but having high density of population. So, relatively the second region is more vulnerable to flood. Social vulnerability helps to explain why different communities can experience the same hazard event differently (Morrow, 2008).

Nobody can stop the hazards, but it is possible to reduce the damage of hazards by improving the socio-economic condition of the society. Social vulnerability identifies the weak points of a society to cope with hazards. This is why social vulnerability assessment is very important to identify the vulnerable region against any type of hazard. The aim of this indicator-based vulnerability assessment is to measure the block level hazard resistance power and causes of vulnerability and try to give some possible remedies to reduce the vulnerability of Purba Medinipur district of West Bengal state of India.

11.3 Methodology

The methodology has seven stages. Figure 11.2 shows the flow chart of the research methodology.

Data Collection

The analysis was performed using data of Census of India (2011), West Bengal, District Human Development Report (Purba Medinipur, 2011), and District Statistical Handbook (2012–2015). Data of Census of India is mostly used in this analysis.

Selection of Indicators

An indicator is a measurable variable which represents an associated factor. In another way, it can be defined as a tool of measurement. In almost all research works, the selection of indicators depends on the objectives of the study. What should be the ideal characteristics of the indicators? The point of views of different authors are different, shown in Table 11.2.

This chapter identifies four characteristics of the indicators. The criteria are listed below:

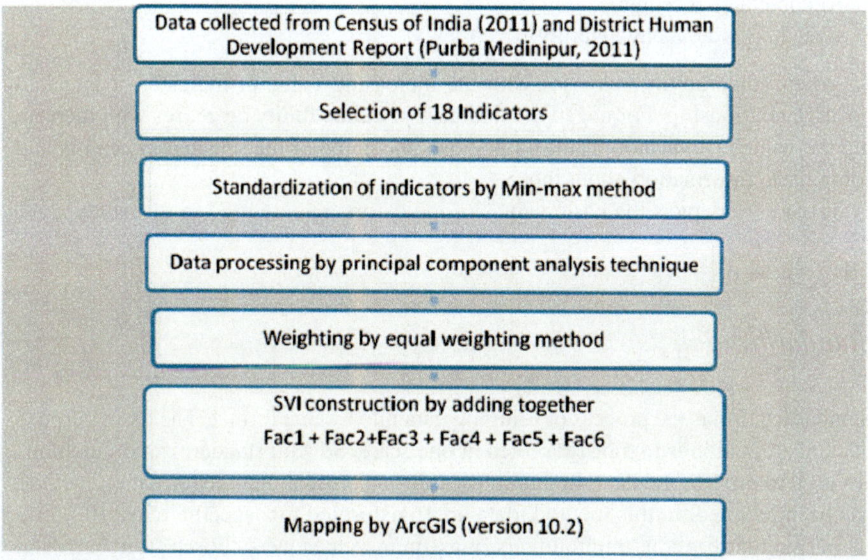

Fig. 11.2 Flow chart of the research methodology

Table 11.2 Characteristics of indicators according to different authors

Schneiderbauer and Ehrlich (2004)	Tapsell et al. (2005)	Birkmann (2006)
• Availability and coverage • Measurability and accuracy • Frequency of update	• Reliable and verifiable • Sensitive to change over time • Simple and easily understood while reflecting complexity of concept • Quantitative-measurable via readily understood model • Recognizable by others • Objective • And ideally, comparable within and between communities	• Measurable • Relevant, represent an issue that is important to the relevant topic • Policy-relevant • Only measure important key elements instead of trying to indicate all aspects • Analytically and statistically sound • Understandable • Easy to interpret • Sensitivity; be sensitive and specific to the underlying phenomenon • Validity/accuracy • Reproducible • Based on available data • Data comparability • Appropriate scope • Cost-effective

- Easily understood and applied by potential users.
- Meaningful and related in research topic.
- Reliable and acceptable.
- Based on the data availability.

When vulnerability increases with the increasing value of indicator, it is called positive relationship. For negative relationship, vulnerability decreases with increasing the value of indicator. Table 11.3 shows the name of the selected indicators and some basic information about those.

Table 11.4 depicts the basic values of indicators that are used in this study.

Standardization

Standardization is the process of bringing data into a same format. The main purpose of standardization is to transfer scores in one scale. Several standardization methods are used to process the data. In this study, min-max rescaling transformation method has been adopted. In this method, data are transformed to a specific range (0–1).

For standardization, minimum and maximum values were chosen from each and every indicator. Then the value of vulnerability for each block of Purba Medinipur

Table 11.3 Selected vulnerability indicators

Sl. No.	Indicators	Name	Description	Relationship with vulnerability	Literature Review
1	Distance from capital	DCA	Capital plays an important role for different administrative services. Economic growth rate is higher in capital compared to other regions. Capital region is well connected by other regions, and also job opportunities are more.	Positive	Gayen et al. (2021)
2	Population density	PDE	High population density indicates more people are vulnerable in any hazard.	Positive	Tascón-González et al. (2020)
3	Population growth rate	PGR	Higher population growth rate increases vulnerability to hazards. Limited space, limited escapes routes, and dense infrastructure lead to the vulnerability.	Positive	Holand and Lujala (2013)
4	Female population	GEN	Women are more vulnerable than men. Women workers are paid less salary than men for the same work.	Positive	Navarro et al. (2020)
5	Child population (0–6 years)	CPO	Children have less physical strength, and they are dependent on others for survival.	Positive	Aksha et al. (2019)
6	Rural population	RPO	Rural peoples have poor medical facilities and higher rates of morbidity due to remoteness from hospital.	Positive	Cutter et al. (2003)
7	Minority population	MPO	Minority population belongs to low levels of education, poor economic condition, poor health, and malnutrition.	Positive	Yang et al. (2019)
8	Literacy rate	LRA	Higher education gives a better job opportunity. Educated society enjoys a greater economic growth.	Negative	Kablan et al. (2017)
9	Households without electricity	HLI	Without electricity no chance to access the television, internet, and modern technologies.	Positive	Mavhura et al. (2017)
10	Households without sanitation	HSA	Sanitation can reduce diarrhea and other diseases and thus reduces health hazards.	Positive	Ge et al. (2017)
11	Households without car	HCA	In case of emergency situation, car of a family member is a need.	Positive	Bergstrand et al. (2015)

(continued)

Table 11.3 (continued)

Sl. No.	Indicators	Name	Description	Relationship with vulnerability	Literature Review
12	Households without kitchen	HKI	Cooking in open space is unhygienic.	Positive	Chen et al. (2013)
13	Households without sewage	HSE	Sewage system helps to keep the environment safe and prevents the spread of waterborne diseases. Poor sewage system also provides less protection against mosquitoes.	Positive	de Mello Rezende (2016)
14	Employment rate	ERA	Higher employment rate indicates healthy food, better medical facilities, and high standard of living.	Negative	Cutter et al. (2008)
15	Households having banking service	HBS	Bank account is an easy way to save money with interest and keep it safe. Online payment of different bills is easy through bank account. Bank also helps to arrange loans for education, home, and business.	Negative	Maiti et al. (2017a)
16	Infant mortality rate	IMR	Infant mortality rate reflects the social, economic, and health-care system of a society. Higher rate of infant mortality indicates poor development of a society.	Positive	de Oliveira Mendes (2009)
17	Pressure of population/hospital (including nursing home)	PHS	If the numbers of hospitals are more, everyone will have the opportunity to get treatment nearby to home. This is very important mainly for patient with serious condition.	Positive	Emrich and Cutter (2011)
18	Pressure of population/bed	PBE	More number of beds in hospital means more people will be able to get treatment.	Positive	Lixin et al. (2014)

district with respect to each indicator has been calculated by using Eqs. (11.1) and (11.2).

For positive relationship,

$$x = \frac{x_i - \min}{\max - \min} \tag{11.1}$$

where x_i = Actual value, min = Minimum value, max = Maximum value.

Table 11.4 Indicators and their range, mean, standard deviation, variance values

Sl. no.	Vulnerability indicators	Range	Mean	Standard deviation	Variance
1	Distance from capital	90.20	44.61	24.10	557.44
2	Population density	1209.71	1319.26	365.10	127962.71
3	Population growth rate	6.03	15.54	1.89	3.41
4	Female population	1.18	48.41	0.25	0.06
5	Child population	3.56	11.57	0.79	0.61
6	Rural population	53.97	89.96	14.80	210.18
7	Minority population	51.63	15.81	9.83	92.80
8	Literacy rate	12.00	80.20	2.65	6.73
9	Households without electricity	42.97	54.40	11.09	118.12
10	Households without sanitation	18.33	10.52	6.11	35.89
11	Households without car	1.44	98.90	0.40	0.15
12	Households without kitchen	27.81	33.85	7.79	58.31
13	Households without sewage	8.27	92.23	2.23	4.76
14	Employment rate	12.31	37.26	3.13	9.43
15	Households having banking service	20.97	38.91	5.46	28.59
16	Infant mortality rate	51.43	20.32	11.61	129.39
17	Pressure of population/hospital	96704.82	38314.72	23159.21	514,895,265
18	Pressure of population/bed	14355.02	5428.80	3236.44	10,055,567

For negative relationship,

$$x = \frac{\max - x_i}{\max - \min} \tag{11.2}$$

where max = Maximum value, x_i = Actual value, min = Minimum value.

After standardization, for positive relationship minimum value of any indicator gets transformed to 0 and maximum value to 1, and for negative relationship it is vice versa. The other values of indicator will change in decimal and stay between 0 and 1.

Data Processing

For data processing and statistical analysis, principal component analysis technique has been employed. Principal component analysis is one of the most extended and successful methodologies (Contreras et al., 2020). It is a technique that used to reduce the dimension of data and help to explain data. The criteria to retain all the factors were selected based on eigenvalues greater than 1 (Kaiser Criterion). The varimax rotation method has been applied in this study. Generally varimax rotation has been considered as the best orthogonal rotation (Fabrigar et al., 1999). Orthogonal rotation generates factors that are uncorrelated. The Statistical Package for the Social Sciences (SPSS, Version 26) was used to calculate the principal component analysis.

Weighting

Weighting is a relative degree which indicates importance of an indicator. Maximum vulnerability assessment studies do not use weights for vulnerability indicators, and the indicators are generally considered as independent and equally important variables (Yoon, 2012). Rygel et al. (2006) showed that vulnerability assessment is possible without practice of weighting for various indicators. An equal weighting was used to avoid the priority of any factor, and each factor has an equal contribution to the county's overall vulnerability (Cutter et al., 2003). In this study equal weight was applied for all the indicators. Except Cutter et al. (2003), Aksha et al. (2019), Mavhura et al. (2017), and Tate, 2012 also followed equal weighting technique for analysis of indicators.

Social Vulnerability Index (SVI) Construction

The SVI was developed by adding together the six component scores for every block.

Equation of SVI = Fac1 + Fac2 + Fac3 + Fac4 + Fac5 + Fac6.

Mapping

Social vulnerability mapping is a method to communicate which areas are more vulnerable to the impacts of disasters (Chen et al., 2013). The maps have been prepared by using ArcGIS (Version 10.2) based on standard deviation method. Standard deviation score is divided into very low, low, moderate, high, and very high vulnerability categories based on the same difference.

Some of the literatures, where same methods for standardization, data processing, and mapping have been used, are listed in Table 11.5.

Table 11.5 Literature review of standardization, data processing, and mapping methodology

Process	Method	Authors
Standardization	Min-max rescaling transformation	Cutter et al. (2010), Khan and Salman (2012), Lixin et al. (2014), Sajjad and Jain (2014), Tali et al. (2016), Žurovec et al. (2017), Ge et al. (2019), Lianxiao and Morimoto (2019), Gayen et al. (2020)
Data processing	Principal component analysis	Boruff et al. (2005), Fekete (2009), de Oliveira Mendes (2009), Bergstrand et al. (2015), Guillard-Gonçalves et al. (2015), Kirby (2015), Letsie and Grab (2015), Frigerio and Amicis (2016), Mavhura et al. (2017), Frigerio et al. (2018), Zhang et al. (2018), Aksha et al. (2019), Dintwa et al. (2019), Ge et al. (2019), Yang et al. (2019)
Mapping	ArcGIS	Letsie and Grab (2015), Gautam (2017), Mavhura et al. (2017), Žurovec et al. (2017), Alizadeh et al. (2018), Lianxiao and Morimoto (2019), Yang et al. (2019)

11.4 Results and Discussion

This block wise social vulnerability assessment identifies vulnerable blocks in each category. Eighteen indicators were standardized by min-max rescaling transformation method, and then principal component analysis was used for data analysis. Principal component analysis was the most common method in the last 10 years for the reduction of variables (Contreras et al., 2020).

If communalities of one variable are low (between 0.0 and 0.4), then that variable may struggle to load significantly on any factor. Higher values of communalities are better. Here all the communalities values are higher than 0.55 (Table 11.6) which indicates that the extracted components represent the variables quite well.

The total variance explained in Table 11.7 shows 18 components. The first component explains the higher variance, and the last component signifies the least variance. Eigenvalue represents the total variance explained by each factor. According to the Kaiser criterion (Kaiser, 1960), an appropriate threshold for component extraction includes those components having an eigenvalue greater than 1.00. Six components have eigenvalues greater than 1. So, only six factors can be retained for further analysis. 27.18% variance is explained by the first component. First six components explained 82.51 percent of the total cumulative variance, and the rest of the 12 components capture only 17.49 percent of the cumulative variance.

The values of "Initial eigenvalues" and "Extraction sums of squared loadings" columns are the same. The "Rotation sums of squared loadings" represent the distribution of the variance after the varimax rotation (here varimax rotation has been used). Total percent of variance explained is the same (cumulative value for factor 6 is 82.51%) for all of three (initial eigenvalues, extraction sums of squared loadings, rotation sums of squared loadings). But rotation changes the extracted

Table 11.6 Communalities

	Initial	Extraction
Distance from capital	1.000	0.898
Population density	1.000	0.938
Population growth rate	1.000	0.829
Female population	1.000	0.922
Child population	1.000	0.947
Rural population	1.000	0.875
Minority population	1.000	0.604
Literacy rate	1.000	0.809
Households without electricity	1.000	0.856
Households without sanitation	1.000	0.868
Households without car	1.000	0.740
Households without kitchen	1.000	0.906
Households without sewage	1.000	0.861
Employment rate	1.000	0.784
Households having banking service	1.000	0.863
Infant mortality rate	1.000	0.770
Pressure of population/hospital	1.000	0.555
Pressure of population/bed	1.000	0.825

factors of eigenvalues. So, after rotation each extracted factor counts for a different percentage of variance explained.

Scree plot (Fig. 11.3) is a line of the eigenvalues against components in an analysis. It always displays the eigenvalues in a downward curve. It helps to determine the number of factors remaining and the relative importance of the factors. The x-axis contains component number and y-axis represents the eigenvalue. From scree plot, the components having score above one (total six) have been considered. At the end of the graph, line is almost flat, which signifies that on that portion each successive component is accounting smaller amounts of the total variance.

After analysis of total variance explained, only six components have been retained, which have been shown in Table 11.8. The rotated component matrix referred to as the loadings for each item on each component as a high positive loading or negative loading. The highest loading for each component is highlighted. Here correlations of less than 0.3 or 0.4 are regarded as being insignificant.

Six factors have been derived from the principal component analysisSocial vulnerability assessmentprincipal component analysis and contribute to social vulnerability assessment in Purba Medinipur district (Table 11.9).

Following Cutter et al. (2003) methodology, all the six factors summed up to obtain final or SVI score (Table 11.10).

SVI scores were classified using the standard deviation by the ArcMap10.2. The block wise standard deviation (SD) score is divided into very low, low, moderate, high, and very high vulnerability categories based on the same difference. Higher

Table 11.7 Total variance explained

Component	Initial eigenvalues			Extraction sums of squared loadings			Rotation sums of squared loadings		
	Total	% of variance	Cumulative %	Total	% of variance	Cumulative %	Total	% of variance	Cumulative %
1	4.893	27.181	27.181	4.893	27.181	27.181	3.445	19.137	19.137
2	3.183	17.685	44.866	3.183	17.685	44.866	2.854	15.857	34.995
3	2.386	13.256	58.123	2.386	13.256	58.123	2.426	13.479	48.473
4	1.818	10.102	68.224	1.818	10.102	68.224	2.415	13.419	61.892
5	1.448	8.046	76.271	1.448	8.046	76.271	1.955	10.862	72.754
6	1.123	6.239	82.510	1.123	6.239	82.510	1.756	9.755	82.510
7	0.921	5.118	87.627						
8	0.645	3.582	91.209						
9	0.429	2.383	93.592						
10	0.324	1.799	95.391						
11	0.263	1.463	96.854						
12	0.211	1.170	98.023						
13	0.161	0.892	98.916						
14	0.085	0.473	99.388						
15	0.063	0.350	99.739						
16	0.034	0.189	99.928						
17	0.009	0.051	99.979						
18	0.004	0.021	100.000						

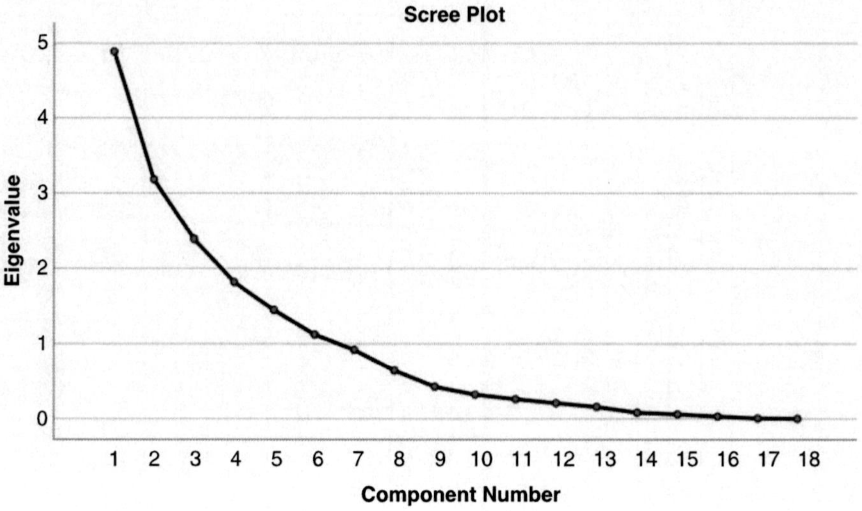

Fig. 11.3 Scree plot

Table 11.8 Rotated component matrix

	Component					
	1	2	3	4	5	6
Distance from capital	0.906	−0.026	−0.085	−0.112	−0.132	−0.196
Population density	−0.901	−0.201	0.135	−0.219	−0.121	−0.073
Households without sanitation	−0.854	0.228	−0.186	−0.199	−0.101	−0.048
Households without electricity	0.655	0.358	0.128	0.184	0.346	0.359
Literacy rate	−0.126	0.859	−0.215	0.065	−0.039	0.054
Minority population	0.211	0.697	0.188	−0.056	0.152	0.109
Child population	−0.050	0.694	0.419	0.389	0.251	0.271
Employment rate	0.015	0.070	0.864	−0.104	−0.142	0.045
Households without kitchen	0.528	0.180	−0.677	−0.089	−0.009	0.359
Population growth rate	0.234	0.471	0.612	−0.118	0.282	0.291
Infant mortality rate	0.005	0.069	0.058	−0.856	0.174	−0.006
Rural population	0.373	0.133	−0.318	0.767	0.151	0.079
Population/hospital	0.095	0.262	0.130	0.506	0.425	0.153
Households without sewage	0.072	0.165	−0.027	−0.174	0.891	−0.066
Population/bed	−0.076	−0.054	−0.536	0.295	0.577	0.330
Households without car	0.192	−0.427	−0.306	0.364	0.491	0.232
Households having banking service	0.081	0.322	0.011	0.199	0.042	0.843
Female population	0.232	0.510	0.101	0.473	−0.056	−0.608

Table 11.9 Dimension of six factors

Name	% of variance explained	Dominant indicator	Component loading
Factor 1	27.181	Distance from capital	0.906
		Population density	−0.901
		Households without sanitation	−0.854
		Households without electricity	0.655
Factor 2	17.685	Literacy rate	0.859
		Minority population	0.697
		Child population	0.694
Factor 3	13.256	Employment rate	0.864
		Households without kitchen	−0.677
		Population growth rate	0.612
Factor 4	10.102	Infant mortality rate	−0.856
		Rural population	0.767
		Population/hospital	0.506
Factor 5	8.046	Households without sewage	0.577
		Population/bed	0.491
		Households without car	0.042
Factor 6	6.239	Households having banking service	0.843
		Female population	−0.608

value indicates higher level of vulnerability and vice versa. SVI values <−0.1.5 SD are considered as a low vulnerability category. −1.5 to −0.50 SD values are assigned under low category. Values from −0.50 to 0.50 SD are termed as moderate vulnerability. High vulnerability values lie between 0.50 and 1.5 SD. Values >1.5 SD are marked as a very high vulnerability. Figure 11.4 shows different zones of social vulnerability in Purba Medinipur district. This study reveals that most vulnerable blocks are lying on the central, east, and western part of the district.

The result shows that the most vulnerable block is Moyna. The block is located near to the district capital Tamluk. But rural population is high (97.09%) in Moyna. Social parameters like households without electricity, no latrine, no kitchen in house, and no sewage system make Moyna the most vulnerable block in the district. Social parameters of vulnerability like households having no sanitation, no sewage system, and no kitchen are related to health problems. Sewage system and sanitation facilities are important to reduce health related problems. Almost one in two people in the developing countries don't have improved sanitation (Human Development Report, 2006). After a decade of twenty-first century more than 18.29% households have no sanitation facility in Moyna block. Even some people do not use latrine, though they have latrine. Sewage systems carry waste water by pipes. From the dumped sewage water, ground water could be contaminated, and water-borne diseases such as hepatitis, cholera, and typhoid could spread.

Table 11.10 SVI scores and vulnerable zones for 25 blocks of Purba Medinipur district

Blocks	Fac1	Fac2	Fac3	Fac4	Fac5	Fac6	SVI score	Rank	Level of vulnerability
Tamluk	−1.86849	−0.20082	0.16243	0.04979	−0.4291	−1.1474	−3.43353	22	Low
Sahid Matangini	−1.82438	−0.63505	0.19564	0.08275	−1.2901	0.0469	−3.42426	21	Low
Panskura	−1.37659	0.76426	1.64011	0.34084	0.0381	−0.5808	0.82593	13	Medium
Kolaghat	−1.52912	−0.42288	1.00205	0.19139	−1.0346	−0.2039	−1.99703	19	Low
Moyna	−0.80505	−0.10743	1.03917	−0.13885	1.4736	2.0857	3.54718	1	Very high
Nandakumar	−1.03044	0.79712	0.5931	0.02964	0.9916	−0.1686	1.21239	10	High
Chandipur	−0.44171	−0.15019	−0.76743	0.74738	0.5707	0.958	0.9168	12	Medium
Mahishadal	−0.61012	−0.54197	−0.68288	1.23374	−0.0127	−0.5481	−1.16198	18	Medium
Nandigram-I	0.07469	0.76035	−1.63292	1.55552	−0.096	1.1707	1.8323	7	High
Nandigram-II	0.24488	−0.39963	−1.63729	1.05733	0.2808	−0.312	−0.76591	17	Medium
Sutahata	−0.72797	0.7796	−1.25186	−1.7638	−0.8517	0.1226	−3.69318	23	Very low
Haldia	−0.50471	−0.22737	−1.77458	−3.11207	1.3926	−0.41	−4.63614	25	Very low
Potashpur-I	0.85323	−0.44343	1.55536	−0.50193	2.0969	−0.7477	2.81241	3	High
Potashpur-II	1.20059	0.46007	0.69389	−0.52517	0.1459	0.5286	2.50384	4	High
Bhagawanpur-I	−0.18615	−0.15798	−0.2148	0.63091	1.4437	0.612	2.12763	5	High
Egra-I	0.96283	1.11313	1.47748	−1.03348	−2.0222	0.1886	0.68633	14	Medium
Egra-II	1.17336	−0.3313	0.58555	−0.04488	−0.4642	0.9249	1.84346	6	High
Khejuri-I	0.65751	0.02055	−0.43742	0.96441	−0.7632	0.7786	1.22052	9	High
Khejuri-II	0.71749	3.39769	−0.72671	0.07372	0.156	−0.3328	3.28542	2	High
Bhagawanpur-II	0.71635	−0.81693	0.04907	0.56674	−0.5456	0.6354	0.60508	15	Medium
Ramnagar-I	0.83607	−0.05633	−0.55242	0.59694	−0.4429	−3.0018	−2.62043	20	Low
Ramnagar-II	1.18157	−0.85295	0.39648	0.07395	0.0646	−1.1586	−0.29495	16	Medium
Contai-I	0.67953	−2.18189	−0.72098	−1.20303	−1.2198	0.7513	−3.89485	24	Very low
Deshapran	0.75834	0.34394	0.55576	−0.34431	−0.5515	0.5728	1.33508	8	High
Contai-III	0.84828	−0.91056	0.45323	0.47245	1.0688	−0.7643	1.16789	11	Medium

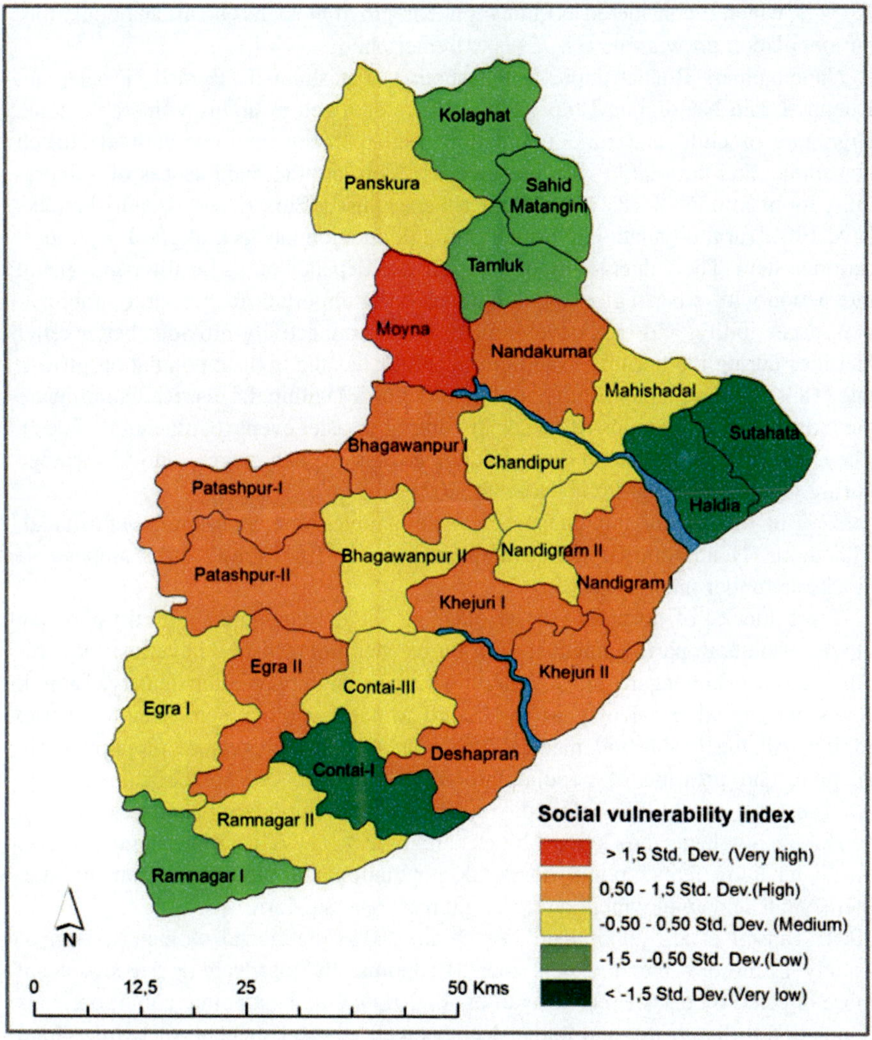

Fig. 11.4 Social vulnerability map of Purba Medinipur district

Mainly middle part of the district is highly vulnerable. Maximum blocks of Purba Medinipur district (36%) come under high vulnerable category. Khejuri-II block belongs to this category and is the second most vulnerable block in the district. Here population density is quite low, but values of other indicators are high. Khejuri-II has the highest rural population, minority population, illiteracy rate, and households without electricity within the district. Another important cause of vulnerability is high population growth rate. Population growth is considered as increase of vulnerability (Cutter et al., 2003). Decadal population growth of Purba Medinipur district is

15.54%, which is considered as quite high. Khejuri-II block has no urban population; but population growth rate (18.75) is extremely high.

Nandakumar, Bhagawanpur-I, Patashpur-I, Patashpur-II, Egra-II, Deshapran, Khejuri-I, and Nandigram-I blocks are also categorized in highly vulnerable zone. High rate of child and rural population, high percentage of without electrified household, and household with no sewage system are the main causes of vulnerability for maximum blocks. Nandakumar, Patashpur-II, Egra-II, and Khejuri-I blocks have 100% rural population. Normally rural population has lack of good communication system. The vulnerability of a region also depends on its location and spatial interaction with other regions. Communication is important for resource mobilization, accessibility, and area development. Good connectivity provides better relief facilities during the hazards. Nandigram-I block has the highest population growth rate (18.97%) and highest child population (13.66%) within the district. Children are the most vulnerable groups in the society during disaster events (Cutter et al., 2003). They are vulnerable due to lack of their education, experience, and knowledge during the time of hazard (Gayen et al., 2021).

32% of blocks fall into the medium vulnerable zone. Panskura, Mahishadal, Chandipur, Nandigram-II, Bhagawanpur-II, Contai-III, Egra-I, and Ramnagar-II blocks are under this category.

Three blocks of northern part (Kolaghat, Sahid Matangini, Tamluk) and one block of southern part (Ramnagar-I) in Purba Medinipur district have low vulnerability. Except Ramnagar-I block, other blocks have high population density. Tamluk is district capital and Kolaghat and Sahid Matangini blocks are near to district capital. All the health and medical indicators like IMR, pressure of population/hospital, and pressure of population/bed contribute to reduce the vulnerability. After disaster, accessibility of medical service is very important (Chen et al., 2013).

Haldia, Sutahata, and Contai-I blocks are in very low vulnerable category. These blocks have low female population, child population, and rural population and high literacy rate and employment rate. Several researchers (Morrow, 1999; Dwyer et al., 2004; Tapsell et al., 2005; Kuhlicke et al., 2011) mentioned women section of society is more vulnerable than men. Economically disadvantaged peoples are more affected by disasters (Flanagan et al., 2011). Good economic condition leads population to live better and longer lives as well as good standard of living. Good economic condition means high rate of employment that translated into the consumption of higher quantity and better quality nutrients. As well, it helps to cope with natural hazard and to a faster recovery from those. Besides these conditions, a number of hospitals and nursing homes and bed availability are quite good than other blocks of the district. In Contai-I block female population and child populations are lowest and literacy rate is highest in the district, but distance from capital is far. Haldia block is least vulnerable within the district. Good demographic, social parameter, and health conditions are the main contributors of low vulnerability for Haldia block. Another important thing is Haldia is the most industrialized block of the district. So, employment rate of Haldia block is high. Though Haldia block has high population growth rate, overall vulnerability index is less.

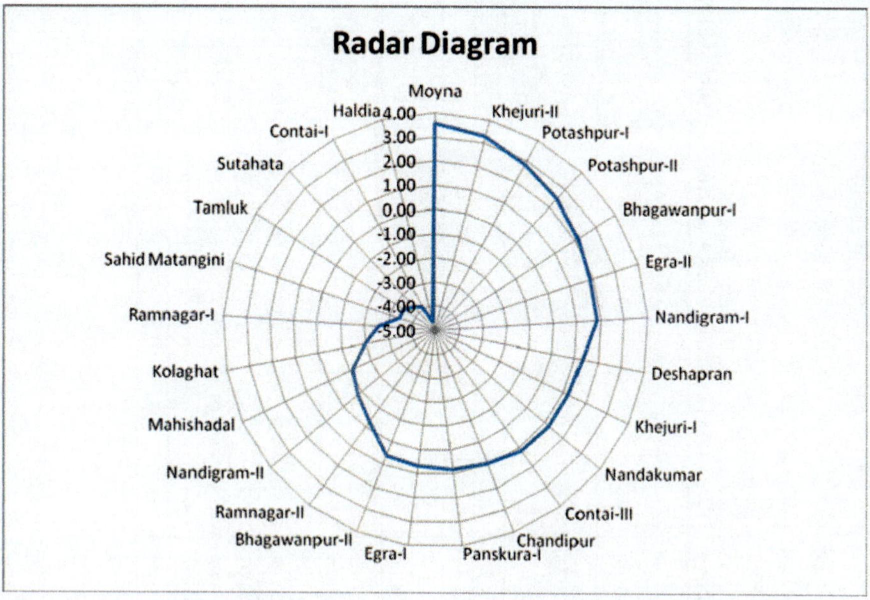

Fig. 11.5 Social vulnerability index of different blocks within the district in radar diagram

Purba Medinipur district comprises of five municipalities: Tamluk, Panskura, Haldia, Egra, and Contai. All the municipalities included blocks that are situated between very low (Contai, Haldia), low (Tamluk), and medium (Panskura, Egra) vulnerable zones, because municipalities are facilitated with different possibilities (communication, more job opportunities, hospital facilities, etc.) than other regions.

The radar chart (Fig. 11.5) shows position of every block within the district in a diagram

The relationship between 18 indicators is presented in Table 11.11. A strong relationship between child population and population growth is not surprising. Population density also decreases with the increasing distance from capital. So relationship between population density and distance from capital is negatively strong. Moderate negative association has been found between rural population and population density. A fair relationship has been found between minority population and population growth and also between minority population and child population.

Recommendations

Purba Medinipur district is a multi-hazard-prone district, specially for hazards like flood and cyclone (District Disaster Management Plan, Purba Medinipur, 2019–20). According to hazard (earthquake, flood, wind and cyclone, landslide, and

Table 11.11 Correlation coefficient between different indicators

	DCA	PDE	PGR	GEN	CPO	RPO	MPO	LRA	HLI	HSA	HCA	HKI	HSE	ERA	HBS	IMR	PHS	PBE
DCA	1.00																	
PDE	−0.77	1.00																
PGR	0.13	−0.28	1.00															
GEN	0.26	−0.38	0.15	1.00														
CPO	−0.21	−0.19	0.74	0.42	1.00													
RPO	0.22	−0.57	−0.14	0.41	0.27	1.00												
MPO	0.11	−0.25	0.46	0.28	0.51	0.10	1.00											
LRA	−0.10	−0.09	0.29	0.34	0.55	0.17	0.40	1.00										
HLI	0.37	−0.72	0.54	0.21	0.52	0.49	0.49	0.18	1.00									
HSA	−0.64	0.76	−0.20	−0.13	0.02	−0.40	−0.03	0.26	−0.60	1.00								
HCA	0.13	−0.25	−0.09	−0.18	−0.08	0.41	−0.20	−0.20	0.22	−0.30	1.00							
HKI	0.50	−0.56	−0.13	−0.14	−0.13	0.40	0.22	0.21	0.43	−0.24	0.28	1.00						
HSE	−0.01	−0.18	0.35	0.03	0.23	0.05	0.23	0.07	0.34	−0.04	0.31	0.07	1.00					
ERA	0.03	0.18	0.47	0.02	0.30	−0.29	0.31	−0.13	0.06	−0.05	−0.36	−0.45	−0.15	1.00				
HBS	−0.14	−0.30	0.43	−0.17	0.55	0.29	0.26	0.29	0.50	−0.07	0.14	0.32	0.10	−0.02	1.00			
IMR	0.01	0.14	0.14	−0.38	−0.22	−0.54	0.05	0.01	−0.03	0.08	−0.37	0.01	0.21	0.14	−0.17	1.00		
PHS	−0.12	−0.28	0.15	0.16	0.52	0.55	0.31	0.24	0.45	−0.29	0.13	0.08	0.22	0.12	0.22	−0.16	1.00	
PBE	−0.15	−0.16	−0.13	−0.18	0.09	0.49	0.00	0.04	0.21	0.06	0.60	0.41	0.35	−0.49	0.28	−0.15	0.37	1.00

subsidence) classification map, the district comes under highly vulnerable category (Nath et al., 2008). To prevent destruction from hazards, good socio-economic construction as well as natural hazard management strategy is needed.

Demographic problems like high population growth rate and high percentage of child population could be reduced only by awareness. Awareness on population control should not be confined to only for a day in the year. It should be conducted at regular intervals from the village level.

To improve economic condition, job opportunity is needed. Purba Medinipur district is the most literate district of West Bengal, having literacy rate of 87%, whereas the literacy rate of West Bengal is 76.26% (Census of India, 2011). Even the literacy rate of Purba Medinipur district is higher than the state capital Kolkata (86.31%). Higher education gives a better job opportunity. People of coastal blocks have a great opportunity to engage themselves in pisciculture and fish-related works. Establishment of small and medium industries can also create work opportunities.

The main cause of vulnerability for maximum blocks of Purba Medinipur district is inferior health condition. Overall health picture in Purba Medinipur district is very poor. Infant mortality rate is the mirror of general public health (Lee, 2014). Average infant mortality rate is 20.32. Institutional birth is considered as an important powerful determinant for infant mortality rate. It reduces the probability of death of newborn baby and mother. But institutional birth rate is quite low, mainly in the rural areas of the district. Goli and Jaleel (2014) showed that institutional deliveries reduce infant mortality rate and maternal mortality rate. The number of government hospitals is also very less. Private hospitals and nursing homes are mainly concentrated in municipality-related blocks like Tamluk, Haldia, Egra-I, and Contai-I. Poor people are unable to get access of these private health facilities because of high cost. They are dependent on existing public health-care center. After any natural hazard, the most important necessity is hospitals with bed capacity. Dependency rate of average number of people/hospital and nursing home is 38314.72, and average number of people/bed is 5428.80. These both factors are in very severe condition in Purba Medinipur district. The national norm of India is 1000 persons per bed (District Human Development Report, Paschim Medinipur, 2011). To improve these conditions, it is essential to establish new hospitals and increase number of beds in existing hospitals. Also the condition of primary health centers (PHCs) needs to improve. These health centers are suffering from basic infrastructural facilities such as beds, toilets, and drinking water facility.

To prevent the natural hazards, hazard management and development planning is needed. In present time geographic information system (GIS) is used as a necessary tool for natural hazard management almost in all countries of the world. GIS data can be used to explain past floods, and to predict future flood-prone area. By using satellite imagery, it is possible to identify flood-prone areas, duration of flood, depth of flood, and progress of flood. Satellite-based information is the major source for cyclone. It is helpful to locate origin, tracking, and movement of cyclone.

During the time of hazard, geo information can help to identify the locations where people may be trapped or injured and to provide medical or food supply.

During the recovery phase, GIS can help damage measurement and impact of environment measurement. Mainly satellite images are very useful to identify the changes of landscape after hazard.

It is not possible to stop natural hazards, but their impacts can be minimized using satellite images, like rescue people with weather forecast.

11.5 Conclusion

Natural hazards are natural phenomenon. But it turns into disaster when destructions are added with it. The destruction is multiplied when combined with demographic, social parameters, economic, and health factors. The impact of hazards is significantly higher in densely populated region. Human cannot reduce the severity of natural hazards. Vulnerability to hazard is highly correlated with the level of development of a society. Human Development Report (2014) highlighted the need of human development to reduce vulnerabilities. Every community is exposed to natural hazards. But those have low levels of resistance power, suffer more damages, and recover less quickly than others. Indicators are one part of disaster risk assessments and provide crucial information necessary for supplementing hazard mitigation assessments. Social vulnerability assessment in Purba Medinipur district is effected due to the lack of availability of data and no updation of data. For future studies, some data of indicators in different categories like income, birth rate, mortality rate, elderly population, dependency ratio, crime rate, suicide cases, etc. could also be added for better explanation. In the present study, an equal weighting technique was accomplished which is a simple aggregation technique. Also future vulnerability studies could be focused on a specific hazard and individual indicator index weighting technique that will help to improve the interpretation of vulnerability results.

This social vulnerability assessment can improve a block's ability to promote hazard reduction, thereby protecting communities, infrastructures, and properties. Social and economic capacity for adaptation, development and application of policies, strategies, and practices should be designed in district level to minimize vulnerabilities from natural hazards.

References

Ahmad, H. F., Bhat, M. S., Alam, A., & Ahmad, S. (2016). Flood hazard zonation and vulnerability assessment of greater Srinagar, J&K India. *International Journal of Advanced Research, 4*(12), 1679–1690.

Aksha, S. K., Juran, L., Resler, L. M., & Zhang, Y. (2019). An analysis of social vulnerability to natural hazards in Nepal using a modified social vulnerability index. *International Journal of Disaster Risk Science, 10*, 103–116.

Alizadeh, M., Alizadeh, E., Kotenaee, S. A., Shahabi, H., Pour, A. B., Panahi, M., Ahmad, B. B., & Saro, L. (2018). Social vulnerability assessment using artificial neural network (ANN) model for earthquake hazard in Tabriz City, Iran. *Sustainability, 10*, 3376.

Bergstrand, K., Mayer, B., Brumback, B., & Zhang, Y. (2015). Assessing the relationship between social vulnerability and community resilience to hazards. *Social Indicators Research, 122*, 391–409.

Birkmann, J. (2006). *Measuring vulnerability to promote disaster-resilient societies: Conceptual frameworks and definitions* (pp. 9–54). United Nations University Press.

Boruff, B. J., Emrich, C., & Cutter, S. L. (2005). Erosion hazard vulnerability of US coastal counties. *Journal of Coastal Research, 21*(5), 932–942.

Cannon, T., Twigg, J., & Rowell, J. (2003). Social vulnerability, sustainable livelihoods and disasters. *Report to DFID conflict and humanitarian assistance department (CHAD) and sustainable livelihoods support office*. Retrieved from https://www.eldis.org/document/A21628

Census of India. (2011). District census handbook, Purba Medinipur. *Village and town wise primary census abstract, Directorate of Census Operations, West Bengal*. Series-20, Part XII-B. Retrieved from http://censusindia.gov.in/2011census/dchb/1919_PART_B_DCHB_PURBA%20MEDINIPUR.pdf

Chen, W., Cutter, S. L., Emrich, C. T., & Shi, P. (2013). Measuring social vulnerability to natural hazards in the Yangtze River Delta region, China. *International Journal of Disaster Risk Science, 4*(4), 169–181.

Contreras, D., Chamorro, A., & Wilkinson, S. (2020). Review article: The spatial dimension in the assessment of urban socio-economic vulnerability related to geohazards. *Natural Hazards and Earth System Sciences, 20*, 1663–1687.

Cutter, S. L., Barnes, L., Berry, M., Burton, C., Evans, E., Tate, E., & Webb, J. (2008). A place-based model for understanding community resilience to natural disasters. *Global Environmental Change, 18*, 598–606.

Cutter, S. L., Boruff, B. J., & Shirley, W. L. (2003). Social vulnerability to environmental hazards. *Social Science Quarterly, 84*(2), 242–261.

Cutter, S. L., Burton, C. G., & Emrich, C. T. (2010). Disaster resilience indicators for benchmarking baseline conditions. *Journal of Homeland Security and Emergency Management, 7*(1), 51.

de Mello Rezende, G. B. (2016). Social vulnerability index: A methodological proposal for application in the cities of Barra do Garcas—MT, Pontal Do Araguaia—MT and Aragarcas—GO, Brazil. *Open Journal of Social Sciences, 4*, 32–45.

de Oliveira Mendes, J. M. (2009). Social vulnerability indexes as planning tools: Beyond the preparedness paradigm. *Journal of Risk Research, 12*(1), 43–58.

Dintwa, K. F., Letamo, G., & Navaneetham, K. (2019). Measuring social vulnerability to natural hazards at the district level in Botswana, Jàmbá. *Journal of Disaster Risk Studies, 11*(1), a447.

District Disaster Management Plan, Purba Medinipur. (2019–20). *District disaster management section, Purba Medinipur*. Retrieved from http://wbdmd.gov.in/pages/district_dm_plan.aspx

District Human Development Report, Paschim Medinipur. (2011). *Development & planning department, government of West Bengal*. Retrieved from http://www.wbpspm.gov.in/publications/District%20Human%20Development%20Report

District Human Development Report, Purba Medinipur. (2011). *Development & planning department, government of West Bengal*. Retrieved from https://www.wbpspm.gov.in/publications/District%20Human%20Development%20Report

Dwyer, A., Zoppou, C., Nielsen, O., Day, S, & Roberts, S. (2004). Quantifying social vulnerability: A methodology for identifying those at risk to natural hazards. *Geoscience Australia Record, 2004/14*.

Emrich, C. T., & Cutter, S. L. (2011). Social vulnerability to climate-sensitive hazards in the southern United States. *Weather, Climate, and Society, 3*(3), 193–208.

Fabrigar, L. R., Wegener, D. T., MacCallum, R. C., & Strahan, E. J. (1999). Evaluating the use of exploratory factor analysis in psychological research. *Psychological Methods, 4*(3), 272–299.

Fekete, A. (2009). Validation of a social vulnerability index in context to river-floods in Germany. *Natural Hazards and Earth System Sciences, 9*, 393–403.

Flanagan, B. E., Gregory, E. W., Hallisey, E. J., Heitgerd, J. L., & Lewis, B. (2011). A social vulnerability index for disaster management. *Journal of Homeland Security and Emergency Management, 8*(1), 3.

Frigerio, I., & Amicis, M. D. (2016). Mapping social vulnerability to natural hazards in Italy: A suitable tool for risk mitigation strategies. *Environmental Science & Policy, 63*, 187–196.

Frigerio, I., Carnelli, F., Cabinio, M., & Amicis, M. D. (2018). Spatiotemporal pattern of social vulnerability in Italy. *International Journal of Disaster Risk Science, 9*, 249–262.

Gautam, D. (2017). Assessment of social vulnerability to natural hazards in Nepal. *Natural Hazards and Earth System Sciences, 17*, 2313–2320.

Gayen, S., Villalta, I. V., & Haque, S. M. (2020). Comparative social vulnerability assessment in Purba Medinipur district, West Bengal, India. *European Journal of Geography, 11*(1), 093–107.

Gayen, S., Villalta, I. V., & Haque, S. M. (2021). Assessment of social vulnerability in Malaga Province, Spain: A comparison of indicator standardization techniques. *Revista de Estudios Andaluces, 41*, 87–108.

Ge, Y., Dou, W., & Dai, J. (2017). A new approach to identify social vulnerability to climate change in the Yangtze River delta. *Sustainability, 9*, 2236.

Ge, Y., Yang, G., Chen, Y., & Dou, W. (2019). Examining social vulnerability and inequality: A joint analysis through a connectivity lens in the urban agglomerations of China. *Sustainability, 11*, 1042.

Goli, S., & Jaleel, A. C. P. (2014). What is the cause of the decline in maternal mortality in India? Evidence from time series and cross-sectional analyses. *Journal of Biosocial Science, 46*(3), 351–365.

Guillard-Gonçalves, C., Cutter, S. L., Emrich, C. T., & Zêzere, J. L. (2015). Application of social vulnerability index (SoVI) and delineation of natural risk zones in greater Lisbon, Portugal. *Journal of Risk Research, 18*(5), 651–674.

Hazard Assessment and Disaster Mitigation for West Bengal due to Tropical Cyclones. (2006). *Department of disaster management, government of West Bengal*. Retrieved from http://www.iczmpwb.org/main/pdf/ebooks/Disaster%20Management%20Report_West%20Bengal.pdf

Holand, I. S., & Lujala, P. (2013). Replicating and adapting an index of social vulnerability to a new context: A comparison study for Norway. *The Professional Geographer, 65*(2), 312–328.

Holand, I. S., Lujala, P., & Rød, J. K. (2011). Social vulnerability assessment for Norway: A quantitative approach. *Norsk Geografisk Tidsskrift - Norwegian Journal of Geography, 65*(1), 1–17.

Human Development Report. (2006). *United Nations development programme, New York, USA*. Retrieved from http://hdr.undp.org/sites/default/files/reports/267/hdr06-complete.pdf

Human Development Report. (2014). *Sustaining human progress: Reducing vulnerabilities and building resilience*. United Nations Development Programme. Retrieved from http://hdr.undp.org/sites/default/files/hdr14-report-en-1.pdf

ISDR. (2004). *Living with risk: A global review of disaster reduction initiatives*. United Nations, New York and Geneva (Vol. 1). Retrieved from https://www.undrr.org/publication/living-risk-global-review-disaster-reduction-initiatives

Jaman, T., Dharanirajan, K., & Sharma, S. V. S. (2021). Assessment of impact of cyclone hazard on social vulnerability of Bhadrak District of Odisha State during Phailin Cyclone in 2013 and Titli Cyclone in 2018 using multi-criteria analysis and geospatial techniques. *International Journal of Disaster Risk Reduction, 53*, 101997.

Kablan, M. K. A., Dongo, K., & Coulibaly, M. (2017). Assessment of social vulnerability to flood in urban Côte d'Ivoire using the MOVE framework. *Water, 9*, 292.

Kaiser, H. F. (1960). The application of electronic computers to factor analysis. *Educational and Psychological Measurement, 20*(1), 141–151.

Khan, F. A., & Salman, A. (2012). A simple human vulnerability index to climate change hazards for Pakistan. *International Journal of Disaster Risk Science, 3*(3), 163–176.

Kirby, R. H. (2015). *Measuring social vulnerability to environmental hazards in the Dutch Province of Zeeland*. LSU Master's Theses. 3800. Retrieved from https://digitalcommons.lsu.edu/gradschool_theses/3800

Kuhlicke, C., Scolobig, A., Tapsell, S., Steinfuhrer, A., & Marchi, B. D. (2011). Contextualizing social vulnerability: Findings from case studies across Europe. *Natural Hazards, 58*, 789–810.

Kumar, D., & Bhattacharjya, R. K. (2020). Study of integrated social vulnerability index SoVIint of hilly region of Uttarakhand, India. *Environmental and Climate Technologies, 24*(1), 105–122.

Lee, Y. J. (2014). Social vulnerability indicators as a sustainable planning tool. *Environmental Impact Assessment Review, 44*, 31–42.

Letsie, M. M., & Grab, S. W. (2015). Assessment of social vulnerability to natural hazards in the mountain Kingdom of Lesotho. *Mountain Research and Development, 35*(2), 115–125.

Lianxiao, & Morimoto, T. (2019). Spatial analysis of social vulnerability to floods based on the MOVE framework and information entropy method: Case study of Katsushika Ward, Tokyo. *Sustainability, 11*, 529.

Lixin, Y., Xi, Z., Lingling, G., & Dong, Z. (2014). Analysis of social vulnerability to hazards in China. *Environment and Earth Science, 71*, 3109–3117.

Maiti, S., Jha, S. K., Garai, S., Nag, A., Bera, A. K., Paul, V., Upadhaya, R. C., & Deb, S. M. (2017a). An assessment of social vulnerability to climate change among the districts of Arunachal Pradesh, India. *Ecological Indicators, 77*, 105–113.

Maiti, S., Jha, S. K., Garai, S., Nag, A., Chakravarty, R., Kadian, K. S., Chandel, B. S., Datta, K. K., & Upadhyay, R. C. (2015). Assessment of social vulnerability to climate change in the eastern coast of India. *Climatic Change, 131*, 287–306.

Maiti, S. K., Nath, S. K., Adhikari, M. D., Srivastava, N., Sengupta, P., & Gupta, A. K. (2017b). Probabilistic seismic hazard model of West Bengal, India. *Journal of Earthquake Engineering, 21*(7), 1113–1157.

Mavhura, E., Manyena, B., & Collins, A. E. (2017). An approach for measuring social vulnerability in context: The case of flood hazards in Muzarabani district, Zimbabwe. *Geoforum, 86*, 103–117.

McBride, J. L. (1995). Tropical cyclone formation. In *Global perspectives on tropical cyclones*. World meteorological organisation, Geneva, WMO/TD-No. 693.

Mondal, D. (2013). Assessing social vulnerability to coastal hazards: An examination on Sagar Island of Sundarban Delta. *Research Journal of Humanities and Social Sciences, 4*, 210–215.

Mondal, M., Paul, S., Bhattacharya, S., & Biswas, A. (2020). Micro-level assessment of rural societal vulnerability of coastal regions: An insight into Sagar Island, West Bengal, India. *Asia-Pacific Journal of Rural Development, 30*, 55–88.

Morrow, B. H. (1999). Identifying and mapping community vulnerability. *Disasters, 23*(1), 1–18.

Morrow, B. H. (2008). Community resilience: A social justice perspective. *CARRI research report 4, community and regional resilience initiative*. doi: https://doi.org/10.13140/RG.2.1.1278.9604

Mukherjee, N., Siddique, G., Basak, A., Roy, A., & Mandal, M. H. (2019). Climate change and livelihood vulnerability of the local population on Sagar Island, India. *Chinese Geographical Science, 29*(3), 417–436.

Müller, A., Reiter, J., & Weiland, U. (2011). Assessment of urban vulnerability towards floods using an indicator-based approach—A case study for Santiago de Chile. *Natural Hazards and Earth System Sciences, 11*, 2107–2123.

Nath, S. K., Roy, D., & Thingbaijam, K. K. S. (2008). Disaster mitigation and management for West Bengal, India—An appraisal. *Current Science, 94*, 7.

Navarro, D., Vallejo, I., & Navarro, M. (2020). Análisis de la vulnerabilidad social a los riesgos naturales mediante técnicas estadísticas multivariantes. *Investigaciones Geográficas, 74*, 29–49.

Rygel, L., O'sullivan, D., & Yarnal, B. (2006). A method for constructing a social vulnerability index: An application to hurricane storm surges in a developed country. *Mitigation and Adaptation Strategies for Global Change, 11*, 741–764.

Sajjad, H., & Jain, P. (2014). Assessment of socio-economic vulnerabilities among urban migrants in south-East Delhi, India. *Journal of Studies in Social Sciences, 7*(1), 65–81.

Schneiderbauer, S., & Ehrlich, D. (2004). Risk, hazard and people's vulnerability to natural hazards: A review of definitions, concepts and data. *Eur. Comm. Jt. Res. Centre., EUR 21410 EN.*

Tali, M. G., Naeimi, A., & Esfandiary, M. (2016). Physical development of Arak City applying natural indicators. *European Journal of Geography, 7*(3), 99–110.

Tapsell, S., Tunstall, S., Green, C., & Fernandez-Bilbao, A. (2005). *Task 11 social indicator set. FLOODsite project report*, T11-5-01. Retrieved from http://www.floodsite.net/html/partner_area/project_docs/M11.1_Indicator-set_7-05.pdf

Tascón-González, L., Ferrer-Julià, M., Ruiz, M., & García-Meléndez, E. (2020). Social vulnerability assessment for flood risk analysis. *Water, 12*(2), 558.

Tate, E. (2012). Social vulnerability indices: A comparative assessment using uncertainty and sensitivity analysis. *Natural Hazards, 63*(2).

Vittal, H., & Karmakar, S. (2019). A comprehensive social vulnerability analysis at a national scale. *Climate Change Signals and Response*, 163–176.

Weichselgartner, J. (2001). Disaster mitigation: The concept of vulnerability revisited. *Disaster Prevention and Management, 10*(2), 85–94.

Yang, X., Lin, L., Zhang, Y., Ye, T., Chen, Q., Jin, C., & Ye, G. (2019). Spatially explicit assessment of social vulnerability in coastal China. *Sustainability, 11*, 5075.

Yoon, D. K. (2012). Assessment of social vulnerability to natural disasters: A comparative study. *Natural Hazards, 63*, 823–843.

Zhang, M., Xiang, W., Chen, M., & Mao, Z. (2018). Measuring social vulnerability to flood disasters in China. *Sustainability, 10*, 2676.

Žurovec, O., Cadro, S., & Sitaula, B. K. (2017). Quantitative assessment of vulnerability to climate change in rural municipalities of Bosnia and Herzegovina. *Sustainability, 9*, 1208.

Chapter 12
Understanding the Development and Progress of Extremely Severe Cyclonic Storm "Fani" Over the Bay of Bengal

Pankaj Bhardwaj and Omvir Singh

Abstract The Bay of Bengal (BoB) experiences the occurrence of tropical cyclones (TCs) almost throughout the year. However, the extremely severe cyclonic storm (ESCS) Fani has formed in April 2019 has shown uniqueness in terms of its location of origin, direction of track, and landfall location. Therefore, in this study, an attempt has been made to examine the development and progress of ESCS Fani over the BoB. The analyses have shown that a low pressure area has formed near equator (approximately 2.7°N latitude) over the southern BoB on 25 April 2019 and strengthened into depression on 26 April at the same location. This depression has further strengthened into cyclonic, severe cyclonic, very severe cyclonic, and extremely severe cyclonic storm and moved northwestward. Then, it has recurved and moved northeastward and make landfall over Orissa coast. It has been reported among the long-lasting cyclones of BoB as it travelled the distance of about 3030 km. The total accumulated cyclone energy and power dissipation index generated by the ESCS Fani have been found higher than their long-term mean (1972–2017). The analyses of large-scale dynamic and thermodynamic conditions have shown favorable environment for the development of cyclone over the southern BoB. The consistent strong convective activity, high SST (approximately 30 °C), more relative humidity, strong vertical motion, low level cyclonic vorticity, and less vertical wind shear have supported for further intensification of cyclonic system. The cyclone Fani has followed the recurving track which has been chiefly steered by an upper tropospheric level anticyclonic circulation.

Keywords Tropical cyclone · Fani · Development · Environmental conditions · Bay of Bengal

P. Bhardwaj
Department of Geography, Government College, Bahu, Jhajjar, India

O. Singh (✉)
Department of Geography, Kurukshetra University, Kurukshetra, India

P. K. Shit et al. (eds.), *Geospatial Technology for Environmental Hazards*, Advances in Geographic Information Science, https://doi.org/10.1007/978-3-030-75197-5_12

12.1 Introduction

Tropical cyclones (TCs) are among the most disastrous events, resulting immense damage to infrastructure and property, and human lives loss at the time of landfall (Peduzzi et al., 2012). Many tropical and subtropical coastal regions are frequently devastated by the occurrence of these TCs (Li & Li, 2013). However, vulnerability and their impact significantly vary from one region to another (Bhardwaj et al., 2020). It is well-known that the development and progress of these TCs is largely controlled by a few thermodynamic and dynamic environmental factors such as high sea surface temperature (SST) (>26.5 °C), reduced vertical wind shear, high low level vorticity, substantial Coriolis force, and high mid-tropospheric relative humidity (Gray, 1968, 1975; Webster et al., 2005). Therefore, these factors have been frequently used as the key predictors for their development and progress (Chan & Liu, 2004; Aiyyer & Thorncroft, 2006).

The Bay of Bengal (BoB) accounts nearly 80 per cent of the total TCs of the North Indian Ocean (IMD, 2011). These TCs cause huge loss of lives in the rim countries, i.e., India, Bangladesh, and Myanmar (Alam & Dominey-Howes, 2015). For example, two extreme cyclones originated in BoB in the years 1970 and 1991 have caused a loss of about 3,00,000 and 1,40,000 human lives, respectively, in Bangladesh (Choudhury, 2001). Likewise, in the year 1999 Orissa super cyclone and in 2008 cyclone Nargis have killed about 10,000 and 1,38,000 people in India and Myanmar, respectively (Chittibabu et al., 2004; Fritz et al., 2009). Bhardwaj and Singh (2020) have shown that majority of TCs in the BoB occurs in two seasons. The post-monsoon season accounts for about 64 per cent, and pre-monsoon season accounts for about 21 per cent of the total annual TCs. However, the conversion rate of cyclonic storms into intense cyclonic storms is greater during pre-monsoon than post-monsoon season in the BoB. Besides, most of pre-monsoon's TCs usually follow the northward tracks or recurved towards northeast and make landfall over Bangladesh and Myanmar coasts. During 1891–2017, 14 TCs have formed in April in the BoB; however, only one TC made landfall over the Indian mainland coast (Sangomla, 2019). Balasubramanian and Chalamalla (2020) have examined the dynamics which led the rapid intensification of Amphan cyclone in the BoB.

Recently, the TCs of BoB have shown an unusual behavior in terms of timings of occurrence and intensification rates. For instance, cyclones Ockhi of 2017 and Titli of 2018 have rapidly intensified just before the landfall. This cyclonic system has formed near the equator (2.7°N latitude). In such lower latitudes, the cyclones' formation is rare. The ESCS Fani of 2019 is second in the past 118 years that has formed in the month of April in BoB and crossed the Indian mainland (Sangomla, 2019). It is the most intense cyclone during satellite era (1965 onwards) which formed during the pre-monsoon season and crossed the Orissa coast. Also, it is the10th most severe TC in Indian subcontinent in the last 52 years. It has affected the large parts of eastern and northeastern Indian states and resulted in about 64 deaths and property loss of about 9000 crores. Moreover, Kumar et al. (2020) have examined the impacts of the Fani cyclone and noticed that it has severely affected the agricultural and built-up area over the eastern India and Bangladesh. Therefore, a detailed study of cyclone Fani is of very much importance. In this study, an attempt has been made to examine the development and progress of ESCS Fani over the BoB. It is believed that the finding pertaining to the formation and intensification of Fani cyclone will be helpful for meteorologists, forecasters, scientific community, and disaster managers to reduce the impacts of future cyclones.

12.2 Data and Methods

For North Indian Ocean, Regional Specialized Meteorological Centre (RSMC), New Delhi, provides the TCs' best track data. Therefore, the best track data of RSMC, New Delhi has been used to examine the various characteristics of Fani cyclone (RSMC, 2019a). The TC Fani is chiefly monitored with satellite supported by meteorological buoys, coastal and inland observations, and Doppler Weather Radars. The dataset comprised a detailed information of Fani TC including the latitudinal and longitudinal position, current intensity number, estimated maximum sustained wind (MSW) (kt), estimated central pressure (hPa), estimated pressure drop at the center (hPa), and grade at 3-hourly time (0000, 0300, 0600, 0900, 1200, 1500, 1800, 2100 UTC) (Table 12.1).

Additionally, to examine the development and progress of Fani cyclone, several dynamic and thermodynamic environmental conditions have been examined. The daily SST data has been obtained from Advanced Very High Resolution Radiometer-SST from National Oceanic and Atmospheric Administration. Besides, the data pertaining to relative humidity, outgoing longwave radiation (OLR), precipitable water, vertical velocity (omega), lower (850 hPa) and upper (200 hPa) winds, and vertical wind shear have been acquired from the National Centers for Environmental Prediction-National Center for Atmospheric Research (NCEP-NCAR) reanalysis (Kalnay et al., 1996). The composite maps of these environmental parameters have been prepared to examine their daily pattern during the cyclone period. The composite maps have been prepared by means of GrADS software. Besides, track of Fani cyclone has been prepared by using the ArcGIS 10.1 software.

12.3 Results and Discussion

A Brief Life History of ESCS Fani

A low-pressure area has formed near equator (approximately 2.7°N latitude) over the southeast BoB in the early morning (0530 IST) on 25 April 2019 (Fig. 12.1). It has deepened over the same region due to favorable environmental conditions and classified as depression on 26 April. Then it has moved northwestward and strengthened into a deep depression on 27 April. It has further intensified into a cyclonic

Table 12.1 Best track positions and other parameters of the ESCS Fani over BoB during 26 April–04 May, 2019

Date	Time (UTC)	Center latitude	Center longitude	Current intensity No.	MSW (kt)	Estimated central pressure (hPa)	Estimated pressure drop at the center (hPa)	Intensity
26-04-19	3	2.7	89.7	1.5	25	998	4	D
	6	3.0	89.4	1.5	25	998	4	D
	12	3.2	89.2	1.5	25	998	4	D
	18	3.7	88.8	1.5	25	998	4	D
27-04-19	0	4.5	88.8	2.0	30	997	5	DD
	3	4.9	88.7	2.0	30	996	6	DD
	6	5.2	88.6	2.5	35	995	7	CS
	9	5.4	88.5	2.5	40	994	8	CS
	12	5.9	88.5	3.0	45	992	10	CS
	15	6.3	88.5	3.0	45	992	10	CS
	18	6.6	88.2	3.0	45	992	10	CS
	21	6.9	87.9	3.0	45	992	10	CS
28-04-19	0	7.3	87.9	3.0	45	992	10	CS
	3	7.3	87.9	3.0	45	992	10	CS
	6	7.4	87.8	3.0	45	992	10	CS
	9	7.7	87.5	3.0	45	992	10	CS
	12	8.2	87.0	3.0	45	992	10	CS
	15	8.3	86.9	3.0	45	992	10	CS
	18	8.4	86.9	3.0	45	992	10	CS
	21	8.5	86.9	3.0	45	992	10	CS
29-04-19	0	8.6	86.9	3.0	45	992	10	CS
	3	8.7	86.9	3.0	45	992	10	CS
	6	9.2	86.9	3.0	45	992	10	CS
	9	9.7	86.8	3.0	45	992	10	CS
	12	10.1	86.7	3.5	55	986	16	SCS
	15	10.4	86.7	3.5	55	986	16	SCS
	18	10.8	86.6	3.5	55	986	16	SCS
	21	11.1	86.5	3.5	60	986	16	SCS
30-04-19	0	11.7	86.5	4.0	65	980	22	VSCS
	3	12.3	86.2	4.5	75	974	28	VSCS
	6	12.6	85.7	4.5	80	970	32	VSCS
	9	13.0	85.3	4.5	85	966	36	VSCS
	12	13.3	84.7	5.0	90	962	40	ESCS
	15	13.4	84.5	5.0	95	957	45	ESCS
	18	13.5	84.4	5.0	95	957	45	ESCS
	21	13.6	84.2	5.0	95	957	45	ESCS
	0	13.9	84.0	5.0	95	957	45	ESCS
	3	14.1	83.9	5.0	95	957	45	ESCS

(continued)

Table 12.1 (continued)

Date	Time (UTC)	Center latitude	Center longitude	Current intensity No.	MSW (kt)	Estimated central pressure (hPa)	Estimated pressure drop at the center (hPa)	Intensity
01-05-19	6	14.2	83.9	5.0	95	957	45	ESCS
	9	14.5	84.1	5.0	95	955	45	ESCS
	12	14.9	84.1	5.5	100	950	50	ESCS
	15	15.1	84.1	5.5	100	950	50	ESCS
	18	15.2	84.1	5.5	100	950	50	ESCS
	21	15.5	84.2	5.5	100	950	50	ESCS
02-05-19	0	15.9	84.5	5.5	105	945	55	ESCS
	3	16.2	84.6	5.5	105	945	55	ESCS
	6	16.7	84.8	5.5	110	940	60	ESCS
	9	17.1	84.8	6.0	115	932	66	ESCS
	12	17.5	84.8	6.0	115	932	66	ESCS
	15	17.8	84.9	6.0	115	934	66	ESCS
	18	18.2	85.0	6.0	115	934	66	ESCS
	21	18.6	85.2	6.0	115	934	66	ESCS
03-05-19	0	19.1	85.5	6.0	105	945	55	ESCS
	3	19.6	85.7	5.5	100	950	50	ESCS
	6	20.0	85.9	.	85	952	46	VSCS
	9	20.5	86.0	.	75	970	28	VSCS
	12	21.1	86.5	.	70	976	22	VSCS
	15	21.5	86.7	.	60	980	18	SCS
	18	21.9	87.1	.	55	986	16	SCS
	21	22.5	87.9	.	50	990	12	SCS
04-05-19	0	23.1	88.2	.	40	994	8	CS
	3	23.6	88.8	.	30	996	6	DD
	6	24.3	89.3	.	25	998	5	D
	12	25.2	90.7	.	20	1000	4	D

storm named as "Fani" nearby noon on 27 April. Then it has moved towards northwest and further strengthened into severe, very severe, and extremely severe cyclonic storm. It has started to recurve towards north and northeastward from 1 May and reached at its peak (115 kt) on 2 May. It has crossed the Orissa coast as an ESCS with MSW about 110 kt in the morning on 3 May. Then it has moved towards northeast and crossed the West Bengal, Bangladesh, and finally diminished nearby central Assam on 4 May. The total accumulated cyclone energy and power dissipation index generated by the ESCS Fani are approximately 16.7 × 10^4 kt^2 and 15.1 × 10^6 kt^3, respectively, which are higher than their long-term mean (1972–2017), i.e., 13.0 × 10^4 kt^2 and 12.4 × 10^6 kt^3 (Bhardwaj & Singh, 2020). The 3-hourly detailed information cyclone of Fani has been presented in Table 12.1, and the track and daily development and progress of ESCS Fani have been displayed

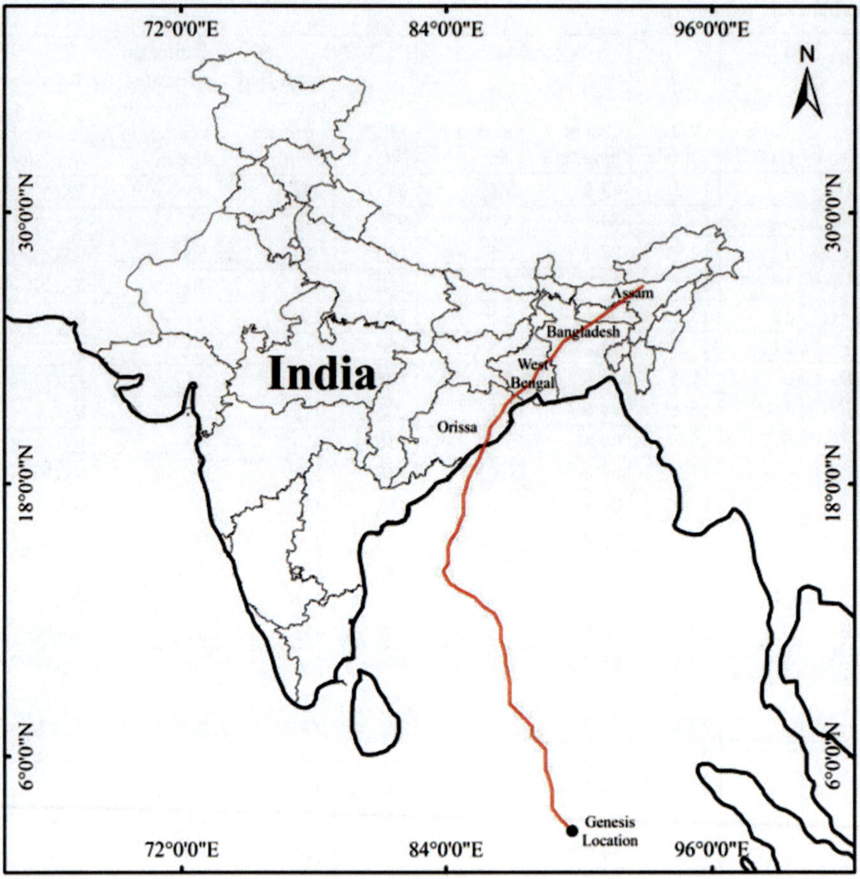

Fig. 12.1 Genesis location and track of ESCS Fani over the Bay of Bengal during April 26 to May 04, 2019

in Figs. 12.1 and 12.2. It is among the long-lasting cyclones of BoB as it travelled the distance of about 3030 km. Hence, all these facts suggest the need for a detailed analysis of the large-scale environmental conditions linked with the development and progress of ESCS Fani. These have been discussed in the subsequent section:

Large-Scale Environmental Conditions

Outgoing Longwave Radiation (OLR)

OLR is a proxy indicator of convective clouds. Strong convective activity over a region supports the formation of TCs. The higher OLR values indicate less convective activity and vice-versa. Figure 12.3 demonstrates the daily mean OLR pattern

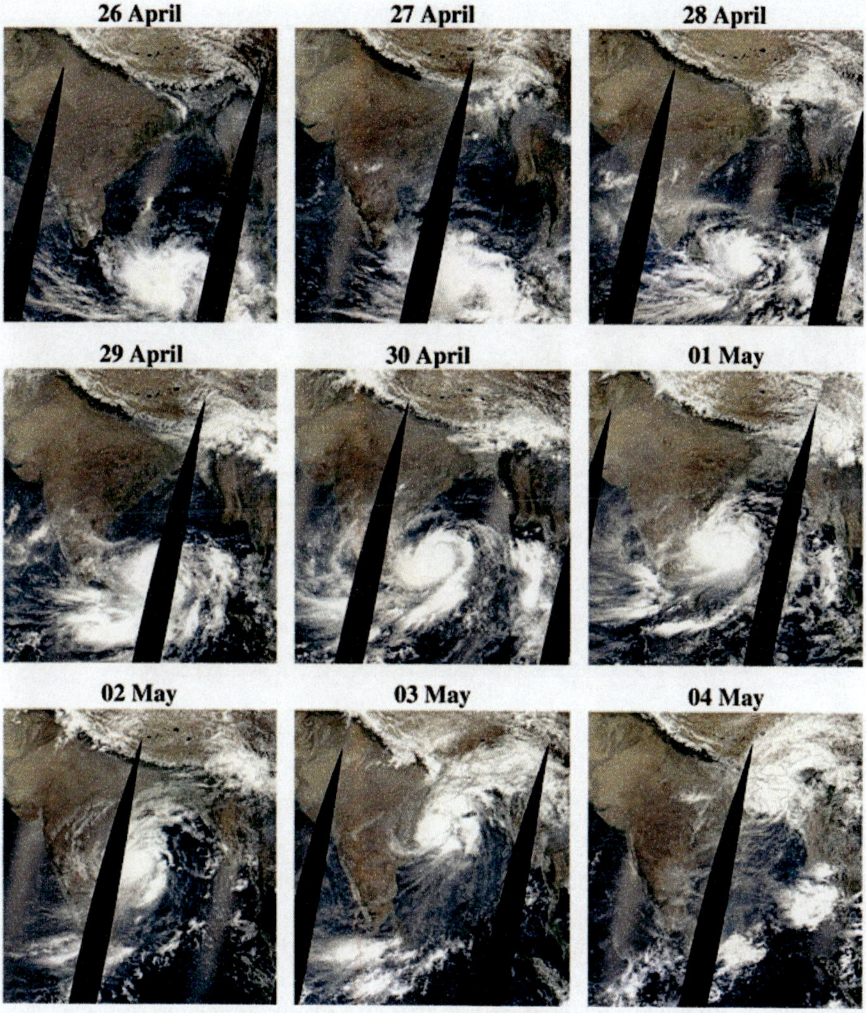

Fig. 12.2 Progress of ESCS Fani over the Bay of Bengal during April 26 to May 04, 2019.Images are of moderate-resolution imaging spectroradiometer (MODIS) Terra sensor

during April 26 to May 04, 2019.The figure clearly shows that the convective activity is higher in the region where the Fani has initially developed as low pressure area. The convectively active phase of Madden-Julian Oscillation (MJO) with amplitude greater than 1 has also located continuously over the Bay of Bengal which provided the favorable conditions for the enhancement of convection and strengthening of Fani (Bhardwaj et al., 2019a; RSMC, 2019b). Then this region of strong convective activity has moved towards Orissa coast. However, the convective activity has weakened near the coast and reduced significantly after the landfall of ESCS Fani.

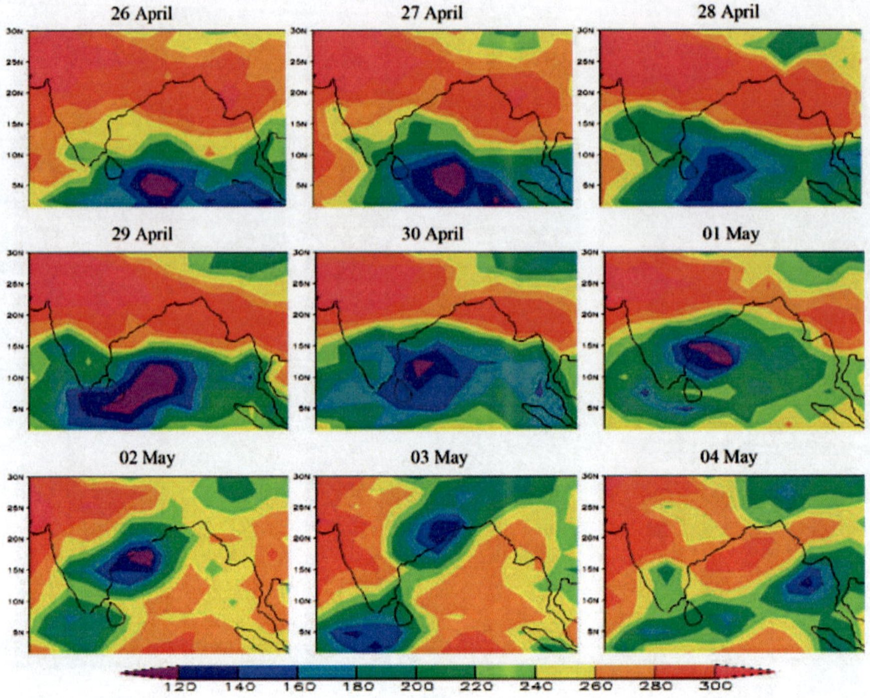

Fig. 12.3 Composites of OLR (Wm^{-2}) during the lifetime of ESCS Fani during 26 April to 04 May, 2019

Sea Surface Temperature (SST)

SST is the most important factor which controls the development and strengthening of TCs (Sebastian & Behera, 2015). Figure 12.4 displays the daily SST pattern over the BoB during April 26–May 04, 2019. The figure clearly exhibits that SST is near about 30 °C over the large parts of BoB during the cyclone period, which is very high than the minimum required SST (>26.5 °C) for the formation of a cyclone (Gray, 1968). The consistent high SST is suitable for the formation and strengthening of cyclone Fani over the BoB. Also, a cooling of SST can be seen over the central parts of the BoB from the 1 May, as the Fani has intensified further. This cooling of SST is mainly ascribed to the evaporation produced by the strong winds and blockage of solar radiations due to presence of large-scale convective clouds associated with Fani (Bhardwaj et al., 2019b). Similarly, a rapid reduction in SST can be seen towards the coastal areas of Orissa where Fani reached at its peak.

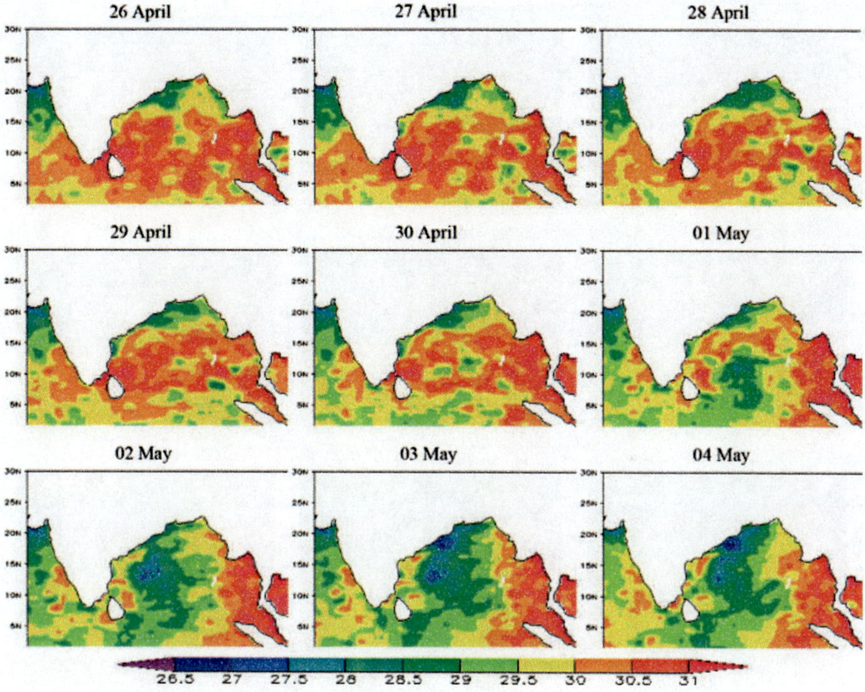

Fig. 12.4 As in Fig. 12.3, but for SST (°C)

Relative Humidity

The sufficient amount of relative humidity (at least 40%) at mid-tropospheric level is required for the intensification of TCs (Gray, 1968). Figure 12.5 exhibits that relative humidity is approximately 80 per cent at mid-tropospheric level where the Fani has developed as depression on 26 April in the southern BoB. Then, this region of high humidity has migrated towards north with the movement of cyclone over the BoB. The presence of higher relative humidity at mid-tropospheric level has released high latent heat which energized and intensified the system. However, after the landfall of the Fani, the relative humidity has decreased rapidly.

Vertical Velocity (Omega)

Figure 12.6 shows the spatial pattern of vertical velocity over the BoB during April 26 to May 04, 2019, at mid-tropospheric level. A negative omega indicates upward motion and vice-versa. Small negative omega values indicate slightly weak vertical motion over the areas where cyclone has initially developed as depression on 26 April over the southern BoB. After that, a strong vertical motion has taken place over the central to northern BoB on 29–30 April. On 30 April, again small

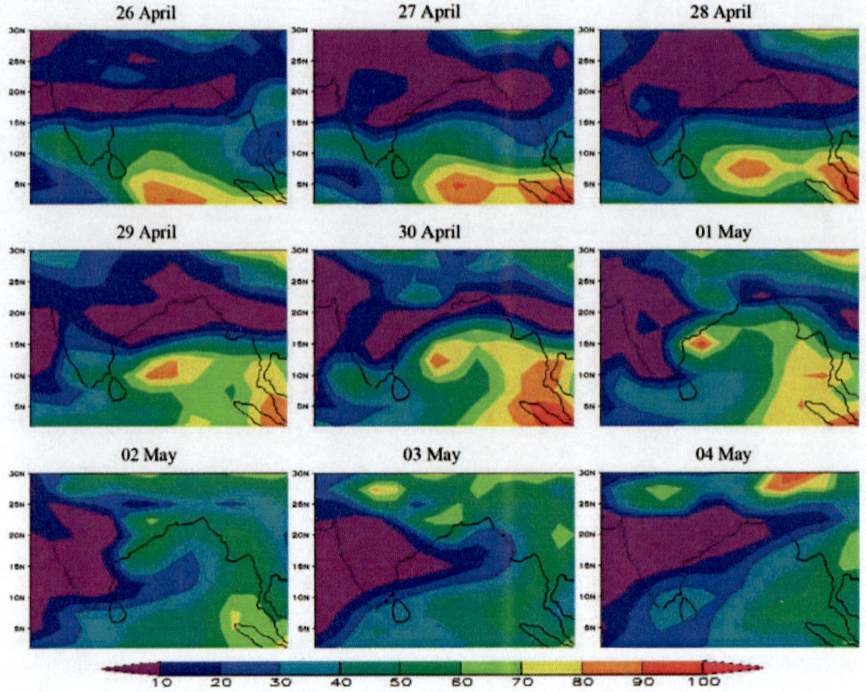

Fig. 12.5 As in Fig. 12.3, but for relative humidity (%)

negative omega values indicate the fast surface motion of cyclone towards Orissa coasts. After 30 April, again strong vertical motion has occurred over the Orissa coast and continued after the landfall. The presence of strong vertical motion over warm oceanic water has transported the moisture to mid-tropospheric level and helped in intensification of cyclone.

Low Level Winds (850 hPa)

Figure 12.7 shows the pattern of low-level winds during April 26 to May 04, 2019, over the BoB. The figure clearly shows that a low level cyclonic circulation is present over the southern BoB on 26 April, which assisted in formation of cyclone. On 27–28 April, the cyclonic circulation has moved towards the central BoB. Later, this cyclonic circulation has intensified with movement towards Orissa coast between April 29 and May 02. On 3 May, the cyclonic circulation has mostly crossed the Orissa coast. The strong winds can be seen around the eye of cyclonic circulation, whereas winds are weak within eye and nearby region. The cyclonic circulation has weakened rapidly on 04 May after the landfall on 3 May.

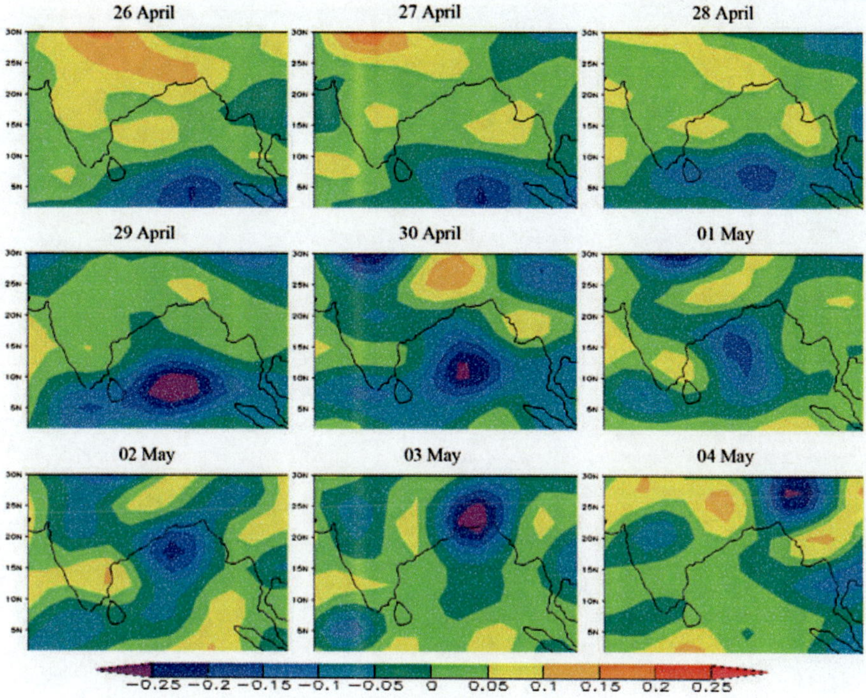

Fig. 12.6 As in Fig. 12.3, but for vertical velocity (Omega; Pa/s)

Upper Level Winds (200 hPa)

It is well-known that upper level winds control the track direction of cyclones. Figure 12.8 displays the pattern of upper level winds during April 26 to May 04, 2019, over the BoB. On 26 and 27 April, strong easterly (westerly) winds can be seen in the upper troposphere over the southern (northern) BoB when the cyclone Fani is in initial stage. An anticyclonic circulation lies over South Thailand and adjoining South Andaman Sea in the middle and upper tropospheric levels (RSMC, 2019b). This anticyclonic circulation has steered the system northwestward. Then, this anticyclonic system has embedded with strong upper level westerly winds and steered the cyclone northeastward. After landfall, due to the impact of upper level westerly winds, the cyclonic system has moved rapidly with an average speed of 24.0 km/h.

Vertical Wind Shear (200–850 hPa)

Vertical wind shear is a vital factor which controls the TCs. The less amount of vertical wind shear is favorable for the development and strengthening of TCs and vice-versa. Figure 12.9 demonstrates the pattern of vertical wind shear during April

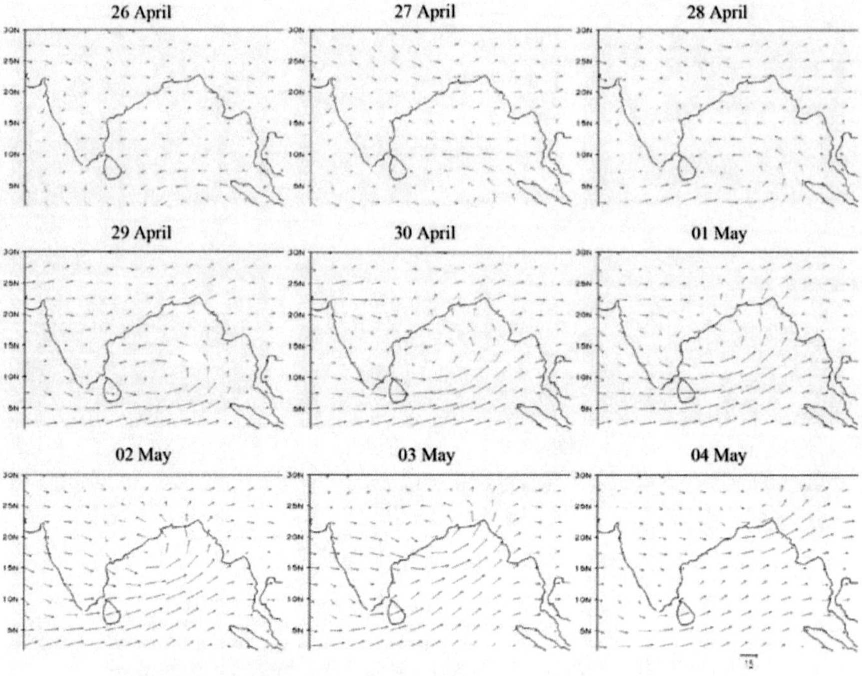

Fig. 12.7 As in Fig. 12.3, but for low level winds (850 hPa; ms^{-1})

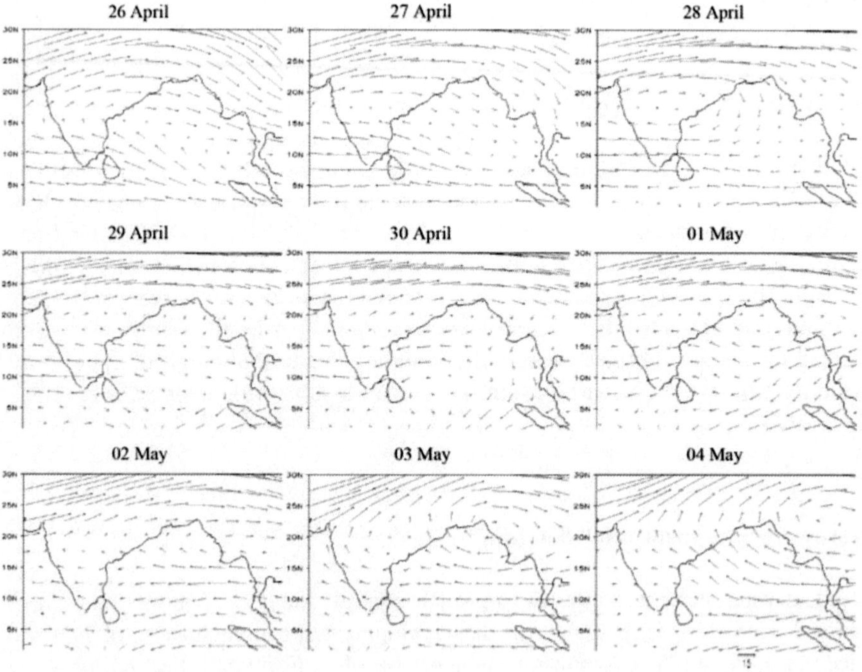

Fig. 12.8 As in Fig. 12.3, but for upper level winds (200 hPa; ms^{-1})

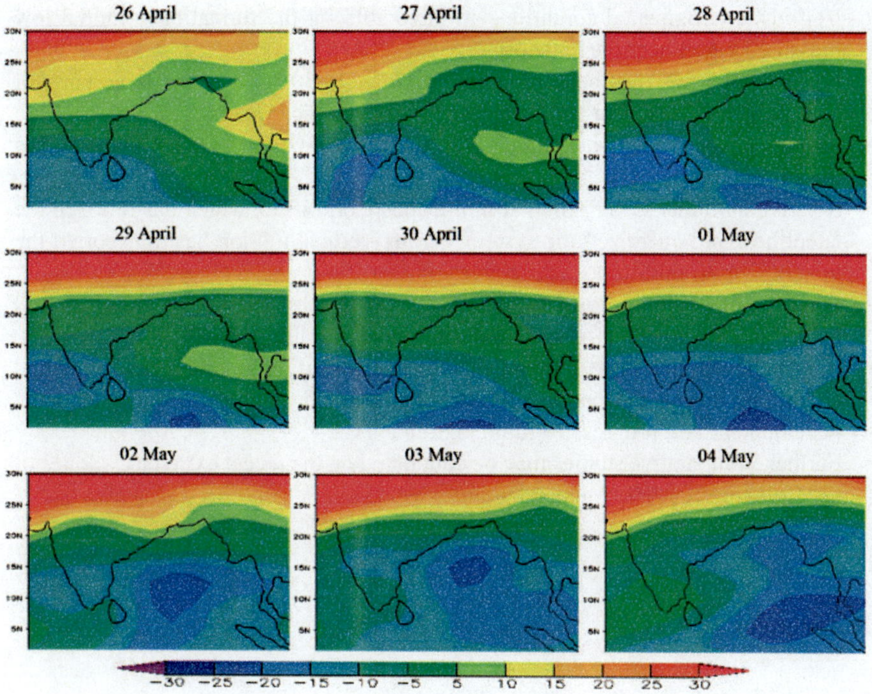

Fig. 12.9 As in Fig. 12.3, but for vertical wind shear (200–850 U; ms^{-1})

26 to May 04, 2019, over the BoB. The vertical wind shear is less on 26 April when initial low pressure system has developed as depression over the southern BoB. Later, the region of less vertical wind shear has spread over all the parts of BoB during the cyclone period. This less vertical wind shear has continuously provided the favorable condition for the intensification of cyclone Fani over the BoB.

12.4 Conclusions

The major conclusions of this study are as follows:

- The cyclone Fani of April 2019 is among the most intense cyclones which formed over the southern BoB and crossed the Indian mainland.
- It is distinctive in terms of its location of origin, direction of track, and landfall location.
- It is the most intense cyclone during the satellite era (1965 onwards), which has formed during the pre-monsoon season and crossed the Orissa state of Indian mainland.

- Various environmental conditions are favorable for the formation of initial low pressure system near the equator.
- For example, SST has been found approximately 30 °C over the large parts of BoB, which is sufficient for development of cyclonic system.
- Strong convective activities along with the active MJO have taken place during the storm period.
- Relative humidity is very high at mid-tropospheric level, which has released the latent heat and energized the system. Strong vertical motion has transported the moisture up to mid-tropospheric level.
- Presence of cyclonic vorticity at lower level and less vertical wind shear has helped in strengthening of system during the entire life period.
- It is believed that this comprehensive analysis of large-scale environmental conditions during the period of ESCS Fani will be helpful to understand the development and progress of future extreme TCs.
- Further, comparative studies may be conducted on the recent extreme cyclones of the BoB for their better understanding.

References

Aiyyer, A. R., & Thorncroft, C. (2006). Climatology of vertical wind shear over the tropical Atlantic. *Journal of Climate, 19*, 2969–2983.

Alam, E., & Dominey-Howes, D. (2015). A new catalogue of tropical cyclones of the northern Bay of Bengal and the distribution and effects of selected landfalling events in Bangladesh. *International Journal of Climatology, 35*, 801–835.

Balasubramanian, S., & Chalamalla, V. K. (2020). Super cyclone Amphan: A dynamical case study. *Atmospheric and Oceanic Physics*. https://arxiv.org/abs/2007.02982.

Bhardwaj, P., & Singh, O. (2020). Climatological characteristics of bay of Bengal tropical cyclones: 1972-2017. *Theoretical and Applied Climatology, 139*, 615–629.

Bhardwaj, P., Pattanaik, D. R., & Singh, O. (2019b). Tropical cyclone activity over Bay of Bengal in relation to El Niño-Southern Oscillation. *International Journal of Climatology, 39*, 5452–5469.

Bhardwaj, P., Singh, O., & Yadav, R. B. S. (2020). Probabilistic assessment of tropical cyclones' extreme wind speed in the Bay of Bengal: Implications for future cyclonic hazard. *Natural Hazards, 101*, 275–295.

Bhardwaj, P., Singh, O., Pattanaik, D. R., & Klotzbach, P. J. (2019a). Modulation of Bay of Bengal tropical cyclone activity by the Madden-Julian oscillation. *Atmospheric Research, 229*, 23–38.

Chan, J. C. L., & Liu, K. S. (2004). Global warming and western North Pacific typhoon activity from an observational perspective. *Journal of Climate, 17*, 4590–4602.

Chittibabu, P. S., Dube, K., Macnabb, J. B., Murty, T. S., Rao, A. D., Mohanty, U. C., & Sinha, P. C. (2004). Mitigation of flooding and cyclone hazard in Orissa, India. *Natural Hazards, 31*, 455–485.

Choudhury, A. M. (2001). Cyclones in Bangladesh. In K. Nizamuddin (Ed.), *Disaster in Bangladesh: Selected readings* (pp. 61–76). Department of Geography and Environment, University of Dhaka, Dhaka.

Fritz, H. M., Blount, C. D., Thwin, S., Thu, M. K., & Chan, N. (2009). Cyclone Nargis storm surge in Myanmar. *Nature Geoscience, 2*, 448–449.

Gray, W. M. (1968). Global view of the origin of tropical disturbances and storms. *Monthly Weather Review, 96*, 669–700.

Gray, W. M. (1975). *Tropical cyclone genesis*. Department of Atmospheric Science. Paper No. 234, Colorado State University, Fort Collins.

IMD. (2011). *Tracks of cyclones and depressions over North Indian Ocean (from 1891 onwards) (Cyclone eAtlas—IMD, Version 2.0)*. Cyclone Warning and Research Centre, India Meteorological Department, Regional Meteorological Centre, Chennai.

Kalnay, E., Kanamitsu, M., Kistler, R., Collins, W., Deaven, D., Gandin, L., Iredell, M., Saha, S., White, G., Woollen, J., Zhu, Y., Chelliah, M., Ebisuzaki, W., Higgins, W., Janowiak, J., Mo, K. C., Ropelewski, C., Wang, J., Leetmaa, A., . . . Joseph, D. (1996). The NCEP/NCAR 40-year reanalysis project. *Bulletin of the American Meteorological Society, 77*, 437–471.

Kumar, S., Lal, P., & Kumar, A. (2020). Turbulence of tropical cyclone 'Fani' in the Bay of Bengal and Indian subcontinent. *Natural Hazards, 103*, 1613–1622.

Li, K., & Li, G. S. (2013). Risk assessment on storm surges in the coastal area of Guangdong province. *Natural Hazards, 68*, 1129–1139.

Peduzzi, P., Chatenoux, B., Dao, H., Bono, A. D., Herold, C., Kossin, J., Mouton, F., & Nordbeck, O. (2012). Global trends in tropical cyclone risk. *Nature Climate Change, 2*, 289–294.

RSMC. (2019a). *Extremely severe cyclonic storm "FANI" over east central equatorial Indian Ocean and adjoining southeast Bay of Bengal (26 April—04 May, 2019): Summary*. Retrieved June 8, 2019, from https://reliefweb.int/sites/reliefweb.int/files/resources/fani.pdf

RSMC. (2019b). *Special tropical weather outlook*. Regional Specialised Meteorological Centre-Tropical Cyclones, New Delhi. Retrieved June 8, 2019, from http://www.rsmcnewdelhi.imd.gov.in/images/pdf/archive/bulletins/2019/rfani.pdf

Sangomla, A. (2019). *Fani to be second severe April cyclone to make landfall in India in 118 years*. Retrieved June 14, 2019, from https://www.downtoearth.org.in/news/natural-disasters/fani-to-be-second-severe-april-cyclone-to-make-landfall-in-india-in-118-years-64278

Sebastian, M., & Behera, M. R. (2015). Impact of SST on tropical cyclones in North Indian Ocean. *Procedia Engineering, 116*, 1072–1077.

Webster, P. J., Holland, G. J., Curry, J. A., & Chang, H. R. (2005). Changes in tropical cyclone number, duration and intensity in a warming environment. *Science, 309*, 1844–1846.

Chapter 13
AHP-Based Spatial Composite Impact Assessment Model (SCIAM) of Highway Broadening in Sikkim Himalaya

Polash Banerjee, Mrinal K. Ghose, and Ratika Pradhan

Abstract The environmental degradation associated with a highway project can sometimes outweigh its benefits. Limited studies on the aggregate spatial impacts of highway projects in remote areas on varied environmental criteria have been done. SCIAM assesses the spatial distribution of impacts of the highway broadening project in East Sikkim on environmental criteria like air, water, noise, biodiversity, socioeconomy, and landslide susceptibility. The impacts predicted under the "with"- and "without"- project scenarios are based on mathematical models, landscape metrics, and socioeconomic survey. Spatially Explicit Sensitivity Analysis (SESA), model validation, and cross-validation criteria suggest that the spatial model is robust. SCIAM indicates that areas near the highway had a severe adverse impact compared to areas away from the highway. Moreover, Spatial Composite Impact Index (SCII) shows that the highway broadening is not viable for the local environment. These outcomes are primarily explained by the outweighing of loss of biodiversity, landslide susceptibility, and water pollution over the gains of socio-economic benefits and decline in air and noise pollution due to the project. SCIAM can be a geovisualization and decision support tool for environmental managers.

Keywords Expert opinion · Aggregate Impact Assessment · Environmental Impact Assessment · Spatial analysis · Highway · Sensitivity analysis · Mountain

---

Supplementary Information The online version of this chapter (https://doi.org/10.1007/978-3-030-75197-5_13) contains supplementary material, which is available to authorized users.

P. Banerjee (✉)
Independent researcher, Gangtok, India

M. K. Ghose
Computer Engineering and Applications, Institute of Engineering & Technology, GLA University, Mathura, UP, India
e-mail: mrinalkanti.ghose@gla.ac.in

R. Pradhan
Department of Computer Engineering, Sikkim Institute of Science and Technology, Sikkim, India

P. K. Shit et al. (eds.), *Geospatial Technology for Environmental Hazards*, Advances in Geographic Information Science, https://doi.org/10.1007/978-3-030-75197-5_13

13.1 Introduction

Remote areas enjoy substantial benefits through highway projects in the form of greater security, development, and connectivity (Brown, 2003; Rudiarto & Handayani, 2011). However, environmental concerns should be weighed against such benefits to assess the viability of a highway project. Conventional Environmental Impact Assessment (EIA) often ignores the composite and spatial impacts of a project (Glasson et al., 2005; Takyl, 2012). GIS and Multi-Criteria Decision Making (MCDM) method based SCIAM can overcome these limitations. Spatial analysis can predict the spatial impacts of a project, while a composite impact analysis based on these predictions can facilitate prompt decision-making by the stakeholders. GIS can be integrated with mathematical models to assess the spatio-temporal impacts of development projects and act as a decision support tool (Atkinson & Canter, 2011). Presently, such integrations are predominantly confined to address spatial impacts concerning one or a few environmental criteria, like air, water, noise, or biodiversity. SCIAM presents a methodology to combine and assess the spatial impacts of a highway broadening project on a wider set of environmental criteria in the East district of Sikkim.

Limited studies have been conducted so far on the application of spatial analysis in the composite impact assessment of development projects in mountainous areas (Geneletti, 2008; Geneletti & Dawa, 2009; Warner & Diab, 2002). To be more specific, applications of spatial analysis of highway projects in mountainous areas remain largely unexplored (Banerjee & Ghose, 2016). GIS-based EIA of highway projects can identify environmental issues that sometimes conventional EIA fails to capture (Atkinson & Canter, 2011). Unfortunately, much of the spatial impact assessments on highway projects focus on either physical impacts due to pollution (Ahmed et al., 2017; Amin et al., 2017; Banerjee et al., 2016; Cai et al., 2015), impacts on biodiversity (Bennett, 2017; Geneletti, 2004; Gülci & Akay, 2015), socioeconomic impacts (Banerjee & Ghose, 2017), or natural hazards associated with highways (Wang et al., 2017; Zhao et al., 2017). The data constraints associated with EIA of remote areas often limit the use of spatial analysis in EIA studies. However, a fair part of such limitations can be overcome with the use of mathematical modelling and expert opinion-based impact assessment (Banerjee & Ghose, 2016; Therivel & Wood, 2017). Use of SESA (H. Chen et al., 2011; Y. Chen et al., 2010; Crosetto et al., 2000; Feizizadeh et al., 2014; Lilburne & Tarantola, 2009; Qi et al., 2013; Xu & Zhang, 2013), model validation criteria (Paliwal & Srivastava, 2014), and spatial cross-validation criteria (Chang, 2017; Lloyd, 2009) can assess the robustness and effectiveness of such impact assessment models (Lo & Yeung, 2006; Longley et al., 2010). Recent attempts to capture the sustainability issues of highway projects have been addressed by very few studies (El-Kholy & Akal, 2020). However, such studies lack the spatial dimensions of environmental impacts.

In the majority of spatial analysis-based highway-related impact studies, sensitivity analysis is often ignored due to its high computation cost, at the expense of ignoring the assessment of the robustness of the model. Moreover, socioeconomic impact assessments of highway projects are overlooked (Burdge, 2002; Geurs et al., 2009). SCIAM intends to address these literature gaps. SCIAM attempts to perform spatial analysis of composite impact assessment of highway projects in a remote area and a developing country. It is a Weighted Linear Combination (WLC) model of environmental criteria maps. Environmental criteria like air quality, water quality, noise, biodiversity value, socioeconomic conditions, and landslide susceptibility and their sub-criteria were estimated either using a mathematical model or through survey methods. Analytic Hierarchy Process (AHP) was used in weighing the environmental criteria and their sub-criteria. AHP is an MCDM method that uses expert opinion-based criteria weighing (Saaty, 1980, 1990; Saaty & Vargas, 1994). In comparison to data-intensive statistical methods, AHP relies on experts' opinion for criteria weighing (Arriaza & Nekhay, 2008). Moreover, unlike non-expert opinion based MCDM, AHP can capture the study area-specific peculiarities of environmental criteria through experts' valuation (Karbassi et al., 2011). Experts used in AHP of environmental criteria should ideally have sufficient knowledge of the local environment of the study area and understanding of the plausible impacts of the concerned development project on the local environment.

In this study, a methodology is presented to construct a spatial model that predicts the magnitude, distribution, and impacts of the project on the local environment. The overarching objectives of this study are:

- Integrate spatial analysis in the environmental impact evaluation of the considered development project.
- Quantification of various impacts of the project on the environmental criteria and present a composite impact map.
- Assess the viability of the project based on spatial impacts.
- Assess the robustness and reliability of the model.

The study showed that the areas near the highway had the highest adverse impacts. Also, SESA suggested that areas near the highway were most sensitive to changes in environmental criteria weight. Model validation, cross-validation, and SESA indicated that SCIAM was robust. The present study is most likely the first attempt to integrate a wide range of environmental criteria and sub-criteria encompassing the physico-chemical, ecological, socioeconomic, and disaster-related impacts of a highway project that affects the mountainous and remote study area in a developing country.

13.2 Materials and Methods

Study Area

Sikkim is a small mountainous state of India in the North-Eastern Hills of Himalaya. Firstly, the state is geopolitically critical, as it shares its borders with Nepal, China, and Bhutan. Secondly, it is culturally unique due to its high proportion of tribal population predominantly following Buddhism. Thirdly, Sikkim is a part of the North-Eastern Himalayan biodiversity hotspot. These features make it a very attractive tourist spot. To address the tourism and defense needs of the state, a project has been undertaken by the government of India in 2007 to broaden the sole national highway of Sikkim NH10 from 7 to 12 m width. The project extends from the border of West Bengal to the Indo-China border in North Sikkim. The project has its fair share of impacts on the local environment. To analyze the spatial impacts of the project, a road corridor of 2 km radius along a stretch of 27 km of NH 10 was considered in the East district of Sikkim between the towns of Ranipool (27°17′28.74″ N, 88°35′31.11″ E, Elevation 847 m) and Rangpo (27°10′31.26″ N, 88°31′44.43″ E, Elevation 300 m) (Banerjee et al., 2016; Banerjee & Ghose, 2017). The likely change in traffic volume and composition for *with-project* and *without project* alternatives were projected (Supplement Table 13.S1). The rationale of considering a small study area was primarily due to the ease of data collection. The study area is mainly covered with dense vegetation, interrupted by agrarian land, villages, and small townships like Rangpo, Singtam, and Ranipool. NH 10 from Ranipool closely follows the river Ranikhola till Singtam, and then river Teesta till Rangpo and beyond.

Overview of SCIAM

To study the impacts of the project on the environment, depending on the nature of environmental criteria, varied extent and shapes of impact areas were considered based on expert opinions and available literature (Antunes et al., 2001; Geneletti, 2004) (Fig. 13.1). The road corridor was considered as an impact area for *air quality*, *noise level*, and *socioeconomic conditions* as well as *biodiversity representative area* (Banerjee et al., 2020; Geneletti, 2003; Stoms, 2000). For estimating the impact of highway broadening on the landslide susceptibility, the highway side mountain slope area was only taken into account. It spanned from the edge of the river to 500 m buffer of the highway (Banerjee et al., 2018a). Accessibility of humans and wildlife to rivers and road runoff were also considered while demarcating the impact area. Hence, the 50 m buffer along the rivers and the highway were merged to construct the water impact area (Banerjee et al., 2018b).

Several submodels were used to assess the impacts of the "with- and without-project" scenarios on the individual environmental criterion. For each environmental criterion, ancillary databases and geodatabase were created to run individual submodels. Maps of each subcriterium under individual environmental criterion (like nitrogen dioxide under air quality criterion, Chemical Oxygen Demand (COD) as water quality criterion, etc.) were generated from the submodels, and WLC was used to prepare Composite Impact Assessment map for each environmental criterion. The weights of the environmental criteria and their sub-criteria were estimated by AHP, an expert opinion-based method.

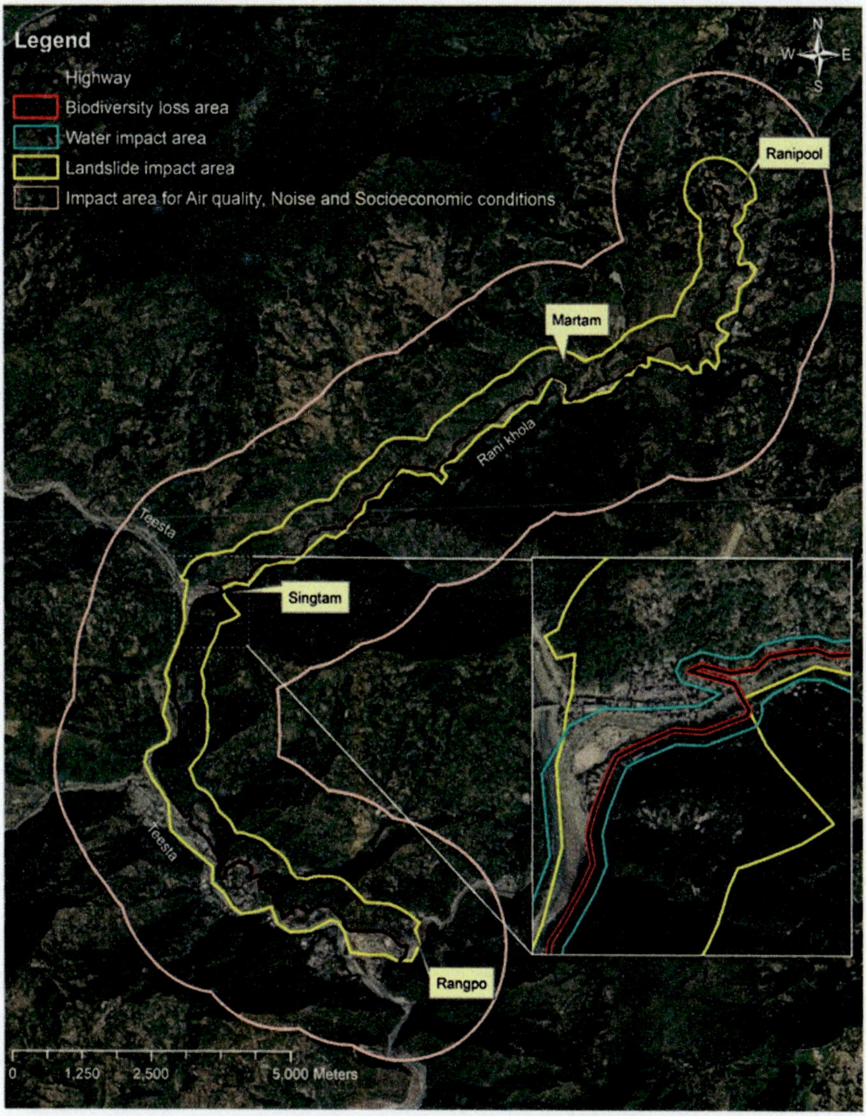

Fig. 13.1 Impact areas under various environmental criteria. (Courtesy of ESRI)

Air quality impact assessment due to traffic was done using Indian Institute of Technology Line Source (IITLS) pollutant dispersal model (Banerjee & Ghose, 2017). Federal Highway Authority Traffic Noise Model (FHWA TNM 2.5) was used to assess the noise level in the impact area (Banerjee et al., 2016). Water quality impact assessment was done using a combination of Soil Conservation Service-Curve Number (SCS-CN) method, an empirical model of traffic-induced highway runoff composition, and mass balance model to assess the contribution of highway runoff on the pollution load of the nearby rivers, viz., Ranikhola and Teesta (Banerjee et al., 2018b). The biodiversity impact assessment was performed using a set of landscape metrics and experts' opinion-based spatial model (Banerjee et al., 2020). For socioeconomic impact assessment, a GPS tagged structured questionnaire method was used (Banerjee & Ghose, 2017). Landslide susceptibility assessment along the highway was based on the information value method (Banerjee et al., 2018a). Spatial composite impact assessment map was prepared by using WLC of normalized impact assessment maps of environmental criteria (Chang, 2017). Furthermore, SESA indices, viz., Mean Absolute change rate (MACR) and Impact Category Change Rate (ICCR), were used to assess the robustness of SCIAM (Banerjee et al., 2018b). Also, the change in spatial impacts of "with-project" scenario from the "without-project" scenario was used in the Spatial Composite Impact Index (SCII) to assess the viability of highway broadening Fig. 13.2). All the GIS-related analysis was performed in ArcGIS 10.3 environment.

Data Collection

Two sets of databases were prepared to describe the environmental conditions of the year 2004 as "without-project" alternative and 2014 as "with-project" alternative. In the "without-project" alternative, the highway width was 7 m, and the road pavement was made up of a non-bitumen emulsion of 300 mm thickness. In contrast, in "with-project" alternative, the highway width was 12 m, and the road pavement was made up of bitumen emulsion of 580 mm thickness. Other model inputs like highway traffic volume and composition and land use and land cover (LULC) were also considered under the two alternatives. A wide set of spatially referred and ancillary data was prepared to perform spatial impact analysis (Table 13.1). These inputs were fed into the submodels for project impact predictions. For instance, during socio-economic impact assessment GPS-tagged responses by the project-affected people were converted into point feature maps for the preparation of socioeconomic rasters.

AHP Model

To find the weights or relative importance of the environmental criteria and their sub-criteria, AHP was performed. The weights estimated by AHP were used in WLC of the criteria rasters for the preparation of impact prediction maps. For this, the project was decomposed into a hierarchical structure of project alternatives and criteria. According to Saaty (1980), based on experts' opinion a comparison matrix of the relative importance of the various criteria and alternatives of the project was constructed on a numerical scale, as shown in Table 13.2. It was used to calculate the weight or importance of the various environmental criteria and project alternatives considered in SCIAM. In AHP, the elements of the comparison matrix, $a_{ij} > 0$,

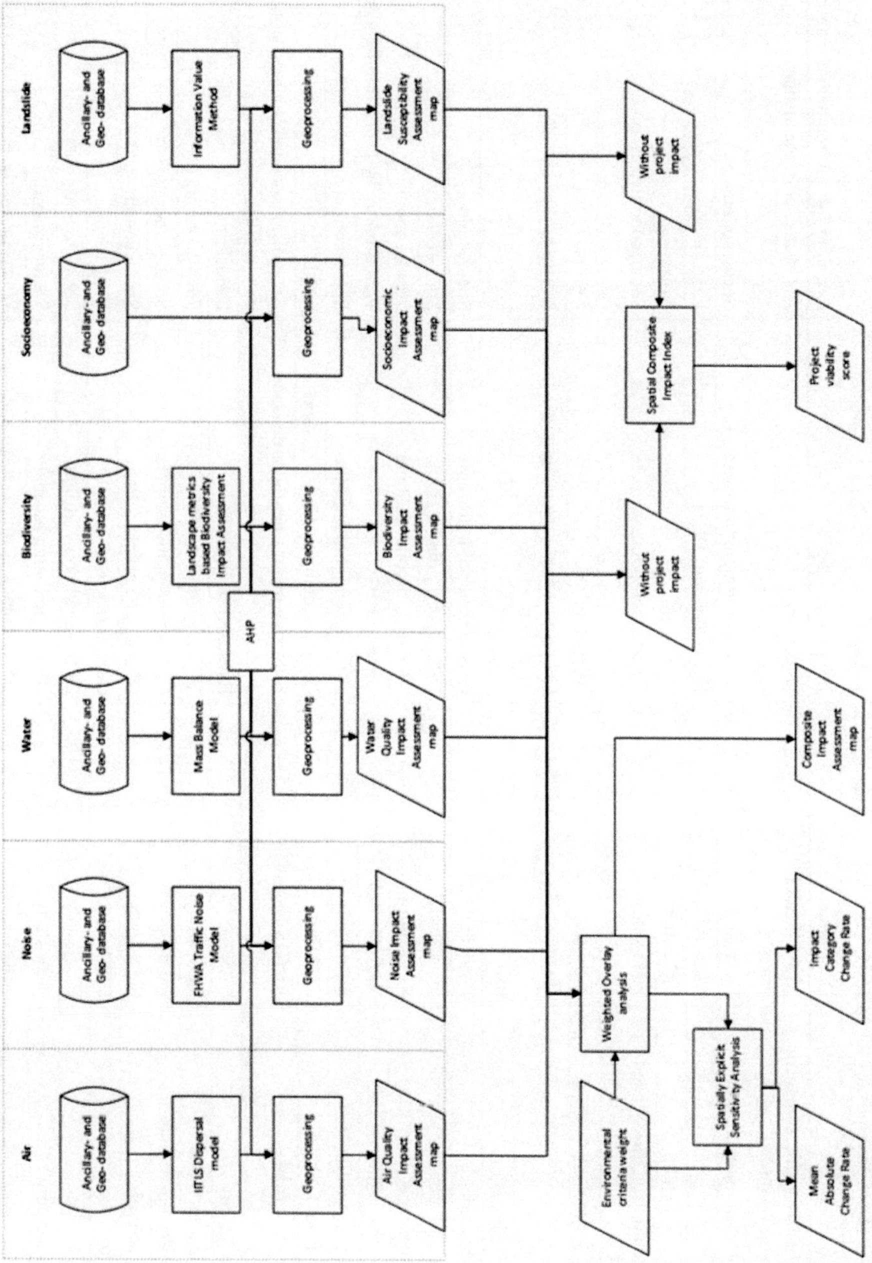

Fig. 13.2 Schematic representation of SCIAM

Table 13.1 Data prerequisites for SCIAM

Data type	Data source	Data processing	Required for environmental criteria
AADT, hourly traffic composition	BRO[a], MoRTH[b]	Direct input	Air, water, noise, landslide
Ambient air quality	SPCB[c]	Direct input	Air
Archaeological monuments	Field study	GPS tagged point attribute	Socioeconomy
Aspect	NRSC/ISRO[d]	Extracted from DEM	Landslide
Average tree height	Field study	Averaging	Noise
Banking and insurance-related activities	Field study	GPS tagged point attribute	Socioeconomy
Building percentage	Field study	Averaging	Noise
Building row coordinate	Google earth	Georeferencing, image merging, vectorization	Noise
Building row height	Field study	Averaging	Noise
Commodity prices	Field study	GPS tagged point attribute	Socioeconomy
Criminal and criminal activities	Field study	GPS tagged point attribute	Socioeconomy
Cultural heritage	Field study	GPS tagged point attribute	Socioeconomy
DEM	NRSC/ISRO	Direct input	Water, noise, landslide, biodiversity
Demographic profile	Field study	GPS tagged point attribute	Socioeconomy
Distance from fault	CISMHE (2008a)	Georeferencing and vectorization	Landslide
Distance from road	ArcGIS Basemap	Buffering nearest neighbor distance	Air, noise, landslide, socioeconomy, biodiversity
Drainage density	NRSC/ISRO	Extracted from DEM	Landslide
Educational facilities	Field study	GPS tagged point attribute	Socioeconomy
Employment opportunities	Field study	GPS tagged point attribute	Socioeconomy
Farming activities	Field study	GPS tagged point attribute	Socioeconomy
Forest types	DEFWMCC[e]	Georeferencing	Biodiversity
Health facilities	Field study	GPS tagged point attribute	Socioeconomy
Industrial activity	Field study	GPS tagged point attribute	Socioeconomy
LULC	SSRSAC[f]	Georeferencing, image classification	Water, landslide, noise, biodiversity
Lithology	CISMHE (2008a)	Georeferencing, rasterization	Landslide
Meteorological parameters	Rahman et al. (2012)	Direct input	Air

Pavement type	BRO	Direct input	Noise
Plan curvature	NRSC/ISRO	Extracted from DEM	Landslide
Rainfall	SSRSAC	Georeferencing and rasterization	Landslide
Receiver coordinate	Google earth, DEM[g]	Georeferencing, image merging, vectorization	Air, noise, water
Receiver height	Field study	Averaging	Noise
Road coordinates	Google earth, DEM	Georeferencing, image merging, vectorization	Air, noise
Road width	Field study, BRO	Direct input	Noise
Slope	NRSC/ISRO	Extracted from DEM	Landslide
Soil depth	CISMHE (2008b)	Georeferencing, rasterization	Landslide
Soil texture	CISMHE (2008b)	Georeferencing, vectorization, rasterization	Landslide
Traffic speed (km/hr)	BRO	Direct input	Air, noise, landslide
Tree zone coordinates	Google earth, LISS III[h], DEM	Georeferencing, image merging, vectorization	Noise
Water quality profile, the year 2004	CISMHE (2008a)	Direct input	Water
Water quality profile, the year 2014	Bhutia (2015), Gurung et al. (2015)	Direct input	Water

[a]BRO, Traffic data of NH 31A (nearest town, Rangpo) collected from 29/06/2012 to 06/07/2012 by Border Roads Organization, Govt. of India. The ADT was projected for 2014 and 2039 based on annual growth rates (in percent) for traffic estimated by BRO
[b]MoRTH, Traffic data of NH 31A (nearest town, Singtam) collected on July 2004 and December 2004 by Ministry of Road Transport and Highways, Govt. of India. Downloaded from http://morth.nic.in/writereaddata/sublinkimages/sikkim2987772247.htm. Accessed on 18/05/2014
[c]Sikkim State Pollution Control Board
[d]National Remote Sensing Centre/Indian Space Research Organization. DEM accessed from http://bhuvan.nrsc.gov.in/data/download/index.php under CartoDEM- all versions as the CARTOSAT-1 satellite image on 11/04/2012
[e]Department of Environment, Forest, Wildlife Management and Climate Change, Govt. of Sikkim
[f]Sikkim State Remote Sensing Applications Centre
[g]DEM accessed from http://bhuvan.nrsc.gov.in/data/download/index.php under CartoDEM- all versions as the CARTOSAT-1 satellite image on 11/04/2012
[h]LISS III accessed from http://bhuvan.nrsc.gov.in/data/download/index.php under Resourcesat-I satellite image on 18/12/2014
[i]Google Earth image data accessed during 10/05/2011 to 21/03/2012

Table 13.2 Importance scale used in AHP

Scale of importance	Description
1	Both decision elements are equally important
3	First element is slightly more influenced than the second
5	First element is stronger than the second
7	First element is significantly stronger than the second
9	First element is extremely significant than the second
2, 4, 6, 8	Judgment values between equally, slightly, strongly, very strongly, and extremely
Reciprocals	When *i-th* criterion is compared to *j-th* criterion, a_{ij}, then $1/a_{ij}$ is the judgment value when *j-th* criterion is compared with *i-th*, i.e., $a_{ij} = 1/a_{ij}$

expresses the expert's preference of the *i*-th criterion in comparison to the *j*-th. For $i = j$, $a_{ij} = 1$,otherwise $a_{ij} = 1/aij$. The total number of pairwise comparisons by an expert is $n(n - 1)/2$, where "*n*" is the total number of criteria under consideration. The eigenvector, **w**, satisfying the maximum eigenvalue, $\lambda_{\max}$, of the comparison matrix is the solution of the AHP model (Eqs. (13.1) and (13.2)):

$$\boldsymbol{Aw} = \lambda_{\max}\boldsymbol{w} \tag{13.1}$$

In another form,

$$\begin{pmatrix} a_{11} & \cdots & a_{1n} \\ \vdots & \ddots & \vdots \\ a_{n1} & \cdots & a_{nn} \end{pmatrix}\begin{pmatrix} w_1 \\ \vdots \\ w_n \end{pmatrix} = \lambda_{\max}\begin{pmatrix} w_1 \\ \vdots \\ w_n \end{pmatrix} \tag{13.2}$$

where **A** is the comparison matrix. The elements of **w** are subject to $\sum_{i=1}^{n} w_i = 1$ and $\lambda_{max} = n$. The reliability of the AHP model is assessed by Consistency Ratio, $CR = CI/RI$, where Consistency Index, $CI = (\lambda_{max} - n)/(n - 1)$, and Random Consistency Index, RI, are obtained by a large number of simulation runs. It varies upon the order of the comparison matrix (Saaty, 2000; Taha, 2010). An inconsistency value ≤ 0.1 is acceptable for an AHP model.

Once the weights were known, they were used in the modelling of impact predictions of the environmental criteria and sub-criteria.

Air Quality Impact Assessment

Air pollutants, viz., CO, NO_2, SO_2, and SPM, were only considered in this study, as other pollutants were in negligible level in the impact area. To estimate the ground-level air pollution, a point-feature map was prepared considering villages, towns, isolated houses, and randomly selected points in the impact area ($n = 873$). IITLS, a Gaussian line source air dispersal model, was selected as it outperformed other dispersal models in mountainous areas of India. Such dispersal models predict the air pollution level at a point in space away from the source based on traffic and meteorological parameters (Agrawal et al., 2007; Goyal et al., 2006; Goyal and Rama Krishna, 1999) (Eqs. (13.3) and (13.4)):

$$C(x,0) = \frac{2q}{\sqrt{2\pi}\sigma_z u} \exp\left[-\frac{1}{2}\left(\frac{H}{\sigma_z}\right)^2\right] \tag{13.3}$$

where

$$q = \sum_{i=1}^{4} e_i v_i N_i \tag{13.4}$$

Model inputs like traffic volume (N), vehicle speed (v), vehicle composition (i), proximity of the receiver from the highway (x), average wind speed (u) (Goyal, 2008), vertical dispersal parameter (σ_z) (Gifford, 1961), and emission factor (q) (CPCB, 2007; Ramachandra & Shwetmala, 2009) were used to estimate the downwind ground-level ($H = 0$) pollutant concentration (C) of the receivers (Table 13.1). The pollution dataset generated by IITLS model was fed into the GIS framework as known points. The known points were interpolated using Empirical Bayesian Kriging (EBK) and clipped with the impact area to create air pollutant maps of CO, NO_2, SO_2, and SPM. EBK is a robust and unsupervised machine-learning spatial interpolation technique. It predicts values in the unsampled areas of a region of interest with the known points with the help of geostatistical algorithms. In comparison to other forms of kriging, EBK includes stochasticity in spatial parameters. It divides the known points into subsets. For each subset, semivariograms are simulated repeatedly and compared with the known point locations. This results in a distribution of several local semivariograms based interpolated surfaces. These local surfaces are put together to get the final output map (Banerjee et al., 2016; Cui et al., 1995; Krivoruchko, 2012; Pilz & Spöck, 2008). Model validation was done by comparing the estimated value with the observed at three well-spaced locations in the impact area. The air pollutant maps were reclassified using Indian Air Quality Index (IND-AQI) (CPCB, 2015).The composite air quality impact map was prepared by WLC of reclassified maps for both the with- and without-project scenarios (Banerjee et al., 2018c).

Water Quality Impact Assessment

An empirical model of highway runoff composition estimation constructed by Kayhanian et al. (2003) in combination with the mass-balance model was used to assess the impact of traffic-induced water pollution. The set of water quality parameters was confined to Chemical Oxygen Demand (COD), pH, Total Dissolved Solids (TDS), Total Suspended Solids (TSS), and Heavy metal (Zn), mainly due to lack of historical data and innate constraints of the empirical model. The microcatchment drainages that feed the highway runoff and rivers along the highway were estimated by SCS-CN method using HEC-GeoHMS, a geospatial hydrological extension of ArcGIS (Vojtek & Vojteková, 2016). Highway runoff pollutant level (C_i^H) was estimated from inputs (x_i), viz., event rainfall, maximum intensity rainfall, antecedent dry period, cumulative seasonal rainfall, watershed area, and Average Annual Daily Traffic (AADT) (Table 13.1) (Eq. (13.5)):

$$C_i^H = \exp\left(b_i + \sum_{j=1}^{6} a_j x_j\right) \tag{13.5}$$

where b_i is the y-intercept of the i-th water quality parameter, a_j is the proportionality coefficient of j-th predictor variable, and H is the highway runoff.

Mass balance model was used to estimate the downstream concentration of i-th water pollution in the r-th river (C_i^R) by considering the j-th upstream discharge rate (Q_j) and the concentration of the water quality parameters in it (C_{ij}), the runoff from the D-th micro-catchments (Q_D), and the concentration of the water quality parameter in the highway runoff (C_i^H) (Barthwal, 2012; Davie, 2008) (Eq. (13.6)):

$$C_i^R = \frac{Q_D C_i^H + \sum_{j=1}^{n} Q_j C_{ij}}{Q_D + \sum_{j=1}^{n} Q_j} \tag{13.6}$$

The estimated concentration of water quality parameters at the sink of the impact area was compared with the observed data using model validation criteria (Paliwal & Srivastava, 2014). The estimated concentration of water quality parameters due to traffic in the rivers along the highway were attributed to randomly selected points ($n = 100$) in the impact area and interpolated using EBK. The interpolated surfaces of water quality parameters were standardized[1] to *Single Factor Pollution Index* (SFPI) maps and reclassified (Li et al., 2009; Yan et al., 2015). The reclassified SFPI maps were linearly combined using AHP weight and normalized to prepare a water quality impact assessment map for both the project scenarios (Mushtaq et al., 2015). A detailed account of the method is available with Banerjee et al. (2018b).

---

[1]Considering the ecological and cultural importance of the study area, US Public Health Service (USPH,1962) and Bureau of Indian Standards (BIS, 2012) water quality standards were used for the local water bodies.

Noise Impact Assessment

Two noise level indices were estimated, viz., Leq(H) (Hourly A-Weighted Equivalent Sound Level) and Ldn (Day-Night Average Sound Level) using FHWA TNM 2.5. It is one of the most popular computer-based traffic noise measurement models, and it is recommended by the Ministry of Environment and Forest (MoEF), Govt. of India, for EIA of highway projects. The TNM algorithm performs a series of adjustments to the basic noise level called reference sound level (L_0) of a stream of vehicles by considering several factors, namely, volume and speed correction (A_{VS}), distance correction (A_D), barrier correction (A_B), flow correction (A_F), gradient correction (A_G), and ground cover correction (A_S)(FHWA, 1998) (Eq. (13.7)):

$$Leq(H) = L_0 + A_{VS} + A_D + A_B + A_F + A_G + A_S \tag{13.7}$$

$Leq(H)$ is more appropriate for the assessment of noise in human habitations, while *Ldn* is better for nonresidential areas like ecologically sensitive areas. Based on the need of TNM, several spatial inputs were created for the estimation of noise level indices. Noise level at various receiver locations ($n = 872$) was fed into the GIS framework and interpolated using EBK. Model validation was done by comparing the estimated value with the observed at three well-spaced locations in the impact area. A composite noise impact assessment map was prepared by a WLC of the noise indices maps for both project scenarios. A detailed account of the method is available in Banerjee et al. (2016).

Biodiversity Impact Assessment

Five landscape metric maps were created from the georeferenced forest type map (FSI, 2011) of the study area, viz., core area, disturbance, patch size index, proximity, and rarity (Geneletti, 2003, 2004). The core area is a measure of the fraction of an undisturbed area of habitat. It was estimated by removing the disturbance zone from a forest patch using the buffer method. The disturbance is the measure of the level of degradation of a habitat. Usually, the proximity of a habitat to human settlement or an asset like roadways is a good measure of disturbance. It was considered as the inverse of the shortest Euclidian distance between the edges of the forest patch and disturbance polygon (like human settlement, road corridor, etc.). Patch size indicates the geometry of a habitat. Ideally, a circular patch is considered least degraded as the core of the habitat is far more than its edges. A patch size greater than one indicates the deformation and deviation of the shape of a forest patch from a perfect circle due to human disturbance. The proximity between two like habitat provides corridors for the species to migrate and disperse. Hence, proximity is a measure of connectivity between habitats. It was estimated as the Shortest Euclidian distance between the edges of polygons of the same habitat type. The rarity of a habitat is a measure of the degree of scarcity of a habitat patch. The rarity of a forest patch was measured by estimating the percentage change in the forest patch from its potential size in the absence of any disturbance (Geneletti, 2003). The potential forest map of the study area was created by reclassification method based on the criteria proposed by Champion and Seth (1968) like elevation, precipitation, and soil type, to name a few.

The range of values in each landscape metrics map was converted into expert opinion value using AHP. These values were normalized, and the curve fitting method (MyCurveFit, 2017) was used to create a normalization equation for each landscape metric. For this study, the proportionate rate growth model showed the best curve fitting (Hobbie & Roth, 2015). The normalization equation was used in converting landscape metric maps into normalized landscape metric maps. The biodiversity value map was made by a WLC of the normalized landscape metric maps (Banerjee et al., 2020).

Socioeconomic Impact Assessment

A structured score-based questionnaire was administered to a set of randomly selected people ($n = 100$) from the local population at various distances and elevations from the highway. Based on expert opinion and pilot survey, 15 socio-economic attributes, viz., archaeological monuments, banking and insurance-related activities, commodity prices, criminal activities, cultural heritage, demographic profile, educational facilities, employment opportunities, farming activities, health facilities, industrial activity, land use and land value, per capita income, tourism, and traditional local values, were considered for the study (Table 13.1). The questionnaire was based on the perception of the extent of change in the socioeconomic criteria under "with-project" scenario compared to the "without-project" scenario. Hence, the assessment was focused on the alterations due to the project. The GPS location of the respondent was also included in the questionnaire. The GPS locations of the respondents were converted into a point map. The points were populated with socioeconomic attributes and interpolated using EBK and clipped to the impact area to create socioeconomic maps of all sub-criteria. Socioeconomic impact assessment map was prepared by WLC of the socioeconomic attribute maps. A detailed account of the method is available in Banerjee and Ghose (2017).

Landslide Susceptibility Assessment

Eleven landslide inducing factors, viz., Aspect, Distance from Fault, Distance from Road, Drainage Density, Land Use and Land Cover, Lithology, Plan Curvature, Rainfall, Slope, Soil Depth, and Soil Texture, were identified as the most relevant factors concerning highway projects in mountainous areas (Table 13.1) (Das & Raja, 2015; Devkota et al., 2013; Pourghasemi et al., 2012). A landslide inventory map of the impact area was prepared using ArcGIS basemaps, Google Earth, Wikimapia, and GPS-based ground-truthing. Each landslide inducing factor raster was reclassified into intervals, and the landslide inventory map was overlaid on them to estimate the information values of the intervals using the information value method (Cao et al., 2016; Pardeshi et al., 2013) (Eq. (13.8)):

$$I(x_i, H) = \ln\left(\frac{N_i/N}{S_i/S}\right) \tag{13.8}$$

where S is the total number of pixels in the study area, N is the total landslide pixels in the study area, S_i is the pixels count of the LIF in the study area, and N_i is the landslide pixels count of the LIF. A higher information value indicates a greater possibility of landslides associated with the concerned landslide inducing factor. The information values estimated for the landslide inducing factors were used in creating their information value maps. Landslide susceptibility assessment map was prepared using WLC of the information value maps of landslide inducing factors (Banerjee et al., 2018b). Validation of landslide susceptibility assessment model was done using the Evaluation Accuracy Analysis method (Cao et al., 2016).

Model Validation and Cross-Validation

Validation of the estimated values of the sub-criteria under air, water, and noise was done by comparing them with the observed values at appropriate locations within the study area. The *correlation coefficient* (COR), *Fractional Bias* (FB), *Normalized Mean Square Error* (NMRE), and *Index of Agreement* (IOA) were estimated as model validation criteria (Paliwal & Srivastava, 2014). Cross-validation of the EBK interpolation was done using *Mean Standardized Error* (MSE) and *Standardized Root Mean Square Error* (SRME) (Chang, 2017; Lloyd, 2009).

Spatial Composite Impact Assessment Map

The environmental criteria impact assessment maps prepared using various submodels were normalized (Chang, 2017) (Eq. (13.9)):

$$\eta_{ij} = \frac{I_{ij} - \min(I_{ij})}{\max(I_{ij}) - \min(I_{ij})} \tag{13.9}$$

where η_{ij} is the normalized impact value of the *i*-th environmental criteria at *j-th* location. $\min(I_{ij})$ and $\max(I_{ij})$ are the minimum and maximum impact value, respectively, in the set of all impact values under the impact area of *i*-th environmental criteria. Spatial composite impact assessment map is a WLC spatial model of normalized environmental criteria maps (Chang, 2017) (Eq. (13.10)):

$$C_j = \sum_{i=1}^{n} w_i \eta_{ij} \tag{13.10}$$

where C_j is the composite impact value at j-th location and w_i is the weight of i-th environmental criteria.

Spatially Explicit Sensitivity Analysis

Sensitivity analysis of SCIAM was performed by changing the environmental criteria weights. The analysis is based on One factor At a Time (OAT) method, which mainly involves assessing the impact of model input change on the model output, considering one input at a time (Murphy et al., 2004). A total of 100 spatial composite impact assessment maps were prepared by manipulating the weight of all environmental criteria in a range of −20% to +20% with a step size of ±2%. The run maps were generated using Eq. (13.11):

$$C_{i\alpha} = w_i \eta_{it} + \sum_{j \neq i}^{n} (1 - w_i) \frac{w_i}{\sum_{j \neq i}^{n} w_j} \eta_{it} \tag{13.11}$$

subject to the condition (Eq. (13.12)):

$$\sum_{i=1}^{n} w_i = 1 \tag{13.12}$$

where $C_{i\alpha}$ is dependent on the normalized value of *i-th* environmental criteria and step size, $\alpha.w_i$, is the changed weight and $(1 - w_i)\frac{W_j}{\sum_{j \neq i}^{n} W_j}$ is the adjusted weight for *j-th* environmental criteria (H. Chen et al., 2011; Xu & Zhang, 2013). *Impact Category Change Rate* (ICCR) (Banerjee et al., 2018a) was used to visualize the role of criteria weight on area coverage by individual impact categories (Eq. (13.13)):

$$ICCR_{i\alpha\in} = \frac{1}{|\alpha|} \times \left| \frac{C_{i\alpha\in} - C_{i0\in}}{C_{i0\in}} \right| \times 100 \tag{13.13}$$

$ICCR_{i\alpha\epsilon} \geq 1$ indicates that the *ε-th* impact category is sensitive to *i-th* environmental criteria weight at *α-th* step size, while $ICCR_{i\alpha\epsilon} < 1$ implies an insensitivity and

ICCR at $\alpha = 0$ is undefined. In other words, SCIAM impact category will be considered sensitive, if a change in the criterion weight causes equivalent or greater change in the model output (Longley et al., 2010). To evaluate the change in the composite impact value per pixel per change in the weight, a change rate function was used (Eq. (13.14)):

$$CR_{ij\alpha} = \frac{C_{ij\alpha} - C_{ij0}}{C_{ij0}} \times 100 \tag{13.14}$$

where $CR_{ij\alpha}$ is the change rate of composite impact value due to *i-th* environmental criteria weight for *j-th* pixel at *α-th* step size in comparison to the without-project scenario, CR_{ij0} *Mean Absolute Change Rate* (MACR), a summary sensitivity index, was used to assess the overall sensitivity of the entire study area with the change in environmental criteria weights (Xu & Zhang, 2013) (Eq. (13.15)):

$$MACR_{i\alpha} = \frac{1}{N} \sum_{i=1}^{N} \left| \frac{C_{ij\alpha} - C_{ij0}}{C_{ij0}} \right| \times 100 \tag{13.15}$$

where $MACR_{i\alpha}$ is the mean absolute value of change rate of composite impact value due to change in the weight of environmental criteria and N is the total number of pixels. $MACR_{i\alpha} \geq \alpha$ indicates that the SBIAM is sensitive to *i-th* environmental criteria weight at *α-th* step size, while $MACR_{i\alpha} < \alpha$ implies an insensitivity (Longley et al., 2010).

Spatial Composite Impact Index

A suite of indices was developed to assess the viability of the project. In case of air quality, water quality, and noise impact assessments, the mean change of the impact values from "without- to with- project" scenarios was considered as the impact scores of the *k-th* environmental criterion (Eq. (13.16)):

$$\overline{I}_k = \frac{1}{N} \sum_{i=1}^{N} \left(\frac{I_{jp}}{I_{jw}} - 1 \right) \tag{13.16}$$

where I_{jp} and I_{jw} are the impact values under with- and without-project scenarios at *j*-th pixel. N is the total number of pixels. For noise, a weighted average of impact scores of $Leq(H)$ and Ldn was taken as the noise impact score.

Regarding the impact due to biodiversity loss, a 50 m buffer along the highway was created as the *biodiversity loss area*, for the road width of 7 m and 12 m representing "without- and with-project" scenarios (Geneletti, 2004). The biodiversity value map was clipped by these buffers. Weighted area loss of biodiversity was estimated by summation of the biodiversity value times area per biodiversity value. This was done for both the scenarios. The difference between the mean values of the weighed area loss under the two scenarios was estimated and considered as the biodiversity impact score (Eq. (13.17)):

$$\bar{I}_k = \frac{\sum_{i=1}^{N_p} A_{jp} B_{jp} \times}{\sum_{i=1}^{N_w} A_{lw} B_{lw}} \frac{\sum_{i=1}^{N_w} A_{lw}}{\sum_{i=1}^{N_p} A_{jp}} - 1 \tag{13.17}$$

where B is biodiversity value and A is the area under the biodiversity value. j and l are the index number of the biodiversity value under "with-" (p) and "without-" (w) project scenarios. N is the number of patch areas.

The mean value of the socioeconomic impact assessment map represented socioeconomic impact score (Eq. (13.18)):

$$\bar{I}_k = \frac{\sum_{i=1}^{N} \Delta S_i}{N} \tag{13.18}$$

where Δ represents the people's perception-based change in socioeconomic criteria S due to the project and N is the total pixel count.

To assess the impact of landslide susceptibility under the two scenarios, a vehicle vulnerability assessment was done, as proposed by Das et al. (2011) (Banerjee et al., 2018a). For this, the hazard areas were demarcated based on two conditions that (a) the area belonged to high to very high landslide susceptibility zone and (b) also had a history of at least one landslide. The expected number of vehicles along a highway section that belonged to a hazard area was estimated from the average daily hourly traffic volume, the average speed of the vehicle type, and the time taken by the vehicles to cover the highway section (Table 13.1) (Guzzetti, 2005). The vehicle vulnerability in a hazard area was estimated from (a) expected number of vehicles of the two scenarios and (b) the coefficient of damage due to landslides (Jaiswal et al., 2009) using a Poisson distribution method (Liu, 2006). Change in the mean vulnerability value under the two scenarios was considered as the landslide impact score (Eq. (13.19)):

$$\bar{I}_k = \frac{\sum_{i=1}^{N_p} x_{i,p} V_{i,p} \times}{\sum_{i=1}^{N_w} x_{i,w} V_{i,w}} \frac{\sum_{i}^{N_w} x_{i,w}}{\sum_{i}^{N_p} x_{i,p}} - 1 \tag{13.19}$$

where x is the road section and V is the vulnerability associated with the i-th road section. The landslide vulnerability score is the weighted mean of the impact scores estimated for the vulnerabilities of light and heavy vehicles.

The impact score can be negative or positive in value. The impact score is negative if the "with-project" scenario leads to a lesser adverse impact on the concerned environmental criterion compared to "without-project" scenario, implying an *environmental gain* due to the project implementation. In contrast, a positive impact score will occur when the mean adverse impact of "with-project" scenario is greater than "without-project" scenario, implying *environmental loss*.

The SCII is the ratio of the weighted negative impact scores to the weighted positive impact scores (Eq. (13.20)):

$$SCII = \frac{\left(\sum_{k=1}^{l} \left|w_k \bar{I}_k\right| \middle| \forall \bar{I}_k < 0\right)}{\left(\sum_{k=1}^{p} \left|w_k \bar{I}_k\right| \middle| \forall \bar{I}_k > 0\right)} \tag{13.20}$$

SCII can hold any value from zero to infinity. A value of SCII more than 1.0 (i.e., the total environmental gain is more than the total environmental loss) is preferred, while SCII < 0.75 is critical. For the present work, the minimum permissible value of SCII for a project alternative was considered as 0.75, based on the information available in the literature (Agrawal, 2005; Leopold et al., 1971; Modak & Biswas, 1999).

13.3 Results

AHP Model

Expert opinion-based AHP model gave higher priority to the "with-project" scenario as compared to the "without-project" scenario. Among environmental criteria, biodiversity and landslide susceptibility had the highest weight. Under air quality, CO and NO_2 had the highest weight. The disturbance subcriterion was the predominant factor in the biodiversity impact assessment. Distance from the road and slope received higher weight under landslide susceptibility. Equal weight was given to the noise indices. Under socioeconomic changes, employment opportunities and changes to LULC were given higher weight (Table 13.3).

Model Validation and Cross-Validation

For air quality modelling, the predicted value of CO significantly deviated from the observed value. This was mainly due to the high variation of the observed CO concentration at various locations in the impact area. Otherwise, IITLS performed satisfactorily in predicting the ambient air concentrations of the remaining pollutants. Water quality predicted by the mass balance model performed satisfactorily, while noise quality predicted by FHWA TNM did well, except under the "Index of Agreement" criterion. The AHP-Information Value Method based landslide susceptibility assessment model performed satisfactorily with 85% predictability of

Table 13.3 AHP model output

Environmental criteria	Environmental criteria weight	Inconsistency	Priority of the project alternatives		Sub-criteria		Sub-criteria weights	Inconsistency	Priority of the project alternatives	
			Without project	With project					Without project	With project
Air quality	0.135	0.04	0.238	0.762	Air	CO	0.361	0.04	0.192	0.808
						NO_2	0.297			
						SO_2	0.212			
						PM	0.129			
						Total	1			
Biodiversity value	0.298				Biodiversity	Core area	0.192	0.06	0.398	0.602
						Disturbance	0.416			
						Proximity	0.218			
						Rarity	0.100			
						Patch shape	0.074			
						Total	1			
Landslide susceptibility	0.201				Landslide susceptibility	Aspect	0.060	0.08	NA	NA
						Distance from fault	0.036			
						Distance from road	0.307			
						Drainage density	0.032			
						Land use and land cover	0.082			
						Lithology	0.035			
						Plan curvature	0.128			
						Rainfall	0.015			

						Shipdepth	0.0510			
						Soil texture	0.014			
						Total	1			
Noise	0.095				Noise	*Leq*(*H*)	0.500	0.01	0.375	0.625
						Ldn	0.500			
						Total	1			
Socioeconomic conditions	0.134				Socioeconomic conditions	Archaeological monuments	0.026	0.04	0.353	0.647
						Banking and insurance-related activities	0.039			
						Commodity prices	0.053			
						Criminal and criminal activities	0.022			
						Cultural heritage	0.024			
						Demographic profile	0.068			
						Educational facilities	0.096			
						Employment opportunities	0.129			
						Farming activities	0.044			
						Health facilities	0.072			
							0.105			

(continued)

Table 13.3 (continued)

Environmental criteria	Environmental criteria weight	Inconsistency	Priority of the project alternatives		Sub-criteria		Sub-criteria weights	Inconsistency	Priority of the project alternatives	
			Without project	With project					Without project	With project
						Industrial activity				
						Land use and land value	0.116			
						Traditional local values	0.025			
						Per capita income	0.096			
						Tourism	0.085			
						Total	1			
Water	0.135				Water	COD	0.197	0.01	0.273	0.727
						TDS	0.150			
						TSS	0.163			
						Heavy metals	0.329			
						pH	0.161			
Total	1					Total	1			

Table 13.4 Model validation criteria

Criteria and sub-criteria (↓)	Fractional bias	Normalized mean square error	Correlation coefficient	Index of agreement
Ideal value (→)	0	0	1	1
CO	0.978	4.597	0.425	0.571
NO_2	0.129	0.031	0.969	0.982
SO_2	−0.058	0.349	0.288	0.871
SPM	0.207	0.082	0.663	0.977
Water	−0.042	0.035	0.997	0.997
Noise	0.230	0.076	0.990	0.291
	Evaluation accuracy analysis method			
Ideal value	1			
Landslide	0.851			

landslides within the impact area. Cross-validation criteria showed satisfactory results for all EBK interpolations (Tables 13.3, 13.4 and 13.5, Supplement Table 13.S2).

Impact Assessment Maps of Environmental Criteria and Spatial Composite Impact Assessment Map

Under air quality impact assessment, high pollution level was estimated near the highway (Fig. 13.3a). Except for a single large forest patch, a very low to low biodiversity value was estimated in the forest patches near the highway (Fig. 13.3b). Landslide susceptibility was found to be high near the highway and along the steep slopes (Fig. 13.3c). The noise level was estimated to be high along the highway and in areas with gentler elevations (Fig. 13.3d). Moderate socioeconomic benefits were expected in the northern part of the impact area, while slight socioeconomic benefits were felt in the remaining impact area (Fig. 13.3e). Water pollution was high throughout the impact area. However, it was highest in the Teesta area (Fig. 13.3f).

The spatial composite impact assessment map showed that the high to severe adverse impacts were confined within a 500 m buffer distance along the highway. The moderate adverse impact was distributed in the entire impact area except for some pockets of slight to low adverse impact (Fig. 13.4).

Table 13.5 Cross-validation of Empirical Bayesian Kriging

Criteria and sub-criteria (↓)	Mean standardized error	Standardized root mean square error
Ideal value (→)	0	1
Air	0.024	0.896
COD	0.003	0.769
pH	0.012	0.884
TSS	0.002	0.850
TDS	0.016	0.963
Zn	0.007	0.819
Leq(H)	0.017	0.965
Ldn	0.024	0.980
Archaeological monuments	−0.045	0.996
Banking and insurance-related activities	−0.029	1.010
Commodity prices	0.008	1.013
Criminal and criminal activities	0.015	0.994
Cultural heritage	−0.072	0.986
Demographic profile	−0.012	0.976
Educational facilities	−0.002	0.991
Employment opportunities	0.036	0.987
Farming activities	−0.003	0.969
Health facilities	−0.037	0.964
Industrial activity	−0.032	0.986
Land use and land value	−0.027	1.008
Per capita income	0.021	0.996
Tourism	−0.051	0.955
Traditional local values	−0.015	0.966

Spatially Explicit Sensitivity Analysis

Impact Category Change Rate

All impact categories of SCIAM showed insensitivity to changes in the weights of air, noise, and water. Furthermore, the weight change of landslide susceptibility and socioeconomy led to marginal sensitivity. ICCR due to a change of air weight showed higher changes in severe adverse impact category confined mainly near the highway. ICCR of SCIAM due to changes in biodiversity weight showed a high sensitivity for all impact categories. At lower biodiversity weight, the moderate adverse impact category was the most sensitive, while at higher biodiversity weight, high and severe adverse impact categories showed the highest sensitivity. Impact categories remained insensitive to changes in noise weight. The rise in

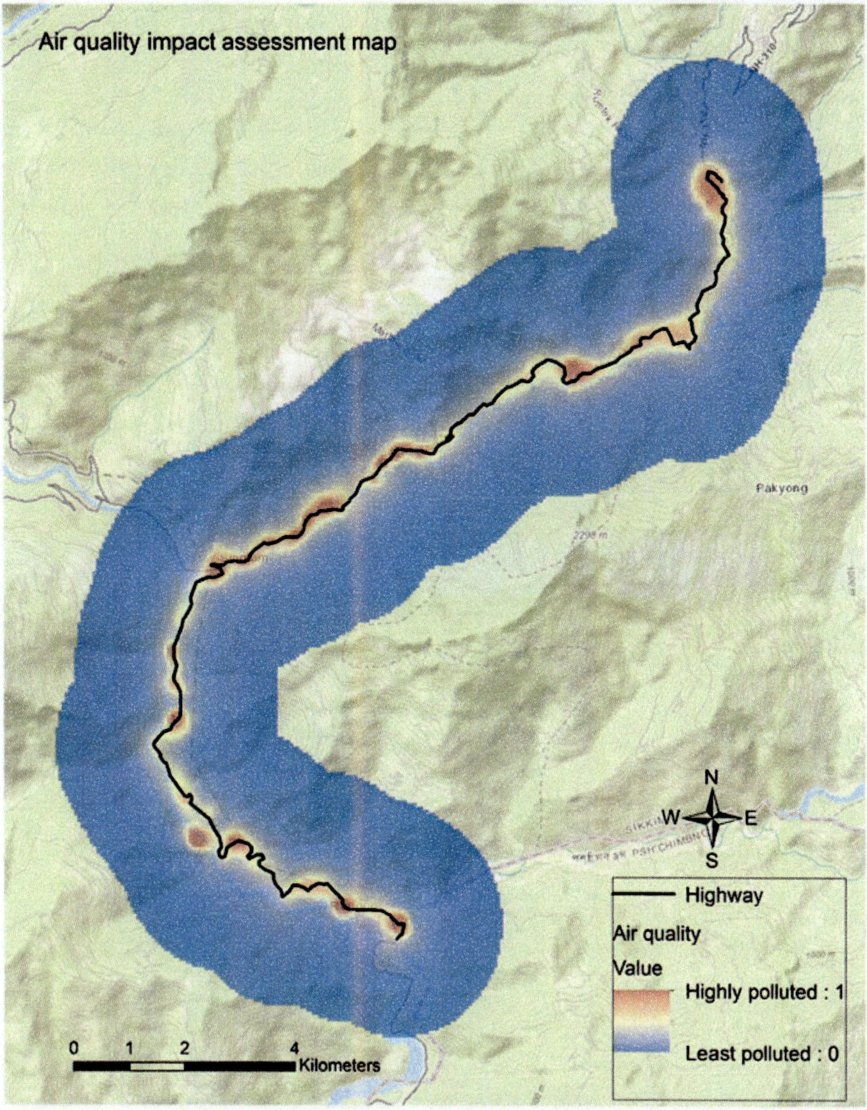

Fig. 13.3 (**a**) Air quality impact assessment. (**b**) Biodiversity impact assessment map. (**c**) Landslide susceptibility map. (**d**) Noise impact assessment map. (**e**) Socioeconomic impact assessment map. (**f**) Water quality impact assessment map

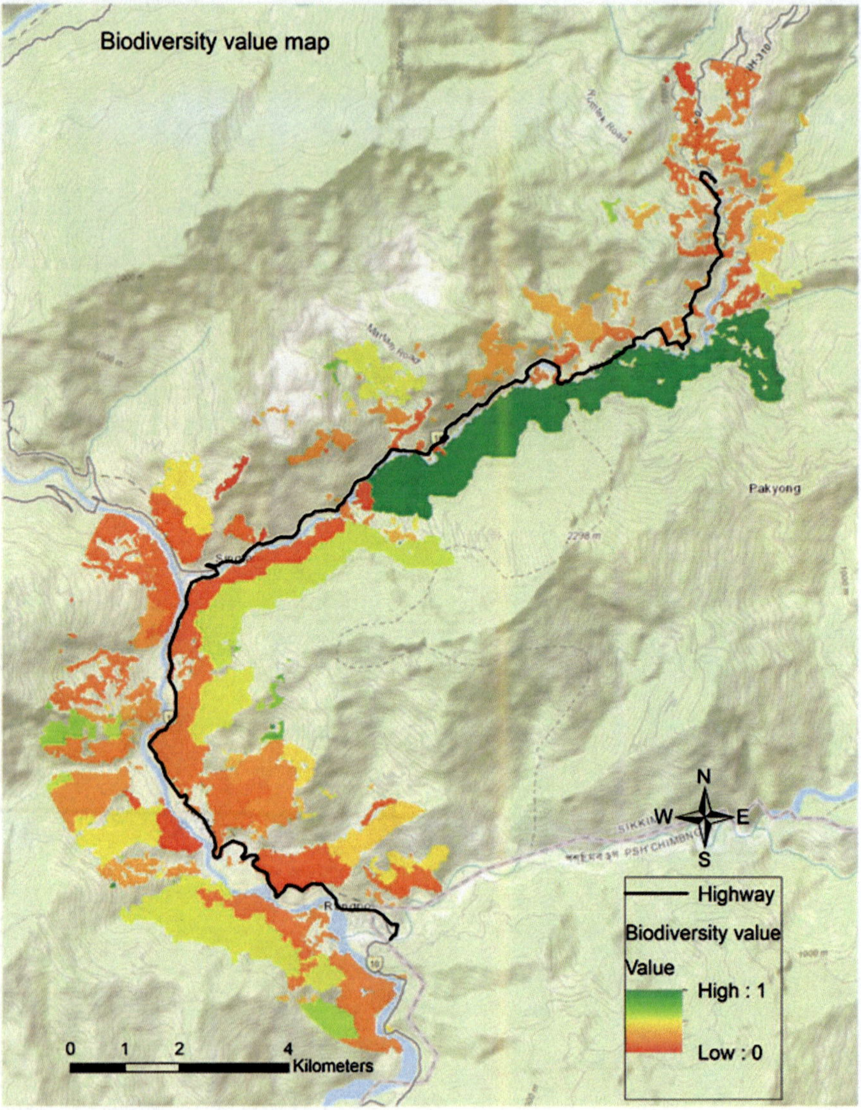

Fig. 13.3 (continued)

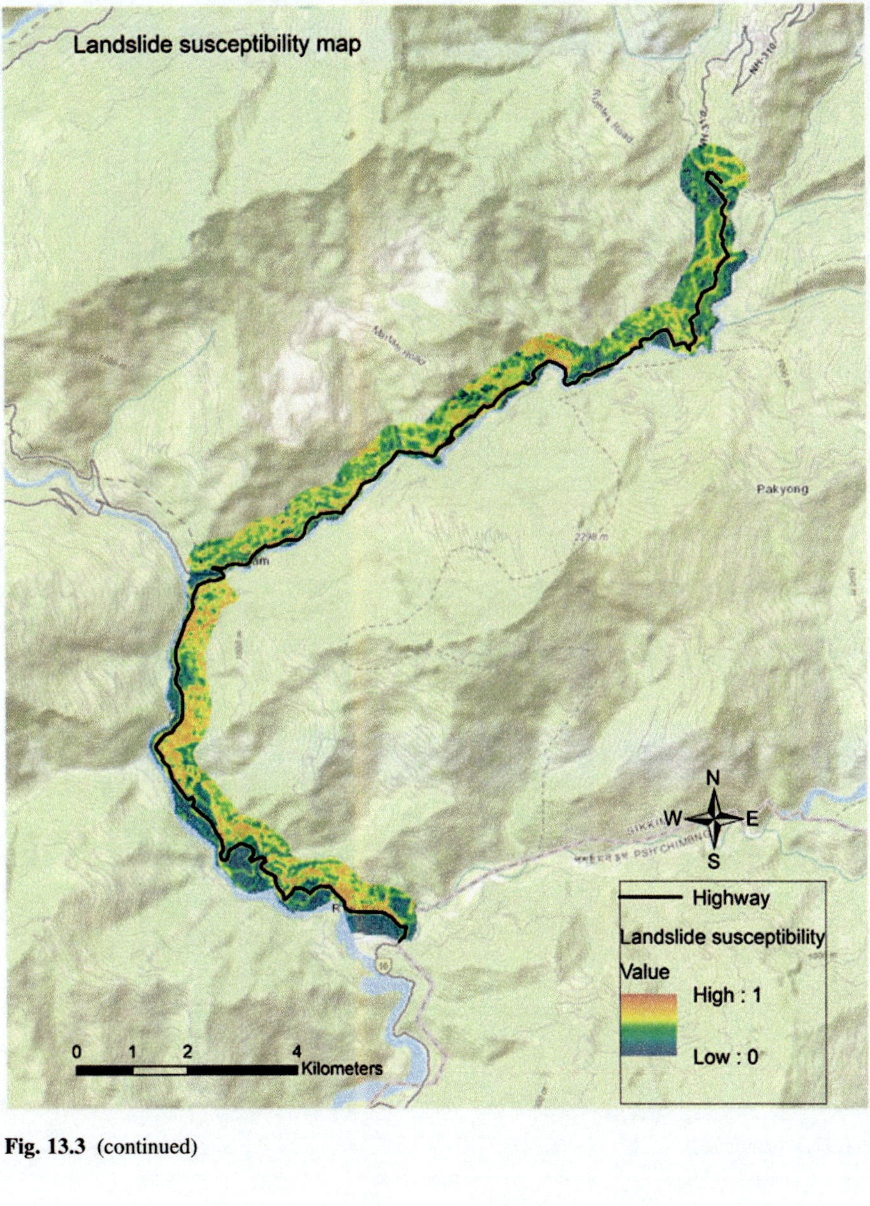

Fig. 13.3 (continued)

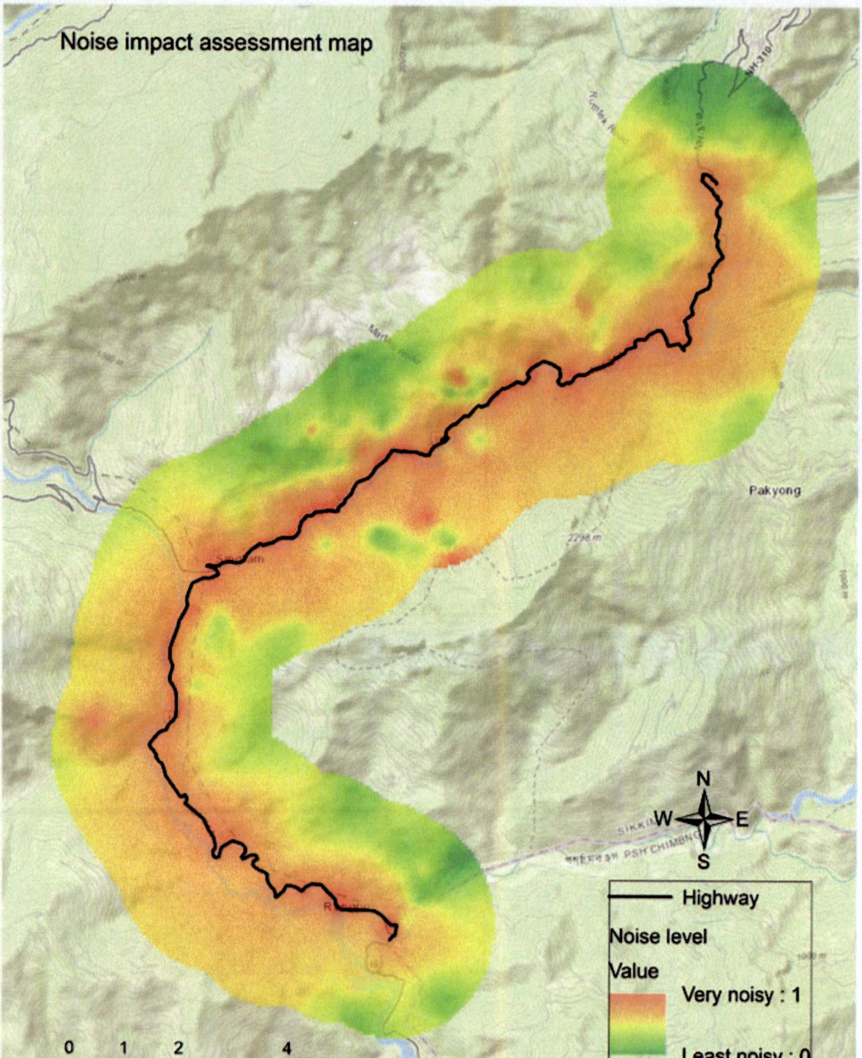

Fig. 13.3 (continued)

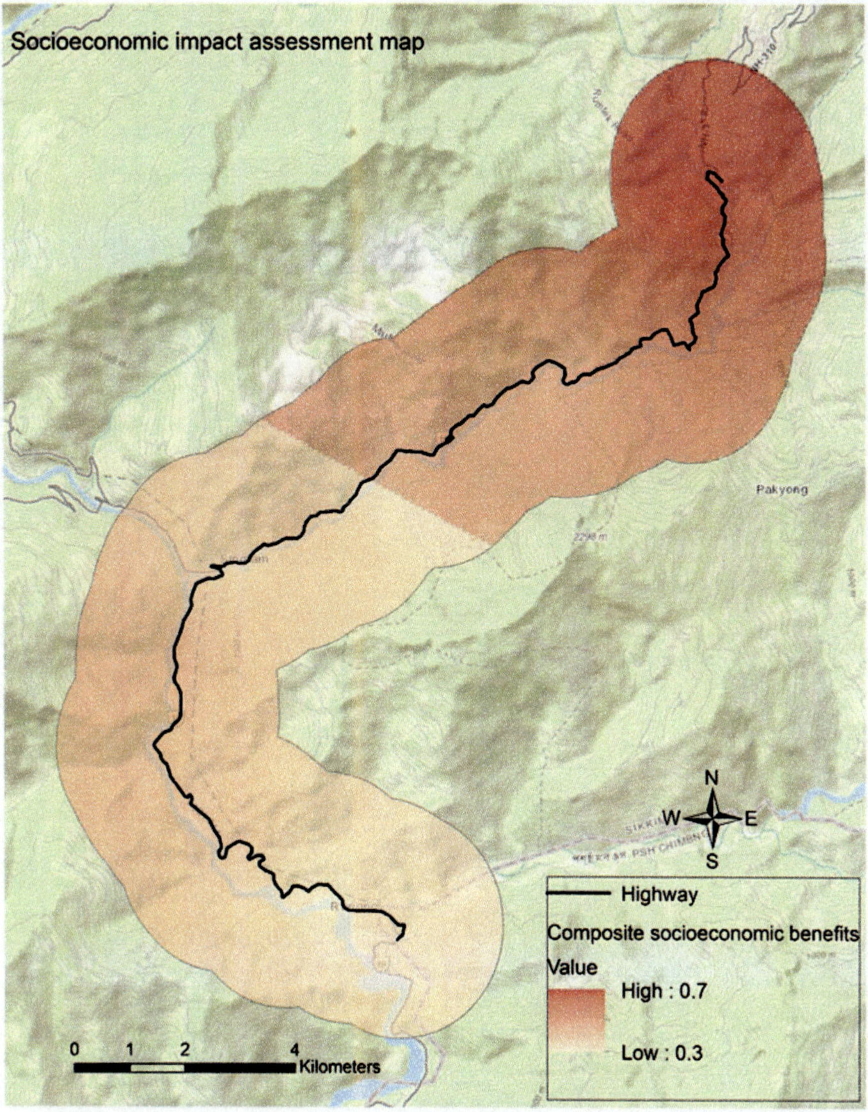

Fig. 13.3 (continued)

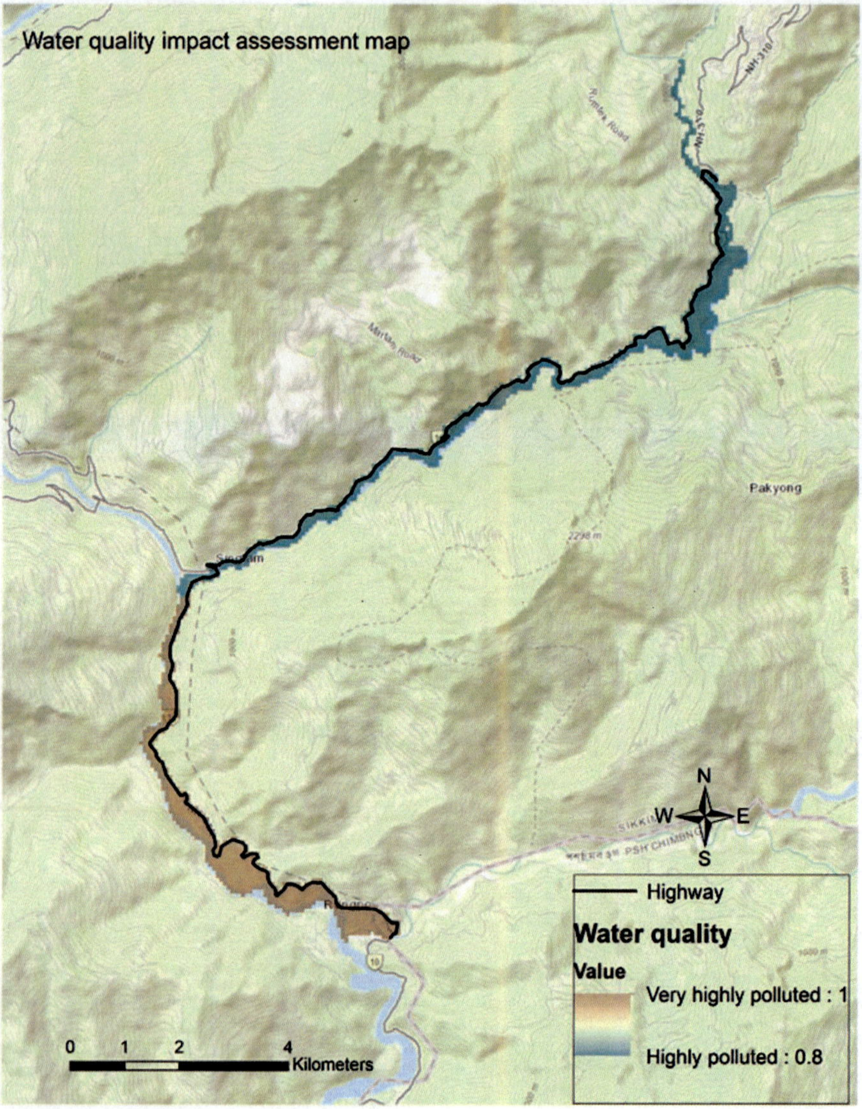

Fig. 13.3 (continued)

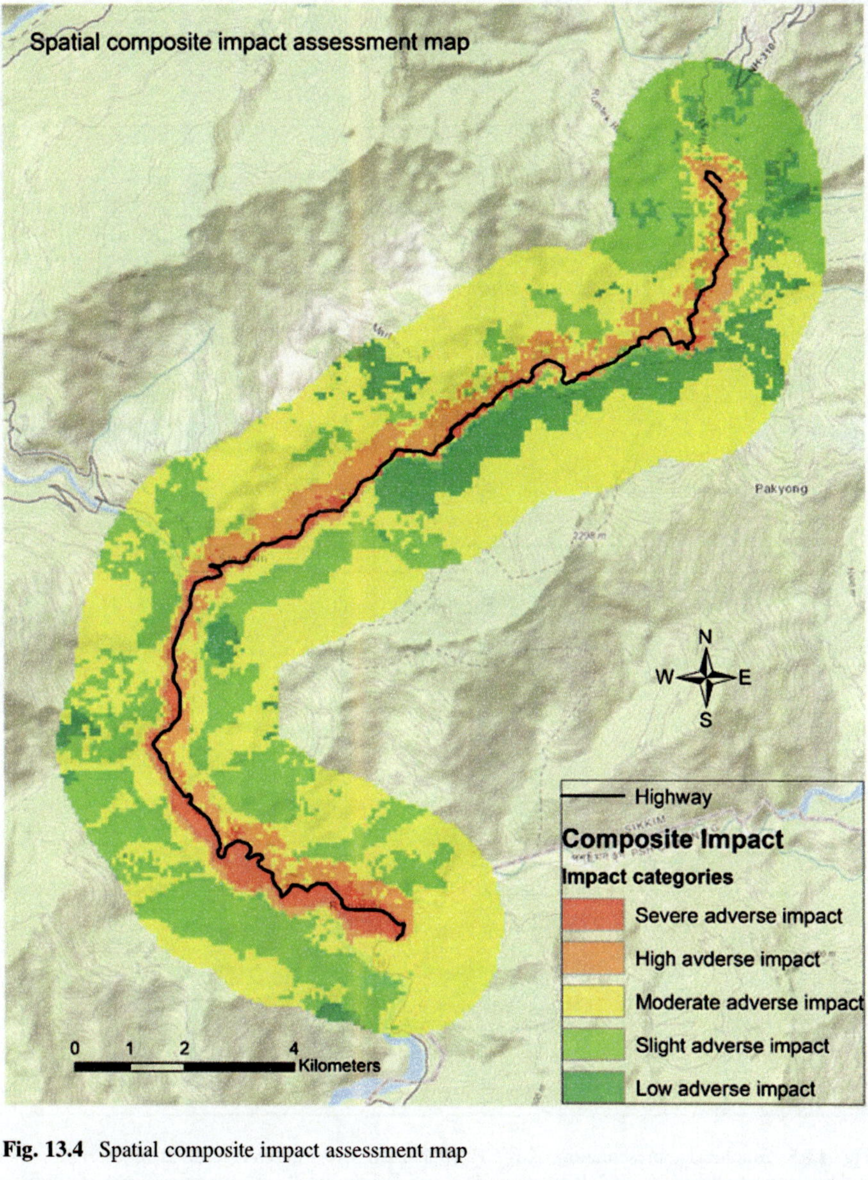

Fig. 13.4 Spatial composite impact assessment map

socioeconomic weight led to a steady decline in the change rate of low and slight adverse impact categories. The high adverse impact category showed marginal change at the higher socioeconomic weight. High adverse impact category showed a greater change to change in water weight (Fig. 13.5).

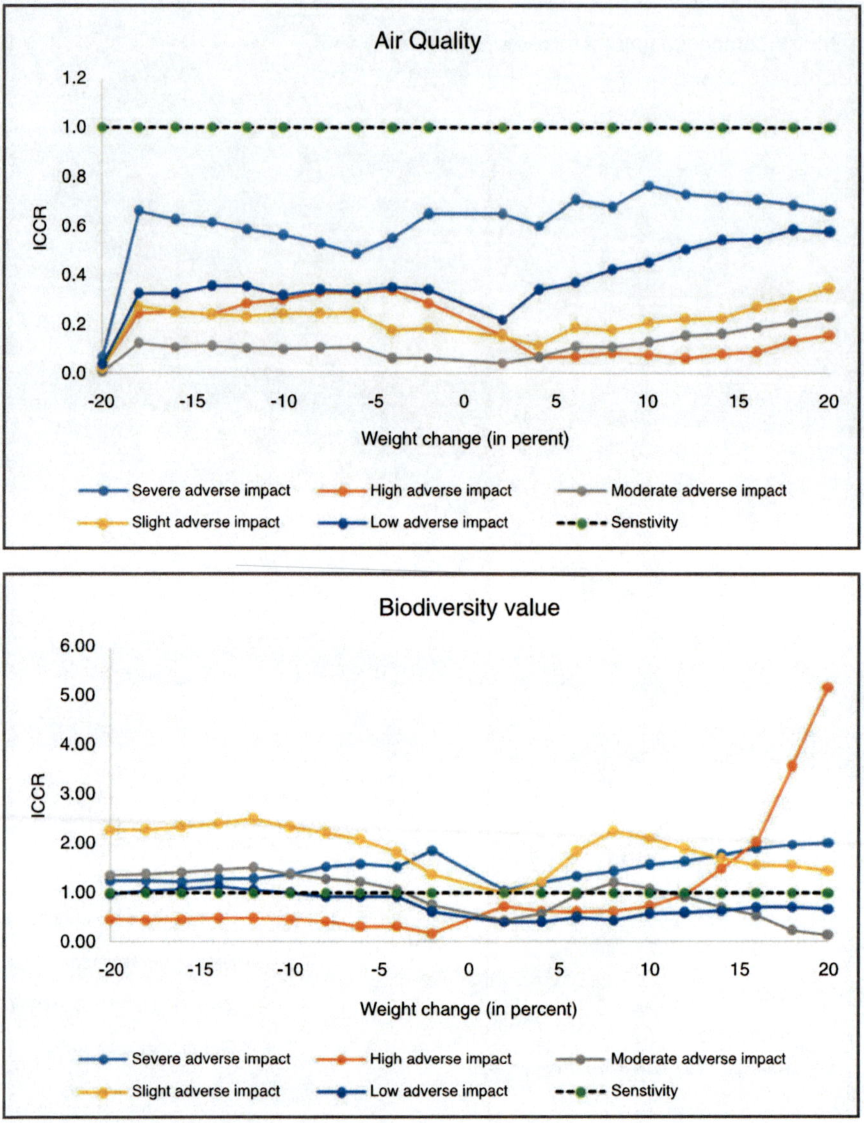

Fig. 13.5 Graphical representation of ICCR of SCIAM over the change in (**a**) air weight. (**b**) Biodiversity weight. (**c**) Landslide weight. (**d**) Noise weight. (**e**) Socioeconomy weight. (**f**) Water weight

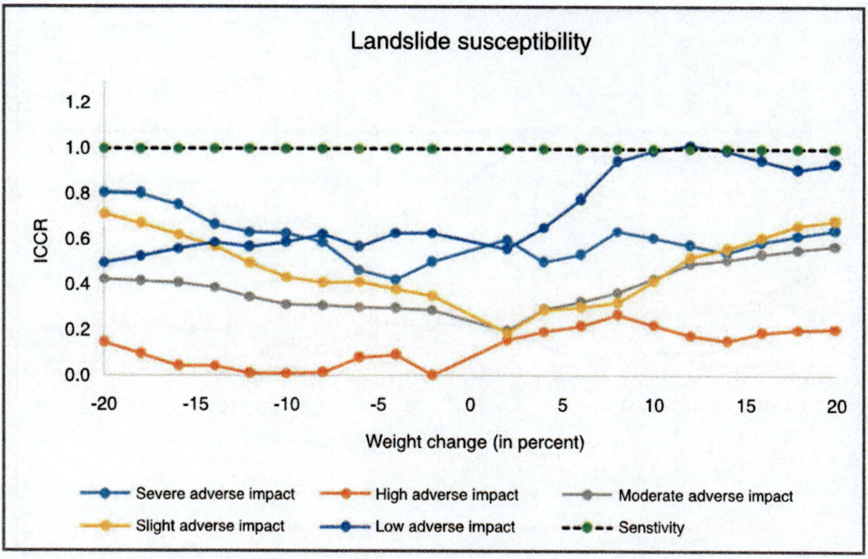

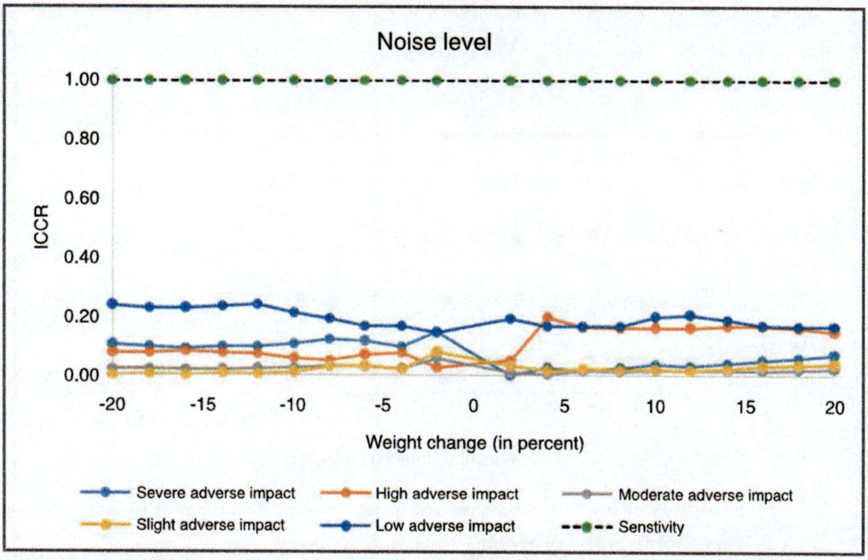

Fig. 13.5 (continued)

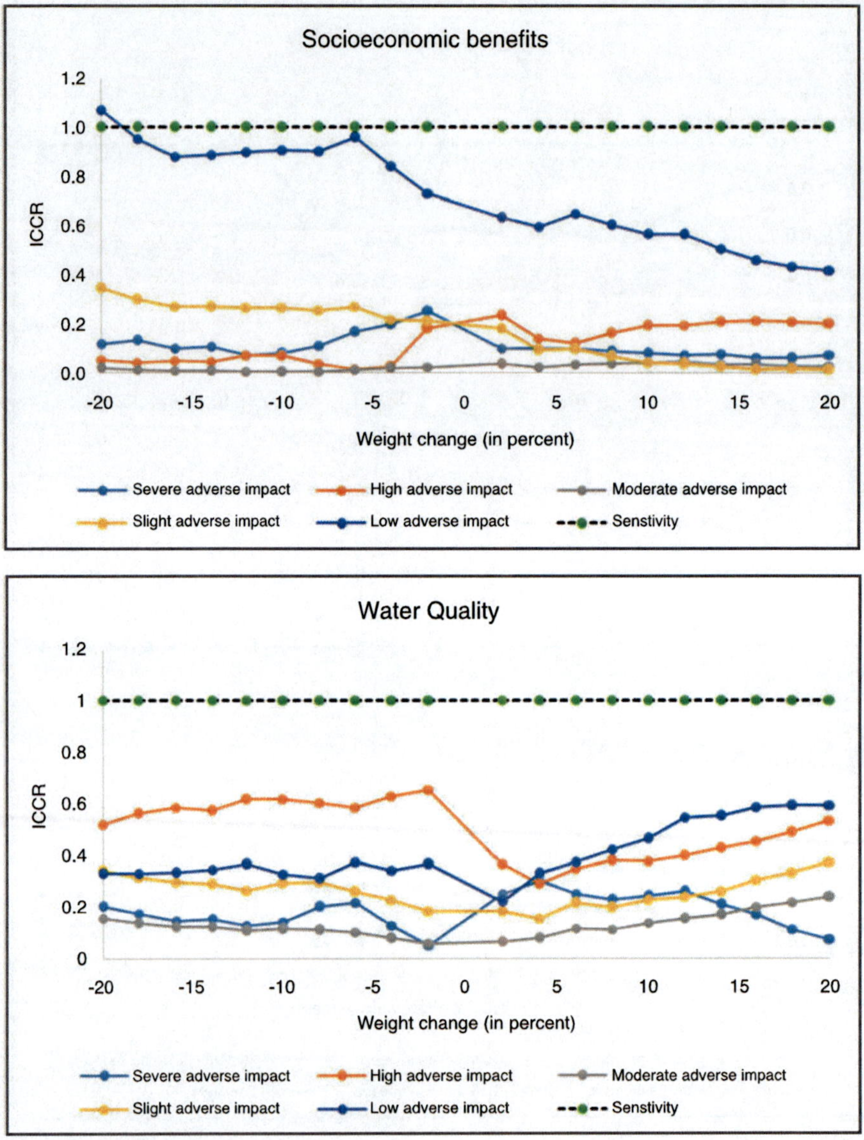

Fig. 13.5 (continued)

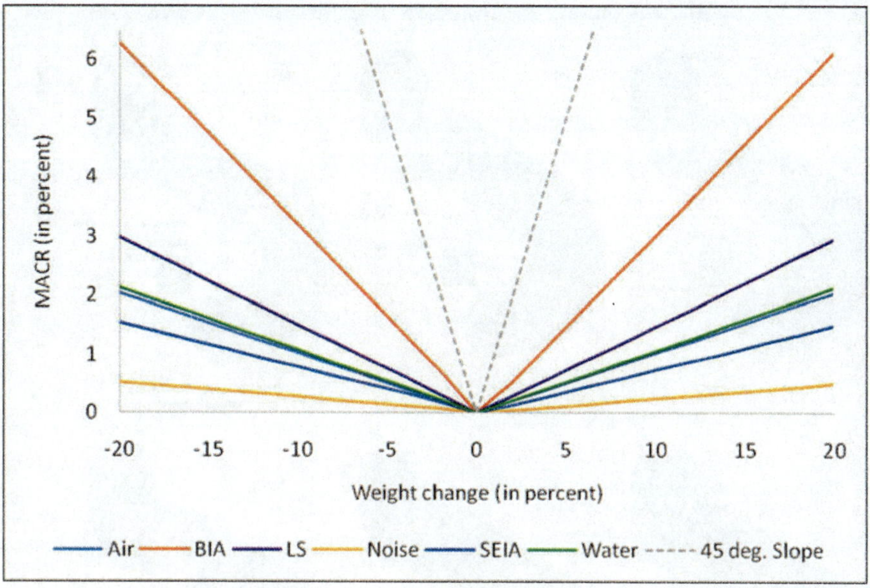

Fig. 13.6 Graphical representation of MACR of SCIAM over the change in environmental criteria weights

Mean Absolute Change Rate

MACR curves of SCIAM showed a linear, symmetrical, and radial distribution of varied slopes around the origin due to equal and opposite changes in environmental criteria weight (Fig. 13.6). Furthermore, MACR slopes were much steeper in the case of biodiversity and landslide susceptibility compared to other environmental criteria weights (Fig. 13.6).

Spatial Change Rate

Figure 13.7 illustrates the percentage change in the composite impact value with 16 percentage rises in the environmental criteria weight. Areas with a high concentration of air pollutants showed the greatest adverse impacts, while the impact value declined in the areas with low air pollution. A similar trend was observed with biodiversity and socioeconomic weight changes. The rise in the landslide weight led to a fall in the impact value in areas with higher landslide susceptibility, while a rise in the impact value was observed in areas with low susceptibility. A similar trend was observed for the changes in water weight too. In the case of noise pollution, the impact value was high near the highway and areas with low noise pollution. Overall, the adverse impact was high near the highway in the cases. An almost equal and

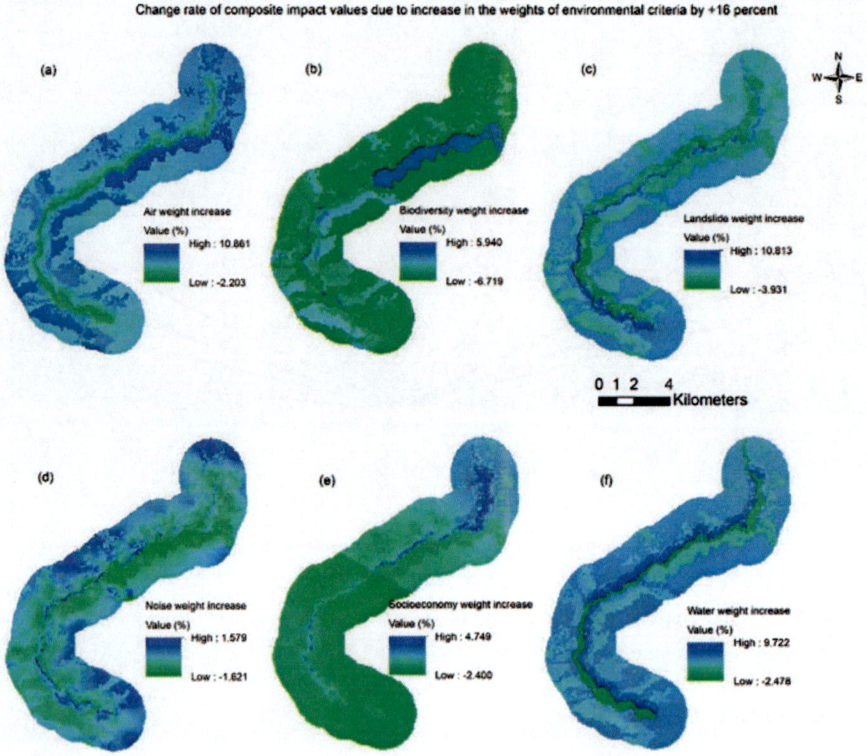

Fig. 13.7 Change rate map of composite impact value due to an increase in environmental criteria weight by +16 percentage

opposite change rate in the impact value was observed with an equal and opposite change in environmental criteria weights (Fig. 13.8).

Spatial Composite Impact Index

SCIAM estimated that broadening of the highway will decrease the air and noise pollution while increasing the socioeconomic benefits. Thereby the project at least initially will cause environmental gains for these criteria (Table 13.6, column 8). In contrast, an increase in adverse impact was estimated for vehicle vulnerability to landslides, a decline in biodiversity value, and an increase in water pollution. Thereby, the project will cause environmental losses for these criteria (Table 13.6, column 7). However, the overall environmental losses outweighed the environmental gains. Based on these arguments and the impact scores estimated, SCII was 0.65 (<0.75), implying that the project was not environmentally viable (Table 13.6, column 9).

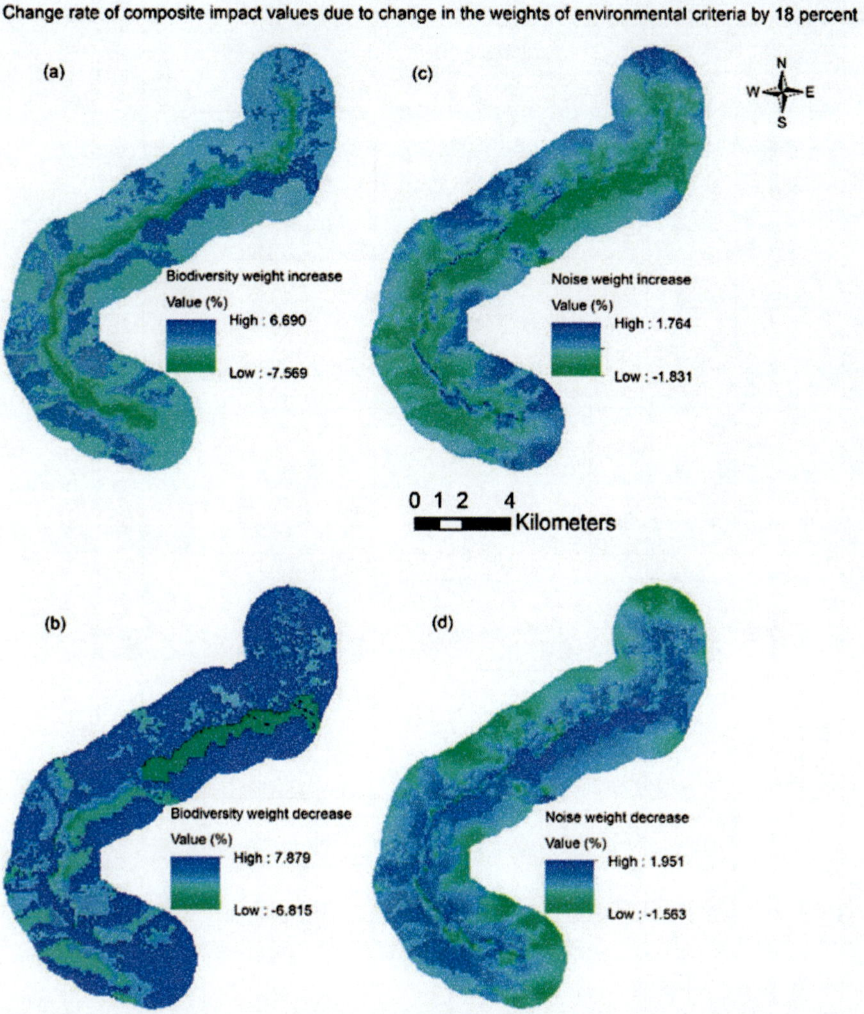

Fig. 13.8 Change rate map of composite impact value due to change in the weight of biodiversity and noise by ±18 percentage

13.4 Discussion

The goal of the study was to construct, implement, and analyze SCIAM in the impact assessment of highway broadening in the East Sikkim. With few exceptions, model validation criteria and cross-validation criteria showed promising results regarding the effectiveness of SCIAM in impact prediction. In addition, SESA showed that the model output, namely, the composite impact value, did not perturb significantly with a substantial change in environmental criteria weight. This can be verified by the

Table 13.6 Estimation of SCII value

	Mean pixel value under without-project scenario (year 2004)	Mean pixel value with-project scenario (year 2014)	Mean change fraction of impact	Adjusted mean change fraction of impact[a]	AHP weight (see Table 10.1)	Weighted change	Environmental gains	Environmental losses	SCII
Environmental criteria	(1)	(2)	$(3)=\frac{(2)-(1)}{(1)}$	(4)	(5)	$(6)=(4)\times(5)$	$(7)=(6)<0$	$(8)=(6)>0$	$(9)=\frac{\lvert\sum(7)\rvert}{\lvert\sum(8)\rvert}$
Air quality	0.245	0.127	−0.479	−0.479	0.135	−0.065	−0.065	0	0.651
Biodiversity value	0.253	0.288	0.138	0.138	0.298	0.041	0	0.041	
Vehicle vulnerability to landslide of light vehicles	0.140	0.319	1.288	0.953	0.201	0.191	0	0.191	
Vehicle vulnerability to landslide of heavy vehicles	0.102	0.165	0.618						
Noise level in *Ldn*	49.549	39.420	−0.204	−0.299	0.095	−0.028	−0.028	0	
Noise level in *Leq*(*H*)	52.924	32.086	−0.394						
Socioeconomic condition	NA	NA	−0.452	−0.452	0.134	−0.061	−0.061	0	
Water quality	0.889	0.910	0.024	0.024	0.135	0.003	0	0.003	
Sum total							−0.154	0.236	

[a]Except for landslide vulnerability and noise level, the mean change fraction of impact (column 3) was brought forward to the adjusted mean change fraction of impact (column 4). It was done since the spatial impact assessment values for these criteria have already been estimated by weighted overlay analysis. However, in the case of vehicle vulnerability to landslide and noise level criteria, adjusted mean change of impact (column 4) was derived by giving equal weights of 0.5 to their sub-criteria (column 3), followed by adding them to get a weight adjusted combined mean change of the criteria. NA is not applicable

flatter slopes of MACR (<45°) and small perturbations of ICCR curves in most of the cases (<1). These observations suggested that SCIAM was robust.

The spatial composite impact assessment map and change rate maps clearly showed that areas near the highway (≤500 m) were most vulnerable to the adverse impacts of the project. Under the AHP model, the high endemicity of the local biodiversity and frequent incidents of landslides, especially during monsoon, was reflected in experts' opinion propensity to giving higher weights to the two criteria. Furthermore, a higher priority to "with-project" scenario indicated the experts' perception of greater adverse impacts due to highway broadening than otherwise. A decline in pollution was estimated in the case of air and noise impact assessments under the "with-project" scenario, mainly due to the decline in the traffic volume of heavy trucks. The socioeconomic impact assessment also showed an overall improvement in socioeconomy due to the project. However, these benefits were outweighed by the loss of biodiversity area, highway runoff induced water pollution, and greater vehicle vulnerability to landslides. This led to a low SCII value indicating non-viability of the project (<0.75).

Socioeconomic benefits caused due to highway broadening were in harmony with earlier studies where highway projects have benefited the local people (Brown, 2003; Rudiarto & Handayani, 2011). Moreover, our study was in line with the claim of the predominant literature that highway runoff is a major pollution source to local waterbodies (Gan et al., 2008) and contradicts the exceptions (Hwang & Weng, 2015). Furthermore, our study acknowledges that highway projects can be a major threat to biodiversity (Atkinson & Canter, 2011; Geneletti, 2003, 2004). However, in contrast to the prevailing literature, air and noise pollution declines in the with-project scenario (Amin et al., 2017; Monazzam et al., 2015).

Some limitations of SCIAM include a limited number of sub-criteria were taken for air and water quality impact assessment. This was mainly due to the lack of a wider set of historical data of these environmental criteria. Next, highway runoff was estimated only for the storm runoff with the onset of the monsoon. The yearlong contribution of the water pollutants generated from the highway runoff was not considered due to the lack of high-resolution meteorological data of the study area. Then, only patch level landscape metrics were considered for biodiversity impact assessment, ignoring the impacts associated with habitat or higher-level ecological complexes. Moreover, the vulnerability assessment due to landslide was only estimated for vehicles. Other elements like property and mortality were not included as it was difficult to predict their temporal trend. Also, instead of a detailed socioeconomic survey, a rapid people's perception-based survey was done, which may not have captured the true nature of the socioeconomic impacts. Finally, SCIAM requires a reliable database, expertise in the use of spatial analysis, and impact prediction software.

SCIAM is probably the first attempt to integrate the biophysical, socioeconomic, and hazard-related impacts due to a highway project in mountainous or remote areas in a developing country. It provides a single impact map and a single impact score that can be easily interpreted by the stakeholders to assess the viability of the project. Also, the present study uses SESA to assess the robustness of the model and highlights the role of criteria weight in overall spatial impact assessment. Possibly, this is the first attempt to include SESA in highway-related spatial impact assessment. The MACR outcomes were in harmony with the SESA method suggested by Xu and Zhang (2013). The effectiveness of SCIAM needs to be assessed in varied terrains, spatial extent, meteorological conditions, project types, and with the integration of submodels. The SESA of SCIAM is deterministic. It can be improved by introducing stochastic methods like Monte Carlo simulation (Xu and Zhang, 2013; Qi et al., 2013). Subject to high-quality data availability, non-expert opinion based MCDM methods like artificial neural networking or multiple regression analysis can be used instead of AHP for SCIAM. Use of habitat connectivity analysis in biodiversity impact assessment and cost-distance analysis in socioeconomic impact assessment can be included in the future. Consideration of the principal component analysis may reduce data redundancy, if any, in SCIAM. The SCII can be pushed towards a viable score ($\geq$0.75) by mitigation measures like slope stabilization, compensatory afforestation, preservation of ecological corridors, and diversion of highway runoff to local water treatment facilities. Overall, SCIAM can be a reliable geovisualization and decision support tool to the project stakeholders.

Geospatial modelling is progressively becoming indispensable in natural as well as anthropogenic hazard mitigation and preparedness. Hazards like landslides, earthquakes, floods, and coastline hurricanes are some major threats to highway infrastructure. On the other hand, roadways constructed on unstable slopes are a major triggering factor of landslides. The likelihood of such natural hazards can be effectively mapped using geospatial modelling. Open access GIS software like QGIS and GRASS, cloud computing-based platforms like Google Earth Engine and Collect Earth, and geocomputation-based predictive mapping in R as well as Python are becoming the cornerstone of geospatial hazard modelling. In the same spirit, SCIAM has been used to assess the susceptibility of landslides in highway NH 10 and the associated vulnerability of such landslides to vehicular traffic.

13.5 Conclusion

SCIAM is a robust model that integrates physico-chemical, ecological, socioeconomic, and hazard susceptibility impacts in spatial impact assessment. It showed that the areas near the highway ($\leq$ 500 m) were most affected by the highway broadening project in the East Sikkim. However, SCIAM requires a comprehensive and reliable database and expertise in the use of spatial analysis and impact modelling. Then again, with the affordability of GIS software, the growing human resources in GIS, and an ever-increasing Internet-based spatially referred data warehouse, SCIAM can be a viable option in developing countries. It can be further improved by the inclusion of stochasticity in MCDM and submodelling of environmental criteria. SCII aids in decision-making by providing a single score to assess the viability of a project, although it fails to identify the factors responsible for it. Both SCIAM and SCII heavily depend on the experts' opinions. Hence a large set of experts specialized in both specific and holistic aspects of the project is a prerequisite. The reliability of SCIAM under varied geographic conditions, spatial extent, and project scenarios needs to be assessed in the future. Consideration of GIS in the mainstream EIA studies is still far from a practice, primarily due to the lack of GIS experts and as it is used mainly in the preparation of the baseline database. In contrast, SCIAM indicates that GIS can be a useful tool for the geovisualization and decision support process during EIA. SCIAM can further help in the prediction of geohazards associated with the highway construction and planning of mitigation measures.

Compliance with Ethical Standards

The authors abide by the ethical standards of the journal.

Funding (Optional) None

Conflict of Interest The authors declare that they have no conflict of interest.

Ethical Approval The manuscript is in abidance with the academic and publication ethics.

Informed Consent The authors have taken consent from the competent authorities for the preparation and communication of this study.

Data Repository https://doi.org/10.6084/m9.figshare.8970335.v1

References

Agrawal, M. L. (2005). *A spatial quantitative approach for environmental impact assessment of highway projects*. IIT.

Agrawal, M. L., Maitra, B., & Ghose, M. K. (2007). A spatial assessment of air pollution impact of a highway project using GIS. In *Information technologies in environmental engineering* (pp. 11–20). Springer. https://doi.org/10.1007/978-3-540-71335-7_5.

Ahmed, F., Fakhruddin, A. N. M., Imam, M. T., Khan, N., Abdullah, A. T. M., Khan, T. A., et al. (2017). Assessment of roadside surface water quality of Savar, Dhaka, Bangladesh using GIS and multivariate statistical techniques. *Applied Water Science, 7*(7), 3511–3525. https://doi.org/10.1007/s13201-017-0619-0

Amin, M. S. R., Tamima, U., & Amador Jimenez, L. (2017). Understanding air pollution from induced traffic during and after the construction of a new highway: Case study of highway 25 in Montreal. *Journal of Advanced Transportation*. https://doi.org/10.1155/2017/5161308

Antunes, P., Santos, R., & Jordão, L. (2001). The application of geographical information systems to determine environmental impact significance. *Environmental Impact Assessment Review, 21* (6), 511–535. https://doi.org/10.1016/S0195-9255(01)00090-7

Arriaza, M., & Nekhay, O. (2008). Combining AHP and GIS modelling to evaluate the suitability of agricultural lands for restoration. 107th EAAE seminar. In: *Modelling agricultural and rural development polices, Sevilla*, January 29 to February 1.

Atkinson, S. F., & Canter, L. W. (2011). Assessing the cumulative effects of projects using geographic information systems. *Environmental Impact Assessment Review, 31*(5), 457–464. https://doi.org/10.1016/j.eiar.2011.01.008

Banerjee, P., & Ghose, M. K. (2016). Spatial analysis of environmental impacts of highway projects with special emphasis on mountainous area: An overview. *Impact Assessment and Project Appraisal, 34*(4), 279–293. https://doi.org/10.1080/14615517.2016.1176403

Banerjee, P., & Ghose, M. K. (2017). A geographic information system-based socioeconomic impact assessment of the broadening of national highway in Sikkim Himalayas: A case study. *Environment, Development and Sustainability, 19*(6), 2333–2354. https://doi.org/10.1007/s10668-016-9859-7

Banerjee, P., Ghose, M. K., & Pradhan, R. (2016). GIS based spatial noise impact analysis (SNIA) of the broadening of national highway in Sikkim Himalayas: A case study. *AIMS Environmental Science, 3*, 714–738. https://doi.org/10.3934/environsci.2016.4.714

Banerjee, P., Ghose, M. K., & Pradhan, R. (2018a). Analytic Hierarchy Process and Information Value Method based Landslide Susceptibility Mapping and Vehicle Vulnerability Assessment along a highway in Sikkim Himalaya. *Arabian Journal of Geosciences, 11*: 139, https://dx.doi.org/10.1007/s12517-018-3488-4

Banerjee, P., Ghose, M. K., & Pradhan, R. (2018b). AHP-based spatial analysis of water quality impact assessment due to the change in vehicular traffic caused by highway broadening in Sikkim Himalaya. *Applied Water Science, 8*(2), 72. https://doi.org/10.1007/s13201-018-0699-5

Banerjee, P., Ghose, M. K., & Pradhan, R. (2018c). AHP based Spatial Air Quality Impact Assessment Model (SAQIAM) of vehicular traffic change due to highway broadening in Sikkim Himalaya. *Annals of GIS, 24*:4, 287–302 https://doi.org/10.1080/19475683.2018.1534889

Banerjee, P., Ghose, M. K., & Pradhan, R. (2020). Analytic Hierarchy Process (AHP) based Spatial Biodiversity Impact Assessment Model (SBIAM) of highway broadening in Sikkim Himalaya. *Geocarto International, 35*(5): 470–493, https://doi.org/10.1080/10106049.2018.1520924

Barthwal, R. R. (2012). *Environmental impact assessment* (2nd ed.). New Age International Private Limited.

Bennett, V. J. (2017). Effects of road density and pattern on the conservation of species and biodiversity. *Current Landscape Ecology Reports, 2*(1), 1–11. https://doi.org/10.1007/s40823-017-0020-6

Bhutia, T. Y. (2015). Biochemical properties of Teesta river system in Sikkim [M.Phil]. Dept. of Chemistry, School of Physical Sciences, Sikkim University.

BIS. (2012). *Indian standard drinking water—specification (Second Revision) IS 10500: 2012*. Bureau of Indian Standards.

Brown, S. (2003). Spatial analysis of socioeconomic issues: Gender and GIS in Nepal. *Mountain Research and Development, 23*(4), 338–344. https://doi.org/10.1659/0276-4741(2003)023[0338:SAOSIG]2.0.CO;2

Burdge, R. J. (2002). Why is social impact assessment the orphan of the assessment process? *Impact Assessment and Project Appraisal, 20*(1), 3–9. https://doi.org/10.3152/147154602781766799

Cai, M., Zou, J., Xie, J., & Ma, X. (2015). Road traffic noise mapping in Guangzhou using GIS and GPS. *Applied Acoustics, 87*, 94–102. https://doi.org/10.1016/j.apacoust.2014.06.005

Cao, C., Wang, Q., Chen, J., Ruan, Y., Zheng, L., Song, S., & Niu, C. (2016). Landslide susceptibility mapping in vertical distribution law of precipitation area: Case of the Xulong Hydropower Station reservoir, southwestern China. *Water, 8*(7), 270.

Champion, H. G., & Seth, S. K. (1968). *A revised survey of the forest types of India*. Manager of Publications. Retrieved from https://dds.crl.edu/crldelivery/23005

Chang, K.-T. (2017). *Introduction to geographic information systems* (4th ed.). McGraw Hill Education.

Chen, H., Wood, M. D., Linstead, C., & Maltby, E. (2011). Uncertainty analysis in a GIS-based multi-criteria analysis tool for river catchment management. *Environmental Modelling & Software, 26*(4), 395–405. https://doi.org/10.1016/j.envsoft.2010.09.005

Chen, Y., Yu, J., & Khan, S. (2010). Spatial sensitivity analysis of multi-criteria weights in GIS-based land suitability evaluation. *Environmental Modelling & Software, 25*(12), 1582–1591. https://doi.org/10.1016/j.envsoft.2010.06.001

CISMHE. (2008a). Aquatic Environment and Water Quality in Carrying capacity of study of Teesta Basin in Sikkim-Volume VI, Biological Environment- Terrestrial and Aquatic Resources. Centre for Inter-Disciplinary Studies of Mountain & Hill Environment, University of Delhi.

CISMHE. (2008b). Carrying Capacity Study of Teesta Basin in Sikkim, Volume – II, land environment—Geophysical environment. Centre for Inter-Disciplinary Studies of Mountain & Hill Environment.

CPCB. (2007). *Transport fuel quality for the year 2005*. Central Pollution Control Board, Government of India.

CPCB. (2015). *National air quality index. CUPS/82/2014-15*. Central Pollution Control Board, Government of India.

Crosetto, M., Tarantola, S., & Saltelli, A. (2000). Sensitivity and uncertainty analysis in spatial modelling based on GIS. *Agriculture, Ecosystems & Environment, 81*(1), 71–79. https://doi.org/10.1016/S0167-8809(00)00169-9

Cui, H., Stein, A., & Myers, D. E. (1995). Extension of spatial information, bayesian kriging and updating of prior variogram parameters. *Environmetrics, 6*(4), 373–384. https://doi.org/10.1002/env.3170060406

Das, I., Kumar, G., Stein, A., Bagchi, A., & Dadhwal, V. K. (2011). Stochastic landslide vulnerability modeling in space and time in a part of the northern Himalayas, India. *Environmental Monitoring and Assessment, 178*, 25–37. https://doi.org/10.1007/s10661-010-1668-0

Das, S., & Raja, D. R. (2015). Susceptibility analysis of landslide in Chittagong City Corporation area, Bangladesh. *International Journal of Environment, 4*(2), 157–181. https://doi.org/10.3126/ije.v4i2.12635

Davie, T. (2008). *Fundamentals of hydrology* (2nd ed.). Routledge.

Devkota, K. C., Regmi, A. D., Pourghasemi, H. R., Yoshida, K., Pradhan, B., Ryu, I. C., et al. (2013). Landslide susceptibility mapping using certainty factor, index of entropy and logistic regression models in GIS and their comparison at Mugling–Narayanghat road section in Nepal Himalaya. *Natural Hazards, 65*(1), 135–165. https://doi.org/10.1007/s11069-012-0347-6

El-Kholy, A. M., & Akal, A. Y. (2020). Proposed Sustainability Composite Index of Highway Infrastructure Projects and Its Practical Implications. *Arabian Journal for Science and Engineering, 45*(5), 3635–3655. https://doi.org/10.1007/s13369-019-04201-1

Feizizadeh, B., Jankowski, P., & Blaschke, T. (2014). A GIS based spatially-explicit sensitivity and uncertainty analysis approach for multi-criteria decision analysis. *Computers & Geosciences, 64* (Supplement C), 81–95. https://doi.org/10.1016/j.cageo.2013.11.009

FHWA. (1998). *FHWA traffic noise model-technical manual*. Federal Highway Administration, FHWA-10-96-010. U.S. Department of Transportation Office of Environment and Planning Research and Special Programs Administration, John A. Volpe National Transportation Systems Centre, Acoustics Facility, Cambridge, MA 02142-1093.

FSI. (2011). *Sikkim in Indian state of forest report 2011* (pp. 214–218). Forest Survey of India. Retrieved from http://fsi.nic.in/details.php?pgID=sb_16

Gan, H., Zhuo, M., Li, D., & Zhou, Y. (2008). Quality characterization and impact assessment of highway runoff in urban and rural area of Guangzhou, China. *Environmental Monitoring and Assessment, 140*(1–3), 147–159. https://doi.org/10.1007/s10661-007-9856-2

Geneletti, D. (2003). Biodiversity impact assessment of roads: An approach based on ecosystem rarity. *Environmental Impact Assessment Review, 23*(3), 343–365. https://doi.org/10.1016/S0195-9255(02)00099-9.

Geneletti, D. (2004). Using spatial indicators and value functions to assess ecosystem fragmentation caused by linear infrastructures. *International Journal of Applied Earth Observation and Geoinformation, 5*(1), 1–15. https://doi.org/10.1016/j.jag.2003.08.004

Geneletti, D. (2008). Impact assessment of proposed ski areas: A GIS approach integrating biological, physical and landscape indicators. *Environmental Impact Assessment Review, 28* (2), 116–130. https://doi.org/10.1016/j.eiar.2007.05.011

Geneletti, D., & Dawa, D. (2009). Environmental impact assessment of mountain tourism in developing regions: A study in Ladakh, Indian Himalaya. *Environmental Impact Assessment Review, 29*(4), 229–242. https://doi.org/10.1016/j.eiar.2009.01.003

Geurs, K. T., Boon, W., & Wee, B. V. (2009). Social impacts of transport: Literature review and the state of the practice of transport appraisal in the Netherlands and the United Kingdom. *Transport Reviews, 29*(1), 69–90. https://doi.org/10.1080/01441640802130490

Gifford, F. A. (1961). Use of routine meteorological observations for estimating atmospheric dispersion. *Nuclear Safety, 2*(4), 47.

Glasson, J., Therivel, R., & Chadwick, A. (2005). *Introduction to environmental impact assessment* (3rd ed.). Routledge.

Goyal, P. (2008). *Carrying capacity study of Teesta Basin in Sikkim—Air environment*. Centre for Atmospheric Sciences, Indian Institute of Technology.

Goyal, P., & Rama Krishna, T. V. B. P. S. (1999). A line source model for Delhi. *Transportation Research Part D: Transport and Environment, 4*(4), 241–249. https://doi.org/10.1016/S1361-9209(99)00007-3

Goyal, P., Anand, S., & Gera, B. S. (2006). Assimilative capacity and pollutant dispersion studies for Gangtok city. *Atmospheric Environment, 40*(9), 1671–1682. https://doi.org/10.1016/j.atmosenv.2005.10.057

Gülci, S., & Akay, A. E. (2015). Assessment of ecological passages along road networks within the Mediterranean forest using GIS-based multi criteria evaluation approach. *Environmental Monitoring and Assessment, 187*(12), 779. https://doi.org/10.1007/s10661-015-5009-1

Gurung, S., Subba, S., Jha, S., & Pandey, M. (2015). Analysis of physic-chemical variations in water samples of river Teesta of Sikkim. International Journal of Engineering, Technology and Research, 2(2), 49–60.

Guzzetti, F. (2005). Landslide hazard and risk assessment. In: *PhD thesis, Mathematics—Scientific faculty*, University of Bonn, Bonn, 389 pp.

Hobbie, R. K., & Roth, B. J. (2015). Exponential growth and decay. In *Intermediate physics for medicine and biology* (pp. 33–51). Springer. https://doi.org/10.1007/978-3-319-12682-1_2.

Hwang, C.-C., & Weng, C.-H. (2015). Effects of rainfall patterns on highway runoff pollution and its control. *Water and Environment Journal, 29*(2), 214–220. https://doi.org/10.1111/wej.12109

Jaiswal, P., van Westen, C. J., & Jetten, V. (2009). *Quantitative landslide risk assessment along the transport lines in South India*. ITC, 27 pp.

Karbassi, A. R., Mir Mohammad Hosseini, F., Baghvand, A., & Nazariha, M. (2011). Development of water quality index (WQI) for Gorganrood River. *International Journal of Environmental Research, 5*(4), 1041–1046. https://doi.org/10.22059/ijer.2011.461

Kayhanian, M., Singh, A., Suverkropp, C., & Borroum, S. (2003). Impact of annual average daily traffic on highway runoff pollutant concentrations. *Journal of Environmental Engineering, 129* (11), 975–990. https://doi.org/10.1061/(ASCE)0733-9372(2003)129:11(975)

Krivoruchko, K. (2012). *Empirical Bayesian kriging—implemented in ArcGIS geostatistical analyst*. ArcUser. Retrieved from http://www.esri.com/news/arcuser/1012/empirical-byesian-kriging.html

Leopold, L. B., Clarke, F. E., Hanshaw, B. B., & Balsley, J. R. (1971). *A procedure for evaluating environmental impact (USGS numbered series no. 645)* (p. 19). U.S. Geological Survey. Retrieved from http://pubs.er.usgs.gov/publication/cir645

Li, Z., Fang, Y., Zeng, G., Li, J., Zhang, Q., Yuan, Q., et al. (2009). Temporal and spatial characteristics of surface water quality by an improved universal pollution index in red soil hilly region of South China: A case study in Liuyanghe River watershed. *Environmental Geology, 58*(1), 101–107. https://doi.org/10.1007/s00254-008-1497-4

Lilburne, L., & Tarantola, S. (2009). Sensitivity analysis of spatial models. *International Journal of Geographical Information Science, 23*(2), 151–168. https://doi.org/10.1080/13658810802094995

Liu, X. (2006). Site-specific vulnerability assessment for debris flows: Two case studies. *Journal of Mountain Science, 3*, 20–27. https://doi.org/10.1007/s11629-006-0020-1

Lloyd, C. (2009). Spatial data analysis: An introduction for GIS users. OUP Oxford.

Lo, C. P., & Yeung, A. K. W. (2006). *Concepts and techniques of geographic information systems* (2nd ed.). Pearson.

Longley, P. A., Goodchild, M., Maguire, D. J., & Rhind, D. W. (2010). *Geographic information systems and science* (3rd ed.). John Wiley & Sons.

Modak, P., & Biswas, A. K. (1999). *Conducting environmental impact assessment for developing countries*. CABI.

Monazzam, M. R., Karimi, E., Abbaspour, M., Nassiri, P., & Taghavi, L. (2015). Spatial traffic noise pollution assessment—a case study. *International Journal of Occupational Medicine and Environmental Health, 28*(3), 625–634. https://doi.org/10.13075/ijomeh.1896.00103

Murphy, J. M., Sexton, D. M. H., Barnett, D. N., Jones, G. S., Webb, M. J., Collins, M., & Stainforth, D. A. (2004). Quantification of modelling uncertainties in a large ensemble of climate change simulations. *Nature, 430*(7001), 768–772. https://doi.org/10.1038/nature02771

Mushtaq, F., Lala, M. G. N., & Pandey, A. C. (2015). Assessment of pollution level in a Himalayan Lake, Kashmir, using geomatics approach. *International Journal of Environmental Analytical Chemistry, 95*(11), 1001–1013. https://doi.org/10.1080/03067319.2015.1077517

MyCurveFit. (2017). Online curve-fitting at mycurvefit.com. Retrieved October 28, 2017, from https://www.mycurvefit.com

Paliwal, R., & Srivastava, L. (2014). *Policy intervention analysis: Environmental impact assessment*. The Energy and Resources Institute, TERI.

Pardeshi, S. D., Autade, S. E., & Pardeshi, S. S. (2013). Landslide hazard assessment: Recent trends and techniques. *Springerplus, 2*, 523. https://doi.org/10.1186/2193-1801-2-523

Pilz, J., & Spöck, G. (2008). Why do we need and how should we implement Bayesian kriging methods. *Stochastic Environmental Research and Risk Assessment, 22*(5), 621–632. https://doi.org/10.1007/s00477-007-0165-7

Pourghasemi, H. R., Pradhan, B., Gokceoglu, C., & Moezzi, K. D. (2012). Landslide susceptibility mapping using a spatial multi criteria evaluation model at Haraz watershed, Iran. In *Terrigenous mass movements* (pp. 23–49). Springer. https://doi.org/10.1007/978-3-642-25495-6_2

Qi, H., Qi, P., & Altinakar, M. S. (2013). GIS-based spatial Monte Carlo analysis for integrated flood management with two dimensional flood simulation. *Water Resources Management, 27* (10), 3631–3645. https://doi.org/10.1007/s11269-013-0370-8

Rahman, H., Karuppaiyan, R., Senapati, P. C., Ngachan, S. V., & Kumar, A. (2012). An analysis of past threedecade weather phenomenon in the mid-hills of Sikkim and strategies for mitigating possible impact of climate change on agriculture in Climate Change in Sikkim Patterns, Impacts and Initiatives. Information and Public Relations Department, Government of Sikkim.

Ramachandra, T. V., & Shwetmala. (2009). Emissions from India's transport sector: State wise synthesis. *Atmospheric Environment*. https://doi.org/10.1016/j.atmosenv.2009.07.015

Rudiarto, I., & Handayani, W. (2011). *Spatial differentiation of socioeconomic and infrastructural development in rural mountain area. In international seminar on urban and regional planning in the era of global change*. Hasanuddin University. Retrieved from https://core.ac.uk/download/pdf/11734194.pdf

Saaty, T. L. (1980). The Analytic Hierarchy Process: Planning, Priority Setting, Resource Allocation. Mcgraw-Hill.

Saaty, T. L. (1990). How to make a decision: The analytic hierarchy process. *European Journal of Operational Research, 48*(1), 9–26. https://doi.org/10.1016/0377-2217(90)90057-I

Saaty, T. L. (2000). Fundamentals of decision making and priority theory with the analytic hierarchy process. RWS Publications, Pittsburg.

Saaty, T. L., & Vargas, L. (1994). Fundamentals of Decision Making and Priority Theory with the Analytic Hierarchy Process: 6 (1st edition). RWS Publications, Pittsburg.

Stoms, D. M. (2000). GAP management status and regional indicators of threats to biodiversity. *Landscape Ecology, 15*(1), 21–33. https://doi.org/10.1023/A:1008105026956

Taha, H. A. (2010). *Operations research: An introduction* (9th ed.). Pearson.

Takyl, S. A. (2012). *Review of environment impact assessment- approach, process and challenges*. Retrieved from https://www.academia.edu/3703629/Review_of_Environmental_Impact_Assessment_EIA_Approach_Process_and_Challenges.

Therivel, R., & Wood, G. (2017). *Methods of environmental and social impact assessment* (4th ed.). Routledge. Retrieved October 13, 2017, from https://www.routledge.com/Methods-of-Environmental-and-Social-Impact-Assessment-4th-Edition/Therivel-Wood/p/book/9781138647671

Vojtek, M., & Vojteková, J. (2016). GIS-based approach to estimate surface runoff in small catchments: A case study. *Quaestiones Geographicae, 35*(3), 97–116. https://doi.org/10.1515/quageo-2016-0030

Wang, F., Xu, P., Wang, C., Wang, N., & Jiang, N. (2017). Application of a GIS-based slope unit method for landslide susceptibility mapping along the Longzi River, southeastern Tibetan plateau, China. *ISPRS International Journal of Geo-Information, 6*(6), 172. https://doi.org/10.3390/ijgi6060172

Warner, L. L., & Diab, R. D. (2002). Use of geographic information systems in an environmental impact assessment of an overhead power line. *Impact Assessment and Project Appraisal, 20*(1), 39–47. https://doi.org/10.3152/147154602781766807

Xu, E., & Zhang, H. (2013). Spatially-explicit sensitivity analysis for land suitability evaluation. *Applied Geography, 45*(Supplement C), 1–9. https://doi.org/10.1016/j.apgeog.2013.08.005

Yan, C.-A., Zhang, W., Zhang, Z., Liu, Y., Deng, C., & Nie, N. (2015). Assessment of water quality and identification of polluted risky regions based on field observations & GIS in the Honghe River Watershed, China. *PLoS One, 10*(3), e0119130. https://doi.org/10.1371/journal.pone.0119130

Zhao, H., Tian, W.-P., Li, J., & Ma, B.-C. (2017). Hazard zoning of trunk highway slope disasters: A case study in northern Shaanxi, China. *Bulletin of Engineering Geology and the Environment*, 1–10. https://doi.org/10.1007/s10064-017-1178-1

Chapter 14
Evaluation of Post-Seismic Ground Deformation Using the D-InSAR Technique

Bijay Halder, Veera Sri Naga Sai, Satiprasad Sahoo, and Pulakesh Das

Abstract Tamenglong, the city of Manipur, and its surrounding areas, were hit by an earthquake at 6.7M_w on 4 January 2016. The epicenter is located at 24.834°N, 93.656°E (Noney in the Tamenglong district), at a depth of 55 km, and caused injuries of 200 people with 11 life losses and huge economic loss. In the present study, the Differential Interferometric Synthetic Aperture Radar (D-InSAR) technique is used to estimate the ground deformation due to this earthquake. A pair of complex C-band Sentinel-1 SAR data, acquired during the pre-seismic and post-seismic events were used to generate the phase change image, indicates the ground deformation due to the earthquake. The use of SAR data for two closest times assumes similar atmospheric attenuation removed employing the metadata information. The SRTM DEM data was used for Doppler terrain correction accounts for the local topographic discrepancies in the SAR images and to derive the deformation map. The D-InSAR technique estimated deformation along the line-of-sight (LoS) indicated a significant deformation along the various cross-sections. Moreover, the comparison between the SRTM DEM and phase image derived DEM estimated the observed deformations. Six regions were highlighted along the western part of the Manipur state to understand the deformations, where the majority of the deformation was observed around 2.3 mm, and minimum deformation was 0.72 mm. The estimated deformation is essential in various related studies, e.g., structural and geodetic observations. Moreover, long-term deformation studies will help in hazard mapping, which will be beneficial in many ways including preparedness and hazard management activities.

Keywords Sentinel-1 · D-InSAR · Manipur earthquake · Ground deformation

B. Halder (✉) · P. Das
Department of Remote Sensing and GIS, Vidyasagar University, West Midnapore, India

V. S. N. Sai
Centre for Ocean, Rivers, Atmosphere and Land Sciences, Indian Institute of Technology Kharagpur, Kharagpur, India

S. Sahoo
Department of Civil Engineering, Indian Institute of Technology Kharagpur, Kharagpur, India

P. K. Shit et al. (eds.), *Geospatial Technology for Environmental Hazards*, Advances in Geographic Information Science, https://doi.org/10.1007/978-3-030-75197-5_14

14.1 Introduction

Hazard assessment of a region poses many challenges and it is a very complex task. Estimating the effects of hazard is a very sophisticated process. Many countries have formed hazard or disaster management departments that continuously monitor various disasters and develop mitigation and action plans. In India, the National Center for Seismology (NCS) monitors earthquake events using 115 stations spread all across the country. Since -•50 Ma, the Indian subcontinent has faced a quick collision with Eurasia, which results in a wide convergence zone about 1500 km and nearly 3500 km in the west and east, respectively, where the corresponding convergence rate is about 43 mm/year and 52 mm/year (DeMets et al., 1994). Recurrent earthquakes are observed in the Himalayan ranges causing significant crustal deformation and ground subsidence. An earthquake produces seismic waves that propagate outward and cause ground shaking and permanent displacement of the earth's surface. Such disaster adversely affects resulting in the loss of life, havoc damages, and huge economic loss. Although there is significant reduction in the fatality rate due to natural disasters, the mortality rate due to earthquake remains persistent (Elliott, 2020). The recent (in the year 2021) earthquake incidents map published by NCS, India, indicates the majority of the events in and around the Himalayan range. Although the Indo-Gangetic Plains and Peninsular India is seismically less active, experienced a number of earthquake events in the past, indicated >60% of the area in India total area under threat (Jain, 2016) (Fig. 14.1).

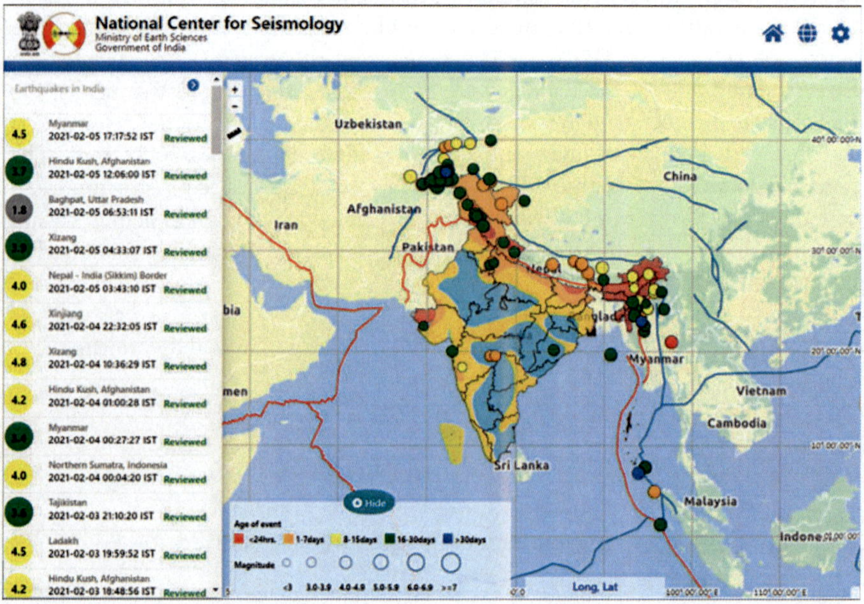

Fig. 14.1 National Center for Seismology portal shows the location of recent (in the year 2021) earthquakes events (https://seismo.gov.in/MIS/riseq/earthquake)

The satellite observations, including imaging and non-imaging systems are an essential source of data for monitoring and assessing the hazards. The highly accurate global position systems are widely used for studying crustal velocity, plate movement, and tectonic deformations (Jade et al., 2004; Bisht et al., 2020). Alternatively, the Synthetic Aperture Radars (SAR) imaging systems operate in the microwave region of the electromagnetic spectrum and enables the generation of the spatial maps on ground deformation. The active microwave systems can operate both in day and night-time, and the microwave signals have the capacity of penetrating the clouds and enable all weather imaging system. The microwave data contains the backscatter microwave signals (amplitude) and the phase information. The microwave signals are transmitted and recorded in two polarizations as horizontal (H) and vertical (V) and produce four data combinations as VV, VH, HV, and HH. The VV data indicates vertically transmitted and vertically received, and VH indicates vertically transmitted and horizontally received, and so on. The interferometric SAR (InSAR) is a widely used technique for ground deformation estimation. It uses two symmetric radar images, which enable the generation of a 3D surface or Digital Elevation Model (DEM). The change analysis of the DEM or ground deformation between two acquisitions is known as Differential InSAR (DInSAR). The InSAR technique derives the information on ground deformation developing the interferogram formed by analyzing the phase differences between two complex SAR images. Moreover, the prerequisite of an interferogram SAR image pair for deformation study is image coherence, i.e., acquired for the same area from two slightly different positions. The two coherent images are acquired either from two antennas on the same platform and separated perpendicularly along the flight direction (single pass SAR interferometry), or from different passes of the same SAR antenna (repeat-pass interferometry). With the movement of any point on the earth surface, the distance between the sensor and the point changes, which modifies the recorded phase information. The coherence between a pair of co-registered complex SAR images decreases or leads to temporal de-correlation due to system noise, volume scattering, temporal changes in vegetation, weathering, change in dielectric property, etc. (Zhou et al. 2009). The phase shift of 2π is equivalent to the distance variation by half of the wavelength ($\lambda/2$; λ is the operating wavelength of the microwave system). The phase change is also caused by the scattering due to the atmospheric constituents. Such attenuation can be minimized by considering a pair of SAR images (master and slave images) having similar observational geometry assuming they have similar scattering properties (Raucoules et al., 2007). The interferogram develops fringe pattern which indicates the lines of equal phase, where the number of fringes surrogates the alteration in the surface or topography. The potential factors that cause the fringe pattern include ground deformation due to earthquake, mining subsidence, land cover changes, change in water levels, land uplift, snow accumulation, etc. The selection of a SAR image pair depends on the coherence and the sensitivity of the interferogram, formed by the two co-registered images, which is expressed by the altitude of ambiguity (h_a). The altitude of ambiguity (h_a) is estimated using the orbital separation between the two image acquisitions, and it is equal to the size of a DEM that would produce one artifactual fringe (Zhou et al.

2009). However, the topographic effects can be ignored when ha is significantly higher compared to the estimated vertical accuracy of a DEM. The h_a is expressed as:

$$h_a = \lambda R \ \sin\theta/2B$$

where θ is local incidence angle, R is slant range between the ground point and antenna in the master image, and B is the distance between the two acquisitions.

The change in signal phase can be expressed as follows:

$$\Delta\varphi = \frac{4\pi}{\lambda}\Delta R + \alpha + \varepsilon + \text{noise}$$

where ΔR is displacement or ground deformation along the satellite line of sight (LoS), α is phase shift due to atmospheric attenuation, and ϵ is the possible errors in the DEM that was used to remove the topographic effects. With a lower noise (less de-correlation) and the phase contribution due to the local topography is precisely measured (e is low), the phase change can be simplified as:

$$\Delta\varphi = \frac{4\pi}{\lambda}\Delta R + \alpha$$

The Sentinel-1 SAR sensor operating in C-band (6 cm, 4–8 GHz) is a freely available SAR data facilitated since 2014 by the Copernicus Programme of the European Space Agency (ESA). The Sentinel-1 mission comprises a constellation of two polar-orbiting satellites which have capabilities for rapid data dissemination at short revisit cycles (12 days for single sensor and 6 days for Sentinel-1A and Sentinel-1B). The interferometric capabilities of Sentinel-1C enable geohazard monitoring including land subsidence assessment. Salvi et al. (2012) reviewed the existing and past SAR system for ground deformation studies comparing with the Sentinel-1 data. They have highlighted the advantages of Sentinel-1 data in estimating the surface displacement. Polcari et al. (2017) employed the COSMO-skyMAP X- and Sentinel-1 C-band InSAR data to estimate the local deformation caused by the 2016–2017 Central Italy seismic sequence. They estimated a deformation ~2–3 cm along the LoS, where the results from X- and C-band indicated similar results (cross-validated each other). Raspini et al. (2018) employed the Sentinel-1 to develop a system for continuous and systematic tracking of the ground deformation with a case study in the Tuscany region (Central Italy). They indicated the operational use of Sentinel-1 data in facilitating the retrieval of crustal deformation at high temporal frequency, which is particularly important for risk mitigation. Neelmeijer et al. (2018) analyzed the ground deformation due to water level variations in the Toktogul Reservoir, Kyrgyzstan, using the Envisat ASAR and Sentinel-1 data for the period 2004–2009 and 2014–2016, respectively. The Sentinel-1 data derived analysis indicated a mean vertical subsidence rate of 2.5 cm/year for the period of March 2015–November 2016 well verified with the results obtained from elastic modelling (TEA12 Earth model). Imamoglu et al. (2019) estimated ground deformation using the multitemporal Sentinel-1 data employing the InSAR technique in Bolvadin, Turkey. The highest subsidence in this region is estimated as 35 mm/year, characterized by the presence of soft alluvial deposits, where the primary cause of subsidence could be due to overexploitation of groundwater and hydrological changes. Ramirez et al. (2020) employed the Sentinel-1 data for long-term monitoring of the ground deformation using the InSAR technique in a few selected sites in South Korea. They have reiterated the applicability of the InSAR technique and SAR data for geohazard assessment. The present study was conducted in the Tamenglong city and its surrounding areas, Manipur state, which was hit by an earthquake at 6.7Mw on 4 January 2016. The Sentinel-1 SAR data was used to estimate the ground deformation employing the InSAR technique.

14.2 Study Area

This state of Manipur is one of the states in the eastern Himalayan region bounded between 23°50′N to 25°42′N latitude and 92°58′E to 94°45′E longitude, and was considered as the study area. It covers an area of 22,327 km^2 and occupies 0.68% of the total geographic area of India. The state is neighboring by Nagaland in the north, Mizoram in the south, Assam in the west, and sharing an international boundary with Burma/Myanmar in the east. Geologically, this state belongs at the young folded with various rocks from the upper Cretaceous to the present alluvium. Two major broad soil types are observed, the red ferruginous soil in the hills and alluvium in the valleys (https://investinmanipur.nic.in/). According to Census India, 2011, the total population of Manipur is 2,721,756. The state belongs to the tropical climate zone with an annual precipitation raging between 1200 mm and 2700 mm, and average annual temperature varies between 14.5 °C and 38 °C. The dominant land cover of the state is forest (~80%) followed by cropland, a biodiversity-rich landscape. Additionally, the current study was focused on the earthquake observed in the Tamenglong city and its surrounding region, hit by an earthquake with magnitude 6.7 on 4 January 2016, caused loss of a few lives and many injuries with significant land deformation (Fig. 14.2).

14.3 Data Used and Methodology

Data Used

The Sentinel-1 C-band (wavelength: ~5.6 cm) SAR data was used to estimate the ground deformation, where the IW products were used before and after the land subsidence event as 11 December 2015 and 28 January 2016, respectively. The Sentinel-2 multi-spectral and Shuttle Radar Topographic Mission (SRTM) DEM were also employed in the current study.

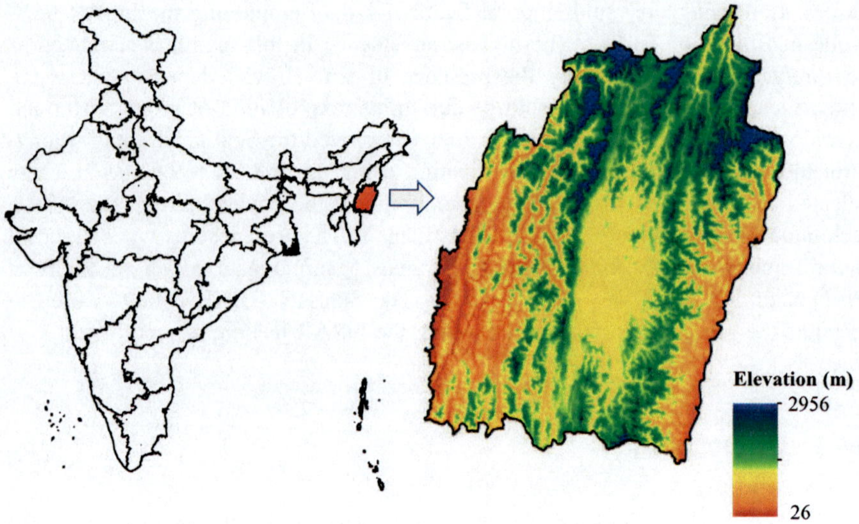

Fig. 14.2 Study area map: elevation (ASTER DEM) map of Manipur state

The SNAP Sentinel-1 toolbox data processing platform is used for data analysis and ArcGIS software for map preparation. The processing of the SAR data for interferogram generation is briefly stated as follows: the image co-registration was performed to remove the relative mis-registration errors between the two interferometric SAR images. The enhanced spectral diversity (ESD) method was used, and the pixel by pixel phase information was retrieved from the different spectral looks. Thereafter, all the patch pairs with good cross-correlation were chosen. The average peak coordinates were extracted to estimate the range and azimuth offsets, which were then applied to the slave image by shifting with the estimated offset values. The interferogram was generated integrating the master and refined slave image, where the coherence for image pair was also estimated. The pixel by pixel phase difference image was obtained by multiplying two images. The metadata and orbit information for the interferometric image pair were integrated to subtract the flat-earth phase (indicates the phase present in the interferometric signal owing to the reference surface curvature) from the generated interferogram. The TOPSAR IW SLC products consist of one image per swath per polarization with multiple sub-swaths. It consists of a series of bursts, where each burst was processed separately and merged to a single image according to the azimuth-time order. The individually focused complex burst images are included into a single sub-swath image, with black-fill demarcation in between. Thereafter, the topographic phase removal was conducted employing the DEM data, which estimates and subtracts the topographic phase from the generated interferogram.

The Goldstein adaptive radar interferogram phase filtering (proposed by Goldstein and Werner, 1998) was applied to enhance the SNR, which fundamentally uses the Fourier spectrum. Where the difference image between the filtered and unfiltered interferogram generates a residual systematic phase trend denotes the loss of resolution due to filtering. The range Doppler terrain correction was then performed to compensate for the topographic distortions and to generate the orthorectified image. These distortions are introduced due to topographic variations in a scene and tilting of satellite sensor. To derive the precise geolocation information, the terrain correction process uses the available orbit state vector information in the metadata, the radar timing annotations, the slant to ground range conversion parameters, and the reference DEM. In the two-pass method, the DEM is used as a synthetic interferogram simulating topographic phase. Although it does not require the unwrapping process, the accuracy depends on the DEM accuracy, which produces phase error of (Error in DEM/h_a) cycles in the interferogram. The overall methodology flow diagram is shown in Fig. 14.3.

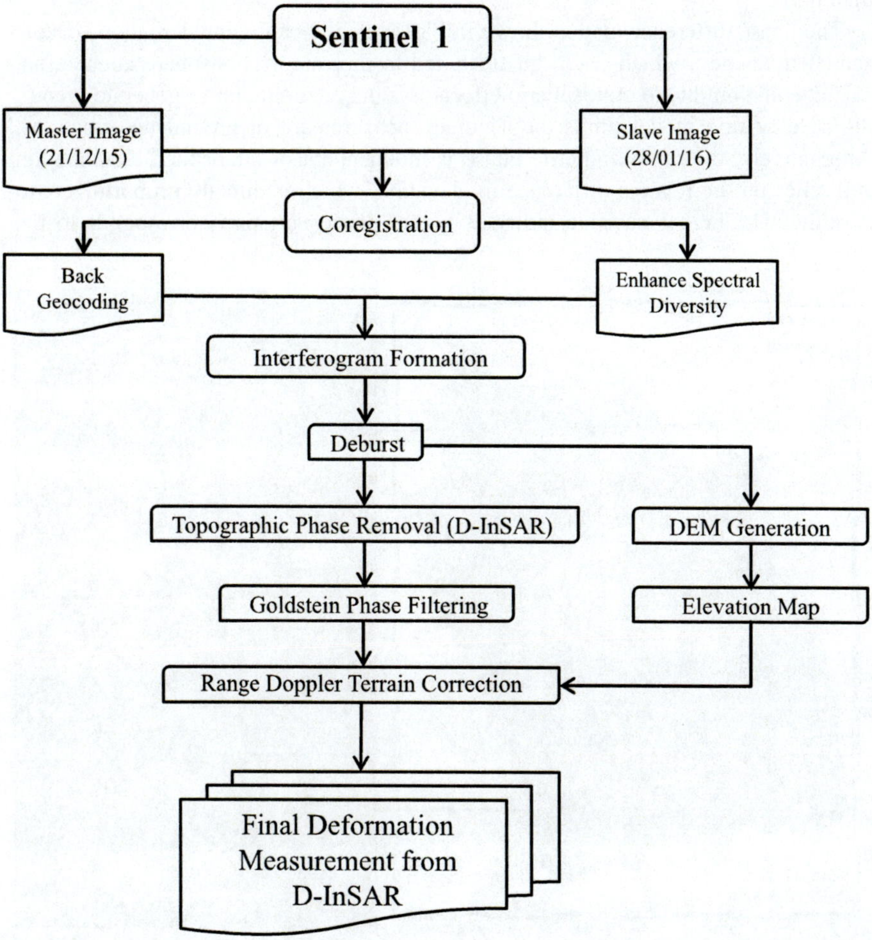

Fig. 14.3 Overall methodology flow diagram

14.4 Result and Discussion

Using the InSAR technique, the interferogram was generated employing two-pass approach with the Sentinel-1 SAR acquired on 11 December 2015 and 28 January 2016. The Sentinel-1 data derived coherence image is shown in Fig. 14.4, which indicated a significant coherence in the majority of the region in Manipur state. The interferometric coherence is an essential factor that indicates the suitability of the SAR image pair to assess the ground deformation. The coherence factor is computed as the absolute value of the correlation coefficient obtained from the radar image pair. The coherence values varied between 0.995 and 0.000015, with an average value of 0.4975. High coherence is observed in the majority of the area with some places with lower coherence values. The observed average coherence in the study indicated much higher coherence reported by Ramirez et al. (2020) estimated as less than 0.3.

The phase difference map is shown in Fig. 14.5, indicating low variations or low spatial frequency, which could be attributed to the minor atmosphere attenuation. The attenuation due to atmospheric effects is a major limitation to generate precise digital elevation model limits the accurate measurement of ground deformations. Alternatively, the interferometric phase is independent of atmospheric attenuation and relies on the relative difference in elevation, which is directly proportionate to baseline. The higher baseline indicates higher phase variation corresponds to the

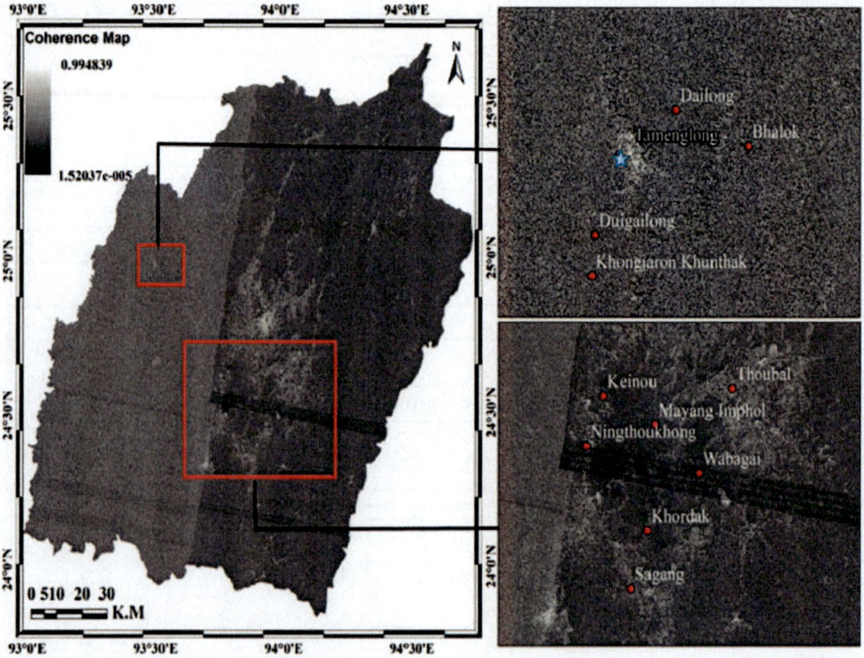

Fig. 14.4 Coherence image generated using the Sentinel-1 SAR data

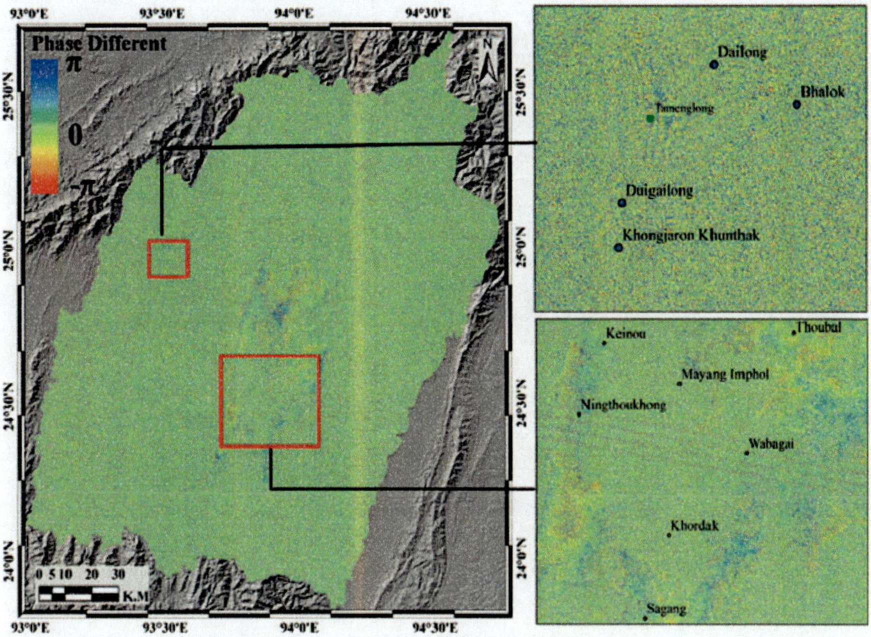

Fig. 14.5 Phase images derived using the InSAR technique

same topography; wherein, the higher baseline of the interferometric image pair indicates lower topographic error.

A differential interferogram was generated employing an external DEM data to nullify the effects due to topography. The fringes in the differential interferometric are clearly visible enabling proper analysis. The result indicated that fringes are identical to those of the real topographic interferogram. Area with higher fringes indicates ground deformation (Fig. 14.6). Three cross-section profiles were drawn on the differential interferogram indicating gradual phase changes along the distance. The red and black lines show the phase before and after the ground deformation caused by the earthquake, respectively (Fig. 14.7).

The DEM developed integrating the SRTM DEM and D-InSAR data exhibits the topographic deformation generated. The SRTM DEM and D-InSAR data derived DEM are shown in Fig. 14.8. Six regions in the image area are shown along with the cross-section to visualize the nature of changes occurred in topography due to the earthquake (Fig. 14.9). The highlighted area clearly indicates the changes in topography compared with the SRTM DEM data. The curve generated between the distance and deformation also exhibits the nature of deviation compared to pre-event topographic variations. The maximum deformation was observed as 2.3 mm, and the lower deformations were observed as 0.72 mm in the highlighted regions.

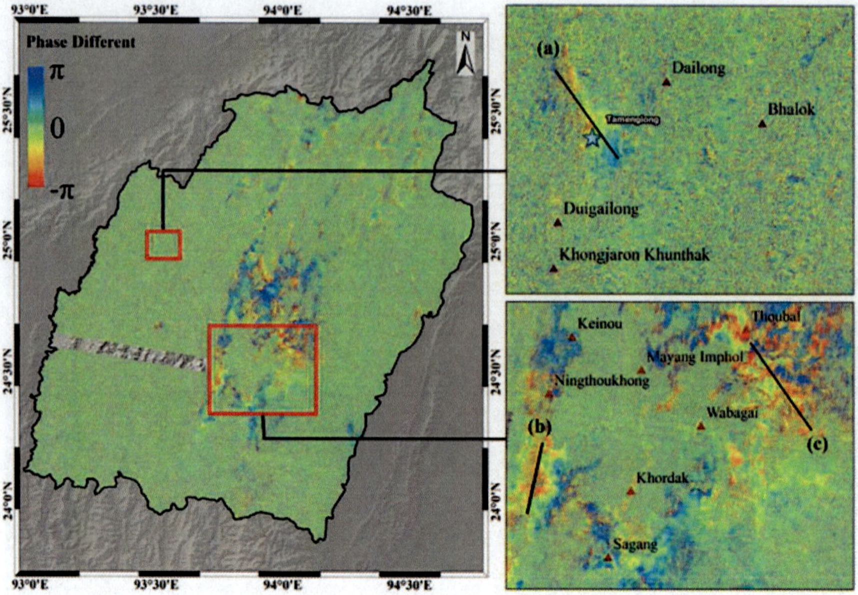

Fig. 14.6 Differential interferogram generated using the D-InSAR technique

14.5 Conclusion

The study highlights the applicability of InSAR and D-InSAR technique for ground deformation assessment employing the freely available Sentinel-1 C-band SAR data. The pre-seismic and post-seismic complex SAR images were acquired for the Manipur state, hit by an earthquake of magnitude 6.7 on 4 January 2016. The obtained land deformation map indicated a realistic value that corroborates the observations reported in previous studies. Significant ground deformation has been identified in various regions around the Tamenglong city and neighboring regions. The high average coherence between the two complex SAR images indicates the reliability of the obtained results. The distance vs deformation curve generated for the cross-sections in different regions clearly exhibits the nature of observed ground deformation. Regular monitoring of the surface deformation can be performed to generate earthquake hazard map, which is essential for identifying the risk zone area over a large region. This will help the managers and planners of various domains in improved planning and management activities. More importantly, the resultant maps will bear significant importance in preparedness and to reduce the damage due to unwanted hazards. The earthquake hazard maps can also be utilized for planning of land use, road and building construction, and many more application areas.

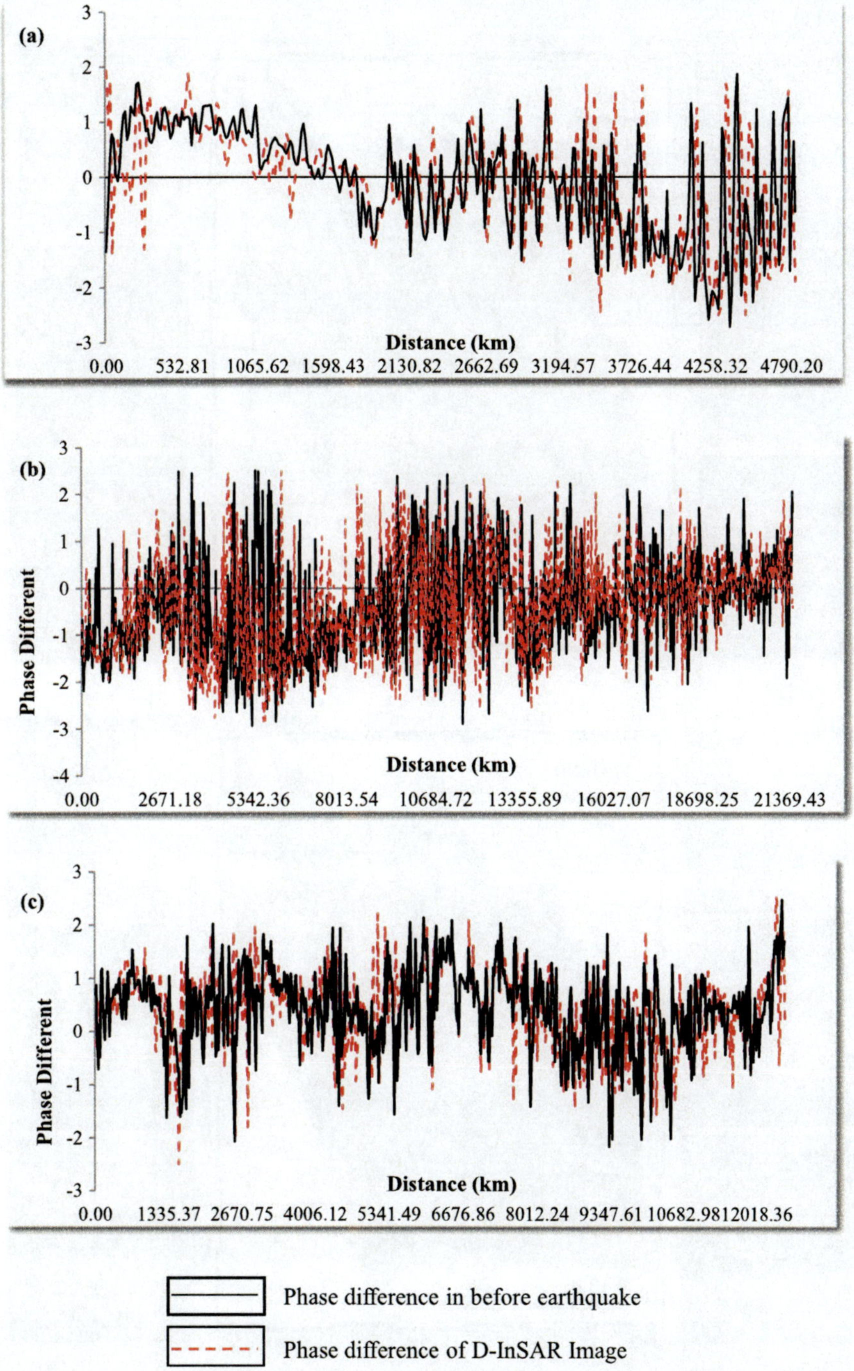

Fig. 14.7 Change in phase due to ground deformation along the three cross-sections

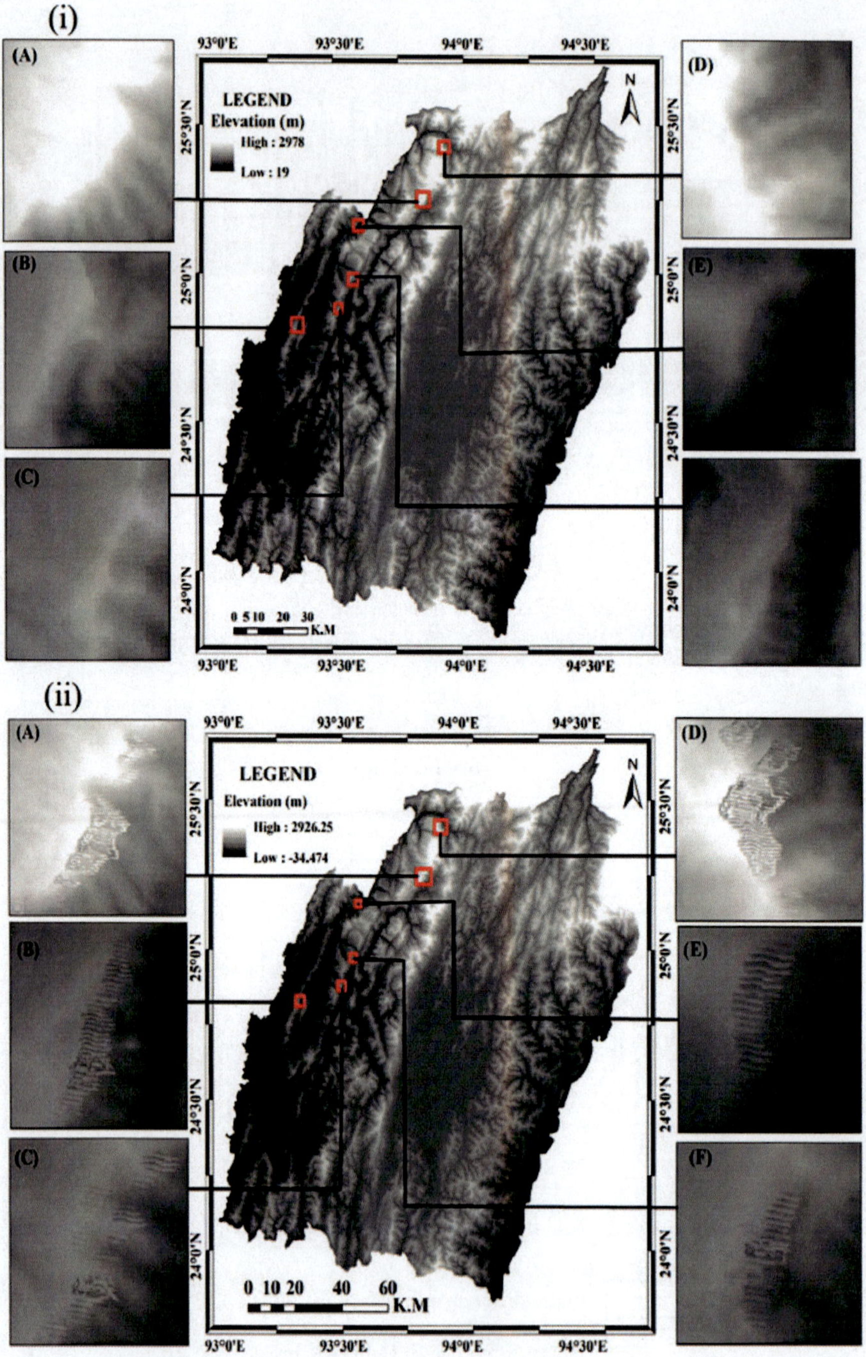

Fig. 14.8 (**a**) SRTM DEM and (**b**) D-InSAR data derived DEM

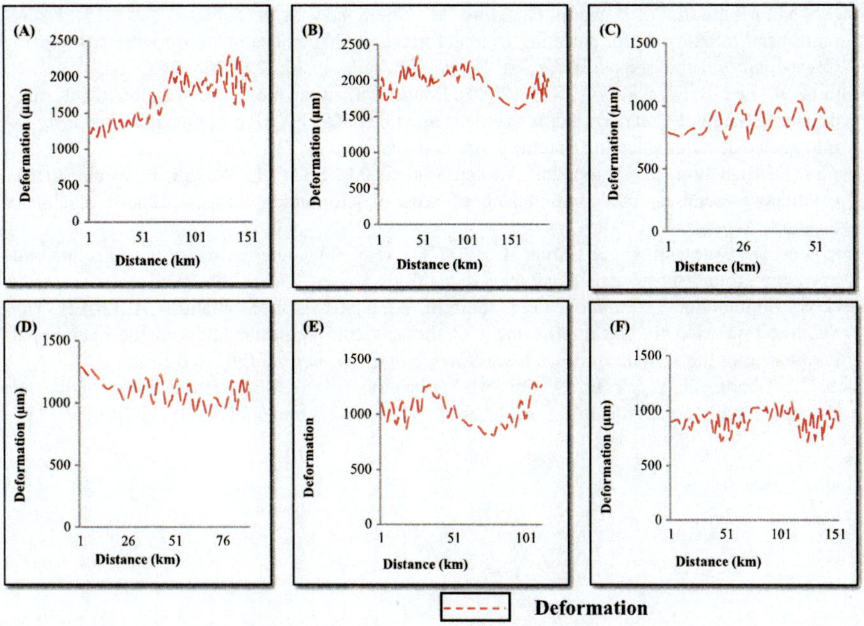

Fig. 14.9 Displacements along the selected cross-section after the earthquake

References

Bisht, H., Kotlia, B. S., Kumar, K., Dumka, R. K., Taloor, A. K., & Upadhyay, R. (2020). GPS derived crustal velocity, tectonic deformation and strain in the Indian Himalayan arc. *Quaternary International, 575–576*, 141–152.

DeMets, C., Gordon, R. G., Argus, D. F., & Stein, S. (1994). Effect of recent revisions to the geomagnetic reversal time scale on estimates of current plate motions. *Geophysical Research Letters, 21*(20), 2191–2194.

Elliott, J. R. (2020). Earth observation for the assessment of earthquake hazard, risk and disaster management. *Surveys in Geophysics, 41*(6), 1323–1354.

Goldstein, R. M., & Werner, C. L. (1998). Radar interferogram filtering for geophysical applications. *Geophysical Research Letters, 25*(21), 4035–4038.

Imamoglu, M., Kahraman, F., Cakir, Z., & Sanli, F. B. (2019). Ground deformation analysis of Bolvadin (W. Turkey) by means of multi-temporal InSAR techniques and Sentinel-1 data. *Remote Sensing, 11*(9), 1069.

Jade, S., Bhatt, B. C., Yang, Z., Bendick, R., Gaur, V. K., Molnar, P., et al. (2004). GPS measurements from the Ladakh Himalaya, India: Preliminary tests of plate-like or continuous deformation in Tibet. *Geological Society of America Bulletin, 116*(11–12), 1385–1391.

Jain, S. K. (2016). Earthquake safety in India: Achievements, challenges and opportunities. *Bulletin of Earthquake Engineering, 14*(5), 1337–1436.

Neelmeijer, J., Schöne, T., Dill, R., Klemann, V., & Motagh, M. (2018). Ground deformations around the Toktogul reservoir, Kyrgyzstan, from Envisat ASAR and Sentinel-1 data—A case study about the impact of atmospheric corrections on InSAR time series. *Remote Sensing, 10*(3), 462.

Polcari, M., Montuori, A., Bignami, C., Moro, M., Stramondo, S., & Tolomei, C. (2017). Using multi-band InSAR data for detecting local deformation phenomena induced by the 2016–2017 Central Italy seismic sequence. *Remote Sensing of Environment, 201*, 234–242.

Ramirez, R., Lee, S. R., & Kwon, T. H. (2020). Long-term remote monitoring of ground deformation using Sentinel-1 interferometric synthetic aperture radar (InSAR): Applications and insights into geotechnical engineering practices. *Applied Sciences, 10*(21), 7447.

Raspini, F., Bianchini, S., Ciampalini, A., Del Soldato, M., Solari, L., Novali, F., et al. (2018). Continuous, semi-automatic monitoring of ground deformation using Sentinel-1 satellites. *Scientific Reports, 8*(1), 1–11.

Raucoules, D., Colesanti, C., & Carnec, C. (2007). Use of SAR interferometry for detecting and assessing ground subsidence. *Comptes Rendus Geoscience, 339*(5), 289–302.

Salvi, S., Stramondo, S., Funning, G. J., Ferretti, A., Sarti, F., & Mouratidis, A. (2012). The Sentinel-1 mission for the improvement of the scientific understanding and the operational monitoring of the seismic cycle. *Remote Sensing of Environment, 120*, 164–174.

Zhou, X., Chang, N. B., & Li, S. (2009). Applications of SAR interferometry in earth and environmental science research. *Sensors, 9*(3), 1876–1912. https://doi.org/10.3390/s9030187

Chapter 15
Spatial Clustering of *P. falciparum* Malaria Epidemiology in Murshidabad District

Poly Patra and Gouri Sankar Bhunia

Abstract The distribution of malaria in low-incidence settings is patchy, with local transfer points being an ongoing source of infection. This research work is based on data on the prevalence of malaria in the Murshidabad district of West Bengal (India), to identify species-specific clusters. *P. falciparum* incidence data was collected in 2009–2016 from Murshidabad, West Bengal, and West Bengal State Health Department. The spatio-temporal cluster of malaria incidences has been analyzed through *Anselin Local Moran's I* statistics. The statistical data of the local *Getis-Ord G* statistics (G_i^*) were calculated for each block based on the spatial weight of disease incidence. *Moran's I* was the highest in 2013 (0.94), while *P. falciparum* malaria was the lowest in *Moran's I* in 2012 (0.53). In 2009, high *P. falciparum* clusters in the central part were observed, and a low-low cluster in the south and east parts were observed. In 2015 and 2016 the high cluster in the western part of the district was observed, and in the south and east the low cluster was identified. Identifying the risk factors of specific malarial species provides key insights into the low transmission epidemiology of malaria that can guide the targeting of additional interventions.

Keywords Malaria · *P. falciparum* · *Moran's I* · Spatial cluster · Disease control

15.1 Introduction

Globally, the incidence of malaria has dropped by 29% between 2000 and 2012, and the mortality rates of severe malaria have fallen by 45% over the same period (WHO, 2017). In 2016, malaria has caused 216 million cases and 445,000 deaths worldwide (WHO, 2017). The malaria map is through with over 35 malaria free-standing approved countries and another 21 indigenous transmission-zero countries (Dhiman et al., 2018). Malaria is transmitted by five species of *Plasmodium*: *Plasmodium ovale (Po)*, *Plasmodium falciparum (Pf)*, *Plasmodium malariae (Pm)*, *Plasmodium*

P. Patra (✉) · G. S. Bhunia
Department of Geography, Seacom Skills University, Kendual, Birbhum, West Bengal, India

P. K. Shit et al. (eds.), *Geospatial Technology for Environmental Hazards*, Advances in Geographic Information Science, https://doi.org/10.1007/978-3-030-75197-5_15

vivax (Pv), and *Plasmodium knowlesi (Pk)* (Cox, 1982; Hundessa et al., 2016). In India *Pv* and *Pf* are traditionally prevalent, and the proportion fluctuates throughout India (Sharma, 1998). Mixed *Plasmodium* infection is another major malaria case typically underreported due to over-reporting by traditional diagnostic methods of the more pronounced *P. falciparum* or *P. vivax* (Ginouves et al., 2015).

In tropical and subtropical areas where the climate is ideal for infection, malaria is a life-threatening infectious disease that seriously affects susceptible members of society. In 2017, India reported 0.84 million malaria cases, and almost all 36 states/UTs are regularly subsidizing malaria cases (NVBDCP, 2018). 80% of the total malaria encumbrance having deliberation of cases (Annual Parasitic Index >10) are donated through north-eastern, eastern, and central Indian states, and these areas are associated with enormous forest asylum, racial tribes, paucity, and the huge amount of rainfall (Sharma, 2012; Sharma et al., 2015). However, there was a severe diminution in the malaria incidence. The death of malaria was caused due to subsequent overview of artemisinin-based combination therapy (ACT) initiating 2010 combined with insecticide-treated netting materials (ITNs). Nevertheless, reports of malaria incidence and death in the research area may be under-reported. The under-reporting factors include administration in the private sector of 2/3 of patients, lack of regular death certification in rural areas, and inadequate malaria diagnosis infrastructure in public healthcare facilities.

Earlier research suggested that the spatial location of malaria is not evenly disseminated in space (Valle & Lima, 2014; Bhunia et al., 2016). Incidence of malaria exhibits spatio-temporal differences with respect to suitable environmental settings (Ikeda et al., 2017; Bi et al., 2013) and immigration of infectious hosts (Bousema et al., 2012) that encourage disease propagation and tenacity. Moreover, the inconsistency of disease pattern due to natural chauffeurs can be further intricate by patterns in host insusceptibility which is a conceivable clarification for intra-annual and inter-annual dissimilarities (Hay et al., 2001; Clements et al., 2009; Wu et al., 2018). Therefore, understanding the interface between connectivity demarcated by the geographical circumstances in disease propagation via spatial statistical models has the prospective to enhance update control and annihilation policies of malaria in real-word situations (Acevedo et al., 2015).

However, studies about the spatio-temporal clustering pattern of the disease have been supportive in effective design and execution of malaria control program and anticipation (Hay & Snow, 2006; Umer et al., 2018). In addition to enhancing control and the allocation of health resources, better understanding of spatio-temporal changes in the distribution of diseases is crucial. Several studies have been performed using space and time scans to detect malaria clusters (Wen et al., 2011; Hundessa et al., 2016). In this regard, GISs are very obliging to detect and understand the epidemiological behavior of developing and reemerging infectious diseases (Franke et al., 2015; Anvikar et al., 2016). In order to inform monitoring and the distribution of health resources, a better understanding of the spatio-temporal amendment in the circulation of diseases is essential. Several studies have taken geographical and spatial scan statistics to identify malaria groupings (Umer et al., 2018) and other infectious diseases in space and time (Lessler et al., 2016; Chowell

& Rothenberg, 2018). This chapter analyzed the spatio-temporal clustering pattern of *Plasmodium falciparum* malaria incidences during the period between 2009 and 2016.

15.2 Study Area

District Murshidabad is located on the river Ganges, which covers an area of 5341 square kilometers and has a population of 7103 million in 2011 Census. The center of West Bengal is Murshidabad between latitudes 23°43'N and 24°52'N and longitude 87°49′E and 88°44′E. The Bhagirathi and Jalangi Rivers and their affluent are flooding the district. The average annual temperature is about 27 °C; the average monthly temperature is between 17 °C and 35 °C. During the Monsoon, approximately 300 mm rainfall occurs in August. Massive cultivation in the west of mulberry is carried out with rice, jute, legumes, oil seeds, wheat, barley, and mangoes. The Scheduled Tribe (ST) is estimated as 12.63% of the population of the district of Murshidabad and the Scheduled Tribe (ST) population was calculated as 1.28%. The population density of the region is 1334 per square kilometer. The district's sexual relationship is 958, and the district's literacy rate is 66.6%. Cultivators and agricultural workers account for 14.7% and 32.5% of the total workers, respectively.

15.3 Materials and Methods

Malaria Epidemic Data Acquisition and Database Creation

In a febrile patient with *P. falciparum* malaria infection microscopically confirmed infection, who lived in Murshidabad between 2009 and 2016. Malaria data from Murshidabad, West Bengal, and the State Department of Health of West Bengal was obtained in 2009–2016. In order to obtain these tests, successful survivors of *P. falciparum* infection were identified during the analysis in various healthcare facilities. The population and demographic figures are measured by age and sex by the Indian government (Fig. 15.1).

Cluster-Outlier Analysis

The spatial cluster among malaria incidence blocks of the district of Murshidabad has been analyzed on the GIS platform (Mitchell, 2005). The research took place using fixed remote bad methods using *Anselin Local Moran's I* statistics, where each block is analyzed in the sense of the adjoining characteristics. The critical distance is

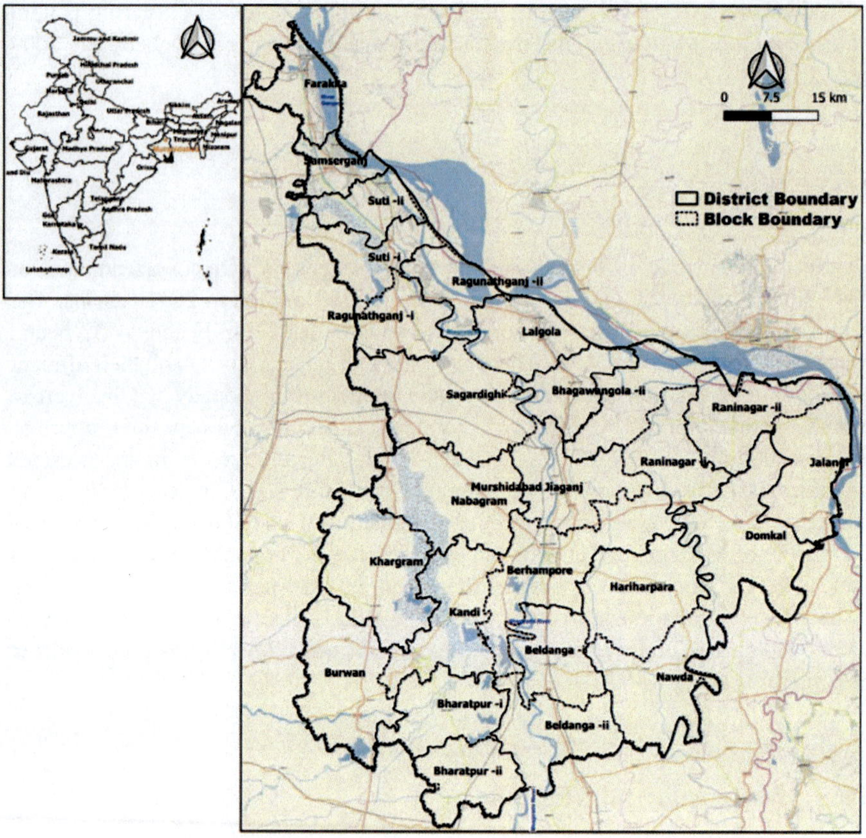

Fig. 15.1 Location map of Murshidabad district in West Bengal, India

less than 500 m in this study. The neighboring block outside the critical distance has a weight of "0" and has no effect on the calculation of blocks affected by malaria. The study has established four cluster types, such as high-high (cluster of high malaria incidence block), low-low (cluster of low malaria incidence block), high-low (outlier in which higher malaria incidence block is surrounded by lower malaria incidence block), and low-high (outlier in which low malaria incidence block are surrounded by high malaria incidence block). With a 95% meaning level, the cluster-outlier analysis was carried out (Anselin, 1995).

Malaria Hotspot and Cold Spot Analysis

Every block was designed for 2009–2016 with the incidence of the spatial cluster existence. The statistical data of the local *Getis-Ord G statistics* (G_i^*) were calculated for each block based on the spatial weight with the following different threshold spaces (Eq. (15.1)):

$$G_i^*(d) = \sum_j \frac{w_{ij}(d)x_j}{\sum_j x_j} \tag{15.1}$$

where W_{ij} is a spatial weight matrix at expected proximity lag in kilometers (d) ($W_{ij}(d)$ is 1 when the proximity from block j to i is within d, or else $W_{ij}(d)$ is 0). The incidence of local malaria clustering was performed depending upon z-score values in the study blocks. The higher positive Z-score is >1.96, suggesting that block "*i*" is surrounded by relatively high malaria incident blocks while block "*i*" is limited to a rather small malaria incidence block with a very high but not very important Z-score value. Z-score values ≥ -1.96 and ≤ 1.96 indicate incidence rate of a malaria regular, random, or cluster pattern (Getis & Ord, 1992).

15.4 Results

Distribution of Malaria Spatial Auto-Correlation

To comprehend the spatial auto-correlation quantitatively, global *Moran's I* statistical analysis was applied. The spatial auto-correlation analysis for *P. falciparum* malaria occurrence showed statistically significant *Moran's I* value (Table 15.1). However, the values of *Moran's I* for *P. falciparum* malaria were varied strangely. Calculated value of *Moran's I* was the highest in 2013 (0.94), while *Moran's I* lowest in 2012 (0.53) was reported for *P. falciparum* malaria. In the period 2009–2016, *Global Moran's I* value of the overall incidence of malaria in Murshidabad differed from 0.64 to 0.80 the *Global Moran's I* of *P. falciparum* varying from 0.63 to 0.94. The spatial auto-correlation of *P. falciparum* is graphically depicted in Fig. 15.2. The results showed that changes in the annual incidence of malaria and the incidence of *P. falciparum* during the entire study period were relatively stable.

Table 15.1 Estimated *Moran's I* value of malaria incidences in Murshidabad district (2009–2016)

Year	*Moran's I*	Z-Score	*P-value*
2009	0.74	4.36	0.00013
2010	0.65	3.81	0.0001
2011	0.63	3.76	0.0004
2012	0.52	3.15	0.0001
2013	0.94	3.49	0.000
2014	0.84	4.79	0.0002
2015	0.74	4.29	0.0001
2016	0.91	5.39	0.000

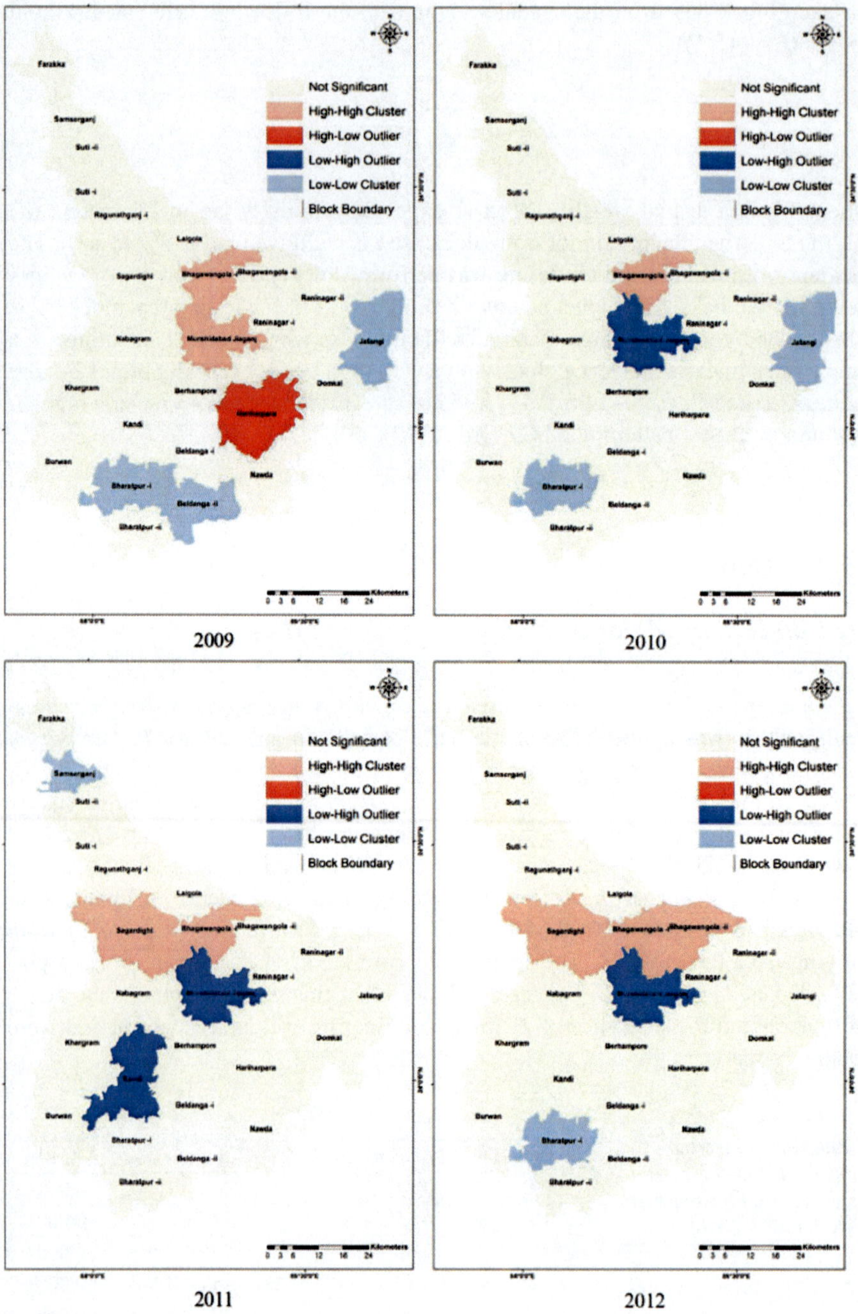

Fig. 15.2 Cluster-Outlier analysis of Plasmodium falciparum in Murshidabad district in 2009 and 2012

Cluster and Outlier Analysis of Malaria Distribution in Murshidabad District

Throughout 2009 there was a strong space cluster in the central part of the district, whereas in the southeast and south of the district, a small cluster of annual incidence blocks of malaria was found. Mapping of *P. falciparum* malaria clusters at the Murshidabad district during the period between 2009 and 2016 is represented in Fig. 15.2. In 2009, a high-high cluster of *P. falciparum* was observed in the central part, and the low-low cluster was observed in the south and eastern part. In the south part of the district, high-low outlier was also observed (Fig. 15.2). Although a small low-high outlier was seen in the central part of the district in 2010, the clustering pattern of *P. falciparum* malaria is more or less the same. High cluster in the central part of the district was observed in 2011 and 2012, while in the central and southern areas the low outlier was observed (Fig. 15.2). A small low-low cluster was in the north in 2011, whereas the small low-low cluster was observed in the south in 2012. In 2013, a high cluster in the north and low outliers in the east of the district were observed. The low-low cluster was observed in the south of the study area (Fig. 15.3). In 2014, the pattern of clustering *P. falciparum* was different as the high-level cluster in the western portion and the low-level cluster in the south and west were observed. A small high-low outlier was observed in the extreme north of the study area. The high cluster in the west of the district was seen in 2015 and 2016, while the low cluster in the south and east of the district was found (Fig. 15.3).

Hotspot and Cold Spot Analysis of P. falciparum Malaria Incidence

Within the null hypothesis, the outcomes of the estimations were analyzed. The incidence rates for the *P. falciparum* malaria in the block level were measured each year (based on *Getis-Ord G* statistics). As the *P-value* is less than 0.5 for *P. falciparum* malaria and the Z-score is positive, the high incidence block spatial distribution was shown to be geographically more clustered than would be projected if the spatial processes were simply random.

The spatial clustering of the blocks in Fig. 15.4 showed the incidence of malaria and major changes over time (2009–2016). The hotspots lead to similar high incidences of malaria and the cold spots to similar values of low blocks of malaria. Maps show hotspots and high-risk areas for important variables with the use of *Getis-Ord* G_i^* and cold spot figures as low-risk areas for major variables. It provides real malaria clustering, which reveals heterogeneousness and hotspots/cold spots at risk of the district of Murshidabad (Table 15.2).

Figure 15.5 demonstrated the study of the hotspot and cold spot of *P. falciparum* malaria in Murshidabad district. The hotspot regions in the middle and east of the district were found in 2009, and there was a small cold spot in the West. The hotspots

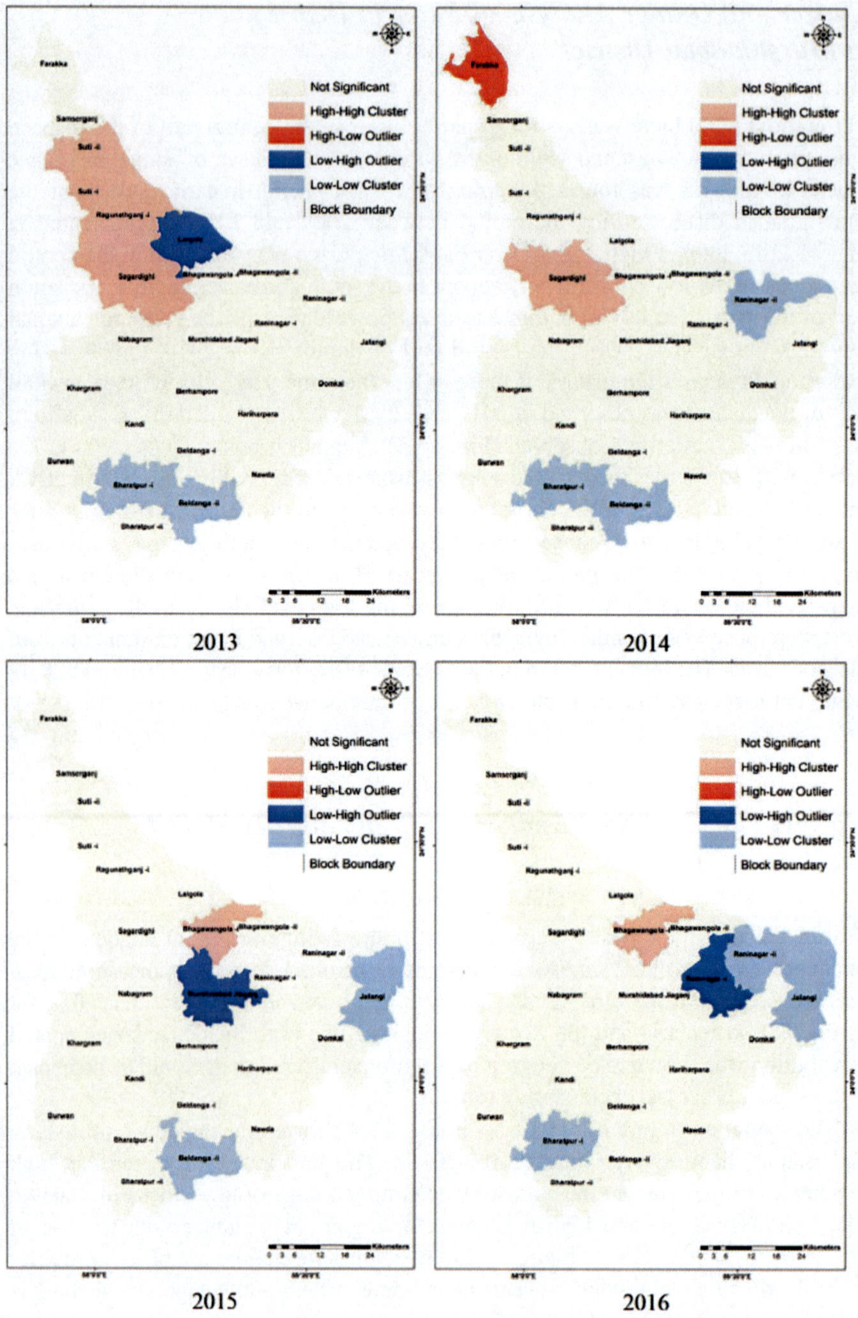

Fig. 15.3 Cluster-Outlier analysis of *Plasmodium falciparum* in Murshidabad district in 2015 and 2016

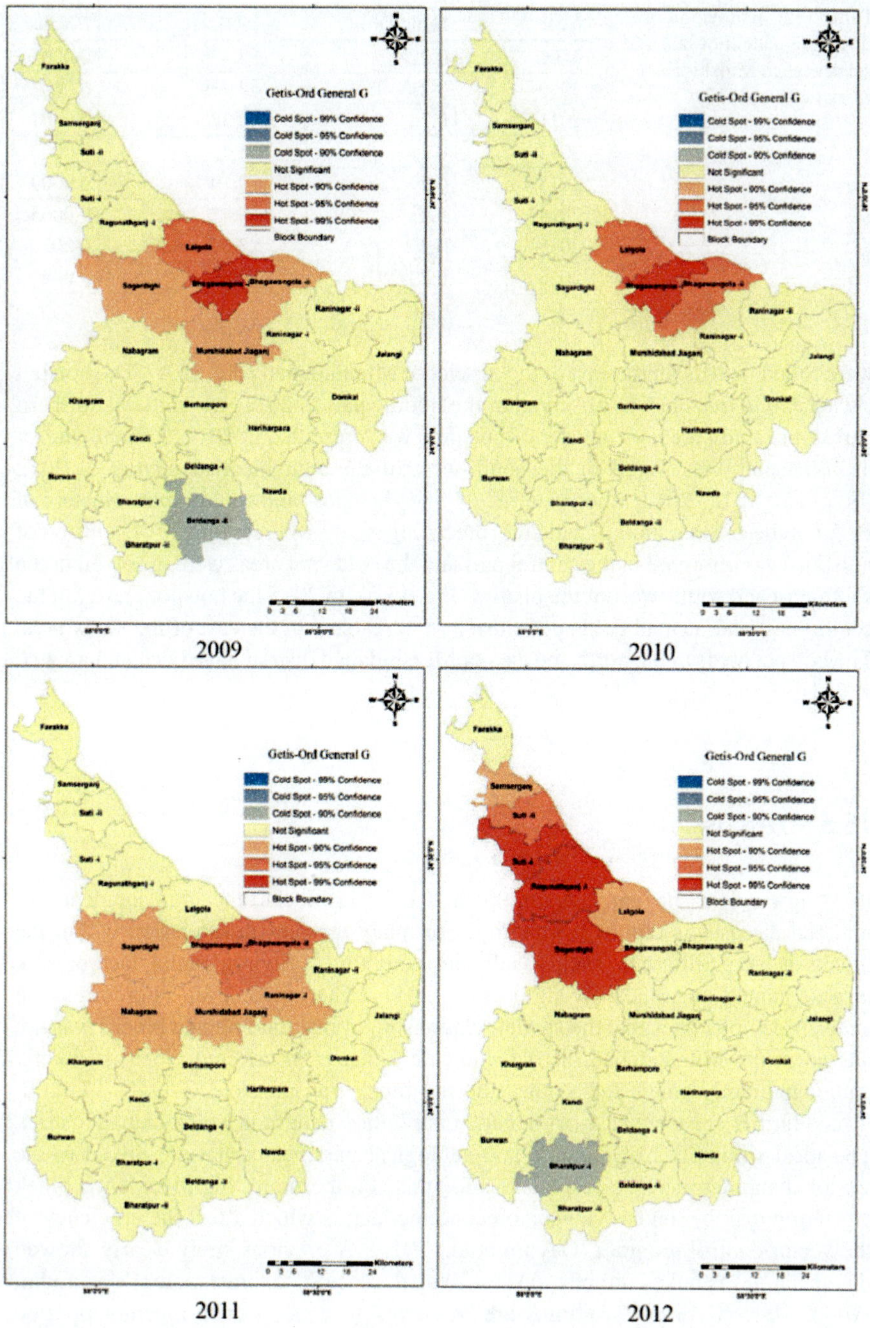

Fig. 15.4 Identification of *Plasmodium falciparum* malaria hotspots and cold spots in Murshidabad district in 2009 and 2012

Table 15.2 Estimation of clustering pattern of malaria incidences in Murshidabad district (2009–2016)

Year	Getis-Ord G	Z-Score	*P*-value
2009	0.22	3.475	0.0005
2010	0.21	3.719	0.0002
2011	0.26	4.38	0.0001
2012	0.35	5.49	0.0000
2013	0.29	3.89	0.0009
2014	0.23	3.47	0.0005
2015	0.27	3.07	0.002
2016	0.23	2.82	0.004

were found in 2010 in the east of the district of Murshidabad (Fig. 15.4). The hotspot cluster *P. falciparum* was observed in the middle part in 2011, while in the northern part of the study area, the cluster of hotspots was presented in 2012. A small cluster of cold spots was found in the south-west of the Murshidabad district in 2012 (Fig. 15.5). In 2013, the pattern of the hotspot and cold spot cluster of *P. falciparum* was similar with the preceding year, whereas, in 2014, cluster of hotspots was observed in the central part and the cold spot areas were observed in the south-east and south-west of the district (Fig. 15.5). In 2015 the hotspot cluster in the central-east and a small cold spot cluster were realized in the east of the study area. The hotspot areas in the north and east of Murshidabad district have been demarcated in 2016.

15.5 Discussion

In Bhagwangola, Jiaganj, Sagardighi, and Lalgola blocks, the high incidence of malarial diseases is reported primarily. For many reasons, transmission within the blocks is very different, such as climate conditions, environmental factors, and disease control measures (Wangdi et al., 2011). The outcome of auto-correlation analysis demonstrates that the spatial relationship of malaria-affected blocks was not altered in the Murshidabad district, indicating that each affected block is directly and/or indirectly connected to the adjacent block transmission. Getis-Ord G confirmed further clustered cases of malaria distribution pattern in Murshidabad district. The local similarities of climatic and ecological parameters that are linked to the vector dynamics may be responsible for this. Furthermore, high-incidence block clustering may be linked with socio-economic factors which affect the efficiency of the vector control program (Qayum et al., 2015). A previous study clearly showed that malaria was linked directly to West Bengal's socio-economic factor (Mazumdar, 2011). Malaria control programs are perceived to contain government-supported public health centers and local vector control efforts in some high-incidence areas (Al-Amin et al., 2015), and in rural and border areas the epidemic continues, underlining the need for a continued and improved malaria control strategy.

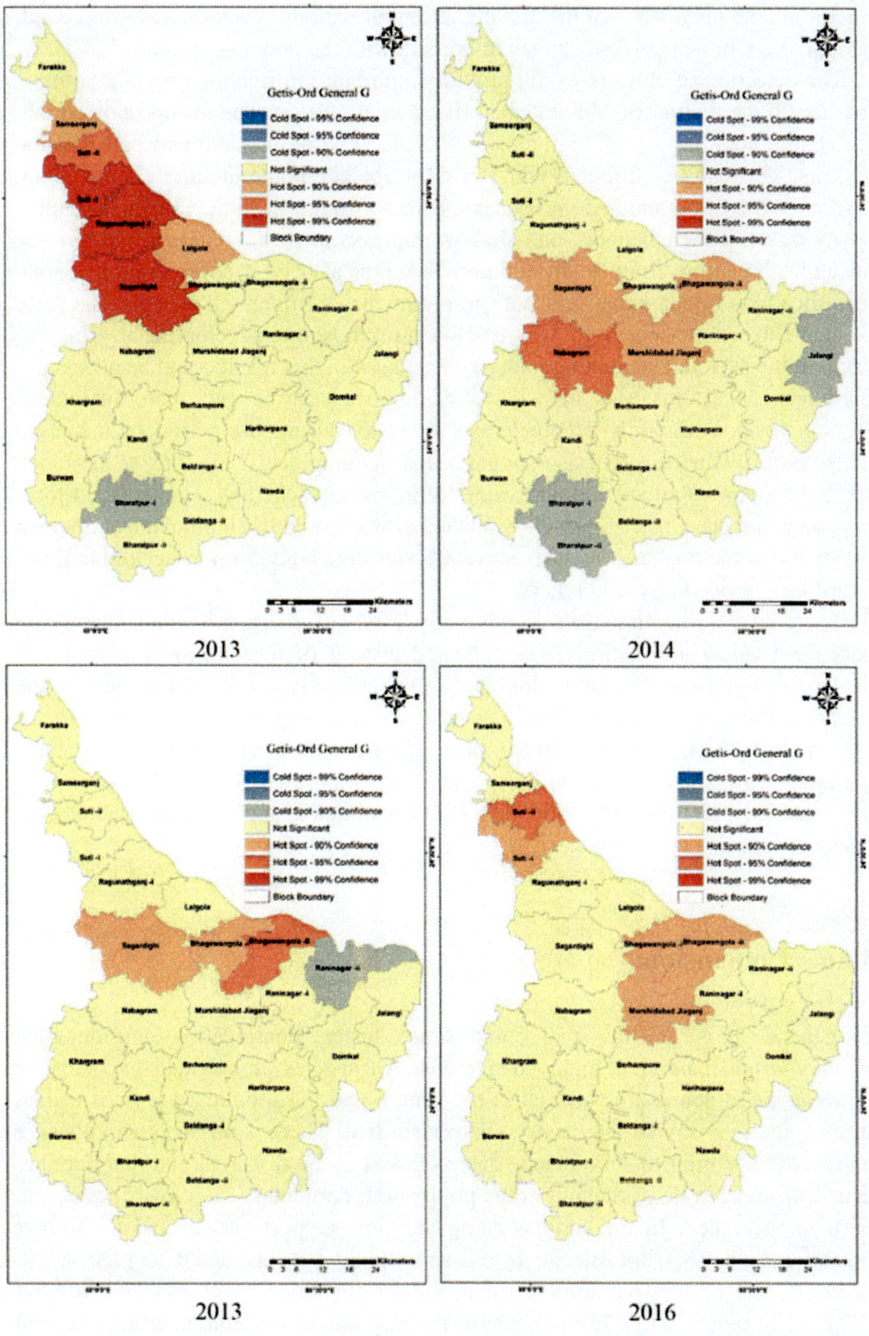

Fig. 15.5 Identification of *Plasmodium falciparum* malaria hotspots and cold spots in Murshidabad district in 2015 and 2016

Therefore, the prevalence of malaria in the central regions is significantly improved, and there are prospects for changes in borders with the endemic regions.

The detection of clusters would provide important information on malaria transmission in the district of Murshidabad (Patra et al., 2017). The results showed that the central area has a very high cluster of malaria while the southern part has low clusters. This can be attributed with the sub-tropical climate to an environmentally favorable condition and also facilitates the vector multiplication. Conversely, most paddy fields, palaeochannels, and shallow depressions in the region could provide mosquito breeding grounds. In addition, it might also be a consideration for the identification of malaria clusters that community fish farming will take place in these blocks. The aquatic ecosystem was developed for mosquito development by this fish farm. In the system extenuating malaria transmission, this result could be necessary. Significant local clustering of malaria incidence ensues between pairs of blocks. *P. falciparum* commonly preponderates *P. vivax* during the period of increased transmission. Our research also suggests that geographical variations at the block level are more beneficial and add value to prevention schemes. The detected local malaria incidence clustering in Murshidabad district could be significantly explained by local risk factors, such as surface waterbodies, the admission to health facilities, monitoring procedures, and others.

There were a few limitations to this study. First, information on clinical and care data from cases and controls was collected instead of forward-looking from the available documents. Second, this study covered only the case reported to the medical officer of the district. The incidence of malaria reported in the private sector was likely lower than in the public sector. The research work could therefore have underrated the share of private sector incidence reporting. Such limitation would have caused the true odds ratio to be underestimated and did not promote the research capacity to conclude.

15.6 Conclusion

In general, data on the spatial and time cluster identification consider that *P. falciparum* malaria is highly dangerous. All species are divided into clusters between the endemic area blocks. In the central and transboundary regions of that district, the majority of the clusters were identified. There is an apparent coverage and credibility limit to surveillance data assessed in the research, and therefore the actual incidence rate of malaria in the population, particularly at a micro-level, can be underestimated. In the early warning decision support system and for further operational analysis, the blocks in clusters should be considered to precede. In addition, this finding recommends that malaria transmission clusters not usually occur in certain regions. Investments in primary data collection at local level will continue to be significant priorities for malaria control and elimination in the next decade, as the landscape of malaria diversifies, evolving and continuing use of comprehensive modelling methods. Furthermore, data should be guided and scientifically established in variable selection procedures to optimize the predictable precision of the risk mapping of malaria.

References

Acevedo, M. A., Prosper, O., Lopiano, K., Ruktanonchai, N., Caughlin, T. T., Martcheva, M., Osenberg, C. W., & Smith, D. L. (2015). Spatial heterogeneity, host movement and mosquito-borne disease transmission. *PLoS One, 10*(6), e0127552.

Al-Amin, H. M., Elahi, R., Mohon, A. N., Kafi, M. A. H., Chakma, S., Lord, J. S., Khan, W. A., Haque, R., Norris, D. E., & Alam, M. S. (2015). Role of underappreciated vectors in malaria transmission in an endemic region of Bangladesh-India border. *Parasites & Vectors, 8*, 195. https://doi.org/10.1186/s13071-015-0803-8

Anselin, L. U. C. (1995). Local indicators of spatial association—LISA. *Geographical Analysis, 27* (2), 93–115.

Anvikar, A. R., Shah, N., Dhariwal, A. C., Anvikar, A. R., Shah, N., Dhariwal, A. C., Sonal, G. S., Pradhan, M. M., Ghosh, S. K., & Valecha, N. (2016). Epidemiology of Plasmodium vivax malaria in India. *The American Journal of Tropical Medicine and Hygiene, 95*(6 Suppl), 108–120.

Bhunia, G. S., Siddiqui, N. A., Shit, P. K., et al. (2016). Spatial clustering of Plasmodium falciparum in Bihar (India) from 2007 to 2015. *Spatial Information Research, 24*, 639–648.

Bi, Y., Yu, W., Hu, W., Lin, H., Guo, Y., Zhou, X. N., & Tong, S. (2013). Impact of climate variability on Plasmodium vivax and Plasmodium falciparum malaria in Yunnan Province, China. *Parasites & Vectors, 6*, 357. https://doi.org/10.1186/1756-3305-6-357

Bousema, T., Griffin, J. T., Sauerwein, R. W., Smith, D. L., Churcher, T. S., Takken, W., Ghani, A., Drakeley, C., Gosling, R. (2012). Hitting hotspots: Spatial targeting of malaria for control and elimination. *PLoS Medicine*, 9: e1001165. https://doi.org/10.1371/journal.pmed.1001165.

Chowell, G., & Rothenberg, R. (2018). Spatial infectious disease epidemiology: On the cusp. *BMC Medicine, 16*(1), 192. https://doi.org/10.1186/s12916-018-1184-6

Clements, A. C., Barnett, A. G., Cheng, Z. W., Snow, R. W., & Zhou, H. N. (2009). Space-time variation of malaria incidence in Yunnan province. China. *Malaria Journal, 8*, 180. https://doi.org/10.1186/1475-2875-8-180

Cox, F. E. (1982). *Modern parasitology. A textbook of parasitology*. Blackwell Scientific Publications.

Dhiman, S., Veer, V., & Dev, V. (2018). *Declining transmission of malaria in India: Accelerating towards elimination* (pp. 257–280). IntechOpen, Chapter 11. Available at https://www.intechopen.com/books/towards-malaria-elimination-a-leap-forward/declining-transmission-of-malaria-in-india-accelerating-towards-elimination

Franke, J., Gebreslasie, M., Bauwens, I., Deleu, J., & Siegert, F. (2015). Earth observation in support of malaria control and epidemiology: Malareo monitoring approaches. *Geospatial Health, 10*(1), 335. https://doi.org/10.4081/gh.2015.335

Getis, A., & Ord, J. K. (1992). The analysis of spatial association by use of distance statistics. *Geographical Analysis, 24*, 189–206.

Ginouves, M., Veron, V., Musset, L., Legrand, E., Stefani, A., Prevot, G., Demar, M., Djossou, F., Brousse, P., Nacher, M., & Carme, B. (2015). Frequency and distribution of mixed Plasmodium falciparum-vivax infections in French Guiana between 2000 and 2008. *Malaria Journal, 14*, 446.

Hay, S. I., Rogers, D. J., Shanks, G. D., Myers, M. F., & Snow, R. W. (2001). Malaria early warning in Kenya. *Trends in Parasitology, 17*, 95–99. https://doi.org/10.1016/S1471-4922(00)01763-3

Hay, S. I., & Snow, R. W. (2006). The malaria atlas project: Developing global maps of malaria risk. *PLoS Medicine, 3*(12), e473.

Hundessa, S. H., Williams, G., Li, S., Guo, J., Chen, L., Zhang, W., & Guo, Y. (2016). Spatial and space–time distribution of *Plasmodium vivax* and *Plasmodium falciparum* malaria in China, 2005–2014. *Malaria Journal, 15*, 595. https://doi.org/10.1186/s12936-016-1646-2

Ikeda, T., Behera, S. K., Morioka, Y., Minakawa, N., Hashizume, M., Tsuzuki, A., Maharaj, R., & Kruger, P. (2017). Seasonally lagged effects of climatic factors on malaria incidence in South Africa. *Scientific Reports, 7*(1), 2458.

Lessler, J., Salje, H., Grabowski, M. K., & Cummings, D. A. (2016). Measuring spatial dependence for infectious disease epidemiology. *PLoS One, 11*(5), e0155249.

Mazumdar, S. (2011). Prevalence, risk factors and treatment-seeking behaviour for malaria: The results of a case study from the Terai region of West Bengal, India. *Annals of Tropical Medicine and Parasitology, 105*(3), 197–208. https://doi.org/10.1179/136485911X12987676649548

Mitchell, A. (2005). *The ESRI guide to GIS analysis, volume 2*. ESRI Press.

Nvbdcp. (2018). Malaria annual report 2018. directorate of national vector-borne diseases control programme,22 sham nath marg, delhi-110054directorate of national vector borne diseases control programme,22 sham nath marg, delhi-110054; Available at: https://nvbdcp.gov.in/Doc/Annual-Report-2018.pdfhttps://nvbdcp.gov.in/Doc/Annual-Report-2018.pdf

Patra, P., Mandal, A. C., Majumder, R., Ghosh, D., & Bhunia, G. S. (2017). Assessing and monitoring malaria epidemiology using remote sensing and GIS in Murshidabad District, West Bengal (India). *Journal of Geography, Environment and Earth Science International, 12*(2), 1–16.

Qayum, A., Arya, R., Kumar, P., & Lynn, A. M. (2015). Socio-economic, epidemiological and geographic features based on GIS-integrated mapping to identify malarial hotspots. *Malaria Journal, 14*, 192.

Sharma, R. K., Singh, M. P., Saha, K. B., Bharti, P. K., Jain, V., Singh, P. P., Silawat, N., Patel, R., Hussain, M., Chand, S. K., Pandey, A., & Singh, N. (2015). Socio-economic & household risk factors of malaria in tribal areas of Madhya Pradesh, central India. *The Indian Journal of Medical Research, 141*(5), 567–575. https://doi.org/10.4103/0971-5916.159515

Sharma, V. P. (1998). Fighting malaria in India. *Current Science, 75*, 1127–1140.

Sharma, V. P. (2012). Continuing challenge of malaria in India. *Current Science, 102*, 678–682.

Umer, M. F., Zofeen, S., Majeed, A., Hu, W., Qi, X., & Zhuang, G. (2018). Spatiotemporal clustering analysis of malaria infection in Pakistan. *International Journal of Environmental Research and Public Health, 15*(6), 1202. https://doi.org/10.3390/ijerph15061202

Valle, D., & Lima, J. M. (2014). Large-scale drivers of malaria and priority areas for prevention and control in the Brazilian Amazon region using a novel multi-pathogen geospatial model. *Malaria Journal, 13*, 443. https://doi.org/10.1186/1475-2875-13-443

Wen, L., Li, C., Lin, M., Yuan, Z., Huo, D., Li, S., et al. (2011). Spatio-temporal analysis of malaria incidence at the village level in a malaria-endemic area in Hainan China. *Malaria Journal, 10*, 88.

World Health Organization (WHO). (2017). *World malaria report 2017* (pp. 33–41). WHO Press.

Wu, D. F., Löhrich, T., Sachse, A., et al. (2018). Seasonal and inter-annual variation of malaria parasite detection in wild chimpanzees. *Malaria Journal, 17*, 38. https://doi.org/10.1186/s12936-018-2187-7

Wu, D. F., Löhrich, T., Sachse, A., et al. (2018). Seasonal and inter-annual variation of malaria parasite detection in wild chimpanzees. *Malaria Journal, 17*, 38. https://doi.org/10.1186/s12936-018-2187-7

Chapter 16
Mapping, Measuring and Modelling Common Fluvial Hazards in Riparian Zones: A Brief Review of Relevant Concepts and Methods

Sayoni Mondal and Priyank Pravin Patel

Abstract The channel and adjacent floodplain tracts of rivers comprise the riparian zone, which is sensitive to a variety of natural/environmental and human-induced hazards. The degradation of this riparian zone leads to the loss of landscape connectivity, interruption of biogeochemical cycles and material fluxes and engenders adverse impacts on the flora and fauna occupying it, both physiologically and through habitat loss and fragmentation. This brief review paper examines the salient characteristics of the riparian zone and the common hazards that afflict it. It also provides insights into the various mapping, measurement and modelling methods construed over time that are in vogue to investigate and characterise such hazards, gauge their impacts and devise possible ameliorative frameworks for the same. The natural hazards considered here are annual floods and high stream flows, soil loss occurring due to overbank flow and runoff, river erosion and bankline failure. Alongside this, the marked degradational impacts engendered by sand mining within the river channel (both on the bed and from in-channel deposits) and from the adjacent floodplain on the local environment are detailed. Word clouds have been used to highlight the most oft-repeated or used catchphrases and terms while undertaking the above researches in the respective hazard domains. We also provide a temporal overview of how these concepts and concerns have come more and more into the fluvial geomorphologic and riparian ecosystem subject purviews, reflecting the rising focus in these areas and the need for further research on the discussed aspects.

S. Mondal · P. P. Patel (✉)
Department of Geography, Presidency University, Kolkata, India
e-mail: priyank.geog@presiuniv.ac.in

P. K. Shit et al. (eds.), *Geospatial Technology for Environmental Hazards*, Advances in Geographic Information Science, https://doi.org/10.1007/978-3-030-75197-5_16

16.1 Introduction

Riparian zone management has been considered as an essential component of river rehabilitation and an integral part of ecological restoration (Gregory et al., 1991; Trimble, 2004; Gurnell et al., 2016). In its most literal meaning, the word *riparian* means or pertains to the bank of a river. It has been further defined as an *undisturbed, naturally vegetated strip of land lying just adjacent to the river that provides critical and valuable ecosystem functions in the form of protecting water quality, supporting wildlife and augmenting flood storage* (Verry et al., 2000). Thus, it is a vegetative zone linking the stream to its adjacent floodplain and upland areas and serves as the connecting link between the aquatic and the terrestrial ecosystem (Agouridis et al., 2010; Cavaille et al., 2013; Fierro et al., 2017). Such stream corridors are characterised by a high level of interaction between the ambient soil, water and vegetation aspects (Naiman & Decamps, 1997; Covino, 2017; McMillan & Noe, 2017) and thus have high physical, structural and biotic functionality (Acuna et al., 2016). This zone is also affected by frequent perturbations including annual inundation, transportation and deposition of sediments and the erosive forces of water (Banerji & Patel, 2019) that in turn create a habitat diversity which is essential in building up an ecologically versatile and self-sustaining ecosystem (Verry et al., 2000; Ramirez et al., 2012). Thus, quite some research has concentrated on examining this particular zone, to understand the factors that govern it and the possible hazards that affect it (López-Baucells et al., 2017). A number of mapping, measurement and modelling methods have accordingly been developed to simulate such hazards (mostly of fluvial origin), and this paper proves a brief overview of the same.

16.2 Ecological Functions of Riparian Zones in Promoting Ecosystem Stability

A typical riparian buffer zone is composed of three main sectors: Zone I which is located just beside the river and is composed of fast and slow-growing water-tolerant species; Zone II is located next to the trees and contains shrubs; and Zone III is the zone of grasses and forbs and is positioned next to the shrubs (Agouridis et al., 2010). Riparian buffers perform a host of functions ranging from the physical and ecological to the economic and social (Nakamura, 1995; Mondal & Patel, 2018). They are also known to stabilise eroding stream banks (Hughes, 2016; Mondal & Patel, 2020). These buffers dissipate stream energy and hydraulic action (Zhang et al., 2019), thus reducing channel scouring. The roots of the vegetative buffers hold soil particles together (Scott, 2016) and thus have greater tensile strength, fostering higher resistance against erosive water action (Eiseltova, 2010; Smith & Wynn-Thompson, 2018). Riparian buffers also act as effective flood storages by retarding the rate of surface runoff (Blackwell et al., 2006; Keeton et al., 2017), thereby retaining and enhancing infiltration of the excess overbank flow that can subsequently recharge the local groundwater table up to a considerable extent (Parkyn, 2004; Grabowski et al., 2014). The above ground biomass of the riparian buffers increases the surface roughness coefficient, and this retards flow velocities of high flood events (Keesstra et al., 2012), in the event diminishing their potentiality to damage life and property (Docker & Hubble, 2008; Maffra et al., 2017). By trapping and filtering excess sediments and contaminants and preventing them from flowing into the channel, it enhances the ambient water quality (Kumar et al., 1992; Patel et al., 2020a), thereby sustaining many aquatic fauna, like fishes (Angermeier & Karr, 1984; Rayner, 2001). They provide food to different species in the form of leaves, twigs, litter and debris (Miserendino et al., 2011), while large woody debris provide suitable habitats where aquatic organisms thrive (Carolyn et al., 2001). The shade created by riparian vegetation helps in reducing in-stream water temperature (Hawes & Smith, 2005) and elevating dissolved oxygen levels, creating better habitats for fishes to thrive in (D'Ambrosio et al., 2014). Deep-rooted vegetation strengthens channel boundary conditions (Shit & Maiti, 2012; Mulyono et al., 2018), thereby providing hydraulic conditions that are ideal for the development of aquatic habitats. These buffers also allow the stream to maintain its base-flow during low flow seasons (Schlosser & Karr, 1981), which directly affects the aquatic habitat extent and quality (O'Brien et al., 2017).

Stream buffers ideally consist of native tree species preserved in their natural form for much of the stream length and provide easy travel and dispersal routes for wildlife (Lees & Peres, 2008), thereby helping preserve the biodiversity of the region (O'Donnell et al., 2015). Forested riparian buffers serve as excellent sediment filters (Naiman et al., 2013) that effectively allow the absorption and retention of sediments by slowing down the overland flow (Wolter et al., 2016). They also act as buffers and filters for excess nutrients and contaminants (Craig et al., 2008; Cockburn et al., 2016). Approximately 80–85% of phosphorus is retained by the plants, and as much as 50–100% of sediments can be retained by riparian zones (Hawes & Smith, 2005; Covino, 2017). Riparian buffers have economic worth, and healthy stream buffers influence property values while also promoting hiking, camping and other recreational activities (Junker & Buchecker, 2008; Vollmer et al., 2015), with these being some of the many other additional ecosystem services it provides (Watson et al., 2016). Large forested buffers allow environmental enthusiasts to enjoy the visual aesthetic appeal provided by undisturbed continuous tracts of diverse native plant species (Zhao et al., 2016; Lopez-Rodriguez et al., 2018; Saha et al., 2020).

16.3 Human Interventions into Riparian Zones and Ambient Fluvial Hazards

These riparian zones that maintain a chain of ecological functions are, however, amongst the most degraded ecosystems in the world, mainly as a result of anthropogenic interventions into fluvial systems (Behbahani et al., 2017). Almost every riparian zone has seen accelerated human activities, which speeds up and negatively impacts their natural functions (Gregory, 2006). In most cases, the degree of change caused by such externalities is greater than the natural resilience and healing rate of such buffer strips, leading to an overall degradation of their natural functions (Chatterjee & Patel, 2016; Patel et al., 2020c; Sahana et al., 2020). In most river basins across the world, the present riparian area bears little resemblance to its original natural character. While small streams have usually been silted up, the larger ones are most often channelized and tamed, and their riparian zones quite entirely cleared of vegetation to support human needs (Behbahani et al., 2017).

Riparian areas are mostly affected by infrastructural and agricultural developments which induce instabilities in the river bed, making them erode their banks quickly and accentuate annual flooding, leading to the loss of otherwise productive areas (Mondal et al., 2016; Olokeogun et al., 2020; Olokeogun & Kumar, 2020). Channels and riparian zones, although different in their respective resilience to floods, suffer from frequent and large flood events due to anthropogenic disturbances having modified the riparian zone, which in turn influences and modifies local livelihoods and land use (Patel & Mondal, 2019; Patel & Dasgupta, 2009). Almost 90% of North American and European floodplains have been severely altered due to agriculture and other intensive floodplain uses for infrastructural developments (Entwistle et al., 2019). Similar scenarios also prevail in Asia, with the situation herein being more grievous due to the already very high and increasing population pressure over the last few decades, engendering marked land cover alterations within foothill-piedmont zones and along river valleys (Sarkar & Patel, 2016). Thus, riparian habitats remain some of the most severely at-risk and threatened ecosystems worldwide (Tockner & Stanford, 2002).

The increasing population pressure and its concentration along rivers for sustenance induces urbanisation, which causes the water table to lower therein (Wakode et al., 2018) due to enhanced extraction and diversion of surface flows, whereas greater tracts of impervious surfaces amplify peak flows, exacerbating floods (Coleman et al., 2005; Du et al., 2015). Increased peak flows also enhance downcutting and erosion of bank sediments, which further leads to hydrologic disconnection of the riparian sub-surface from the stream channel. Hydrologic disconnections between the channel and riparian floodplains in the form of embankments, structural crossings and dams degrade riparian areas by separating a substantial part of the riparian zone from active fluvial processes (Nilsson & Berggren, 2000; New & Xie, 2008). Bank stabilisation structures also disrupt riparian processes by restricting water availability along channels (Baird et al., 2015). Cleared riparian zones for agriculture often raise nutrient inflow into streams, which have serious consequences on the river biota (Dosskey et al., 2010).

Riparian areas are also subjected to mining activities in addition to sand excavation from the river bed. Sand mining not only causes bed degradation (Pitchaiah, 2017) and engenders morphological and hydrological changes within the channel (Padmalal et al., 2008), but also affects floodplains and riparian zones and directly impacts the riparian biota. Complete removal of riparian vegetation, in some instances, in order to create access roads for tractors and hydraulic excavators to enter into the river bed causes habitat fragmentation and destruction of floodplain species, in addition to accelerating bank erosion and loss of valuable arable lands (Asraf et al., 2011). Researches worldwide have asserted that sand mining alone has caused severe destruction of riparian forested lands, along with the depletion of native vegetation colonies and water pollution (Behbahani et al., 2017).

This paper discusses the various fluvial hazards that occur in a riparian zone and provides a very brief review of the various methods available to investigate and measure the consequences of such hazards in the channel-riparian zone interaction.

16.4 Flood Management, Modelling and Mitigation: Considerations and Methods

Riparian zones witness a number of fluvial hazards. Significant amongst them are flooding and bank erosion, soil erosion and land degradation and sand excavation practices, all of which have far-reaching implications on the fluvial system. A brief review of these hazards is presented below.

Floods have been identified as the single most harmful natural disaster which caused almost 31% of worldwide economic losses and 55% of lives lost within a single decade of 1986–1995 (Borrows & Bruin, 2006). Yearly, about 170 million people worldwide get afflicted by floods (Das, 2019), with the United Nations reporting that nearly 2.3 billion were affected with 157,000 deaths from flooding during 1995–2015 (Hoque et al., 2019). In India, out of 3290 lakh hectare of geographical land, 40 mha have been declared as flood prone, with an annual average of 75 lakh hectares being affected, either directly or indirectly (Gangwar, 2013). Causative natural factors relating to hydrology, geology and geo-physical attributes like the general elevation and slope of the region, soil type, precipitation received and land use pattern determine the susceptibility level of a region to floods in general (Blistanova et al., 2016). Although the average frequency of major flood events is 5 years, such catastrophic events have severely increased in recent years due to global issues like climate change, rapid urbanisation and associated developments, river management programmes and unscientific agricultural practices within the riparian zone (Das, 2019). Such hazards have the potential of setting back development goals and hamper the process of economic development in any region, and thus flood hazard preparedness and management has been the most obvious ways of combating such disasters (Behanzin ct al., 2015).

Comprehensive flood management requires detailed information regarding all aspects of flooding, i.e. hydrological, physiographical, geotechnical, economic, social as well as political (Tehrany et al., 2014). Concepts like flood risk management and flood hazard preparedness have become particularly important in view of the rising concern regarding the mitigation of this natural disaster (Rimba et al., 2017). Forecasting and early prevention being the only strategy to reduce its ill effects, flood management programmes have shifted focus towards assessing the vulnerability and susceptibility of regions prone to flooding rather than just evaluating damages already caused by it (Hoque et al., 2019). Integrating the physical and social vulnerability factors along with the coping capacity of the concerned community is an important indicator in assessing the flood risk (Rakib et al., 2017).

A most crucial component within flood risk management is the identification of areas that are more susceptible to flooding so that early forecasting can help provide early warning facilities that would significantly reduce the chaos caused due to it (Sahana & Patel, 2019). Defining flood-prone areas in a particular region based on detailed physical and socio-economic data can be really helpful in formulating quick evacuation plans once the disaster starts hitting. This not only reduces flood fatalities and damages caused, but also helps in building up a more resilient society through stronger community participation and collective cooperation, thus promoting sustainable land planning and preserving valuable ecological corridors along most rivers (Cao et al., 2016). This can also aid in identifying areas that are comparatively distant and thus safer from the flood effect, especially in urban areas, so that further commercial and residential expansions can take place in these zones at minimal risk from such events. Such a process would also help determine the suitability of urban commercial and residential developments (Rimba et al., 2017).

The use of Multi-Criteria Decision-Making techniques (MCDM) with geospatial data have greatly aided flood susceptibility studies. MCDMs are used in the process of decision-making to arrive at the best possible alternative for a particular event. These methods use numeric techniques to choose amongst discrete sets of alternative decisions (Triantaphyllou et al., 1998). MCDMs are basically optimization approaches used for setting ranks, especially if multidimensional criteria are considered (Jahan et al., 2016). The most common MCDM technique used in flood susceptibility studies is the Analytical Hierarchy Process (AHP) which is a decision-support tool and uses a multi-level structural hierarchy to prioritise alternatives, using a set of pair-wise comparison matrices (Triantaphyllou & Mann, 1995). The comparison matrix is constructed based on a numerical scale where each number denotes the magnitude of dominance of one particular aspect over the other on the basis of the comparison criterion (Saaty, 2008). A considerable pool of literature has used AHP and its variants in susceptibility studies (Lawal et al., 2012; Dahri & Abida, 2017; Kaur et al., 2017; Santos et al., 2018; Sozer et al., 2018; Hammani et al., 2019; Talha et al., 2019; Chakraborty & Mukhopadhyay, 2019; Nourani & Andaryani, 2019; Ghezelsofloo & Hajibigloo, 2020; Sozer, 2020).

However, MCDMs have often been criticised for being too qualitative in nature and therefore give rise to uncertainty that arises from the selection, comparison and ranking of criteria, which invariably involves expert knowledge in assigning the weights that would ultimately define the susceptible zones (Tehrany et al., 2014). Thus, flood susceptibility studies have further used machine learning algorithms like Artificial Neural Network (ANN), decision trees, Support Vector Machines (SVM), Fuzzy Logic, Frequency Ratio (FR), Multivariate Regression analysis (MR) and a multitude of other models which seemingly provide a better reliability and accuracy (Valizadeh et al., 2017; Das, 2019; Malik et al., 2020). A combination of bivariate probability methods together with multivariate statistical techniques allows better precision and thus enhances performances of individual methods (Tehrany et al., 2014). Hydraulic risk mapping is another integral part of flood susceptibility assessment that is more focussed on understanding the flow and velocity dynamics like flood hydrographs, peak discharge curves and flow depth and extent of inundation (Cao et al., 2016).

Another aspect of flood management and mitigation deals with gauging the inherent vulnerability of the populace exposed to the hazard. The report of the World Conference on Disaster Reduction denotes vulnerability as the situation/condition as determined by the economic, social, physical and environmental factors that can possibly augment the at-risk status and susceptibility of a societal group (UN/ISDR, 2004; Behanzin et al., 2015). It has often been taken to encapsulate an overall measure of the ambient human welfare of a societal group as reflected in the institutional, economic and political capabilities of its members (Blistanova et al., 2016; Proske, 2008).

The flood risk has therefore been defined as hazard multiplied by vulnerability, i.e. it considers both the susceptibility of a region to flooding and the vulnerability of the local community affected by it (Kron, 2005). Vulnerability has been viewed from three aspects: a community's exposure to a crisis and its associated stress, its inadequate coping capacity (often termed as resilience) and the consequent risks associated with slow recovery from the event (sensitivity). This perspective thus highlights that the most vulnerable communities are often those who are most exposed to the event, have highest sensitivity, have the least coping capacity and are the weakest in terms of recovery from the event (Gangwar, 2013). Vulnerability assessment studies therefore start with understanding the interaction between the physical, social and human systems and how these factors vary locally, i.e. the differential exposure of people and places to natural hazards (Singh et al., 2014). These help identify regions and communities that are at risk when hit by a hazard and thus form part of the disaster preparedness plan. This can also help improve the flood map inventory by providing valuable information regarding areas at high risk (Schwarz et al., 2018).

Social vulnerability has been considered to be the most important factor that possibly governs the resilience of the residents in overcoming a disaster's impacts and is measured at various scales-regional, community and individual levels, based on indicator-specific approaches (Behanzin et al., 2015; Lixin et al., 2017). As vulnerability mainly deals with the risk associated with the hazard, community-based risk assessment has been the prime focus of vulnerability assessment studies, with social vulnerability being the sole indicator in such methods. Just as flooding depth, magnitude and duration have been considered as most important factors affecting vulnerability in a region (Li et al., 2016), other factors like the population distribution pattern and density of the resident community, their economic condition, social diversity and level of innovative technology and preparedness also condition the extent of hazard exposure and the associated vulnerability (Lixin et al., 2017). Questions like what are the exact estimates of flood damages from previous floods and how are they linked to hydrologic flood characteristics, whether such results can be compared and simulated for upcoming floods of similar or higher magnitude, are crucial and should be properly addressed for accurate estimation of the vulnerability of a region to flooding (Wijayanti et al., 2017).

Many studies have analysed and measured the social vulnerability to flood hazards (Zheng et al., 2009; Rufat et al., 2015; Torok, 2018; De Silva & Kawasaki, 2018; Kirby et al., 2019; Munyai et al., 2019; Tascon-Gonzalez et al., 2020) based on the Social Flood Vulnerability Index (FVI_{social}) that is computed using various demographic, economic and social parameters. Concepts like community resilience and household-level preparedness have subsequently become important terms in flood management studies. Community resilience has been denoted as the capability of a community to cope with a disaster, the time taken by the community to come back to normalcy after the event has passed and also the ability to learn and adapt skilfully from the impacts of the disaster. An important aspect of community resilience has been social responsibility (SR). FEMA, in the USA, has also denoted personal responsibility as one of the key elements in building up a resilient community (Soetanto et al., 2017), while household-level preparedness to combat disasters has been considered as an important criterion in hazard preparedness research. The consideration of the individuals' psychological and cognitive factors, their ability to take joint decisions in emergency situations, their economic, social and ecological factors, their perception of a hazard and their mobility and awareness all end up cumulatively affecting the overall hazard preparedness and can be considered crucial in preparing management plans (Hung, 2017).

Efforts to improve flood prediction methods have led to the development of flood simulation models that have greatly facilitated flood risk and damage assessment and mapping, real-time flood forecasting, simulation of inundation extents and depths and the effects of hydraulic structures in causing floods. These methods have further moved on to incorporate floodplain and channel sediment transport mechanisms, together with channel and catchment hydrology dynamics (Teng et al., 2017). Flood simulation models have been widely used in disaster management, with special emphasis on engineering designs relating to floodplain management in urban scenarios. These models often include rainfall-runoff estimations, modelling of the hydrodynamics of channel and overland flow, with special inputs for sediment transport rates and groundwater movements (Li et al., 2016). The use of hydrodynamic models involving 1D, 2D and coupled 1D-2D methodologies to simulate flow characteristics within the channel and the floodplains has gained popularity within the scientific community. A critical review of the various available hydrodynamic models used in flood modelling studies has been given by Teng et al. (2017). Numerous studies have dealt with various hydrodynamic models like HEC-RAS (Hussain, 2017; Ongdas et al., 2020; Ogras & Onen, 2020), LISFLOOD-FP (Sanyal et al., 2014; Amarnath et al., 2015; Neal et al., 2018; Shustikova et al., 2019), SRH-2D (Lavoie & Mahdi, 2017) and CCHE2D-Flow (Hasan et al., 2007; ShahiriParsa et al., 2016; Salunkhe et al., 2018; Jia & Hunt, 2016; Ying & Wang, 2005). The HAZUS-MH flood model employs a variety of topographic and flow-discharge data to model flood frequencies and inundation extents of big floods, especially for wide rivers having extensive floodplains. This model is also equipped with flood loss and damage estimation capabilities that effectively support management decisions. It also provides scope to calculate both direct and indirect damages and includes special modules that can calculate both the physical and economic losses caused due to such hazards (Scawthorn et al., 2006a). Further, this model uses flood-depth data and depth-damage function curves which permit rapid analysis of possible damages that can be caused to buildings and infrastructures situated close to rivers and thus allows the formulation of better flood warnings/ preparedness/insurance that can help in minimising flood vulnerabilities to a great extent (Scawthorn et al., 2006b).

16.5 Soil Loss and Land Degradation: Estimation and Analysis

Besides flooding, riparian zones are affected by topsoil loss and concomitant land degradation. This can be through rill or sheet erosion or from denudation of unstable river banks and has been considered as a most critical threat to the ecological stability of a landscape.

Overland flow induced soil loss is the wearing away of the productive uppermost layer of the soil that has direct implications on soil productivity, and is thus considered as a serious cause of concern all over the world (Bekele et al., 2019). Natural soil erosion occurs at about 0.5 ton/ha/year, which is generally quite similar to the average soil formation rate (Roy, 2018). However, most of the present-day soils are considered to be either in poor or in fair condition, with estimates of around 56% of them showing traces of acute soil erosion, mainly by water. The conversion of forested lands into cultivable plots contributes significantly to soil loss with almost 40% of the Earth's land now devoted to agriculture (Alewell et al., 2019). At an estimate, the total river-borne sediments debouched into oceans has risen considerably from about 9.9 billion tons/year to almost 265 billion tons/year, mainly as a result of agricultural and grazing activities and related anthropogenic pressures on the land (Roy, 2018). The global average annual soil erosion rate is 12–15 tons/ ha, a number which is quite alarming (Susanti et al., 2019). This has direct bearing on crop productivity. Approximately 90% of the nitrogen and phosphorus loss from agricultural watersheds occurs as a result of soil erosion, which directly affects soil fertility. Studies have shown that in the USA, the total productivity loss is around $25 billion, with 90% of the croplands losing soil at unsustainable rates. Iowa, a major agricultural region, has lost half of its fertile topsoil in the last 150 years, which is still continuing at a rate of about 30 ton/ha/year (Zhang et al., 2009). Another estimate says that around 17% reduction in crop productivity has been noticed all over the world due to topsoil loss, particularly in African and Asian countries, induced by the surmounting population pressure on land and its consequent lack of soil conservation measures by small or marginal landholders (Bekele et al., 2019) who farm most of the available lands. In India, soil erosion results in further deterioration of almost 87% of the nation's total degraded lands (Roy, 2018).

Apart from climatic factors and the intrinsic soil characteristics, such accelerated soil erosion is largely a product of anthropogenic interventions into natural landscapes through reckless vegetation clearance to promote agricultural practices, even on poorly maintained marginal slopes, overambitious infrastructural development projects and unscientific agricultural practices (Sobral et al., 2014). The most erosive soils have been found in the vicinity of agricultural fields and settlement zones, while the naturally forested soils have shown lower signs of erosion, once again indicating the fact that most of the modern-day erosion activities have been further accentuated by intense anthropogenic interventions in natural processes (Ergodan et al., 2007).

Soil loss and the consequent land degradation being a grievous problem, numerous studies have tried to quantify the total actual soil loss and investigate the causes behind such occurrences. Spatial soil loss and soil erosion susceptibility maps provide a very good basis for prioritising erosion-prone zones to allow primary land conservation measures to be properly implemented, especially in agricultural watersheds (Gayen et al., 2019). The driving forces operative behind soil erosion have been much researched using soil erosion models that tend to measure the amount of soil loss along with identifying zones that have an enhanced susceptibility to soil erosion (Benavidez et al., 2018).

Although calculating soil loss from agricultural fields has been in vogue since the 1940s, through studies in the US Corn Belt using a technique called the slope-practice method, the classic Universal Soil Loss Equation (USLE) of Wischmeier and Smith (1978) is the most frequently used model in predicting soil loss across landscapes (Majhi et al., 2021). USLE estimates the long-term average soil eroded from arable lands based on those factors that can influence erosion via rainfall events and its consequent overland flow (Simanton et al., 1980) and uses alternative cropping combinations and management practices based on the specific soil type, rainfall pattern and topographic setting of the region. The USLE equation also groups the various interrelated physical, crop management and conservation factors that tend to influence the rate of erosion where site-specific values of each parameter can be expressed in numeric terms. The whole database for this model has been generated from specific farm fields and small erosion-prone regions of eastern USA and has been designed to model rill and sheet erosion mainly from agricultural plots under specific conditions. Soil loss tolerance values are generated, which depict the highest allowable limit of soil erosion that may still sustain enhanced crop productivity, both economically and ecologically (Wischmeier & Smith, 1978). A number of authors have used this model to provide an estimate of the average annual soil loss in their examined regions (Kitahara et al., 2000; Karamage et al., 2016; Jazouli et al., 2017; Gao et al., 2017; Bera, 2017; Roy, 2018; Botelho et al., 2018; Liu et al., 2018; Pham et al., 2018; Marques et al., 2019).

However, this equation was primarily grounded in small agricultural plot-wise information generated and collected from eastern USA. It was particularly framed to estimate the possible yield of sediment from a single storm event, as measured at the watershed outlet, based on that catchment's runoff characteristics. It thus aimed to make the sediment yield prediction more accurate and eliminate the requirement of delivery ratios (Arekhi et al., 2012). Thus its application in other regions needs case-specific modifications to the original parameters stated and relationships used. This becomes one of the major drawbacks of the model, especially where it has failed to properly estimate soil loss rates for larger watersheds. The basic premises of the original model assumed that from small watersheds the soil loss would match the sediment yield and thus the sediment delivery ratio would be unity, which seldom happens in real cases. The inconsistencies related to the prediction of sediment yield as obtained from the sediment delivery ratios often vary spatially and can lead to erroneous results in some scenarios. It was also stated that the soil erodibility nomograph would provide representative K values for every soil type, which again does not always fit properly in every region (Simanton et al., 1980), as was shown by testing its validity for estimating erosion amounts from ultisols in Brazil in comparison to that measured by direct field methods (Cassol et al., 2018). This model also cannot predict deposition rates or sediment delivery rates and yields separately resulting from gully, streambed and stream bank erosion (Wischmeier & Smith, 1978).

The Revised Universal Soil Loss Equation (RUSLE) was a modification of the USLE model to incorporate methods developed after 1978 and to address sediment transport and deposition rates associated with concentrated flows and ponded areas along with inter-rill and rill erosion (Lin & Wang, 2006; Guabi et al., 2016; Gianinetto et al., 2019). The RUSLE also specially addresses the R-factor (rainfall-runoff erosivity) and the K-factor (soil erodibility) (McCool et al., 1995). MUSLE is a further modification of the RUSLE which substitutes the rainfall energy using a runoff factor to make the sediment yield prediction more accurate by eliminating delivery ratios. This then allows the equation to be used to gauge the effects of singular storms (Sadeghi et al., 2014). This runoff coefficient is calculated from the Curve Number (CN) method while the peak discharge is computed from the graphical peak discharge calculation method. The ArcMUSLE toolset, an extension of ArcGIS, suitably predicts soil loss from a particular rainfall event and also prioritises suitable soil erosion areas so that better soil conservation methods can be employed for this purpose (Zhang et al., 2009).

The revaluation of factors like soil factors and cropping practices and land management norms/methods on runoff characteristics have led to the construction of the USLE-M model, a refined form of the USLE, through explicit consideration of the parameters like rainfall erosivity, soil erodibility and crop management. The new model has refined the K and C values by 1.4–3.9 times and 1.0–32.3 times, respectively, with the result that the final model can now predict soil loss more precisely than the original version. This model can further effectively measure soil loss from impervious conditions (Kinnell & Risse, 1998). MUSLE has successfully and accurately predicted sediment yield from numerous diverse watersheds across the world, e.g. from the Kengir watershed in Iran (Arekhi et al., 2012), with high precision results being obtained by the authors when they compared the model generated values with the actual observed ones. Annual soil loss estimates were calculated for the Dwarka and Brahmani river interfluves near the Rajmahal Traps using a combination of USLE and the MMF (Morgan and Morgan and Finney) method to judge the efficacy of both methods in predicting soil loss. While the traditional USLE has been considered as a benchmark technique in estimating soil loss, the MMF method addresses more detailed internal and external soil loss factors, while incorporating separately the water erosion and sediment removal phases (Ghosh & Guchhait, 2012).

Similar other models estimating soil loss include Water Erosion Prediction Project (WEPP, Flanagan et al., 2001; Amore et al., 2004; Shen et al., 2009; Landi et al., 2011; Han et al., 2016; Effendy et al., 2019), Chemicals, Runoff and Erosion from Agricultural Management Systems (CREAMS) Model (Knisel, 1980) and the Soil and Water Analysis Tool (SWAT, Tibebe & Bewket, 2011; Moshabi et al., 2012; Dutta & Sen, 2017; Melaku et al., 2018; Abbaspour et al., 2019; Hussain et al., 2019; Briak et al., 2019). A very succinct review of the USLE and its various revised and modified versions such as RUSLE, MUSLE and other soil erosion prediction models has been provided in the previously/aforementioned studies, while critically analysing the various sub-factors used in the equations, their strengths and limitations in their applicability in various climatic regions and the uncertainties associated with them (for the USLE's applicability and use in the Indian context, see Majhi et al., 2021). Several recommendations have also been provided by the authors indicating their future potentials in using such models (Avwunudiogba & Hudson, 2014; Igwe et al., 2017; Benavidez et al., 2018).

16.6 Riverbank Erosion Hazard: Measurement and Simulation

Soil removal due to bank erosion/failure is another natural geomorphic process by which rivers tend to shape their morphological pattern and behaviour and transport sediments from their upper catchments towards their lower courses (Bordoloi et al., 2020). It has been considered as a desirable attribute in rivers that creates dynamic habitats for riparian communities by promoting riparian vegetation succession and is thus crucial for river ecosystem functions. The various desirable attributes of bank erosion as a natural process have been succinctly explained by Florsheim et al. (2008). However, certain river management programmes, while apparently promoting landscape and economic stability, have accelerated this natural process to an extent where it has been increasingly termed as an environmental hazard associated with accelerated vertical bank instability and active meander migration (Thorne, 1992).

Eroding streambanks are the single main provider of sediments in channels, approximately contributing around 80% of the total load (Mandal, 2017), although this rate varies as a function of channel geometry and time (Thorne, 1992). Streams with unstable banklines and profiles and beside farmlands that have limited or no riparian vegetation often become the prime sites for erosion, which become more active during the monsoon season. Increased streambank erosion is considered as a major cause of non-point pollution in rivers, contributing to an increased sediment load into the channel and thereby decreasing the water quality. Rosgen (2001) further identified that accelerated stream bank erosion often destabilised the stream by changing its major stream type (Ramirez-Avila et al., 2010), resulting in its planform metamorphosing.

Over the years, understanding the processes and mechanisms involved in bank erosion and the rates at which they occur has become fundamental to any bank erosion assessment study. The processes of sediment accretion and degradation and lateral shifting of bank lines as a result of erosion have also been studied because these processes contribute to channel instability in addition to contributing sediment into the channel while causing the destruction of arable lands and valuable floodplain resources along major streams (Lovric & Tosic, 2016). Bank erosion has thus been treated as a hazard which is not easily diagnosed in the field, yet its analysis and prediction are considered vital towards predicting long-term channel stability (NRCS USDA, 2009). Identifying the factors of vertical and lateral channel instability, their relation to bank retreat, the active dominant erosive processes, the bank geotechnical and morphological characteristics and the major bank failure mechanisms are the key concerns that need to be addressed in any bank erosion assessment study (Thorne, 1992).

Although instability in some river channels can be inherent due to the basic nature of the fluvial system or tectonic history of the region, the stability of a streambank is said to be a direct function of channel and flow characteristics and the composition of the bank material. A number of natural factors have been identified as affecting streambank erosion, most important amongst which are the pre-storm soil moisture conditions and nature of storm event, flow characteristics of the channel, bank and channel geometry, bank material composition, stream power and the curvature of the meander bend (Islam, 2008). Recent bank failure mechanisms have however seen a heavy impact of human-induced causes like removal of riparian vegetation (Heeren et al., 2012), increasing imperviousness leading to soil compaction and lesser infiltration and trampling by livestock and vehicles (Watson & Basher, 2006). Globally relevant bank erosion coefficients framed by Hasegawa (1989) identified this process as a function of cross-sectional mean velocity while others have associated bank erosion rate with the geotechnical nature of the bank substrate and the effect of channel hydraulics. Duan (2005) used a deterministic method to quantify the basal erosion rate by predicting bank failure, which varies according to the effects of hydraulic force and the bank geometry and material, through derived analytical equations.

Bank failure occurs by the dual acts of entrainment and basal erosion coupled with mass wasting, the former as a result of hydraulic force, while the latter is a result of gravity (Coryat, 2014). Hydraulic forces, at or below the water surface, scour and entrain sediments and directly cause erosion of mostly non-cohesive materials through the processes of toe undercutting, basal clean out and bed degradation (Watson & Basher, 2006). The hydraulic tractive force on exceeding the critical shear stress of the bank material causes it to fail, and this is usually accelerated by a lack of riparian vegetation and persistent high boundary velocities (NRCS USDA, 2009). Thus the erodibility of the bank material and its shear strength, i.e. the ability of the material to withstand gravitational forces, are considered as important parameters while assessing bank erosion rates (Thorne, 1981).

The second process (i.e. mass wasting) includes mass failure mechanisms which tend to detach sediments mostly from cohesive banks under the influence of gravity and make those sediments available for fluvial transport. These processes mainly include shallow and rotational failures, cantilever and piping failures along the vertical bank face (Watson & Basher, 2006). This is related to the bank geotechnical character and depends on the ambient or antecedent soil moisture condition. The soil moisture amount affects the bank's ability to withstand stresses, and thus geotechnical failures occur, often as a result of bank toe scouring and undercutting and the consequent mass failure (NRCS USDA, 2009). Seepage or piping, a lesser known process, also sometimes contributes to bank erosion. Seepage at the bank toe results in toe undercutting, which makes the entire vertical section of the bank prone to mass failure (Coryat, 2014). The above two processes are frequently interlinked and often work in conjunction, with the incident hydraulic forces being the precursor to gravitational failures (Watson & Basher, 2006).

Quantification methods of bank erosion rates are classified into two broad groups: in situ methods and remote sensing methods. Traditional in situ methods are mostly field intensive techniques requiring detailed data on channel morphology and banklines, typically involving the use of survey instruments. These are short-term studies involving a smaller portion of a stream reach that can be investigated by repeat topographic profiling with much more precision and accuracy but is not always time and cost-effective (Watson & Basher, 2006). Also, since erosive banks are not evenly distributed in space and largely vary with respect to time, channel planform repeat surveys are not usually adequate in picking up the temporal and spatial variations, and thus erosion pins have been widely used as an alternative to this. First used by Wolman (1959), several studies have used erosion pins to successfully estimate bank erosion rates. However, in rivers where the bank is not dominated by alluvial material, erosion pins do not always provide the desired results (Thorne, 1981). Moreover, these are comparatively difficult to install at sites that have high banks and often tend to loosen and impact vertical bank profiles wherever there is coarse material. Complete loss or burial of erosion pins can also occur at sites where rotational mass failure processes are dominant (Watson & Basher, 2006).

Sedimentological and botanical evidences have also been employed to quantify rates of bank erosion in long-term studies (decadal or longer). Chronologies of historical alluvial deposition help in reconstructing the history of river activity based on the types of river deposits, their age and spatial distribution. This process is mainly done to record histories dating back to more than 50 years and can be effectively used to reconstruct the major channel pattern and floodplain evolution and the changing nature of land use in the recent past (Watson & Basher, 2006). A continuous monitoring of turbidity and event sampling of suspended sediment in a particular river reach can also provide valuable insights into the volume of material lost to bank failure.

Geospatial techniques, using remote sensing, on the other hand are a comparatively recent method which has contributed markedly to bank assessment studies. The ability of remote sensing techniques to capture large amounts of information, over a large temporal extent, has further helped in precisely quantifying the rates of erosion and the volume of eroded materials (Lovric & Tosic, 2016). Satellite remote sensing also has the advantage of providing a synoptic view of a large area which allows easy detection of changes occurring along channel banklines (Bordoloi et al., 2020). These methods have four main characteristics, frequency, extent, precision and resolution, and thus a diverse variety of aim-specific methods have been formulated (Duro et al., 2018). Aerial photographs and historical imageries available over a greater period of time can provide reliable information on the extent of bank migration and bank profile development. However, it has often been argued that historical maps and aerial photographs do not always provide the detailed measurements and insight into the actual fluvial processes which are at play, as obtained from on-site surveys and thus these two methods have been considered as complimentary and not always necessarily alternatives. Rather an integration of these two methods would provide a better insight into the actual bank erosion process and rate under consideration (Thorne, 1981). A much more recent technique, i.e. Structure-from-Motion photogrammetry (SfM), is now being used quite frequently to measure bank erosion rates and surface erosion (Patel et al., 2020b). This has evolved as a cost-effective measure of quantifying bank retreat and for identifying the major processes acting behind erosion, although the failure surface needs to be devoid of vegetation and the channel must have low water levels to properly capture the photographs (Duro et al., 2018).

Channel stability assessment protocols are often used to assess bank condition and stability, because they are rapid, are easy to use and provide useful information regarding bank conditions. Rapid Geomorphic Assessment Protocols (RGA) are used to characterise stream reaches with homogenous streambank characteristics, based on the Channel Stability Index (CSI), which is a metric score based on a number of parameters like bed material, bank geometry, channel evolution phase, percentage of channel constriction, meander geometry and fluvial erosion processes. Stream reaches are easily categorised into suitable classes of instability based on their metric score. Such methodologies have been adopted for the Barren Fork and Spavinaw Creeks and later been collated into the Oklahoma Streambank Erosion Potential Index (OSEPI) to fix the existing protocol into ecoregion-specific terms for large streams of the region (Heeren et al., 2012). The Australian River Assessment System (AUSRIVAS) has a special module and scoring matrices for bank erosion and stability module. The SEDNET framework (also developed for Australia) predicts the bank retreat rate based on bankfull discharge and uses empirical equations to calculate erosion rates (Watson & Basher, 2006). Conversely, the Stream Bank and Bed Stability Assessment Protocol assesses channel stability by identifying areas along channel reaches that are most severely affected by erosion and are at the greatest risk of future failure based on parameters like bed and bank material, bank slope and vegetation coverage and the amount of active bed and bank erosion, providing a final score that would necessarily be the initial step for restoration projects (Sholtes & Giordanengo, 2014).

The Bank Assessment for Non-point source Consequences of Sediment (BANCS), developed by Rosgen (2001), uses the Bank Erosion Hazard Index (BEHI) and Near Bank Stress (NBS) to demarcate potential erosion zones (Kwan & Swanson, 2014). This process-integration approach using rational estimation quantifies and predicts stream bank erosion rates through intensive field surveys and empirically derived relationships (Rosgen, 2001). It is a multimetric semi-quantitative index for quantifying the potentiality of bank erosion of unstable streams (Simpson et al., 2014). The BEHI parameters are a combination of visual assessments and measured field indices that together determine a bank's susceptibility to erosion (Sass & Keane, 2012). The estimated variables include bank-height ratio, weighted root density, root-depth ratio, surface protection, stream bank angle, stream bank material and its stratification, whereas, for the quantification of NBS, there are five methods for general prediction, and two methods for detailed prediction (Rosgen, 1996). Some examples of places where this technique has been employed are the Stony Clove watershed in the Catskills Mountains (Coryat, 2014) and along the River Ganga upstream of the Farakka Barrage in West Bengal (Mandal, 2017).

Another most commonly used model for bank failure assessment prediction and stabilization is the Bank Stability and Toe Erosion Model (BSTEM), from USDA-ARS (US Department of Agriculture-Agricultural Research Service) (Simon et al., 1999). It contains two components: modules for Bank Stability and Toe Erosion. To assess bank stability, it combines and simulates three Limit Equilibrium Methods: horizontal layers, vertical slices for failure blocks having tension crack and cantilever failure. The horizontal layer method accounts for the vertical heterogeneity of bank materials and divides the bank profile into layers of distinct geotechnical character. These are user-input layers depending on the riverbank characteristics. Along the vertical slices, the shear and normal forces active in the failure block segments are evaluated. Failure blocks having tension cracks are generally analysed by this method. The cantilever failure algorithm is a further modification of the CONCEPTS model (Langendoen, 2011; Langendoen, 2011), simulating incised stream evolution and reduces sediment yield for long-term stream rehabilitation and management. Combining these three algorithms, the model generates a numerical value called the Factor of Safety (FoS). The value is a ratio of the driving and the resisting forces along a potential failure plane, where the driving forces include tall bank height, steep bank angle and non-cohesive materials while the resisting forces are low bank height, surface protection, gentle bank angle and presence of cohesive materials. The factors which influence bank stability are effective internal angle of friction, matric suction force or pore water pressure, effective cohesion and the bank slope. This model also includes a RipRoot component, which explains the effectiveness of vegetation in providing slope stability. The BSTEM model has been used to evaluate soil erosion in the sub-watersheds of the Ganetti station in Sudan using the Sediment Transport Index (STI) (Kheiralla & Siddeg, 2015).

The use of modelling techniques has further refined such studies, with the devised algorithms typically ranging from empirical equations to process-based approaches. Prevalent are some catchment-based erosion models like RIVERMorph (Klavon et al., 2016), SRH-2D (Lai, 2017) and MIKE-21C(Nikhil et al., 2019), which have special modules and are trained to estimate and predict sediment routing and transportation rates, bed and bank erosion and soil loss from watersheds. The use of 2D mathematical models to predict bank erosion rates depending on local channel geometry, flow and sediment pattern and distribution has also become common in the last decade (Islam, 2008). However, more process-based models are focused on incorporating both gravitational and hydraulic failure mechanisms into the analysis to arrive at the erosion rate. The GRAF model predicts erosion based on historical channel change and flood information where the probability of bank erosion is directly related to the flood magnitude and inversely so to the upstream and lateral distance from a channel (Watson & Basher, 2006).

16.7 Sand Mining and Channel Instabilities: Mapping and Impacts

Another human-made hazard having far reaching implications on the channel and its riparian areas is sand excavation and mining. Sand mining is the removal of sands (loose or consolidated) from the foreshore regions of rivers, lakes and oceans and including from their beds (Mukerji, 2016). It is practised both mechanically and manually. Manual methods of mining are considered to be more eco-friendly as these are more localised phenomena, carried out on a much smaller-scale and the volume of extracted sand is also quite low. This type of mining is essentially practised in the smaller river basins where the quantity of available sand is often low. Mechanical mining, on the other hand, especially when practised on a large-scale commercially, is what causes environmental degradation by altering riverine ecosystem functions and balance. High-powered jet pumps, draglines and hydraulic excavators are generally used in this type of mining.

Commercial extraction of sand from rivers and riparian zones occurs worldwide and is one of the most profitable enterprises because this industry seldom projects/gauges itself in any EIA or cost-benefit scenario (Kondolf, 1994). Worldwide, sand is the next most extracted natural resource after freshwater and is the most traded commodity in terms of weight. 85% of the global mining activity has been sand and gravel, which is largely used in the construction industry and glass, ceramics and electronic industries. Globally, the total volume of sand extracted per year is estimated to be around 40 billion tonnes, out of which only 15 billion has been stated as legally traded (UNEP, 2019). China's Poyang Lake has been the world's largest sand mine with a staggering 236 million cubic meters per year yield. Singapore has emerged as the biggest importer of sand in recent times to meet up its over-ambitious land reclamation projects (Mahadevan, 2019).

Sand mining in India also has a long history, with the industry booming since the 1990s (in the post-economic reforms phase that has ushered in marked growth in real estate and urbanisation in the country) and has been widespread across all the major river basins in India (Padmalal & Maya, 2014; Mukerji, 2016). It has been considered as a disorganised crime in a growing economy like India and is thought to be a problem of large but unknown proportions. With a tripling in the demand for sand from 2000 to 2017 and a total market worth of 150 billion rupees, India houses the third-biggest construction sector globally, accounting for 9% of its economy and employing over 35 million people. Extraction of sand from river beds did not pose a problem in the past since it was considered that natural process would replenish the voids created by mining. But the real problem is the scale at which it occurs. To meet up the growing demand, illicit sand mining often occurs at rates that are beyond the capacity of the natural process to replenish the same. In recent times, there is possibly no river in India that has not been affected by illegal sand mining activities (Padmalal & Maya, 2014). It is said to be practised in more or less every state in India, with 22 out of 29 provinces reporting illegal sand mining and the prevalence of sand mafias that run this industry. According to one estimate, a single river in Kerala was reported to have been mined for sand at a rate 40 times greater than the natural rate of replenishment, while the highest rates of extraction have been seen to be in the vicinity of Kochi city, one of the most industrialised and urbanised cities in the country (Mahadevan, 2019). The Palar and Vaigai rivers in Tamil Nadu; the Periyar and Bharatapuzha rivers in Kerala; the Yamuna, Ganga and Ghaghar rivers in the National Capital Territory; the Narmada in Madhya Pradesh; and the Damodar, Mahanadi, Darakeswar and Kangsabati in West Bengal are examples where extensive illegal sand mining activities occur. The Tawa River, which is Narmada's tributary, has been known to change its course due to extensive mining activities, frequently breaching its banks and causing massive floods (Mukerji, 2016). The Irrigation and Waterways Department in West Bengal has identified around 248 mining sites around the state that are operated by sand mafias practising illegal mining along the major rivers of Rarh Bengal (Ghosh et al., 2016). A detailed state-wise report on the mining activities in different rivers in India has been documented by Kelkar (2016), along with the mentioning of different government acts of omissions and commissions to control and regulate illegal excavation of sand from riverbeds.

River aggregate mining (Koehnken et al., 2020) has been categorised into three groups: in-stream or wet mining, i.e. excavating directly from the flowing channel with materials collected from either the main channel or from the edges of the rivers; dry mining, i.e. collection of sand from either the floodplains, terraces or from the dry exposed channel margin or from ephemeral streams that are dry for a greater part of the year; and bar skimming or bar scalping which involves raking away of sand from the upper layers of the sand bars, often up to varying depths (Koehnken & Rintoul, 2018). In-stream sand mining occurs from a number of geomorphic river units like sand bars, point bars and the active channel itself. It requires less processing and is the most common type of mining activity prevalent in recent times. Bar skimming or scalping refers to the process of removing the top layer of sand to smoothen the irregularities present on the uppermost layers of the bar, without hollowing out the surface to beneath the water level in summer. To maintain the hydraulic control prevailing in the upstream of the bar, the most preferred bar skimming technique involves scalping the remaining two-thirds portion of the bar along its downstream end and leaving the uppermost one-third portion intact (Kondolf, 1994). Wet pit channel mining involves sand extraction from dry ephemeral stream beds below the alluvial ground water table. Such type of mining activities often require the use of draglines or hydraulic excavators to dig out sand from below the water line (Ghosh et al., 2016). Both pit excavation and bar skimming affect the channel morphology, with the first one leading to an increase in the channel depth and the latter causing an increase in the channel width. Both these processes lead to slower flow velocities and consequent slow energy gradients at the mining sites, creating a situation where the normal flow becomes devoid of required sediments and therefore tends to pick up more sediment from the downstream segments causing a "hungry water effect" (Kondolf, 1997; Graviletea, 2017).

Sand mining is known to disturb channel equilibrium through the interception of the sediment load that moves along such a dynamic system and thus triggers morphologic responses, inducing channel instability. Changes in bed forms are by far the most obvious and direct alterations caused by massive sand mining activities. Even though it remains confined to limited areas within the channel bed, it triggers bed and bank erosion which increases the sediment load in channels (DID, 2009). Bed degradation is the most distinguishable effect of sand mining and occurs primarily through the processes of headward cutting and hungry water effect. Sand excavation from within the channel increases the channel depth, which creates local knick points that steepen channel slopes and increase erosion. Such knick points during high flows migrate upstream through headward erosion, thereby increasing the sediment volume in the downstream sections. The channel cross-sectional characteristics and planform geometry also undergo modification due to sand mining. Bed deepening from channel incision becomes apparent through cross-sectional modification of the channel. Furthermore, rivers have "long memories," with channel adjustments due to sand extractions being realised at a much later date even after the activity has ceased to occur (Kondolf, 1998). Large-scale sand mining from alluvial rivers has led to irrevocable changes in fluvial systems, and this has still continued unabated, as the geomorphic and hydrologic effects of such activities often manifest itself at a later stage during events of high flows. Impacts of sand mining have been classified as on-site and off-site impacts (Madyise, 2013). While the on-site impacts are directly related to channel and floodplain modifications, the off-site changes are mostly related to the transportation of the excavated materials and pollution-related aspects. Although sand mining is sometimes said to have positive impacts, which reduces river flooding by deepening and widening channel courses that provides more lateral space to a channel and also increases its carrying capacity, the negative impacts associated with channel instability outnumbers its positive impacts (Nandakumaran et al., 2014).

Indiscriminate and unscientific sand mining activities in the alluvial reaches of the River Damodar have brought about changes in thalweg dynamics and river planform, instability of river bars, bank erosion, changes in riffle-pool sequences and channel bifurcation (Ghosh et al., 2016). Mergers of nearby pits during the monsoonal flows often lower the bed at a much faster rate, thereby causing channel instability and bank erosion problems. Studies from across the world have highlighted how such channel instability has been engendered and perpetuated. The Cheney Creek in the US experienced a bed lowering of almost 4.9 m while a Russian channel near Healdsburg incised on average over 3 m within a stretch of only 11 km, with this rising up to 6 m in some spots. The seven rivers flowing into Kerala's Vembanad Lake have reported around 12 million tonnes of sand being removed in a single year, resulting in the average lowering of the river beds by 7–15 cm/year (Padmalal et al., 2008). The Periyar River, considered as the lifeline of Kochi city, has witnessed a maximum bed lowering rate of 19 cm/year and is still now exhibiting continuous bed lowering due to relentless mining activities (Mukerji, 2016). The Balason River in the vicinity of the Siliguri in West Bengal has seen a steady rise in the volume of sand extracted from the river during 2002–2010, with clear signs of channel incision and fluctuations in water levels measured across the Matigara Bridge on this river. Government reports have suggested that within 20 years the river bed level has been lowered by approximately 1.3 m, i.e. 7 cm/year on an average (Wiejaczka et al., 2018). In certain instances, the removal of the coarser layer from the bar top had exposed the finer sediment layer below, which was then mobilised even at lower flows, causing destruction of the entire sand bar and resulting in altering the hydrodynamic regime of the river. Creation of new pools, elongation of existing pools and destruction of riffles are amongst the common morphological impacts of such mining (Koehnken & Rintoul, 2018).

Another noted impact of such sand mining is the loss of in-stream and riparian species and the overall degradation of the riparian habitat. Natural rivers are blessed with a unique assemblage of in-stream and floodplain biota that helps in maintaining the ecological balance in riverine ecosystems (Sheeba, 2009). Sand mining affects stream biota through sequential changes in the biological, physical and chemical constituents of flowing water. The selective removal of finer sand particles and their deposition in other sections of the channel also alter the natural habitat of certain in-stream organisms, which greatly affect their taxonomic diversity and bring about ecological changes in the entire system. Sand mining induces channel widening, which shallows down the river forming braided river flows that further affect the movement of fishes in the pool sections of the channel (Pitchaiah, 2017). Indiscriminate sand mining from alluvial reaches also alters sediment size characteristics in rivers, as seen in the Manimala and Muvattupuzha Rivers in South India (Padmalal & Maya, 2014).

Physical habitat degradation and impairment of food-web sequential functions have also been reported for some rivers due to this anthropogenic activity. Changes in channel morphology and sediment characteristics adversely affect fish and other benthic communities as well as the three-tier vegetation gradient in rivers. A high level of turbidity due to excess sediments in the stagnated pools prevents photosynthetic activities of in-stream organisms and adversely affects planktonic and benthic growth in channels, causing destabilisation of the food and energy transfer mechanisms that are important for their survival (as was reported by Prabhakar et al. (2019) from the River Ganga in Bihar). The environmental and ecological ramifications of sand mining along the Ithikkara River in Kerala were a considerable decline in its fish population due to loss of breeding and feeding grounds, due to depletion of the detritus which serves as the common food for such organisms (Sheeba, 2009). It also provides suitable bottom substrate for various benthic organisms along with facilitating favourable environmental conditions for spawning, breeding and movement of fishes across river channels. Sand excavation often to the extent of exposing the underlying hard rock has also led to the complete disappearance of *Wallago attu*, a common species that once flourished abundantly in that river (Sheeba, 2009).

A significant alteration in the stream biota along with the water quality has been noticed along the middle course of the Kangsabati River in the vicinity of its sand mining sites as compared to the upper and lower courses, where mining activities are rare. Increased turbidity and reduction in fish species (as much as 80% for some species) have been reported by Bhattacharya (2018) for this area. The Habitat Suitability Index (HSI) was used to quantify habitat fragmentation and degradation and loss of species abundance and diversity for this river (Bhattacharya & Chatterjee, 2016; Bhattacharya et al., 2019a). Indicators like pH, TDS, DO, salinity, turbidity and electrical conductivity were used to assess the water quality, and then its impact on the in-stream biota was measured using the Simpson's Index of Diversity, Simpson's Reciprocal Index, Shannon's Diversity Index and Pielou's Evenness Index, all of which indicated that a deterioration in the water quality due to increased sand mining activities engendered changes in species richness, diversity and overall ecological instability in the landscape (Bhattacharya et al., 2019b).

Creation of access roads to support mining activities has caused fragmentation of forested riparian zones along stretches of France's Lower Eygues River (Kondolf et al., 2007). Kondolf (1994) also observed an increase in the water temperature in pockets where flow velocities were lower as a result of channel widening. Such increased water temperatures often cause changes in the habitat structure due to lower DO concentrations and an increase in toxic pollutants like heavy metals, insecticides, pesticides and other natural toxicants (Koehnken & Rintoul, 2018). A decline in the growth rate of a perennial grass species which usually thrives on the exposed dry bed of River Darakeshwar in West Bengal during the post-monsoon season was observed followed by a changing channel pattern in terms of river deepening due to excessive wet mining activities directly from the channel bed (Sinha, 2016).

Another notable impact of sand mining is the lowering of ground water table (Padmalal et al., 2008). Not only do they act as aquifer systems storing groundwater, but also act as filters to various kinds of pollutants, thus improving water quality and maintaining ecosystem health. In alluvial channels, as the banks are drained to the river level, increased channel incision would cause lowering of ground water tables and affect the water supply in nearby wells and tubewells (Sreebha & Padmalal,

2011). Another widely documented impact of a lowered water table is the degradation of the adjacent riparian/floodplain vegetation as and when the water level goes beneath their root zone. A lowered water table also induces ecological changes through the loss of hyporheic habitats through dewatering of adjacent river banks (Koehnken et al., 2020). Baseflow in channels is also lowered as a result of decreased water table that brings about instability in the hydrologic and ecologic regime of a stream (Oude Essink, 2001).

Measuring and quantifying the above-noted impacts of sand mining are quite difficult, while the total volume of sediment that is actually being mined seldom gets documented. River cross-sectional profiling and morphological changes are quantified using different field survey techniques, while the areas affected by mining are measured by geospatial techniques. In recent times, the development of hydraulic and sediment transport models can gauge the possible sand deposition rate and their particular flow direction. This has been done in the downstream segments of the Selangor River in Malaysia using the HEC-RAS 1D model. It could also predict the maximum depth up to which sand excavation should be permitted and allowed in a river to maintain channel stability. The maximum allowable mining depth as predicted by the model was up to 1.5 m above the natural channel thalweg (Asraf et al., 2011).

16.8 Overview of Methods in Riparian Zone Hazard Assessment

The above sections have outlined in detail some of the major hazards (both natural/environmental and societal/human induced) that affect riparian zones along streams. We have also delved into some of the methods that are used to examine these hazards. Taking a selection of the papers considered in this study, word clouds (Fig.16.1) have been prepared for each of the four main hazard types—riverbank erosion, soil erosion from overland flow/runoff, floods (including the flood susceptibility and flood vulnerability aspect) and sand mining—taking the selected paper titles and their respectively mentioned keywords. In each case, the preponderance of geospatial methods to gauge the extent and severity of these hazards is quite apparent, with these methods being more extensively used for the analysis of floods (quite a few studies in this domain have used the AHP method) and soil loss mapping/estimation (USLE (along with its variations) and SWAT models) than in other cases. The riverbank erosion aspect is mostly examined based on developing combined field and geospatial data-based indices to enumerate the channel bankline changes (e.g. the BEHI-NBS-BANCS method) while a dearth of any such overwhelmingly apparent technical investigative method was seen in case of the sand mining hazard, denoting the relatively lesser amount of studies that have examined this aspect. Thus, possibly there is much greater scope for developing holistic investigative methods in this domain that considers the physical alteration of the stream's morphological aspects along with the changes wrought about in the ambient riparian ecological character and attributes.

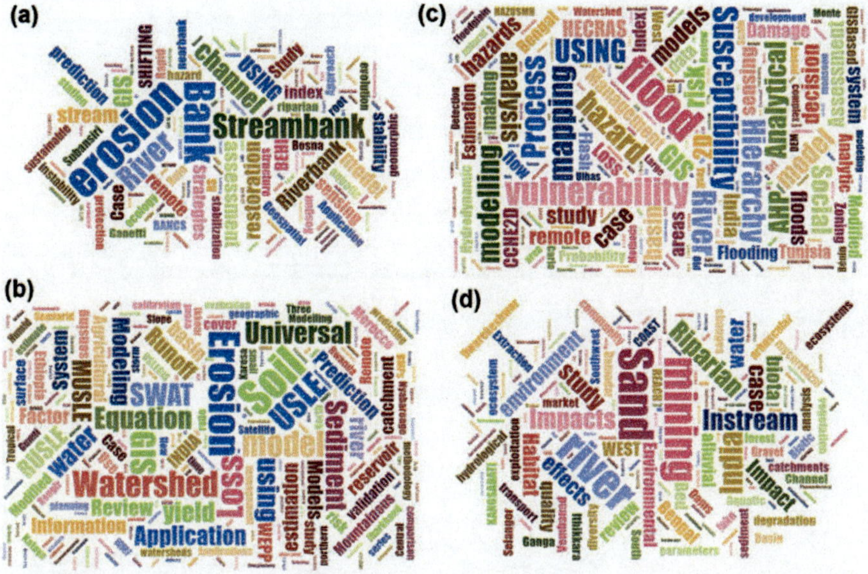

Fig. 16.1 Word clouds prepared for the different hazards examined based on select references in this paper and their keywords, for (**a**) riverbank erosion, (**b**) soil loss and erosion, (**c**) flood hazard, including flood susceptibility and vulnerability, and (**d**) sand mining effect on river channels and riparian zones

Our other investigation into the temporal prevalence of studies that have examined the above four main hazards (Fig. 16.2) also reveals the over-dominance of studies looking to soil loss and soil erosion, followed by the studies of the impact of floods (flood hazard and its related aspects of inundation preparedness and flood risk/vulnerability/susceptibility). This analysis uses the dataset available in the Google Ngram viewer, and as such only the present texts between the years 1900 and 2012 were examined in respect of the frequency of occurrence of some key phrases. It is therefore quite logical that geospatial models developed to gauge the soil loss from catchments (USLE and its modifications—RUSLE, MUSLE) should rank highest followed by flood simulation and bank erosion analysis methods. What the diagrams reveal too is that there has been a growing focus on studies that look into the riparian zone and river ecology and not just into the physical character of the stream.

16.9 Conclusion

With the ever-increasing human pressure on riparian zones, continuous degradation of riparian health has been seen to occur across the major river basins of the world. Indiscriminate felling of riparian vegetation to make space for agricultural lands, extensive river channelization programmes, illegal and massive sand excavation

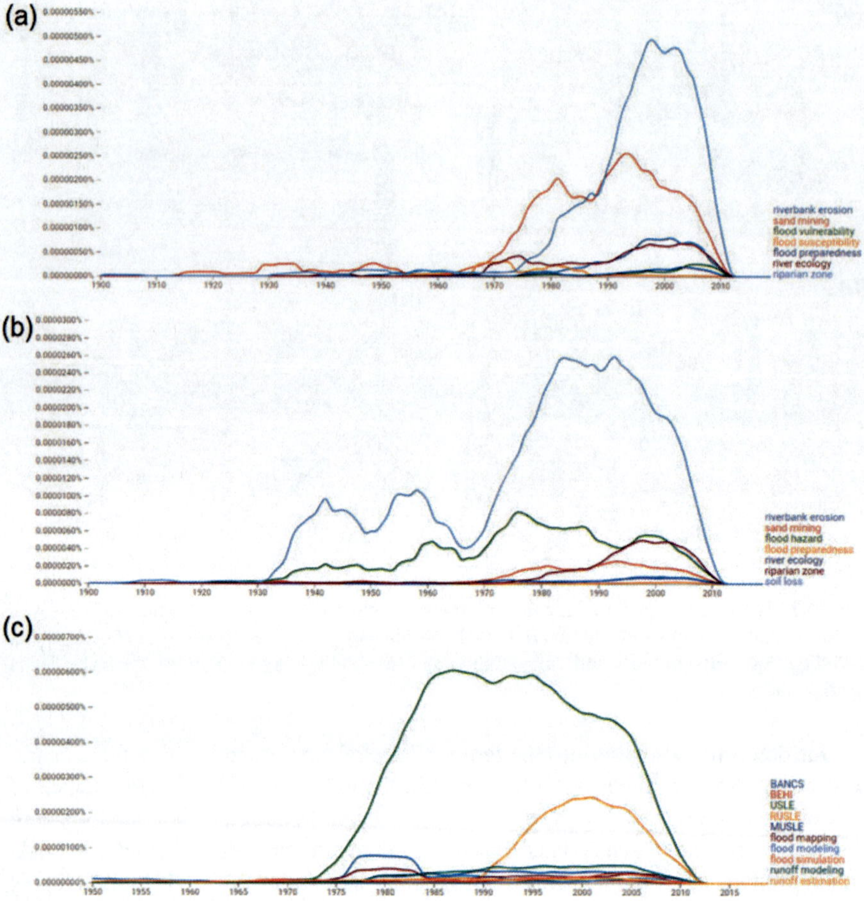

Fig. 16.2 Frequency of key phrase in the examined texts in Google Ngram—(**a**) different hazards and aspects without considering the soil erosion hazard, (**b**) including the soil erosion aspect, which shows its marked dominance in hazard studies, and (**c**) the preponderance of soil loss estimation models in the examined literature

practices, rapid expansion of infrastructural developments mainly along the major river catchments and faulty and unscientific agricultural practices all have led to enhanced channel instability and resultant loss of valuable riparian tracts. With the advancement of scientific technologies aided by a myriad of geospatial techniques, the quantification of such hazards have become easier, but their proper management and mitigation still remains a problem, especially in poorly resourced developing nations wherein these hazards are an annual menace. In developing nations like India, often the riparian health is compensated to make way for hard engineering structures to manage these hazards. The ensuing loss of channel-floodplain linkages and the consequent fragmentation of the riparian habitat and impairment of riparian biodiversity have been identified as some of the most crucial riparian health hazards noticeable amongst Indian rivers. Flooding and increased bank erosion problems that cause loss of arable land is another recognised problem. Especially the latter is markedly accentuated by exploitative sand mining activities that not only make the channel more unstable but also put at risk the entire riparian ecosystem. Compared to the traditional and long undertaken studies on soil loss, riverbank erosion and flood hazards, examinations of sand mining activities have received comparatively lesser attention, with a dearth of investigative frameworks and guidelines on the same. Further studies in this domain are thus pertinent. It is also paramount that in all such hazard modelling investigations of riparian zones and river components, the ecological aspect and connectivity issues be incorporated, in order to better gauge the impacts of such disturbances on the entirety of the lotic environment.

Acknowledgement This research has been funded by the Department of Science and Technology and Biotechnology, Government of West Bengal Grant (Memo No. 256(Sanc.)/ST/P/S&T/10G-13/2017 dated 25/03/2018) awarded to Priyank Pravin Patel. The UGC-SRF Award of Sayoni Mondal is also acknowledged.

References

Abbaspour, K. C., Vaghefi, S. A., Yang, H., & Srinivasan, R. (2019). Global soil, landuse, evapotranspiration historical and future weather databases for SWAT applications. *Scientific Data, 6*, 263.

Acuna, V., Hunter, M., & Ruhi, A. (2016). Managing temporary streams and rivers as unique rather than second-class ecosystems. *Biological Conservation, 211*, 12–19.

Agouridis, C. T., Wrightman, S. J., Barton, C. D., & Gumbert, A. A. (2010). *Planting a riparian buffer*. University of Kentucky, Cooperative Extension Service: ID-185.

Alewell, C., Borrelli, P., Meusburger, K., & Panagos, P. (2019). Using the USLE: Chances, challenges and limitations of soil erosion modelling. *International Soil and Water Conservation Research, 7*, 203–225.

Amarnath, G., Umer, Y. M., Alahacoon, N., & Inada, Y. (2015). Modelling the flood-risk extent using LISSFLOOD-FP in a complex watershed: Case study of Munderi Aru River basin, Sri Lanka. *Proceedings of the International Association of Hydrological Sciences, 370*, 131–138.

Amore, E., Modica, C., Nearing, M. A., & Santoro, V. C. (2004). Scale effect in USLE and WEPP application for soil erosion computation from three Sicilian basins. *Journal of Hydrology, 293*, 100–114.

Angermeier, P. L., & Karr, J. R. (1984). Relationships between woody debris and fish habitat in a small warmwater stream. *Transactions of the American Fisheries Society, 113*(6), 716–726.

Arekhi, S., Shabani, A., & Rostamizad, G. (2012). Application of the modified universal soil loss equation (MUSLE) in prediction of sediment yield (Case study: Kengir Watershed, Iran). *Arabian Journal of Geosciences, 5*, 1259–1267.

Asraf, M. A., Maah, M. J., Yusoff, I., Wajid, A., & Mahmood, K. (2011). Sand mining effects, causes and concerns: A case study from Bestari Jaya, Selangor, Peninsular Malaysia. *Scientific Research and Essays, 6*(6), 1216–1231.

Avwunudiogba, A., & Hudson, P. F. (2014). A review of soil erosion models with special reference to the needs of humid tropical mountainous environments. *European Journal of Sustainable Development, 3*(4), 299–310.

Banerji, D., & Patel, P. P. (2019). Morphological aspects of the Bakreshwar River Corridor, West Bengal, India. In B. Das, S. Ghosh, A. Islam (eds.) *Advances in micro geomorphology of lower Ganga basin – Part I: Fluvial geomorphology* (pp. 155–189). Springer International Publishing, Cham.

Baird, D.C., Fotherby, L., Klumpp, C.C. & Sculock, S.M. (2015) *Bank stabilization design guidelines. Report No. SRH-2015-25*, U.S. Department of the Interior Bureau of Reclamation, .

Behanzin, I. D., Thiel, M., Szarzynski, J., & Boko, M. (2015). GIS-based mapping of flood vulnerability and risk in the Benin Niger River valley. *International Journal of Geomatics and Geosciences, 3*(6), 1653–1669.

Behbahani, S. M., Moradi, M., Basiri, R., & Mirzaei, J. (2017). Sand mining disturbances and their effects on the diversity of arbuscular mycorrhizal fungi in a riparian forest of Iran. *Journal of Arid Land, 9*(6), 837–849.

Bekele, B., Muluneh, A., & Wondrade, N. (2019). Geographic information system (GIS) based soil loss estimation using universal soil loss equation model (USLE) for soil conservation planning in Karesa watershed, Dawuro zone, south West Ethiopia. *International Journal of Water Resources and Environmental Engineering, 11*(8), 143–158.

Benavidez, R., Jackson, B., Maxwell, D., & Norton, K. (2018). A review of the (revised) universal soil loss equation ((R)USLE): With a view to increase its global applicability and improving soil loss estimates. *Hydrology and Earth System Sciences, 22*, 6059–6086.

Bera, A. (2017). Assessment of soil loss by universal soil loss equation (USLE) using GIS techniques: A case study of Gumti River Basin, Tripura, India. *Modelling Earth Systems and Environments, 3*(29).

Bhattacharya, R. K., & Chatterjee, N. D. (2016). Sand mining impact on river biota: A case study of Kangsabati River in West Bengal. *Proceedings of the Conference on Dirology.*

Bhattacharya, R. K. (2018). Instream sand mining impact on water quality and benthos community in an alluvial reach: A case study on river Kangsabati, West Bengal. *International Journal of Current Research in Life Sciences, 7*(8), 2613–12617.

Bhattacharya, R. K., Chatterjee, N. D., & Das, K. (2019a). Impact of instream sand mining on habitat destruction or transformation using coupling models of HSI and MLR. *Spatial Information Research, 28*, 67–85.

Bhattacharya, R.K., Chatterjee, N.D. & Dolui, G. (2019b) Consequences of sand mining on water quality and instream biota in alluvial stream: A case-specific study in South-Bengal River, India. Sustainable Water Resources Management. 4/2019.

Blackwell, M. S. A., Maltby, E., & Gerritsen, A. L. (2006). *Ecoflood guidelines: How to use floodplains for flood risk reduction*. Publications Office of the European Union. Retrieved from https://publications.europa.eu/en/publication-detail/-/publication/9444a5bd-8f06-4692-a9d3-2ffa29164e02/language-en.

Blistanova, M., Zelenakova, M., Blistan, P., & Ferencz, V. (2016). Assessment of flood vulnerability in Bodva river basin, Slovakia. *Acta Montanistica Slovaca, 21*(1), 19–28.

Bordoloi, K., Nikam, B. R., Srivastav, S. K., & Sahariah, D. (2020). Assessment of river bank erosion and erosion probability using geospatial approach: a case study of the Subansiri River, Assam, India. *Applied Geomatics, 12*, 265–280.

Borrows, P., & Bruin, D. (2006). The management of riverine flood risk. *Irrigation and Drainage, 55*(S1), 151–157.

Botelho, T. H. A., Jacomo, S. A., Almeida, R. T. S., & Greibeler, N. P. (2018). Use of USLE/GIS technology for identifying criteria for monitoring soil erosion losses in agricultural areas. *Journal of Brazilian Association of Agricultural Engineering, 38*(1), 13–21.

Briak, M., Mrabet, R., Moussadek, R., & Aboumaria, K. (2019). Use of a calibrated SWAT model to evaluate the effects of agricultural BMPs on sediments of the Kalaya river basin (north of Morocco). *International Soil and Water Conservation Research, 7*(2), 176–183.

Cao, C., Xu, P., Wang, Y., Chen, J., Zheng, L., & Niu, C. (2016). Flash flood hazard susceptibility mapping using frequency ratio and statistical index methods in coalmine subsidence areas. *Sustainability, 8*, 948.

Carolyn, A., et al. (2001). *Stream corridor restoration: Principles, processes and practices. Adopted as part 653 of the National Engineering Handbook*. USDA-Natural Resources Conservation Service.

Cassol, E. A., da Silva, T. S., Eltz, F. L. F., & Levien, R. (2018). Soil erodibility under natural rainfall conditions as the K factor of the universal soil loss equation and application of the nomograph for the subtropical Ultisol. *Revista Brasileira de Ciência do Solo, 42*.

Cavaille, P., Dommanget, F., Daumergue, N., Loucougaray, G., Spiegelberger, T., Tabacchi, E., & Evette, A. (2013). Biodiversity assessment following a naturality gradient of riverbank protection structures in French Prealps Rivers. *Ecological Engineering, 53*, 23–30.

Chakraborty, S., & Mukhopadhyay, S. (2019). Assessing flood risk using analytical hierarchy process (AHP) and geographical information system (GIS): Application in Coochbehar District of West Bengal, India. *Natural Hazards, 99*, 247–274.

Chatterjee, S., & Patel, P. P. (2016). Quantifying landscape structure and ecological risk analysis in subarnarekha sub-watershed, Ranchi. In D. K. Mondol (ed.) *Application of geospatial technology for sustainable development*, (pp. 54–76). North Bengal University Press, Raja Rammohunpur.

Cockburn, J. M. H., Villard, P. V., & Hutton, C. (2016). Assessing instream habitat suitability and hydraulic signatures of geomorphic units in a reconstructed single thread meandering channel. *Ecohydrology, 9*(6), 1094–1104.

Coleman, D., MacRae, C. & Stein, E.D. (2005) Effect of increases in peak flows and imperviousness on the morphology of Southern California streams. *Technical report 450, Southern California coastal water research project*, Westminster, CA.

Coryat, M. (2014). *Analysis of the bank assessment for non-point source consequence of sediment (BANCS) Approach for the Prediction of Strembank Stability and Erosion along Stony Clove Creek in the Catskills*. PhD Thesis. Syracuse University. Retrieved December 20, 2020 from https://surface.syr.edu/thesis/21

Covino, T. (2017). Hydrologic connectivity as a framework for understanding biogeochemical flux through watersheds and along fluvial networks. *Geomorphology, 277*, 133–144.

Craig, L. S., Palmer, M. A., Richardson, D. C., Filoso, S., Bernhardt, E. S., Bledsoe, B. P., Doyle, M. W., Groffman, P. M., Hassett, B. A., Kaushal, S. S., Mayer, P. M., Smith, S. M., & Wilcock, P. R. (2008). Stream restoration strategies for reducing river nitrogen loads. *Frontiers in Ecology and the Environment, 6*(10), 529–538.

D'Ambrosio, J. L., Williams, L. R., Williams, M. G., Witter, J. D., & Ward, A. D. (2014). Geomorphology, habitat and spatial location influences on fish and macroinvertebrate communities in modified channels of an agriculturally-dominated watershed in Ohio, USA. *Ecological Engineering, 68*, 32–46.

Dahri, N., & Abida, H. (2017). Monte Carlo simulation-aided analytical hierarchy process (AHP) for flood susceptibility mapping in Gabes Basin (southeastern Tunisia). *Environmental Earth Sciences, 76*, 302.

Das, S. (2019). Geospatial mapping of flood susceptibility and hydro-geomorphic response to the floods in Ulhas basin, India. *Remote Sensing Applications: Society and Environment, 14*, 60–74.

De Silva, M. M. G. T., & Kawasaki, A. (2018). Socioeconomic vulnerability to disaster risk: A case study of flood and drought impact in a rural Sri Lankan community. *Ecological Economics, 152*, 131–140.

DID (2009). Riparian Strategy: Sustaining Healthy Rivers and Communities. River Sand Mining Management Guideline. *Report prepared by the Department of Irrigation and Drainage, Ministry of Natural resources and Environment*, Malaysia.

Docker, B. B., & Hubble, T. C. T. (2008). Quantifying root-reinforcement of river bank soils by four Australian tree species. *Geomorphology, 100*(3–4), 401–418.

Dosskey, M. G., Vidon, P., Gurwick, N. P., Allan, K. J., Duval, T. P., & Lowrance, R. (2010). The role of riparian vegetation in protecting and improving chemical water quality in streams. *Journal of the American Water Resources Association, 46*(2), 261–277.

Du, S., Shi, P., Rompaey, A. V., & Wen, J. (2015). Quantifying the impact of impervious surface location on flood peak discharges in urban areas. *Natural Hazards, 76*, 1457–1471.

Duan, J. G. (2005). Analytical approach to calculate rate of bank erosion. *Journal of Hydraulic Engineering, 131*, 980–990.

Duro, G., Crosato, A, Kleinhans, M.G. & Uijttewaal, W.S.J. (2018) A low-cost technique to measure bank erosion processes along middle-size river reaches. Earth Surface Dynamics.

Dutta, S., & Sen, D. (2017). Application of SWAT model for predicting soil erosion and sediment yield. Sustainable. *Water Resources Management, 4*, 447–468.

Effendy, Z., Setiawan, M. A., & Mardiatno, D. (2019). Geospatial-interface water erosion prediction project (GeoWEPP) application for the planning of Bompon watershed conservation-prioritized area, Magelang, Central Java, Indonesia. *Proceeding of the International Conference on Environmental Resources Management in Global Region, 256*(012017).

Eiseltova, M. (Ed.). (2010). *Restoration of lakes, streams, floodplains and bogs in Europe: Principles and case studies*. Springer.

Entwistle, N. S., Heritage, G. L., Schofield, L. A., & Williamson, R. J. (2019). Recent changes to floodplain character and functionality in England. *Catena, 174*, 490–498.

Ergodan, E. H., Erpul, G., & Bayramin, I. (2007). Use of USLE/GIS methodology for predicting soil loss in a semiarid agricultural watershed. *Environmental Monitoring Assessment, 131*, 153–161.

Fierro, P., Bertrán, C., Tapia, J., Hauenstein, E., Peña-Cortés, F., Vergara, C., Cerna, C., & Vargas-Chacoff, L. (2017). Effects of local land-use on riparian vegetation, water quality, and the functional organization of macroinvertebrate assemblages. *Science of the Total Environment, 609*, 724–734.

Flanagan, D. C., Ascough, J. C., II, Nearing, M. A., & Laflen, J. M. (2001). The water erosion prediction project (WEPP) model. In R. S. Harmon & W. W. Doe (Eds.), *Landscape erosion and evolution modeling*. Springer.

Florsheim, J. L., Mount, J. F., & Chin, A. (2008). Bank erosion as a desirable attribute of rivers. *Bioscience, 58*(6), 519–529.

Gangwar, S. (2013). Flood vulnerability in India: A remote sensing and GIS approach for warning, mitigation and management. *International Journal of Environmental Science: Development and Monitoring, 4*(2), 77–79.

Gao, F., Wang, Y., & Yang, J. (2017). Assessing soil erosion using USLE model and the MODIS data in the Guangdong, China. *Procceding of the IOP Conference Series: Earth and Environmental Science, 74*(012007).

Gayen, A., Saha, S., & Pourghasemi, H. R. (2019). Soil erosion assessment using RUSLE model and its validation by FR probability model. *Geocarto International, 35*(15), 1750–1768.

Ghezelsofloo, A. A., & Hajibigloo, M. (2020). Application of flood Hazard potential zoning by using AHP algorithm. *Civil Engineering Research Journal, 9*(5), 150–159.

Ghosh, P. K., Bandyopadhyay, S., Jana, N. C., & Mukhopadhyay, R. (2016). Sand quarrying activities in an alluvial reach of Damodar River: Towards a geomorphic assessment. *International Journal of River Basin Management, 14*(4), 477–489.

Ghosh, S., & Guchhait, S. K. (2012). Soil loss estimation through USLE and MMF methods in the lateritic tracts of eastern plateau fringe of Rajmahal traps, India. *Ethiopian Journal of Environmental Studies and Management, 5*(4), 529–541.

Gianinetto, C., Aiello, M., Polinelli, F., Frassy, F., Rulli, M. C., Ravazzani, G., Bocchiola, D., Chiarelli, D. D., Soncini, A., & Vezolli, R. (2019). D-RUSLE: A dynamic model to estimate potential soil erosion with satellite time series in the Italian Alps. *European Journal of Remote Sensing, 52*(4), 34–53.

Grabowski, R. C., Surian, N., & Gurnell, A. M. (2014). Characterizing geomorphological change to support sustainable river restoration and management. *Wiley Interdisciplinary Reviews Water, 1* (5), 483–512.

Graviletea, M. D. (2017). Environmental impacts of sand exploration. Analysis of sand market. *Sustainability, 9*(7), 1118.

Gregory, K. J. (2006). The human role in changing river channels. *Geomorphology, 79*, 172–191.

Gregory, S. V., Swanson, F. J., McKee, W. A., & Cummins, K. W. (1991). An ecosystm perspective of riparian zones. *Bioscience, 41*(8), 540–551.

Guabi, I., Chaabani, A., Mammou, A. B., & Hamza, M. H. (2016). A GIS-based soil erosion prediction using the revised universal soil loss equation (RUSLE) (Lebna watershed, cap bon, Tunisia). *Natural Hazards, 86*, 219–239.

Gurnell, A. M., Corenblit, D., Garcia de Jalon, D., Gonzalez del Tango, M., Grabowski, R. C., O'Hare, M. T., & Szewczyk, M. (2016). A conceptual model of vegetation-hydrogeomorphology interactions with river corridors. *River Research and Applications, 32* (2), 142–163.

Hammani, S., Zouhri, L., Souissi, D., Souei, L., Zghibi, A., Marzougui, A., & Dlala, M. (2019). Application of the GIS based multi-criteria decision analysis and analytical hierarchy process (AHP) in the flood susceptibility mapping (Tunisia). *Arabian Journal of Geosciences, 12*, 653.

Han, F., Ren, L., Zhang, X., & Li, Z. (2016). The WEPP model application in a small watershed in the loess plateau. *PLoS One, 11*(3), e0148445.

Hasan, Z. A., Ghani, A. A. B., & Zakaria, N. A. (2007) Application of 2D modelling for Muda River using CCHE2D. In *Proceedings of the 2nd international conference on managing Rivers in the 21st century: Solutions towards Sustainable River basins. Malaysia* (pp 249–253).

Hasegawa, K. (1989). Universal bank erosion coefficient for meandering rivers. *Journal of Hydraulic Engineering 115*(6), 744–765. https://doi.org/10.1061/(ASCE)0733-9429(1989)115:6(744)

Hawes, E., & Smith, M. (2005). Riparian buffer zones: Functions and recommended widths. *Report published by the Yale School of Forestry and Environmental Studies*, New Haven.

Heeren, D. M., Mittelstet, A. R., Fox, G. A., Storm, D. E., Al-Madhhachi, A. T., Migdley, T. L., Stringer, A. F., Stunkel, K. B., & Tejral, R. D. (2012). Using rapid geomorphic assessments to assess streambank stability in Oklahoma Ozark streams. *Transactions of the American Society of Agricultural and Biological Engineers, 55*(3), 957–968.

Hoque, M. A., Tasfia, S., Ahmed, N., & Pradhan, B. (2019). Assessing spatial flood vulnerability at Kalapara Upazila in Bangladesh using an analytical hierarchy process. *Sensors, 19*, 1302.

Hughes, A. O. (2016). Riparian management and stream bank erosion in New Zealand. *New Zealand Journal of Marine and Freshwater Research, 50*(2), 277–290.

Hung, L. S. (2017). Married couples' decision-making about household natural hazard preparedness: A case study of hurricane hazards in Sarasota County, Florida. *Natural Hazards, 87*, 1057–1081.

Hussain, A. (2017). Flood modelling by HEC-RAS. *International Journal of Engineering Trends and Technology, 50*(1), 1–7.

Hussain, F., Nabi, G., Wu, R., Hussain, B., & Abbas, T. (2019). Parameter evaluation for soil erosion estimation on small watersheds using SWAT model. *International Journal of Agricultural and Biological Engineering, 12*(1), 96–108.

Igwe, P. U., Onuigbo, A. A., Chinedu, O. C., Ezeaku, I. I., & Mouneke, M. M. (2017). Soil erosion: A review of models and applications. *International Journal of Advanced Engineering Research and Science, 4*(12), 138–150.

Islam, M.S. (2008) River bank erosion and sustainable protection strategies. In *Proceedings of the fourth international conference on scour and erosion* (pp. 316–323).

Jahan, A., Edwards, K. L., & Bahraminasab, M. (2016). Multi criteria decision making for materials selection. In A. Jahan, K. L. Edwards, & M. Bahraminasab (Eds.), *Multi-criteria decision analysis for supporting the selection of engineering materials in product design* (2nd ed., pp. 63–80). Elsevier.

Jazouli, A. E., Barakat, A., Ghafiri, A., Moutaki, S. E., & Ettaqy, A. (2017). Soil erosion modeled with USLE, GIS, and remote sensing: A case study of Ikkour watershed in middle atlas (Morocco). *Geoscience Letters, 4*(25).

Jia, Y., & Hunt, S. (2016). Development of CCHE2D embankment break model. *American Society of Agricultural and Biological Engineers, 59*(4), 805–814.

Junker, B., & Buchecker, M. (2008). Aesthetic preferences versus ecological objectives in river restoration. *Landscape and Urban Planning, 85*, 141–154.

Karamage, F., Zhang, C., Kayiranga, A., Shao, H., Fang, X., Ndayisaba, F., Nahayo, L., Mupenzi, C., & Tian, G. (2016). USLE-based assessment of soil erosion by water in the Nyabarongo River catchment, Rwanda. *International Journal of Environmental Research and Public Health, 13*(8), 835.

Kaur, H., Gupta, S., Parkash, S., Thapa, R., & Mandal, R. (2017). Geospatial modelling of flood susceptibility pattern in a subtropical area of West Bengal, India. *Environmental Earth Sciences, 76*, 339.

Keesstra, S. D., Kondrlova, E., Czajka, A., Seeger, M., & Maroulis, J. (2012). Assessing riparian zone impacts on water and sediment movement: A new approach. *Netherlands Journal of Geosciences, 91*(1–2), 245–255.

Keeton, W. S., Copeland, E. S., Sullivan, M. P., & Watzin, M. C. (2017). Riparian forest structure and stream geomorphic condition: Implications for flood resilience. *Canadian Journal of Forest Research, 47*(4), 476–487.

Kelkar, N. (2016). Digging our river's graves? A summary analysis of the ecological impacts of the National Waterways Bill (2015). *South Asia Network on Dams, Rivers and People (SANDRP), 14*(1–2).

Kheiralla, K. M., & Siddeg, A. S. (2015). Control over river bank erosion: A case study of Ganetti Station, northern states, Sudan. *Journal of Earth Science and Climatic Change, 6*(7), 287.

Kinnell, P. I. A., & Risse, L. M. (1998). USLE-M: Empirical modeling rainfall erosion through runoff and sediment concentration. *Soil Science Society of America Journal, 62*, 1667–1672.

Kirby, R. H., Reams, M. A., Lam, N. S. N., Zou, L., Dekker, G. G. J., & Fundter, D. Q. P. (2019). Assessing social vulnerability to flood hazards in the Dutch Province of Zeeland. *International Journal of Disaster Risk Science, 10*, 233–243.

Kitahara, H., Okura, Y., Sammori, T., & Kawanami, A. (2000). Application of universal soil loss equation (USLE) in mountainous forests in Japan. *Journal of Forest Research, 5*, 231–236.

Klavon, K., Fox, G., Guertault, L., Langendoen, E., Enlow, H., Miller, R., & Khanal, A. (2016). Evaluating a process-based model for use in streambank stabilization: Insights on the Bank stability and toe Erosion model (BSTEM). *Earth Surface Processes and Landforms, 42*(1), 191–213.

Knisel, W. G. (1980). *CREAMS: A field-scale model for chemicals, runoff, and erosion from agricultural management systems*. United States Department of Agriculture, Conservation Research Report No. 26.

Koehnken, L., & Rintoul, M. (2018). *Impacts of sand mining on ecosystem structure*. Process and Biodiversity in Rivers.

Koehnken, L., Rintoul, M. S., Goichot, M., Tickner, D., Loftus, A. C., & Acreman, M. C. (2020). Impacts of riverine sand mining on freshwater ecosystems: A review of the scientific evidence and guidance for future research. *River Research and Applications, 36*, 362–370.

Kondolf, G. M. (1994). Geomorphic and environmental effects of instream gravel mining. *Landscape and Urban Planning, 28*(2–3), 225–243.

Kondolf, G. M. (1997). Hungry water: Effects of dams and gravel mining on river channels. *Environmental Management, 21*(4), 533–551.

Kondolf, G. M. (1998). Large scale extraction of alluvial deposits from rivers in California: Geomorphic effects and regulatory strategies. In P. C. Klingeman, R. L. Beschta, P. D. Komar, & J. B. Bradley (Eds.), *Gravel bed Rivers in the environment* (pp. 455–470). Water Resources Publication.

Kondolf, G. M., Piegay, H., & Landon, N. (2007). Changes in the riparian zone of the lower Eygues River, France, since 1830. *Landscape Ecology, 22*, 367–384.

Kron, W. (2005). Flood risk = Hazard. Values. Vulnerability. *International Water Resources Association, 30*(1), 58–68.

Kumar, R., Ambasht, R.S. & Srivastava, M.K. (1992) Conservation efficiency of five common riparian weeds in movement of soil, water and phosphorous. Journal of Applied Ecology. 29. 737-744.

Kwan, H., & Swanson, S. (2014). Prediction for annual streambank erosion for Sequoia National Forest, California. *Journal of the American Water Resources Association, 50*(6), 1439–1447.

Lai, Y. G. (2017). Modeling stream bank erosion: Practical stream results and future needs. *Water, 9*, 950.

Landi, A., Barzegar, A. R., Sayadi, J., & Khademalrasoul, A. (2011). Assessment of soil loss using WEPP model and geographical information system. *Journal of Spatial Hydrology, 11*(1), 40–51.

Langendoen, E. J. (2011). Application of the CONCEPTS channel evolution model in stream restoration strategies. *Geophysical Monograph Series, 194*, 487–502.

Lavoie, B., & Mahdi, T. F. (2017). Comparison of two-dimensional flood propagation models: SRH-2D and Hydro_AS-2D. *Natural Hazards, 86*, 1207–1222.

Lawal, D. U., Matori, A. N., Hashim, A. M., Yusof, K. W., & Chandio, I. A. (2012). *Detecting flood susceptible areas using GIS-based analytic hierarchy process*. Proceedings of the International Conference on future Environment and Energy.

Lees, A. C., & Peres, C. A. (2008). Conservation value of remnant riparian forest corridors of varying quality for Amazonian birds and mammals. *Conservation Biology., 22*(2), 439–449.

Li, H.C., Wu, T., Wei, H.P., Shih, H.J. & Chao, Y.C. (2016) Basinwide disaster loss assessments under extreme climatic scenarios: A case study of the Kaoping River basin. Natural hazards. 86. P.1039-1058.

Lin, Q., & Wang, X. (2006). Soil erosion prediction using RUSLE with GIS: A case study in upper Chaobai River basin in China. *Proceeding of the IEEE International Symposium on Geoscience and remote Sensing*. https://doi.org/10.1109/IGARSS.2006.280

Liu, Y., Li, D., Chen, W., Lin, B., Seeboonruang, U., & Tsai, F. (2018). Soil erosion modeling and comparison using slope units and Grid cells in Shihmen reservoir watershed in northern Taiwan, *10*(10), 1387.

Lixin, Y., Ke, C., Xiaoying, C., Yueling, S., Xiaoqing, C., & Ye, H. (2017). Analysis of social vulnerability of residential community to hazards in Tianjin, China. *Natural Hazards, 87*, 1223–1243.

López-Baucells, A., Casanova, L., Puig-Montserrat, X., Espinal, A., Páramo, F., & Flaquer, C. (2017). Evaluating the use of Myotis daubentonii as an ecological indicator in Mediterranean riparian habitats. *Ecological Indicators 74*, 19–27. https://doi.org/10.1016/j.ecolind.2016.11.012

Lopez-Rodriguez, A., Escribano-Bombin, R., Hernandez-Jimenez, V., & Bell, S. (2018). Perceptions of ecological and aesthetic quality by natural resource professionals and local people. A qualitative exploration in a mountainous landscape (La Rioja, Spain). *Landscape Research, 44*, 241–255.

Lovric, N., & Tosic, R. (2016). Assessment of bank erosion, accretion and channel shifting using remote sensing and GIS: Case study- lower course of the Bosna River. *Quaestiones Geographicae, 35*(1), 81–92.

Madyise, D. (2013). *Case studies of environmental impacts of sand mining and gravel extraction for urban development in Gabarone*. Master of Science Thesis, Department of Environmental Sciences, University of South Australia, Australia.

Maffra, C., De Moraes, M. T., Sousa, R., Sutili, F. J., Pinheiro, R. J. B., & Soares, J. M. D. (2017). Evaluation methods of plants influence and contribution on slope stability. *Scientia Agraria Paranaensis, 18*(4), 129–143.

Mahadevan, P. (2019). Sand mafias in India: Disorganized crime in a growing economy. *Report of the global initiative against transnational organized crime*.

Majhi, A., Shaw, R., Mallick, K., & Patel, P. P. (2021). Towards improved USLE-based soil erosion modelling in India: A review of prevalent pitfalls and implementation of exemplar methods. *Earth-Science Reviews 221*, 103786. https://doi.org/10.1016/j.earscirev.2021.103786

Malik, S., Pal, S. C., Chowdhuri, I., Chakraborty, R., Roy, P., & Das, B. (2020). Prediction of highly flood prone areas by GIS based heuristic and statistical model in a monsoon dominated region of Bengal Basin. *Remote Sensing Applications: Society and Environment, 19*, 100,343.

Mandal, S. (2017). Assessing the instability and shifting character of the river Bank ganga in Manikchak Diara of Malda District, West Bengal, using Bank Erosion Hazard Index (BEHI), RS & GIS. *European Journal of Geography, 8*(4), 6–25.

Marques, V. S., Ceddia, M. B., Antunes, M. A. H., Carvalho, D. F., Anache, J. A. A., Rodrigues, D. B. B., & Oliveria, P. T. S. (2019). USLE K-factor method selection for a tropical catchment. *Sustainability, 11*(7), 1840.

McCool, D. K., Foster, G. R., Renard, K. G., Yoder, D. C., & Weesies, G. A. (1995). *The revised universal soil loss equation*. Department of Defense/interagency workshop on technologies to address soil erosion on Department of Defense Lands, San Antonio. Report No. 1132.

McMillan, S. K., & Noe, G. B. (2017). Increasing floodplain connectivity through urban stream restoration increases nutrient and sediment retention. *Ecological Engineering, 108*(A), 284–295.

Melaku, N. D., Renschler, C. S., Holzmann, H., Strohmeier, S., Bayu, W., Zucca, C., Ziadat, F., & Klik, A. (2018). Prediction of soil and water conservation structure impacts on runoff and erosion processes using SWAT model in the northern Ethiopian highlands. *Journal of Soils and Sediments, 18*, 1743–1755.

Miserendino, M. L., Casaux, R., Archangelsky, M., Di Prinzio, C. Y., Brand, C., & Kutschker, A. M. (2011). Assessing land-use effects on water quality, in-stream habitat, riparian ecosystems and biodiversity in Patagonian northwest streams. *Science of Total Environment, 409*(3), 612–624.

Mondal, S., & Patel, P. P. (2020). Implementing vetiver grass-based riverbank protection programmes in rural West Bengal, India. *Natural Hazards, 103*, 1051–1076.

Mondal, S., & Patel, P. P. (2018). Examining the utility of river restoration approaches for flood mitigation and channel stability enhancement: A recent review. *Environmental Earth Sciences, 77*, 195.

Mondal, S., Sarkar, A., & Patel, P. P. (2016). Causes of drainage congestion in the Moyna block, Purba Medinipur, West Bengal. In D. K. Mondol (Ed.), *Application of geospatial technology for sustainable development* (pp. 1–9). University of North Bengal, India, North Bengal University Press.

Moshabi, M., Benabdallah, S., & Boussema, M. R. (2012). Assessment of soil erosion risk using SWAT model. *Arabian Journal of Geosciences, 6*, 4011–4019.

Mukerji, N. (2016). Illegal sand mining- its effect on ecology and environment. *Lex Terrain, 23*, 7–12.

Mulyono, A., Subardja, A., Ekasari, I., Lailati, M., Sudirja, R., & Nirgrum, W. (2018). The hydrodynamics of vegetation for slope stabilization. *IOP Conference Series: Earth and Environmental Science, 118*, 012038.

Munyai, R. B., Musyoki, A., & Nethengwe, N. S. (2019). An assessment of flood vulnerability and adaptation: A case study of Hamutsha Muungamunwe village, Makhado municipality. *Journal of Disaster Risk Studies, 11*(2).

Naiman, R. J., & Decamps, H. (1997). The ecology of interfaces: Riparian zones. *Annual Reviews, 28*, 621–658.

Naiman, R. J., Decamps, H., & McClain, M. E. (2013). Riparian landscapes. In S. A. Levin (Ed.), *Encyclopaedia of biodiversity* (pp. 461–468). Elsevier.

Nakamura, F. (1995). Structure and function of riparian zone and implications for Japanese River management. *Transactions of Japanese Geomorphological Union, 16*(3), 237–256.

Nandakumaran, P., Shyam, T. S. A., Chandran, M., Rani, V. R., Srinath, G., & Chand, A. D. A. (2014). Impact of river sand mining on groundwater regime in Kerala- an overview. *Proceedings of the Mineralia conference, Trivandrum, Kerala.*

Neal, J., Dunne, T., Sampson, C., Smith, A., & Bates, P. (2018). Optimisation of the two-dimensional hydraulic model LISSFLOOD-FP for CPU architecture. *Environmental Modelling and Software, 107*, 148–157.

New, T., & Xie, Z. (2008). Impacts of large dams on riparian vegetation: Applying global experience to the case of China's three gorges dam. *Biodiversity Conservation, 17*, 3149–3163.

Nikhil, K., Kazi, S., Mani, K., & Anil, K. (2019) Short to medium term morphological prediction of river erosion and planform change: Advance knowledge for decision support system in river monitoring and river management. Journal of Indian Society of Remote Sensing.

Nilsson, C., & Berggren, K. (2000). Alterations of riparian ecosystems caused by river regulation. *Bioscience, 50*(9), 783–792.

Nourani, V. & Andaryani, S. (2019) Flood susceptibility mapping in densely populated urban areas using MCDM and fuzzy techniques. *Proceedings of the 5th international conference on modeling and simulation in civil engineering, India.*

O'Brien, G. R., Wheaton, J., Fryirs, K., McHugh, P., Bouwes, N., Brierley, G., & Jordan, C. (2017). A geomorphic assessment to inform strategic stream restoration planning in the middle fork John Day watershed, Oregon, USA. *Journal of Maps, 13*(2), 369–381.

O'Donnell, J., Fryirs, K., & Leishman, M. R. (2015). Can the regeneration of vegetation from riparian seed banks support biogeomorphic succession and the geomorphic recovery of degraded river channels? *River Research and Applications, 31*(7), 834–846.

Ogras, S., & Onen, F. (2020). Flood analysis with HEC-RAS: A case study of Tigris River. *Advances in Civil Engineering*. https://doi.org/10.1155/2020/6131982

Olokeogun, O. S., & Kumar, M. (2020). An indicator based approach for assessing the vulnerability of riparian ecosystem under the influence of urbanization in the Indian Himalayan city, Dehradun. *Ecological Indicators, 19*, 106,796.

Olokeogun, O. S., Ayanlade, A., & Popoola, O. O. (2020). Assessment of riparian zone dynamics and its flood-related implications in Eleyele area of Ibadan, Nigeria. *Environmental Systems Research, 9*(6).

Ongdas, N., Akiyanova, F., Karakulov, Y., Muratbayeva, A., & Zinabdin, N. (2020). Application of HEC-RAS (2D) for flood hazard maps generation for Yesil (Ishim) river in Kazakhstan. *Water, 12*, 2672.

Oude Essink, G. H. P. (2001). Improving fresh groundwater supply—Problems and solutions. *Ocean & Coastal Management, 44*(5–6), 429–449.

Padmalal, D., & Maya, K. (2014). Impacts of river sand mining. In D. Padmalal & K. Maya (Eds.), *Sand mining: Environmental impacts and selected case studies* (pp. 31–56). Springer.

Padmalal, D., Maya, K., Sreebha, S., & Sreeja, R. (2008). Environmental effects of river sand mining: a case from the river catchments of Vembanad Lake, Southwest coast of India. *Environmental Geology, 54*, 879–889.

Parkyn, S. (2004). *Review of Riparian Buffer zone effectiveness. Technical paper No: 2004/05*. Ministry of Agriculture and Forestry, Wellington, New Zealand.

Patel, P. P., & Dasgupta, R. (2009). Flood induced land use change in the Dulung River valley, West Bengal. In: R. B. Singh, S. D. D. Roy, H. D. D. K. Samuel, V. D. Singh, & G. D. Biji (ed.) Geoinformatics for monitoring and modelling land-use, bio-diversity and climate change – contribution towards international year of planet earth (Vol. 1, pp. 103–123). NMCC Publication, Marthandam.

Patel, P. P., Dasgupta, R., & Mondal, S. (2020b). Using ground-based photogrammetry for fine-scale gully morphology studies: Some examples. In P. K. Shit, H. R. Pourghasemi, & G. Sankar (Eds.), *Gully erosion studies from India and surrounding regions* (pp. 207–220). Springer International Publishing.

Patel, P. P., Mondal, S., & Ghosh, K. G. (2020a). Some respite for India's dirtiest river? Examining the Yamuna's water quality at Delhi during the COVID-19 lockdown period. *Science of the Total Environment, 744*, 140,851.

Patel, P. P., Mondal, S., & Prasad, R. (2020c). Modifications of the geomorphic diversity by anthropogenic interventions in the Silabati river basin. In: B. C. Das, S. Ghosh, A. Islam, &

S. Roy (eds.) *Anthropogeomorphology of Bhagirathi-Hooghly river system in India* (pp. 469–490). Routledge.

Patel, P. P., & Mondal, S. (2019). Terrain – Landuse relation in Garbeta-I block, Paschim Medinipur District, West Bengal. In: S. Mukherjee (ed.) *Importance and Utilities of GIS* (pp. 82–101). Avenel Press, Burdwan

Pham, T. G., Degener, J., & Kappas, M. (2018). Integrated universal soil loss equation (USLE) and geographic information system (GIS) for soil Erosion estimation in a Sap Basin, Central Vietnam. *International Soil and Water Conservation Research, 6*(2), 99–110.

Pitchaiah, P. S. (2017). Impacts of sand mining on environment- a review. *International Journal of Geo informatics and Geological Science, 4*(1), 1–5.

Prabhakar, R., Kumari, A., & Neetu. & Sinha, R.K. (2019). Impact of sand mining on zooplankton of river ganga in and around Patna, Bihar, India. *Environment and Ecology, 37*(4A), 1301–1308.

Proske, D. (2008). *Catalogue of risks. Natural, technical, social and health risks*. Springer.

Rakib, M. R., Islam, M. N., & Islam, M. N. (2017). Flood vulnerability mapping to riverine floods: A study on the old Brahmaputra River. *Current Research in Geosciences, 7*(2), 47–58.

Ramirez, A., Engman, A., Rosas, K. G., Perez-Reyes, O., & Martino-Cardona, D. M. (2012). Urban impacts on tropical island streams: Some key aspects influencing ecosystem response. *Urban Ecosystem, 15*(2), 315–325.

Ramirez-Avila, J. J., Langendoen, E. J., McAnally, W. H., Martin, J. L., & Ortega-Achury, S.,L. (2010) Assessment and estimation of streambank Erosion rates in the Southeastern Plains ecoregion of Mississippi. *Proceedings of the 2nd joint interagency conference, Las Vegas.*

Rayner, T. S. (2001). *Importance of large woody debris for fish-habitat associations on sand-bed forest stream of south-eastern Australia*. Bachelor of environmental science thesis. University of New South Wales, Australia.

Rimba, A. B., Setiawati, M. D., Sambah, A. B., & Miura, F. (2017). Physical flood vulnerability mapping applying geospatial techniques in Okazaki city, Aichi prefecture, Japan. *Urban Science, 1*, 7.

Rosgen, D. L. (1996). *Applied river morphology*. Wildland Hydrology.

Rosgen, D. L. (2001). A practical method of computing stream Bank Erosion rate. *Proceeding of the Seventh Federal Interagency Sedimentation Conference, 2*(2), 9–18.

Roy, P. (2018). Application of USLE in a GIS environment to estimate soil erosion in the Irga watershed, Jharkhand, India. *Physical Geography, 40*(4), 361–383.

Rufat, S., Tate, E., Burton, C. G., & Maroof, A. F. (2015). Social vulnerability to floods: Review of case studies and implications for measurement. *International Journal of Disaster Risk Reduction, 14*(4), 470–486.

Saaty, T. L. (2008). Decision making with the analytic hierarchy process. *International Journal of Services Sciences, 1*(1), 83–98.

Sadeghi, S. H. R., Gholami, L., Darvishan, A. K., & Saeidi, P. (2014). A review of the application of the MUSLE model worldwide. *Hydrological Sciences Journal, 59*(2), 365–375.

Saha, D., Das, D., Dasgupta, R., & Patel, P. P. (2020). Application of ecological and aesthetic parameters for riparian quality assessment of a small tropical river in eastern India. *Ecological Indicators, 117*, 106627.

Sahana, M., & Patel, P. P. (2019). A comparison of frequency ratio and fuzzy logic models for flood susceptibility assessment of the lower Kosi River basin in India. *Environmental Earth Sciences, 78*, 289.

Sahana, M., Rihan, Md., Deb, S., Patel, P.P., Ahmad, W.S., & Imdad, K. (2020). Detecting the facets of anthropogenic interventions on the palaeochannels of Saraswati and Jamuna. In: B. C. Das, S. Ghosh, A. Islam, & S. Roy (eds.) *Anthropogeomorphology of Bhagirathi-Hooghly river system in India* (pp. 469–490). Routledge.

Salunkhe, S. S., Rao, S. S., Prabu, I., Venkataraman, V. R., Krishna Murthy, Y. V. N., Sadolikar, C., & Deshpande, S. (2018). Flood inundation hazard modelling using CCHE2D hydrodynamic model and geospatial data for embankment breaching scenario of Brahmaputra River in Assam. *Journal of the Indian Society of Remote Sensing, 46*, 915–925.

Santos, P. P., Pereira, S., Zezere, J. L., Reis, E., Garcia, R. A. C., Oliveira, S. C., & Santos, M. (2018). Flood susceptibility assessment based on Analytical Hierarchy Process: Application in mainland Portugal. *Geophysical Research Abstracts, 20*, EGU2018-4940.

Sanyal, J., Carbonneau, P., & Densmore, A. L. (2014). Low-cost inundation modelling at the reach scale with sparse data in the lower Damodar River basin, India. *Hydrological Sciences Journal, 59*(12), 2086–2102.

Sarkar, A., & Patel, P. P. (2016). Land use – Terrain correlations in the piedmont tract of Eastern India: A case study of the Dulung river basin. In: A. Santra, & S. Mitra (eds.) *Handbook of research on remote sensing applications in earth and environmental studies* (pp. 147–193). IGI Global, USA. https://doi.org/10.4018/978-1-5225-1814-3.ch008

Sass, C. K., & Keane, T. D. (2012). Application of Rosgen's BANCS model for NE Kansas and the development of predictive stream-bank erosion curves. *Journal of American Water Resources Association, 48*(4), 774–787.

Scawthorn, C., Blais, N., Seligson, H., Tate, E., Mifflin, E., Thomas, W., Murphy, J., & Jones, C. (2006a). HAZUS-MH flood loss estimation methodology. I: Overview and flood Hazard characterization. *Natural Hazards Review, 7*(2), 60–71.

Scawthorn, C., Flores, P., Blais, N., Seligson, H., Tate, E., Chang, S., Mifflin, E., Thomas, W., Murphy, J., Jones, C., & Lawrence, M. (2006b). HAZUS-MF flood loss estimation methodology. II: Damage and loss assessment. *Natural Hazards Review, 7*(2), 72–81.

Schlosser, I. J., & Karr, J. R. (1981). Water quality in agricultural watersheds: Impact of riparian vegetation during base flow. *Journal of the American Water Resources Association, 17*(2), 233–240.

Schwarz, B., Pestre, G., Tellman, B., Sullivan, J., Kuhn, C., Mahtta, R., Pandey, B., & Hammett, L. (2018). Mapping floods and assessing flood vulnerability for disaster decision-making: A case study remote sensing application in Senegal. In P. P. Mathieu & C. Aubrecht (Eds.), *Earth observation open science and innovation*. Springer.

Scott, L. (2016). *Assessing statistical instream sediment models as an approach to prioritize riparian restoration in two Virginia river basins*. Master's Thesis, Plymouth State University.

ShahiriParsa, A., Noori, M., Heydari, M. & Rashidi, M. (2016) Floodplain zoning simulation by using HEC-RAS and CCHE2D models in the Sungai Maka River. Air, Soil and Water Research. 9. p. 55–62.

Sheeba, S. (2009). Biotic environment and sand mining—A case study from Ithikkara River, southwest coast of India. *Journal of Industrial Pollution Control, 25*(2), 133–138.

Shen, Z. Y., Gong, Y. W., Li, Y. H., Hong, Q., Xu, L., & Liu, R. M. (2009). A comparison of WEPP and SWAT for modeling soil erosion of the Zhangjiachong watershed in the three gorges reservoir area. *Agricultural Water Management, 96*, 1435–1442.

Shit, P. K., & Maiti, R. (2012). Effects of plant roots on soil anti-scouribility of topsoil during concentrated flow. *International Journal of Engineering Research and Technology, 1*(4), 1–8.

Sholtes, J., & Giordanengo, J. (2014). *Stream Bank and bed stability assessment protocol* (9p). AloTerra Restoration Services, LLC.

Shustikova, I., Domeneghetti, A., Neal, J. C., Bates, P., & Castellarin, A. (2019). Comparing 2D capabilities of HEC-RAS and LISSFLOOD-FP on complex topography. *Hydrological Sciences Journal, 64*(14), 1769–1782.

Simanton, J. R., Osborn, H. B., & Renard, K. G. (1980). Application of the USLE to southwestern rangelands. *Hydrology and Water Resources in Arizona and the Southwest, 10*, 213–220.

Simon, A., Curini, A., Darby, S., & Langendoen, E. J. (1999). Stream-bank mechanics and the role of bank and near-bank processes in incised channels. In S. Darby & A. Simon (Eds.), *Incised river channels* (pp. 123–152). Wiley.

Simpson, A., Turner, I., Brantley, E., & Helms, B. (2014). Bank erosion hazard index as an indicator of near-bank aquatic habitat and community structure in the southeastern Piedmont stream. *Ecological Indicators, 43*, 19–28.

Singh, S. R., Eghdami, M. R., & Singh, S. (2014). The concept of social vulnerability: A review from disasters perspectives. *International Journal of Interdisciplinary and Multidisciplinary Studies, 1*(6), 71–82.

Sinha, M. (2016). Dwarekeshwar River basin and anthropogenic intervention as sand mines. *International Journal of Research in Geography, 2*(2), 48–56.

Smith, D. J., & Wynn-Thompson, T. (2018). Do roots bind soil: Comparing the physical and biological role of roots in fluvial streambank erosion resistance. *Proceedings of the American Geophysical Union, Fall Meeting, abstract #EP41C-2689.*

Sobral, A. C., Peixoto, A. S. P., Nascimento, V. F., Rodgers, J., & Silva, A. M. (2014). Natural and anthropogenic influence on soil erosion in a rural watershed in the Brazilian south eastern region. *Regional Environmental Change, 15*(4), 709–720.

Soetanto, R., Mullins, A., & Achour, N. (2017). The perception for social responsibility for community resilience to flooding: The impact of past experience, age, gender and ethnicity. *Natural Hazards, 86*, 1105–1126.

Sozer, B. (2020). *Flood susceptibility mapping for Ankara using modified analytical hierarchy process (M_AHP).* Master's Thesis. Hacettepe University, Turkey.

Sozer, B., Kocaman, S., Nefeslioglu, H. A., Firat, O., & Gokceoglu, C. (2018). Preliminary investigations on flood susceptibility mapping in Ankara (Turkey) using modified analytical hierarchy process (M-AHP). *Proceedings of the international archives of the photogrammetry, remote sensing and spatial information sciences, Dehradun* (Vol. XLII-5, pp. 361–365).

Sreebha, S., & Padmalal, D. (2011). Environmental impact assessment of sand mining from the small catchment Rivers in the southwestern coast of India: A case study. *Environmental Management, 47*, 130–140.

NRCS USDA (2009). Streambank Erosion Factors, Mechanisms and Causes. (n.d.). *United States Department of Agriculture Natural Resource Conservation Service Companion Document. Report No. 580-4.*

Susanti, Y., Syafrudin, S., & Helmi, M. (2019). Soil erosion modelling at watershed level in Indonesia: a Review. *Proceedings from the fourth international conference on energy, environment, epidemiology and information system (ICENIS).* 125(01008).

Talha, S., Maanan, M., Atika, H., & Rhinane, H. (2019). Prediction of flash flood susceptibility using fuzzy analytical hierarchy process (FAHP) and GIS: A case study of Guelmim region in Southwestern of Morocco. *Proceedings of the international archives of the photogrammetry, remote sensing and spatial information sciences, Philippines* (Vol. XLII-4/W19, pp. 407–414).

Tascon-Gonzalez, L., Ferrer-Julia, M., Ruiz, M., & Garcia-Melendez, E. (2020). Social vulnerability assessment for flood risk analysis. *Water, 12*, 558.

Tehrany, M. S., Lee, M. J., Pradhan, B., Jebur, M. N., & Lee, S. (2014). Flood susceptibility mapping using integrated bivariate and multivariate statistical models. *Environmental Earth Sciences, 72*, 4001–4015.

Teng, J., Jakeman, A. J., Vaze, J., Croke, B. F. W., Dutta, D., & Kim, S. (2017). Flood inundation modelling: A review of methods, recent advances and uncertainty analysis. *Environmental Modelling and Software, 90*, 201–216.

Thorne, C. R. (1981). Field measurements of rates of bank erosion and bank material strength. *Proceedings of the florence symposium* (pp. 503–512).

Thorne, C. R. (1992) .Field assessment techniques for bank erosion modeling. Prepared for US Army, European Research Office, London, England.

Tibebe, D., & Bewket, W. (2011). Surface Runoff and Soil Erosion Estimation using the SWAT model in the Keleta Watershed, Ethiopia. *Land Degradation & Development, 22*, 551–564.

Tockner, K., & Stanford, J. A. (2002). Riverine flood plains: Present state and future trends. *Environmental Conservation, 29*(3), 308–330.

Torok, I. (2018). Qualitative assessment of social vulnerability to flood hazards in Romania. *Sustainability, 10*, 3780.

Triantaphyllou, E., & Mann, S. H. (1995). Using the analytical hierarchy process for decision making in engineering applications: Some challenges. *International Journal of Industrial Engineering: Applications and Practice, 2*(1), 35–44.

Triantaphyllou, E., Shu, B., Sanchez, S. N., & Ray, T. (1998). Multi-criteria decision making: An operations research approach. In J. G. Webster (Ed.), *Encyclopaedia of electrical and electronics engineering* (pp. 175–186). John Wiley & Sons.

Trimble, S. W. (2004). Effects of riparian vegetation on stream channel stability and sediment budgets. In S. J. Bennett & A. Simon (Eds.), *Riparian vegetation and fluvial geomorphology, water science and application* (Vol. 8, pp. 153–170). AGU Books Board.

UNEP. (2019). *Sand and Sustainability: Finding new solutions for environmental governance of global sand resources*. GRID-Geneva, United Nations Environment Programme, Geneva, Switzerland.

UN/ISDR. (2004). *Guidelines for reducing flood losses*. United Nations, Secretariat of the International Strategy for Disaster Reduction, Geneva.

Valizadeh, N., Mirzaei, M., Allawi, M. F., Afan, H. A., Mohd, N. S., Hussain, A., & El-Shafie, A. (2017). Artificial Intelligence and geo-statistical models for streamflow forecasting in ungauged stations: state of the art. *Natural Hazards, 86*, 1377–1392.

Verry, E. S., Hornbeck, J. W., & Dolloff, C. A. (2000). *Riparian management in forests to the continental Eastern United States* (pp. 125–138). Lewis Publishers.

Vollmer, D., Prescott, M. F., Padawangi, R., Girot, C., & Gret-Regamey, A. (2015). Understanding the value of urban riparian corridors: considerations in planning for cultural services along an Indonesian river. *Landscape and Urban Planning, 138*, 144–154.

Wakode, H. B., Baier, K., Jha, R., & Azzam, R. (2018). Imapct of urbanization on groundwater recharge and urban water balance for the city of Hyderabad, India. *International Soil and Water Conservation Research, 6*(1), 51–62.

Watson, K. B., Ricketts, T., Galford, G., Polasky, S., & O'Neil-Dunne, J. (2016). Quantifying flood mitigation services: the economic value of Otter Creek wetlands and floodplain to Middlebury, VT. *Ecological Economics, 130*, 16–24.

Watson, A. J., & Basher, L. R. (2006). Stream bank erosion: a review of processes of bank failure, measurement and assessment techniques and modelling approaches. *Motueka Integrated Catchment Management Programme Report Series. report No. 2005-2996/01.*

Wiejaczka, L., Tamang, L., Pirog, D., & Prokop, P. (2018). Socioenvironmental issues of river bed material extraction in the Himalayan Piedmont (India). *Environmental Earth Sciences, 77*, 718.

Wijayanti, P., Zhu, X., Hellegers, P., Budiyono, Y., & Ierland, E. C. (2017). Estimation of river flood damages in Jakarta, Indonesia. *Natural Hazards, 86*, 1059–1079.

Wischmeier, W. H., & Smith, D. D. (1978). *Predicting rainfall erosion losses—A guide to conservation planning*. U.S. Department of Agriculture. Agriculture Handbook No. 537.

Wolman, M. G. (1959). Factors influencing erosion of cohesive river banks. *American Journal of Science, 257*(3), 204–216.

Wolter, C., Buijse, A. D., & Parasiewicz, P. (2016). Temporal and spatial patterns of fish response to hydromorphological processes. *River Research and Applications, 32*(2), 190–201.

Ying, X., & Wang, S. S. Y. (2005). Modeling flood inundation due to dam and Levee Breach. *Proceedings of the US-China workshop on advanced computational modeling on hydroscience and engineering*. Oxford, Mississippi, USA.

Zhang, Y., Degroote, J., & Wolter, C. (2009). Integration of Modified Universal Soil Loss Equation (MUSLE) in a GIS framework to assess soil erosion risk. *Land Degradation & Development., 20*(1), 84–91.

Zhang, Y., Zhao, Q., Cao, Z., & Ding, S. (2019). Inhibiting effects of vegetation on the characteristics of runoff and sediment yield on riparian slope along the lower Yellow River. *Sustainability, 11*, 3685.

Zhao, J., Wang, R., Luo, P., Xing, L., & Sun, T. (2016). Visual ecology: exploring the relationships between ecological quality and aesthetic preference. *Landscape and Ecological Engineering, 13*, 107–118.

Zheng, N., Takara, K., Yamashiki, Y., & Tachikawa, Y. (2009). *Large scale quantitative vulnerability analysis for regional flood hazard*. Annuals of Disaster Prevention and Research Institute, Kyoto University. No. 52B, Japan.

Chapter 17
Village Level Landslide Probability Analysis Based on Weighted Sum Method of Multi-Criteria Decision-Making Process of Darjeeling Himalaya, West Bengal, India

Santanu Samanta, Jyotibrata Chakraborty, and Subrata B. Dutta

Abstract The Himalayan mountains are highly prone to landslides, and every year a multitude of slide events take place along its entire ranges. Darjeeling Himalayan region of West Bengal is not an exception and also highly prone to landslides. In the Darjeeling Himalayan extents, landslides are the major threat to the lives and properties of the people and also hamper the connectivity and commerce of the region aggravating lopsided development and poor economic condition. Considering the magnitude of landslide frequency and its extremities, this study aims to develop a landslide probability index (LPI) for the projection of landslide probabilities at the village level. The proposed methodology of the LPI is based on the weighted sum method of multi-criteria decision-making processes that involve eight explicitly localised proxy/indicator quantitative analysis, normalisation and weight assignment. The result of the derived LPI is divided into 4 distinct classes, viz. (a) very high landslide probability class comprising 45 villages, (b) high landslide probability class consists of 138 villages, (c) moderate landslide probability class includes 109 villages and (d) 56 villages are in low landslide probability class. It is evident from the study that the nature of landslide phenomenon is resulting from the interaction of contributing physical indicators, i.e. fragile lithology shares 34.58% in the radar diagram as the prime contributor of landslides in the DHR followed by drainage density at 17.91% and high degree of slope at 14.77%, and identification of such drivers of landslide probability can also be used in developing targeted adaptation planning and interventions.

Key words Darjeeling Himalayan region · Landslide probability index · Multi-criteria decision-making · Village level · Adaptation planning

---

S. Samanta (✉) · J. Chakraborty
Geoinformatics and Remote Sensing Cell, West Bengal State Council of Science and Technology, Government of West Bengal, Kolkata, West Bengal, India

S. B. Dutta
Department of Science and Technology and Biotechnology, Government of West Bengal, Kolkata, West Bengal, India

P. K. Shit et al. (eds.), *Geospatial Technology for Environmental Hazards*, Advances in Geographic Information Science, https://doi.org/10.1007/978-3-030-75197-5_17

17.1 Introduction

The Himalayan mountains are highly prone to landslides. Landslides are simply defined as the movement of the mass of dislodged rock, debris or earth materials down a slope including a broad range of motions whereby falling, sliding and flowing under the influence of gravity (Mandal & Maiti, 2014). Basically, landslide is nothing but a complex and systematic interaction between sets of physical and manmade factors governed by geologic, hydrologic, climatic, geomorphic and land use attributes (Wilson & Keefer, 1985). Every year a multitude of slide events takes place along the entire Himalayan ranges. Darjeeling Himalayan region (DHR) of West Bengal is no exception and is also highly susceptible to landslides. Every consecutive year more than hundreds of new landslide events add to the region's landscape during the monsoon and immediate post-monsoon season (Mondal, 2016).

Regionally, the geomorphology of this terrain represents an intricate interplay of erosional and gravitational processes (e.g. mass wasting) aided by the active tectonic effects of the Himalayas due to proximity of regional tectonic plates (Ghosh et al., 2014). Mallet (1874) opined that in the Darjeeling territory the Gondwana rocks are overlain by metamorphic rocks, which are termed as Darjeeling gneiss (mainly mica-gneisses and schists) and Daling (mainly slates and phyllites). These two lithological units are highly fractured and have been weathered by rainwater to considerable depths below the surface. Thus, in the mountainous region, the ground behind the cliffs is highly decomposed vis-a-vis porous, and during the rains as saturation takes place, the rock face is gradually detached from the ground leading to small and large blocks of rock bodies to slide (Basu & Majumder, 2006). Alongside, the rainwater triggers landslides as it alters the pressure within the slope, which leads to slope instability. Consequently, the heavy water-laden slope materials (soil, rock, etc.) will succumb to the forces of gravity (Conners, 2017). Starkel (1972) for the first time observed the geomorphic effects of an extreme rainfall event in the eastern Himalaya (Darjeeling), India. Froehlich et al. (1990) investigated the same area and found that shallow slides and slumps on steep slope segments occur when 24 h rainfall reaches 130–150 mm or continuous 3 days rainfall totals 180–200 mm. Though the natural factors are very conducive to the occurrence of slides, owing to the regional instability, immature geology and meandering streams, the actual chaos on slopes came when the man entered the scene (Bhandari, 2006). Vast areas of the DHR hilly parts fell to his axe and were robbed of the protective vegetal cover in terms of lopping of trees for fuel or fodder, overgrazing, increased domestic and industrial consumptions of timber, which were chiefly responsible for deforestation. Nevertheless, the spurt of human settlements has increased the number and frequency of landslides, over the period of past few decades (Bhattacharya, 2012). Growing urbanisation and uncontrolled land use are increasing overwhelming pressure on immediately available sloppy mountainous land and making the ecosystem vulnerable with a rising trend of landslide hazards. These quasi-natural events are major threat to the lives and properties of the people and also hamper the connectivity and commerce of the region aggravating lopsided development and poor economic condition.

Earlier studies have analysed landslides of the Darjeeling Himalayas from different perspectives. Many of them begin with geographical characteristics and climate that impacts the landslide configuration followed by some with unexpected consequences of the landslide. Contemporary research works have also tried to introduce new approaches to the landslide studies. Caine and Mool (1982) and Dhakal et al. (1999) focused on landslide risk assessment in Himalayan terrains by analysing physical properties of landslides and debris flows, effects of regional and local geological settings and recommendations for environmental-friendly preventive measures. Some studies attempt to establish the hazard mapping methods that consider natural terrain parameters which play a role in slope failure (Hays, 1981; Kitutu et al., 2009; Sarkar & Kanungo, 2010). Nevertheless, interesting works are also available in which the authors tried to apply intensive methods for landslide vulnerability zonation, and most of them include geologic and geomorphometric analysis of the study area to understand the various terrain parameters responsible for the different magnitudes of landslide susceptibility and field survey (Powde & Saha, 1976; Ghosh, 2011; Bhattacharya, 2014). Basically, the landslide vulnerability zonation contains two vital parts starting with the prediction of spatial probability of the occurrence of landslides in various magnitude zones over an area (Sharma et al., 2009, 2011; Luan et al., 2010; Mukhopadhyay et al., 2012) and ending with their use in the planning of risk mitigation as well as understanding their relation with the elements at risk, e.g. the population, buildings, civil engineering works, economic activities, public services, utilities, infrastructure, etc. (Govt. of India, 2009). Thus, it is quite clear that the landslide phenomenon of hilly regions has been examined from different viewpoints in the past, but this present study aims to develop a probability index of landslide occurrence at village level in the DHR. Based on the available data and geospatial techniques used, this particular study tries to identify possible sites of landslide occurrence on a microscale prior to the events within the selected geographical location.

The increasing trend of landslide events demands a better understanding of what causes landslides and how to mitigate future damage. The field of landslide risk assessment is growing rapidly, and many spatio-temporal modelling tools are addressing how to predict landslide frequency and severity. In this context, considering the magnitude of landslide frequency of the DHR and its extremities, the extended part of this study prepares a detailed inventory of landslide events with spatial information. The geoinformation technology played a formidable role in quantitative model building for village-wise landslide probability index (LPI) involving region-specific significant indicators and will empower us to identify villages with a higher probability of landslide occurrence as well as suggests strategic intervention mechanism. The outcomes of this study will enable the local administrators and stakeholders in proper decision-making and enforcement of better management strategies to combat with the recurring quasi-natural hazard owing to the regional settings within the concerned region and considered to be necessary input for effective area planning in near future.

17.2 Materials and Method

Study Area

The Darjeeling Himalayas hilly parts are the portion of tertiary young fold mountain system, characterised by lofty ridges, numerous river valleys, mountain spurs and foothills (Terai) with its ramifications comprising a total area about 2390 km^2, extending from 27°13'N to 26°44'N latitude and 88°53′E to 87°59′E longitude, bordered by Mechi and Jaldhaka rivers in the western and eastern sides, respectively, and bounded by the state of Sikkim in the north, Nepal in the west, Bhutan on the north-east and Bangladesh in the south-east directions (Bhattacharya, 2017). More specifically the present study area deals with 6 hilly community development (C.D) blocks of Darjeeling district and 3 C.D blocks of Kalimpong district that includes 117 gram panchayats in which 348 villages/census and towns/municipalities have been well demarcated (Fig. 17.1).

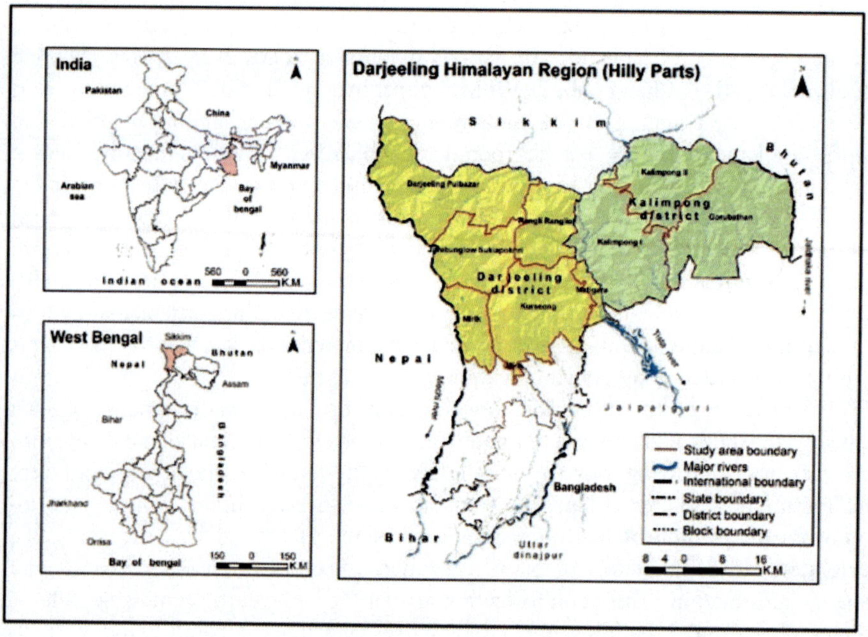

Fig. 17.1 Location map of the study area

Collection of Dataset and Processing

To generate the base database for this study, topographical sheets bearing numbers 78A/4, 78A/8, 78A/12, 78A/16, 78B/1, 78B/2, 78B/3, 78B/5, 78B/6, 78B/7, 78B/9 and 78B/13 on 1:50,000, 1:63,360 and 1:2,50,000 have been collected earlier from Survey of India and National Atlas and Thematic Mapping Organisation Kolkata. Initially, toposheets have been geo-referenced using ESRI ArcGIS v10.3 software in Universal Traverse Mercator projection system. During the geo-referencing process, four ground control points were taken over the whole area from each topographical sheet whose latitude and longitude are identified. Afterwards, known coordinate values have been manually put to create the link, and a first-order transformation has been used for registration with root mean square error of 0.5 which is good. Resampling is performed by nearest neighbour technique to create a final geo-referenced map.

Pre-processed Linear Imaging Self Scanning-IV multispectral image (spatial resolution 5.8 m, path/row 584-587/272-273) of 2015–2016 fused with Cartosat-1 panchromatic data (spatial resolution 2.5 m) of 2015–2016 that were obtained from National Remote Sensing Centre, India, images and eventually exercised to obtain Standard False Colour Composite (SFCC) images in ERDAS Imagine v14 software. The Advanced Land Observing Satellite, PALSAR, orthorectified digital elevation model (DEM) data (spatial resolution 12.5 m) of 2015 were retrieved from the Alaska Satellite Facility repository to determine the spatial distribution of slope and aspects. Secondary data sources like geo-referenced digital lithology and mineral occurrence map on scale 1:2,50,000 have been collected from the Geological Survey of India.

Inventorisation of Landslide Events

The most crucial step in this landslide probability study is to make a landslide inventory for past events. Geospatial inventory of landslides helps to draw the spatial relationship not only between landslides and biophysical factors, but it also enables us to enlighten the role of human interventions in the occurrence of the landslide. A landslide inventory is usually a collection of points and polygon shapes which represent the boundary of a single event or a complex landslide, generated by multiple events (van Westen, 1993; Guzzetti et al., 2012). Nowadays landslides are delineated digitally in a geographical information system environment (Carrara et al., 1991; Van Den Eeckhaut & Hervás, 2012) and even in several complex environments (forested areas with frequent debris or earth flows triggered by earthquakes) where certain geospatial data are available (high-resolution imagery, DEM) (Scaioni et al., 2014). Hence in this work, initially based on the SFCC images, landslides were marked in and around the study area by placing a single point on each landslide event that has occurred between 2005 and 2020. Consequently, the village level spatial location of single landslide points has been precisely validated in the Google Earth Pro platform with limited field checks. For the sake of a detailed study, the derived landslide points are being analysed with every indicator data that are taken to develop the LPI to figure out the maximum correlation between them. Such case study approach that has been taken here for a transparent understanding in a substantial proportion for landslide inventories based on topographic maps and DEM could also be used for checking landslide probabilities in respect to the geomorphological-topological inconsistencies.

Development of LPI

Selection of Region-Specific Indicators

Landslide susceptibility mapping using digital spatial information has been conducted extensively since the early 1980s (van Westen et al., 2008). Most susceptibility mapping techniques in landslide study the potential use of previous landslide records in predicting future landslides with the reflection of certain regional factors (Abdulwahid & Pradhan, 2016). Therefore, in case of developing LPI for the projected landslide probabilities at the village level, a set of conditioning factors of the DHR have been taken into serious consideration. The proposed methodology of the village-wise LPI is based on the weighted sum method of multi-criteria decision-making (MCDM) processes involving eight explicitly localised proxy/indicators, namely, (a) fragile lithology (phyllite, schist, sandstone dominated), (b)more than 30° slope, (c) drainage density, (d) the number of landslide events that occurred, (e) forest cover, (f) pucca road density, (g) south and south-east aspect and (h) settlement.

All of these indicators were found to be an active contributor in the landsliding process specifically in the hilly terrain, and the rationales behind their involvement are well articulated in previous research studies. A vast amount of geotechnical information has been already collected as a result of demand-oriented studies. These studies reveal that the lithology of the DHR in most of the places is fragile in nature, and it is observed that phyllite-dominated Proterozoic litho-unit Reyang Formation of Daling group and another schist-dominated Proterozoic litho-unit Gorubathan Formation of the same group are highly weathered and foliated with maximum landslide probability (Basu, 1969–70). The slope is a constant influential acclimatising agent in landslide occurrence. This factor directly affects landslide occurrence and is typically considered in landslide susceptibility analysis (Alimohammadlou et al., 2013; Biswajit & Sameen, 2017). As the slope becomes steep, the vertical component of gravity increases; hence, most of the landslides occur on more than 30° slope (Tournadour et al., 2015; Abdulwahid & Pradhan, 2016). The slope aspect denotes downslope direction in respect to the alignment of slope facet, is an important casuative element for landslides. This factor stimulates weathering process by receiving direct hit of precipitation, wind and sunlight

(Budimir et al., 2015). In this study south and south-east aspects have been taken into account. Due to its windward face, south-west monsoon hits directly throughout the monsoon season resulting in the maximum number of landslide events. The drainage density is often used to estimate the surface flow and potential instability of fragile slope (Mondal & Maiti, 2013), along with higher drainage density of an area which indicates the higher magnitude of headward erosion on steep slope with fragile lithological ground results in slope failure-type landslide events (Roy & Sensharma, 1967). Among the selected indicators, only the forest cover maintains a negative functional relationship with the landslide, as the root system of forested vegetation stabilises hillslope by reinforcing soil shear strength. The more the roots penetrate a potential a shear plane, the greater is the chance that the vegetation will increase slope stability (Abe & Ziemer, 1990). However, changes in the forest landscape or large-scale logging can alter the soil infiltration and ground evapotranspiration rates, thus indirectly affecting the water contents in soil and reducing slope stability (Vanacker et al., 2005; van Noordwijk, 2005; Chuang & Shiu, 2017).

Anthropogenic interventions like the extensive spread of human habitation in form of settlement are highly site- and situation-oriented entity of land use practice, but the availability of suitable land for settlement in the mountains is quite challenging. At present, people of high mountains are living in risky and hazardous terrain and facing serious threats. The replacement of forests by agriculture and settlements is thought to cause severe erosion and landslides (Glade, 2003; Soini, 2005; Chuang & Shiu, 2017). The growing population of the hills demand better communication and mobility that in turn give rise to metalled road construction that are generally believed to compromise slope stability and trigger landslides as well as debris flows (Burton & Bathurst, 1998; Chuang & Shiu, 2017).

Indicators Quantification

Defined indicators are correlated with landslide by their functional relationship. It may be described in terms of *sensitivity* and *adaptive capacity*. Indicators for sensitivity reflect to what extent a system is sensitive or responding to the exposure from an external stress or hazard such landslide, whereas the adaptive capacity indicators represent the ability or capacity of systems, institutions and infrastructure to adjust to potential damage or change due to external stresses or hazards, including climate (Esteves et al., 2016; Sharma et al., 2019). In the case of LPI, the indicators having higher sensitivity will be of higher landslide probability and vice versa. Furthermore, to quantify all considered indicators in terms of numerical values and to fit them in the ultimate equation of LPI, mathematical operations have been performed accordingly (Table 17.1) (Fig. 17.2).

Table 17.1 Village level selected indicators and their data extraction method for LPI development

Village level indicators	Methods of data extraction
Percentage of area covered by the fragile lithological classes (phyllite, schist, sandstone dominated)	Geo-referenced lithology and mineral occurrence map that has been collected from geological survey of India vectorised precisely and intersected to village layer in GIS platform to determine spatial distribution
Area percentage of more than 30° slope	Using the slope algorithm of spatial analyst tool in ArcGIS, pixel-based calculation has been performed to figure out slope value as the maximum rate of change in value from that cell to its neighbours. After obtaining the downslope values, the output raster has been reclassified for vectorisation and spatially joined with the village layer
Percentage area under south and south-east aspect	Performed using spatial analyst aspect tool that identifies the downslope direction of the maximum rate of change in value from each cell to its neighbours. The values of each cell in the output raster indicate the compass direction that the surface faces at that location. Furthermore, spatially joined to villages to retrieve desired results
Drainage density in km/km^2	Based on the SFCC images drainage, forest cover, settlement area and road network up to village level were extracted in and around the study area by on-screen digitisation on scale 1:10,000
Percentage of area covered with forest	
Percentage of settlement area	
Pucca road density in km/km^2	
Number of landslide events that occurred	Remote sensing methods based on image properties of both SFCC and high-resolution Google earth imageries were used to obtain historical records of the landslides over the past 15 (2005–2020) years. Field validations have been conducted using geographical positioning system device

Normalisation of Indicators

The normalisation of indicators is an essential part to compare among the indicators in which information can be combined in a logical way. Indicators are in different units, that is why it cannot be simply added to imply the index value (Esteves et al., 2016). The normalisation procedure enables aggregation of indicators with different units, by removing the units and converting all the values into dimensionless units. The normalised values of indicators lie between 0 and 1 and thus could be aggregated (Esteves et al., 2016; Sharma et al., 2019). Thus, in this study, normalisation has been done for landslide probability indexing and ranking that enable comparison and prioritisation among landslide-probable villages. Before starting the normalisation processes, it is essential to establish the functional relationship between the indicators and probability. There are basically two types of possible functional relationship: positive and negative relationship (Ravindranath et al., 2011).

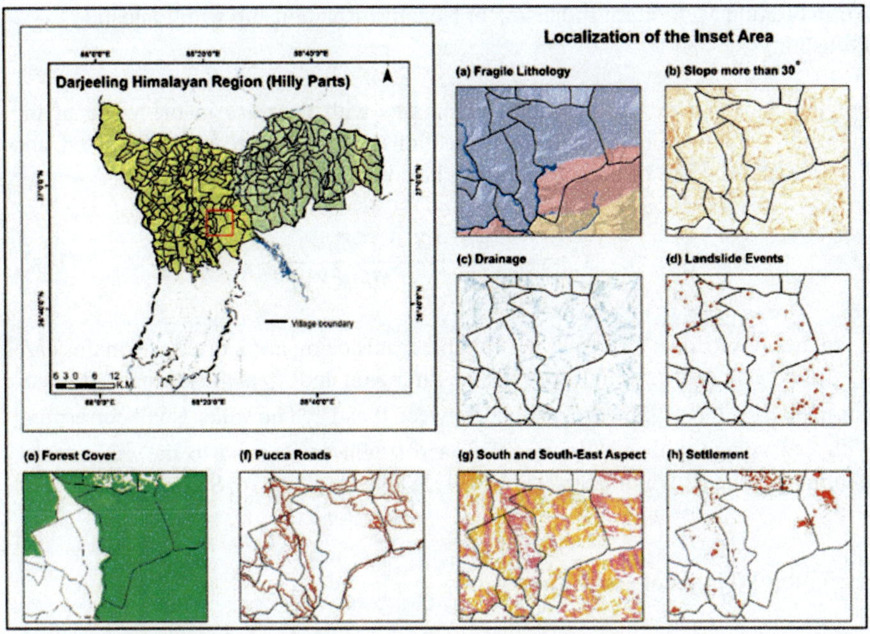

Fig. 17.2 Example of the village level selected indicators data extraction method to develop partial and full landslide probability indices. (**a**) Percentage of area covered by the fragile lithological units (phyllite, schist, sandstone) represented by different colours, (**b**) area percentage of more than 30° slope, (**c**) estimation of drainage density (length of drainage/unit area), (**d**) number of landslide events that occurred, (**e**) percentage of area covered with forest, (**f**) measurement of pucca road density in km/unit area, (**g**) percentage area under south and south-east aspect and (**h**) percentage of settlement area

Normalisation Method for Indicators of Positive Relationship with Landslide Probability

In positive relationship cases, the higher the value of the indicator, the higher will be the probability. As the values of these indicators increase, greater will be the vulnerability of the community. In such cases the variables have direct and positive functional relationship with probability, and normalisation is done using Eq. (17.1):

$$X_{ij}^{p} = \frac{X_{ij} - Min_i\{X_{ij}\}}{Max_i\{X_{ij}\} - Min_i\{X_{ij}\}} \tag{17.1}$$

where X_{ij} is the variable that is being normalised, i.e. in this case X_{ij} is the value of *j*th indicator for the *i*th region and X_{ij}^{p} is the normalised value. Normalised value of X_{ij}^{p} scores will lie between 0 and 1. The value 1 will correspond to that village with maximum value, and 0 will correspond to the village with minimum value (Esteves et al., 2016; Sharma et al., 2019).

Normalisation Method for Indicators of Negative Relationship with Landslide Probability

For indicators where the probability increases with decrease in the value of the indicator, the following normalisation equation (Eq. (17.2)) needs to be adopted, and the normalised score of the indicator can be computed accordingly:

$$X_{ij}^{n} = \frac{Max_i\{X_{ij}\} - X_{ij}}{Max_i\{X_{ij}\} - Min_i\{X_{ij}\}} \tag{17.2}$$

In the above Eq. (17.2), X_{ij} is the variable that is being normalised, i.e. in this case X_{ij} is the value of the *j*th indicator for the *i*th region and X_{ij}^{n} is the normalised value. Normalised value of X_{ij}^{n} scores will lie between 0 and 1. The value 1 will correspond with that village with maximum value, and 0 will correspond to the village with minimum value (Ravindranath et al., 2011; Esteves et al., 2016; Sharma et al., 2019).

Assigning Weights of Indicators

Assignment of proper weights to indicators is practically crucial job to imply reliable results. The problem of choosing an appropriate method of determining weights in MCDM process is very important which complies with the decision-making process (Pamučar et al., 2018). There are different methods of weight assignment, but in this specific study, expert judgement method was adopted, and the opinion of experts was considered to attribute weights for an individual indicator on 0 to 1 rating scale. Regarding the set of experts, who are knowledgeable about the region, their views on selected factors responsible for landslides and relationship with the process in the DHR are seriously taken into account for arranging indicator's weight (Table 17.2).

Aggregation of Indicators and Indexing

Aggregation of different indicators with appropriate weights is necessary to obtain a composite index or aggregated value (Esteves et al., 2016, Sharma et al., 2019). The weights are multiplied with the normalised indicator value and aggregated to obtain the LPI or the ranking of the systems. Probability index is normally developed to assist the policymakers, development administrators, NGOs and planners in prioritising the villages. It may also suggest (a) a comparative index value of different villages, (b) the spatial distribution of the high or low probability units and (c) helps to identify the drivers of landslide hazard, so that adaptation investments can be focused on dominant drivers (Sharma et al., 2015; Sharma et al., 2019). The result of the probability assessment can be classified on a scale of very high to very low vulnerability and represented spatially through profile maps. The results indicate the vulnerability ranking of landslide-prone villages on the basis of LPI

Table 17.2 Village level selected indicators and their functional relationship with landslide and assignment of weights to develop LPI

Sl No.	Village level indicators	Functional relationship with landslide (positive/negative)	Weights
a	Percentage of area covered by the fragile lithological classes (phyllite, schist, sandstone dominated)	Positive	0.21
b	Area percentage of more than 30° slope	Positive	0.17
c	Drainage density in km/km^2	Positive	0.14
d	Number of landslide events that occurred	Positive	0.12
e	Percentage of area covered with forest	Negative	0.12
f	Pucca road density in km/km^2	Positive	0.10
g	Percentage area under south and south-east aspect	Positive	0.08
h	Percentage of settlement area	Positive	0.06
Total			**1**

values, and a critical utility of this probability assessment is the identification of the drivers of landslides along with their proportional contribution. Information about the indicators or factors that contribute most to the aggregate probability is useful in prioritising development and implementation of adaptation interventions.

17.3 Results and Discussion

Quantitative Analysis of the Regional Indicators

This part of the study is categorised into some major dimensions, as the exploration of diversified region-specific varied indicators data and their spatial components that provide efficient support for intensive landscape hazard analysis. The combination of lithology and slope involving primary rock strata and structural control of lithology on the fluvial characteristics influences the local terrain as well as climatic characteristics and impacts the micro-level land use planning, whereas the drainage is similarly one of the most important factors that contribute to the landslide mechanism. Other influencing regional characteristics that are basically a result of modern-day man-environment interaction are also considered here for landslide probability assessment.

Lithological Setup and Landslides Occurrences: An Association

The DHR can be subdivided laterally from south to north as (a) the sub-Himalaya or the Siwalik belt composed of Upper Tertiary Siwalik sediments; (b) the lower or lesser Himalayas composed of Gondwana, Lower Tertiaries and Achaean rocks

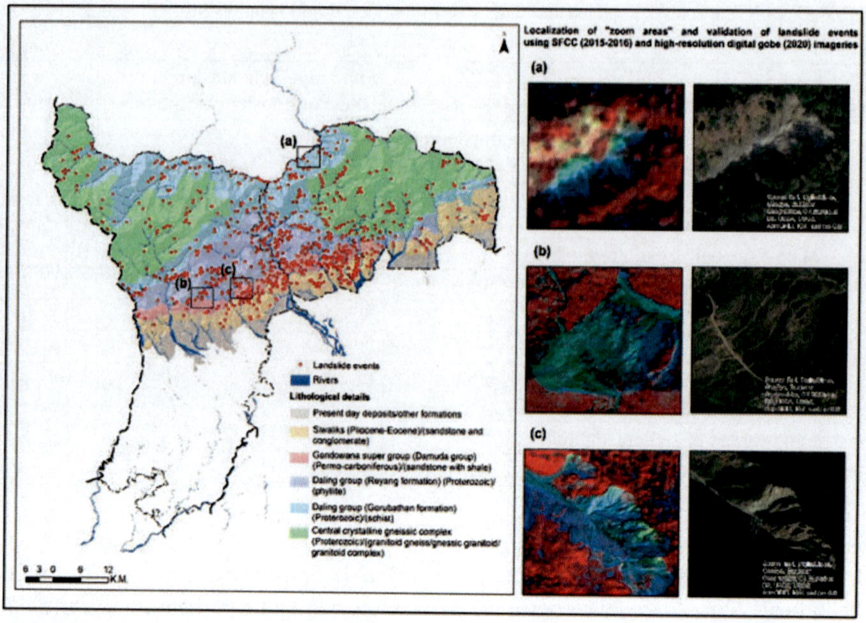

Fig. 17.3 Fragile lithological details and its association with landslides. As demonstrated in (**a**) landslide over Daling group (Gorubathan Formation) in Sangser village of Kalimpong II block, (**b**) landslide of the South Shibkhola Tea Garden village of Kurseong block on Daling group (Reyang Formation) and (**c**) another landslide located on Gondwana supergroup (Damuda group) in Latpanchar forest village

along with some intrusive; and (c) the higher Himalayas, composed of intrusive Tertiary granites and other igneous and metamorphic rocks. The Siwalik belt is subdivided into an upper pebbly sandstone/conglomerate facies and lower sandstone facies, and it comprises of soft greyish sandstone, mudstone, shales and conglomerates that cover about 209.18 km^2 area in the studied region. During the monsoon season, a high amount of rainfall seeping through the pore spaces of these coarse-textured sandstone decomposed the feldspar to form kaolin which acts as a lubricant for sliding of rock materials. Hence, an amalgamation of this triggering factor makes the Siwalik a landslide regime of the region; a total of 196 landslide events have been identified here, and these are concentrated mainly in deforested Lish and Gish basin area of the Kalimpong district. Subsequently, upper Carboniferous to Permian feldspathic and micaceous quartzite, sandstone and carbonaceous slates with thin seams of crushed coal contain the Damuda group of Gondwana supergroup, which is highly landslide prone, as 149 past landslide events have been demarcated within 114.63 km^2 area of this group (Fig. 17.3).

Further to the north, low-grade meta-psammopelitic sequence of Precambrian Daling group's Gorubathan and Reyang formations cover 16% area of the study area, are taken into account due to its extensive geographical coverage with the selected studied reach. The Gorubathan Formation comprises green slate, phyllite, phyllonite, cherty chlorite quartzite and green tuffaceous wake with basic metavolcanic materials, while the Reyang Formation consists of variegated slates, phyllite and low-graded schists. Highly metamorphosed argillaceous rocks in both the formations of the Daling group, where 136 landslides in Gorubathan and 261 landslides in Reyang formation have occurred respectively, are extremely fragile and most landslide prone litho-units of this region.

On the other hand, the Central Crystalline Gneissic Complex (CCGC) is dominantly present in the DHR. Fundamentally this group is made up with schists and Darjeeling gneisses of higher rank with subordinate quartzites and lenticular isolated bodies of calcsilicate rocks. It is spatially distributed over the entire north-western part of Darjeeling and north-eastern corner of the Kalimpong district with an area coverage about 28% comprising 163 landslide events. Considering the geographical area of this group of formation, the number of landslide events that occurred is inconsequential to contribute sensitivity, as the rock types of this formation are harder in nature and make it less vulnerable to landslide. Although variability in precipitation, altitudinal effects and aspects are all responsible for a landslide, the diversified lithological characteristics, as slight recrystallised and coarse-grained sandstone, are characterised by cataclastic deformation which destroy the clastic texture with intense granulation along ramifying narrow zones of fractures, which initiate landslide hazards (Basu & Sarkar, 1985; Basu & Ghosh, 1993). Furthermore, intense metamorphism of Daling is referred to as the classical ground of the low-grade metamorphosed rocks, influencing landslides (Acharya, 1972).

Characterisation of Slope and Aspect with Landslides

The high degrees of slope, specifically more than 30° with fragile lithology, are the most favourable element of slope failure in the selected region. This category of slopes covers 27.05% area to the total area of the study. More than 30° slopes are often regarded as steep slopes, mainly concentrated on escarpments, and the source region of first-order streams and coupled with the fragile lithology range is witnessing greater percentage of landslide events. A total of 498 out of 1045 number of landslide events have occurred in this slope category which is 47.66% of total observed landslides and the rest of landslide events under other lower degree of slopes, which covers 72.95% of the total study area. The south- and south-east-facing slopes are in windward direction and get a high amount of sunlight, respectively, compared to north-facing slopes, which have steeper inclinations and receive a higher amount of rainfall as the branch of south-west monsoon hits directly from the south during monsoons. This group of aspects covers 652.22 km^2 area that denotes only 27.29%, but 503 out of the total number of landslide events are accountable over it (Fig. 17.4a, b).

Owing to the unique tectonic setting of the Himalayan thrust front failures, earth materials along the slope are common. The strength of hillslope materials is a function of geologic composition and stress state and is modified by past movement,

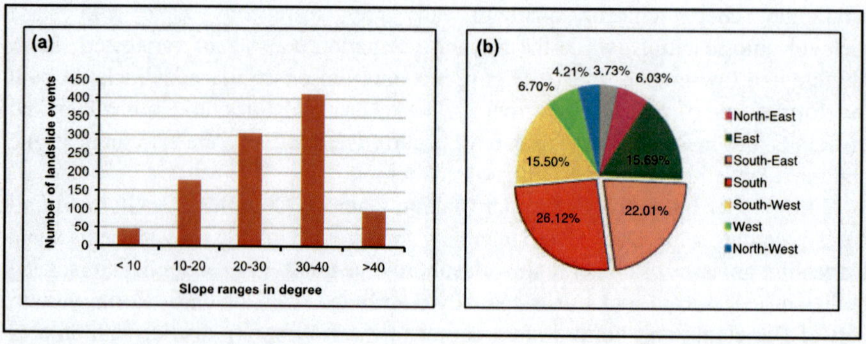

Fig. 17.4 Characterisation of slope and aspect with landslides as stated in (**a**) slope ranges and corresponding landslides events that occurred on it and (**b**) percentage of landslide occurrences in different aspects

weathering, vegetation and hydrologic processes (Lu & Godt, 2008), and here south-facing slopes exhibit more landslide activity and are thus an important predictor of susceptibility to translational landslide (Ghosh et al., 2011).

Drainage Density Estimation

The complex interplay of rock types and its structure in the Himalaya produces varied hydrogeological environments that often lie in proximity to each other. Further, variability in precipitation, altitudinal effects and aspects are all responsible for the formation of intricate drainage network in the DHR. The hilly Darjeeling Himalaya gave rise to unaccountable streams and numerous rivers and have been divided into two parts by the Tista River. In this work, village level drainage density has been calculated and used as a significant proxy that alone could address the litho-structural effect active on the region participating in the landsliding process. Village-wise drainage density of the DHR shows a diverse characteristic as the values range from 0 to 12.75 km/km^2, and the maximum density was observed in the north-western part which is composed of central crystalline rock group as well as in parts of Kalimpong districts where the CCGC and Daling rock groups are prevalent (Fig. 17.5a, b). Here, high densities can also indicate a greater probability of landslide (Mandal & Mandal, 2016). Drainage density is also determined as a good susceptibility index for rainfall-induced landslides, and study of drainage density is a practical approach for disaster management (Hasegawa et al., 2009).

Forest Cover Depletion and Spread of Settlement and Transport Network

The forest cover of the study area is considered as an adaptive proxy in the LPI study. Usually, the preventive nature of forest vegetation root system helps to arrest soil erosion and slope failures in a hillslope. From the historical perspective, the

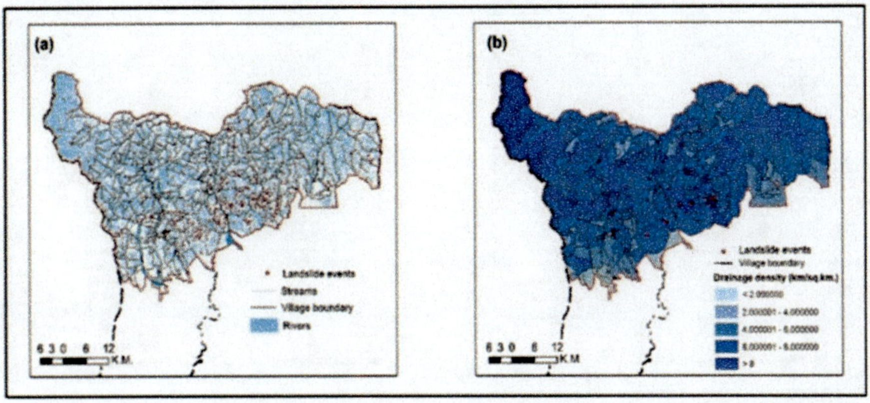

Fig. 17.5 Village level drainage density assessment. (**a**) Drainage network and landslides events of the study area and (**b**) computed drainage density of villages that ranges from 0 to 12.75 km/km^2

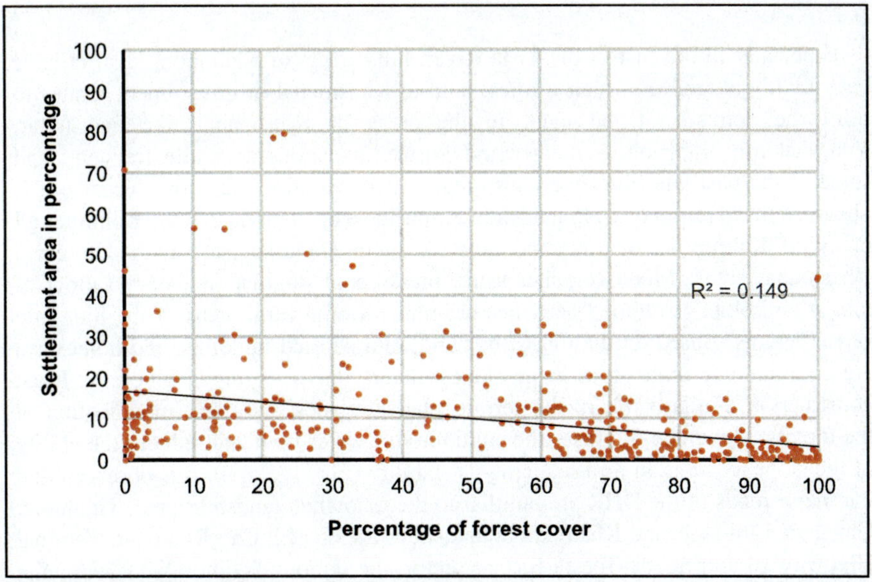

Fig. 17.6 Bivariate analysis shows correlation between percentage forest cover and settlement areas percentage at village level of the study area

entire hilly tract of Darjeeling Himalaya was completely clothed with dense forests from the top of the hills to the very bottom of the valleys, but timely demand of land for settlements, cultivation, tea (*Cinchona*) plantation and road constructions leads to deforestation activities. Figure 17.6 establishes this fact, as it shows there is a negative correlation between the percentage of forest cover and settlement areas in villages within the defined study area.

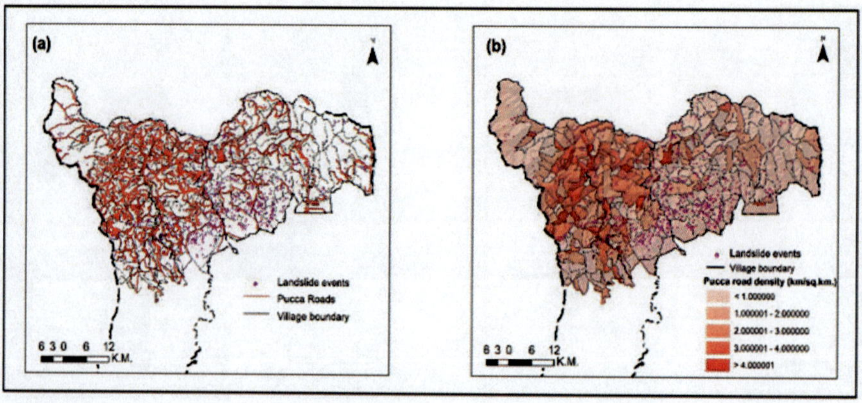

Fig. 17.7 Village level pucca road density analysis. (**a**) Landslides events located along the pucca road corridor in the study area and (**b**) calculated pucca road density of villages that ranges from 0 to 18.395 km/km^2

Especially in the east of the Tista River, hilly tracts of Kalimpong district have been highly altered from dense forest land to Khasmahal or government estates to fulfil the demand of the land. In the areas of Khasmahal like Pabringtar, Yangmakum, Nobgaon, Samether and Suruk, landslides are quite frequent, and some of the past landslides were also reactivated within the last few decades. It is observed that 415 out of 1045 landslide events have been marked in the Kalimpong I block of Kalimpong district, and most of them are associated with deforested Khasmahal areas. Moreover, increasing pressure of population, heavy inflow of tourist and other livelihood activities are transforming rural areas of the hills into urban centres. Based on the tourist demand, multistoried buildings and homestays are continuously built on unstable slopes underlying with fragile lithology. Road construction similarly alters the terrain characteristics, as slope modification at the toe of a slope compromises the continuity of water flow and reduces the ability of the slope to drain, in turn reducing its stability (Fig. 17.7a, b). It is observed that the major roads of the DHR are familiar to the renowned landslides, e.g. Tindharia, Gayabari 14th Mile and Khairekhola landslides are on Hill Cart Road and National Highway 10 and Sevok-Teesta Bazaar sector, the famous Setijhora and 29th Mile landslides are on State Highway 12 (SH 12) and Algarah-Gorubathan sector and the Chibo, Bong, Chisopani, Bhameygaon and Kolbong landslides of the Kalimpong district are located nearby SH 12.

Khasmahals are severely affected by extensive deforestation, tremendous agricultural pressure and overgrazing; as a result, most of the precipitated water goes down the deforested slopes causing soil erosion and sliding (Basu & Ghosh, 1993). Besides, the vulnerability of roads is not only delaying overall development by impacting the livelihood of the people and delaying administrative activities and accessibility to daily needs of goods and medicine, but also heavy vehicular congestion along the narrow passage of hilly road increases expenses from longer driving distances due to road blockages (Ze^zere et al., 2007; Chuang & Shiu, 2017). This indirect damage entails the effects on society from disruptions to utility services and local businesses, which result in a loss of revenue and tourism (van Westen et al., 2008).

Interpretation of LPI

Village level quantification of differential contributing data in a precise manner further leads to normalisation of indicators in accordance with the stated methodology. Therefore, functional relationships of the selected proxies have been rationally attributed to indicators in compliance with weight assignment to them, and finally, aggregation was done for different indicator value to obtain the LPI. The prepared LPI is divided into four distinct classes by maintaining equal intervals of LPI values. The (a) very high landslide probability class comprises 45 villages with LPI value more than 4.33, (b) high landslide probability consists of 138 villages having the LPI range from 4.32 to 3.03, (c) moderate landslide probability zone includes 109 villages where the values LPI ranging between 3.03 and 1.73, and 56 villages are in (d) low landslide probability zone of LPI value less than 1.73 (Fig. 17.8).

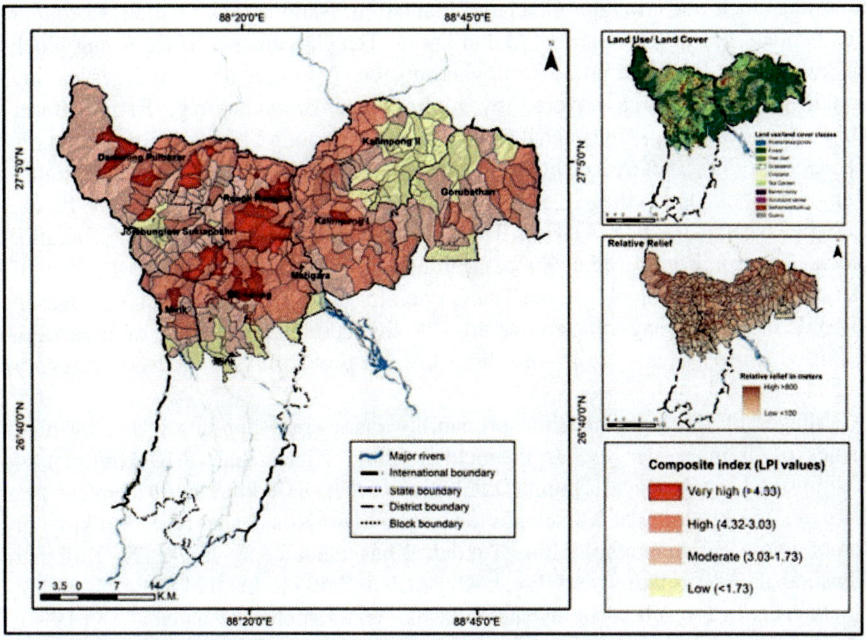

Fig. 17.8 Landslide probability map showing landslide-prone villages of different categories, e.g. very high landslide probability, high landslide probability, moderate landslide probability and low landslide probability villages

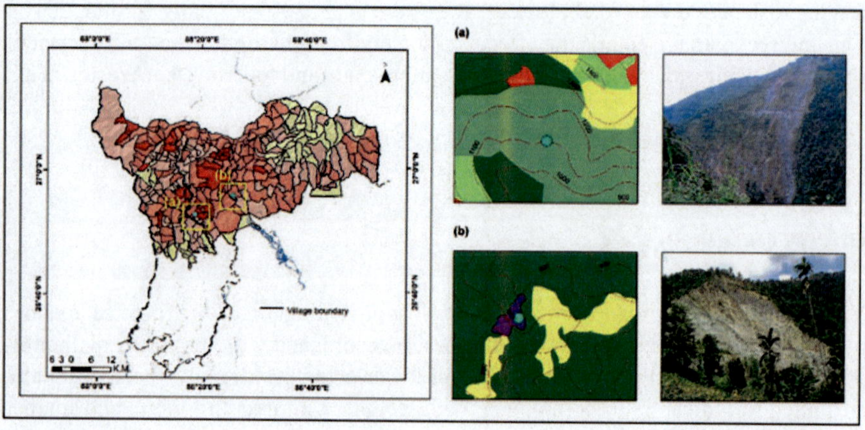

Fig. 17.9 Validation of LPI results as with limited field checks as exemplified in (**a**) the Paglajhora, one of the famous landslides of the DHR falls in the very high landslide probability class and (**b**) the devastating lower Karmat landslide is under the high landslide probability class

Village-wise lithological configuration shows that very high and high probability of landslide occurrence villages like Kolbong Khasmahal, Lingding Khasmahal, Chegra Khasmahal, Hum Tukdah Khasmahal, Labda Khasmahal, etc. of Rangli Rangliot block and Mangpu Cinchona Plantation, Malotar Tea Garden, Lizziepur Tea Garden, Majua Tea Garden, Mahal Diram Tea Garden, etc. of Kurseong block are lying over landslide-prone Daling group. On the other hand, villages having certain percentage area covered by high degree of slope, e.g. Rambi Bazar D.I.F. (97.84%), Singi Khasmahal (40.61%) of Kalimpong I block, North Shibkhola Tea Garden (44.23%) of Kurseong block, Kolbong Khasmahal (34.97%) of Rangli Rangliot block, and villages, namely, Phuguri Forest (82.58%) of Mirik block, Chegra Khasmahal (75.96%), Hum Tukdah Khasmahal (74.39%) of Rangli Rangliot block, Turzam Forest (65.67%) of Kalimpong I block, Lodhama (61.49%) of Darjeeling Pulbazar block, Majua Tea Garden (57.92%) and Jungpana Tea Garden (57.35%) of Kurseong block covered with by certain percentage of area of a specified slope aspects, are in very high to high probability of landslide category (Fig. 17.9a, b).

Villages of the high landslide probability class correspondingly display high values of drainage density, as identical in Singi Khasmahal (8.10 km/km^2) of Kalimpong I block, Dayal Thong (D.R.) Tea Garden (8.08 km/km^2), Edenvale Tea Garden (7.51 km/km^2) of Kurseong block, Parmaguri Khasmahal (7.92 km/km^2) of Jorebunglow Sukhiapokhri, Hum Tukdah Khasmahal (7.89 km/km2), Kolbong Khasmahal (7.80 km/km^2) and Rayak Khasmahal (7.40 km/km^2) of Rangli Rangliot block. The LPI result also reveals villages, viz. Lapchu Khasmahal (36.19%), Soriang Khasmahal (35.85%) of Rangli Rangliot block, Saurinibasti (29.05%) of Mirik block, North Shibkhola Tea Garden (25.44%) of Kurseong block, Patliabas Forest (27.33%) and Lodhama (25.75%) of Darjeeling Pulbazar block having maximum settlement covers, are also included very high to high chances of landslide occurrence.

In contrast, villages characterised by harder rocks type of CCGC, a lesser area covered with high degree of slope and determinant aspect, low drainage density along with the high percentage of forest cover and insignificant settlement transport distribution are in moderate to low landslide probability category. Most of the villages of these classes are sporadically distributed over the concerned study area; the majority of them are situated near Neora Valley National Park of Kalimpong II block, while others are close to the foothills zone. So the weakest lithologies with cohesion, consolidation or interlayering of materials having highly varied permeabilities can lead to high degrees of fracturing, jointing or erosional dissection (Cruden & Varnes, 1996). As landslide risk is fundamentally a product of hazard and vulnerability; these two phenomena can be managed in mutually varying proportions (Alexander, 2012), and, thus, the assessment of landslide probability is very essential as it predicts the places associated with hazard at a microscale prior to its occurrence.

Identification of Drivers of Landslide Probability for Adaptation Planning

The LPI assessments are often designed to support and improve adaptation planning, with the overall objective of reducing vulnerability in the region under consideration. It can also help to substantiate decision-making when it comes to selecting adaptation measures, based on the assessment of drivers of landslide occurrences with their index value. Thus, this LPI can be designated to assess the drivers of probability for developing targeted adaptation planning. Here in this study, the contribution of a single indicator to the LPI was obtained as a product of its weight and measured value. In this study, the different drivers of landslides and their contribution to it are shown in the radar diagram (Fig. 17.10) which depicts the

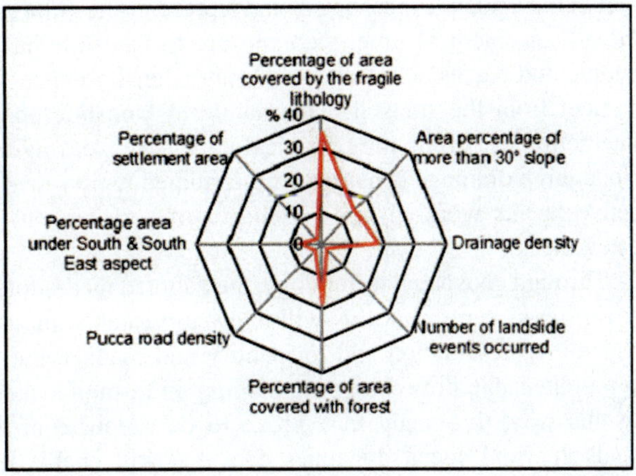

Fig. 17.10 Drivers of landslide probability and their contribution in the LPI

fragile lithology shares 34.58% as the prime contributor of landslides in the DHR followed by drainage density (17.91%) and high degree of slope (14.77%). The study reveals that the physical indicators like fragile lithology, drainage density and high degree of slope are main causative factors to the occurrence of landslide events of the region, but most significantly the forest cover holds 18.39% in the radar plot and that stands for the adaptive capacity measures to minimise landslide hazards.

Contrastingly, indicators like south and south-east aspects and the number of landslide events contribute 6.92% and 1.98%, respectively, in the landslide probability. Landslide events identified on south and south-east aspects are mainly monsoon driven in nature. In case of the proportional contribution of landslide events, it is necessary here to mention that this particular proxy is result oriented while others are causative. Eventually, reactivation nature of previously occurred major landslides locally increases high chances of slope failure which is not reflected throughout the region. It is also observed from the present study that the frequency of landslides under forested area is less, while other parts of the study area are highly infested with such hazards as the percentage of settlement area and road density share 2.18% and 3.25% correspondingly. Here, the contemporary ground scenario is an evidence that reveals that the combination of anthropogenic activities like deforestation, slope cutting, digging for settlements and road construction is highly impacting the recurring hazard phenomenon in the entire DHR. As the Himalaya represents one of the most fragile mountain ecosystems of the world, systematic planning is a must for successful implementation of developmental schemes (Anbalagan et al., 2008); thereafter, the LPI and its results are truly a practical approach that might be taken into account for both the inherent and external parameters responsible for slope instability and future landslides.

17.4 Conclusion

The proposed index has analysed all the significant indicators that are actively taking part in the landslide mechanism and represents the vulnerable state of the DHR. With the advancement of time, the exposure to landslide hazards may increase to both people and places as the elements may intensify depending on situations. It is also evident from the study that the nature of landslide phenomenon results from the interaction of contributing physical indicators, i.e. fragile lithology, high degree of slope, high drainage density over the studied region, which are considered to be the active agents working in 183 villages of very high to high landslide probability categories.

Through geospatial technology, the adopted methodology can be applicable to all territorial systems, and this will allow to visualise the results, through maps, as a realistic representation and to identify and manage the process through an easily comprehensible flow chart programming in form of spatial decision support system. In this way, this study may prove to be adequate in the integration of climate, landscape and social dynamics data that will enable us not only in developing landslide early warning system but also considered as the base input to the planners and policymakers for strategic intervention and good governance approach to minimise the landslide vulnerability, which is directly related to the human life and livelihood of the region.

Acknowledgements Sincere thanks to Sri. Anil Verma, IAS, Additional Chief Secretary, Department of Science and Technology and Biotechnology, Government of West Bengal (DSTBT, GoWB), for extending all necessary support in accomplishment of this study. The authors would like to appreciate the initiatives that were taken by the Department of Science and Technology, Government of India (DST, GoI), to support this research work as it has been carried out under the National Mission for Sustaining the Himalayan Ecosystem (NMSHE) project, Climate Change Programme Division funded by DST, GoI. The authors are grateful to Dr. P.B. Hazra, Principal Scientist, DSTBT, GoWB, for his continuous encouragement to this work and also indebted to Dr. B. Samanta, Senior Scientist, DSTBT, GoWB, for his valuable suggestions and recommendation.

Declaration

Compliance with Ethical Standards This research work is carried out in compliance with transparency, moral values, honesty and hard work. No human or animal experiments are involved in this research work.

Funding This research work has been carried out under the NMSHE project, Climate Change Programme Division, funded by the Department of Science and Technology, Government of India.

Conflict of Interest There is no question of conflicts of interest for any co-authors of this work directly or indirectly.

Ethical Approval The methodology, findings and conclusions made here belong to original research work as per our knowledge and belief.

Informed Consent Everything is carried out with collective decision and consent.

References

Abdulwahid, W. M., & Pradhan, B. (2016). Landslide vulnerability and risk assessment for multi-Hazard scenarios using airborne laser scanning data (Lidar). *Landslides, 14*, 1057–1076. https://doi.org/10.1007/s10346-016-0744-0

Abe, K., &Ziemer, R. R. (1990). Effect of tree roots on shallow-seated landslides. *Paper presented at the technical session on geomorphic hazards in managed forests, XIX world congress, IUFRO*, August 5–11, 1990, Montreal, Canada.

Acharya, S. K. (1972). Geology of Darjeeling coalfield, with a reference to its intrusive records. *Geological Survey of India, 99*(2), 80–100.

Alexander, D. (2012). Vulnerability to landslides. In T. Glade, M. Anderson, & M. J. Crozier (Eds.), *Landslide hazard and risk* (pp. 175–198). John Wiley & Sons. https://doi.org/10.1002/9780470012659.ch5.

Alimohammadlou, Y., Najafi, A., & Yalcin, A. (2013). Landslide process and impacts: A proposed classification method. *Catena, 104*, 219–232. https://doi.org/10.1016/j.catena.2012.11.013

Anbalagan, R., Chakraborty, D., & Kohli, A. (2008). Landslide hazard zonation (lhz) mapping on meso-scale for systematic town planning in mountainous terrain. *Journal of Scientific and Industrial Research, 67*, 486–497.

Basu, S. R., & Ghosh, L. (1993). A comprehensive study of landslides and floods in the Lish basin of Darjeeling Himalaya. *Indian Journal of Power and River Valley Development, XLIII*, I 96–I203.

Basu, A. R. (1969–70). *A geotechnical note on investigations of certain stretches of border roads under project Swastik in Sikkim and Darjeeling district, West Bengal.* Unpublished G.S.I. Report.

Basu, S. R., & Majumder, P. (2006). Landslide scenario of the Darjeeling Himalayas in west Beengal, India. *Geographical Review of India, 68*(2), 161–173.

Basu, S. R., & Sarkar, S. (1985). Some considerations on recent landslides at Tindharia and their control. *Indian Journal of Power and River Valley Development*, I 86–I192.

Bhandari, R. K. (2006). The Indian landslide scenario, strategic issues and action points. *Keynote Address*, 19.The First India Disaster Management Congress, New Delhi, 29–30 November 2006.

Bhattacharya, A. (2017). *Current Science, 8*, 149–154. https://doi.org/10.19071/cb.2017.v8.3256

Bhattacharya, S. K. (2014). A calibration technique for better estimation of soil loss in the Terai and adjacent hills of West Bengal, India. *Research Forum International Journal of Social Sciences, 2*(3), 1–6.

Bhattacharya, S. K. (2012). Zonation mapping techniques of the landslide Hazard in Darjeeling Hills, West Bengal, India. *Geo-Analyst, 2*, 12–17.

Biswajit, P., & Sameen, M. (2017). *Laser scanning applications in landslide assessment* (pp. 3–19). Springer International Publishing. Part I.

Budimir, M. E. A., Atkinson, P. M., & Lewis, H. G. (2015). A systematic review of landslide probability mapping using logistic regression. *Landslides, 12*(3), 419–436. [online]. https://doi.org/10.1007/s10346-014-0550-5

Burton, A., & Bathurst, J. C. (1998). Physically based modelling of shallow landslide sediment yield at a catchment scale. *Environmental Geology, 35*(2), 89–99. https://doi.org/10.1007/s002540050296

Caine, N., & Mool, P. (1982). Landslides in the Kolpukhola drainage, Middle Mountains, Nepal. *Mountain Research and Development, 2*(2), 157–173. https://doi.org/10.2307/3672961

Carrara, A., Cardinali, M., Detti, R., Guzzetti, F., Pasqui, V., & Reichenbach, P. (1991). Gis techniques and statistical models in evaluating landslide Hazard. *Earth Surface Processes and Landforms, 16*(5), 427–445. https://doi.org/10.1002/esp.3290160505

Chuang, C. Y., & Shiu, S. Y. (2017). Relationship between landslides and mountain development-integrating geospatial statistics and a new long-term database. *Science of the Total Environment*, 622–623. https://doi.org/10.1016/j.scitotenv.2017.12.039

Conners, D. (2017). What causes landslides and mudslides? *Earth-Human World*. online document-https://earthsky.org/

Cruden, D. M., &Varnes, D. J. (1996). Landslide types and processes. In *Landslides: Investigation and mitigation* (pp. 36–75). Special report 247, Transportation Research Board, National Research Council, Chapter 3.

Dhakal, A. S., Amada, T. K., & Aniya, M. (1999). Landslide hazard mapping and application of GIS in the Kulekhani watershed, Nepal. *Mountain Research and Development, 19*(1), 3–16.

Esteves, T, Ravindranath, D, Bedamatta, S, Raju,K, Sharma, J, Bala, G and Murthy, I. (2016). Multi-scale vulnerability assessment for adaptation planning. Current Science, 110(7), pp. 1225–1239.

Froehlich, W., Gil, E., Kasza, I., & Starkel, L. (1990). Thresholds in the transformation of slopes and river channels in the Darjeeling Himalaya, India. *Mountain Research and Development, 10* (4), 301–312.

Ghosh, S. (2011). Knowledge *guided empirical prediction of landslide hazard* (pp. 7–9). Ph.D. Dissertation, University of Twente.

Ghosh, S., Bhattacharjee, N. R., Chakraborty, I., & Nath, S. (2014). A report on the site-specific geotechnical investigations of Tindharia landslide on National Highway - 55, Darjeeling District, West Bengal, GSI, 2013. pp. 36.

Ghosh, S., Carranza, E. J. M., van Westen, C. J., Jetten, V. G., & Bhattacharya, D. N. (2011). Selecting and weighting spatial predictors for empirical modeling of landslide susceptibility in the Darjeeling Himalayas (India). *Geomorphology, 131*(1), 35–56. https://doi.org/10.1016/j.geomorph.2011.04.019

Glade, T. (2003). Landslide occurrence as a response to land use change: A review of evidence from New Zealand. *Catena, 51*, 297–314. https://doi.org/10.1016/S0341-8162(02)00170-4

Govt. of India. (2009). *National disaster management guidelines: Management of landslides and snow avalanches*. National Disaster Management Authority.

Guzzetti, F., Mondini, A. C., Cardinali, M., Fiorucci, F., Santangelo, M., & Chang, K.-T. (2012). Landslide inventory maps: New tools for an old problem. *Earth-Science Reviews, 112*(1), 42–66. https://doi.org/10.1016/j.earscirev.2012.02.001

Hasegawa S., Yamanaka M., Mimura T., Dahal R.K., Nonomura A. (2009). Drainage density as rainfall induced landslides susceptibility index. *International Seminar on Hazard Management for Sustainable Development in Kathmandu, Nepal* (pp. 72–75).

Hays, W. W. (Ed.). (1981). *Facing geologic and hydrologic hazards; earth-science considerations* (report no. 1240B; professional paper). USGS Publications Warehouse. https://doi.org/10.3133/pp1240B

Kitutu, M., Muwanga, A., Poesen, J., & Deckers, J. (2009). Influence of soil properties on landslide occurrences in Bududa District, eastern Uganda. *African Journal of Agricultural Research, 4*, 611–620.

Lu, N., & Godt, J. (2008). Infinite slope stability under steady unsaturated seepage conditions. *Water Resources Research, 44*(11). https://doi.org/10.1029/2008WR006976

Luan, T. X., Shibayama, M., Cannata, M., & Long, N. H. (2010). Area informatics and TRIGRS model for study shallow landslide vulnerability assessment. A case study in Bavi area, Hanoi region, Vietnam. In: *International symposium on geoinformatics for spatial infrastructure development in Earth and Allied Sciences*.

Mallet, F. R. (1874). On the geology and mineral resources of the Darjeeling District and the Western Duars. *Memoir GSI, 11*(1), 72.

Mandal, B., & Mandal, S. (2016). Assessment of mountain slope instability in the Lish River basin of eastern Darjeeling Himalaya using frequency ratio model (FRM). *Modelling Earth Systems and Environment, 2*, 121. https://doi.org/10.1007/s40808-016-0169-8

Mandal, S., & Maiti, R. (2014). Role of lithological composition and lineaments in Landsliding: A case study of Shivkhola watershed, Darjeeling Himalaya. *International Journal of Geology, Earth and Environmental Sciences, 4*, 126–132.

Mondal, S. (2016). Geomorphic threshold and landsliding in Paglajhora Sinking Zone, Darjiling Himalaya. *International Journal of Research in Geography, 2*. https://doi.org/10.20431/2454-8685.0201001

Mondal, S., & Maiti, R. (2013). Integrating the analytical hierarchy process (ahp) and the frequency ratio (fr) model in landslide susceptibility mapping of shiv-Khola watershed, Darjeeling Himalaya. *International Journal of Disaster Risk Science, 4*, 200–212. https://doi.org/10.1007/s13753-013-0021-y

Mukhopadhyay, B. P., Roy, S., Chaudhuri, S., & Mitra, S. (2012). Influence of geological parameters on landslide vulnerability zonation of Darjeeling town, in eastern Himalayas. *Asian Journal of Environment and Disaster Management, 4*(2), 27–46.

Pamučar, D., Stević, Ž., & Sremac, S. (2018). A new model for determining weight coefficients of criteria in Mcdm models: Full consistency method (Fucom). *Symmetry, 10*(9), 393. https://doi.org/10.3390/sym10090393

Powde, M. B., & Saha, S. S. (1976). Geology of the Darjeeling Himalayas. *Geological Survey of India Miscellaneous Publication, 41*, 50–55.

Ravindranath, N., Rao, S., Sharma, N., Nair, M., Gopalakrishnan, R., Rao, A., Malaviya, S., Tiwari, R., Sagadevan, A., Munsi, M., Krishna, N., & Govindasamy, B. (2011). Climate change vulnerability profiles for north East India. *Current Science, 101*, 384–394.

Roy, S., & Sensharma, S. B. (1967). *Geological report on the stability of hill slopes in and around Darjeeling town, Darjeeling Dist. S.W.* Unpublished Geological Survey of India Report.

Sarkar, S., & Kanungo, D. P. (2010). Landslides in relation to terrain parameters—A remote sensing and GIS approach (pp. 1–6). *Natural Hazard Management*, GISdevelopment.net.

Scaioni, M., Longoni, L., Melillo, V., & Papini, M. (2014). Remote sensing for landslide investigations: An overview of recent achievements and perspectives. *Remote Sensing, 6*(10), 9600–9652.

Sharma, L. P., Debnath, P., Patel, N., & Ghose, M. K. (2009). *Landslide vulnerability*. Retrieved from http://mycoordinates.org/landslide-vulnerability

Sharma, L. P., Patel, N., Ghose, M. K., & Debnath, P. (2011). Landslide vulnerability assessment and zonation through ranking of causative parameters based on landslide density-derived statistical indicators. *Geocarto International, 26*(6), 491–504.

Sharma, J., Chaturvedi, R. K., Bala, G., & Ravindranath, N. H. (2015). Assessing "inherent vulnerability" of forests: A methodological approach and a case study from Western Ghats, India. *Mitigation and Adaptation Strategies for Global Change, 20*(4), 573–590. https://doi.org/10.1007/s11027-013-9508-5

Sharma, J., Murthy, K.I., Esteves, T., Negi, P., Sushma, S., Dasgupta, S., Barua, A., Bala, G., Ravindranath, N.H., (2019). *Climate vulnerability and risk assessment: Framework, methods and guidelines for the Indian Himalayan Region*, Report.

Soini, E. (2005). Land use change patterns and livelihood dynamics on the slopes of Mt. Kilimanjaro, Tanzania. *Agricultural Systems, 85*, 306–323. https://doi.org/10.1016/j.agsy.2005.06.013

Starkel, L. (1972). The role of catastrophic rainfall in the shaping of the relief of the lower Himalaya (Darjeeling Hills). *Geographia Polonica, 21*, 103–147.

Tournadour, E., Mulder, T., Borgomano, J., Hanquiez, V., Ducassou, E., & Gillet, H. (2015). Origin and architecture of a mass transport complex on the northwest slope of little Bahama Bank (Bahamas): Relations between off-Bank transport, bottom current sedimentation and submarine landslides. *Sedimentary Geology, 317*, 9–26. https://doi.org/10.1016/j.sedgeo.2014.10.003

Van Den Eeckhaut, M., & Hervás, J. (2012). Landslide inventories in Europe and policy recommendations for their interoperability and harmonization. *A JRC contribution to the EU-FP7 Safe Land project, JRC Science and Policy Reports, EUR 25666EN*. http://doi.org/10.2788/75587.

van Noordwijk, M. (2005). *RUPES typology of environmental services worthy of reward* (p. 46). World Agroforestry Centre—ICRAF, SEA Regional Office.

van Westen, C. J. (1993). *Application of geographic information systems to landslide hazard zonation*. International Institute for Geo-Information Science and Earth Observation. Retrieved from http://www.itc.nl/library/Papers_1993/phd/vanwesten.pdf

van Westen, C. J., Castellanos, E., & Kuriakose, S. L. (2008). Spatial data for landslide susceptibility, Hazard, and vulnerability assessment: An overview. *Engineering Geology, 102*, 112–131. https://doi.org/10.1016/j.enggeo.2008.03.010

Vanacker, V., Molina, A., Govers, G., Poesen, J., Dercon, G., & Deckers, S. (2005). River channel response to short-term human-induced change in landscape connectivity in Andean ecosystems. *Geomorphology, 72*, 340–353.

Wilson, R., & Keefer, D. (1985). Predicting areal limits of earthquake-induced landsliding. In: Earthquake hazards in the Los Angeles region-an earth science perspective. Ziony J (ed), U.S. *Geological Survey Professional Paper, 1360*, 317–345.

Ze^zere, J. L., Oliveira, S. C., Garcia, R. A. C., & Reis, E. (2007). Landslide risk analysis in the area north of Lisbon (Portugal): Evaluation of direct and indirect costs resulting from a motorway disruption by slope movements. *Landslides, 4*, 123–136.

Chapter 18
Vulnerability Assessment of Landslide with the Help of Geospatial Approach in Western Himalayas, Upper Basin of River Sutlej, India

Amit Jamwal and Vikram Sharma

Abstract Vulnerability assessment is an important part of environmental management, and this approach is used for the identification of hazards and its potential risk in the upper basin of the river Sutlej. The geospatial tool was used to analyse, monitor and map landslide vulnerability. Rockslide, rock fall, slump, earthflow, and subsidence types of landslides were identified in the field. High summital convexity (1), rectilinear (0.8), high relative relief (>1000 m), high dissection ratio (>0.97), less forest cover (8%), slope aspects; southeast (1), south (0.9), the fine texture of soil (1), subhumid region (1), limestone-based lithology (0.9), high earthquakes magnitudes (1), and hydropower construction (1) were the major factors that indicated a high degree of vulnerability (0.68) and a high weighted score (0.58). The major finding of the vulnerability assessment indicated that 27% (1812 km^2) area of the basin had a high vulnerability of landslide; however, 39% (2617 km^2) area of the basin exists with low vulnerability. In the future, if anthropogenic activities increase in this basin, then the impacts of landslides and their loss of physical environment shall be increased.

Keywords Landslide · Vulnerability assessment · Geospatial approach

A. Jamwal (✉)
Aryabhatta Geo-informatics & Space Application Centre (AGiSAC), Shimla, Himachal Pradesh, India

Department of Geography, D.S.B. Campus, Kumaun University, Nainital, Uttarakhand, India

V. Sharma
Department of Geography, Faculty of Science, Banaras Hindu University, Varanasi, India

P. K. Shit et al. (eds.), *Geospatial Technology for Environmental Hazards*, Advances in Geographic Information Science, https://doi.org/10.1007/978-3-030-75197-5_18

18.1 Introduction

Hazards are natural phenomena, which also include geological and meteorological hazards such as earthquakes, volcanic eruptions, wildfires, cyclones, floods, droughts, avalanches, and landslides. Landslide refers to the terms of a different form of mass wasting that include rock fall, deep-seated slope failures, mudflow, and debris flow (Rosi et al., 2018). Lithology, rainfall, nature of the slope, status of land use, land cover, snowfall, aquifer recharge, increasing hydrostatic pressure in rocks cracks, soil structure, and forest fire are all factors that control frequency and magnitude of the mass movement (Wei et al., 2018) Physical and chemical weathering also weaken the rocks and also increase the incidences of landslides (Di Maio et al., 2017). In recent times landslide incidences identification is a very simple method. Remote sensing and geographic information system are an applied science tool, which is used in many fields. It is more helpful in tuff terrain where field observation is not possible. The vulnerability assessment of hazards depends on the presence of threat and its impact factor. The Sutlej Basin of district Kinnaur is known for its geohazards such as landslides, cloud bursts, earthquakes, and avalanches. Landslide is one of the major responsible incidences for the loss of physical landscape in the basin (Jamwal et al., 2019). The anthropogenic activities such as hydropower construction are one major triggering factor for the loss of physical and social environment. Slope failure is very common in the basin because of construction activities (Blasting and mining) (Fan et al., 2019).

The climate of the Sutlej Basin is subhumid to temperate type. The lower region has high rainfall, and the upper region has maximum rainfall in the form of snowfall. The snowfall region has arid condition, and incidences of shooting stone are very common (IPCC, 2007a, 2007b). The lower region (Rampur, Noli) suffered very much because of its varied climatic condition. Debris flow, earthflow, debris slide, rock fall, rock slide, slump, and subsidence were some common types of mass movement. When the slope material becomes saturated with water, it becomes debris flow or mudflow (Di Maio et al., 2017). Very fine grain clay, fine sand silt, and fine-grained pyroclastic material were found in the basin areas like Karcham Wangtoo, Shuda Rang, Dhakhu, Barang, and Tapri. Earthflow incidences were recorded in the area of Tapri, Shongtong, and Powari during the rainy season from June to August. This also increased pour pressure and decreases the shared stress of material (David et al., 1984).

Debris slide incidences were commonly observed in the lower portion of the basin where rainfall intensity was high and soil texture was coarse. The debris includes the small fragments of rocks, trees, and coarse grain of soil. Doris slides also occurred in the upper region of the basin where the impact of snowfall was high. The fine particles of soil and small pieces of rock were common with the rolling snow (Hutchinson, 1968). Small and shallow landslides were observed in the area like Kasang, Barang, Apka, Sumdo, Khab, and Reckong Peo. High permeable soil was on the top, and low permeable soil was in the bottom. But road cutting, dam construction, and blasting caused deep-seated landslides. The Urni landslide occurred after the high-intensity rainfall of 2014. The recorded average debris flow was 23.5 m/s (Kumar et al., 2019). The Urni villages (elevation, 2300 m) agriculture and horticulture land was damage. This landslide was one of the deep-seated landslides that were associated with slope failure in the form of rock fall, rockslide, rotational, transitional, and complex movement. The Sutlej Basin River was highly affected under the incidences of landslides from Khab to Tapri. The natural setting and land use status of this basin indicate the vulnerability of the physical landscape. The study revealed that incidences of landslides were increased because of haphazard development of hydropower project in the basin (Kuniyal et al., 2019). The vulnerability assessment includes the phenomena identification, monitoring, mapping, buffering, risk, and threat. It could be highly valuable on ground level if the development process considers the vulnerability map. But in reality, the landslide safety majors are not considered in the Indian Himalayas region. Management information system (MIS) is more effective for the real time monitoring of hazards (Chen & Wang, 2007). The landslide vulnerability, susceptibility, and risk depend on the topographical characteristics and human activities in that region. The selected indicators such as anthropogenic activities, curvature, geological structure, slope, slope profile, relative relief, land use and land cover, lithology, soil texture, and precipitation were taken to complete this process (Jamwal et al., 2019).

The landslide hazard analysis was done with factors such as geomorphology, geology, land use, and hydrology. High-resolution satellite imageries were very helpful to study topography very precisely (De La Ville et al., 2002). Along the National Highway 5, a number of landslides incidences were observed. The incidences of slope failure were very common on the convexity of the basin such as Dublin landslide, Spillow landslide, Khadra Dhaank landslide, Lippa landslide, Pangi Nala landslide, Powari landslide, Sapni landslide, Brua landslide, Kuppa landslide, Urni landslides, Sholding landslide, and Nathpa landslide (Fig. 18.1).

Landslide analysis through the geospatial approach provides us a valuable strategy to control the potential risk. Geospatial technology is also suitable for the vulnerability assessment of the physical landscape, and low vulnerable areas can be considered for development. High vulnerable areas can be put under the management.

18.2 Study Area

The landslide vulnerability assessment was done in the upper basin of Sutlej River, Himachal Pradesh, India. Geographically this area was extended from 31°30′12"N to 32°22′16"N and 77°40′16″E to 79°13′16″E. The total area covered under this region was 6401 km^2 and elevated from 2320 m to 6816 m. This region was covered by Tibet and China in the east direction and Shimla in the west. This basin was known for its geographical complexities. The MCT passing through this basin and a number of faults were found in this area. Main central thrust (MCT) was defined as the boundary between quartzite and phyllite, from the Lesser Himalayan sequence, and the orthogenesis biotitic-rich schist, which belongs to the Greater Himalayan

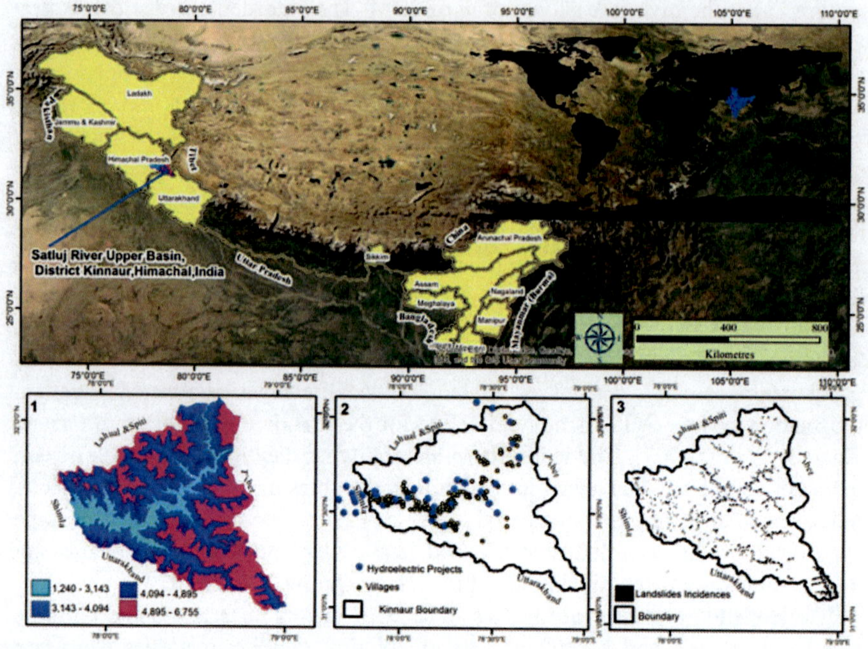

Fig. 18.1 Study area, upper basin, district Kinnaur, Himachal Pradesh. (1) Elevation status of upper basin. (2) Habitations and locations of HEPs. (3) Landslide incidences

crystalline complex (Daniel, 2003). The 80% (5483.92 km^2) basin area was found under the categories of high steep slope (>30°) and 74% area under the high altitude (>3000 m) (Jamwal et al., 2019). The basin had subhumid to arid temperate climate. This basin is known for its geohazards, and unique culture was related to tribal population of Mongoloid as well as Mediterranean. This region was known for its hydropower development and many social issues related to its development (Kuniyal et al., 2017). The upper basin of river Sutlej had the temperature type of subhumid temperate alpine highland and frigid aridic type. The low-altitude region of the basin had high rainfall during the monsoon season, and rainfall varies from 600 mm to 1400 mm. The average annual temperature of the basin was 13 °C. The total population (84,181) habituated in the 10% area of the basin and suffered huge losses during the incidences of hazards (Fig. 18.1).

18.3 Methodology

The vulnerability assessment of landslide was done on the basis of the selected parameters. All parameters were elected on the basis of geophysical aspects such as relative relief, anthropogenic activities, curvature, geological structure, slope aspect, slope profile, land use and land cover, lithology, precipitation, and soil texture. The slope, slope aspect, relative relief, and curvature were generated from the ASTER DEM (USGS, 2004) of 30 m resolution. The geological map was prepared on the basis of the Geological Survey of India (GSI). The soil texture map was prepared with the help of a second map of the National Bureau of Soil Science (NBSS). The slope profile map was prepared on the basis of the geomorphological map, and later it was corrected and analysed through a field survey (SOI). The precipitation map of the study area was prepared by using weather station data. The landslide points (GPS) were collected from the field along the Sutlej Valley, Tapri to Khab, and also along the Baspa Valley. The terrain was very complex, and it was difficult to collect all landslide incidences, so these landslide points were extracted from Google Earth. Google Earth images of WGS84 were used to extract the affected area under the landslides at the resolution of 15 m. Then, the shape files of landslide area were overlapped on the raster data set with other parameters. The extracted area was calculated on every subclass of parameters. The selected parameters were classified, and the affected areas were identified on these subclasses. The highest affected area was scored with the highest number from the total number of subclasses. The values were normalized from 0 to 1 to check the impacts on parameter classes.

The formula was used to normalized the value:

$$P_{\mathrm{N}}\ (\text{parameters normalized}) = P_{\mathrm{s}}\ (\text{Parameters Score}) \\ /N_{\mathrm{p}}\ (\text{Total the number of parameters}) \tag{18.1}$$

Then vulnerability was also identified in every subclass. The vulnerability scored on the basis of impact factors such as 0 for no vulnerability and 1 for vulnerability.

The risk was also identified on the basis of normalizing value and impact (threat value), and the formula is as follows:

$$\text{Risk} = \text{Impacts (Threat area)} \times \text{weight rank}\ (W_n) \tag{18.2}$$

Equation (18.2) was used to analyse the risk on every parameter. Then the average vulnerability and risk score were calculated. The vulnerability and risk score index was generated. Finally, the overlay analysis was done, and one vulnerability score map was prepared. This indicated the region with the high, medium, and low vulnerability of landslides.

18.4 Types of Landslides

Landslides were classified on the basis of its nature of occurrences and its geometrical shape. Rock fall, rock slide, debris slide, complex slide, and transitional debris were noticed as mass movement in the field. Crown cracks slide and transitional debris slide were commonly noticed in the field. Minor scarps were observed on the rectilinear section of the slope profile. Debris fall was noticed where soil structure belongs to coarse grain soil. Earthflow incidences were highly noticed in the upper region of Sutlej like that Akpa, Spillow, Khas, Pangi, and Kwangi. Debris flow was noticed in the upper basin of Sutlej and known for its devastating impact. Pagal Nall received a huge amount of rocks, fine soil, and clay with water with rapid movement (Varnes, 1978). During the rainfall in the basin, the incidences of debris were very commonly found in the place of Akpa, Ribba, Purbani, Powari, Shongtong, Karcham, and Ghanvi (Jamwal et al., 2019). Geological elements had the dominant impact on such types of landslides which include rotational failure, flow, falls, and debris slides. The basin was also known for its earthquake incidences. The vibration-based landslides occurred in places like Shoultu, Pangi, Kang, Purbani, Sudharang Dhaku, and Malling. The human-induced landslides were noticed in and around of hydroelectric project affected area of Khab, Jangi-Thopan-Powari, Shongtong Karcham, and Karcham Wangtoo. These landslides were caused due to the slope excavation, mining, and blasting (Fig. 18.2).

Fig. 18.2 Landslides incidences; (**a**) Tapri, (**b**) Near Pangi, (**c**) Near Powari, and (**d**) Near Barang

18.5 Result and Discussion

Identification and Status of Landslides as per the Selected Parameters

The Sutlej Basin in district Kinnaur is known for its complex topography. The basin had received a number of landslides of different scale; some were very large, and some were very small. The triggering factors were such as relative relief, slope, slope aspect, lithology, curvature, soil texture, lithology, effective land use, and anthropogenic activities. Geographic information system indicated that the landslide's occurrence value was higher in and around of hydroelectric project affected area of Khab, Jangi-Thopan-Powari, Shongtong Karcham, and Karcham Wangtoo.

Relative Relief (R_R)

Relative relief is one of the important topographic factors for the analysis of the landscape. The high degree of relative relief (R_R = 1244–6755 m) and high dissection indicate the high vulnerability of landslide and soil erosion (Singh, 2004a). In the Sutlej upper basin, maximum landslides occurred under the high relative relief (Muthukumar et al., 2009). The human activities in this elevation were the major cause of this environmental loss. The whole basin had high relative relief >3000 m which had high vulnerability (1) and risk and for agriculture and horticulture. Human properties and lives lost were observed along the river Sutlej Rampur to Khab with a risk score of 1 (Table 18.1).

Slopes

The slope was classified on the basis of A. Young classification (Young, 1964). The sloping nature of the region decided the vulnerability of landslide and erosion (Webb et al., 2011). The region had a high degree of slope, and maximum degraded area was observed under the vertical slope (17.01 km^2) and under the moderately steep slope (6.42 km^2). The vertical slope had a high degree of risk (17.01 km^2) and vulnerability scored as 1. The risk factor was increased on the steep slope (Table 18.1).

Slope Profile

The slope profiles were one of the deciding factors for the acceleration of eroded material and slope segments classified on the basis of geometrical shape (Singh, 2004b). Summit convexity, rectilinear section, free face, and basal connectivity were the main identified slope profiles in the basin. About 8.1 km^2 area was degraded

Table 18.1 Landslide impacts and status on different parameters. (W_n) *W*, Weighted; *n*, number of classes of parameter

		Area degradation (mass movements)			
S. No. parameters (Weight)	Classification	Identification and weight rank (W_n)		Vulnerability	Risk
1. Anthropogenic (W_4)	Under construction	5.47	1	1	5.47
	Commissioned	2.41	0.7	1	1.687
	Obtaining clearance	1.14	0.2	0	0.228
	Under investigation	2.03	0.5	0	1.015
2. Curvature (W_2)	Concavity	7.3 to −7.3	1	1	1
	Convexity	8.12 to −8.12	0.5	0	0.5
3. Geological aspects (W_2)	Geological structure	2 fault 1 thrust	1	1	1
	Magnitudes	5.5	0.5	1	2.75
4. Slope (W_6)	Gentle slope (0°–5°)	0.31	0.1	0	0.031
	Moderate slope (5°–10°)	3.31	0.3	0	0.993
	Moderate steep slope (10°–18°)	6.42	0.8	0	5.136
	Steep slope (18°–30°)	6.12	0.6	1	3.672
	Very steep slope (30°–45°)	4.45	0.5	1	2.225
	Vertical slope (45°–90°)	17.01	1	1	17.01
5. Slope profile (W_4)	Summital convexity	4.6	1	1	4.6
	Rectilinear section	1.9	0.7	1	1.33
	Free face	1.8	0.5	1	0.9
	Basal concavity	0.1	0.2	0	0.02
6. Relative relief W_2	R_R 1244–3000	D_I (0.93)	0.5	0	0.5
	$R_R > 3000$	D_I (0.97)	1	1	1
7. Land use and land cover (LULC) (W_8)	Settlements	0.15	0.2	1	0.03
	Agricultural	0.7	0.3	1	0.21
	Forest cover	0.8	0.5	0	0.4
	Wasteland	8.46	1	1	8.46
	Grass/grazing	8.39	0.8	1	6.712
	Scrubland	1.97	0.7	1	1.379
	Water bodies	1.17	0.6	0	0.702
	Snow/glacier	–	0.1	0	0
8. Slope aspect	Flat	0.85	0.2	0	0.17
	North	1	0.5	0	0.5
	North-east	0.71	0.3	1	0.213
	East	3.24	0.7	1	2.268
	South-east	9.47	0.8	1	7.576

(continued)

Table 18.1 (continued)

S. No. parameters (Weight)	Classification	Area degradation (mass movements) Identification and weight rank (W_n)		Vulnerability	Risk
	South	**14.12**	**1**	**1**	**14.12**
	South-west	2.93	0.6	1	1.758
	West	0.22	0.1	0	0.022
	North-west	0.91	0.4	1	0.364
9. Lithology	P_{t1} regionally metamorphosed	11.46	1	0	11.46
	P_{t3e} greenish grey sandstone	5.7	0.8	1	4.56
	Y granite and granitoid	0.91	0.2	1	0.182
	P_{g3o} Boulder conglomerate, sandstone, shale, clay	4.24	0.7	1	2.968
	OC limestone, siltstone, shale	2.31	0.4	1	0.924
	P_{t23} slate, phyllite, quartzite, grey shale	3.28	0.5	1	1.64
	P_{t2} Ortho-quartzite, basic volcanic	0	0.1	1	0
10. Soil texture	Coarse texture	4.3	0.5	1 0.5	2.15
	Fine texture	15.1	1	1	15.1
	Medium texture	3.8	0.2	1	0.76
	Rocky/badland	4.5	0.7	1	3.15
11. Precipitation	Subhumid	16.1	1	1	16.1
	Subtropical	4.9	1	1	4.9
	Subhumid temperate	8.9	0.5	1	4.45
			0.58	0.7	164.2

under all different segments. The summital convexity had high vulnerability (1) and high risk (4.6). The free face segments of the slope were highly affected due to construction activities (Table 18.1). Rock fall incidences were highly noticed under this section of the slope. The basal concavity segments of the slope profile were highly affected due to flood and soil erosion (Fig. 18.3).

Slope Aspects

Solar energy is the main driving force that affects the slope aspects (Dearman, 1974). In general, the slope aspect can influence the distribution and density of mass

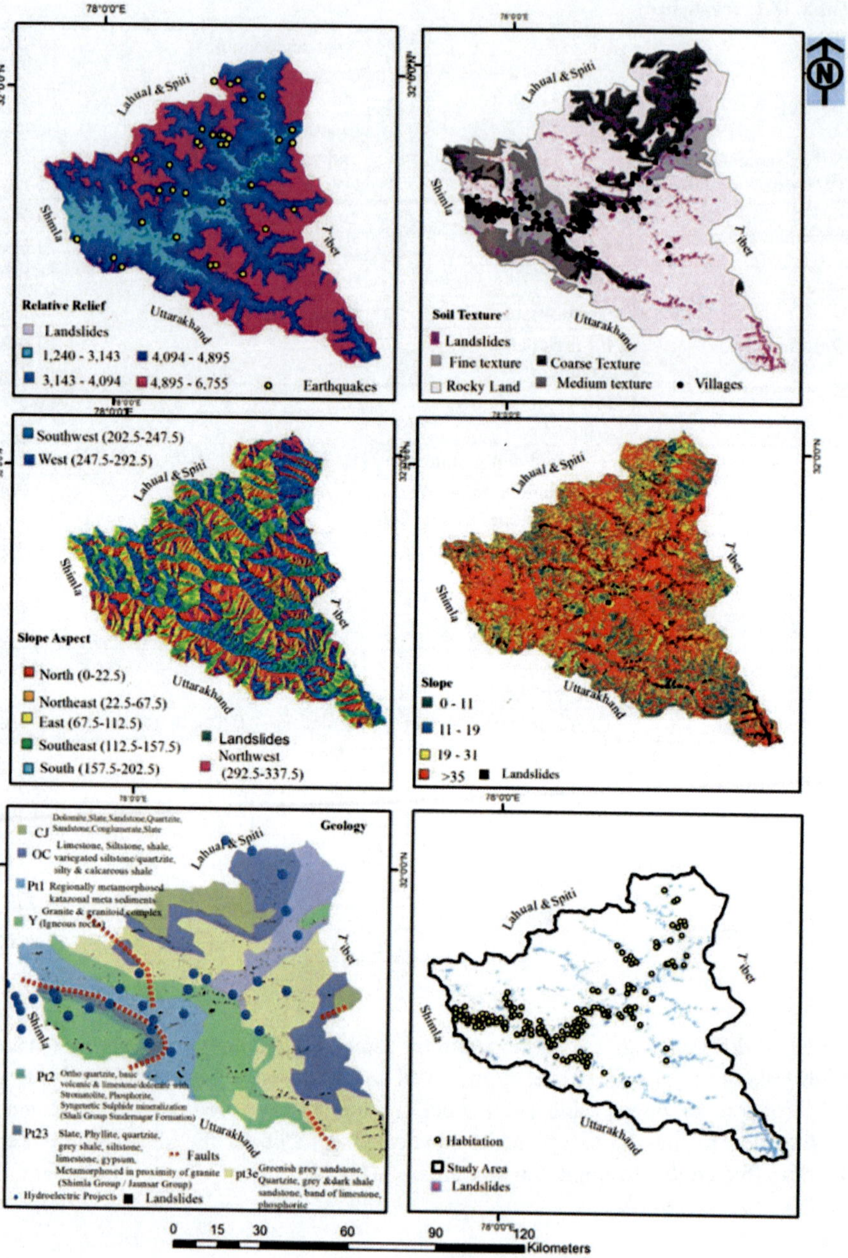

Fig. 18.3 Parameter-based analysis

movement by controlling the concentration of soil moisture or orientation of tectonic fracture (Wieczorek et al., 1997). In the northern hemisphere above the 33° latitudes, the maximum sunny area was found under the south, south-west, and south-east aspect of a slope. The high risk was found on the south (1), south-east (0.8), east (0.7), and south-west (0.6) aspect of slope. High vulnerability (1) and high risk (14.2 km^2) were observed under the slope aspect of the south.

Curvatures

The acceleration and deceleration of flow across on earth's surface depend on the concavity and convexity of a slope. A negative value (−10.4) indicates that the surface is upwardly convex at that cell, and flow will be decelerated. A positive profile (10.4) indicates that the surface is upwardly concave at that cell, and the flow will be accelerated (Rautelal & Lakhera, 2000). The incidences of soil erosion and landslide were noticed both curvatures of the slope. But affected pixels had a high value on the concavity of slope. The vulnerability (1) and risk (1) were high on the concavity of slope. The risk factors of human settlements were very high at the convexity segments of the slope, and the concave segments of the slope were highly affected due to anthropogenic activities (road and dam construction). The convexities of slopes were highly scored as 1, and concavities of the slope were scored as 0.5 (Table 18.1).

Soil Texture

Soil texture and rainfall are two important factors which affect the movement of landslides. Coarse texture, fine texture, medium texture, and very fine texture were classified categories of soil in the study region. Very fine textures were commonly observed in the upper basin and coarse texture in the lower basin. In the lower basin of the study region, the coarse texture gives birth to debris slides and debris fall. The very fine texture of soil generates the mudflow with tiny particles in the presence of rainfall. This was very common in the upper region of district Kinnaur, where climate types are arid and very cold (subhumid temperate). Here maximum rainfall in the form of snowfall and vegetation belongs to alpine type. The rocky surface of the region was not very much affected but un-stabilized by haphazard human activities.

High vulnerability and risk were found under the fine texture of the soil (1). This indicates a high risk (15.1 km^2) of the physical landscape. Fine textures were scored as 1, the coarse texture was scored as 0.5, a medium-fine was scored as 0.2, and rocky badland was scored as 0.7 (Table 18.1). The soil textures of the region were also determined by the climatic condition of the region. But it's very fine, and the fine texture of the region had sparse vegetation, which was more susceptible to landslides and soil erosion (Jenny, 1941; NRSC, 2010).

Lithology

The lithology of the study region is a type of sandstone, shale, clay, limestone, siltstone, and regionally metamorphism. The lithology is one of the dominant factors which control the rate of surface erosion. Lithology consist the rock types such as alluvial, sedimentry, limestone, siltstone, shale, and dolomite in the lower regions, were affected with mass movement during the season of rainfall. Below the Shongtong and Tapri area region, most of the rock strata were loosed because of the high rainfall intensity. The highest affected area was found under the categories of regionally metamorphosed rocks type Pt1 and Pg3o and scored as 1 with high risk and a high risk of 11.46 km^2. Boulder conglomerate, sandstone, shale, and clay were scored as 0.7, P_{t23} is scored as 0.8, limestone, siltstone, and shale OC were scored as 0.4, greenish grey sandstone p_{t3e} was scored as 0.8, granite granitoid base was scored as 0.2, and orthoquartzite basic volcanic rocks were scored with the lowest 0.1 (Gupta, 2003).

The average magnitude of the basin varies from M5.5 to 6.0, and this magnitude was sufficient to generate small- and large-scale landslides (Fig. 18.4). The slopes cutting process were including the vibration of slopes area which later becomes the causes of slopes failure. These continuous vibrations of magnitude were generating a small number of landslides. The Sudharang Dakhu landslide occurred due to high rainfall in June 2014 but still continuous sliding one of possible causes was earthquakes vibrations (Keefer, 1994). Shallow-focus earthquakes were crustal earthquakes, and they existed in the crustal layer. Deep-focus earthquakes were generated because of the collision of two plates and generated from the depth 300–700 km. Deep-focus earthquakes generate high energy, which had a high vulnerability of destruction (Meunier et al., 2013).

Land Use and Land Cover (LULC)

The study region has covered only 8% forest cover, the maximum area of the study region falls under snow-covered 2490 km^2 (39%), and the wasteland was 2030.63 km^2 (31%). Barren and sparsely vegetated areas were more prone to weathering and erosion (Majumder et al., 2019). According to forest policy, the minimum 33% area should be covered under the forest cover for a healthy landscape.

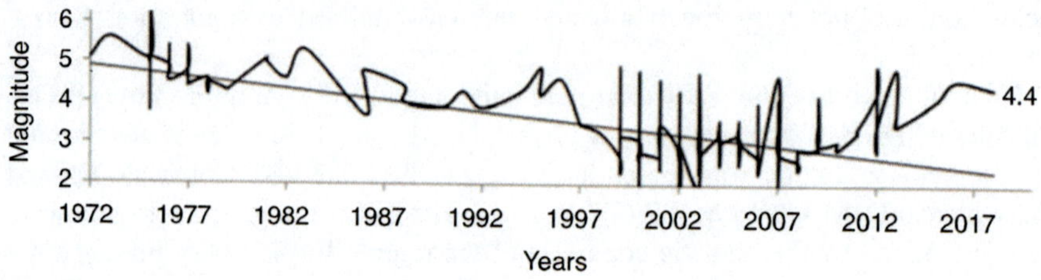

Fig. 18.4 Magnitude of earthquakes

The percentage of wasteland and snow-covered land was very high, which was 70%. The 1.89 km^2 of agriculture land was very low and is about 1.2%, which indicates the tuff terrain and low possibility of agriculture (0.7 km^2) and settlement (0.15 km^2) development in this region.

The increase of wasteland is a very serious problem in the basin, and wasteland has high vulnerability (1) and high risk of 8.46 km^2. It was evident from the field survey that the other land use type's agriculture (0.7 km^2), settlement (0.15 km^2) types were highly affected by mass movement and agriculture was scored as 0.7 and settlements were scored as 0.2. The scrubland (1.97 km^2) was scored as 0.7. A forest of the region was also affected (0.8 km^2) under the mass movement and scored as 0.5 (Table 18.1).

Anthropogenic Activities

The Sutlej Basin of district Kinnaur is known for its development of hydroelectric projects. Different types of hydroelectric projects such as under construction, commissioned, obtaining clearance and under investigation were identified in study region. Under construction types, hydroelectric projects have a number of landslide incidences. Tunnelling and blasting had adverse impacts on human settlements and the physical landscape (Kuniyal et al., 2015; Kuniyal et al., 2017). Slope instability is found under the surrounding area of hydropower projects (CEIA, 2014). Increased landslide incidences, flash floods, river morphological changes, water quality deterioration, reduction in agricultural/horticultural production, forest degradation, land degradations, inadequate compensation due to construction activities, damage to human health due to dust, and cracks in houses are major adverse impacts noticed within the project-affected area (Lata et al., 2017) (Fig. 18.5). The impact value was high under the construction types of HEPs, which have a threat of mass movement and scored with high vulnerability and risk (1). The susceptibility score (21.88 km^2) was high under these categories. But the other categories of the hydroelectric projects have a low vulnerability and less susceptibility. Land degradation and loss of agricultural land are found under the construction hydroelectric projects. A high vulnerability is found within the buffer of under-construction hydroelectric projects.

Precipitation

Precipitation is one of the dominant factors for the generation of landslides. About 80% of landslide is related to rainfall, and the remaining 20% belongs to the other responsible factors. According to the intergovermental panel of climate change (IPCC) 2014 model projection relative to 1986–2005, RCP2.6 (left) and RCP8.5 (right), the dotted pattern shows the projected change and variability in rainfall. This IPCC model also indicated the sign of the change (IPCC, 2014) (Fig. 18.6). The precipitation pattern was different in the study region which was influenced by

Fig. 18.5 Glimpse of study area; (**a**) degraded river valley, (**b**) landslides at Sudharang Dakhu, (**c**) interaction with respondent, (**d**) River valley at Powari

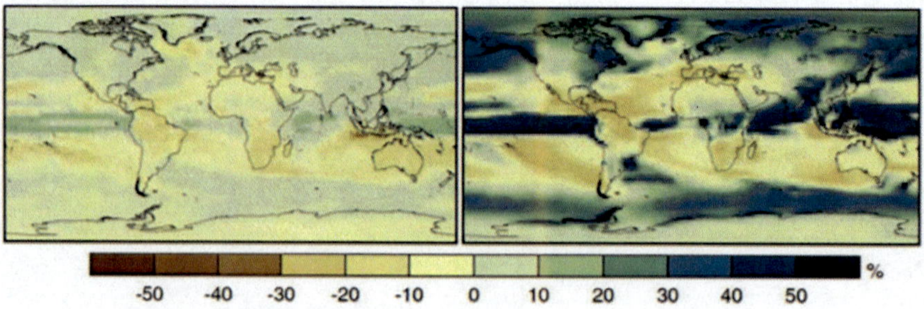

Fig. 18.6 Change in average precipitation (1986–2005 to 2081–2100) after: IPCC, 2014

topographic variance (434–6448 m). Isohyets study made it clear that the annual isohyets varied from 100 to 1400 mm. Annual rainfall in this basin was decreased from the Lesser to the Greater Himalayan range. Subhumid subtropical region (below Tapri) received maximum incidences of the landslide which was scored as

1, and above Tapri, the region was found under the sub-temperate which received fewer incidences of landslides which were scored as 0.5 (Table 18.1).

18.6 Comprehensive Vulnerability

No vulnerability region had no any anthropogenic activities and highly covered with forest. The low vulnerability region had a vulnerability score of 0.3–0.5 and sparsely covered with the vegetation. The incidences of landslides were also recorded in this region. Moderate vulnerability score was varies from 0.5 to 0.7, and human settlements were also recorded in this region. The incidences if earthquakes were also recorded in this zone. This zone was not much affected by hydropower development but highly affected by the incidences of slope failures because of road cutting. The region of high vulnerability had most incidences of landslides, floods, and earthquakes. The region of high vulnerability had a high vulnerability score >9 with high losses of the landscape. High incidences of landslides were also recorded in this region (Fig. 18.7). This region recorded along the river Sutlej from Tapir to Khab. The numbers of under-construction hydroelectric projects were developed in this

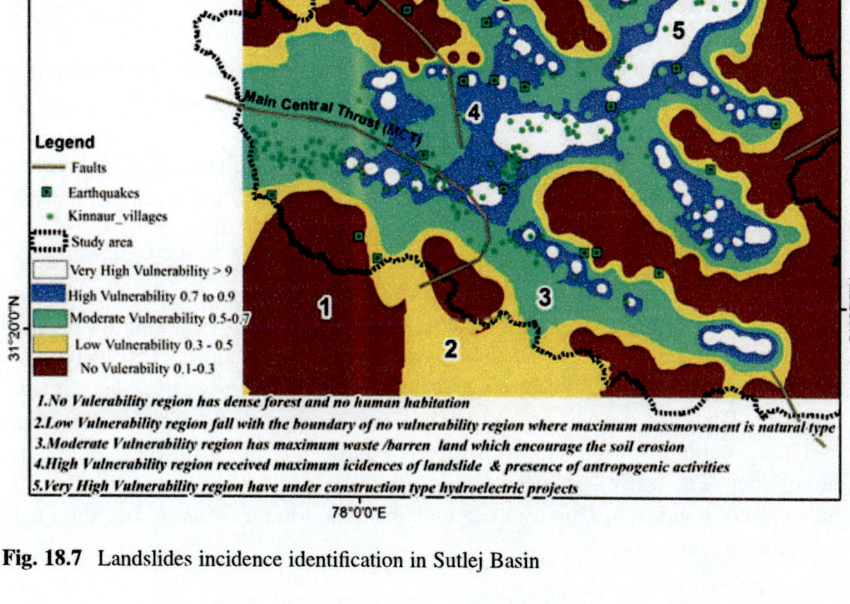

Fig. 18.7 Landslides incidence identification in Sutlej Basin

region. Constructed hydroelectric projects were more responsible for the number of landslides. The development of hydroelectric projects and any other developmental activities should be a sustainable way, and all suggestive approaches and methods should be adopted to control the incidences of landslides.

18.7 Suggestive Approach to Control the Incidences of Landslides

- In the Sutlej Basin, structural managements were very poor and not properly followed. Channel linings are another method for stabilizing a stream. The boulders could be deposited near the bank of the river. It could be helpful to protect the riverbank and settlements/road near the river. This can be used near the riverbank of Karcham Wangtoo. Linings are not costly like the check dam.
- Basically check dams are small, sediment storage dams which are made of dressed and undressed stone. This can be implemented in the upper Sutlej Basin from Khab to Rampur. In Himachal Pradesh the check dam formation process was completed under the forest department. This technique reduces the soil erosion, reduces the steep gradient of the river, and offers toe support to small stream slopes.
- Ditches and drains method could be implemented in the convex types of slopes, where the soil of the lower slope segment has a fine to medium texture. The lower trench could be little excavated to the base of the shallow soil. The eroded material will be deposited on the lower section of the slope. The stability of the concave slope could be attained. The ditches and drains methods could be implemented in Pangi and Powari villages.
- Horizontal drain piping is a commonly used method to reduce the risk of landslide. In this method, PVC pipes were installed in the wall. This hydraulic method reduces the water table and also reduces the risk of slope failure. This method will be very effective near the Powari region where the soil is fine and sandy and the water table fluctuates with rainfall. In the rainfall, the water table raises, and vulnerability of landslide is increased. The drain pipes reduce the intensity of storing water in sandy soil and reduce the risk of landslides.
- The gabion wall-based wire mesh technique is widely used in the basin because this technique required less engineering knowledge. Most of constructed works of gabion walls in Tapri, Kwangi, Karcham, Powari, Brang, Shongtong, Sudharang Dhaku, Khab, Sangla, and Chitkul were done by local labourers. This technique became helpful to cut soil erosion and landslide incidences but not very much effective. There is a need to strengthen the base foundation of gabion walls. Large and long gabion walls were constructed on the left bank of river Sutlej in Shongtong. The army settlements were protected by this gabion wall from a massive flood in 2014. This wall can be constructed in the Powari village on the left bank of river Sutlej.

- It is a very serious matter that in the Sutlej Basin 8% area covers green cover and 36% falls under barren or wasteland. The slope stabilized through the growth of vegetation. The shrub is enough to control soil erosion and landslide incidences. The shrub-covered area was less prone to landslides and reduces the surface runoff. Bioengineering techniques can be used to stabilize the slope in the area of Urni and Apka landslides. But here slope was very steep, and hydraulic seedling can be used for the reduction of slope failure incidences. However, this region was very complex but not impossible.
- In the valley along the roadside, >65% area had the steep slope, and the problem of rock fall and shooting slope was very high from the Tapri to Khab road and another link road. Wire mesh can be placed on the rocky surface of the slope. It can be helpful to protect the people from the falling rocks. This mesh can be helpful to prevent small rocks less than 0.75 m from falling. The rock curtain can be used in and surrounding of highly vulnerable area (Apka) of the Sutlej Basin, where wind speed is very high and rolls out the rock fragment on the road. An open rock shed can be constructed where the slope was very steep and the problem of shooting stone was very high (Apka and Khab).
- This is best to avoid the blasting and mining, according to the people of the study region, as they are the most destructive of the region due to the anthropogenic activities. The blasting and mining activities lead to incidences of slope failures and landslide incidences. A hydraulic rock hammer could be used to down the rock from the slope.
- If we see the future perspective of this research, this validates the way for physical vulnerability. However when we go to resource management and their sustainable development, then we have to go through geospatial technology. The real implementation of vulnerability assessment required micro-level study and also requires a better budget.

Acknowledgement The authors would like to thank the cooperation of local people during the time of the field survey. I am also thankful to the Kumaun University Nainital for giving me the opportunity of research in the upper basin of Sutlej River and also thankful to the Aryabhatta Geo-informatics & Space Application Centre (AGiSAC), Himachal Pradesh, for support of research and development.

References

Chen, Z., & Wang, J. (2007). Landslide hazard mapping using logistic regression model in Mackenzie Valley, Canada. *Natural Hazards, 42*, 75–89. https://doi.org/10.1007/s11069-006-9061-6.22

Daniel, C. G. (2003). Exhumation of the Main central thrust from lower crustal depths, eastern Bhutan Himalaya. *Journal of Metamorphic Geology, 21*(4), 317–334.

David, K., Keefer and Arvid, M. (1984) *Earth flows: Morphology, mobilization, and movement, geological survey professional paper geological Society of America, Rock-colour chart* (pp. 1–55).

De La Ville, N., Diaz, A. C., & Ramirez, D. (2002). Remote sensing and GIS technologies as tools to support sustainable management of areas devastated by landslides, environment. *Development and Sustainability, 4*(2), 221–229.

Dearman, W. R. (1974). Weathering classification in the characterization of rock for engineering purposes in British practice. *Bulletin. International Association of Engineering Geologists, 9*, 33–42.

Di Maio, C., Vassallo, R., Scaringi, G., De Rosa, J., Pontolillo, D. M., & Grimaldi, G. M. (2017). Monitoring and analysis of an earthflow in tectonized clay shale and study of a remedial intervention by KCl wells. *RivistaItaliana di Geotecnica, 51*, 48–63. https://doi.org/10.19199/2017.3.0557-1405.048

Fan, X., Qiang, X., & Scaringi, G. (2019). The long run out rock avalanche in Pusa, China, on august 28, 2017: A preliminary report. *Landslides*, 1–16. https://doi.org/10.1007/s10346-018-1084-z

Gupta, R. P. (2003). *Remote sensing of geology* (pp. 498–524). Springer.

Hutchinson, J. N. (1968). Mass movement. In Fairbridge (Ed.), *Encyclopedia of geomorphology* (pp. 688–695). Reinhold Publishers.

Intergovernmental Panel on Climate Change. (2014). *Climate Change 2014: Synthesis report. Contribution of working groups I, II and III to the fifth assessment, Report of the Intergovernmental Panel on Climate Change*, Geneva, Switzerland, p. 151.

IPCC. (2007a). A: Climate change 2007: Synthesis report, Contribution of working groups I, II and III to the fourth assessment report of the intergovernmental panel on climate change (pp. 1–104), Core Writing Team, R.K. Pachauri, and A. Reisinger (eds.), IPCC, Geneva, Switzerland.

IPCC. (2007b). Synthesis report—An assessment of the intergovernmental panel on climate change. Retrieved April 1, 2019, from https://portals.iucn.org/library/sites/library/files/documents/IO-UN-IPCC-2007-004pdf

Jamwal, A., Kanwar, N., & Kuniyal, J. C. (2019). Use of geographic information system for the vulnerability assessment of landscape in upper Satluj basin of district Kinnaur, Himachal Pradesh, India. *Geology, Ecology, and Landscapes*. https://doi.org/10.1080/24749508.2019.1608410

Jenny, H. (1941). *Factors of soil formation* (pp. 13–134). McGraw-Hill.

Keefer, D. K. (1994). The importance of earthquake-induced landslides to long-term slope erosion and slope-failure hazards in seismically active regions. *Geomorphology and Natural Hazards, 12*, 265–284. https://doi.org/10.1016/B978-0-444-82012-9.50022-0

Kumar, V., Gupta, V., Jamir, I., & Chattoraj, S. L. (2019). Evaluation of potential landslide damming: Case study of Urni landslide, Kinnaur, Satluj valley, India. *Geoscience Frontiers, 10*(2), 753–767. https://doi.org/10.1016/j.gsf.2018.05.004

Kuniyal, J. C., Jamwal, A., & Kanwar, N. (2019). Vulnerability assessment of the Satluj catchment for sustainable development of hydroelectric projects in the northwestern Himalaya. *Journal of Mountain Science, 16*, 2714–2738. https://doi.org/10.1007/s11629-017-4653-z

Kuniyal, J. C., Lata, R., & Kumar, A. (2017). Strategic environmental assessment (SEA) of hydropower projects. *Current Science, 113*(12), 2239–2240.

Kuniyal, J. C., Shashni, S., & Kumar, A. (2015). Strategic environmental assessment. *Current Science, 108*(4), 480–481.

Lata, R., Herojeet, R., & Dolma, K. (2017). Environmental and social impact assessment: A study of hydroelectric power projects in Satluj Basin in district Kinnaur, Himachal Pradesh, India. *International Journal of Earth Science and Engineering, 10*(02), 270–280. https://doi.org/10.21276/ijese.2017.10.0219

Majumder, A., Kingra, P. K., Setia, R., Singh, S. P., & Brijendra, P. (2019). Influence of land use/land cover changes on surface temperature and its effect on crop yield in different agro-climatic regions of Indian Punjab. *Geocarto International*. https://doi.org/10.1080/10106049.2018.1520927

Meunier, P., Uchida, T., & Hovius, N. (2013). Landslide patterns reveal the sources of large earthquakes. *Earth and Planetary Science Letters, 363*, 27–33. https://doi.org/10.1016/j.epsl.2012.12.018

Muthukumar, M., Ramasamy, S. M., Mohammad SartajBasha, S. K., & Kumanan, C. J. (2009). Geomorphic control of landslides, The Nilgiris Mountains, South India. *Journal of Indian Landslides, 2*, 35–40.

NRSC. (2010). *Evaluation of IRS-1C data for mapping soil resources and degraded lands*, Project report (pp. 81–107).

Rautelal, P., & Lakhera, R. C. (2000). Landslide risk analysis between Giri and tons Rivers in Himachal Himalaya, India. *International Journal of Applied Earth Observation and Geo-information, 2*, 153–160.

Rosi, V., Tofani, L., Tanteri, C., Tacconi, S. A., Agostini, F., Catani, & Casagli, N. (2018). The new landslide inventory of Tuscany (Italy) updated with PS-InSAR: geomorphological features and landslide distribution. *Landslides, 15*, 5–19. https://doi.org/10.1007/s10346-017-0861-4

Singh, S. (2004a). *Geomorphology* (4th ed., pp. 381–382). Kalyan Publication.

Singh, S. (2004b). *Geomorphology* (5th ed., pp. 267–296). Kalyan Publication.

USGS. (2004). Lynn highland: Landslide types and processes. Retrieved from http://pubs.usgs.gov/fs/2004/3072/fs-30-72.

Varnes, D. J. (1978). Slope movement types and processes, in Schuster, R.L., Krizek, R.J., eds., Landslides—Analysis and control: National Research Council, Washington, D.-C. *Transportation Research Board, Special Report, 176*, 11–33.

Webb, A., Yin, A. G., & Harrison, A. (2011). Cenozoic tectonic history of the Himachal Himalaya (northwestern India) and its constraints on the formation mechanism of the Himalayan Orogen. *Geosphere, 7*, 1013–1061. https://doi.org/10.1130/ges00627.1

Wei, H., Scaringi, G., Xu, Q., Van, A., & Theo, W. J. (2018). Suction and rate-dependent behavior of a shear-zone soil from a landslide in a gently-inclined mudstone-sandstone sequence in the Sichuan basin, China. *Engineering Geology, 237*, 1–11. https://doi.org/10.1016/j.enggeo.2018.02.005. ISSN: 0013-7952.

Wieczorek, G. F., Mandrone, G., & Colla, L. (1997). The influence of hill-slope shape on debris flow initiation. In C. L. Chen (Ed.), *Debris-flow hazards mitigation: Mechanics prediction and assessment* (pp. 21–31). American Society of Civil Engineers.

Young, A. (1964). Deductive models of slope evolution. *International Geographical Union Slopes Commission 3*, 45–66.

Chapter 19
Application of Fractal Dimension Technique on a Badland Topography in Tapi Basin, Deccan Trap Region, India

Veena Joshi

Abstract A fractal is a fragmented geometric shape which is characterized by scale invariance under contractions or dilations. Fractal is statistically a self-similar body, which implies that some aspect of a process or phenomena is invariant under scale-changing transformation. Fractal dimension is applied in geomorphology in wide range of topics, such as tectonics, coastal configuration, river basin geometry, landslides, soil studies, karst features, etc. With the availability of high-resolution digital elevation data and operating GIS tools, further new interests have arisen in the technique. In the present study, fractal dimension has been applied to an alluvial badland topography along the banks of Tapi River, in the Deccan Trap region, India. The area is characterized by semi-arid climate. A newly developed software 'Viz-Morphotec' was used to calculate fractal dimension (*D*) for the entire area of badlands which yielded that '*D*' values between 2.9 and above clearly coincide with the location of badlands in the basin. Two sample catchments were selected for determining fractal properties of these badlands and to understand microprocesses operating in this topography. Fractal dimensions were calculated at three levels, namely, linear, perimeter and surface. Variograms were computed for both the catchments also. Results indicate a multifractal topography, where two or more processes are operating in the landscape. The curves of the variograms indicate possible influence of diffusional and erosional processes operating on the topography or could be a result of tectonics or changes in the climatic conditions that are still manifested in the landscape. Hence, these badland areas indicate multifractal topography where more than one process are operating within it. The results also reveal that rivers are actively eroding, and linear erosion is predominant in the whole region.

Keywords Fractal dimension · Viz-Morphotec · Variogram · Badlands · Deccan Trap region · Cartosat DEM

V. Joshi (✉)
Department of Geography, S.P Pune University, Pune, India
e-mail: vujoshi@unipune.ac.in

P. K. Shit et al. (eds.), *Geospatial Technology for Environmental Hazards*, Advances in Geographic Information Science, https://doi.org/10.1007/978-3-030-75197-5_19

19.1 Introduction

A fractal is a fragmented geometric rough shape that can be delineated into parts, each of which is a reduced-size copy of the whole structure. The characteristics, so-called symmetries, in a fractal signify invariance under contractions or dilations. The roughness and fragmentation of a natural fractal shape neither fluctuate up and down or vanish but stay unchanged at any zoom level. The key to the whole structure is still attached to the structure of every piece in a fractal. Fractals possess self-similarity in a structure across all scales. As we zoom in or zoom out a feature, the geometry and appearance of a fractal surface remain unchanged or self-similar (Mandelbrot, 1989). Self-similarity has been used under two contexts. It suggests that each part of the body/structure of a fractal is very much like a tree branch or a piece of a broccoli which is simply a linear geometric reduction of the whole body. It also expresses that not only is the reduction linearly geometric but also the ratio of reduction is the same in all directions. Self-similarity indicates that some aspect of a process or phenomena is invariant under scale-changing transformation, such as simple zooming in or out. In short, a self-similar structure repeats a unit pattern on different scale or size. Self-affine shapes and self-affinity are recent terms that are mostly used in geomorphology and relief analysis, which suggests that though the reduction is still linear, it is different in a way that, as one goes from a large piece to small piece, we must contract the vertical and horizontal coordinates in different ratios (Burrough, 1981). If a natural scene is self-similar, determination of its scale is not possible. The simplest visual test of self-similarity is to see whether any enlargement of any part of a structure remains indistinguishable from the whole or from any other part of the same body.

Since the term fractal was first coined by Benoit Mandelbrot in 1975, fractal models and related analysis techniques were matters for speculative coffee break discussions between sessions at geoscience conferences. For the first time, fractal geometry proposed by Mandelbrot (1975) has provided the possibility of precisely simulating and describing landscapes by employing a mathematical model. Fractal analysis without any question has shown to look beyond the traditional techniques to contain 'new' information in a phenomenon. If a landscape possesses fractal character, it should reveal statistically self-similar or statistically self-affine nature. Simulating landforms processes using Mandelbrot's fractional Brownian motion (fBm) has gained enormous popularity in the last three decades. The application of fractal technique in geomorphology started sprouting only by the 1980s, though it was suggested that the application of fractal was always important to geomorphology, even before Mandelbrot's coining the word 'fractal'. Reviewing all the papers of fractal in geomorphology is beyond the scope of this paper; however, some of the most cited landmark papers that focused on different themes of geomorphology are presented here.

Goodchild (1982) presented fractal Brownian process as a terrain simulation model. Mark and Aronson (1984) studied scale-dependent fractal dimension of topographic surface: with special application in geomorphology. 'Self-similar' profiles of deep-sea topography were investigated by Fox (1986). Culling and Datko (1987) applied fractal geometry to soil-covered landscapes. Goodchild and Mark (1987) conducted a comprehensive review of the relevance of fractals to geography under three headings: self-similarity, the response of measure to scale and the recursive subdivision of space. The fractal geometry of the landscape was measured by Milne (1988). Gilbert, 1989 evaluated whether the topographic data sets have fractal properties or otherwise. Fractal mapping of digital images of Arizona were conducted by Huang and Turcotte (1989). Fractal sinuosity of stream channel was determined by Snow (1989). Unwin (1989) introduced fractals and geosciences to computer geosciences. The surface roughness of talus slope was studied by applying fractal techniques by Andrle and Abrahams (1989). Fingerprints and fractal terrain were assessed by Piech & Piech, 1990. Polidori et al. (1991) applied fractal technique to assess quality of digital elevation model.

In 1992, a special issue was published in the journal *Geomorphology*, devoted to the application of fractal geometry on landform analysis, where there are nine articles in the theme, ranging in topic including catchment evolution model, fluvial land sculpturing, tectonic, climate and lithology control, measurement of self-affinity, fractal significance of drainage basin parameters, fractal dimension of sinkholes, desert storm, etc. (Willgoose et al., 1991a; Willgoose et al., 1991b; Tarboton et al., 1992; Chase, 1992; Klinkengerg, 1992; Nathaniel & Chase, 1992; Ouchi & Matsushita, 1992; Breyer & Snow, 1992; Reams, 1992; Mayer, 1992; Snow, 1992). Whether there exists any relationship between fractals and morphometric measures has been investigated by Klinkengerg (1992) in the same issue. A review article on fractals, fractal dimensions and landscapes came out in 1993 by Xu et al. The next year, another review article was published by Jie & Haosheng (1994) on fractal geomorphology, where the authors focused on the issues of geomorphic fractals and geomorphic conditions, fractal dimensions and geomorphic processes, digital modelling of fractal landscape, range scale of geomorphic fractals and fractal characteristics of geomorphic phenomena in space and time. Gao & Xia published fractals in *Physical Geography* in 1996 (Gao & Xia, 1996). Chaos, fractals and self-organization in coastal geomorphology were investigated by Baas (2002). The author simulated dune landscapes in vegetated coastal environments. Hagerhalla et al. (2004) studied fractal dimension of landscape silhouette outlines as a predictor of landscape preference. Taud and Parrot (2005) measured roughness of DEM applying the local fractal dimension. Another review paper came out by Sun et al. (2006) where the scholars took a survey of several methods for fractal dimension calculation which are commonly used in many studies. Turcotte (2007) related fundamental statistical properties of landform and drainage networks that have been developed in statistical physics. Fractal dimension of a badland topography at Deccan was investigated by Joshi et al. (2009). Relationships between the fractal dimension of the drainage basins were assessed by Khanbabaei et al. (2013). A review article on the methods of fractal geometry used in the study of complex geomorphic networks was published by Kusák, 2014, where the focus was on the comparisons of the basic terms used in fractal geometry. Tectonic and lithological control on topography and their reflection in fractal dimension were presented by

Liucci & Melelli (2017). Pardo-Iguzquiza et al. (2019) took a complete review of fractals in karst to demonstrate fractal behaviour of karst topography. A case study of fractal-based modelling and spatial analysis of urban form and growth of Shenzhen in China was conducted by Man & Chen, 2020. Patuano & Tara (2020) presented the summary of a literature review of the methods and interpretations of fractal geometry, currently used in landscape architecture. Over 40 studies were examined for their use of fractal concepts within the analysis of landscape-related elements.

A brief traverse through the fractal journey of the landscape from the 1980s till the present, as outlined in the previous paragraphs, revealed that the application field is widely variable, ranging from tectonics, coastal configuration, river basin geometry, landslides, soil studies and hordes of other applications. With the availability of high-resolution digital elevation data and operating GIS tools, further new interests have arisen in the technique.

Fractal geomorphology emerges as a new discipline to evaluate the origin, process and distribution of relief on the earth. Though landforms widely exhibit fractal nature, it is complex to asses it due to the non-homogeneity of relief geometry in space. The main objective of the present study is to assess the applicability of fractal dimension to identify badland locations and to understand microprocesses operating in badland topography. The significance of the study is that fractal dimension technique addresses spatial characteristics of the landform features and therefore represents a powerful method to investigate the relationships between landforms and their underlying processes.

19.2 Study Area

The study area selected for the study is the Tapi Basin which is the second largest west flowing peninsular river in India (Fig. 19.1). It flows through the states of Madhya Pradesh, Maharashtra and Gujarat. Its length is 724 km, and it drains an area of 64,750 km^2. It rises from Satpura Ranges at an elevation of 762 m and drains into Arabian Sea. It flows through a rift valley, and its trough accommodates considerable deposits of alluvium. The area extends between 75°15′E to 75°45′E longitude and 20°25′ to 22°35’N latitude. The climate is semi-arid with the annual average rainfall between the range of 650 and 780 mm. December is the coldest month with the mean daily minimum temperature at 11.9 °C and the mean daily maximum at 29.8 °C. Temperatures begin to rise steadily from the beginning of the March, reaching its peak in May. Mean daily maximum temperature reaches 45–48 °C on the hottest day in May. Natural vegetation is mostly absent except in the form of acacia thorny plants. Deccan Trap region as a whole is rocky landscape where sediment deposits are restricted and thin, if at all they are present. River Tapi flows for a large part within a rift valley, and its trough accommodates considerable deposits of alluvium. On the right-hand bank of the river, numerous tributaries, collecting their headwaters from foot of Satpura systems, have dissected the alluvium-covered pediment surface and have caused badland formations along a

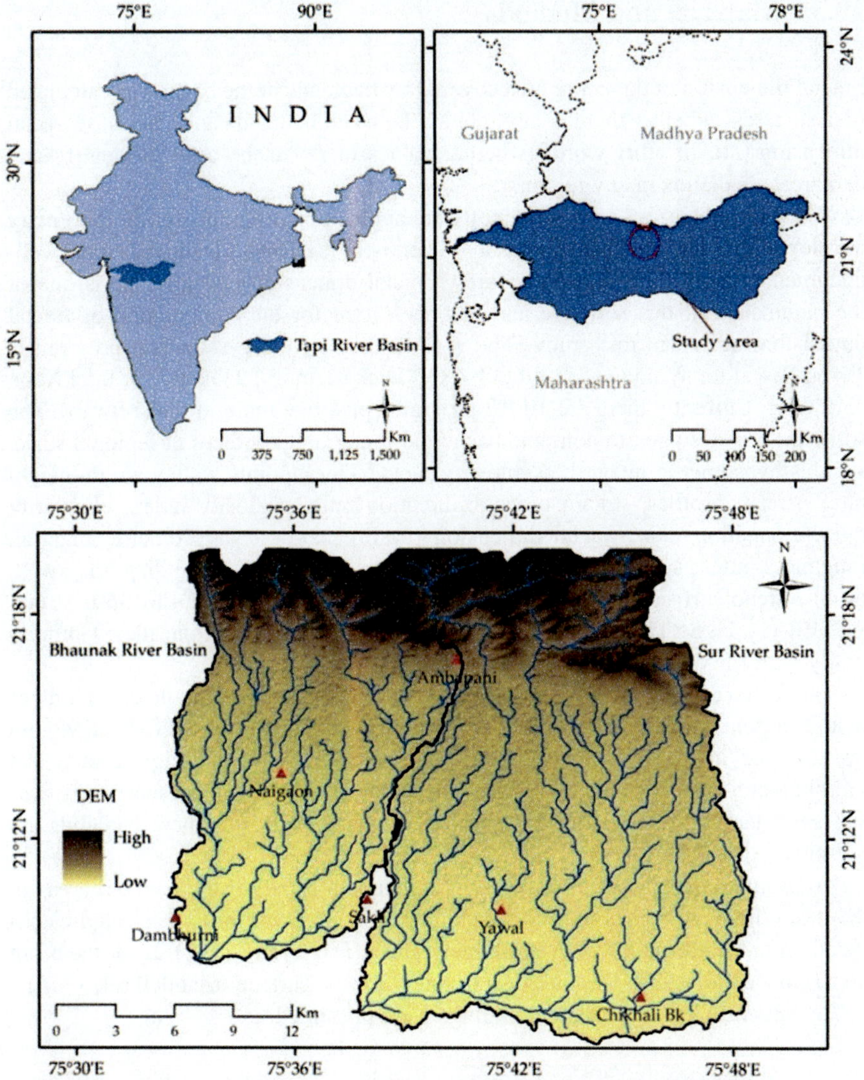

Fig. 19.1 Location map of the study area in general and also showing two sample basins

stretch of 60–70 km. There has not been any document/report or research article on the origin of these badlands, though tectonics as well as climate forcing during the LGM are speculated during the scientific discussions. The focus of the present study is these badland areas.

19.3 Material and Methods

Fractal dimension of the entire area covered by badlands in the basin was calculated first, to assess whether there is any relation between badlands and values of fractal dimension (*D*). In other words, whether values of *D* can be used to identify and demarcate badlands in any region.

'Viz-Morphotec' is a new set of software application programs for morphometry developed by Dr. Prakash Joglekar (Scientist F, DRDO, Delhi) as per well-documented algorithms and procedures. Fractal dimension of landscape is one of the parameters in this software and has been used for the computation of fractal dimension indices in this study. The specification of the system and program is Turbo Pascal for Windows, 32-bit OS, 4 GB Disk Memory, 2 GB RAM, Intel Xeon 3.40 GHz. Currently there are 10–12 programs plus few more in the software. The software was designed to compute basin and topographic indices at regional scale, such as hypsometric integral, asymmetry factor, knick points, valley width–height ratio, stream profiles, stream order, bifurcation ratio, sinuosity index, circularity ratio, elongation ratio, fractal dimension (basin), iso-base, hypsometric integral, roughness index, surface index, mountain front sinuosity and rose diagram, swath profiles trend surface and fractal dimension (topography). Input DEM 3601×3601 ASTER (1×1), SRTM (3×3), Re-sampled SRTM (15×15). Output files: Compatibility with ENVI, ERDAS, Geomatica/.bmp.

The fractal dimension program in the software was used to compute the D indices in the present study. For morphometric analysis, ASTER and SRTM DEMs are mostly used in the software, since the software was mainly designed to detect morphometry at a regional scale. But the input DEM for the present study was self-generated Cartosat (10 m) for the part of the Tapi Basin where badlands are present.

To evaluate the fractal property (self-similar/self-affine) of the badland areas in the basin, fractal dimensions were calculated at micro level for two selected catchments from the area at three levels, namely, linear *D* (Laverty 1987), *D* for the basin area (Goodchild, 1982; Turcotte, 1992) and *D* for the surface (Mandelbrot, 1975).

Results were synthesized, and findings were presented.

DEM Extraction Using Cartosat-1 Stereo Imagery

Cartosat-1, which was launched on May 5, 2005 (carrying two panchromatic sensors with 2.5-m spatial resolution and having fore-aft stereo capability), has been designed to generate DEMs and ortho-images for terrain modelling and is widely used nowadays, especially in India. Cartosat DEM of 30-m resolution is available on Bhuvan portal for free download. But the study needs a higher-resolution DEM than that; hence a DEM was self-generated for the present study using 8 Cartosat-1 stereo

Table 19.1 Details of each scene of the eight Cartosat images used to create DEM

Sr. No.	Path	Row	Date
1	524	299	12 Jan 2012
2	524	300	12 Jan 2012
3	525	299	31 Dec 2011
4	525	300	31 Dec 2011
5	526	299	18 Dec 2011
6	526	300	18 Dec 2011
7	527	299	30 March 2011
8	527	300	30 March 2011

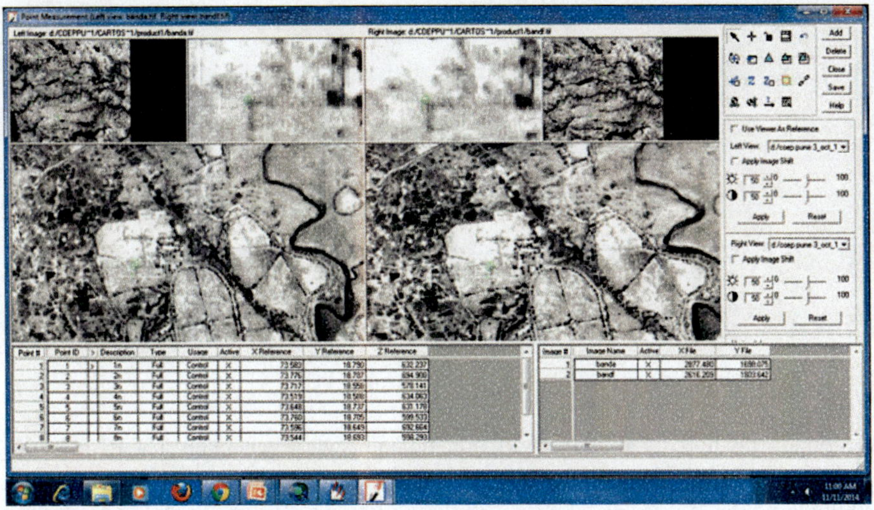

Fig. 19.2 Input GCP, classic point measurement tool in LPS

images taken over the badlands along Tapi River. The details of each scene of the eight Cartosat images are displayed in Table 19.1.

Extraction of DEM from stereo images can be treated as a semi-automatic procedure. Leica Photogrammetry Suite (LPS) 9.2 was used for DEM generation. The first step is to identify the GCPs for each image, and for that, both the Google image and the Cartosat images were opened in ArcGIS. After the points have been identified, the images PAN-A and PAN-F were imported in Leica LPS as indicated in Fig. 19.2. Cartosat-1 images are provided with 'rational polynomial coefficients' (RPC). These RPCs are computed by using the available information, i.e. sensor model, sensor position and attitude data, ellipsoid parameters and map projection. Further, rational polynomial coefficients (RPC) file is attached in the block file as a part of image orientation. GCPs were collected directly from the field by using differential Global Positioning System (dGPS). Five GCPs were taken for each image. Figure 19.3 indicates the GCPs of one image actually in the field. Twenty-five GCPs were used as control points to refine orientation results (Fig. 19.4). Using

Fig. 19.3 Measuring ground control points in the field using dGPS

the 'bundle method' of aerial triangulation, tie points were generated. The RMSEs of the residuals were obtained after block triangulation. Once aerial triangulation with optimum RMSE is done, the data is exported in ArcGIS 10, and DEM was extracted at 10-m resolution. Standard WGS-84 projection and datum were assigned for the Cartosat 1 photogrammetry model. The ortho-rectified image is indicated in Fig. 19.5. For accuracy checking, the Survey of India topographical maps of the area were used which revealed error-free matching with great accuracy. The flow chart in Fig. 19.6 demonstrates the entire steps of DEM creation from IRS Cartosat stereo images using LPS and ArcGIS.

Figure 19.7 demonstrates the Carto DEM of 10-m resolution which became the input DEM for all the further calculations.

Calculation of Fractal Dimension

Calculations of the fractals were computed for 137 × 77 grids cell of 100 × 100 pixel dimension. The box counting was performed by varying grid dimension from 4 to 50 cells in *x*, *y* and *z* axes. The fractal dimension (*D*) was computed for each pixel based on the slope of regression of graph of log (*N*) against log (*r*). *N* is the number of boxes that cover the pattern, and r is the magnification or the inverse of

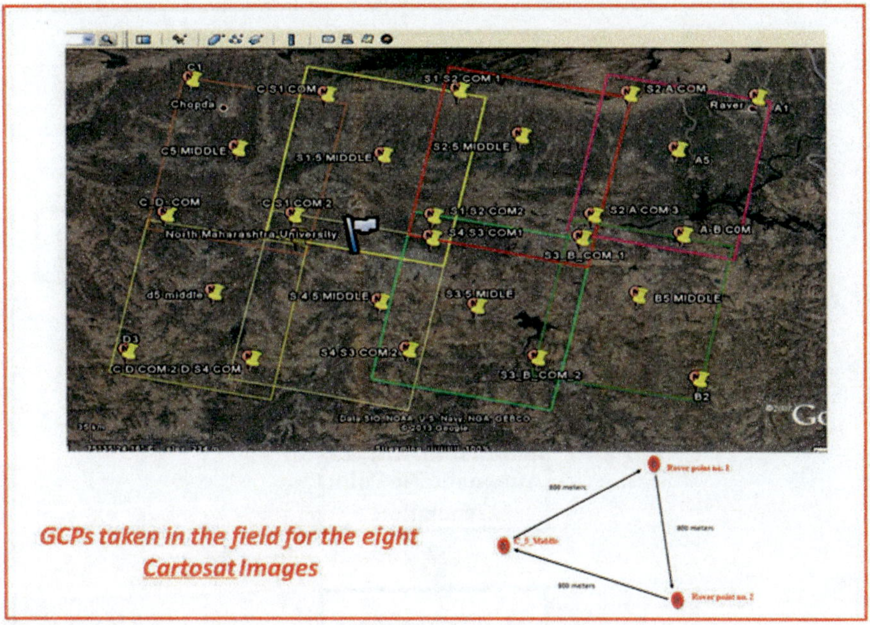

Fig. 19.4 Depicting the positions of the 25 GCPs as control points to refine orientation results

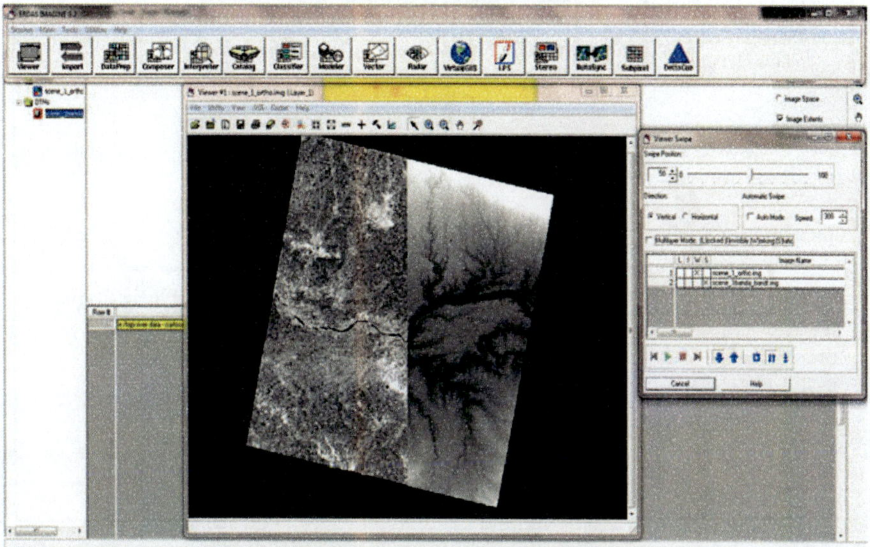

Fig. 19.5 DEM and the ortho-rectified image

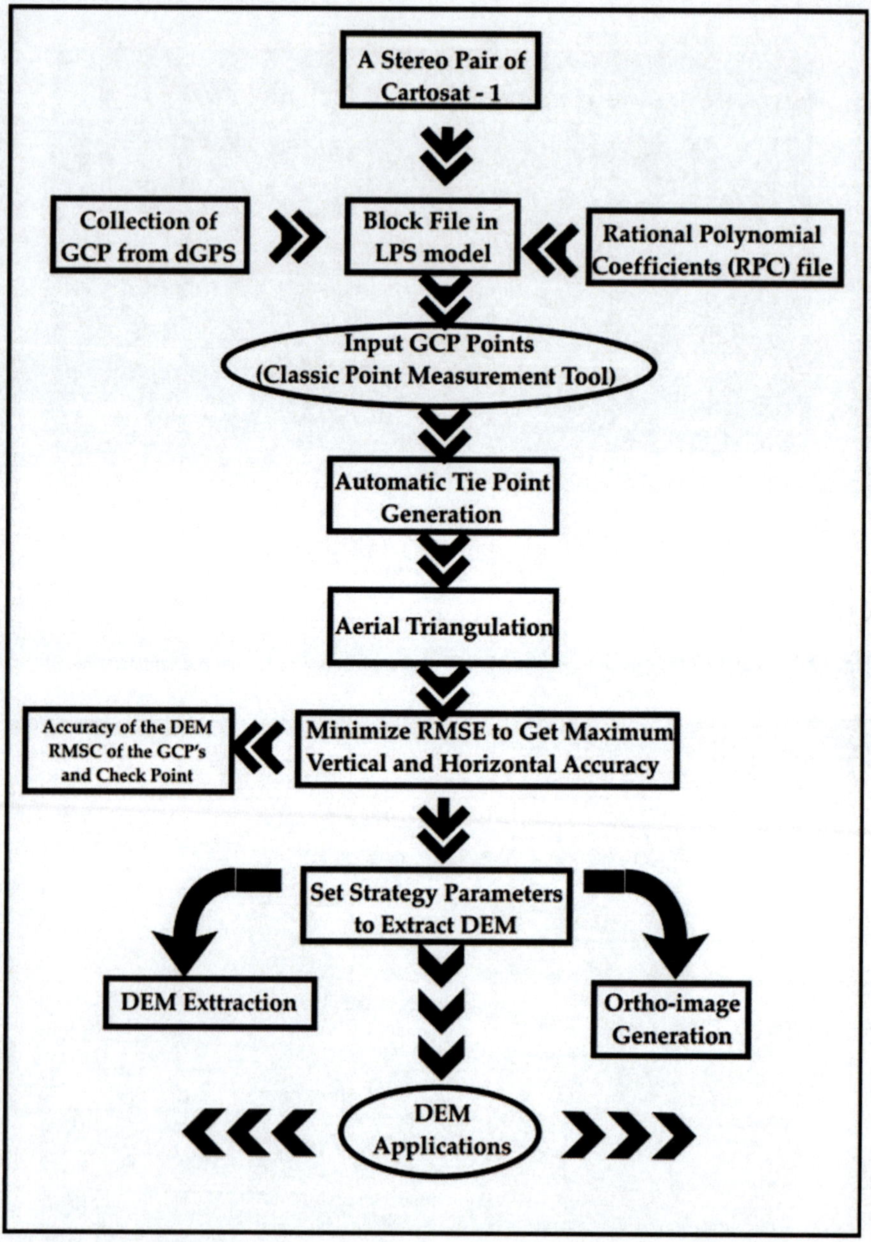

Fig. 19.6 Flow chart demonstrating the entire steps of DEM creation from IRS Cartosat stereo images using LPS and ArcGIS

the box size. Density-sliced image for local D (D range 2.8–3.0) is presented in Fig. 19.8. The density-sliced image of D has been draped on Google Earth image of the area (Fig. 19.9) which reveals a remarkable match of the areas with D values

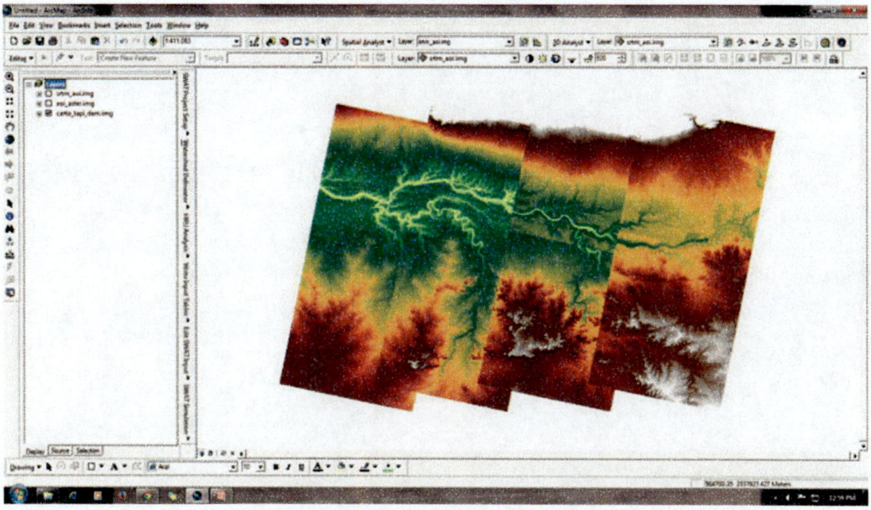

Fig. 19.7 Carto DEM of 10-m resolution which is the input DEM for Viz-Morphotec software

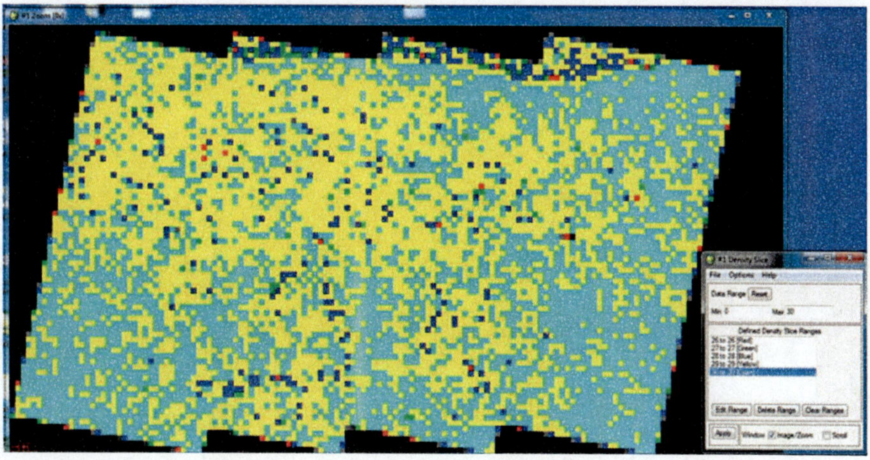

Fig. 19.8 Density-sliced image for local *D*, Calculated using Viz-Morphotec software (*D* ranges from 2.8 to 3.0)

above 2.8 with the location of badlands in the basin. The percentage area against the *D* values were obtained (Fig. 19.10) which indicates clearly that *D* values around 2.90–2.95 were observed over the badlands in the region. The result is showing that fractal dimension can be effectively used to delineate badlands in any basin. The software proved to be useful in determining fractal dimension on a broad scale in addition to other morphometric parameters.

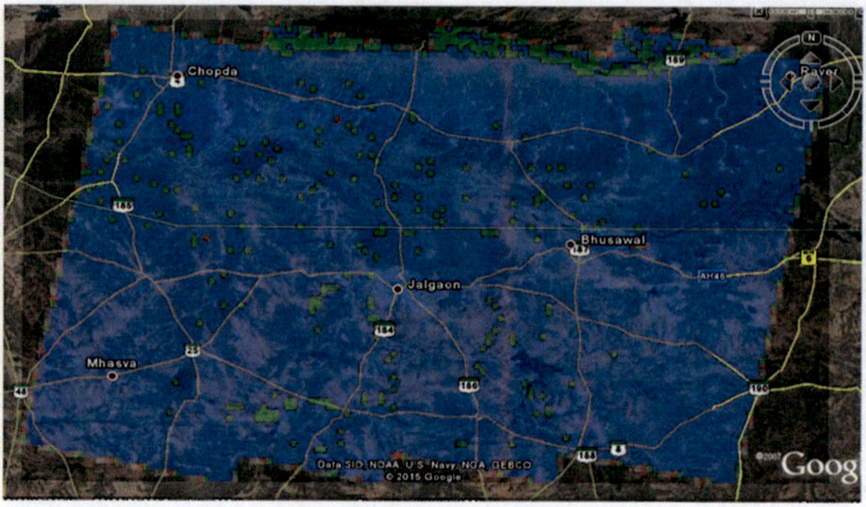

Fig. 19.9 Density-sliced image of *D* has been draped on Google Earth image of the area, depicting the perfect coincidence of badlands with *D* values between 2.8 and 3

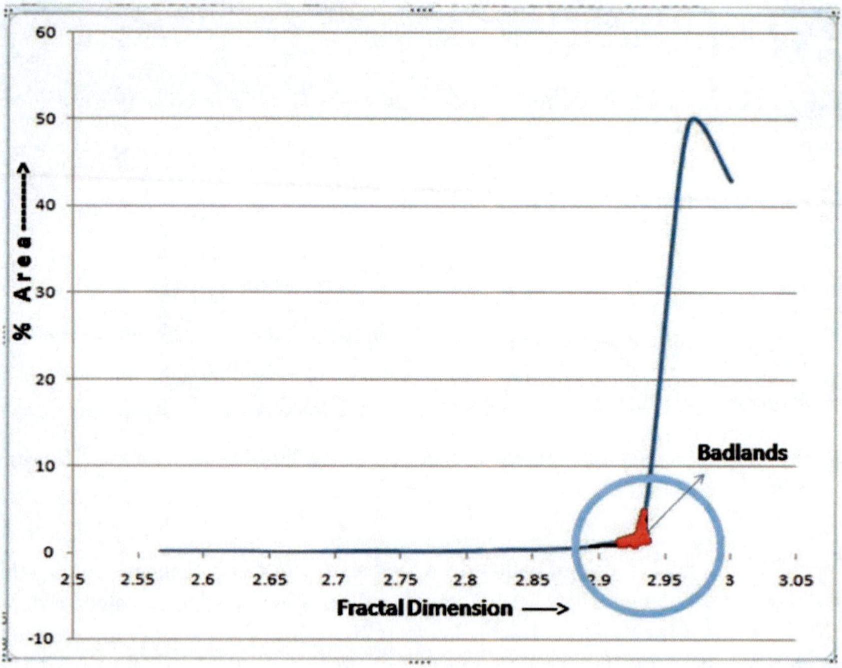

Fig. 19.10 Percentage area against the *D* values, revealing that values above 2.9 are clustered within the badlands in the area, as shown by red in the circle

The next step is to test whether we can meaningfully apply the technique to evaluate the fractal properties of the badlands. Badlands are deeply dissected topography, with dense drainage density. Relief within a badland can be variable, such as in areas like Chambal badlands of India where relief of more than 100 m is found, but the badlands that are developed over the Deccan region are usually of a lot smaller dimension in terms of vertical relief and horizontal extent (Joshi et al., 2009). The Deccan Trap region as a whole is rocky, and sediments and soils have been formed at only restricted patches along some riverbanks and few foot slope pediment zones. The badlands in the study area have maximum relief of 10–25 m, so landform processes are operating at a much finer resolution. Hence, Cartosat DEM of 10-m resolution is believed to be a fair match with the landscape operating scale here.

Linear Fractal Dimension of the Channel

In order to evaluate the linear fractal dimension of the streams in the basin, a field survey was conducted to select a few badland catchments using the locations of badlands identified by the *D* values of the area (Fig. 19.9). Two sample catchments were selected in the field and demarcated them in the DEM. The locations of these two sample catchments are indicated in the location map (Fig. 19.1), and field views are shown in Fig. 19.11. They are Bhaunak Basin and Sur Basin. General

Fig. 19.11 The field views of the two sample catchments, where (**a**) and (**b**) depict parts of Bhaunak Basin and (**c**) and (**d**) capture parts of Sur Basin. What is seen in the pictures are the true badland parts of the basins

Table 19.2 Topographical parameters of the sample basins

Parameters	Bhaunak Basin	Sur basin
Basin area	209.093 km^2	347.409 km^2
Basin length	22.329 km	26.85 km
Basin width	12.743 km	17.84 km
Relative relief	911 m	896 m
Dissection index	85.78%	84.44%
Absolute relief	1062 m	1061 m
Slope	2.64°	2.02°

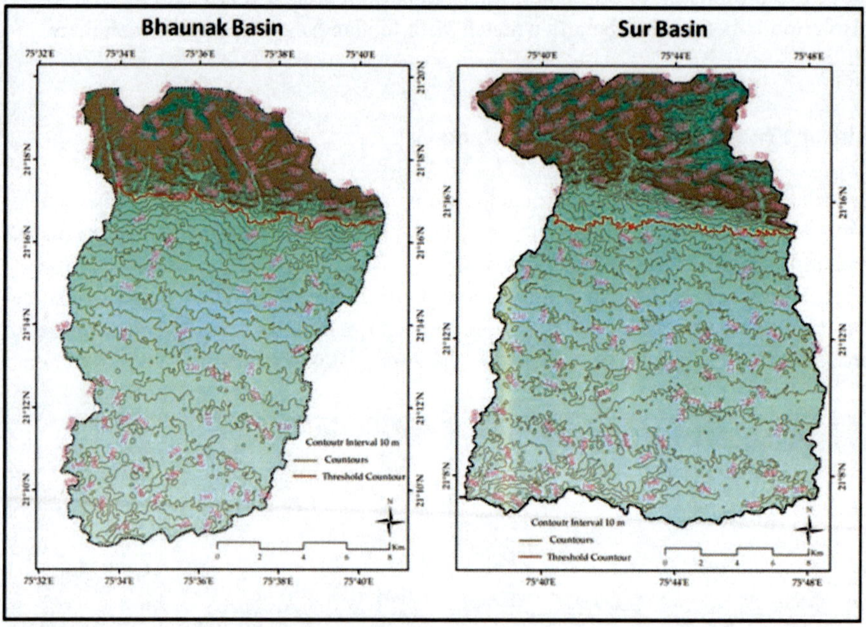

Fig. 19.12 Contour maps of Sur and Bhaunak basins and the threshold contour (in red colour), which demarcates upstream and downstream reaches

geomorphometry of these two sample basins is presented in Table 19.2. There are two distinct geomorphic units within both the basins, namely, pediments and badlands. It is clearly visible in Fig. 19.12 that there is a sharp break of slope, indicated by a red line in the diagram. Figures 19.13 and 19.14 display surface and slope maps of the basins. As we can see in these maps, the upstream part is the hillslope pediment zone, and the downstream reach is the true badland alluvial zone. The fractal dimension (*D*), in general, ranges between 1 (indicating almost straight) and 2 (nearly filling the plane). Over a range of scales, statistically self-similar lines demonstrate constant values of *D* (Mandelbrot, 1967). The fractal value (*D*) of a curve is calculated by measuring the entire length of the curve using various step sizes. When the curve is irregular, the step size increases, leading to increase in the total length of the curve. Two stream channels were selected each from the two

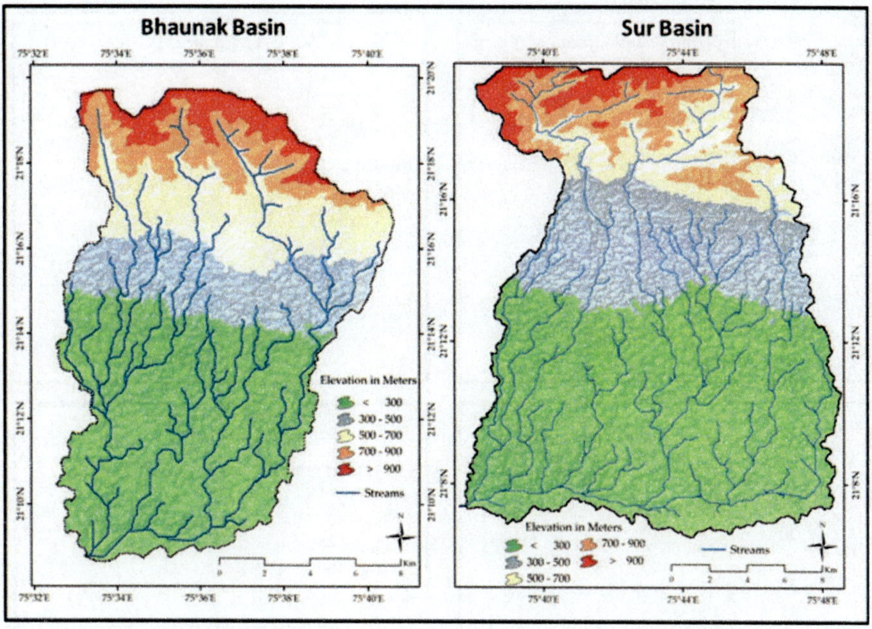

Fig. 19.13 Surface maps of the two sample basins, namely, Sur and Bhaunak

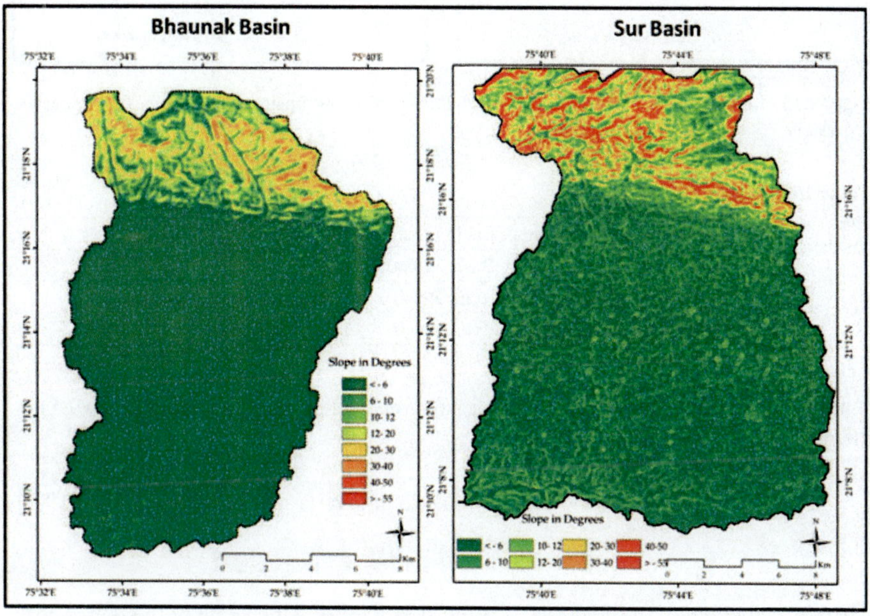

Fig. 19.14 Slope maps of the two sample basins, namely, Sur and Bhaunak

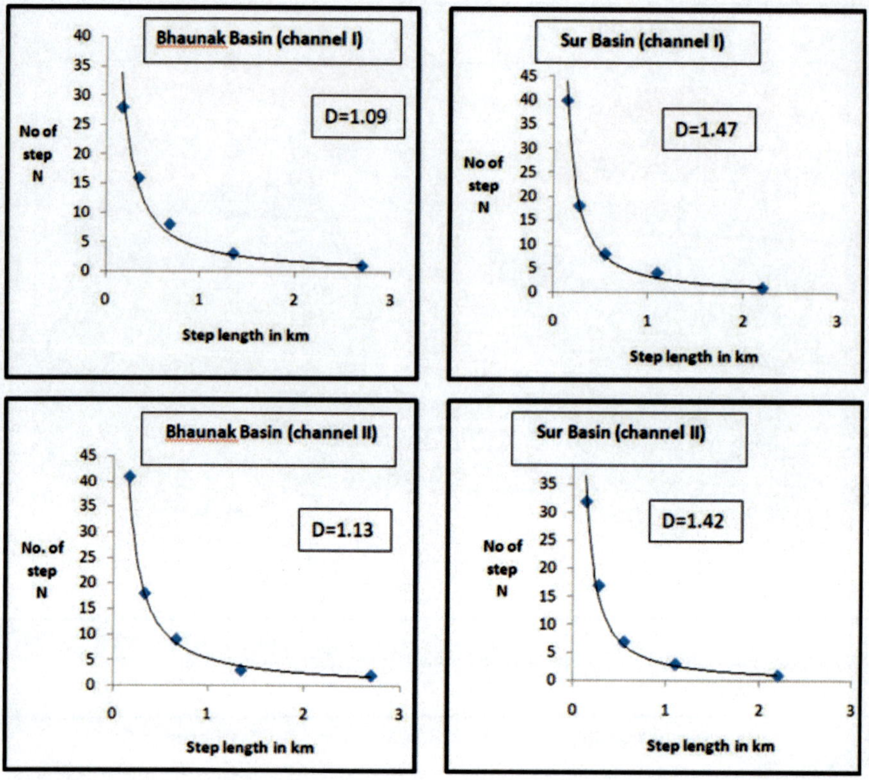

Fig. 19.15 Linear fractal dimensions of the channels, showing upstream and downstream curves separately

Table 19.3 Linear fractal dimension for the sample basins

Name	Fractal value (*D*)
Bhaunak Basin (channel I)	1.09
Bhaunak Basin (channel II)	1.13
Sur Basin (channel I)	1.47
Sur Basin (channel II)	1.42

sample catchments, and linear fractal dimensions were computed using Eq. (19.1) in the program:

$$\log L = K + B \log d \tag{19.1}$$

$$D = 1 - B$$

where L is the length of the curve, d is the step size, B is the slope of the regression, K is a constant, and D is function of the regression slope B. The steeper the negative slope (B is negative value), the higher the fractal dimension. The results of the calculation are present in Fig. 19.15 and Table 19.3. It is clear from the table that the values are closer to 1 than 2, indicating low sinuosity of the channels in question. Variations in fractal dimensions are observed between Bhaunak and the Sur basins, but there is no significant variation within the individual basins. Bhaunak shows lower *D* than the Sur Basin.

Fractal Dimensions of the Basin Area

The catchment boundaries of Bhaunak Nadi and Sur Nadi were demarcated with the help of hydrology tool in ArcGIS 10. The fractal dimension of the basin area for the sample basins was calculated using box-counting method (Goodchild, 1982; Turcotte, 1992), for that grids of different scales were plotted on area as well as sample basin boundary, and fractal has been calculated with the help of Eq. (19.2):

$$(D) = \frac{\ln\left(N_n + \frac{1}{N_n}\right)}{\ln\left(\frac{r_n}{r_n}\right) + 1} \tag{19.2}$$

where (*r*) is the size of the grid and (*N*) signifies the number of boxes to cover the entire area for each grid size. The computed *D* for the perimeter for the two basins is demonstrated in Table 19.4. The box-counting dimension is much more widely used than the self-similarity dimension since the box-counting dimension can measure pictures that are not self-similar (and most real-life applications are not self-similar). Due to the nearly box-shaped nature of the demarcated basins, the values are close to 2, indicating near space filling.

Variograms and the Surface Fractal Dimensions

The roughness or spatial continuity in a data set is indicated by variograms. Of all the varieties of methods of computation of *D* for surface, variogram technique is the most widely used one. The variogram of a surface is constructed by considering the variance of its elevation as a function of its horizontal distance. For a pair of points

Table 19.4 Box sizes and the number of boxes (for box-counting method) and fractal values for the two basins

Basin name	Size of each box (*r*)	No. of required boxes (*N*)	Fractal values (*D*) of the whole catchment
Bhaunak Basin			
Channel I	1.0 km	147	1.971
Channel II	0.5 km	333	
Sur Basin			
Channel I	1.0 km	203	1.989
Channel II	0.5 km	429	

x_1, y_1, z_1 and x_2, y_2, z_2 on a grid of digital topography, with x and y horizontal coordinates and 'z' being the elevation, the contribution to the variance is expressed as

$$(Z_1 - Z_2)^2 = (\Delta z)^2 \tag{19.3}$$

and the horizontal distance is depicted as:

$$[(x_1 - x_2) + (y_1 - y_2)]^{½} = \Delta x \tag{19.4}$$

It calculates these for every pair of points on the grid and plots the logarithm of the standard deviation over a binned distance interval, against the logarithm of the distance at the logarithmic midpoint of that interval. Manually it is computationally intensive operation, because n by n grid contains $n^2(n^2 - 1)/2$ points. The following equation (Eq. (19.5), Carr, 1995) has been employed in the program to calculate variograms of the sample basins, which is expressed as:

$$y(h) = (1/2n) \sum \left[Z(x_i, y_i) - Z\left(x_{i+h}, y_{i+h}\right)\right]^2 \tag{19.5}$$

where $\gamma(h)$ = semivariance at lag distance h; $Z(x_i,y_i)$ = data value at location i; $Z(x_{i+h}, y_{i+h})$ = data value at location plus distance h; and n = number of points in the data set.

Estimation of Fractal Dimension from Variograms

Fractal dimension of the landscape can be directly obtained from the variogram, assuming that land surfaces have statistical properties to those of fractional Brownian surface (Mandelbrot, 1975). The two random function of $F(t)$ and $F(rt)/rH$, when properly rescaled, are statistically similar. For a surface, the single variable t is replaced by point coordinated x and y on a plane to give $F(x,y)$ as the surface altitude z at position (x,y). The surface that consists of these $F(x.y)$ points is usually called a fractional Brownian surface (fBm) (Mandelbrot, 1975).

On fractional Brownian surface, the variogram is described by what is expressed in Eq. (19.6), such as

$$E\Big[F(x,y) - F(x + \Delta x, y + \Delta y)^2 = \left(\Delta x^2 + \Delta y^2\right)^{2H} \tag{19.6}$$

The variogram takes on the form of a power function in which H (Hurst exponent) should range between 0 and 1. In case of a fractional Brownian surface, $D = 3 - H$. As H increases toward its upper limit (i.e. small D), the variability of the surface is locally small but rises rapidly with distance; whereas, when H is small (i.e. large D), the surface shows high local variability but a slow increase at large distance (Mandelbrot, 1975).

The slope and the trend of a variogram can be directly used to interpret the nature of fractal of the landscape. On a log plot of standard deviation *v/s* distance, the slope of a variogram will be indicated by $(3 - D)/2$, resulting into a steep variogram representing low fractal dimension, while a gentle slope implies high fractal dimension. Straight variogram represents a true fractal (self-similar). If the variogram has breaks in the slope, it indicates that each break is associated with changes in the process or lithology and represents a 'multifractal' landscape. In a topography where there are variations in either lithology and/or processes, a variogram shows generally a multifractal topography (Voss, 1988).

Relation Between *H* and Self-Similarity of the Landscape

If the relief (or variance of elevation) in a small area resembles that of the entire area, when the relief is magnified by the area factor (entire area/small area ratio), the landform is considered as self-similar, where $H = 1$. If relief increases, on the other hand, then the landform is considered self-affine ($0 < H < 1$). This is a more probable situation in real landscapes. With similar total relief, the greater the local relief, the lesser the value of H. With similar local relief, the greater the value of H, the lesser the total relief.

Texture is a word used in landform studies to indicate arrangements of topographic heights and the frequency of changes, and surface fractal dimension is simply the measure of it. The calculated D values fall within the general range of 2 (flat) to 3 (completely space filling).Using the equations cited above, variograms and the fractal dimensions were computed for the two sample basins. Both Bhaunak and Sur basins were demarcated into two sectors, namely, upstream pediment section and downstream badland section. Variograms were separately generated for all the four sections, and results are depicted in Fig. 19.16 and Table 19.5. Hurst components (H) were also calculated for the sample basins, and they are also included in Table 19.5.

19.4 Discussions

In the present study, demarcations of badlands were done using a new software, Viz-Morphotec, which clearly reveals a strong association of D values with the occurrences of badlands. At a micro-level, the fractal calculations were done at three levels, namely, linear, perimeter and surface, for two sample basins. The data presented above reveals that in case of linear D, the low values for both the channels in Bhaunak suggest that the rivers are actively eroding, and linear erosion is predominant in the entire section being included in the analysis. They are also close to the source areas. Sinuosity will always indicate low under such conditions.

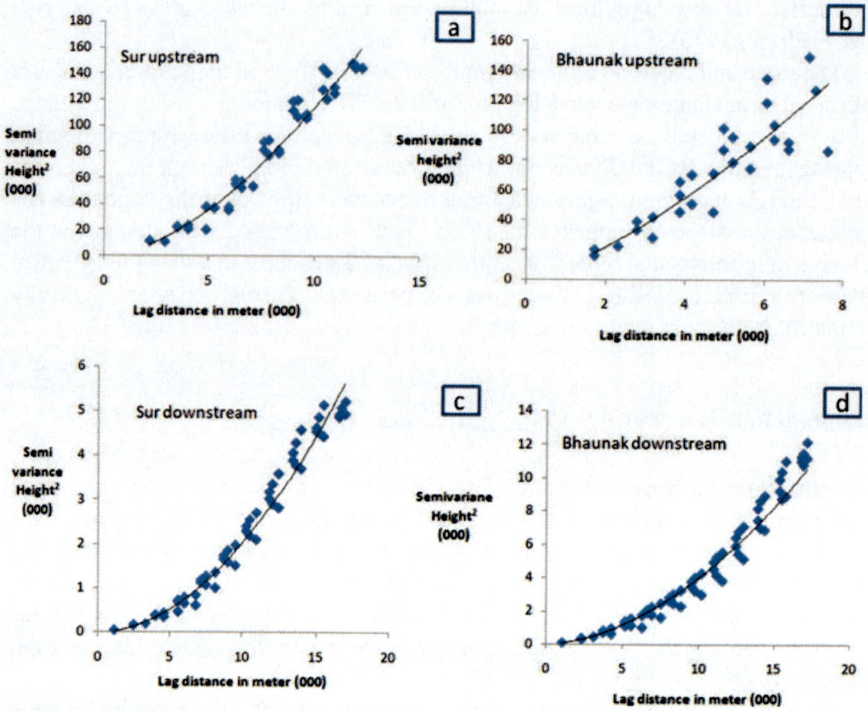

Fig. 19.16 Variograms for both the upstream and downstream sections of the sample basins. Upstream reaches of both the basins as depicted in (**a**) and (**b**) are straight lines, implying self-similarity, whereas, downstream reaches shown in (**c**) and (**d**) show slight concavity, that is deviation from self-similarity but more of self-affinity

Table 19.5 Fractal dimension of the surface and 'H' values of the sample basins

River name	Fractal dimension (*D*)	Hurst exponent (*H*)
Bhaunak Basin (upstream)	1.56	1.44
Bhaunak Basin (downstream)	2.36	0.64
Sur Basin (upstream)	1.82	1.18
Sur Basin (downstream)	2.77	0.23

The gradients of these badland gullies/streams are high that directly drain into the main streams. Hence there is rapid linear erosion resulting in low values of *D*. Sur is revealing higher values of *D*, showing more sinuosity. Minor variations prevail in the values of *D* between the two catchments, which could be due to the variations in their textural and stratigraphic characteristics as well as gradients of the longitudinal profiles.

The fractal dimension of the basin perimeters conducted using box-counting method yielded in the high values of *D* for both the basins, reaching almost 2. This is due to the nearly box-shaped nature of the demarcated basins.

In surface *D*, values above 2.5 are considered to be very rough surface, and the values for both the basins are more than that. Values for upstream basins are below 2. Generally, surface fractal values range between 2 and 3, but in this calculation, for the upstream reaches, the values are less than 2. This is very unusual. The surface shows high local variability. There is a slow increase in roughness with distance for the badland areas, while the hillslope areas demonstrate exactly the opposite trend. The fractal dimension (*D*) of a surface is a relief-independent parameter. High fractal does not necessarily mean a rough surface but rapid change in the relief in a small local area (but gradual change in a large area), and low fractal suggests slow changes in a small area but large variation over a long distance (Sung and Chen, 2004). High values of *D* and small *H* (Table 19.5) as shown in these areas are in accordance with the normal trend. Hurst exponent is useful to determine the self-similarity of the landscape and even more useful when comparisons are made of the minor variations in the surface morphology between the two sites. The upstream areas are very small in aerial extent, and hence variation in the surface relief is not significantly reflected in the analysis.

A straight variogram suggest self-similar landscape. The fractal dimension of a topography is controlled by the changes in its variability with distance but not the amplitude of that variability (Joshi et al., 2009). A topography which is uncorrelated at all length scales will indicate high fractal value, while topography that is strongly correlated at short wavelengths (but less so at long wavelengths) will reveal low fractal dimension (Mandelbrot, 1989). A true fractal must reveal a straight variogram. Departure from the straightness indicates multifractal topography, more so at lower fractal dimension, at lesser wavelength scales (Voss, 1988; Mandelbrot, 1989). Multifractal can be interpreted in two ways: that the landscape is not self-similar and also that more than one process is operating in the region.

The trend lines of the two sample basins (Fig. 19.15) do not reveal noteworthy difference with each other. The variograms of the two sample basins are smooth and slightly concave, not straight, indicating multifractal topographic distribution for this range. Sur Upper course (Fig. 19.16a) shows almost smooth and straight trend, indicating a fractal self-similar topography for a short range. Even the upstream trend of Bhaunak (Fig. 19.16b) also is straight. However, downstream for both the basins show smooth but slightly concave profile. Breaks in the variogram slopes are indicative of multifractal topography and change in the processes operating within the area. The variograms show steeper slopes at shorter length scales with *D* clustering around 2.2/2.3 and gentler slopes at wavelengths longer than 1 km, representing *D*, in the vicinity of 2.4 and above. The interpretation of the variogram slopes suggest that 'statistical self-similarity' is not indicated here and also that two processes are dominantly operating in the badland areas. There is no sharp break in the slope of the variograms, but smooth concave slope, indicating that one process merges into the other.

19.5 Conclusions

Along the banks of Tapi River within Maharashtra, badlands have been developed along a stretch of approximately 70 km. The exact locations of these badlands have been delineated using fractal values calculated in a new software, Viz-Morphotec. The microprocesses operating at these badlands were further attempted at finer landscape operation scale at linear, perimeter and surface levels by selecting two badland catchments and performing the fractal calculations.

Within the two catchments under review, two distinct geomorphic units could be identified: hillslope and pediment zone. The pediment zone is deeply overlain by alluvium that continues till the riverbanks of the main stream. These alluviums are severely dissected to form badlands. Fractal values of the streams (linear) show low to moderate values, indicating less sinuosity of the streams and high competence. The *D* values for basin perimeters are close to 2, which is nearly space filling. The trend lines of the variograms show slightly concave curve, indicating multifractal topography (not self-similar in true sense), and two processes are dominantly operating, with one merging smoothly into another. Though fractal dimension does not highlight the actual process in the area, based on the understanding of the general processes operating on the badlands and the slope of the variograms, it is very likely that in the upstream reach, which is the hillslope units of these basins, unusually low *D* values are found as against high values downslope. It is probably because high *D* values do not indicate necessarily high relief or rougher surface, but it indicates that the rate of change in relief over a small area is high but gradual at the long distance and vice versa. Surface fractal dimension indicates values higher than 2.5 implying a rough surface, which is very typical of badland topography. Marginal variations occur in the two sample basins, but not significant enough to warrant attention. The curves of the variograms indicate possible influence of erosional and diffusional processes operating on the topography, or they could be the result of tectonics or changes in the climatic conditions, which are still manifested in the landscape. In future, more data need to be generated from similar badland watersheds to make comparison and for better understanding of the landscape processes operating at micro-scale within the badland watersheds in the region.

Acknowledgements The paper is a part of a research project funded by the Department of Science and Technology, India. The authors would like to thank the commission for the financial support. Author would also like to acknowledge the help from many students of the department during the fieldwork. They are Sandeep Pawar, Gaurav Gamre, Sadashiv Tukanvar and Nilesh Susware. I offer special thanks to Mr. Nilesh Susware for helping me to prepare the figures for this paper. Thanks are also due to National Remote Sensing Centre (NRSC), India, for helping us to choose the right scenes of Cartosat I stereo images and also making them available for the study. My sincere thanks and acknowledgement to Dr. Prakash Joglekar, the then Scientist 'F' Defence Research and Development Organization (DRDO), New Delhi, India, for allowing me to use the software Viz-Morphotec, which was used for computation of the fractal indices in the study.

References

Andrle, R., & Abrahams, A. D. (1989). Fractal techniques and the surface roughness of talus slopes. *Earth Surface, Process and Landforms, 14*, 197–209.

Baas, A. C. W. (2002). Chaos, fractals and self-organisation in coastal geomorphology: Simulating dune landscapes in vegetated environments. *Geomorphology, 48*, 309–328.

Breyer, S. P., & Snow, R. S. (1992). Drainage basin perimeters: A fractal significance. *Geomorphology, 5*, 141–157.

Burrough, P. A. (1981). Fractal dimensions of landscapes and other environmental data. *Nature, 294*(5838), 240–242.

Carr, J. R. (1995). *Numerical analysis for the geological sciences* (p. 592). Pentice Hall.

Chase, C. G. (1992). Fluvial landsculpturing and fractal dimension of topography. *Geomorphology, 5*, 39–57.

Culling, W. E. H., & Datko, M. (1987). The fractal geometry of the soil covered landscape. *Earth Surface Processes and Landforms, 8*, 369–385.

Fox, C. (1986). Self-similar profiles of deep-sea topography (abstr.) EOS. *Transactions of the American Geophysical Union, 67*, 871.

Gao, J., & Xia, Z. (1996). Fractals in physical geography. *Progress In Physical Geography: Earth and Environment, 20*(2), 178–191.

Gilbert, L. E. (1989). Are topographic datasets fractal? *Pure and Applied Geophysics, 131*, 241–254.

Goodchild, M. F. (1982). The fractal Brownian process as a terrain simulation model. *Modelling and Simulation, 13*, 1133–1137.

Goodchild, M. F., & Mark, D. M. (1987). The fractal nature of geographic phenomena. *Annals of the Association of American Geographers, 77*(2), 265–278.

Hagerhalla, M. C., Purcella, T., Taylor, R. (2004). Fractal dimension of landscape silhouette outlines as a predictor of landscape preference , *Journal of Environmental Psychology 24*, 247–255.

Huang, J., & Turcotte, D. L. (1989). Fractal mapping of digitized images: Application to the topography of Arizona and comparisons with synthetic images. *Journal of Geophysical Research, 94*(B6), 7491–7495.

Jie, Z., & Haosheng, B. (1994). Fractal geomorphology—Review and prospect. *Geographical Research, 13*(3), 104–112.

Joshi, V. U., Tambe, D., Dhawade, G., et al. (2009). Geomorphometry and fractal dimension of a riverine badland in Maharshtra, India. *Journal Geological Society of India, 73*(3), 355–370.

Khanbabaei, Z., Karam, A., Rostamizad, G., et al. (2013). Studying relationships between the fractal dimension of the drainage basins and some of their geomorphological characteristics. *International Journal of Geosciences, 4*(3), 636–642.

Klinkengerg, B. (1992). Fractals and morphometric measures: Is there a relationship? *Geomorphology, 5*, 5–20.

Kusák, M. (2014). Review article: Methods of fractal geometry used in the study of complex geomorphic networks. *Acta Universitatis Carolinae Geographica Univerzita Karlova, 49*(2), 99–110.

Laverty, M. (1987). Fractals in Karst: *Earth Surf. Proc. and Land., 12*, 475–480.

Liucci, L., & Melelli, L. (2017). The fractal properties of topography as controlled by the interactions of tectonic, lithological, and geomorphological processes. *Earth Surface Processes and Landforms, 42*, 2585–2598.

Man, X., & Chen, Y. (2020). Fractal-based modeling and spatial analysis of urban form and growth: A case study of Shenzhen in China. *ISPRS International Journal of Geo-Information, 9*(11), 672. https://doi.org/10.3390/ijgi9110672

Mandelbrot, B. (1967). How long is the coast of Britain? Statistical self-similarity and fractional dimension. *Science, 156*(3775), 636–638.

Mandelbrot, B. (1975). On the geometry of homogeneous turbulence, with stress on the fractal dimension of the iso-surfaces of scalars. *Journal of Fluid Mechanics, 72*(3), 401–416.

Mandelbrot, B. (1989). Multifractal measures, especially for the geophysicist. *Pageoph, 131*, 5–42.

Mark, D. M., & Aronson, P. B. (1984). Scale dependent fractal dimensions of topographic surfaces: An empirical investigation with applications in geomorphology and computer mapping. *Mathematical Geology, 16*(7), 671–683.

Mayer, L. (1992). Fractal characteristics of desert storm sequences and implications for geomorphic studies. *Geomorphology, 5*, 167–183.

Milne, B. T. (1988). Measuring the fractal geometry of landscapes. *Applied Mathematics and Computation, 27*, 67–79.

Nathaniel, A. L., & Chase, C. G. (1992). Tectonic, climatic and lithologic influences on landscape fractal dimension and hypsometry: Implications for landscape evolution in the San Gabriel Mountains, California. *Geomorphology, 5*, 77–114.

Ouchi, S., & Matsushita, M. (1992). Measurment of self affinity on surfaces as a trial application of fractal geometry to landform analysis. *Geomorphology, 5*, 115–130.

Patuano, A., & Tara, A. (2020). Fractal geometry for landscape architecture: Review of methodologies and interpretation. *Landscape Architecture, 5*, 72–80.

Pardo-Iguzquiza, E., Dowd, P. A., Duran, J. J., & Robledo-Ardila, P. (2019). A review of fractals in karst. *International Journal of Speleology, 48*(1), 11–20.

Polidori, L., Chorowicz, J., Guillande, R., et al. (1991). Description of terrain as a fractal surface and application to digital elevation model quality assessment. *Photogrammetric Engineering and Remote Sensing, 57*(10), 1329–1332.

Piech, M. A., & Piech, K. R. (1990). Fingerprints and fractal terrain. *Mathematical Geology, 22*, 457–485.

Reams, M. W. (1992). Fractal dimensions of sinkholes. *Geomorphology, 5*, 159–165.

Snow, R. S. (1989). Fractal sinuosity of stream channels. *Pure and Applied Geophysics, 11*, 99–109.

Snow, R. S. (1992). The cantor dust model for discontinuity in geomorphic process rates. *Geomorphology, 5*, 185–194.

Sun, W., Xu, G., Gong, P., Liang, S., et al. (2006). Fractal analysis of remotely sensed images: A review of methods and applications. *International Journal of Remote Sensing, 27*(22), 4963–4990.

Sung, Q. C., Chen, Y. C. (2004). Self-affinity dimension of topography and its implications in morphotectonics: an example from Taiwan. *Geomorphology, 62*, 181–198.

Tarboton, D. G., Bras, R. L., Rodriguez-Iturbe, I., et al. (1992). A physical basis for drainage density. *Geomorphology, 5*, 59–76.

Taud, H., & Parrot, J. (2005). Measurement of DEM roughness using the local fractal dimension. *Géomorphologie: Relief, Processus, Environnement, 4*, 327–338.

Turcotte, D. L. (1992). *Fractals and Chaos in geology and geophysics*. Cambrige University Press.

Turcotte, D. L. (2007). Self-organised complexity in geomorphology: Observations and models. *Geomorphology, 91*, 302–310.

Unwin, D. (1989). Fractals and geosciences, introduction. *Computers and Geosciences, 15*, 163–165.

Voss, R. F. (1988). Fractals in nature: From characterization to simulation. In H. O. Petigen & D. Saupe (Eds.), *The science of fractal images* (pp. 21–70). Springer-Verlag.

Willgoose, G., Bras, R. L., Rodriguez-Iturbe, I., et al. (1991a). A coupled channel network growth and hillslope evolution model, I, Theory. *Water Resources Research, 27*, 1671–1684.

Willgoose, G., Bras, R. L., Rodriguez-Iturbe, I., et al. (1991b). A coupled channel network growth and hill slope evolution model, 2, Applications. *Water Resources Research, 27*, 1685–1696.

Xu, T., Moore, I. D., Gallant, C., et al. (1993). Fractals, fractal dimensions and landscapes—A review. *Geomorphology, 8*(4), 245–226.

Chapter 20
Flood Hazard Mapping in Assam Using Sentinel-1 SAR Data

Sujoy Mudi, Jaya Prakash A, and Pulakesh Das

Abstract Floods are one of the most devastating natural hazards, which lead to fatal outcomes to socio-economic and human life for several weeks in vast areas. Floods mostly occur in the monsoon season when cloud-free optical data are hardly available; whereas, microwave data has advantages to penetrate the cloud and enables real-time flood mapping. The current study focuses using Sentinel-1 synthetic aperture radar (SAR) images for flood or water inundation mapping in Assam during July – September in 2018, 2019 and 2020. Otsu's automatic thresholding method was applied on the vertically transmitted vertically received (VV) polarization band Sentinel-1 SAR data to identify the water inundated areas. The Google Earth Engine (GEE) platform was used for data processing, which requires lesser data processing capabilities in the user system and less time than other traditional approaches. The comparison with the Sentinel-2 optical data available for few regions indicated high mapping accuracy ($\geq$87%). The total water inundated area was identified as identified as 1453, 2081 and 3634 km^2 in 2018, 2019 and 2020, respectively. All the districts in the floodplain of the Brahmaputra River and three districts in the southern region of Assam experienced three-time flood events in the last 3 years. Overlaying with vegetation map indicated that Vthe flood events severely affected vast cropland, grassland and fores. Overlaying with vegetation map indicated that the flood events severely affected vast cropland, grassland and forest. The Sentinel-1 SAR data and GEE platform allowed rapid flood assessment and enabled real-time mapping, which is particularly important for the decision-makers to develop flood controlling measures, mitigation, relief,and rescue planning.

Keywords Flood mapping · Brahmaputra River · Sentinel-1 · Google Earth Engine (GEE)

S. Mudi (✉) · Jaya Prakash A
Centre for Ocean, Rivers, Land and Atmospheric Sciences, Indian Institute of Technology Kharagpur, Kharagpur, India

P. Das
World Resources Institute India, New Delhi, India

P. K. Shit et al. (eds.), *Geospatial Technology for Environmental Hazards*, Advances in Geographic Information Science, https://doi.org/10.1007/978-3-030-75197-5_20

20.1 Introduction

Flood is one of the most powerful natural disasters affecting the socio-economic conditions of millions of people, causing loss of life, food, infrastructure, etc. (DeVries et al., 2020). The severity of flood events depends on the location, flood preparedness, mitigation and action plans. Periodic flood mapping and long-term monitoring can help in developing appropriate policies and preparedness, whereas real-time flood mapping will help the decision-makers in disaster management through improved mitigation and action plans. The identification of water inundation through remote sensing techniques is useful for assessing flood hazards, especially in remote areas. Both the permanent and temporary surface water areas can easily be identified using optical (mono- and multispectral) and microwave satellite images (synthetic aperture radar, SAR) due to their unique spectral behaviour and backscattering properties, respectively. The optical data is applicable during daylight and under the cloud-free sky conditions, which has limitations during flood events due to the cloud covers. On the contrary, SAR data has advantages over optical data as it can operate at night and can penetrate through clouds and other aerosols. Past studies have shown that low backscatter values for water-dominated pixels in SAR images are beneficial for flood extent mapping (Sghaier et al., 2018). However, the urban areas consist of various artificial structures that act as scatterer and restrict water inundation area mapping due to the higher incidence angle causing the radar shadow.

The lower surface reflectance for water areas beyond the visible bands allows easy identification in multispectral remote sensing images. Several indices, e.g. normalized difference water index [NDWI = (Green − NIR)/(Green+NIR)], modified normalized difference water index [MNDWI = (Green − SWIR)/(Green + SWIR)] and image classification algorithms (parametric and non-parametric), are used to identify the surface water areas (Das & Pandey, 2019; Behera et al., 2018). On the contrary, in microwave images, the smooth texture and higher dielectric constant of the water body induce significantly lower backscatter values compared to other land surface features, which enables the use of the thresholding method for water area mapping (Das et al., 2021; Manjusree et al., 2012). Several studies have employed different supervised or unsupervised classification and machine learning (ML) techniques for the water inundation area mapping using SAR data (Townsend, 2001; Martinis & Twele, 2010; Giustarini et al., 2016). The use of supervised classification achieves higher accuracy but has limitations in training sample collection in each time (Kussul et al., 2008; Song et al., 2007). Pulvirenti et al. (2011) developed an automatic classifier based on fuzzy logic, which also depends on the human-labelled samples. Several studies were carried out, where the threshold value method was used for surface water area mapping (Kuenzer et al., 2013; Martinis & Rieke, 2015; Schumann et al., 2009). Various ML approaches, e.g. support vector machines (SVM), random forest (RF) and artificial neural networks (ANN), are widely used for various regression and classification studies (Dadhich et al., 2019; Uddin et al., 2019; Kussul et al., 2011). The ML algorithms are compelling non-linear regression methods that are used as an alternative to traditional ways to handle complex and non-linear problems (Breiman, 2001). In ML algorithms, no prior assumptions are made about the nature of the relationships between the response and predictor variables (Prasad et al., 2006). As a result, these algorithms allow for all the possible interactions, including complex non-linear ones, among the predictor variables, and do not rely on the data distribution (Ali et al., 2015).

The RGB colour composite image is a fast and reliable method to enhance selective features in an image depending on the pre- and during-event conditions. The changes in dielectric constant and backscatter values during the pre- and during-flood events allow easy delineation of water spread areas (Lu et al., 2014). Conde and De Mata Muñoz (2019) applied the RGB clustering and thresholding technique for the flood inundation assessment and reported that the Sentinel-1 SAR data has a high potential for flood inundation extent mapping. Shen et al. (2019) highlighted that the SAR data-derived inundation results may comprise errors due to other water-like surfaces, geometric error and speckle noise, and removal of such errors requires additional inputs or use of the semi-automatic approaches. Otsu's method, named after Nobuyuki Otsu, is an automatic image thresholding technique (Sezgin & Sankur, 2004). This algorithm returns a single-intensity threshold computed from the radiometric histogram, which separates the image pixels into two classes as foreground and background. This threshold is determined by minimizing intra-class intensity variance or equivalently by maximizing inter-class variance (Otsu, 1979). Lee et al. (1990) reported an accepted accuracy of this method when object area covers at least 30% of the total image area, and the accuracy declines when the object area is reduced below 10%.

The Sentinel-1 is a freely available C-band SAR data and records the backscatter values in two polarizations as vertically transmitted vertically received (VV) and vertically transmitted horizontally received (VH). Previous studies have indicated a higher accuracy for VV polarization compared to VH polarization for the surface water mapping (Psomiadis, 2016; Agnihotri et al., 2019). Clement et al. (2018) applied Otsu's method for water inundation area mapping employing the Sentinel-1 SAR data. They estimated the accuracy compared with the MNDWI results and reported higher accuracy for VV band. Ruzza et al. (2019) compared the manual threshold approach, Otsu methodology, and K-means clustering for flood mapping using Sentinel-1 data and observed a similar accuracy. The latest Google Earth Engine (GEE) is a cloud-based image-processing platform that consists of various spatial datasets including the Sentinel-1 SAR data (Huang et al., 2017). Vanama et al. (2020) applied Otsu's thresholding algorithm on the Sentinel-1 data in the GEE platform to map the water inundation areas during the Kerala flood event in August 2018 and observed a well-accepted accuracy. The Brahmaputra, Ganga and Meghana River basins occupy about 60% of the total area in India and experience recurrent flood events, almost every year (Shivaprasad Sharma & Roy, 2017). The present study focuses on water inundation area mapping using the Otsu thresholding approach employing the Sentinel-1 SAR data in the state of Assam.

20.2 Study Area

The present study was conducted for flood mapping in Assam, India (Fig. 20.1). The study area is geographically situated between 88.25°E to 96.00°E longitude and 24.50°N to 28.00°N latitude. The major rivers of Assam include Brahmaputra River, Barak River and Manas river. The total area of Assam state is ~78,438, wherein 56,194 and 22,244 km^2 areas are occupied by the Brahmaputra and Barak River basins (Govt. of Assam, n.d. water resources). The Brahmaputra River is both a snow- and rainfed river and has continuous flow throughout the year (Shivaprasad Sharma & Roy, 2017). The state of Assam is affected by recurrent flood events and experiences nearly one severe flood event every year. According to the Rashtriya Barh Ayog, about 31,500 km^2 in Assam is flood-prone area, which is about 39.58% of the total land area of Assam and about 9.40% of the total flood-prone area of the whole country. The total population of Assam is about more than 30 million, who are mostly living along the floodplain of Brahmaputra River and dependent on agriculture and allied sectors. In addition, the study region is home to several important sites, e.g. Kaziranga National Park, a UNESCO World Heritage Site in the Eastern Himalayas, an ecologically important region.

20.3 Materials and Methodology

Materials

A total number of 30 Sentinel-1 C-band image tiles [interferometric large swath (IW) GRD] were used in this study to map the flood inundation areas from 2018 to 2020. The Sentinel-2 optical data was used for reference and accuracy assessment.

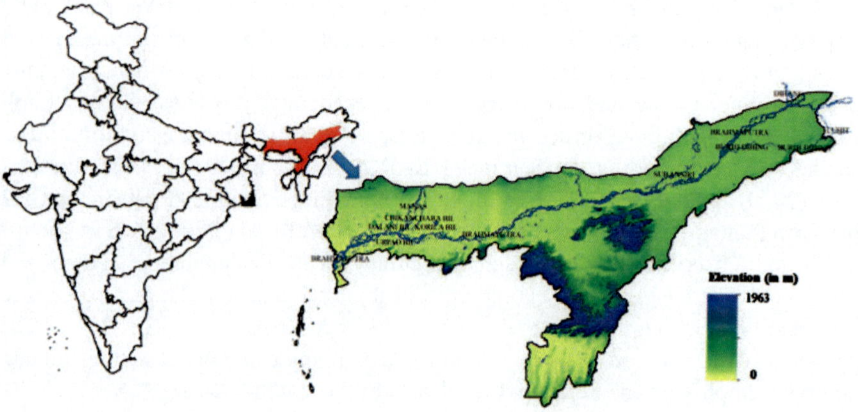

Fig. 20.1 Study area showing an elevation map with river channels

Method

Sentinel-1 Preprocessing

Sentinel-1 SAR images require a number of image corrections as robust geometric, radiometric, thermal and speckle corrections. The ESA Sentinel Application Platform (SNAP) tool is widely used for Sentinel-1 data preprocessing using the following steps: (a) precise geocoding employing the orbit file, (b) radiometric calibration, (c) thermal noise removal, (d) speckle noise removal, (e) the range Doppler terrain correction and (f) deriving the backscattering coefficients (in dB) from the orthorectified sigma band (Dadhich et al., 2019). However, the GEE offers the preprocessed Sentinel-1 Ground Range Detected (GRD) images readily usable for image processing (Filipponi, 2019). In the present study, the Sentinel-1 images for the monsoon season (July to September) were processed for the flood mapping in Assam in 2018, 2019 and 2020.

The preprocessed image data were classified into water and non-water areas employing Otsu's automatic thresholding method. Otsu's algorithm assumes the distribution of image pixel intensities as a bimodal histogram and separates the classes. This method uses the clustering method to generate the threshold value and to separate the classes as the foreground and background. To determine the optimal threshold value, it minimizes the weighted sum of within-class (intra-class) variances for the foreground and background pixels. The mathematical expression for Otsu's algorithm is given as follows:

The pixels in a given image are represented in L grey levels (1, 2, 3, L). n_i represents the number of pixels at level i, and N denotes the total number of pixels:

$$N = n_1 + n_2 n_3 + n_L \tag{20.1}$$

The normalized grey level histogram and probability can be expressed as Eqs. (20.2)–(20.4):

$$P_i = n_i / N \tag{20.2}$$

$$P_i \geq 0 \tag{20.3}$$

$$\sum\nolimits_{i=1}^{L} P_1 = 1 \tag{20.4}$$

Threshold k divides the pixels into two classes (i.e. foreground and background). The optimum value of k can be computed by maximizing the intra-class variance (within the class). The criterion function ρ is introduced and defined as.

$$P(k) = \sigma_B^2(k) / \sigma_{BT}^2 \tag{20.5}$$

σ_B^2 and σ_{BT}^2 can be expressed as

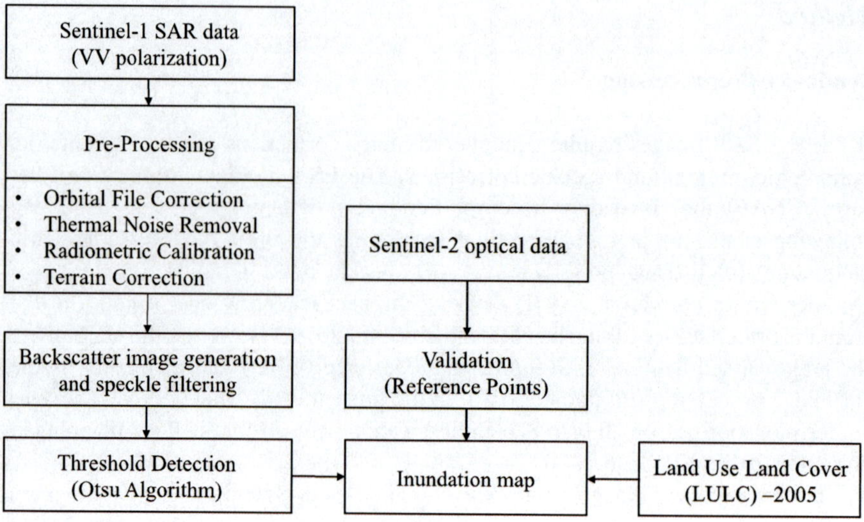

Fig. 20.2 Methodology flow chart diagram

$$\sigma_B^2 = \omega_0\omega_1(\mu_1 - \mu_0)^2 \tag{20.6}$$

$$\sigma_{BT}^2 = \sum\nolimits_{i=1}^{L}(i - \mu_T)^2\eta_i \tag{20.7}$$

where ρ is the criteria function, σ_B^2 is the between class variance, σ_{BT}^2 is the total variance, ω_0 is the probabilities of class occurrence for background, ω_1 is the probabilities of class occurrence for foreground, μ_0 is the class mean of background, μ_1 is the class mean of foreground, μ_T is the total mean grey level of the image, and k is the threshold value.

The probability of class occurrence and the class mean levels are expressed in the Eqs. (20.8) and (20.9):

$$\omega(k) = \sum\nolimits_{i=1}^{k}\eta_i \tag{20.8}$$

$$\mu(k) = \sum\nolimits_{i=1}^{k}i\eta_i \tag{20.9}$$

The Sentinel-2 multispectral data for monsoon was accessed to validate the water inundation map generated by the Sentinel-1 data. The Sentinel-2 images were visually interpreted to collect the reference points for validation. In 2018, 95 reference points were generated, which were 110 and 140 for 2019 and 2020, respectively. The national scale land use and land cover (LULC) map generated by Roy et al. (2016) was downloaded from the ORNL-DAAC website. The LULC map for the year 2005 was the latest available LULC map, which was employed in the current analysis to assess the flood impact. The overall data processing methodology is shown in Fig. 20.2.

20.4 Results and Discussion

The water inundation areas were identified employing Otsu's automatic threshold method using the VV polarization data. The GEE platform was employed for data processing and identifying water inundation in 2018, 2019 and 2020. The bimodal histograms generated for the VV polarization images for 2018, 2019 and 2020 are shown in Fig. 20.3. Otsu's algorithm uses an optimal threshold value to differentiate between the foreground and background, i.e. the water and non-water areas. The identified threshold value was −14.90 for 2018, which were −14.92 and −14.93 for 2019 and 2020, respectively.

The comparison of the reference points visually identified in the Sentinel-2 optical data indicated a well-accepted accuracy for all 3 years. The estimated overall accuracies were 89%, 87% and 91% for 2018, 2019 and 2020, respectively. A few image tiles were created for the cloud-free optical data (in standard false colour composite, SFCC), and the overlaid water inundation layers on the SFCC map (Fig. 20.4) which also indicated high accuracy. Vanama et al. (2020) applied Otsu's thresholding algorithm on the Sentinel-1 data in the GEE platform to map the water inundation area during the Kerala flood (2018) and reported a similar accuracy (overall accuracy of 82% and 78.5% for flood class). A few field photographs were collected for the year 2020 during the flood events in Assam, as shown in Fig. 20.5. These field photographs were collected for the Hatibat and Notun goan in Makum, Tinsukia, Bishnupur and Pohukhowa Gaon region.

The water inundation area maps of 2018, 2019 and 2020 are shown in Fig. 20.6. The LULC areas affected by the flood events in different years are shown in Fig. 20.7. The majority of the flooded regions were identified along the Brahmaputra River, where the floodplains were mostly inundated in the last 3 years (during 2018–2020) (Fig. 20.6). The total inundation area was identified as 1453.82 km^2 in 2018, which increased to 2080.95 km^2 in 2019 and 3633.81 km^2 in 2020 (Table 20.1). The flood hazard map of Assam prepared by the Govt. of Assam through Assam State Disaster Management Authority (ASDMA, n.d.) identified about 28.75% area of Assam was affected by flood during 1998–2015, wherein 17 out of 34 districts were severely affected. The flood hazard map indicated a high similarity with the water inundation area maps developed in the current study. Flood events during 2018–2020 were observed in all the districts in the floodplain of Brahmaputra River and the three districts in the southern region as Karimganj, Hailakandi and Cachar (Fig. 20.6).

The maximum water inundation in the year 2018 was observed for cropland (777.43 km^2) followed by grassland (477.78 km^2) (Table 20.1; Fig. 20.7). In comparison, about 1379.40 and 2670.55 km^2 cropland were flooded in 2019 and 2020, respectively. About 152.83 km^2 forest area was flooded with 15.24 km^2 shrubland in 2018. The forested areas inundated in 2019 and 2020 were 183.91 and 272.33 km^2, respectively, and the shrubland inundated areas were 10.95 and 21.34 km^2, respectively. About 463.08 and 616.81 km^2 grassland were inundated in 2019 and 2020, respectively. Moreover, various settlement areas that were flooded

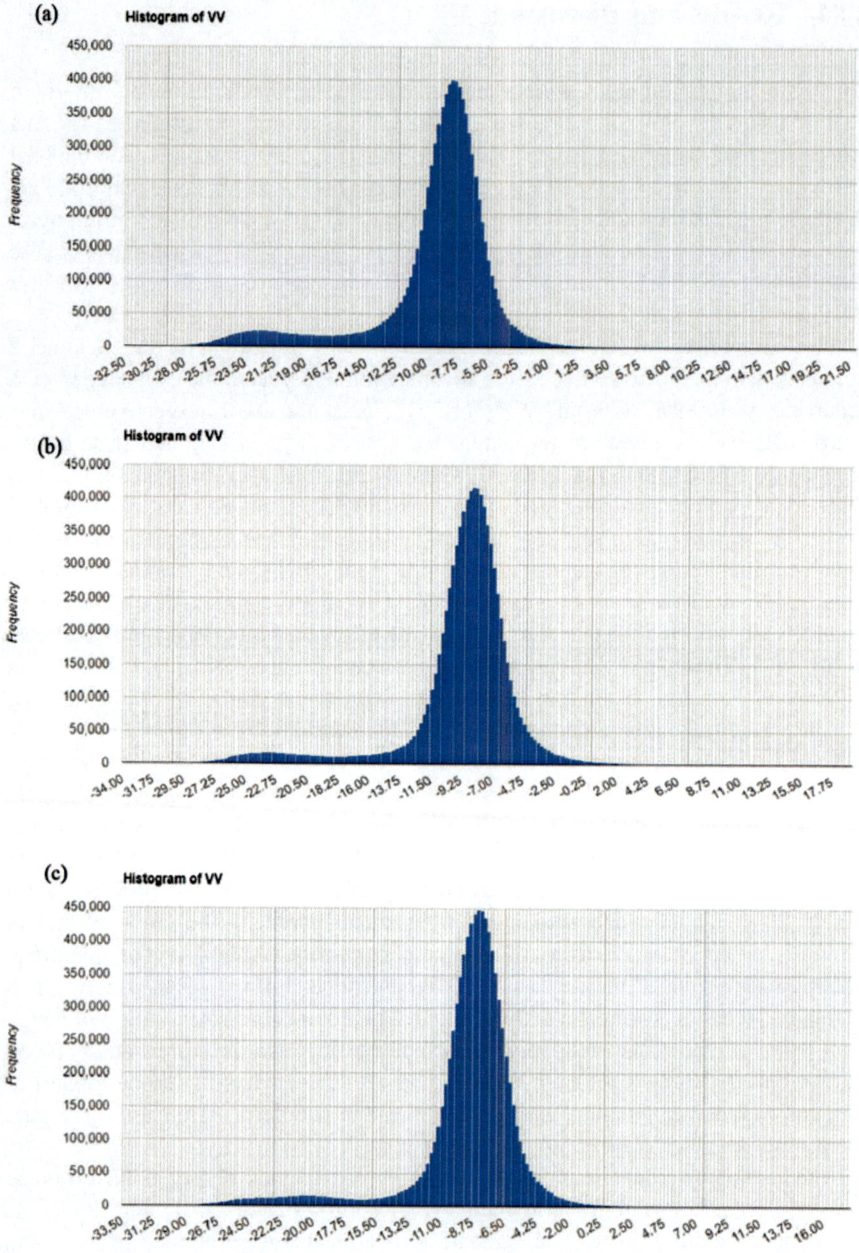

Fig. 20.3 Bimodal histograms of VV images for the year (**a**) 2018, (**b**) 2019 and (**c**) 2020

(a) (c) (b) (d) (e) (g) (f) (h)

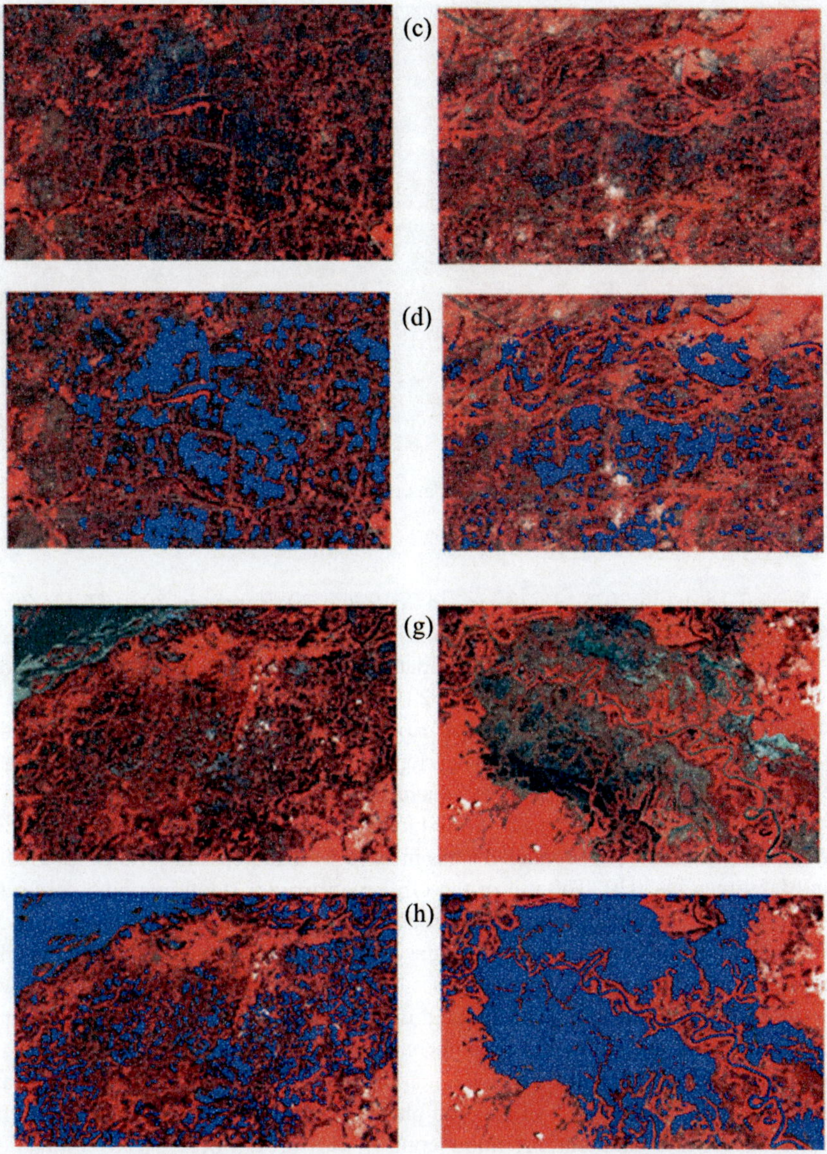

Fig. 20.4 Comparison of optical data and identified flooded areas for the year 2019: (**a**) SFCC image and (**b**) flood map overlaid on SFCC image tile-1; (**c**) SFCC image and (**d**) flood map overlaid on SFCC image tile-2 for the year 2020: (**e**) SFCC image and (**f**) flood map overlaid on SFCC image tile-3; (**g**) SFCC image and (**h**) flood map overlaid on SFCC image tile-4

Fig. 20.5 Field photographs of water inundation in (**a**) Hatibat (**b**) Notun goan in Makum, (**c**) Bishnupur (**d**) Pohukhowa Gaon

indicated 7 km^2 in 2018, 18.70 km^2 in 2019 and 20.61 km^2 in 2020 (Table 20.1). Borah et al. (2018) studied the water inundated areas in the Kaziranga National Park, Assam, employing the Sentinel-1 SAR data and reported that about 35% of the total area was inundated twice in the year 2017. Shivaprasad Sharma and Roy (2017) studied the flood inundation in the Kopili River basin, Assam, and recorded 183 flood events in two decades. They reported that 29% of the Kopili River basin experienced floods during 1977, 1988 and 1998–2015.

The water inundation maps identified for the past 3 years indicated high accuracy (>87%). The total water inundated areas in 2018 were increased by 1.5 times in 2019 and 2.5 times in 2020. The use of the GEE platform for flood mapping solves the issue of data downloading, data preprocessing and good computation facility for Sentinel-1 data processing. Moreover, the computation facility provided by the GEE enables quick data processing and accessibility of the data employing a simple web browser from any location. The adopted method is useful to generate the flood map in a short time and allows to prepare real-time water inundation maps using the Sentinel-1 data. The real-time flood mapping will help the decision-makers and policy developers in developing action plans during flood events, rescue and relief operations, route preparation, flood control measures, flood hazard mitigation, preparedness, assessing the resource loss, etc.

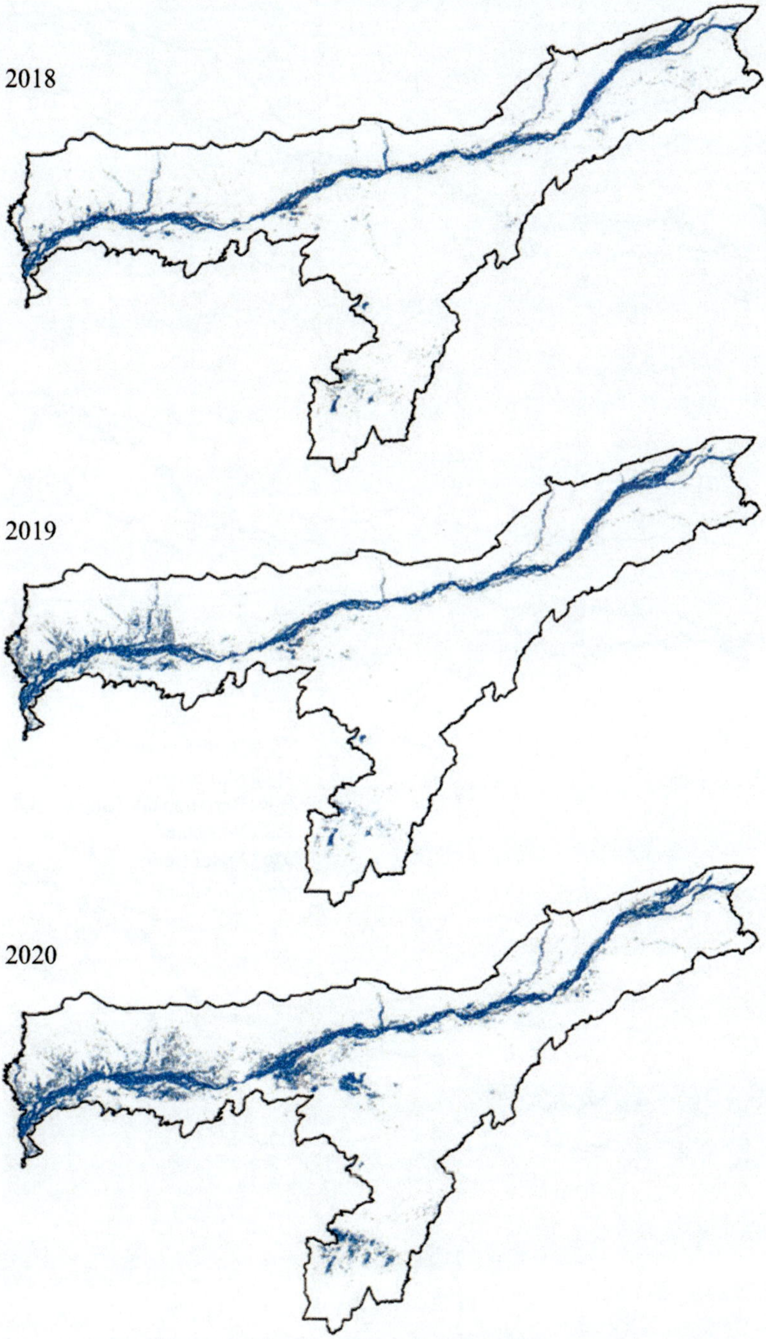

Fig. 20.6 Water inundation map for the year (**a**) 2018, (**b**) 2019, and (**c**) 2020

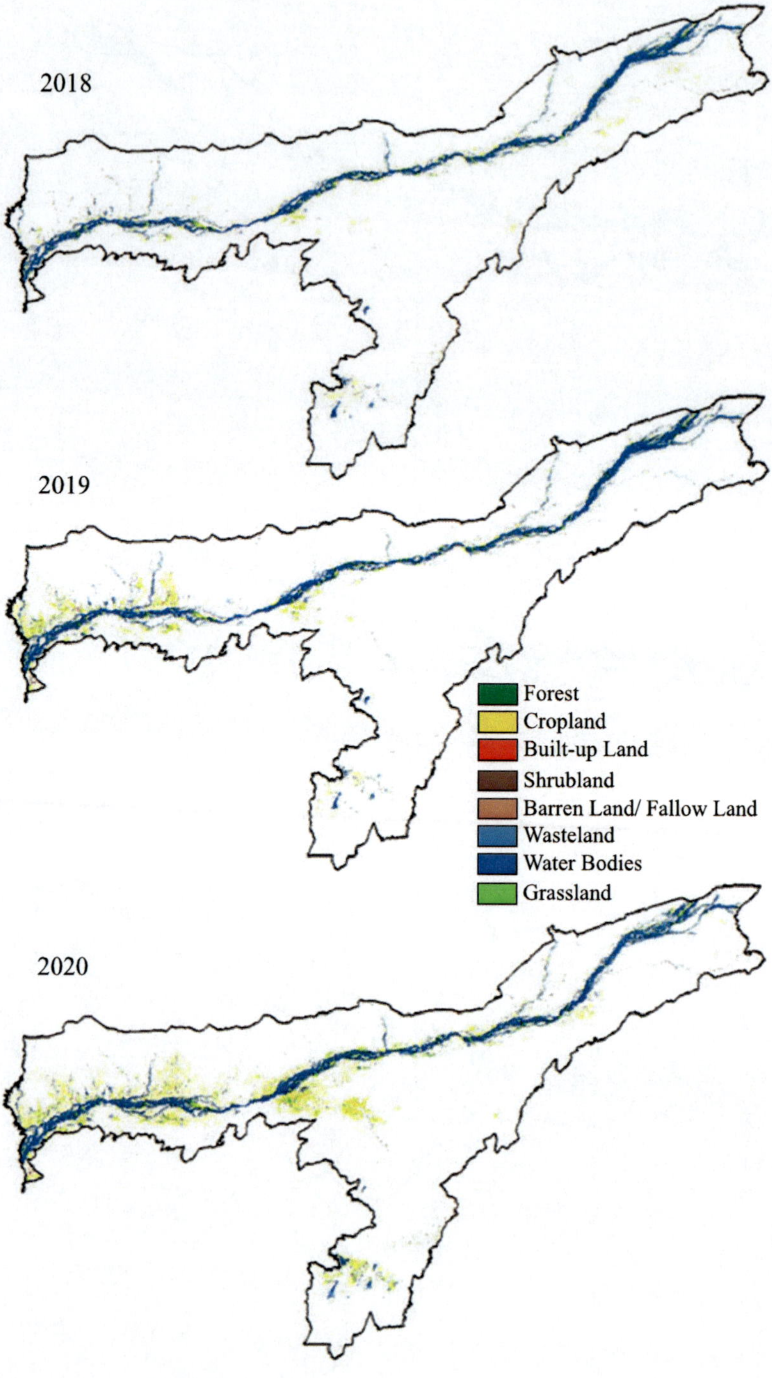

Fig. 20.7 LULC affected by the water inundation

Table 20.1 Flood inundated area (in km^2) and LULC area affected in different years

LULC	2018	2019	2020
Forest	152.83	183.91	272.33
Cropland	777.43	1379.40	2670.55
Built-up land	7.18	18.70	20.61
Shrubland	15.24	10.95	21.34
Barren land	0.17	0.05	0.18
Fallow land	0.99	1.57	3.17
Wasteland	23.20	23.29	28.80
Grassland	476.78	463.08	616.81
Total	1453.82	2080.95	3633.81

20.5 Conclusion

Accurate identification of water inundation areas during a flood event can provide valuable and timely information to the policymakers. Otsu's automatic thresholding method was applied to identify the water inundation areas using the Sentinel-1 SAR data in the GEE platform. The verification with the Sentinel-2 optical data indicated high accuracy ($\geq$87%) for the last 3 years (2018, 2019 and 2020). The adopted method identified the water inundation areas without site-specific training data and local knowledge. The use of the GEE platform allowed us to perform the data processing with minimum system requirement and in a short time. We have observed severe flood-affected areas in Assam for the last 3 years (i.e. 2018, 2019 and 2020). All the districts situated in the floodplain of the Brahmaputra River and three districts in the southern part of Assam experienced water inundation. Extensive cropland, grassland and forest areas were flooded in the last 3 years, which indicating a total area of 1453.82, 2080.95 and 3633.81 km^2 in 2018, 2019 and 2020, respectively. The adopted method and generated maps appeal to increasing interest in real-time flood mapping using Web-GIS platforms like GEE and Sentinel-1 SAR data.

References

Agnihotri, A. K., Ohri, A., Gaur, S., Das, N., & Mishra, S. (2019). Flood inundation mapping and monitoring using SAR data and its impact on Ramganga River in Ganga basin. *Environmental Monitoring and Assessment, 191*(12), 760.

Ali, I., Greifeneder, F., Stamenkovic, J., Neumann, M., & Notarnicola, C. (2015). Review of machine learning approaches for biomass and soil moisture retrievals from remote sensing data. *Remote Sensing*. https://doi.org/10.3390/rs71215841

Behera, M. D., Gupta, A. K., Barik, S. K., Das, P., & Panda, R. M. (2018). Use of satellite remote sensing as a monitoring tool for land and water resources development activities in an Indian tropical site. *Environmental Monitoring and Assessment, 190*(7), 401.

Borah, S. B., Sivasankar, T., Ramya, M. N. S., & Raju, P. L. N. (2018). Flood inundation mapping and monitoring in Kaziranga National Park, Assam using Sentinel-1 SAR data. *Environmental Monitoring and Assessment, 190*(9), 520.

Clement, M. A., Kilsby, C. G., & Moore, P. (2018). Multi-temporal synthetic aperture radar flood mapping using change detection. *Journal of Flood Risk Management, 11*(2), 152–168.

Conde, F. C., & De Mata Muñoz, M. (2019). Flood monitoring based on the study of Sentinel-1 SAR images: The Ebro River case study. *Water (Switzerland), 11*, 1–25. https://doi.org/10.3390/w11122454

Dadhich, G., Miyazaki, H., & Babel, M. (2019). Applications of sentinel-1 synthetic aperture radar imagery for floods damage assessment: A case study of Nakhon Si Thammarat, Thailand. *The International Archives of the Photogrammetry, Remote Sensing and Spatial Information Sciences, 42*, 1927–1931. https://doi.org/10.5194/isprs-archives-XLII-2-W13-1927-2019

Das, P., & Pandey, V. (2019). Use of logistic regression in land-cover classification with moderate-resolution multispectral data. *Journal of the Indian Society of Remote Sensing, 47*(8), 1443–1454.

Das, P., Pandey, V., & Dutta, D. (2021). Land surface water resource monitoring and climate change. In *Mapping, monitoring and modeling land and water resources*. Taylor & Francis. [*accepted*].

DeVries, B., Huang, C., Armston, J., Huang, W., Jones, J. W., & Lang, M. W. (2020). Rapid and robust monitoring of flood events using Sentinel-1 and Landsat data on the Google Earth Engine. *Remote Sensing of Environment, 240*, 111664. https://doi.org/10.1016/j.rse.2020.111664

Filipponi, F. (2019). Sentinel-1 GRD Preprocessing workflow. In Multidisciplinary Digital Publishing Institute Proceedings (18, 1, 11).

Giustarini, L., Hostache, R., Kavetski, D., Chini, M., Corato, G., Schlaffer, S., & Matgen, P. (2016). Probabilistic flood mapping using synthetic aperture radar data. *IEEE Transactions on Geoscience and Remote Sensing, 54*(12), 6958–6969.

Govt. of Assam through Assam State Disaster Management Authority (ASDMA) (n.d.). https://www.isro.gov.in/updated-flood-hazard-atlas-assam-state; Accessed on 15-01-2021.

Govt. of Assam water resources (n.d..): https://waterresources.assam.gov.in/portlets/river-system-of-assam

Huang, H., Chen, Y., Clinton, N., Wang, J., Wang, X., Liu, C., Gong, P., Yang, J., Bai, Y., Zheng, Y., & Zhu, Z. (2017). Mapping major land cover dynamics in Beijing using all Landsat images in Google Earth Engine. *Remote Sensing of Environment, 202*, 166–176. https://doi.org/10.1016/j.rse.2017.02.021

Kuenzer, C., Guo, H., Huth, J., Leinenkugel, P., Li, X., & Dech, S. (2013). Flood mapping and flood dynamics of the mekong delta: ENVISAT-ASAR-WSM based time series analyses. *Remote Sensing, 5*, 687–715. https://doi.org/10.3390/rs5020687

Kussul, N., Shelestov, A., & Skakun, S. (2011). Flood monitoring from SAR data. In *Use of satellite and in-situ data to improve sustainability* (pp. 19–29). Springer.

Kussul, N., Shelestov, A., & Skakun, S. (2008). Grid system for flood extent extraction from satellite images. *Earth Science Informatics, 1*, 105–117. https://doi.org/10.1007/s12145-008-0014-3

Lee, S. U., Chung, S. Y., & Park, R. H. (1990). A comparative performance study of several global thresholding techniques for segmentation. *Computer Vision, Graphics, and Image Processing, 52*(2), 171–190.

Leo, Breiman Machine Learning 45(1), 5–32. https://doi.org/10.1023/A:1010933404324

Lu, J., Giustarini, L., Xiong, B., Zhao, L., Jiang, Y., & Kuang, G. (2014). Automated flood detection with improved robustness and efficiency using multi-temporal SAR data. *Remote Sensing Letters, 5*, 240–248. https://doi.org/10.1080/2150704X.2014.898190

Manjusree, P., Kumar, L. P., Bhatt, C. M., Rao, G. S., & Bhanumurthy, V. (2012). Optimization of threshold ranges for rapid flood inundation mapping by evaluating backscatter profiles of high incidence angle SAR images. *International Journal of Disaster Risk Science, 3*(2), 113–122.

Martinis, S., & Twele, A. (2010). A hierarchical spatio-temporal Markov model for improved flood mapping using multi-temporal X-band SAR data. *Remote Sensing, 2*(9), 2240–2258.

Martinis, S., & Rieke, C. (2015). Backscatter analysis using multi-temporal and multi-frequency SAR data in the context of flood mapping at River Saale, Germany. *Remote Sensing, 7*, 7732–7752. https://doi.org/10.3390/rs70607732

Otsu, N. (1979). A threshold selection method from gray-level histograms. *IEEE Transactions on Systems, Man, and Cybernetics, 9*(1), 62–66. https://doi.org/10.1109/TSMC.1979.4310076

Prasad, A. M., Iverson, L. R., & Liaw, A. (2006). Newer classification and regression tree techniques: Bagging and random forests for ecological prediction. *Ecosystems, 9*, 181–199. https://doi.org/10.1007/s10021-005-0054-1

Psomiadis, E. (2016). Flash flood area mapping utilising SENTINEL-1 radar data. In Earth Resources and Environmental Remote Sensing/GIS Applications VII (Vol. 10005, p. 100051G). International Society for Optics and Photonics.

Pulvirenti, L., Chini, M., Pierdicca, N., Guerriero, L., & Ferrazzoli, P. (2011). Flood monitoring using multi-temporal COSMO-skymed data: Image segmentation and signature interpretation. *Remote Sensing of Environment, 115*, 990–1002. https://doi.org/10.1016/j.rse.2010.12.002

Roy, P. S., Meiyappan, P., Joshi, P. K., Kale, M. P., Srivastav, V. K., Srivasatava, S. K., Behera, M. D., Roy, A., Sharma, Y., Ramachandran, R. M., Bhavani, P., Jain, A. K., & Krishnamurthy, Y. V. N. (2016). *Decadal land use and land cover classifications across India, 1985, 1995, 2005*. ORNL DAAC. https://doi.org/10.3334/ORNLDAAC/1336

Ruzza, G., Guerriero, L., Grelle, G., Guadagno, F. M., & Revellino, P. (2019). Multi-method tracking of monsoon floods using Sentinel-1 imagery. *Water, 11*(11), 2289.

Schumann, G., Di Baldassarre, G., & Bates, P. D. (2009). The utility of spaceborne radar to render flood inundation maps based on multialgorithm ensembles. *IEEE Transactions on Geoscience and Remote Sensing, 47*, 2801–2807. https://doi.org/10.1109/TGRS.2009.2017937

Sezgin, M., & Sankur, B. (2004). Survey over image thresholding techniques and quantitative performance evaluation. *Journal of Electronic Imaging, 13*(1), 146–166. https://doi.org/10.1117/1.1631315

Sghaier, M. O., Hammami, I., Foucher, S., & Lepage, R. (2018). Flood extent mapping from time-series SAR images based on texture analysis and data fusion. *Remote Sensing, 10*, 1–30. https://doi.org/10.3390/rs10020237

Shen, X., Wang, D., Mao, K., Anagnostou, E., & Hong, Y. (2019). Inundation extent mapping by synthetic aperture radar: A review. *Remote Sensing, 11*, 1–17. https://doi.org/10.3390/RS11070879

Song, Y. S., Sohn, H. G., & Park, C. H. (2007). Efficient water area classification using radarsat-1 SAR imagery in a high relief mountainous environment. Photogramm. Eng. *Remote Sensing, 73*, 285–296. https://doi.org/10.14358/PERS.73.3.285

Shivaprasad Sharma, S. V., & Roy, P. S. (2017). Extraction of detailed level flood hazard zones using multi-temporal historical satellite data-sets–a case study of Kopili River basin, Assam, India. *Geomatics, Natural Hazards and Risk, 8*(2), 792–802.

Townsend, P. A. (2001). Mapping seasonal flooding in forested wetlands using multi-temporal Radarsat SAR. *Photogrammetric Engineering and Remote Sensing, 67*(7), 857–864.

Uddin, K., Matin, M. A., & Meyer, F. J. (2019). Operational flood mapping using multi-temporal Sentinel-1 SAR images: A case study from Bangladesh. *Remote Sensing, 11*. https://doi.org/10.3390/rs11131581

Vanama, V. S. K., Mandal, D., & Rao, Y. S. (2020). GEE4FLOOD: Rapid mapping of FLOOD areas using temporal Sentinel-1 SAR images with Google earth engine cloud platform. *Journal of Applied Remote Sensing, 14*(3), 034505.

Chapter 21
Assessment of Socio-Economic Vulnerability in a Forested Region: An Indicator-Based Study in Bankura District of West Bengal, India

Shyamal Dutta and Soumen Chatterjee

Abstract Vulnerability is the degree to which people, resources, economic and cultural environment or activity are exposed to hostile agent and are suffering from the propensity of being harmed from them. Climate change coupled with environmental degradation seems to be the major driver for forest clearance which pushes millions of people on the verge of socio-economic vulnerability. It is basically the endogenous inability that leads the society towards the inequality and social conflict which not only poses problem for local development but also considered to be a matter of serious concern for regional development on a sustainable way. Accordingly several efforts have been made by FAO, USAID and WFP to reduce vulnerability to combat with reducing the inequality, alleviation of poverty and maintain food security and well-being. So study of vulnerability is not only important for resource management but also essential for future planning. The present study addresses this problem over the forested district of Bankura to explore the extent of socio-economic vulnerability by using three major components of vulnerability defined by IPCC, namely, exposure, sensitivity and adaptive capacity. The extent of this component has been computed at subdistrict level using quantitative techniques, and the whole data has been processed in the GIS environment. The result reflects that exposure is closely associated with physiological conditions, whereas economic environment controls the sensitivity factors. It was also found that eastern and southern part are better equipped with adaptive capacity by virtue of their developed infrastructure. Combining all those environmental stress prevailing on the study area has been assessed with adequate measures.

Keywords Vulnerability · Quantitative analysis · Indicator-based approach · Forest ecology, GIS

S. Dutta (✉) · S. Chatterjee
Independent Researcher, Bardhaman, West Bengal, India

P. K. Shit et al. (eds.), *Geospatial Technology for Environmental Hazards*, Advances in Geographic Information Science, https://doi.org/10.1007/978-3-030-75197-5_21

21.1 Introduction

Vulnerability may be considered as a state of diminished capacity of an individual or a group to anticipate with a changing environment which may prove hostile to them while other may not find it as such. Alternatively it may be considered as an extent to which persons, resources, socio-economic environment as well as life and livelihood are liable to damage or deterioration being uncovered to intimidating causes or factor. In the circumstance of changing climate with environmental degradation, many ecosystems are either degraded heavily in some form or other on a regional scale. It not only pushes the species of those ecosystem on the verge of extinction to some extent but also makes most of them vulnerable in the changing face of the environment to a great extent. Hence, the study of vulnerability is considered to be a major issue of scientific research in recent time with special emphasis on forest ecosystem which is reflected from the work carried out by major apex global bodies. For instance, IPCC considers climate change as foreseeable and expected to exacerbate the physiological hassle on green space of environment through temperature rising, prolonged drought and customized rate of occurrence of any farthest events (IPCC, 2012; Sharma et al., 2017). In the twenty-first century, such levels of climatic as well as non-climatic occurrences are likely to have stern impacts for the vulnerability of forests especially in the tropics (IPCC, 2014).

Extending over 30% of the global land cover, forest ecosystems are one of the most biologically rich and genetically assorted ecosystems on the planet (Köhl et al., 2015). Apart from the various ecological functions like moulding the climate, protection of the soil cover and as a host, forest indirectly provides opportunities to sustain millions of livelihood. It is popularly reported that above 400 million people and more than 1 billion people are highly dependent on forests for continuation with income as well as for forest-based products and services partially for their livelihoods, respectively (Munang et al., 2011). As a matter of fact, anthropogenic influences over the forest cover are found to be increasing day by day. Forest continues to serve the ever-increasing demands of man which is increasingly putting stress on the natural resources and making them more vulnerable. Increasing human population pooled with irrational uses of resources, underprivileged management and conservation practices further contributes to their vulnerability (Sharma et al., 2009; Tse-ring et al., 2010). Therefore, forest vulnerability assessment has emerged as a critical prerequisite to ensure forest resource management, conservation and long-term adaptation and/or mitigation under increasing perturbations (Murthy et al., 2011; Ribot, 2011). Seidl et al. (2011) also consider the assessment of vulnerability as urgent due to the time factor needed for any adaptive measures undertaken and development of adaptive capacities to formulate. Vulnerability assessments should be conducted at a local scale (e.g. small forest patch) as well as regional scales like broad landscapes to serve diverse objectives (Sharma et al., 2013; Engström et al., 2020). Assessments at regional scales would support in identification of vulnerable forest areas as well as prioritize them for competent resource allocation (Naess et al.,

2006), where assessment in local level is obligatory for scheming site-specific enhancement measures for forest resilience (Sharma et al., 2013).

Assessment of association between physical and social system using diversified techniques is foremost important for vulnerability assessment in any space as assortment of appropriate site-specific indicators from both the system is mandatory to address versatile issues for vulnerability assessment (Hahn et al., 2009). Malakar and Mishra (2017) point out that this assessment fundamentally depends on the social and economic state of affairs of any social system. That's why such assessment is very crucial for shaping the extent of agony of the relying population and their economic space due to disasters (Sahana et al., 2019). Vulnerability database can be efficiently accomplished using both the data sources, primary (Challinor et al., 2010) and secondary (Sahana & Sajjad, 2019). Several studies are found to be conducted on this particular issue over different ecological setups by applying several techniques to obtain targeted outcome. For instance, Sullivan et al. (2002) applied gap method, Bray et al. (2012) used human development index, composite vulnerability index has been applied by Rygel et al. (2006) with Sajjad and Nasreen (2016) who implied sustainable livelihood security index, whereas fuzzy logic has been used Ahmed et al. (2018) for the assessment of vulnerability. Beside this, several approaches have been found to be applied in different time to assess the level of vulnerability, for example, technical measures (Brooks, 2003), social science-oriented approaches and human-oriented approach involving diverse aspects from ecological and socio-economic as well as institutional parameters (Blaikie et al., 2005) in different time.

However, a holistic and interdisciplinary approach seems to be more appropriate and useful for the assessment of vulnerability as the concept itself consists of heterogeneous components. Use of indicators from several aspects proves to be more useful and applicable in this context as they have the provision to use quantitative as well as qualitative indicators which are informative, analytical and collaborative in nature, so they will yield a more realistic result which may facilitate the research and planning and also useful for decision-making process. Socio-ecological vulnerability which resulted from changing climate as well as unprincipled distribution of resource in the Himalayan region has been highlighted by Pandey and Bardsley (2015), whereas Sahana et al. (2019) highlighted the socio-economic vulnerability in the Indian Sundarban region using pragmatic approach. Thakur et al. (2019) and Sharma et al. (2013) applied a similar technique over the forest areas of western Himalaya and Western Ghat region, respectively. Sharma et al. (2015) in another study assessed the vulnerability on a national scale and categorized different levels of vulnerability. They also suggested minimum human interference and conservation of biodiversity may help to reduce the vulnerability.

The present study has been carried over the dry deciduous forest region of Bankura district. The district has vast forested patches with a large population dependant on the forest. The present study intended to address several aspects of socio-economic vulnerability with reference to livelihood and well-being of the people in the light of traditional conflict assessment. This will surely provide an accurate and prevention-oriented approach to minimize the conflict and inequality and also will be very useful to improve the coordination between the local people and developmental agencies engaged in management and policy implementation. Beside this the assessment will be able to monitor the changing well-being of the environment with reference to the economic and social condition and risk managing apatite of the household. So the study considers being necessary with respect to the need of the time. The present study considers the IPCC definition of vulnerability which is 'the degree to which a system is susceptible to or unable to cope with adverse effects of climate change, including climate variability and extremes'. As per definition, three key components of vulnerability are exposure, sensitivity and adaptive capacity. Exposure (E) is mainly highlighted in the climatic extreme phenomenon as well as ecological factors, which include topographic factors and environmental aspects too. So exposure factor can be an amalgamation of both changing climatic phenomena and ecosystem factors. Sensitivity (S) is a distinguishing feature of the system and represents 'dose-response relationship' between exposure and its associated impacts. Adaptive capacity (A) is a property of the system to fiddle with its uniqueness or behaviour in order to expand its coping range under existing variability in socio-economic as well as ecological system over certain time. The major objective of the present study is to focus on the subdistrict or block level assessment of socio-economic vulnerability in a forested district based on the relative importance of three fundamental components with their constituting subcomponents, i.e. exposure, sensitivity and adaptation capacity. Another objective of the study is to identify the priority blocks based on these parameter-wise responses which will eventually assist for preparation of future planning and management.

21.2 Methods and Database

Study Area

Containing diversified regional identity in physiography, climate and edaphic as well as vegetation characteristics, Bankura is positioned in the western part of the West Bengal which is known as *Rarh* in Bengal bounded between the latitudes of 22°38′ and 23°38′ North and longitudes of 86°36′ and 87°46′ East (Fig.21.1). River Damodar separated this district from Purba and Paschim Bardhaman district in the north. In the south-east, Bankura is bounded by Hugli district, on the south by Paschim Medinipur district and on the west by Puruliya district. The Survey of India (SOI) toposheet covering the districts is 73I, 73 J, 73 M and 73 N. Except Damodar, other significant rivers are *Dwarakeswar*, *Sali*, *Gandheswari*, *Shilabati* and *Kangsabati* which are mostly identified as hill streams, originating in the hills in the western plateau region and flow from the north-east to the south-west direction in courses roughly parallel to one another (O'Malley, 1908). Bankura district is one of the vegetated districts comprising three divisions, viz. Bankura (North) Division, Bankura (South) Division and Panchet Division, which is about 1285.58 km^2

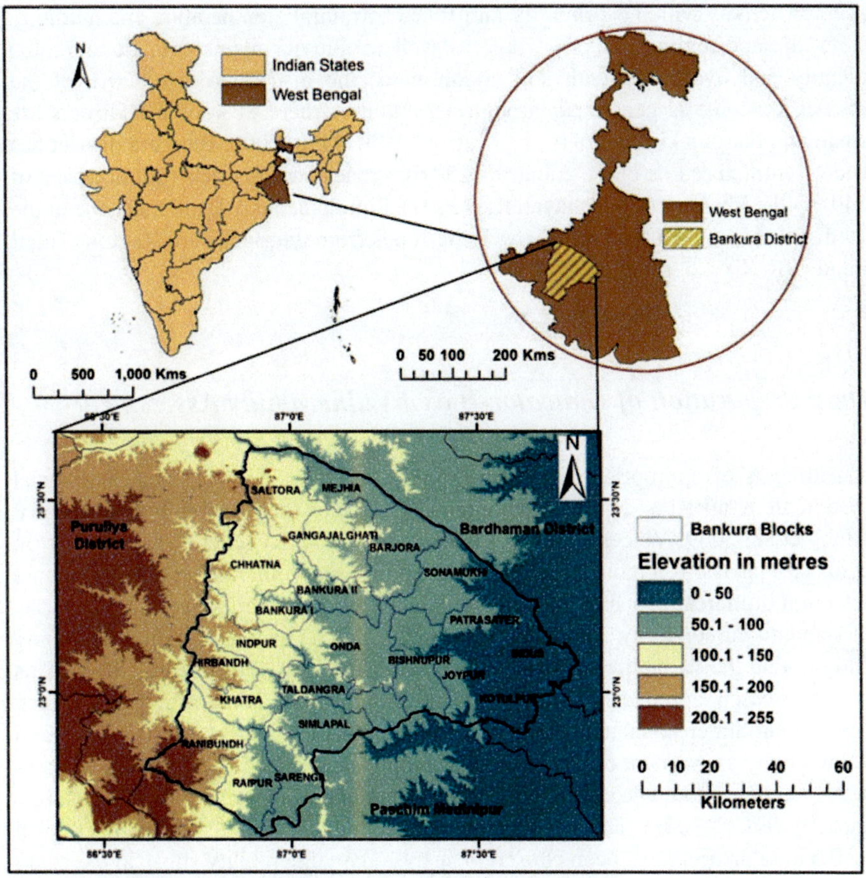

Fig. 21.1 Location of the study area and its geo-environmental setup

territorial forest jurisdiction and covers 18.68% of the total land area of the district. Per capita availability of forest in this district is 0.035 ha which is low than the other south-western districts. The climatic scenario especially in the western upland is much drier than in the eastern or southern part of the district. In pre-monsoon period (March to early June), when the monsoon sets in, hot westerly winds prevail, and the temperature raises up to 45 °C. Winter months of the district are pleasing with temperatures dropping down to below 27 °C at end of the year. The total average rainfall is recorded as 1400 mm, which mostly comes in the months of June to September.

Bankura district has a total population of 3,596,674 with a population density of 523 inhabitants per square kilometre (Census of India, 2011). Population growth rate over the last decade (2001–2011) was recorded as 12.64% with sex ratio of 954 females for every 1000 males as well as a literacy rate of 70.95%. Bankura is economically underdeveloped where agriculture and farming activities are the main primary activity which is primarily influenced by rainfall, temperature and humidity from climatic regime, plateau fringe as well as alluvial plains from topographic identity and hydrology with soil condition of the district. Almost 70% of the district's income is generated through agriculture, where 80% of the farmers are small and marginal in nature (Census of India, 2011). In 2006, Bankura district has been identified as one of the country's 250 most backward districts (out of a total of 640) by the Ministry of Panchayati Raj (2009). Simultaneously Bankura is one of the 19 districts in the state presently receiving funds from the Backward Regions Grant Fund (BRGF) programme.

Data Preparation of Indicator-Based Vulnerability Assessment

Assortments of appropriate site-specific indicators from both physical and social system are required to deal with comprehensive issues for vulnerability assessment (Hahn et al., 2009; Vincent, 2004; O'Brien et al., 2004; Rygel et al., 2006; Preston et al., 2011). The foremost steps of an indicator-based vulnerability assessment are selection of indicators using a defined vulnerability framework covering all aspects of standard vulnerability components, normalization as well as weighing, clubbing and plotting the variability in space context (Preston et al., 2011; Tate, 2012). This indicator-based approach facilitates the development of composite index which covers multidimensional temperament of vulnerability (Wiréhn et al., 2016). Though vulnerability assessment of any region has biophysical, social and economic dimensions, all these should be covered under the trio of exposure, sensitivity and adaptive capacity (IPCC, 2001). The site-specific indicators mainly subdistrict or block level of Bankura district have been chosen to analyse the vulnerability stress in the study area. There is no unbendable conformity on applying indicators to socio-economic vulnerability analysis. All these block level biophysical along with socio-economic indicators for assessment of socio-economic vulnerability also help in evaluating the priority areas where efforts can be made in reducing exposure and sensitivity and increasing adaptation. The following are the block level indicators with their associated role in vulnerability framework to construct socio-economic vulnerability index (Table 21.1).

- **Exposure indicators**: Elevation, slope and aspect form topographic perspective, rainfall and temperature from climatic perspective and per capita availability of forest land as well as agricultural land from environmental and carrying capacity perspectives.
- **Sensitivity indicators**: Total area under forest and agricultural land uses, population density, household density, total work participation, engagement in primary economic activity, engagement as main agricultural labourer, non-working population, engagement of female workers, illiteracy, marginalized population (belongs to Scheduled Caste and Scheduled Tribe population) and total agricultural land under irrigation.

Table 21.1 Framework of indicators with dimensions for socio-economic vulnerability assessment

Domains	ID	Description of indicators	Dimension	Data source
Exposure	E1	Elevation in metres	Physical dimension	SRTM DEM (https://earthexplorer.usgs.gov/)
	E2	Slope in degree	Physical dimension	SRTM DEM (https://earthexplorer.usgs.gov/)
	E3	Aspect	Physical dimension	SRTM DEM (https://earthexplorer.usgs.gov/)
	E4	Average rainfall (mm)	Climatic dimension	Agriculture Meteorologist, Directorate of Agriculture, W.B (2013)
	E5	Average temperature in °C	Climatic dimension	Agriculture Meteorologist, Directorate of Agriculture, W.B (2013)
	E6	Per capita availability of forest (ha/100 person)	Environmental carrying capacity	District Census Handbook, Bankura WB (2011)
	E7	Per capita availability of agricultural land (ha/100 person)	Environmental carrying capacity	District Census Handbook, Bankura WB (2011)
Sensitivity	S1	Total forest area (ha)	Environmental dimensions	District Census Handbook, Bankura WB (2011)
	S2	Total land under agricultural land uses (ha)	Environmental dimensions	District Census Handbook, Bankura WB (2011)
	S3	Population density	Demographic dimensions	District Census Handbook, Bankura WB (2011)
	S4	Household density	Demographic dimensions	District Census Handbook, Bankura WB (2011)
	S5	Engagement in primary economic activity	Economic dimensions	District Census Handbook, Bankura WB (2011)
	S6	Engagement as main agricultural labourer	Economic dimensions	District Census Handbook, Bankura WB (2011)
	S7	Non-working population	Economic dimensions	District Census Handbook, Bankura WB (2011)
	S8	Engagement of female workers	Economic dimensions	District Census Handbook, Bankura WB (2011)
	S9	Total work participation	Economic dimensions	District Census Handbook, Bankura WB (2011)

(continued)

Table 21.1 (continued)

Domains	ID	Description of indicators	Dimension	Data source
	S10	Illiteracy	Social dimensions	District Census Handbook, Bankura WB (2011)
	S11	Total agricultural land under irrigation	Infrastructural dimensions	District Census Handbook, Bankura WB (2011)
	S12	Marginalized population (SC and ST population)	Social dimensions	District Census Handbook, Bankura WB (2011)
Adaptation	A1	Total literate population	Social dimensions	District Census Handbook, Bankura WB (2011)
	A2	Female literate population	Social dimensions	District Census Handbook, Bankura WB (2011)
	A3	Road density per km^2	Infrastructural dimensions	District Statistical Handbook, Bankura (2014)
	A4	Number of banks served per 1000 persons	Infrastructural dimensions	District Statistical Handbook, Bankura (2014)
	A5	Number of cooperative societies	Infrastructural dimensions	District Statistical Handbook, Bankura (2014)
	A6	Distance nearest railway station from the block H.Q. (K.M.)	Infrastructural dimensions	District Statistical Handbook, Bankura (2014)
	A7	Number of primary schools	Infrastructural and social dimensions	District Statistical Handbook, Bankura (2014)
	A8	Number of primary health centres	Infrastructural and health dimensions	District Statistical Handbook, Bankura (2014)
	A9	Number of originating/terminating bus routes	Infrastructural dimensions	District Statistical Handbook, Bankura (2014)

- **Adaptation indicators**: Total literate population with female literate population, road density per km^2, number of banks served per 1000 persons, number of cooperative society, distance nearest railway station from the block headquarter, number of originating/terminating bus routes, number of primary schools and number of primary health centres.

All the respective indicators under three vulnerability components were normalized using universal normalization technique of minimum-maximum rescaling which provides indicators a unique range between zero (0) and one (1). Numerous composite indices (e.g. HDI) were also formulated using such type of normalization technique (UNDP, 2014). Normalization of data has been formulated by using the following equation:

$$N_{iab} = \frac{X_{iab} - \min_{ab(X_i)}}{\max_{ab(X_{iab})} - \min_{ab(X_i)} \qquad (21.1)$$

where N_{ias} indicates normalized indicator value, i denotes to variables (1,2,3...i), 'a' indicates components and 'b' indicates blocks of the district.

Though the impacts of indicators are unknown before putting them under vulnerability assessment, equal weights have been incorporated to assign (Tate, 2012). This weightage is applied when comprehension about indicators is limited. Then, during the clubbing of all these indicators has been normalized and summed up to acquire the arithmetic mean which in a well-known method (Tate, 2012). Simultaneously, all the indicators are assumed to be equally significant to stir up vulnerability in any region (Krishnan et al., 2019). Various researchers have publicized effectiveness of integrated approach where a number of researches have suggested equal weighting approach (e.g. Holsten & Kropp, 2012; Krishnan et al., 2019; Sahana & Sajjad, 2019; Haq, 2003). Human development index (HDI) using three constituents, i.e. education, knowledge and standard of living by UNDP (2015), followed equal weights which justified uses of such weights as indicators direct and easily interpretable. This consistency factor of equal weight techniques is also convinced by different scholars over time (Chowdhury & Squire, 2006; Nguefack-Tsague et al., 2011; Villa & McLeod, 2002).

Selection of the Indicators

Exposure refers to the extent of any system or region which is uncovered to the variable climatic phenomena and change in prime elements of physical environment. It also gives impulse to the economic sector of the region by influencing vegetation growth and crop production. Elevation, slope and aspects all are significant variables of exposure for analysing socio-economic vulnerability as elevation helps to identify vulnerable zones where areas with high elevation received ample amount of sunshine and encourage the growth of vegetations. On the contrary areas with high slope become susceptible to soil erosion (Deb et al., 2019). Propensity to soil erosion which resulted from advanced topographic slope enhances inherent vulnerability of any forested region. Aspects strongly give impact on temperature and soil which has impact on growth of vegetation. It is a causative parameter that is responsible for forest fire too. SRTM-DEM with 30 m spatial resolution has been obtained from Earth Explorer (https://earthexplorer.usgs.gov/) to generate overall as well as block level average slope, aspect and elevation of the study area. Diversity in topographic elevation direct to variability in precipitation and temperature with soil type etc. which ultimately affects the vegetation growth whereas Slope, aspect is indispensable in understanding the vegetation growth and for sustainable forest management (Sinha et al., 2018; Hu et al., 2018). Thus these three physical determinants, i.e. slope, elevation and aspect, are imperative to analyse the vulnerability in forested region. Climate aspects, mainly rainfall and temperature, are central for vulnerability assessment in any forested region as well as area proliferated with agro-based economic activities. Allocation of vegetation and agricultural land as well as cropping pattern is largely exaggerated by the variability in climatic phenomena (Cui et al., 2016).

Rationale of the Selection of the Indicators

In the present study, layers containing average annual temperature and precipitation have been prepared from the collected database from 11 block agricultural meteorologist section and interpolated to the whole district to prepare block level climatic scenario. Population parameter has importance to judge the impact of demography on natural element by various means, i.e. per capita availability and land use pressure. Per capita availability of forest area and agricultural land in terms of net sown area has been calculated by summing up the village level amenities data from District Census Handbook. Both these sub-components have been incorporated as this region is endowed with forest as well as belongs to one of forested districts of South Bengal, and agriculture builds the main economic structure of this region. All these seven indicators have been integrated to construct an exposure index of the study area by using the following formula and categorized using quintile method in GIS platform:

$$Ei = \frac{\sum x_{ie}}{\sum i \sum x_{ie}} \tag{21.2}$$

where Ei indicates to exposure index, $\sum X_{ie}$ depicts summation of variables and $\sum_i \sum X_{ie}$ depicts the number of indicators.

Sensitivity indicates the extent to which the blocks of the district are affected by any stress on socio-economic system. It includes the level of tolerance in an existing social organization. So it has immense importance in vulnerability frameworks. Population attributes, infrastructural availability and population belonging to working categories are the vital components of socio-economic framework (Schmidtlein et al., 2008). Population data as well as data belonging to socio-economic status was gathered from Primary Census Abstract (Census of India, 2011) based on different indicators such as population density, marginalized population who mainly belong to Scheduled Caste and Scheduled Tribe categories, illiteracy, category- and gender-wise working categories of population and availability of resources in terms of different land uses (forest, agricultural land and irrigated area). Population parameter has importance to judge the impact of demography on natural element in terms of carrying capacity (Shukla et al., 2016; Sharma et al., 2013). Percentage of area under forest cover and agricultural land as well as irrigated are to total net sown area are imperative to judge the sensitivity of the different blocks to carry the increasing population pressure. The people engaged in primary sector, i.e. agricultural activities (as cultivators and agricultural labourer), will be more affected by any type of stress arising in socio-economic system, whereas the non-working population already has trends to be vulnerable (Sharma et al., 2013). Agricultural labours works in others' farms on the basis of daily wages have low economic security. Marginalization in social ladder by belonging to Scheduled Caste and Scheduled Tribes categories as well as illiteracy factor has direct sensitivity to overall social system. So they are in corporate in the present framework as this clubbing is being trigger with social weeping as well as paucity of resources. Sensitivity index has been formulated by using the following equation:

$$Si = \frac{\sum x_{is}}{\sum i \sum x_{is}} \tag{21.3}$$

where Si indicates sensitivity index, $\sum X_{is}$ refers to summation of variables and $\sum_i \sum X_{is}$ refers to the number of indicators.

Adaptation indicates adjustment as well as potentiality of the inhabitants of any region to cope when any kind of stress arises in socio-economic system as well as environmental scenario of that particular region. In most of the cases, it depends on socio-economic quality as well as infrastructural facilities and access of basic amenities in a social system. Percentage share of literate population to total population as well as number of female literate to total female population has been obtained from Primary Census Abstract (Census of India, 2011). Road density in terms of length of road per area, availability of bank and cooperative society as well as their serving categories, distance of block headquarter from nearest railway station, number of originating and terminating bus routes, availability of primary school and primary health centres was obtained from District Statistical Handbook of Bankura district. Literate population might have improved perceptive and awareness about adaptation measures. General literate population and female literate population will endow with insights about the level of awareness and education among the communities. High degree of road density, minimum distance from railway and bus connectivity indicate better accessibility of resources which enhances the quality of life and reduces the chance to be vulnerable. Availability of education facility by primary schools and healthcare facilities by primary health centres, banking facility and availability of cooperative society cumulatively have great significance in rural agro-based socio-economic system as they have more adaptation capacity. The adaptation capacity index was formulated as

$$Ai = \frac{\sum x_{ia}}{\sum i \sum x_{ia}} \tag{21.4}$$

where Ai indicates adaptation capacity index, $\sum X_{ia}$ refers to the sum of variables of adaptation indicator and $\sum_i \sum X_{ia}$ refers to the total number indicators.

Finally, the indicators of these three components, i.e. exposure, sensitivity and adaptation, have been incorporated to formulate composite SCVI, i.e. socio-economic vulnerability index, by the following equation:

$$\mathrm{SCVI} = \left(\frac{\sum x_{ie}}{\sum i \sum x_{ie}} + \frac{\sum x_{is}}{\sum i \sum x_{is}}\right) - \frac{\sum x_{ia}}{\sum i \sum x_{ia}} \quad (21.5)$$

21.3 Analysis and Results

Block Level Scenario of Exposure, Sensitivity and Adaptation

Very high level (value >0.47) of exposure is found in the western part of the study area covering the blocks of Ranibandh, Hirbandh, Chhatna and Bankura I. High level of mean elevation, average slope factor along with moderate level of rainfall with high temperature as well as high per capita availability of agricultural land are the probable causes of high level of exposure in the study area (Fig. 21.2). In Ranibandh, mean elevation and average slope recorded as 168.97 metres and 3.48°, respectively, both of which are highest among all blocks in Bankura district. Per capita availability of agricultural land is also highest (i.e. 20.99 ha per 100 person) in the district. Climatic parameters like average rainfall and average temperature are also moderate to high level in this particular block which ultimately depicted the high level of exposure. In rest of the three blocks, mean elevation and slope are also moderate to high. High level of exposure (value 0.40–0.46) was found in the blocks of Saltora followed by Gangajalghati in northern part, Indpur in western part and Taldangra and Sarenga in southern part. Moderate to high level of elevation and slope along with the characteristics under medium rainfall temperature regime with moderate per capita availability of agricultural land are the prime controlling factors of these five blocks to be incorporated in highly exposed zone. Low value of exposure (<0.36) was found in the easternmost blocks in a continuous orientation. These are Mejia, Barjora, Patrasayer, Joypur-, Indus, and Kotulpur- in the eastern part of the study area adjoining to Purba Bardhaman District and in Khatra block (0.36) in the Central part of the study area. This region mainly belongs to flat in topographic expression with lesser surface slope and aspects with high level of rainfall and low to moderate level of temperature. The rest of the blocks like Onda, Simlapal, Raipur and Bankura II fall under the medium level of exposure zone. All determining factors are low to moderate level in this part.

There are no specific spatial similarity in sensitivity in the study area as the very high sensitivity (value >0.56) was found in the blocks of Sonamukhi, Patrasayer and Indus in the eastern part, followed by Ranibandh and Raipur in the south-western part as well as Onda in the middlemost part of the district (Fig. 21.3). High values in socio-economic factors, e.g. high-level participation in primary economic activities,

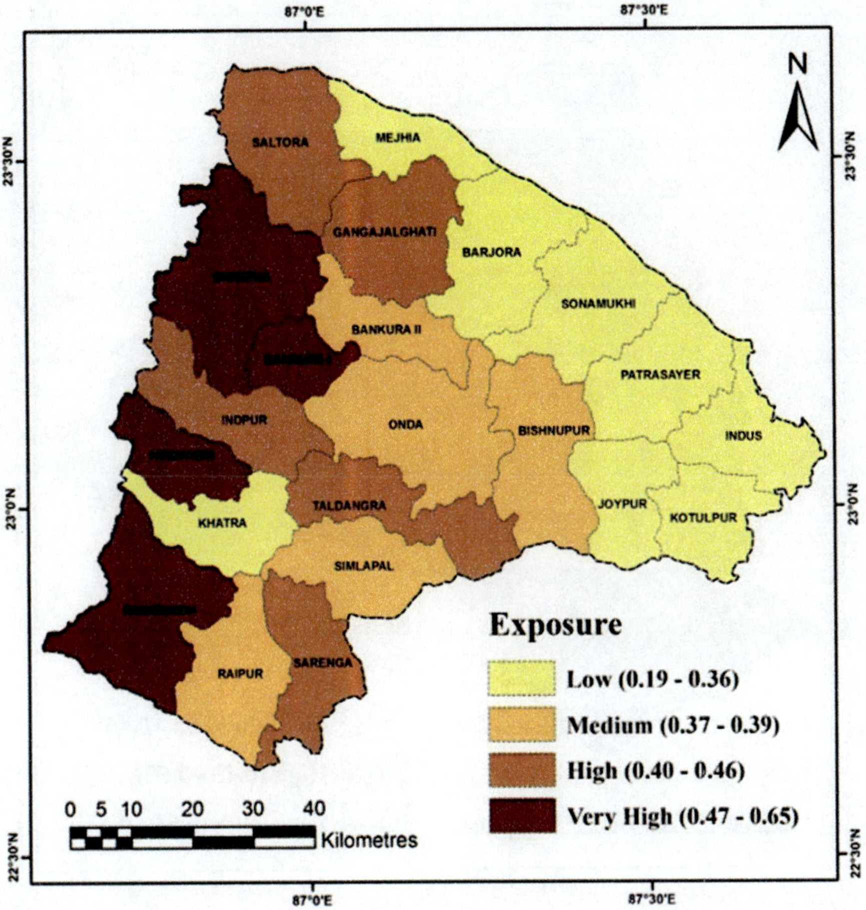

Fig. 21.2 Spatial distribution of degree of exposure in different blocks of Bankura district

moderate-level engagement as main agricultural labourers and low level of literacy along with low-level irrigated area coverage (except Indus) in these six blocks, create very high level of sensitivity in this district. High level of sensitivity (value 0.45 to 0.55) has been found in the blocks of Indpur and Hirbandh in the western part followed by Taldangra, Simlapal and Sarenga in the southern part as well as Bishnupur, Kotulpur and Joypur in a contiguous manner in eastern part of the district. In Sonamukhi and Patrasayer (both have highest value of 0.625), high level of work participation in primary sector as well and main agricultural labourers along with high level of illiteracy pushes them to become more sensitive to any type of stress. In Bishnupur illiteracy (33.7%) and moderate level engagement as main agricultural labourers (38.76%). In Joypur high level of household density (131/km^2), high level of engagement in primary economic activity (70.3%) as well

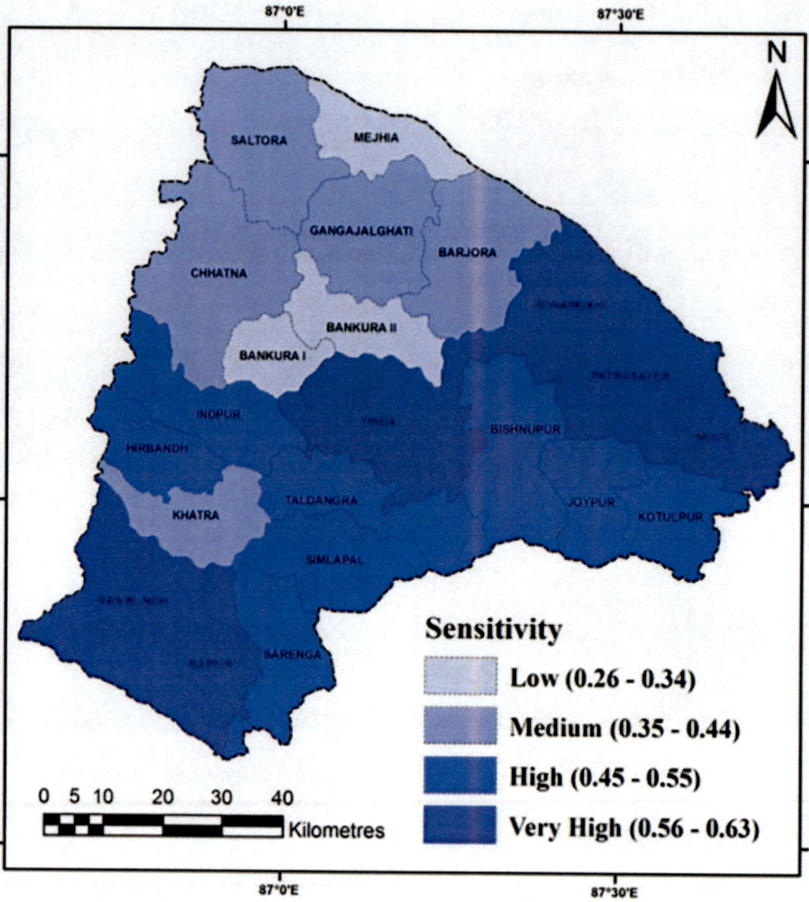

Fig. 21.3 Spatial distribution of degree of sensitivity in different blocks of Bankura district

as high level of irrigated area (above 75% of total NSA) are the controlling factors of high level of sensitivity. High population density (754/km^2, i.e. highest in all blocks) and high household density (164/km^2, i.e. highest in all blocks) played a vital role to enhance the sensitivity. Low level of sensitivity (<0.34) was found only in three blocks of Bankura I and II and Mejia. Medium sensitivity zone (0.35–0.44) was also found around the low sensitive zone in the northern portion of the district containing the four blocks of Saltora, Gangajalghati, Barjora and Chhatna (Fig. 21.3).

In vulnerability frameworks of the district, exposure and sensitivity have better contact on adaptive capacity of the region (Table 21.2). In the present study, area with very high adaptation capacity (value >0.48) was found in a dispersed manner in the blocks of Sonamukhi followed by Kotulpur, Taldangra, Simlapal and Raipur. Very high connectivity parameters like road density (Raipur: 2036 km/km^2), distance from railway station and number of bus routes along with availability of

Table 21.2 Socio-economic vulnerability and its components in different C.D. blocks in Bankura district

Name of the Blocks	Exposure		Sensitivity		Adaptive capacity		Vulnerability Index	
	Score	Category	Score	Category	Score	Category	Score	Category
Bankura I	0.493	Very high	0.336	Low	0.320	Low	0.51	High
Bankura II	0.395	Medium	0.295	Low	0.371	Low	0.32	Low
Barjora	0.357	Low	0.394	Medium	0.471	High	0.28	Low
Bishnupur	0.380	Medium	0.551	High	0.350	Low	0.58	Very high
Chhatna	0.470	Very high	0.435	Medium	0.451	High	0.45	Medium
Gangajalghati	0.438	High	0.408	Medium	0.379	Medium	0.47	Medium
Hirbandh	0.469	Very high	0.514	High	0.302	Low	0.68	Very High
Indpur	0.409	High	0.507	High	0.443	Medium	0.47	Medium
Indus	0.254	Low	0.577	Very high	0.429	Medium	0.40	Medium
Joypur	0.241	Low	0.527	High	0.465	High	0.30	Low
Khatra	0.361	Low	0.437	Medium	0.432	Medium	0.37	Medium
Kotulpur	0.193	Low	0.518	High	0.708	Very high	0.00	Low
Mejia	0.307	Low	0.256	Low	0.229	Low	0.33	Low
Onda	0.385	Medium	0.576	Very high	0.466	High	0.49	High
Patrasayer	0.264	Low	0.625	Very high	0.407	Medium	0.48	High
Raipur	0.393	Medium	0.580	Very high	0.708	Very high	0.27	Low
Ranibandh	0.646	Very high	0.598	Very high	0.414	Medium	0.83	Very high
Saltora	0.456	High	0.422	Medium	0.214	Low	0.66	Very high
Sarenga	0.425	High	0.516	High	0.465	High	0.48	High
Simlapal	0.380	Medium	0.519	High	0.472	Very high	0.43	Medium
Sonamukhi	0.360	Low	0.625	Very high	0.492	Very high	0.49	High
Taldangra	0.430	High	0.522	High	0.535	Very high	0.42	Medium

cooperative society are the dominant indicators behind such promising level of adaptation capacity of these blocks (Fig. 21.4). High-level adaptation capacity (0.45–0.47) was found in the blocks of Joypur, Onda, Chhatna, Sarenga and Barjora where the literacy and availability of primary schools are high. In the blocks of Bishnupur, Bankura I and II along with Hirbandh, low level of adaptation capacity is found due to low level of primary education and lack of primary health facility. Due to excess population pressure in the blocks, the education facility and health and connectivity infrastructure faced serious lacking.

Socio-Economic Vulnerability Index

Block level score value of the three domains of vulnerability has been used in the mentioned formulae to prepare SCVI of the district (Fig. 21.5). Vulnerability assessment based on specific indicators discovered that very high vulnerability (value above 0.52) was found in four blocks of Ranibandh (0.83), Hirbandh (0.68), Saltora (0.66) and Bishnupur (0.58). Very high level of exposure and sensitivity with medium adaptation capacity are the prime causes for Ranibandh to be in very high vulnerability, whereas low level of adaptation capacity with high

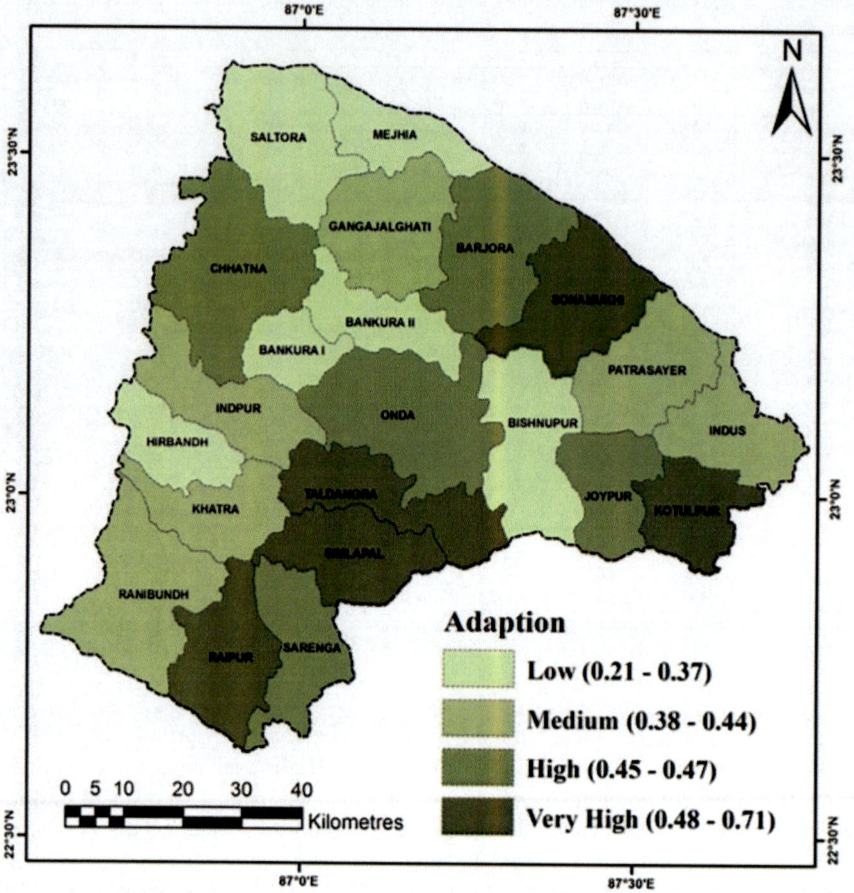

Fig. 21.4 Spatial distribution of degree of adaptation capacity in different blocks of Bankura district

level of sensitivity is responsible for very high vulnerability in case of Hirbandh, Saltora and Bishnupur. High level of vulnerability (0.48–0.51) was found in the blocks of Bankura I, Onda, Patrasayer, Sonamukhi and Sarenga. High to medium level of exposure and sensitivity with low to medium adaptation capacity are the prime reasons behind such scenario except Sarenga where the three domains are in the high category, but the value of adaptation is quite low in comparison to the other two factors.

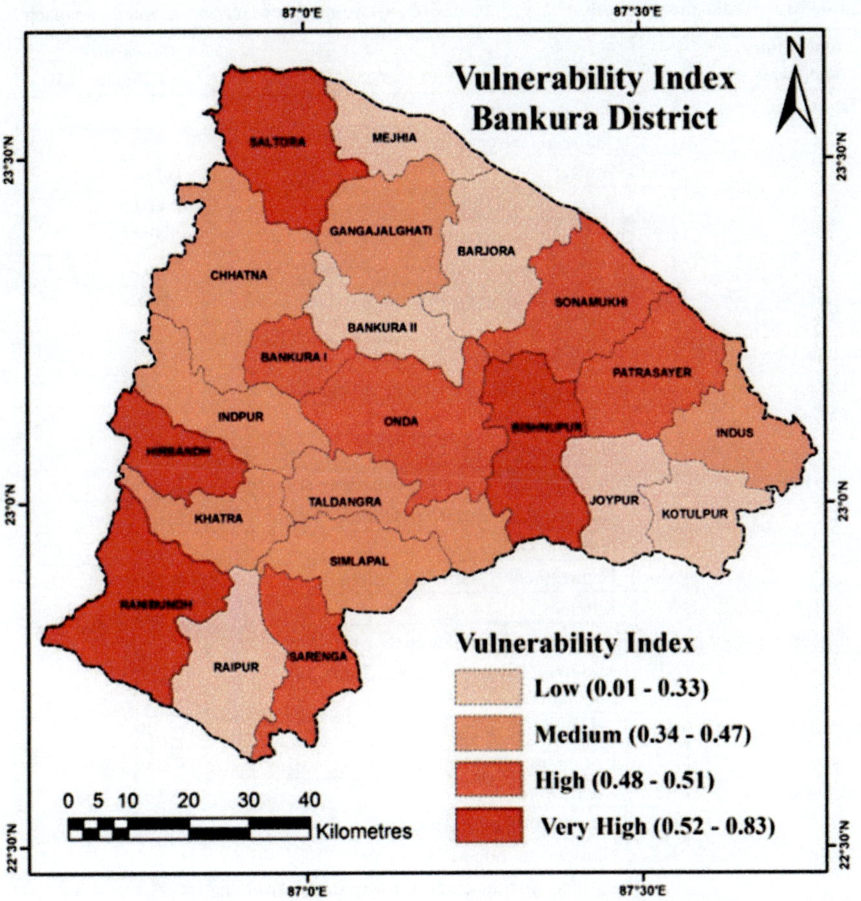

Fig. 21.5 Spatial distribution of socio-economic vulnerability index in different blocks of Bankura district

21.4 Discussions

Block level indicator-based scores have been processed in regression frame by least square methods to judge the relative importance of individual indicator to the overall exposure domain under vulnerability assessment (Table 21.3). Exposure index has been prepared based on the reflection of the seven indicators. From the standardized coefficient values, it is clear that average temperature (0.457), average rainfall (0.403) and mean elevation (0.382) are the more dominant indicators to overall refection of exposure domain (Fig. 21.6). Sensitivity domain of vulnerability framework is composed of 12 indicators from demographic, socio-economic and infrastructural dimensions. In this domain, except the household density (0.077) and

Table 21.3 Indicator-wise reflection of β values of exposure, sensitivity and adaptation capacity

Vulnerability components	Sl. no.	Indicators	Standardized coefficients (β)
Exposure	E1	Elevation in metres	0.382
	E2	Slope in degree	0.299
	E3	Aspect	0.354
	E4	Average rainfall (mm)	0.403
	E5	Average temperature in °C	0.457
	E6	Per capita availability of forest (ha/100 person)	0.291
	E7	Per capita availability of agricultural land (ha/100 person)	0.318
Sensitivity	S1	Total forest area (ha)	0.275
	S2	Total land under agricultural land uses (ha)	0.241
	S3	Population density	0.280
	S4	Household density	0.077
	S5	Engagement in primary economic activity	0.259
	S6	Engagement as main agricultural labourer	0.269
	S7	Non-working population	0.017
	S8	Engagement of female workers	0.152
	S10	Illiteracy	0.182
	S11	Total agricultural land under irrigation	0.270
	S12	Marginalized population (SC and ST population)	0.240
Adaptive capacity	A1	Total literate population	0.198
	A2	Female literate population	0.239
	A3	Road density per km^2	0.287
	A4	Number of banks served per 1000 persons	0.189
	A5	Number of cooperative society	0.236
	A6	Distance to the nearest railway station from the block H.Q. (K.M.)	0.294
	A7	Number of primary schools	0.222
	A8	Number of primary health centres	0.224
	A9	Number of originating/terminating bus routes	0.363

non-working population (0.017), the rest of the indicators respond almost the same as their coefficient values as nearly 0.2. So all these indicators are almost equally relevant to overall sensitivity of the study region. Among them, population density (0.28) from demographic perspectives, engagement in primary workers (0.259), along with engagement as main agricultural labourer (0.269) from socio-economic dimensions and area under irrigation (0.270) from infrastructural dimensions are worth mentioning. Adaptation capacity is composed of nine indicators from socio-economic and infrastructural dimensions with availability of basic services perspectives. Here connectivity indicators like distance of block HQ from railway station (0.294) and the number of originating and terminating bus routes (0.363) and road density (0.287) play the dominant role to enhance the adaptation capacity.

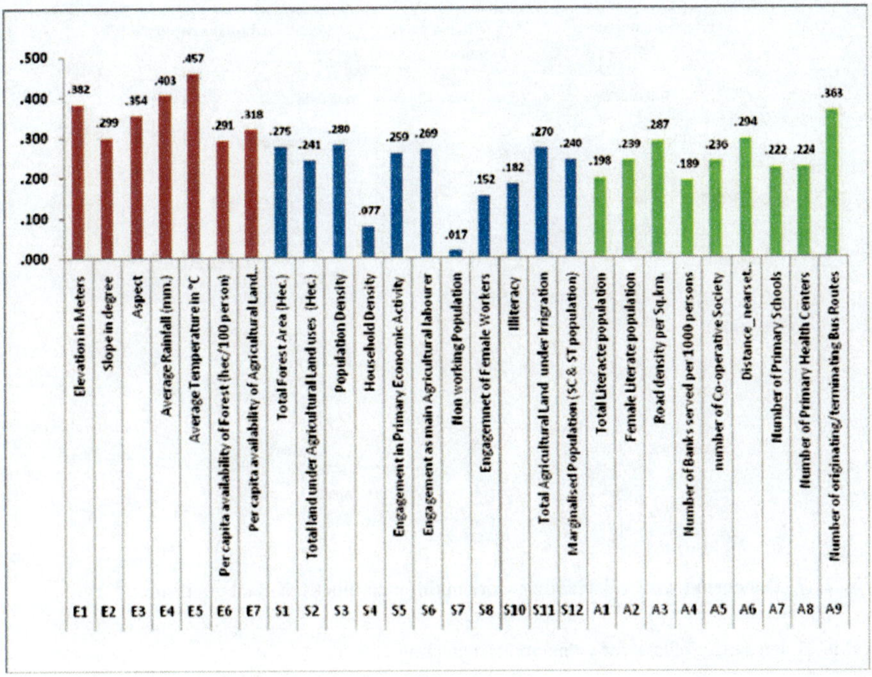

Fig. 21.6 Indicator-wise standardized coefficient (β) of exposure, sensitivity and adaptation capacity

The overall analysis of socio-economic vulnerability revealed that there are three blocks (Saltora, Hirbandh and Ranibandh) mainly located in the westernmost elevated region (elevation above 100 m) with moderate rainfall and high temperature except Bishnupur where the sensitivity factor is most dominant to be incorporated in the very high vulnerability class. So exposure plays the most determining role in the vulnerability class as the correlation coefficient between exposure score and overall SEV score is 0.754 in comparison to sensitivity ($r = 0.21$) and adaptation capacity ($r = -0.60$). The determining factors in high-vulnerable blocks (e.g. Sonamukhi, Patrasayer, Onda, etc.) have shifted from exposure to sensitivity factor highly but partial implication of exposure factor is still present mainly in case of Sarenga and Bankura I where the exposure score is still very high (above 0.47). Elevation, Slope, Temperature factors responses comparatively low in the block of Sonamukhi, Patrasayer, Onda and Sarenga which ultimately impacted on exposure score enormously (Fig. 21.7).

Socio-economic vulnerability assessment has been recognized as helpful implementation for analysing vulnerability at subdistrict rank. It increases the knowledge about any socio-economic as well as environmental stress intervention in any socio-economic strata and support in site-specific vulnerability assessment (Ahsan & Warner, 2014; Sorg et al., 2018). SEVI analysis discovered that blocks of Ranibandh, Saltora, Hirbandh and Bankura I along with Sonamukhi, Patrasayer

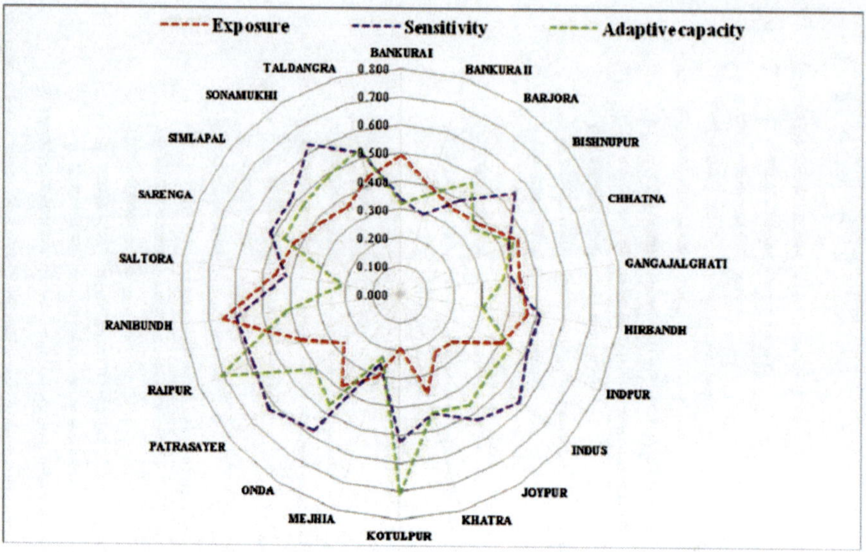

Fig. 21.7 Dimension-wise vulnerability score in different blocks of Bankura district

Table 21.4 Priority blocks for vulnerability reductions

Blocks	Exposure index	Sensitivity index	Adaptation index
Bankura I	√	–	√
Bankura II	–	–	√
Barjora	–	–	–
Bishnupur	–	√	√
Chhatna	√	–	–
Gangajalghati	√	–	√
Hirbandh	√	√	√
Indpur	√	√	√
Indus	–	√	√
Joypur	–	√	–
Khatra	–	–	√
Kotulpur	–	√	–
Mejia	–	–	√
Onda	–	√	–
Patrasayer	–	√	√
Raipur	–	√	–
Ranibandh	√	√	√
Saltora	√	–	√
Sarenga	√	√	–
Simlapal	–	√	–
Sonamukhi	–	√	–
Taldangra	√	√	–

and Sarenga require instantaneous consideration to reduce the level of vulnerability (Table 21.4). These blocks have registered a high degree of sensitivity (first priority) and exposure with negligible adaptation. These priority blocks are mostly located along the highly elevated along with less rainfall with high temperature regime with normally create a physical constrains to overall socio-economic status. As reduction in exposure is not so easy to conduct, thus efforts could be made to increase the adaptive capacity simultaneously with reducing sensitivity by indicator-wise development mainly in primary education, basic health services and infrastructural sector development mainly through connectivity enhancement to reduce the socio-economic vulnerability. Enhancement of adaptation capacity and reduction in sensitivity factors are also needed in case of Chhatna, Indpur, Gangajalghati, Indus and Simlapal to reduce future chances of high vulnerability (Table 21.4).

21.5 Conclusion

The present study thus analyses the level of socio-economic vulnerability in dry deciduous forested district of Bankura. The study uses three major components of vulnerability, namely, exposure, sensitivity and adaptive capacity. The index value of each has been computed by quantitative method. The western blocks, namely, Ranibandh, Hirbandh and Chhatna having areas with greater elevation, average slope, per capita arable land, moderate rainfall and high temperature, were found to have high value of exposure. Moderate exposure value was found to be associated with the northern, central and southern blocks and the low exposure was found to prevail over the eastern blocks. The exposure value shows close association with their physiographical regime. No definite trend was found to exist for the sensitivity value which seems to be associated with work participation rate, irrigation facilities, etc. Some western and eastern blocks have greater sensitivity value, while the northern part has low value zones. Transport, communication and financial aspect are considered important for adaptation capacity value. The eastern and southern part was found to be equipped with higher capacity; contrary to this, western blocks have low value. Eventually, combining all those, Ranibandh, Hirbandh, Saltora and Bishnupur are found to be more vulnerable socio-economically with higher index value. This will help to identify the existing environmental stress prevailing over the subdistricts and also helps to identify the major causes for their poor performance in the basic amenities. Being rigid in nature, the exposure factors hardly change; hence emphasis should be given on adaptive capacity building measures coupled with basic infrastructure development in health, education and communication which may be considered to be helpful for reducing the socio-economic vulnerability on a long-term basis.

References

Agriculture Meteorologist, Directorate of Agriculture, W. B. (2013). A report on Climatic Variation in West Bengal.

Ahmed, R., Sajjad, H., & Husain, I. (2018). Morphometric parameters-based prioritization of sub-watersheds using fuzzy analytical hierarchy process: A case study of lower Barpani Watershed, India. *Natural Resources Research, 27*(1), 67–75.

Ahsan, M. N., & Warner, J. (2014). The socioeconomic vulnerability index: A pragmatic approach for assessing climate change led risks–A case study in the south-western coastal Bangladesh. *International Journal of Disaster Risk Reduction, 8*, 32–49.

Blaikie, P., Cannon, T., Davis, I., & Wisner, B. (2005). *At risk: Natural hazards, people's vulnerability and disasters*. Routledge.

Bray, F., Jemal, A., Grey, N., Ferlay, J., & Forman, D. (2012). Global cancer transitions according to the human development index (2008–2030): A population-based study. *The Lancet Oncology, 13*(8), 790–801.

Brooks, N. (2003). Vulnerability, risk and adaptation: A conceptual framework. *Tyndall Centre for Climate Change Research Working Paper, 38*(38), 1–16.

Census of India. (2011). Primary Census Abstracts. Office of the Registrar General and Census Commissioner, Ministry of Home Affairs, Government of India.

Challinor, A. J., Simelton, E. S., Fraser, E. D. G., Hemming, D., & Collins, M. (2010). Increased crop failure due to climate change: Assessing adaptation options using models and socio-economic data for wheat in China. *Environmental Research Letters, 5*(3), 034012. (8pp).

Chowdhury, S., & Squire, L. (2006). Setting weights for aggregate indices: An application to the commitment to development index and human development index. *Journal of Development Studies, 42*(5), 761–771.

Cui, G., Kwak, H., Choi, S., Kim, M., Lim, C. H., Lee, W. K., & Chae, Y. (2016). Assessing vulnerability of forests to climate change in South Korea. *Journal of Forest Research, 27*(3), 489–503.

Deb, P., Debnath, P., Denis, A. F., & Lepcha, O. T. (2019). Variability of soil physicochemical properties at different agroecological zones of Himalayan region: Sikkim, India. *Environment, Development and Sustainability, 21*, 2321–2339.

District Census Handbook (2011). Bankura District West Bengal https://censusindia.gov.in/2011census/dchb/DCHB_A/19/1913_PART_A_DCHB_BANKURA.pdf

District Statistical Handbook, Bankura, (2014). *Department of Statistics and Programme Implementation. Government of West Bengal.*

Engström, J., Jafarzadegan, K., & Moradkhani, H. (2020). Drought vulnerability in the United States: An integrated assessment. *Water, 12*(7), 2033. https://doi.org/10.3390/w12072033

Hahn, M. B., Riederer, A. M., & Foster, S. O. (2009). The livelihood vulnerability index: A pragmatic approach to assessing risks from climate variability and change—A case study in Mozambique. *Global Environmental Change, 19*(1), 74–88.

Haq, M. (2003). The birth of the human development index. In S. Fukuda-Parr & A. K. Shiva Kumar (Eds.), *Readings inhuman development* (pp. 127–137). Oxford University Press.

Holsten, A., & Kropp, J. P. (2012). An integrated and transferable climate change vulnerability assessment for regional application. *Natural Hazards, 64*(3), 1977–1999.

Hu, S., Ma, J., Shugart, H. H., & Yan, X. (2018). Evaluating the impacts of slope aspect on forest dynamic succession in Northwest China based on FAREAST model. *Environmental Research Letters, 13*(3), 034027.

IPCC. (2012). Managing the risks of extreme events and disasters to advance climate change adaptation. A special report of working groups I and II of the intergovernmental panel on climate change. In C. B. Field, V. Barros, T. F. Stocker, D. Qin, D. J. Dokken, K. L. Ebi, M. D. Mastrandrea, K. J. Mach, G. K. Plattner, S. K. Allen, M. Tignor, & P. M. Midgley (Eds.), (p. 582). Cambridge University Press.

IPCC. (2014.); Field CB, Barros VR, Mach KJ et al.). Technical summary. Climate change 2014: Impacts, adaptation, and vulnerability. Part A: Global and sectoral aspects. Contribution of working group II to the fifth assessment report of the intergovernmental panel on climate change. In C. B. Field, V. R. Barros, D. J. Dokken, K. J. Mach, M. D. Mastrandrea, T. E. Bilir, & Chatterjee (Eds.).

IPCC: Intergovernmental Panel on Climate Change. (2001). *Technical summary: climate change 2001: Impacts,adaptation, and vulnerability*. A Report of Working GroupII of the Intergovernmental Panel on Climate Change. http://www.grida.no/climate/ipcc_tar/wg2/pdf/wg2TARtechsum.pdf. Accessed on 22nd September, 2019.

Köhl, M., Lasco, R., Cifuentes, M., Jonsson, Ö., Korhonen, K. T., Mundhenk, P., de Jesus Navar, J., & Stinson, G. (2015). Changes in forest production, biomass and carbon: Results from the 2015 UN FAO Global Forest Resource Assessment. *Forest Ecology and Management, 352*, 21–34.

Krishnan, P., Ananthan, P. S., Purvaja, R., Jeevamani, J. J. J., Infantina, J. A., Rao, C. S., et al. (2019). Framework for mapping the drivers of coastal vulnerability and spatial decision making for climate-change adaptation: A case study from Maharashtra, India. *Ambio, 48*(2), 192–212.

Malakar, K., & Mishra, T. (2017). Assessing socio-economic vulnerability to climate change: A city-level index-based approach. *Climate and Development, 9*(4), 348–363.

Ministry of Panchayati Raj. (2009). "A Note on the Backward Regions Grant Fund Programme" (PDF). National Institute of Rural Development. Archived from the original (PDF) on 5 April 2012. Retrieved 27 September 2011.

Munang, R., Thiaw, I., Thompson, J., Ganz, D., Girvetz, E., Rivington, M. (2011). Sustaining Forests: Investing in Our Common Future. *UNEP Policy Series, 5*, Nairobi, Kenya.

Murthy, I. K., Tiwari, R., & Ravindranath, N. H. (2011). Climate change and forests in India: Adaptation opportunities and challenges. *Mitigation and Adaptation Strategies of Global Change, 16*, 161–175.

Naess, L. O., Norland, I. T., Lafferty, W. M., & Aall, C. (2006). Data and processes linking vulnerability assessment to adaptation decision-making on climate change in Norway. *Global Environmental Change, 16*(2), 221–233.

Nguefack-Tsague, G., Klasen, S., & Zucchini, W. (2011). On weighting the components of the human development index: A statistical justification. *Journal of Human Development and Capabilities, 12*(2), 183–202.

O'Brien, K., Leichenko, R., Kelkar, U., Venema, H., Aandahl, G., Tompkins, H., ... West, J. (2004). Mapping vulnerability to multiple stressors: Climate change and globalization in India. *Global Environmental Change, 14*(4), 303–313.

O'Malley, L. S. S., ICS, *Bankura*, Bengal District Gazetteers, pp. 1–20, first published 1908, 1995 reprint, Government of West Bengal.

Pandey, R., & Bardsley, D. K. (2015). Social-ecological vulnerability to climate change in the Nepali Himalaya. *Applied Geography, 64*, 74–86.

Preston, B. L., Yuen, E. J., & Westaway, R. M. (2011). Putting vulnerability to climate change on the map: A review of approaches, benefits, and risks. *Sustainability Science, 6*(2), 177–202.

Ribot, J. (2011). Vulnerability before adaption: Towards transformative climate action. *Global Environmental Change, 21*, 1160–1162.

Rygel, L., O'sullivan, D., & Yarnal, B. (2006). A method for constructing a social vulnerability index: An application to hurricane storm surges in a developed country. *Mitigation and Adaptation Strategies for Global Change, 11*(3), 741–764.

Sahana, M., & Sajjad, H. (2019). Vulnerability to storm surge flood using remote sensing and GIS techniques: A study on Sundarban biosphere reserve, India. *Remote Sensing Applications: Society and Environment, 13*, 106–120.

Sahana, M., Rehman, S., Paul, A. K., & Sajjad, H. (2019). Assessing socio-economic vulnerability to climate change-induced disasters: Evidence from Sundarban biosphere reserve, India. *Geology, Ecology, and Landscapes*, 1–13. https://doi.org/10.1080/24749508.2019.1700670

Sajjad, H., & Nasreen, I. (2016). Assessing farm-level agricultural sustainability using site-specific indicators and sustainable livelihood security index: Evidence from Vaishali district, India. *Community Development, 47*(5), 602–619.

Schmidtlein, M. C., Deutsch, R. C., Piegorsch, W. W., & Cutter, S. L. (2008). A sensitivity analysis of the social vulnerability index. *Risk Analysis: An International Journal, 28*(4), 1099–1114.

Seidl, R., Rammer, W., & Lexer, M. J. (2011). Climate change vulnerability of sustainable forest management in the eastern Alps. *Climatic Change, 106*(2), 225–254.

Sharma, J., Chaturvedi, R. K., Bala, G., et al. (2013). Challenges in vulnerability assessment of forests under climate change. *Carbon Management, 4*(4), 403–411.

Sharma, J., Upgupta, S., Kumar, R., et al. (2015). Assessment of inherent vulnerability of forests at landscape level: A case study from Western Ghats in India. *Mitigation and Adaptation Strategies for Global Change*. https://doi.org/10.1007/s11027-015-9659-7

Sharma, E., Chettri, N., Tsering, K., Shrestha, A. B., Jing, F., Mool, P., & Eriksson, M. (2009). *Climate change impacts and vulnerability in the eastern Himalayas*. ICIMOD.

Sharma, J., Upgupta, S., Jayaraman, M., Chaturvedi, R., Bala, G., & Ravindranath, N. H. (2017). Vulnerability of forests in India: A National Scale Assessment. *Environmental Management*, 544–553. https://doi.org/10.1007/s00267-017-0894-4

Shukla, R., Sachdeva, K., & Joshi, P. K. (2016). An indicator-based approach to assess village level social and biophysical vulnerability of agriculture communities in Uttarakhand, India. *Journal of Mountain Science, 13*(12), 2260–2271.

Sinha, S., Badola, H. K., Chhetri, B., Gaira, K. S., Lepcha, J., & Dhyani, P. P. (2018). Effect of altitude and climate in shaping the forest compositions of Singalila National Park in Khangchendzonga landscape, eastern Himalaya, India. *Journal of Asia-Pacific Biodiversity, 11*(2), 267–275.

Sorg, L., Medina, N., Feldmeyer, D., Sanchez, A., Vojinovic, Z., Birkmann, J., & Marchese, A. (2018). Capturing the multifaceted phenomena of socioeconomic vulnerability. *Natural Hazards, 92*(1), 257–282.

Sullivan, C. A., Meigh, J. R., & Fediw, T. S. (2002). Derivation and testing of the water poverty index phase 1. Final report may 2002. Retrieved from http://nora.nerc.ac.uk/id/eprint/503246

Tate, E. (2012). Social vulnerability indices: A comparative assessment using uncertainty and sensitivity analysis. *Natural Hazards, 63*(2), 325–347.

Thakur, S., Negi, V. S., Pathak, R., Dhyani, R., Durgapal, K., & Rawal, R. S. (2019). *Forest Ecology and Management*. https://doi.org/10.1016/j.foreco

Tse-ring, K., Sharma, E., Chettri, N., & Shrestha, A. (Eds.). (2010). *Climate change vulnerability of mountain ecosystems in the eastern Himalayas—Synthesis report*. ICIMOD.

UNDP. United Nations Development Programme (2014). *Human development index (HDI) report 2014*. UNDP.

UNDP-United Nations Development Programme. (2015). Training Material for Producing National Human Development Reports, Occasional paper, March 2015.

Villa, F., & McLeod, H. E. L. E. N. A. (2002). Environmental vulnerability indicators for environmental planning and decision-making: Guidelines and applications. *Environmental Management, 29*(3), 335–348.

Vincent, K. (2004). Creating an index of social vulnerability to climate change for Africa. *Tyndall Center for Climate Change Research Working Paper, 56*(41), 1–50.

Wiréhn, L., Opach, T., & Neset, T.-S. (2016). Assessing agricultural vulnerability to climate change in the Nordic countries – An interactive geovisualization approach. *Journal of Environmental Planning and Management. pp*, 1–20. https://doi.org/10.1080/09640568.2016.1143351

Chapter 22
Assessing River Bank Erosion in the Ganges Using Remote Sensing and GIS

Masjuda Khatun, Sk Mujibar Rahaman, Sanjoy Garai, Pulakesh Das, and Sharad Tiwari

Abstract River bank erosion is one of major issues occurring due to high flow, extra silt charge and river channel shift and causes several physical and socio-economic issues to the dependent population. Remote sensing and geographic information system (GIS) are an idyllic and seemingly tool for monitoring river erosion and its line transfer. The study evaluates spatial and temporal river bank changes, channel shifting and its impact on land use land cover (LULC) change around the Ganges River from the bottom of the Rajmahal Hill to the Farakka Barrage for the period 1980–2020. The Landsat imagery with 30/60 m spatial resolution was employed to river bank demarcation and LULC mapping. The GIS layers on the river banks lines were created at decadal interval and overlaid. The results indicated significant shift of the river bank towards the left bank along the flow direction. The total erosion and accretion of the study zone from 1980 to 2020 were estimated as 250.82 and 236.75 km^2, respectively. Moreover, a significant increase (more than sevenfold) was observed in the island's area and numbers during the study period, which could be attributed due to the construction of Farakka Barrage. The river bank erosion and accretion with the LULC change analysis indicated major transactions between agriculture land, agriculture fallow, river and sand. The finding of the study has policy level implications and useful for agriculture, water resource and landscape planners for improved management and planning.

Keywords River bank erosion · Accretion · LULC change · Ganges

---

M. Khatun · S. M. Rahaman · S. Garai · S. Tiwari (✉)
Institute of Forest Productivity, Ranchi, India

P. Das
World Resources Institute India, New Delhi, India

P. K. Shit et al. (eds.), *Geospatial Technology for Environmental Hazards*, Advances in Geographic Information Science, https://doi.org/10.1007/978-3-030-75197-5_22

22.1 Introduction

The river bank erosion is an erratic natural disaster that causes extensive land damage. River channel migration is a process that can cause catastrophic local or regional changes. The widespread impact of such changes becomes a socio-economic disaster for the people living in the respective rivers' floodplains. River bank erosion causes various physical and socio-economic problems on the river banks. Every year, millions of people are affected by river bank erosion and shift that causes loss of croplands, farmland and settlement areas, destruction of artificial structures, etc. (Rahman & Gain, 2020). The seasonal and periodic changes occur in the floodplain, mostly due to alterations in river discharge and poly load (Guite & Bora, 2016). Most of the world's river banks are suffering from erosion and waterlogging due to both natural and human activities (Majumdar & Mandal, 2019). The river bank erosion is typically associated with high flow, extra silt charge and river channel shift (Das et al., 2014). Much of the river is controlled by areas such as river drainage basins, groundwater, vegetation cover, land use land cover (LULC), tectonic activity, and climate factors (rainfall and temperature) (Das et al., 2018; Debnath et al., 2017). The abrupt increase in river discharge during monsoon enforces pressure on its two banks due to the turbulent water flow condition and initiates lateral erosion and leads to river bank erosion. The changes in river bank lines cause various alterations, e.g. change in the flow direction due to neck cut-off and alteration of the meander bends that account for translational, lateral, rotational and extensional shifts (Sarma et al., 2007). The spatio-temporal river channel assessment is essential for geomorphological studies, improved river, water resource and agriculture management activities and planning.

The seven major rivers and their tributaries form India's dominant river systems (Water Resources Systems Division, 2020). Recurring flood events in India cause severe damages to the land and other resources (Singh & Kumar, 2017). The Ganges and Brahmaputra are the two major rivers experiencing severe erosion for decades (Phukan et al., 2012). The Ganges is one of the most important river systems in India, covering an enormous portion of India. Every year, the Ganges, its tributaries and distributors carry millions of tons of sediment and deposit in the plains. As an alluvial river with low velocity and flat stream, the Ganges deposits large silt and causes many issues like reduction of depth, waterlogging and flood events in various parts of Bihar and West Bengal mostly (Iqbal, 2010). The Ganges system has developed some exclusive features like shifting discharge, sedimentary pressure, tidal interruption, erosion, reduced water flow in the western tributaries and different tectonic structures of delta (Das, 2011). Studies have indicated that the right bank has erosion-resistant basalt rock, whereas the left bank has weak alluvial soil structures that make it more vulnerable to excessive river bank erosion (Majumdar & Mandal, 2019). Several studies have been carried out assessing the river bank erosion and its impact on the Ganges River in recent time. Biswas and Anwaruzzaman (2019) used the pressure and release (PAR) model to measure the vulnerability posed by the erosion of the Ganges in Malda district of West Bengal state. They have highlighted the socio-economic issues villagers face due the river bank erosion. Several other studies have identified the consequences of river bank erosion and flood on people's livelihood and socio-economic condition in various districts in West Bengal state (Majumdar, 2018; Das et al., 2012; Iqbal, 2010).

Remote sensing and geographic information system (GIS) are the idyllic and seemingly tool for identifying and monitoring river erosion and its line transfer. Sarma et al. (2007) studied the sequential changes in Burhi Dihing River's bank lines employing the Survey of India (SOI) toposheets (for the years 1934 and 1972) and the IRS 1D and IRS P4 satellite images (for the years 2001 and 2004). They considered 13 segments along the flow direction (at 5′ longitude equivalent to 15 km intervals) to assess the overall shift and erosion. Thakur et al. (2012) studied the river bank erosion using toposheet (1955) and Landsat satellite images (1977 to 2005) due to the morphometric transformation of the Ganges in between Rajmahal and Farakka Barrage. They observed drastic changes in the morphometric parameters such as sinuosity, beardedness index and the island area percentage. Ghosh and Sahu (2019) assessed the impact of the Ganges-Bhagirathi river bank line shifting on LULC changes during 1980–2010 in the Murshidabad district, West Bengal, India. Debnath et al. (2017) evaluated bank erosion and channel migration along the Khowai River in north-eastern India using remotely sensed data from 1975 to 2014. Akter et al. (2017) assessed the lateral channel movements of the lower reaches of the Jamuna River from 1976 to 2015 using space-based imagery. Hassan et al. (2017) measured the erosion and accretion of spatio-temporal dynamics at different locations on the Jamuna River. Islam and Rashid (2011) studied the Meghna river bank erosion and its impact on the population displacement and socio-economic status in Bangladesh. They concluded that the socio-economic condition, removal of settlement areas, household size changes, educational attainment, labour force participation and occupational status, land holding and income were negatively affected at both individual and household levels.

The current study focuses on assessing the river bank erosion in the part of the Ganges River falling under Malda district of West Bengal, employing the multi-temporal Landsat images. The study evaluates spatial and temporal changes in channel shifting and erosion and its impact on LULC in the study region.

22.2 Study Area

The current study was conducted in the Malda district of West Bengal state, India, on the Ganges between the Rajmahal Hills and Farakka Barrage (Fig. 22.1). Majority of this area covers four blocks as Manikchak, Kaliachak I, Kaliachak II and Kaliachak III of the Malda district. Geographically the study area extended between the latitudes 24°37′41″N to 25°4′48″N and the longitudes 87°45′30″E to 88°08′22″E. The Farakka Barrage was constructed primarily to maintain the water flow at Calcutta Port. However, the construction of barrage has led to significant changes in the upstream and river high sediment deposition and river bank erosion. Previous

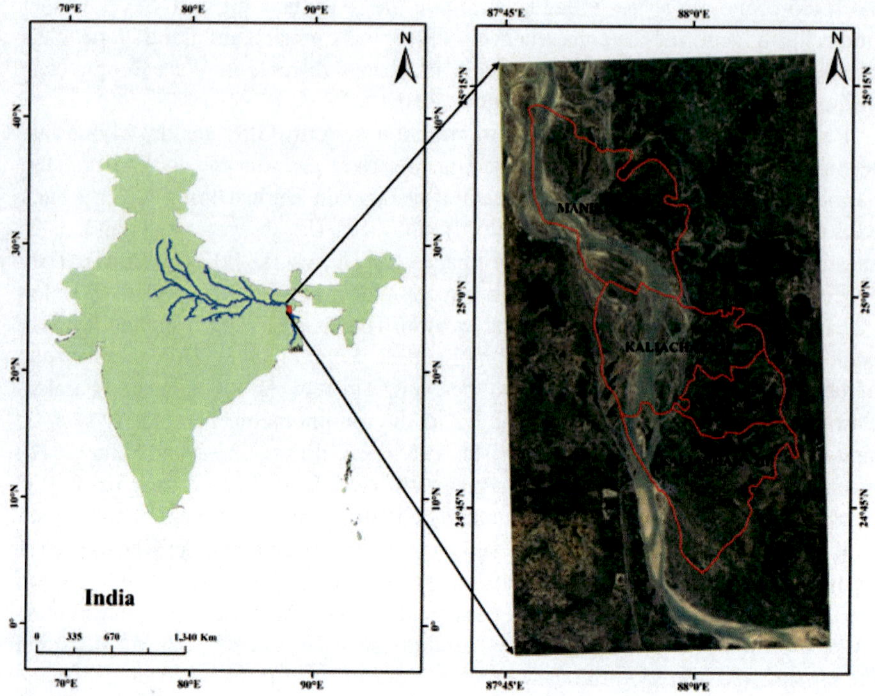

Fig. 22.1 The Ganges River network map and the Google Earth image view of the study area

studies have identified significant loss of river banks, e.g. around 750 km^2 in the Manikchak and Kaliachak blocks in the past 30 years (Banerjee, 1999).

22.3 Data and Methodology

Landsat satellite images were used for the spatio-temporal analysis of river bank shift for the five different time period (at 10 years interval) from 1980 to 2020. The cloud-free surface reflectance images were acquired for the pre-monsoon dry period. The image data were downloaded from the USGS Earth Explorer website and are enlisted in Table 22.1.

The maximum likelihood supervised image classification approach was adopted to generate the LULC maps. The 2000 and 2020 images were classified into six major classes such as agricultural land, agricultural fallow, river, sand, plantation and fallow land. For accuracy assessment, a set of random points was created and compared with the high-resolution multi-temporal Google Earth imagery. The river bank lines and islands were demarcated for five time periods employing the visual

Table 22.1 Details of Landsat data used in the current study

Sensor(Landsat)	Path/row	Date of acquisition	Spatial resolution (m)
MSS	149/43	17/01/1980	60
TM	139/43	06/02/1990	30
ETM+	139/43	29/03/2000	30
TM	139/43	02/04/2010	30
OLI/TIRS	139/43	12/03/2020	30

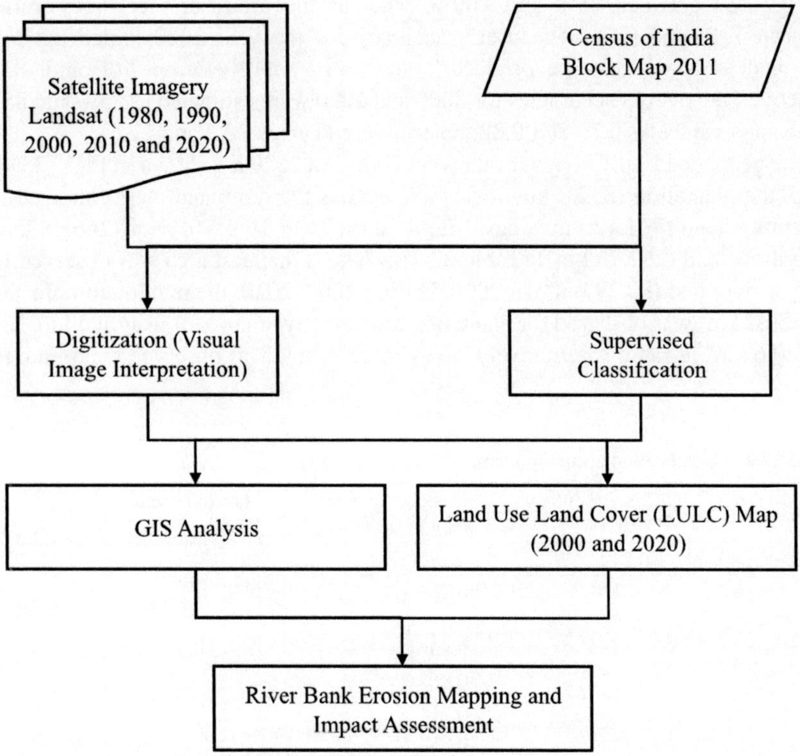

Fig. 22.2 Methodology flow diagram

image interpretation technique. The temporal data layers on river bank and islands were further employed for GIS analysis to assess the various processes like erosion, accretion, river bank line shifting. The methodology used in the present study is shown in Fig. 22.2.

22.4 Result

Land Use Land Cover Change

The LULC maps generated for the years 2000 and 2020 were validated with 84 points randomly generated on the image. The estimated accuracy for both the years indicated well-accepted accuracy, where a higher accuracy was observed for 2020 compared to 2000 (Tables 22.2 and 22.3). The user's accuracy for the year 2000 varied between 75% and 100%, whereas the producer's accuracy varied between 71% and 90%. The user's accuracy for the year 2020 varied between 68% and 94%, whereas the producer's accuracy varied between 77% and 91%. Moreover, the overall accuracies for 2000 and 2020 were estimated as 83% and 85% and Kappa values as 0.79 and 0.82, respectively (Tables 22.2 and 22.3).

The generated LULC maps for the year 2000 and 2020 are shown in Fig. 22.3. In 2000, the plantation (837.8 km^2) was observed as the dominant class followed by agriculture land (761.12 km^2), agriculture fallow (496.81 km^2), river (266.08 km^2) and fallow land (252.73 km^2) (Table 22.4), whereas the least area was observed for sand or river bed (80.29 km^2) in 2000. During 2000–2020, the maximum reduction of 135.27 km^2 was observed for plantation followed by 96.54 km^2 in agriculture land and 20.6 km^2 in sand. Alternatively, an increase in area was observed maximum for

Table 22.2 Classification accuracy metrics for the year 2000

Class	User accuracy (%)	Producer accuracy (%)	Overall accuracy (%)	Kappa
Agricultural land	0.84	0.87	83	0.79
Agricultural fallow	0.83	0.75		
Plantation	0.83	0.90		
Fallow land	0.81	0.86		
River	1	0.71		
Sand	0.75	0.85		

Table 22.3 Classification accuracy metrics for the year 2020

Class	User accuracy (in %)	Producer accuracy (in %)	Overall accuracy (in %)	Kappa
Agricultural land	0.88	0.91	85	0.82
Agricultural fallow	0.94	0.85		
Plantation	0.91	0.84		
Fallow land	0.68	0.84		
River	0.8	0.8		
Sand	0.88	0.77		

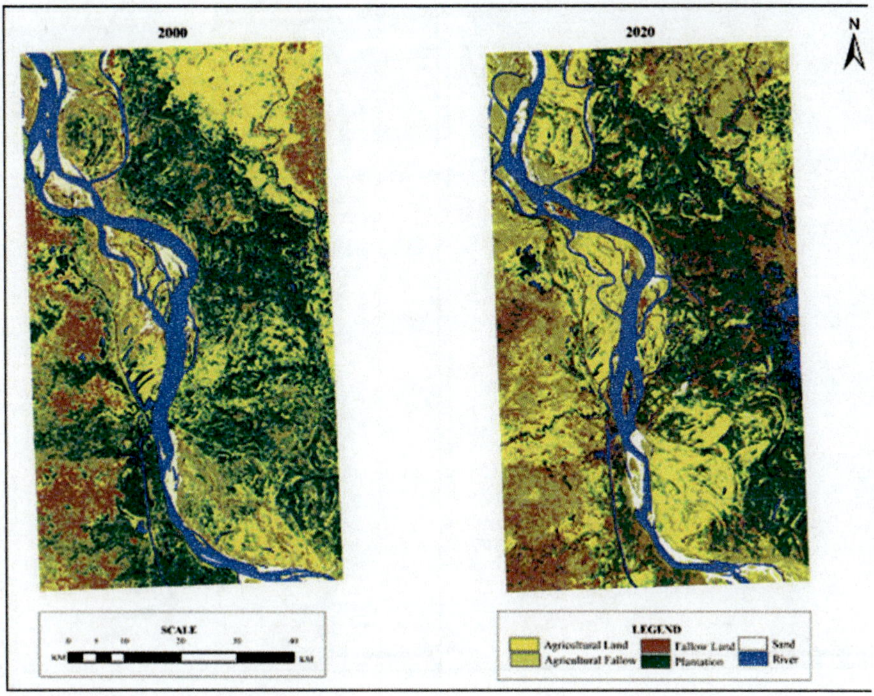

Fig. 22.3 LULC map of the year of 2000 and 2020

Table 22.4 LULC class-wise change in different landscape areas due to erosion during the period between 2000 and 2020

Sl. no.	Class name	Area (km^2)		
		2000	2020	Change
1	Agricultural fallow	496.81	605.14	108.33
2	Agricultural land	761.12	664.58	−96.54
3	Fallow land	252.73	389.38	136.65
4	Plantation	837.80	702.53	−135.27
5	River	266.08	273.51	7.43
6	Sand	80.29	59.69	−20.60

fallow land (136.65 km^2) followed by agriculture fallow (108.33 km^2) and river (7.43 km^2).

River Channel Shifting, Erosion and Accretion

The overall shifts in the river banks identified in the satellite images during 1980 and 2020 are shown in Fig. 22.4. The change analysis indicated that the course of the river has significantly changed, leading to the differential alteration in the river

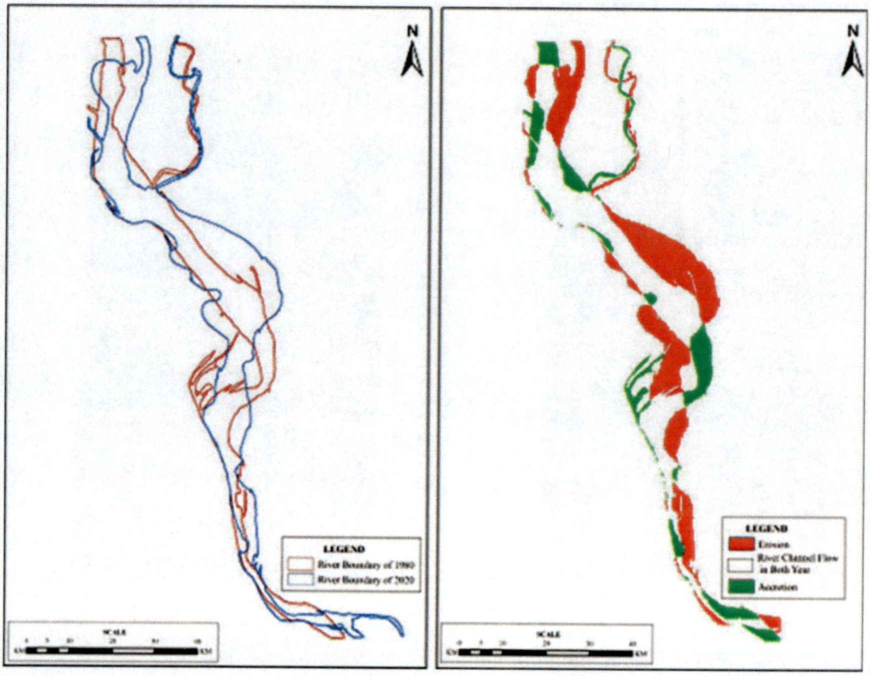

Fig. 22.4 The overall change in river channel, erosion and accretion observed from 1980 to 2020

banks. The maps are inferring the shift of river bank line in both the sides, which indicates a greater shift towards the left bank along the flow direction.

To find the sequential changes of the river bank and channel from 1980 to 2020, the river boundaries identified in the different years were overlaid. The erosion and accretion observed at 10 years interval are shown in Fig. 22.5. The corresponding changes in area due to erosion and accretion are given in Table 22.5. During 1980–1990, the estimated erosion was 71.91 km^2, and accretion was 67.60 km^2, which were 91.95 and 23.85 km^2 during 1990–2000, 38.60 and 74.05 km^2 during 2000–2010 and 48.36 and 71.25 km^2 during 2010–2020 (Table 22.5). The maximum erosion was thus observed for the period 1990–2000, and the minimum during 2000–2010 (Fig. 22.6), whereas the maximum accretion was observed between 2000 and 2010 and the minimum between 1990 and 2000. The common surface water area of the river was observed least (156.75 km^2) during 1980–1990 and maximum (285.11 km^2) during 2000–2010.

Change Analysis of Island (Char)

The islands identified within the study area in the different years are shown in Fig. 22.7, and area statistics is shown in Fig. 22.6. The maps indicated a significant

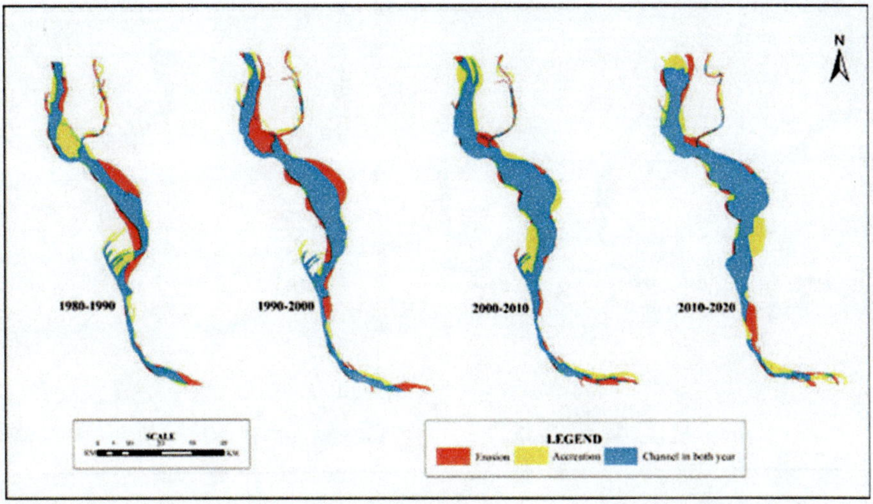

Fig. 22.5 Erosion and accretion maps for the different period from1980 to 2020

Table 22.5 Year-wise erosion and accretion area (km^2)

Duration	Erosion	Accretion	Channel in both year
1980–1990	71.91	67.60	156.75
1990–2000	91.95	23.85	204.81
2000–2010	38.60	74.05	285.11
2010–2020	48.36	71.25	260.90

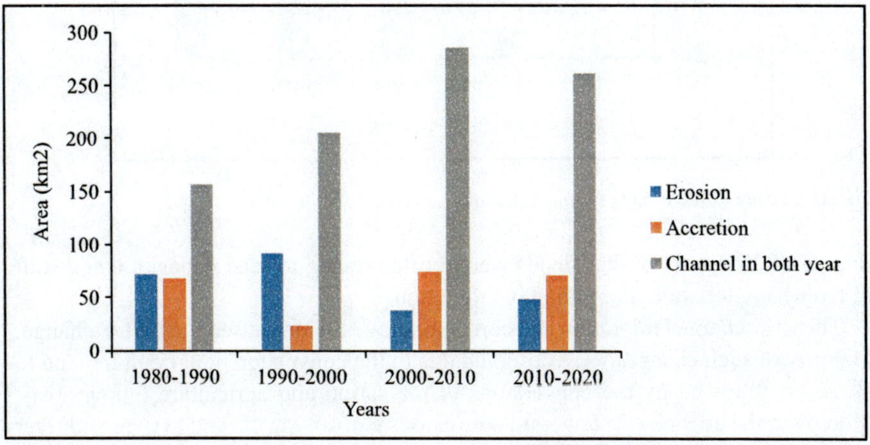

Fig. 22.6 Diagram showing erosion and accretion for the different periods

increase in the numbers and area of islands since 2000. The maximum area was estimated in 2010 (132 km^2) followed by 2020 (102 km^2) and 2000 (89 km^2), whereas the areas were mapped as 14 and 32 km^2 in 1980 and 1990, respectively

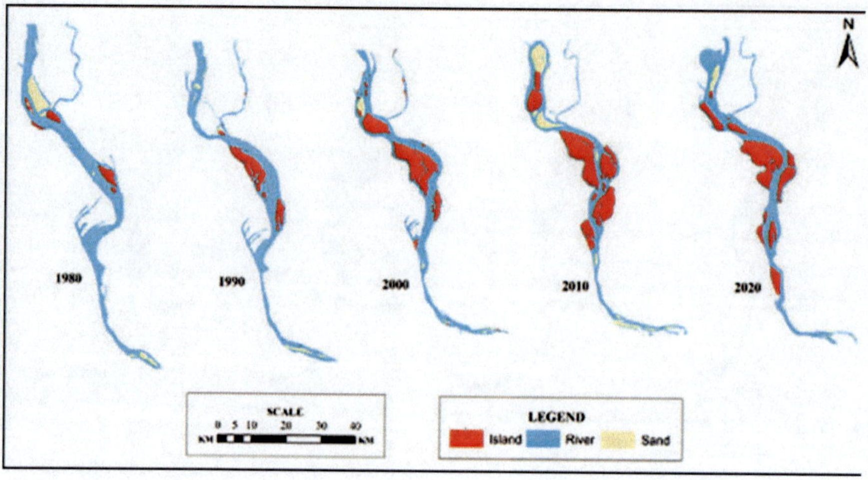

Fig. 22.7 Islands area identified in the different years from 1980 to 2020

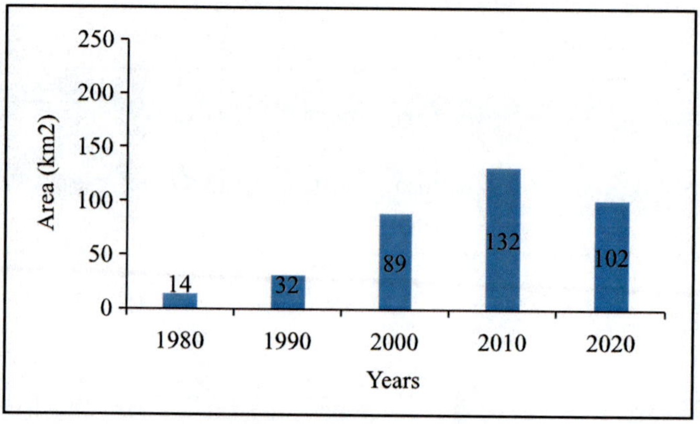

Fig. 22.8 Area of the islands for the different years from 1980 to 2020

(Fig. 22.8). Majority of the islands were created owing to sand deposition and shift of river bank leftward along the flow direction.

The river channel migration and corresponding erosion caused the LULC change. Majority of such changes were observed due to the conversion of agriculture land to the river, followed by the conversions of plantation and agriculture fallow areas. Moreover, the erosion of a few sand bars causes conversion of sand to river and river to sand conversion. Similarly, the reverse conversions took place between agriculture land, agriculture fallow and rivers.

22.5 Discussion

In the present study, the GIS analysis for different time periods ranging from the year 1980 to 2020 has shown the migration of river bank lines of the Ganges within the study area. The results indicated the maximum erosion (91.95 km^2) and the minimum accretion (23.85 km^2) occurred between 1990 and 2000. In contrast, the minimum erosion (38.60 km^2) and the maximum accretion (74.05 km^2) were observed in between 2000 and 2010. The erosion is higher when there is more storage capacity in the river and vice versa (Shahin, 2007). The shift of river bank line indicated an overall leftward shift along the flow direction. A similar change was also reported by Majumdar and Mandal (2019), who observed leftward shift of the river bank owing to weak alluvial soil structures in the left bank compared to erosion-resistant basalt rock in the right bank. The study revealed that the formation of river islands has been steadily increasing from 1980 to 2010, except in 2020 when the river accretion was decreased as compared to 2010. The total island area of 14 km^2 experienced more than ninefold and sevenfold increase, resulting in the total area of 132 and 102 km^2 in 2010 and 2020, respectively. This could be indicating higher sediment accumulation during the period 2000–2020 in comparison to the earlier decades. Moreover, the amount of accretion was maximum in 2010, which led to maximum islands in 2010 (132 km^2). The significant change in the area of curvature, leading to river meandering, infers the high river bank erosion (Zhou & Endreny, 2020).The major drivers of river bank line shifts could be either climatic or anthropogenic. Moreover, the flood events and soil erosion in the upper catchment also influence such changes. The soil erosion in upper catchment leads to sediment accumulation in these places (Mehebub et al., 2015). The swiftness of the river channel has led to significant alterations in the river bed and adjoining land use patterns. Majority of the LULC changes were observed as the conversion between agriculture land, agriculture fallow, river and sand. The rise in the number of sand bars or islands could be attributed to the increasing deposition of sediment due to the construction of the Farakka Barrage, leading to lower river flow. Thakur et al. (2012) also analysed the river bank erosion between Rajmahal and Farakka Barrage from 1955 to 2005. They have reported that the soil stratification of the river bank, presence of hard rocky area around Rajmahal, higher sediment load, lack of dredging and construction of Farakka Barrage which act as an obstruction to the natural river flow could be the major factors of river bank failure in this region. Ghosh and Sahu (2019) studied the impact of bank line shifting on LULC changes for an adjoining Ganga-Bhagirathi River in the Jangipur subdivision of Murshidabad district, West Bengal, and reported much lesser change (erosion, 7.93 km^2; deposition, 3.244 km^2) compared to the current study.

There is a historical outlook on the relationship between human and river ecosystem, which has furnished the regional cultural activities. The versatile human interaction with the river benefits in several ways. Thus, changes in the river modify the local environment, hydrological and climatic conditions, water availability, irrigation and agriculture, socio-economic condition of the dependent population, etc. The noticeable incidents or changes observed during the study period are summarized as follows:

1. Eradication the soil and plants on the river banks.
2. River bank degeneration causes destruction or relocation of habitat settlements.
3. Removal of productive soil and cropland leads to significant loss in crop yield.
4. Reduction in water level due to deepening of the soil bed.
5. Unutilized sedimentation has occurred on the opposite side of decay.
6. River bank erosion and corresponding changes in river's physical and the underground allies.
7. Alteration in the cultural activities.

22.6 Conclusion

The current study employed the satellite remote sensing data-derived GIS layers to assess the river morphology in the Ganges in parts of Malda district in West Bengal, India. The Landsat satellite imageries were used for visual image interpretation to map the erosion and accretion and corresponding changes in the river bank line. The GIS analysis of the past four decades (1980–2020) indicated significant changes in the river bank with an overall leftward shift along the flow direction. The maximum erosion with least accretion was observed during 1990–2000, whereas the minimum erosion with maximum accretion was observed during 2000–2010. Such changes in the river bank mostly reduced the fertile croplands and rural settlements along the river channel. A significant increase was observed in the number of islands and total area. The time series changes observed in the past four decades indicate the river meandering. The total islands area of 11 km^2 in 1980 monotonically increased to 132 km^2 in 2010 and then reduced to 102 km^2 in 2020, which could be indicating significant accumulation of sediment in the river beds. However, the deposition of sediment or developments of chars in the region may lead to flooding during the monsoon. The generated maps are useful in understanding the river processes and local hydrological regimes. Moreover, the study outcomes have multiple usage and implications. These will help the water resource, agriculture, forest and landscape planners and managers for improved management and mitigation planning, policy development, conservation, etc.

References

Akter, S., Zahra, F. T., Sakib, M. N., Sen, D., and Chowdhury, M. (2017). Analysis of river bank erosion and calculation of universal erosion co-efficient using space-borne gis and rs technique: a analysis of river bank erosion and calculation of universal erosion co-efficient using Space-Borne Gis and Rs Technique : a Study O. *International Conference on Disaster Risk Mitigation, April.*

Banerjee, M. (1999). *A report on the impact of Farakka barrage on the human fabric*. South Asian Network on Dams, Rivers and People.

Biswas, R., & Anwaruzzaman, A. K. M. (2019). Measuring hazard vulnerability by bank erosion of the ganga river in Malda district using PAR model. *Journal of Geography, Environment and Earth Science International, 22*(1), 1–15.

Das, B. (2011). Stakeholders' perception in identification of river bank erosion hazard: A case study. *Natural Hazards, 58*(3), 905–928.

Das, B., Mondae, M., & Das, A. (2012). Monitoring of bank line erosion of river ganga, Maida District, and West Bengal: Using RS and GIS compiled with statistical techniques. *International Journal of Geomatics and Geosciences, 3*(1), 239–248.

Das, P., Behera, M. D., Patidar, N., Sahoo, B., Tripathi, P., Behera, P. R., ... Krishnamurthy, Y. V. N. (2018). Impact of LULC change on the runoff, base flow and evapotranspiration dynamics in eastern Indian river basins during 1985–2005 using variable infiltration capacity approach. *Journal of Earth System Science, 127*(2), 19.

Das, T. K., Haldar, S. K., Gupta, I. D., Sen, S., & others. (2014). River bank erosion induced human displacement and its consequences. *Living Review of Landscape Research, 8*(3), 1–35.

Debnath, J., Pan, N. D., Ahmed, I., & Bhowmik, M. (2017). Channel migration and its impact on land use/land cover using RS and GIS: A study on Khowai River of Tripura, North-East India. *The Egyptian Journal of Remote Sensing and Space Science, 20*(2), 197–210.

Ghosh, D., & Sahu, A. S. (2019). Bank line migration and its impact on land use and land cover change: A case study in Jangipur subdivision of Murshidabad District, West Bengal. *Journal of the Indian Society of Remote Sensing, 47*(12), 1969–1988.

Guite, L. T. S., & Bora, A. (2016). Impact of river bank erosion on landcover in lower Subansiri River flood plain. *International Journal of Scientific and Research Publications, 6*(5), 480–486.

Hassan, M. A., Ratna, S. J., Hassan, M., & Tamanna, S. (2017). Remote sensing and GIS for the Spatio-temporal change analysis of the east and the West River Bank erosion and accretion of Jamuna River (1995-2015), Bangladesh. *Journal of Geoscience and Environment Protection, 5* (9), 79–92.

Iqbal, S. (2010). Flood and erosion induced population displacements: A socio-economic case study in the Gangetic riverine tract at Malda district, West Bengal, India. *Journal of Human Ecology, 30*(3), 201–211.

Islam, M. D. F., & Rashid, A. N. M. B. (2011). Riverbank erosion displaces in Bangladesh: Need for institutional response and policy intervention. *Bangladesh Journal of Bioethics, 2*(2), 4–19.

Majumdar, S. (2018). Impact of flood and Bank erosion on human life: A case study of Panchanandapur at Malda, West Bengal, India. *International Journal of Basic and Applied Research, 8*, 9. http://www.pragatipublication.com/assets/uploads/doc/5d892-1038-1056.13825.pdf

Majumdar, S., & Mandal, S. (2019). Identification of suitable human habitation sites through the river bank stability analysis on Ganga—Pagla interfluves area of West Bengal, India. *Spatial Information Research*, 1–15.

Mehebub, S., Raihan, A., Nuhul, H., & Haroon, S. (2015). Assessing flood inundation extent and landscape vulnerability to flood using geospatial technology: a study of Malda district of West Bengal, India. In *Forum geografic* (Vol. 14, No. 2, p. 156). University of Craiova, Department of Geography.

Phukan, A., Goswami, R., Borah, D., Nath, A., & Mailanta, C. (2012). River bank erosion and restoration in the Brahmaputra river in India. *The Clarion-International Multidisciplinary Journal, 1*(1), 1–7.

Rahman, M. S., & Gain, A. (2020). Adaptation to river bank erosion induced displacement in Koyra Upazila of Bangladesh. *Progress in Disaster Science, 5*, 100055.

Sarma, J. N., Borah, D., & Goswami, U. (2007). Change of river channel and bank erosion of the Burhi Dihing river (Assam), assessed using remote sensing data and GIS. *Journal of the Indian Society of Remote Sensing, 35*(1), 93–100.

Shahin, M. (2007). Erosion and sedimentation in drainage basins and in storage reservoirs. In *Water resources and hydrometeorology of the Arab Region. Water science and technology library, vol 59*. Springer. https://doi.org/10.1007/1-4020-5414-9_8

Singh, O., & Kumar, M. (2017). Flood occurrences, damages, and management challenges in India: A geographical perspective. *Arabian Journal of Geosciences, 10*(5), 102.

Thakur, P. K., Laha, C., & Aggarwal, S. P. (2012). River bank erosion hazard study of river Ganga, upstream of Farakka barrage using remote sensing and GIS. *Natural Hazards, 61*(3), 967–987.

Water Resources Systems Division., N. I. of H. (2020). River of India. *Hydrology and water resources information system for India.*

Zhou, T., & Endreny, T. (2020). The straightening of a river meander leads to extensive losses in flow complexity and ecosystem services. *Water, 12*(6), 1680.

Chapter 23
Spatiotemporal Detection and Delineation of Bhagirathi-Hooghly River Bank Erosion Using GIS Analytics, West Bengal, India

Mantu Das and Snehasish Saha

Abstract West Bengal is a riverine state in India. So the floodplain areas in the lower Gangetic plain experience several natural hazards like monsoon floods and bank failure every year due to the extensive river network. The study area is a part of the moribund delta and is formed by alluvial deposits (sand, silt, and clay). Due to regular river course shifting, the Ganga River faces frequent bank erosion hazards. The primary objective of this research is to detect bank erosion sites with concerning spatiotemporal channel migration in between Nabadwip and Katwa (84-km-long reach) in West Bengal, India. In recent time, remote sensing (RS) and geographical information system (GIS) have been considered powerful applications for the river study basically in spatial change identification. In this study, using six selected maps from 1954 to 2020, the river bank erosion patterns and dynamics of the Ganga (Bhagirathi-Hooghly) river resulting from processes for 66 years were analyzed. Therefore, here, multispectral satellite imagery (Landsat 8 OLI/TIRS, Landsat 7 ETM+, Landsat 5 TM, and Landsat 3 MSS) for the year 1980, 1990, 2000, 2010 and 2020, and 1954 have been used. Toposheet (NF-45/3) was used as base information. Temporal river course detection is a studious task. Moreover, here, Toposheet was resampled with WGS 84 UTM projection, and satellite imageries were also resampled and rectified and finally processed through modified Normalized Difference Water Index (NDWI) to detect active river flow more precisely. In the period between 1954 and 2020, most of the sites of the right bank (76% of referenced sites; 24% toward leftward direction) are being shifted toward the rightward direction. But in the case of the left bank, 52% of sites are eroded toward the rightward direction, and 48% of sites are moved toward the leftward. The western part of the river Bhagirathi-Hooghly (Nabadwip to Katwa) is highly vulnerable due to bank erosion. The Bhagirathi-Hooghly river course within the studied periods (1954–2020) normally followed a straight to the sinuous path, and the SSI ranged between 1.006 and 1.031. In 1954–2020, topographic interferences played a great role (TSI gt; HIS) in river course shifting at the Nabadwip-Katwa section. The river shifting rates are based on the differences in 1954–1980, 1980–1990,

M. Das · S. Saha (✉)
Department of Geography and Applied Geography, University of North Bengal, Siliguri, India

P. K. Shit et al. (eds.), *Geospatial Technology for Environmental Hazards*, Advances in Geographic Information Science, https://doi.org/10.1007/978-3-030-75197-5_23

1990–2000, 2000–2010, and 2010–2020. Results show few sites along the river course are highly affected by bank erosion, i.e., near Akandanga (T8), Uttar Srirampur (T12), Bargara (T17), and Sankarpur (T20). Therefore, these sites experienced greater economic loss by degrading land loss, i.e., fertile agricultural land, settlement area, etc., due to the river bank erosion. So bank protection and hazard mitigation fulfil the purpose of sustainable river basin management and will be good practice in order for human to benefit.

Keywords Bank erosion · Bank line shifting · NDWI · Remote sensing · GIS · Bhagirathi-Hooghly River

23.1 Introduction

River bank erosion occurs as a natural threat caused by water and land interactions and is greatly affected during monsoon period due to bankful discharge with astonishing flow dynamics, but sometimes it's called quasi-natural because of human involvement as soil digging, riverine agricultural practices, settlement expansion, etc. Lateral displacement of the channel confers to the spatial shift of a river course as a response of flow dynamics, sediment supply, bank materials (cohesive and non-cohesive materials), and channel configuration/geometry (Bandyopadhyay et al., 2014; Roy et al., 2020). Therefore, channel migration is a process that is continuously amalgamated with river bank erosion to the channel bed and channel wall under dynamic flow situations which may cause baneful local and regional changes. In the present research work, authors have acquired knowledge about the river bank erosion with special reference to lateral shifting of the channel within the Nabadwip-Katwa section of the lower Ganga (Bhagirathi-Hooghly) river.

Rivers are complex systems in dynamic equilibrium within natural environmental settings (Das et al., 2014; Chakraborty & Mukhopadhyay, 2015). Alluvial river courses are a very subtle exponent of channel morphodynamic changes that can readjust through the interaction of flow characteristics, bank bed geometry, and sediment supply and tectonics activity (Schumm et al., 2002; Konsoer et al., 2016). Throughout the river network, diverse bank erosion mechanisms take place from upstream to downstream (base level). River bank erosion and bank line shifting on a spatiotemporal scale are a fluvio-morpho-dynamical phenomena under various flow regimes (Das et al., 2007; Das & Saraf, 2007; Pati et al., 2008; Ghosh & Sahu, 2018; Biswas & Anwaruzzaman, 2019; Dekaraja & Mahanta, 2020).

A review study on the assessment of river bank collapse and lateral channel adjustments was performed by Lawler (1993), who stated that sedimentological data, botanical proof, historical records, repetitive cross profiling, erosion pins, and terrestrial photogrammetry were the methods covered to fulfil the objectives. Muller et al. (1993) and Lovric and Tosic (2016) made a summary of multispectral satellite data with GIS applications to river studies. In this review, all types of remotely sensed multispectral data with the varying spatial resolution are currently available on the free domain (USGS Earth Explorer, Bhuvan-Indian Geo-Platform of ISRO, etc.) and payable mode, and their versatile uses in river studies basically in channel migration, bank erosion, and morphological changes have been highlighted. Remote sensing and GIS applications in bank erosion study have a great role in management study (Das & Saraf, 2007; Das et al., 2012; Ophra et al., 2018).

India is a country of rivers which is comprised of seven major rivers and associated tributaries and distributaries making up the river network in India (Thakur et al., 2012; Das et al., 2014). Every year monsoon floods are a frequent natural hazard in India. Two main river systems, namely, Ganga and Brahmaputra river system, are highly affected by bank erosion activity (Das et al., 2014; Sarma, 2013; Mukherjee et al., 2017; Islam & Guchhait, 2019; Dekaraja & Mahanta, 2020). It is noteworthy that Nadia district is experiencing badly invaded by the bank erosion dynamics along the Bhagirathi river course (Islam and Guchhait 2018; Islam & Guchhait, 2015). The Gangetic plain has immense value in human life as well as environmental perspective also. But every year in the plain region (old stage of the river), the river bears millions of tons of sediment loads and deposits on the river bed, bank, and surface of the base level of erosion causing morphological modification to the channel. (Khan et al., 2018; Parua, 1999). The sediment accumulation (bed materials and suspended particles) creates several problems like water quality issues, channel shrinkage, and the decrease of river depth due to riverbed siltation (Wilkinson et al., 2006). But when the discharge increases or at bankful discharge (during monsoon rainfall), the river exerts pressure on its bank walls and begins lateral expansion by the erosive agents (water, suspended, and bed materials) depending on bank materials, compactness, tension magnitude, etc., which in the floodplain area becomes the source of several floods and river bank erosion hazards (Wu et al., 2008; Xia et al., 2010). Around the same time, anthropogenic interference along the river course (embankment, dam construction, unauthorized encroachment, etc.) disturbs the balance of river dynamics and speeds up the rate of bank erosion (Das et al., 2014). Previous studies show West Bengal, especially Ganga major and Bhagirathi-Hooghly River, has experienced high bank erosion phases, and the Farakka Barrage was the main triggering factor (Rudra, 2010; Chatterjee & Mistri, 2013) upstream Malda. So it is necessary to analyze the behavior of the ever-changing river course on natural settings with human interferences over the spatio-temporal scale and its severity in terms of unhealthy impact to the areas where the concerned authorities can take preventive measures. To accomplish these targets, this research work has been perpetrated with the following subject matters: to quantify the magnitude of bank line migration or changes due to bank erosion on space-time domain and quantification of these changes using the morphometric parameters of the Hooghly River.

Natural hazard like river bank erosion is vulnerable to the riverine environment which includes a rich ecosystem, human society, and valuable earth surface. Moreover, it appears in our society as a great evil toward a sustainable environment during the peak monsoon period because of bankful discharge and dynamic hydraulic nature. Riverbank shifting over time and space due to bank erosion continuously knowing to human society, what is the changing nature? What are the causes and consequences? Through this study we can predict the spatiotemporal changing trend of the bank erosion in the recent future and help us to take necessary precautions to minimize the risk, e.g., economic loss, river ecosystem, social forestry, natural vegetation, and so on. So river bank erosion study has an immense value not only for human society rather in a sustainable physical environment. Several investigations (Das et al., 2017; Chakraborty & Mukhopadhyay, 2015; Mondal et al., 2014; Sarma, 2014; Mandal, 2017) revealed the great significance of this study to manage natural hazards as river bank erosion.

23.2 Brief Description of the Study Area

The studied river course, Hooghly River, passes along the administrative boundary of Nadia (western side) and Burdwan (eastern side) districts of West Bengal (Fig. 23.1). The selected river reach of the Bhagirathi-Hooghly River middle course is 84 km of water track from Katwa (upstream) to Nabadwip (downstream), comprising a portion of the districts of Burdwan and Nadia. This present study is comprised of the bank displacement of the river Hooghly caused by erosional activities. The studied reach is located within the coordinates of 23°45′12.24″ N to 23°21′49.66″ N and 88°11′33.68″ E to 88°21′22.55″ E. The average height of the study area is 8 m above MSL and comes under the lower Ganga plain (Sinha et al.,

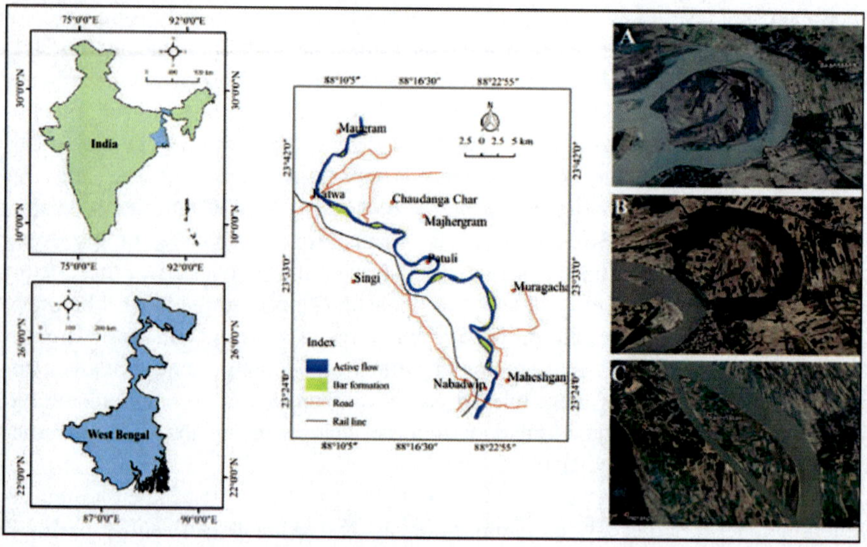

Fig. 23.1 Locational personality of the study area characterized by morphological features emerged as a footprint of erosional and depositional activity of the river. (**a**) Meander loop near Akandanga in 2010, (**b**) oxbow lake near Akandanga in 2020, (**c**) mid channel bar near Kashthasali in 2020

2005) (Fig. 23.2). Three main landforms are accessible, i.e., from the Lower Ganga plains, uplands, old fluvial/deltaic plains, and young fluvial plains are established (Singh et al., 1998), in which the study area is situated within old deltaic plains. This area is situated geologically in the Rarh region, the lower part of the Bengal Basin's dying deltaic section. (Islam & Guchhait, 2015) and is made up of a recent Pleistocene deposit. Initially, along the river course, the region is covered by sandy clay, sand, fine silt, sandy loam, and loamy soil contained in the extensive low gradient land surface of the plain. Moreover, fertile soil of the floodplain areas is suitable for quality or productive cultivation. The studied reach has been divided into a total of 25 cross section lines. These cross section lines are made on the basis of the maximum amount of shifting, analyzing the satellite images through logical perspectives (Fig. 23.3).

23.3 Data Sources and Methodology

Data Used

The following data were considered to study the river bank erosion mapping and channel morphodynamic changes. The overview of the methodology is represented in Fig. 23.2.

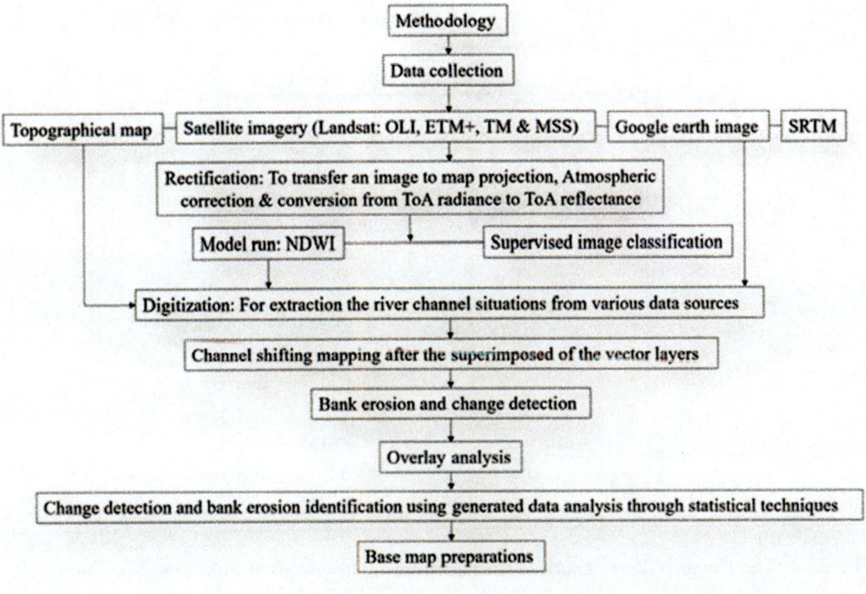

Fig. 23.2 Comprehensive flowchart of the river bank erosion mapping research approach

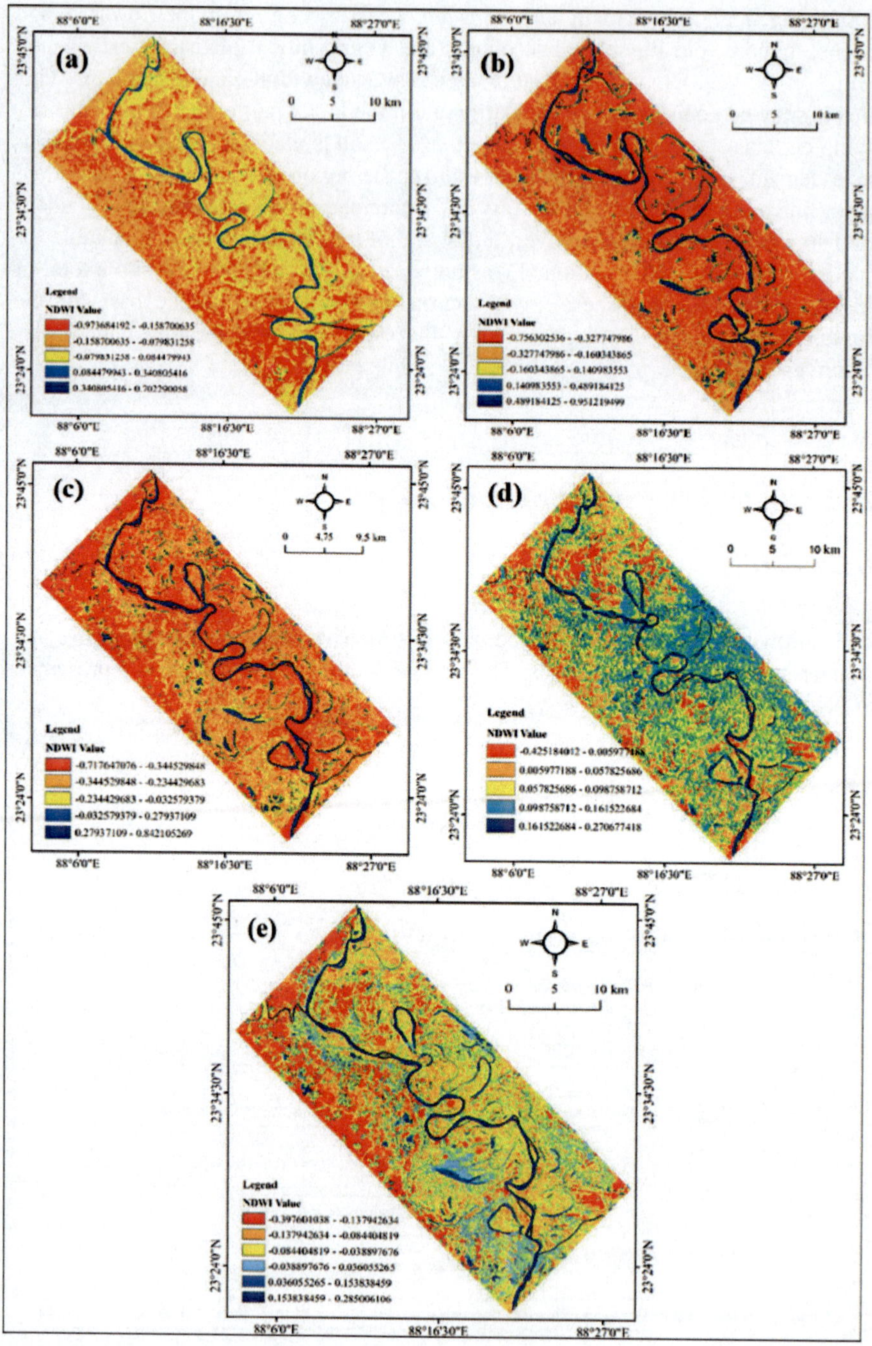

Fig. 23.3 The Hooghly River course extracted from multispectral satellite imageries by using MNDWI model. (**a**) 1980, (**b**) 1990, (**c**) 2000, (**d**) 2010, and (**e**) 2020

Data Processing

Topographical map is on a sufficiently large scale to enable individual features to be shown on the map. The study area's topographical map is 1:250,000 series map which was surveyed in 1954 as found in the NF 45/3 Toposheet. Remote sensing is one of the techniques in which the capturing of spatial information without physical contact with a spatial object of the earth's surface, which is not only essential for comprehensive spatial analysis rather has immense value for precise resource management. In this regard, primarily multispectral satellite imagery for the year 1980, 1990, 2000, 2010, and 2020 has been resampled and rectified in ArcMap software for the present study. So, here, two steps have been developed. At first the pixel value of individual bands was converted into surface reflectance to precisely identify the surface features (Bhatta, 2011; Lillesand et al., 2015). The DN values of a particular band were converted to the top of atmosphere (ToA) radiance by the spectral radiance scaling method and finally converted to the ToA reflectance. Then, the required atmospheric correction (minimized cloud coverage over the study area) has been performed in the GIS environment. The proper findings of river shifting and bank erosion are obtained from various data sources and GIS techniques (Langat et al., 2019; Momin et al., 2020). The base map prepared by software analyses the precision of river shifts over topographical and remote sensing data.

River Course Detection

River channel detection with high precision is a sturdy task to the researcher. Here, satellite imagery and Modified Normalized Difference Water Index (MNDWI) have been accomplished to detect the channel over spatiotemporal scale (Jovanovic et al., 2014; Tiwari et al., 2016; Nijhawan & Jain, 2018). For MNDWI, two bands, namely, green band and short-wave infrared (SWIR) band, are the key components, and aftermath raster calculation results show the difference between water body and land surface (Eq. 23.1). Results, if lower the raster calculated value (−1 to 0) means earth surface covered with water. On the other hand, a greater value (0 to 1) refers to the dry surface (land surface) on earth. The NDWI has been determined by using the following formula (Xu, 2006; Mukherjee et al., 2017):

$$\mathrm{NDWI} = \frac{(\text{Green band} - \text{SWIR band})}{(\text{Green band} + \text{SWIR band})} \quad \text{(Modified NDWI after H.Xu 2006)} \tag{23.1}$$

Bank Erosion Mapping

This research work is accomplished with the help of remote sensing multispectral satellite data, topographic map (NF 45-3), and Google Earth imagery. All these data have individually been processed and analyzed in a GIS environment (ArcMap 10.8 pirated version) to prepare thematic maps on bank erosion. By the topographical map, the situation of 1954 of Hooghly River has been drafted, while the satellite imagery data are showing the conditions of 1980, 1990, 2000, 2010, and 2020 of the river. Two databases were overlaid on one another along with the WGS-84 UTM projections in GIS platform to find out the channel shifting of Hooghly River (Nabadwip to Katwa). This shifted segment of the river is mapped by vectorization in the GIS domain.

Morphometric Measurements

Channel morphometric analysis has been carried out using remote sensing and GIS applications. The sinuosity index (Aswathy et al., 2008; Kumar et al., 2014) has been applied in river study to access the temporal changes of channel braiding nature due to bank erosion. The following equations (Muller 1968) were incorporated here:

$$\text{Hydraulic Sinuosity Index (HSI)} = \frac{(\text{Channel index} - \text{Valley index})}{(\text{Channel index} - 1)} \tag{23.2}$$

$$\text{Topographic Sinuosity Index (TSI)} = \frac{(\text{Valley index} - 1)}{(\text{Channel index} - 1)} \tag{23.3}$$

$$\text{Standard Sinuosity Index (SSI)} = \frac{\text{Channel length (CL)}}{\text{Valley length (VL)}} \tag{23.4}$$

Standard Sinuosity Index (SSI) represents the nature of river course which ranges from a straight to braided course on the basis of ratio between channel length (the length of the channel in the stream under study) and valley length (the valley length along a stream, i.e., the length of a line which is everywhere midway between the base of the valley walls. It will equal channel length wherever the valley walls descend directly to the water's edge and will be less than channel length wherever a floodplain has developed) (Eq. 23.4). According to Vijith and Dodge-Wan (2018), Maurya and Yadav (2016), and Tiwari et al. (2016), SSI value is categorized into five distinct groups, i.e., straight course (<1.05), sinuous (1.05–1.30), moderate meandering (1.30–1.50), meandering (1.50–2.00), and braided course (>2.00).

On the other hand, HSI and TSI reveal that course departure from their straight course is due to hydraulic characteristics and topographic behavior over space and time, respectively (Mueller, 1968; Ghosh & Mistri, 2012; Kumar et al., 2014). The total value of TSI, when added to HSI, should always be 100. It means any one of these two (TSI or HSI) has been calculated and the next one has also been derived through simple subtraction (100-HSI or TSI). Results show (Eqs. (23.2) and (23.3)) higher TSI and HSI means river course is highly deviated from straight path due to geo-topographic interferences (e.g., geology, undulating surface, plain, plateau, foothill, and mountain topography) and ever-changing hydraulic nature which is greatly affected during peak monsoon period.

Statistical Analysis

Statistical tools are considered as an important key for establishing the facts in a precise way. Statistical analysis includes mean, standard deviation (SD), coefficient of variation (CV), skewness, and kurtosis, which have been proceeded to comprehend the changing behavior (space-time domain) of river bank line and their consistency for understanding the future trend (Mandal et al., 2018; Dey & Mandal, 2019). The selected statistical methods have been performed in the IBM SPSS Statistics (trial) software. Spatial mapping of bank erosion as lateral shifting of bank line is a common phenomenon in the natural riverine environment. Standard deviation is a widely used measure of the variability of the bank line migration (Gabet, 1998; Dey & Mandal, 2019). Higher SD value refers to migration rates distributed over the wide range of scales, and low SD value means that migration rates appear to be very normal migration due to erosional activity. The CV of variation of bank line migration is higher which means less consistent migration and vice versa (uniform in nature). Kurtosis is employed to refer to the peakedness of a distribution curve of bank line shifting. The nature of the distribution pattern of the bank line migration is analyzed by kurtosis. And finally, skewness has also been applied to convey the degree of asymmetry of the spatial migration of the temporal bank line due to bank erosion activity.

23.4 Results and Discussion

Spatiotemporal Scenario of River Bank Erosion (1954–2020)

Hooghly River Reach in 1954

River course has been delineated from the 1954 Texas University topographical map (Table 23.1). A total of 25 transects across the river have been considered as a reference site (Fig. 23.4), and this Toposheet is also considered a base map for the

Table 23.1 Description of the topographical map

Toposheet No.	Series	Area	Scale	Survey year	Publisher
NF 45-3	U502	Burdwan	1: 2,50,000	1954	Map room, General Libraries, PCL 1.302, University of Texas, Austin, TX 78713-7330

Source: https://legacy.lib.utexas.edu/maps/ams/india/

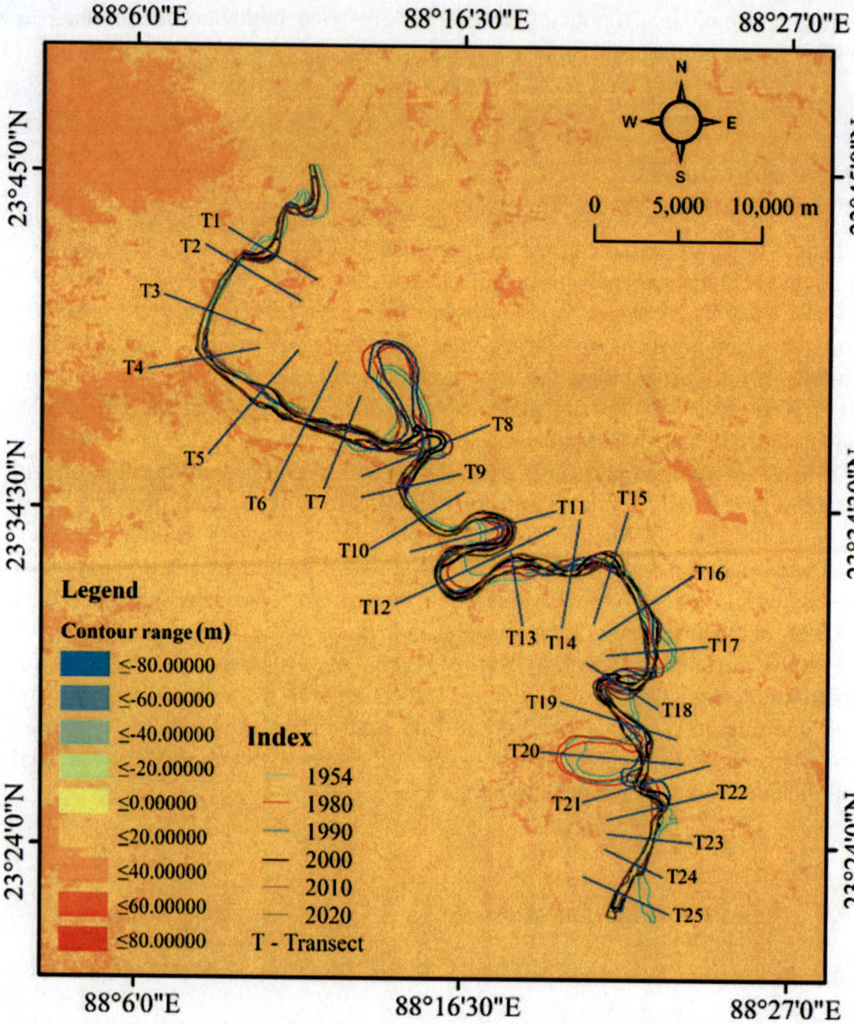

Fig. 23.4 Referenced transects across the Hooghly River course in relation to topographic elevation between Nabadwip and Katwa, West Bengal

Table 23.2 Temporal status of river morphometric parameters

Year	HSI	TSI	SSI
1954	0.029	0.971	1.016
1980	0.008	0.992	1.006
1990	0.027	0.972	1.012
2000	0.043	0.956	1.019
2010	0.005	0.934	1.031
2020	0.034	0.965	1.018

Source: Calculated by Das and Saha (this volume)
HSI Hydraulic Sinuosity Index, *TSI* Topographic Sinuosity Index, *SSI* Standard Sinuosity Index

Table 23.3 Description of satellite imageries

Sl. No	Satellite	Sensor	Path	Row	Acquisition date	Spatial resolution
1	Landsat 3	MSS	149	44	02nd Feb. 1980	30 × 30 m
2	Landsat 5	TM	138	44	14th Nov. 1990	30 × 30 m
3	Landsat 5	TM	138	44	27th Dec. 2000	30 × 30 m
4	Landsat 7	ETM+	138	44	15th Dec. 2010	30 × 30 m
5	Landsat 8	OLI/TIRS	138	44	02nd Dec. 2020	30 × 30 m

Source: USGS Earth explorer (http://earthexplorer.usgs.gov/)

study. In 1954, the river course was near the straight path (Horacio, 2014). The SSI value for the selected study reach from Nabadwip to Katwa was 1.016 (Eq. 23.2) (<1.05 indicates straight course). It is also observed that the calculated HSI and TSI were 0.029 (Eq. 23.3) and 0.971 (Eq. 23.4), respectively (Table 23.2).

Hooghly River Reach in 1980

The river course has been detected and extracted from Landsat 3 MSS multispectral satellite image (Table 23.3). During this time, the study reach was almost a straight course. The calculated SSI, HSI, and TSI were 1.006 (Eq. 23.2), 0.008 (Eq. 23.3), and 0.992 (Eq. 23.4), respectively (Table 23.2). During this period, a total of 10 referenced spots on the left bank and 16 spots on the right bank were affected by the bank erosion (Table 23.4 and Fig. 23.4).

Hooghly River Reach in 1990

The Hooghly River course has been extracted from Landsat 5 TM satellite imagery (Table 23.3). In 1990, the river course followed almost a straight path (SSI-1.012). Temporal changes of sinuosity index refer to the impact of the erosional and depositional activity. Therefore, it was noticed that the measured SSI, HSI, and TSI were about 1.019 (Eq. 23.2), 0.027 (Eq. 23.3), and 0.972 (Eq. 23.4), respectively (Table 23.2). Channel morphometric parameters are the indicators of spatiotemporal

Table 23.4 River bank erosion spots along Hooghly River (1954–2020)

Time interval	Left bank erosion	Right bank erosion
1954–1980	T1, T3, T4, T6, T7, T11, T13, T14, T20, T22	T2, T5, T6, T8, T9, T10, T12, T15, T16, T17, T18, T19, T20, T21, T23, T25
1980–1990	T1, T5, T7, T8, T9, T11, T13, T15, T16, T20, T21, T22, T24	T1, T4, T5, T6, T7, T10, T12, T14, T17, T18, T19, T21, T23, T25
1990–2000	T3, T5, T6, T7, T8, T9, T10, T11, T14, T15, T16, T18, T19, T20, T22, T23, T24, T25	T1, T2, T4, T6, T9, T12, T13, T14, T16, T17, T18, T19, T21, T22, T23, T25
2000–2010	T1, T3, T5, T7, T8, T9, T10, T11, T12, T13, T15, T17, T18, T19, T20, T21, T24	T2, T3, T4, T6, T7, T8, T10, T12, T13, T14, T15, T16, T17, T19, T22, T23, T25
2010–2020	T1, T3, T5, T6, T10, T11, T12, T13, T14, T16, T18, T19, T21, T23	T1, T2, T4, T7, T8, T9, T14, T19, T22, T25

Source: Prepared by Das and Saha (this volume)

river course change. Bank erosion is common in every monsoon season for Hooghly River course. Table 23.4 shows that 13 sites on the left bank and 14 sites on the right bank experienced bank erosion.

Hooghly River Reach in 2000

The river has been digitized from Landsat 5 TM satellite imagery (Table 23.3) after processing the NDWI model (Fig. 23.3). According to Table 23.2, the calculated SSI value was 1.019 which indicates near to straight river course. Moreover, the HSI and TSI were 0.043 (Eq. 23.3) and 0.956 (Eq. 23.4), respectively. During this period, most of the sites were affected by erosion. It is observed that 18 spots on the left bank and 16 spots on the right bank have received identifiable mark of bank erosion.

Hooghly River Reach in 2010

The river has been extracted from Landsat 7 ETM+ multispectral satellite imagery (Table 23.3). In the period between 2000 and 2010, the river Hooghly had experienced a devastating flood which greatly affected the river morphology. The river reach received SSI of 1.031 (Eq. 23.2), and 0.065 (Eq. 23.3) and 0.934 (Eq. 23.4) were the calculated values of HSI and TSI accordingly. Moreover, 17 referenced sites on the left bank and 17 sites on the right bank were affected by bank erosion.

Hooghly River Reach in 2020

The river has been digitized from Google Earth satellite image and verified by overlay analysis using Landsat 8 OLI/TIRS satellite imagery in GIS environment. During monsoon period every year, the river is continuously experiencing bank erosion with the changing nature of bank, and the river readjusts itself by auto-cut-fill process. Therefore, at present time the river course follows almost a straight path. In 2020, the calculated SSI, HSI, and TSI were 1.018 (Eq. 23.2), 0.034 (Eq. 23.3), and 0.965 (Eq. 23.4), respectively (Table 23.2). At the end of 2020, 14 referenced spots on the left bank and 10 spots on the right bank have been identified as erosional sites with reference to 2010 base map.

Temporal Shifts of the Hooghly River Bank

River shifting is the cause of river erosion and depositional activity itself or accelerating by human interventions. In the period between 1954 and 1980, four sites on the right bank vastly shifted, i.e., T1 (735.73 m), T5 (522.90 m), T11 (958.93 m), and T25 (1384.78 m), which are located near Naliapur, Basatpur, Jhaudanga, and Parmedia, respectively. But in left bank, six sites experienced high shifting rate, i.e., T1 (969.68 m) at Naliapur, T11 (1070.47 m) at Jhaudanga, T12 (1028.59 m) at Uttar Srirampur, T17 (1084.39 m) at Mertala, T18 (1262.43 m) at Gopipur, and T20 (1028.22 m) at Sankarpur. In the period between 1980 and 1990, the highly shifted places were T19, T20, T21, and T22 on the right bank; on the other hand, T12, T20, and T22 on the left bank experienced a drastic change of bank line. Year 2000 shows five referenced spots (i.e., T8, T11, T14, T15, and T19) on the right bank, and four spots (i.e., T8, T15, T15, and T21) were higher shifted sites. During 2000–2010, a large amount of shifting was experienced at T8, T15, and T18 on the left bank and at T8 and T22 on the right bank of Hooghly River. But during the last decade (2010–2020), maximum sites experienced a more consistent nature of bank line shifting (Table 23.5). It is observed that only T22 referenced site on the right bank and four sites (i.e., T7, T8, T16, and T17) on the left bank were identified as large amounts of bank line shifting (Table 23.6). Moreover, it is observed that most of the sites were identified as a normal migration within 10 years of temporal scale.

Change of River Width

River width, according to spatiotemporal scale, is changing parameters in natural environmental settings by the river process and man-induced causes. The spatial dimension of channel geometry has been modified due to the high river discharge, velocity during heavy flows, and (the monsoon period) lateral erosion during floods

Table 23.5 Temporal shifting characteristics of right bank line with direction of Hooghly River from Nabadwip to Katwa (1954–2020)

Transects	1954–1980	Direction	1980–1990	Direction	1990–2000	Direction	2000–2010	Direction	2010–2020	Direction	1954–2020	Direction
T1	735.73	Left	58.71	Right	35.18	Right	39.02	Left	54.80	Right	630.79	Left
T2	182.08	Right	28.31	Left	21.47	Right	3.00	Right	14.40	Right	191.93	Right
T3	209.61	Right	11.76	Left	4.45	Left	14.31	Right	1.28	Left	206.64	Right
T4	252.50	Right	16.03	Right	1.29	Right	2.10	Right	2.97	Right	279.36	Right
T5	522.90	Right	134.32	Right	4.84	Right	104.02	Left	22.50	Left	530.20	Right
T6	179.28	Right	40.59	Right	46.00	Right	22.71	Right	5.85	Right	292.24	Right
T7	207.84	Left	113.61	Right	0.64	Right	18.11	Right	28.90	Right	45.41	Left
T8	324.63	Right	10.09	Left	1103.94	Left	1126.07	Right	99.40	Right	421.52	Right
T9	351.13	Right	78.55	Left	18.26	Right	38.95	Left	63.50	Right	314.87	Right
T10	111.86	Right	36.85	Right	4.26	Left	12.40	Right	0.56	Left	158.07	Right
T11	958.93	Left	228.53	Left	324.42	Left	64.09	Left	125.00	Left	1702.42	Left
T12	406.18	Right	316.56	Right	31.21	Right	12.71	Right	1.61	Right	772.96	Right
T13	487.42	Left	228.78	Left	83.21	Right	19.95	Right	12.30	Left	624.10	Left
T14	171.12	Left	260.45	Right	399.03	Right	11.22	Right	13.60	Right	512.79	Right
T15	249.14	Right	56.64	Left	224.55	Left	100.91	Right	252.00	Left	183.55	Left
T16	169.92	Right	38.67	Left	18.10	Right	33.49	Right	7.99	Left	175.12	Right
T17	1236.66	Right	163.67	Right	176.09	Left	54.42	Right	60.80	Left	1243.6	Right
T18	1246.09	Right	147.25	Right	111.13	Right	1253.00	Left	170.00	Left	83.17	Right
T19	297.81	Right	430.07	Right	154.58	Right	68.42	Right	8.63	Left	973.12	Right
T20	385.56	Right	4327.56	Left	258.34	Left	87.06	Left	13.40	Right	4323.41	Left
T21	41.91	Left	360.67	Right	148.00	Right	13.52	Left	24.00	Right	475.98	Right
T22	402.96	Right	435.45	Right	163.46	Left	189.35	Right	9.47	Right	18.92	Right
T23	27.50	Left	1.34	Right	9.97	Right	12.89	Right	11.10	Left	6.83	Right
T24	195.91	Right	1.26	Left	6.04	Left	7.42	Left	4.62	Right	166.42	Right
T25	1384.78	Right	17.63	Left	43.93	Right	46.75	Right	4.67	Right	1469.95	Right

Source: Calculated by Das and Saha (this volume)

Table 23.6 Temporal shifting characteristics of left bank line with direction of Hooghly River from Nabadwip to Katwa (1954–2020)

Transects	1954–1980	Direction	1980–1990	Direction	1990–2000	Direction	2000–2010	Direction	2010–2020	Direction	1954–2020	Direction
T1	969.68	Left	87.03	Left	114.86	Right	25.68	Left	120	Right	1077.77	Left
T2	80.68	Right	28.53	Right	4.41	Right	26.51	Right	1.84	Right	141.09	Right
T3	28.89	Right	18.97	Right	8.03	Left	16.51	Left	12	Left	11.18	Right
T4	57.62	Right	136.07	Right	163.99	Right	13.98	Right	18.4	Right	391.72	Right
T5	121.82	Right	139.22	Left	258.22	Left	22.46	Left	4.95	Left	305.45	Left
T6	29.29	Left	220.35	Right	5.06	Left	100.14	Right	43.5	Left	243.27	Right
T7	363.3	Left	42.02	Left	195.28	Left	124.83	Left	217	Right	512.79	Left
T8	146.45	Right	79.49	Left	923.18	Left	312.02	Left	1441	Right	263.95	Right
T9	316.33	Right	204.4	Left	155.17	Left	14.1	Left	323	Right	264.37	Right
T10	100.29	Right	22.17	Right	11.32	Left	8.19	Left	3.41	Left	99.08	Right
T11	1070.47	Left	235.08	Left	188.42	Left	105.71	Left	73.9	Left	1671.89	Left
T12	1028.59	Right	463.59	Right	0.79	Right	18.72	Left	4.79	Left	1480.85	Right
T13	494.9	Left	266.99	Left	167.65	Left	146.99	Left	35.4	Left	1119.75	Left
T14	209.09	Left	271.41	Right	452.92	Right	1.7	Left	7.54	Left	506.46	Right
T15	125.69	Right	15.99	Left	418.46	Left	266.37	Left	6.38	Right	576.42	Left
T16	11.42	Right	128.41	Left	63.94	Right	4.4	Left	20	Left	205.95	Left
T17	1084.39	Right	108.25	Right	144.46	Left	13.25	Left	18.7	Right	1060.87	Right
T18	1262.43	Right	183.56	Right	107.8	Left	583.03	Left	182	Left	150.02	Left
T19	526.31	Left	311.75	Right	132.41	Left	120.28	Left	0.4	Left	546.96	Left
T20	1028.22	Right	4564.33	Left	202.3	Left	16.74	Left	79.7	Left	3730.19	Left
T21	440.94	Left	116.67	Left	954.9	Right	217.5	Left	66.6	Right	102.41	Right
T22	305.83	Right	590.59	Left	18	Right	72.18	Right	1.74	Left	126.72	Left
T23	142.79	Right	9.43	Right	18.36	Left	0.18	Left	50.1	Right	128.77	Right
T24	188.94	Right	200.66	Left	80.03	Left	9.67	Left	123	Left	53.15	Left
T25	1514.46	Right	34.1	Left	11.95	Left	3.15	Left	122	Left	1589.27	Right

Source: Calculated by Das and Saha (this volume)

Table 23.7 Descriptive statistics on temporal changes of channel width (1954–2020)

Transects	Minimum	Maximum	Mean	SD	CV	Skewness	Kurtosis
T1	231.22	708.60	513.63	162.71	0.32	−0.98	1.68
T2	165.17	273.39	218.14	35.23	0.16	0.14	1.58
T3	161.70	373.67	317.89	78.93	0.25	−2.13	4.77
T4	297.53	634.80	416.45	132.77	0.32	0.98	−0.15
T5	203.05	1183.37	846.74	374.27	0.44	−1.22	0.68
T6	327.10	537.04	389.13	78.77	0.20	1.69	2.99
T7	215.84	883.32	572.72	247.10	0.43	−0.34	−1.00
T8	353.76	1970.02	740.17	609.73	0.82	2.32	5.52
T9	247.09	607.92	406.66	156.55	0.38	0.50	−2.11
T10	220.63	282.74	253.89	24.73	0.10	−0.09	−1.50
T11	284.22	442.73	350.90	69.76	0.20	0.69	−1.88
T12	236.75	1067.37	429.84	316.67	0.74	2.31	5.42
T13	356.86	852.62	577.19	232.15	0.40	0.28	−2.62
T14	292.47	361.20	325.91	26.75	0.08	0.25	−1.47
T15	217.81	878.47	441.38	254.48	0.58	1.29	0.58
T16	288.05	691.61	550.77	156.34	0.28	−1.05	0.28
T17	289.40	578.14	480.70	100.35	0.21	−1.74	3.71
T18	317.31	1251.53	581.78	358.44	0.62	1.64	2.80
T19	231.65	1871.17	1302.07	598.68	0.46	−1.30	1.77
T20	260.29	961.43	459.77	256.80	0.56	2.01	4.30
T21	320.73	1273.12	684.15	336.70	0.49	1.06	1.59
T22	242.48	518.70	379.92	98.06	0.26	0.06	−0.47
T23	288.90	484.03	346.52	71.06	0.21	1.91	4.02
T24	398.82	700.33	576.17	138.91	0.24	−0.74	−1.88
T25	475.66	624.46	546.65	65.48	0.12	0.26	−2.39

Source: Calculated by Das and Saha (this volume)

over time. The measurement of river width over defined cross sections is required to detect the strength of lateral erosion to account for this change of the river bank (Table 23.7; Fig. 23.5). The high variability in channel width is observed near Rajnagar (T5), Akandanga (T8), Kashthasali (T19), and Rudrapara (T21) along the Hooghly River (Fig. 23.5). Moreover, maybe due to the erosive agents with high discharge, the lateral erosion has increased.

In 1954–2020, the Hooghly River course between Nabadwip and Katwa has gone through several natural hazards, i.e., flood (devastating flood in 1978 and 2000), bank erosion, channel shifting, land loss, etc. Clear picture of the site-specific bank erosion and associated aftermath result is very crucial in river management practices. On the right bank side, from statistical point of view, greater variability of bank line shifting is observed at T1 (308.20), T8 (543.73), T11 (359.89), T17 (505.33), T18 (606.56), T20 (1857.83), and T25 (606.92) which points out that bank line shifting rates are distributed over an extensive range of scale during the period between 1954 and 2020 (Table 23.8). It is highlighted that the coefficient of variation (CV) is

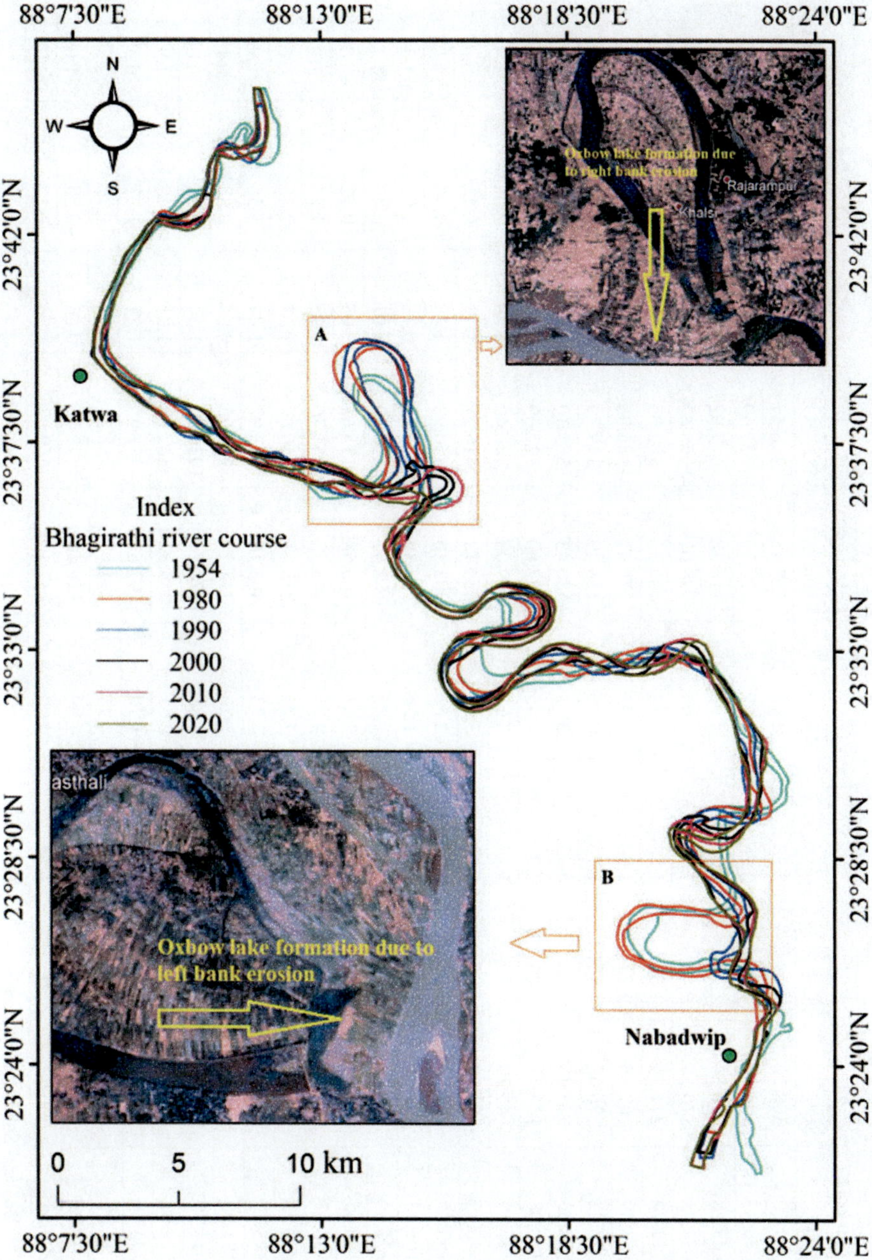

Fig. 23.5 Temporal changes of channel width at various transect points (1954–2020)

Table 23.8 Descriptive statistics of bank line shifting (1954–2020)

Transects	Right bank					Left bank				
	Mean	SD	CV	Skewness	Kurtosis	Mean	SD	CV	Skewness	Kurtosis
T1	184.69	308.20	1.67	2.23	4.98	263.45	396.57	1.51	2.19	4.83
T2	49.85	74.51	1.49	2.15	4.69	28.39	31.69	1.12	1.45	2.29
T3	48.28	90.34	1.87	2.22	4.93	16.88	7.92	0.47	0.78	0.75
T4	54.98	110.58	2.01	2.22	4.94	78.01	68.62	0.88	0.45	−2.61
T5	157.72	211.21	1.34	1.88	3.73	109.33	102.05	0.93	0.60	−0.30
T6	58.89	69.13	1.17	1.94	4.03	79.67	86.06	1.08	1.44	1.82
T7	73.82	86.63	1.17	1.16	0.16	188.49	119.29	0.63	0.48	0.71
T8	532.83	543.73	1.02	0.42	−3.11	580.43	585.10	1.01	0.94	−0.90
T9	110.08	136.71	1.24	2.08	4.45	202.60	127.69	0.63	−0.72	−0.26
T10	33.19	46.20	1.39	1.77	3.07	29.08	40.40	1.39	2.08	4.41
T11	340.19	359.89	1.06	1.83	3.53	334.72	416.27	1.24	2.11	4.53
T12	153.65	192.54	1.25	0.72	−2.61	303.30	450.93	1.49	1.40	1.08
T13	166.33	199.41	1.20	1.37	1.26	222.39	173.13	0.78	1.06	1.44
T14	171.08	166.06	0.97	0.42	−1.37	188.53	190.30	1.01	0.40	−1.11
T15	176.65	91.33	0.52	−0.68	−2.51	166.58	175.65	1.05	0.72	−1.04
T16	53.63	66.14	1.23	2.05	4.35	45.63	51.76	1.13	1.34	1.02
T17	338.33	505.33	1.49	2.17	4.75	273.81	456.66	1.67	2.15	4.69
T18	585.49	606.56	1.04	0.60	−3.33	463.76	483.92	1.04	1.55	1.93
T19	191.90	171.88	0.90	0.54	−1.26	218.23	204.98	0.94	0.88	0.13
T20	1014.38	1857.83	1.83	2.20	4.88	1178.26	1936.26	1.64	2.01	4.06
T21	117.62	146.06	1.24	1.61	2.21	359.32	362.65	1.01	1.48	1.91
T22	240.14	177.71	0.74	−0.08	−1.70	197.67	251.25	1.27	1.23	0.39
T23	12.56	9.46	0.75	0.94	2.25	44.17	58.24	1.32	1.72	2.89
T24	43.05	85.48	1.99	2.23	4.99	120.46	79.12	0.66	−0.50	−1.09
T25	299.55	606.92	2.03	2.23	4.98	337.13	659.83	1.96	2.21	4.90

Source: Calculated by Das and Saha (this volume)
SD, standard deviation, *CV* coefficient of variation

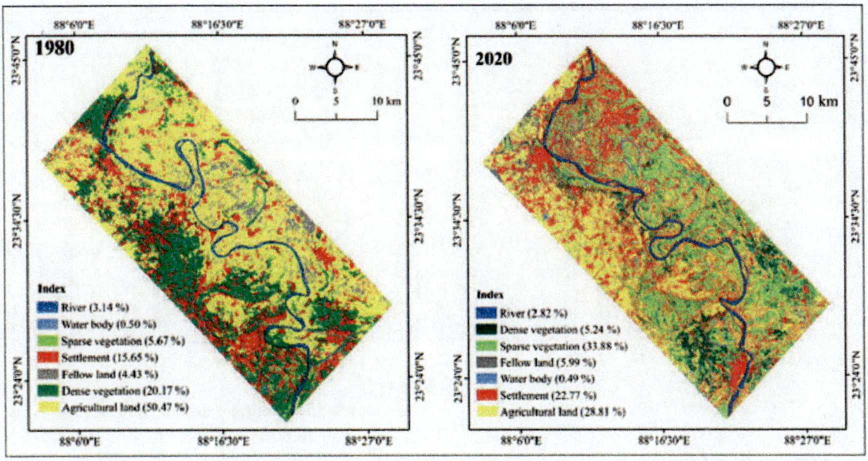

Fig. 23.6 River bank shifting for the period between 1954 and 2020. (**A**) Oxbow lake formation due to extensive right bank erosion at Rajarampur and (**B**) oxbow lake formation due to lateral erosion toward left bank near Purbasthali

comparatively higher in the upper section (e.g., T1, T2, T3, T4, T5) of the study reach which reveals less consistent lateral migration. But most of the cross section sites have experienced a more uniform nature of migration. Moreover, excessive asymmetry (skewness >2.00) of bank line migration has been observed in the entire study area (40%). Contrary to this, for the left bank, extreme variability is calculated at T8 (585.10), T11 (416.27), T12 (450.93), T17 (456.66), T18 (483.92), T20 (1936.26), and T25 (659.83) during the study periods. The coefficient of variation is high in some specific place, i.e., T1, T17, T20, and T25 which indicates more flexible lateral migration. However, higher asymmetry of bank line migration was observed at about 24% referenced site from Nabadwip to Katwa along Hooghly River. Channel cutoff and finally oxbow lakes formation are the landmark of extreme channel migration. In the study area, two oxbow lakes have emerged at Rajarampur and Purbasthali by extensive channel shifting (Fig. 23.6).

The average river width in the study area was 508.36 m. From statistical analysis of the river width, it is clear that greater variability of river width was at T5 (374.27), T8 (609.73), T18 (358.44), and T19 (598.68), which means there was a wide range of variation. Meanwhile, coefficient of variation shows uniform nature of river width variation. Here, most of the sites experienced symmetric form of channel width in order to temporal scales (1954–2020) (Table 23.7).

Land use/land cover (LULC) change over spatiotemporal scale is one of the driving factors of river bank erosion, course change, morphometric change, etc. In this regard, 40 years temporal LULC (1980–2020) in the study is showing various changes which directly or indirectly affect the bank erosion. In 1980, dense vegetation, sparse vegetation, river, water body, agricultural land, fallow land, and settlement covered an area of 20.17%, 5.67%, 3.14%, 0.50%, 50.47%, 4.43%, and 15.65%, respectively (Fig. 23.7). But, in 2020, the dense vegetation was decreased

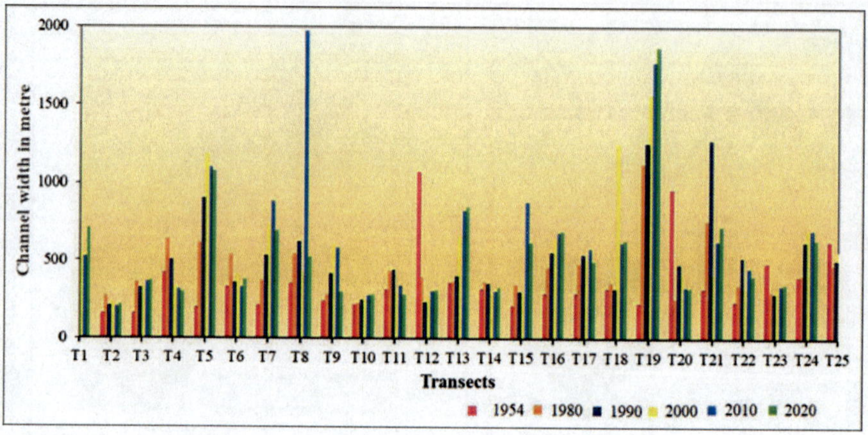

Fig. 23.7 Land use /land cover changes along Hooghly River course (Nabadwip to Katwa) in the period between 1980 and 2020

by an area of 5.24% due to population growth (settlement area 22.77%). River area is gradually decreasing in trend due to human encroachment by practicing agriculture and settlement in *Char* areas. It is notable that vegetation signifies a special alternative to conventional inert material-based preventive measures (Bache & Macaskill, 1981; Polvi et al., 2014; Zanetti et al., 2016). But with flourishing urbanization, vegetation loss along or near the bank has been considered as a bank failure as well as economic loss by destroying the valuable fertile land.

23.5 Bank Erosion Management Plan through GIS Application

Natural hazards like bank erosion are the victims of a sustainable/healthy society. So risk mitigation is the main challenge to diminish the losses, e.g., valuable fertile agricultural land, settlement destruction, river ecology, infrastructural insecurity, and so on. Various initiatives were developed for collaborative mapping of topographic features (spatial attributes of earth surface), also directed to as "crowd sourcing." For example, Open Street Map (OSM) is a free editable map of the whole world, which is made employing collaborative mapping by volunteers/citizens. It authorizes users to collect, view, edit, and use geographical data (spatial changes like bank erosion, land use and land cover alteration, different topographic features, etc.) in a collaborative way from any places any time on the earth (Open Street Map, 2021 https://www.openstreetmap.org/edit#map = 16/22.9865/87.8556). Through this application, we can easily understand the present scenario (during bank erosion) and what makes it in our grip. The Bhagirathi-Hooghly River reach (Nabadwip-Katwa section) passed through eight C.D. blocks (four blocks come under Nadia district and four blocks in

Bardhaman district). The total population was 1,048,699 (Census, 2011a, 2011b), out of which approximately 5% of the population is residing along the river and over the riverine landscape. Therefore, every year they are experiencing river bank erosion and aftermath bad consequences upon rehabilitation, land degradation, destroying the standing crop, etc. So monthly monitoring (depends on bank erosion rate and seasonal behavior) through OSM application can make a good platform to take decisions according to risk magnitude. The Virtual Disaster Viewer is a crowdsourcing tool for collaborative natural disaster impact and damage assessment. Remote sensing experts can be assigned in specific areas of the affected sites to review and provide their assessment by comparing before and after situation using high-resolution satellite images, which became obtainable on Google Earth instantly after the disaster and which served as the groundwork for the collaborative mapping. On the other hand, Internet-based GIS systems have been developed in which all the individual layers are separated (multitier approach), thus allowing many clients to access and visualize the geo-data at the same time (Van Westen, 2013). A WebGIS is a special GIS tool that uses the Internet as means to access and transmit remote data, conduct analysis, and present GIS results. WebGIS applications for risk visualization have been developed for different purposes. Users can visualize, download, or extract data on past hazardous events, human and economic hazard exposure, and risk from natural hazards on a platform compliant with the Open Geospatial Consortium (OGC) Web Services (OWS). Such a collaborative mapping platform and Internet-based GIS applications might become very significant tools in the future.

23.6 Conclusion

This research is primarily depending upon remote sensing and GIS techniques associated with statistical methods corresponding to the spatial approach to degradation of river banks and management structures, with an emphasis on local to focal level operations. Basically, in the field of environmental conservation, hydro-technical studies, and in various areas of environmental protection, the study of river bank erosion has played a very important role in basin management. Bank erosion is the successive collapse of a river banks and is a significant problem in the area of water and land conservation. In this research, 66 years of temporal study provides us with an evolving pattern and trend of bank line shifting in relation to bank erosion, which would be a helpful document for us for improved management practice.

Continuous monitoring of the river bank erosion is a ticklish task, but it gives a sophisticated knowledge toward maintaining the riverine environment. Such kind of study basically has a significant scope in the field of land resource management, agricultural management along riverine space, water and near-bank ecosystem management, river health management (water quality, channel efficiency, etc.), and finally good social practices. Better management practices in river bank erosion can simplify our life in order to minimize the risk.

Acknowledgments The authors heartily appreciate all the secondary sources for literature for information and relevant cognitive ideas to shape up the conceptual framework and all concerned help during the working process. They concomitantly appraise the support of departmental SAP Laboratory of Dept. of Geography and Applied Geography, University of North Bengal, for preprocessing of map works and layout generations. They explicitly want to be thankful to Dr. Pravat Kr. Shit for his encouraging support to carry out the work.

References

Aswathy, M. V., Vijith, H., & Satheesh, R. (2008). Factors influencing the sinuosity of Pannagon River, Kottayam, Kerala, India: An assessment using remote sensing and GIS. *Environmental Monitoring and Assessment, 138*(1-3), 173–180.

Bache, D. H., & Macaskill, I. A. (1981). Vegetation in coastal and stream-bank protection. *Landscape planning, 8*(4), 363–385. https://doi.org/10.1016/0304-3924(81)90003-4

Bandyopadhyay, S., Ghosh, K., & De, S. K. (2014). A proposed method of bank erosion vulnerability zonation and its application on the River Haora, Tripura, India. *Geomorphology, 224*, 111–121. https://doi.org/10.1016/j.geomorph.2014.07.018

Bhatta, B. (2011). *Remote sensing and GIS (second edition)*. Oxford University press.

Biswas, R., & Anwaruzzaman, A. K. M. (2019). Measuring hazard vulnerability by bank erosion of the ganga river in Malda district using PAR model. *Journal of Geography, Environment and Earth Science International, 22*(1), 1–15. https://doi.org/10.9734/JGEESI/2019/v22i130136

Barddhaman District Census Handbook. Census 2011b (https://censusindia.gov.in/).

Chakraborty, S., & Mukhopadhyay, S. (2015). Riverbank erosion and channel width adjustments across a Meandering Channel of North Bengal, India. *Earth Science India, 8*(3), 61.

Chatterjee, S., & Mistri, B. (2013). Impact of river bank erosion on human life: A case study in Shantipur block, Nadia District, West Bengal. *Population, 66*(26.009), 7–17.

Das, M., & Saha, S. (this volume). Spatiotemporal detection and delineation of Bhagirathi-Hooghly river bank erosion using GIS analytics, West Bengal, India. In P. K. Shit, H. R. Pourghasemi, G. S. Bhunia, P. Das, & A. Narsimha (Eds.), *Geospatial technology for environmental hazards: Modeling and management in Asian countries*. Springer.

Das, B., Mondae, M., & Das, A. (2012). Monitoring of bank line erosion of River Ganga, Maida District, and West Bengal: Using RS and GIS compiled with statistical techniques. *International Journal of Geomatics and Geosciences, 3*(1), 239–248.

Das, J. D., Dutta, T., & Saraf, A. K. (2007). Remote sensing and GIS application in change detection of the Barak river channel, NE India. *Journal of the Indian Society of Remote Sensing, 35*(4), 301–312.

Das, J. D., & Saraf, A. K. (2007). Remote sensing in the mapping of the Brahmaputra/Jamuna River channel patterns and its relation to various landforms and tectonic environment. *International Journal of Remote Sensing, 28*(16), 3619–3631. https://doi.org/10.1080/01431160601009664

Das, T. K., Haldar, S. K., Gupta, I. D., & Sen, S. (2014). River bank erosion induced human displacement and its consequences. *Living Review of Landscape Research, 8*(3), 1–35. https://doi.org/10.12942/lrlr-2014-3

Das, T. K., Haldar, S. K., Sarkar, D., Borderon, M., Kienberger, S., Gupta, I. D., & Guha-Sapir, D. (2017). Impact of riverbank erosion: A case study. *Australasian Journal of Disaster and Trauma Studies, 21*(2), 73–81. http://trauma.massey.ac.nz/issues/2017-2/AJDTS_21_2_Das.pdf

Dekaraja, D., & Mahanta, R. (2020). Riverbank Erosion and Migration, Inter-linkage: With Special Focus on Assam, India. https://orcid.org/0000-0002-7810-2846

Dey, S., & Mandal, S. (2019). Assessing channel migration dynamics and vulnerability (1977–2018) of the Torsa River in the Duars and Tal region of eastern Himalayan foothills,

West Bengal, India. *Spatial Information Research, 27*(1), 75–86. https://doi.org/10.1007/s41324-018-0213-z
Gabet, E. J. (1998). Lateral migration and bank erosion in a salt marsh tidal channel in San Francisco Bay, California. *Estuaries, 21*(4), 745–753.
Ghosh, D., & Sahu, A. S. (2018). Problem of river bank failure and the condition of the erosion victims: A case study in Dhulian, West Bengal, India. *Regional Science Inquiry, 10*(2), 205–214.
Ghosh, S., & Mistri, B. (2012). Hydrogeomorphic significance of sinuosity index in relation to river instability: A case study of Damodar River, West Bengal, India. *International Journal of Advances in Earth Sciences, 1*(2), 49–57.
Horacio, J (2014). River sinuosity index: geomorphological characterisation. Technical note 2. *CIREF* and *Wetlands International*, 6 p.
Islam, A., & Guchhait, S. K. (2015). Is severity of River Bank erosion proportional to social vulnerability? A perspective from West Bengal, India. *Life and Living Through Newer Spectrum of Geography, 35*.
Islam, A., & Guchhait, S. K. (2018). Analysis of Social and Psychological Terrain of Bank Erosion Victims: A Study Along the Bhagirathi River, West Bengal, India. *Chinese geographical science, 28*(6). https://doi.org/10.1007/s11769-018-0937-7
Islam, A., & Guchhait, S. K. (2019). Social engineering as shock absorbing mechanism against bank erosion: A study along Bhagirathi River, West Bengal, India. *International Journal of River Basin Management*, 1–14. https://doi.org/10.1080/15715124.2019.1574263
Jovanovic, N., Garcia, C. L., Bugan, R. D., Teich, I., & Rodriguez, C. M. G. (2014). Validation of remotely-sensed evapotranspiration and NDWI using ground measurements at Riverlands, South Africa. *Water SA, 40*(2), 211–220.
Khan, S., Sinha, R., Whitehead, P., Sarkar, S., Jin, L., & Futter, M. N. (2018). Flows and sediment dynamics in the Ganga River under present and future climate scenarios. *Hydrological Sciences Journal, 63*(5), 763–782. https://doi.org/10.1080/02626667.2018.1447113
Konsoer, K. M., Rhoads, B. L., Langendoen, E. J., Best, J. L., Ursic, M. E., Abad, J. D., & Garcia, M. H. (2016). Spatial variability in bank resistance to erosion on a large meandering, mixed bedrock-alluvial river. *Geomorphology, 252*, 80–97. https://doi.org/10.1016/j.geomorph.2015.08.002
Kumar, B. A., Gopinath, G., & Chandran, M. S. (2014). River sinuosity in a humid tropical river basin, south west coast of India. *Arabian Journal of Geosciences, 7*(5), 1763–1772. https://doi.org/10.1007/s12517-013-0864-y
Langat, P. K., Kumar, L., & Koech, R. (2019). Monitoring river channel dynamics using remote sensing and GIS techniques. *Geomorphology, 325*, 92–102. https://doi.org/10.1016/j.geomorph.2018.10.007
Lawler, D. M. (1993). The measurement of river bank erosion and lateral channel change: A review. *Earth Surface Processes and Landforms, 18*(9), 777–821.
Lillesand, T. M., Kiefer, R. W., & Chipman, J. W. (2015). *Remote sensing and image interpretation* (7th ed.). Wiley.
Lovric, N., & Tosic, R. (2016). Assessment of Bank erosion, accretion and channel shifting using remote sensing and GIS: Case study–lower course of the Bosna River. *Quaestiones Geographicae, 35*(1), 81–92. https://doi.org/10.1515/quageo-2016-0008
Mandal, A. C., Patra, P., Majumder, R., Ghosh, D. K., & Bhunia, G. S. (2018). Evaluating meander shifting dynamics (1977–2017) of the Bhagirathi river course in Murshidabad District, West Bengal, India. *Spatial Information Research, 26*(1), 33–45. https://doi.org/10.1007/s41324-017-0153-z
Mandal, S. (2017). Assessing the instability and shifting character of the river bank Ganga in Manikchak Diara of Malda district, West Bengal using bank erosion hazard index (BEHI), RS & GIS. *European Journal of Geography, 8*(4), 6–25.

Maurya, S. P., & Yadav, A. K. (2016). Evaluations of course change detection of Ramganga river using remote sensing and GIS, India. *Weather and Climate Extremes, 13*, 68–72. https://doi.org/10.1016/j.wace.2016.08.001

Momin, H., Biswas, R., & Tamang, C. (2020). Morphological analysis and channel shifting of the Fulahar river in Malda district, West Bengal, India using remote sensing and GIS techniques. *Geo Journal*, 1–17. https://doi.org/10.1007/s10708-020-10248-7

Mondal, J., Debanshi, S., & Mandal, S. (2014). Dynamicity of the river Ganga and Bank erosion induced land loss in Manikchak Diara of Malda district of West Bengal, India: A RS and GIS based Geo-spatial approach. *International Journal of Applied Remote Sensing GIS, 3*(1), 43–56.

Mueller, J. E. (1968). An introduction to the hydraulic and topographic sinuosity indexes. *Annals of the Association of American Geographers, 58*(2), 371–385.

Mukherjee, R., Bilas, R., Biswas, S. S., & Pal, R. (2017). Bank erosion and accretion dynamics explored by GIS techniques in lower Ramganga River, Western Uttar Pradesh, India. *Spatial Information Research, 25*(1), 23–38. https://doi.org/10.1007/s41324-016-0074-2

Muller, E., Décamps, H., & DOBSON, M. K. (1993). Contribution of space remote sensing to river studies. *Freshwater Biology, 29*(2), 301–312.

Nadia District Census Handbook. Census 2011a (https://censusindia.gov.in/).

Nijhawan, R., and Jain, K. (2018). Glacier terminus position monitoring and modelling using remote sensing data. In International Conference on Advances in Computing and Data Sciences (pp. 11-23). Springer, Singapore. https://doi.org/10.1007/978-981-13-1813-9_2.

Ophra, S. J., Begum, S., Islam, R., & Islam, M. N. (2018). Assessment of bank erosion and channel shifting of Padma River in Bangladesh using RS and GIS techniques. *Spatial Information Research, 26*(6), 599–605. https://doi.org/10.1007/s41324-018-0202-2

Parua, P. K. (1999). Erosion problems of the river ganga in the districts of Malda and Murshidabad in West Bengal. *Civil Engineering Today, ASCE: Calcutta, 13*(2), 3–20.

Pati, J. K., Lal, J., Prakash, K., & Bhusan, R. (2008). Spatio-temporal shift of western bank of the Ganga river, Allahabad city and its implications. *Journal of the Indian Society of Remote Sensing, 36*(3), 289–297.

Polvi, L. E., Wohl, E., & Merritt, D. M. (2014). Modeling the functional influence of vegetation type on streambank cohesion. *Earth Surface Processes and Landforms, 39*(9), 1245–1258. https://doi.org/10.1002/esp.3577

Roy, S., Barman, K., Das, V. K., Debnath, K., & Mazumder, B. S. (2020). Experimental investigation of undercut mechanisms of River Bank erosion based on 3D turbulence characteristics. *Environmental Processes, 7*(1), 341–366. https://doi.org/10.1007/s40710-019-00417-3

Rudra, K. (2010). Dynamics of the Ganga in West Bengal, India (1764–2007): Implications for science–policy interaction. *Quaternary International, 227*(2), 161–169. https://doi.org/10.1016/j.quaint.2009.10.043

Sarma, A. (2014). Landscape degradation of river island Majuli, Assam (India) due to flood and erosion by river Brahmaputra and its restoration. *Journal of Medical and Bioengineering, 3*, 4. https://doi.org/10.12720/jomb.3.4.272-276

Sarma, D. I. P. I. M. A. (2013). Rural risk assessment due to flooding and riverbank erosion in Majuli, Assam, India. University of Twente Faculty of Geo-Information and Earth Observation (ITC).

Schumm, S. A., Schumm, S. A., Dumont, J. F., & Holbrook, J. M. (2002). *Active tectonics and alluvial rivers*. Cambridge University Press.

Singh, L. P., Parkash, B., & Singhvi, A. K. (1998). Evolution of the lower Gangetic plain landforms and soils in West Bengal, India. *Catena, 33*(2), 75–104. https://doi.org/10.1016/S0341-8162(98)00066-6

Sinha, R., Tandon, S. K., Gibling, M. R., Bhattacharjee, P. S., & Dasgupta, A. S. (2005). Late quaternary geology and alluvial stratigraphy of the Ganga basin. *Himalayan Geology, 26*(1), 223–240.

Thakur, P. K., Laha, C., & Aggarwal, S. P. (2012). River bank erosion hazard study of river Ganga, upstream of Farakka barrage using remote sensing and GIS. *Natural Hazards, 61*(3), 967–987. https://doi.org/10.1007/s11069-011-9944-z

Tiwari, H., Rai, S. P., & Shivangi, K. (2016). Bridging the gap or broadening the problem? *Natural Hazards, 84*(1), 351–366. https://doi.org/10.1007/s11069-016-2422-x

Van Westen, C. J. (2013). Remote sensing and GIS for natural hazards assessment and disaster risk management. *Treatise on Geomorphology, 3*, 259–298.

Vijith, H., & Dodge-Wan, D. (2018). Morphology and channel characteristics of an equatorial tropical river in Malaysian Borneo: A detailed evaluation through spatially explicit geomorphometric modelling. *Model Earth System Environment, 4*, 325–337. https://doi.org/10.1007/s40808-017-0407-8

Wilkinson, S. N., Prosser, I. P., & Hughes, A. O. (2006). Predicting the distribution of bed material accumulation using river network sediment budgets. *Water Resources Research, 42*, 10. https://doi.org/10.1029/2006WR004958

Wu, B., Wang, G., Xia, J., Fu, X., & Zhang, Y. (2008). Response of bankful discharge to discharge and sediment load in the lower Yellow River. *Geomorphology, 100*(3-4), 366–376. https://doi.org/10.1016/j.geomorph.2008.01.007

Xia, J., Wu, B., Wang, G., & Wang, Y. (2010). Estimation of bankful discharge in the lower Yellow River using different approaches. *Geomorphology, 117*(1-2), 66–77. https://doi.org/10.1016/j.geomorph.2009.11.007

Xu, H. (2006). Modification of normalised difference water index (NDWI) to enhance open water features in remotely sensed imagery. *International Journal of Remote Sensing, 27*(14), 3025–3033. https://doi.org/10.1080/01431160600589179

Zanetti, C., Macia, J., Liency, N., Vennetier, M., Mériaux, P., & Provensal, M. (2016). Roles of the riparian vegetation: the antagonism between flooding risk and the protection of environment. *3rd European Conference on Flood Risk Management*. https://doi.org/10.1051/e3sconf/20160713015.

Chapter 24
Landslide Susceptibility Mapping in Gangtok, Sikkim Himalaya

Arnab Sengupta and Sankar Kumar Nath

Abstract Landslides and other mass movements are serious geo-environmental hazards in the Greater Gangtok region. It is intensified manifolded by human interferences in terrain. Massive landslides kill thousands of people with catastrophic damages in this region. Therefore, in the present study, a methodology for landslide susceptibility mapping has been developed through an integrated GIS technique. Thus, four major causative factors, i.e. morphometric, geological, environmental and triggering for landslides, have been prepared on the GIS platform. A frequency ratio technique has been used for landslide susceptibility zonation mapping. Thereafter, the entire Sikkim Himalaya has been categorized into five landslide susceptible classes, viz. low, moderate, high, very high and severe. The susceptibility map has been validated by correlating with the landslide inventory, and it is found that more than 80% of the landslide has occurred in the severe to high susceptibility zone. Also, it is observed that the major urban clusters like Chandmari, Bhurtuk, Tadong and the northwestern part of the terrain are most vulnerable to the landslide.

Keywords Landslide Susceptibility Zones · Greater Gangtok · Landslide Inventory

24.1 Introduction

Landslide, mass movements and slope instability are a common and serious geo-environmental problem in highland ecosystems, particularly in seismically active regions like the Himalayas. In fact, wherever the mountain slopes are steep, there is always a possibility of the occurrence of disastrous landslides due to gravity sliding of saturated rock debris. Widespread landsliding is triggered there by earthquakes in the Sikkim Himalaya region. The rapid movement of a mass of rock, debris or earth down a slope separated from stationary part of highlands produces

A. Sengupta (✉) · S. K. Nath
Department of Geology and Geophysics, Indian Institute of Technology Kharagpur, Kharagpur, India

P. K. Shit et al. (eds.), *Geospatial Technology for Environmental Hazards*, Advances in Geographic Information Science, https://doi.org/10.1007/978-3-030-75197-5_24

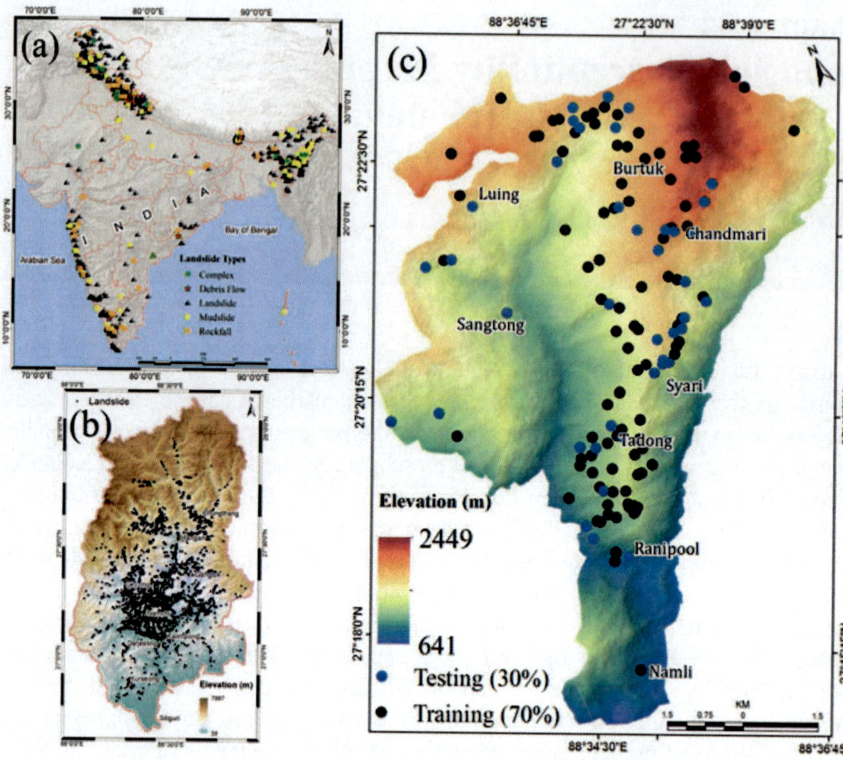

Fig. 24.1 Location map of the study region. (**a**) Landslide inventory of Indian peninsula, (**b**) Darjeeling-Sikkim Himalaya overlaid with landslide inventory, (**c**) Gangtok, the capital city of Sikkim-Himalaya, located in East Sikkim district. The black circle represents the training (70%) landslide dataset, and the blue circle represents the testing (30%) landslide dataset

disastrous effects on both the natural environment and man-made structures, weakening infrastructural facilities, rendering people homeless and also disrupting the productive bases. Although landslide is a local phenomenon, the loss of life and property due to its impact is particularly seen more frequently in recent times. Figure 24.1(b) depicts a landslide inventory map of the Sikkim Himalaya containing about 1200 landslides.

Gangtok, the state capital of Sikkim, is situated within a structurally complex and perennially landslide-prone fold-thrust belt in Sikkim Himalaya with an elevation of 678–2482 m from mean sea level, as is depicted in Fig. 24.1(c). The city has a demographic growth of 82% with 98,658 inhabitants (Chandramouli & General, 2011), compared to 2001 which contains only 43,711 inhabitants and is characterized by steep to moderate topographic gradient, fragile geologic base, active tectonics, accelerated erosion, rapid channel changes, widespread deforestation and intense land use pressure. The region was exaggerated by several significant active/palaeo-landslides illustrated in Table 24.1, among which the most disastrous causing innumerable destruction are situated at the ninth Mile slide, a perennial landsliding

Table 24.1 Inventory of major landslide in and around Gangtok that occurred during 1957–2007 (Sharma, 2008)

Slide name	Lithology	Nature of mass movement	Causative factors	Current status
9th Mile (1957 to present)	Biotite schist and highly fractured slates	Rotational slide, circular failure with subsidence	Debris/soil slump due to pore pressure increase	Active slide
Ranipool (24th September 2005)	Phyllitic quartzites	Debris slide	High-intensity rainfall	Palaeo-slide
Samdur (24th September 2005)	Highly weathered schist			
6th Mile (1997 to present)	Schist/phyllite			Active slide
Amdo Golai (13th June 2007)	Schist/gneiss	Rockslide	Vast boulder broke loose following heavy rains	
Tathangchen (ninth June 1997)	Schist	Composite schistosity controlled in the upper part; slumping in the lower part	High-intensity rainfall	Palaeo-slide
Ganesh Tok (12th July 2007)	Schist/granite	Debris slide		
Chandmari (1960 to present)	Gneiss			Active slide
Bakthang (sixth July 2005)	Schist		Heavy rain, presence of highly weathered realize material and water seepage	Active slide
Majadhari	Gneiss boulders, no orientation		Perennial water flow, steep slope, road cutting	Palaeo-slide
Panchmil I	Hornblende gneiss with mica flakes	Rock-cum-debris slide	Weathering of micaceous material, steep slope	Active slide
Panchmil II	Fine-grained hornblende gneiss	Rockslide (planar failure)	Weathering of gneiss, toe erosion, unfavourable discontinuities	
Panchmil III	Quartzite and gneiss	Rock-cum-debris slide	Gully erosion debris slide due to high pore pressure	
Lumsey Basty (1995)	Schist	Debris slide	High-intensity rainfall	Palaeo-slide

(continued)

Table 24.1 (continued)

Slide name	Lithology	Nature of mass movement	Causative factors	Current status
Tadong (eighth June 1997)	Schist/ phyllite			
Syari (16th September 1990)	Schist/ phyllite/ gneiss			
Upper Sichey (1997)	Schist/ quartzite			
5th Mile (third June 2005)	Schist/ phyllite			
Arithang (2005)	Schist/ gneiss			

zone with recurrence for over 50 years undergoing differential subsidence and sinking both across and along the slope since 1957. The entire distressed zone is covered by overburden material with a few exposures of highly weathered and decomposed schistose rock mass and is characterized by debris slide in which the toe of the slide extends up to the Rongni Chu River. The 6th Mile sinking zone is located in southern Gangtok, tranquiled with steep to moderate slope characterized by debris slide under the metamorphic terrain within the lower grade, and is composed of organic and red soil, fractured saturated zone in the weathered rock mass and sliding surface discontinuity. Chandmari slide is a palaeo-landsliding zone in the eastern part of Gangtok and is experiencing both rainfall and earthquake-induced landslides since 1960 in which the slope movement involves a combination of earth slide and debris flow as the maximum topsoil in the area is comprised of medium-grained sandy soil which is mixed with boulders and weathered mica gneiss.

Landslide susceptibility mapping (LSM) is a continuously progressing part of the research and a broad review of susceptibility models. The consistency of LSM depends mostly on the availability of spatial/aspatial data, scale and the selection of the appropriate methodology of analysis. Comprehensive studies have been conducted in and around the Greater Gangtok region for landslide susceptibility mapping during the last decades using a heuristic, deterministic, statistical and data-driven techniques by Kaur et al. (2018), Mandal and Mandal (2018), Kaur et al. (2019), Sarkar et al. (2008), Gupta and Shukla (2018), Bhasin et al. (2002) and Sharma et al. (2011). Thus, in the present study, our main motivation is to classify the entire terrain into various landslide susceptibility zones by using the frequency ratio technique on GIS environment for the appropriate mitigational measures in highly vulnerable areas in the view of the socio-economic aspect and will also bridge the gap between the contemporary levels of landslide experts.

24.2 Data Acquisition and Methodology

There are various interrelated landslide causative factors, based on the data availability listed in Table 24.2. In the present study, elevation, slope, Terrain Ruggedness Index, aspect, drainage density and Topographic Wetness Index are morphometric causative factors, while surface geology, geomorphology, hydrologic soil group, landform and lineament density are used as geological causative factors, whereas land use/land cover, Normalized Differences Vegetation Index (NDVI) and road density thematic layers are used as an environmental factors; however, average annual rainfall, surface-level peak ground acceleration (PGA) and epicentre proximity are used as a landslide triggering factors.

Frequency Ratio (Probability Approach)

The framework of the methodology embraced in the present study is depicted in Fig. 24.2 which includes data acquisition and causative factors generation; thereafter, all the causative factors have been converted to a raster grid of 12.5 m × 12.5 m cell size and integrated through a probabilistic approach using frequency ratio method on GIS environment.

The frequency ratio model is a probabilistic approach that is based on the pragmatic liaison between the landslide distribution and each landslide-dependent factors. It is the ratio between the landslides in the class as a percentage of all

Table 24.2 List of spatial/aspatial data used in the present study

Spatial/aspatial data	Data format	Source
Landslide inventory	Vector	Hazard, risk and vulnerability analysis (HRVA)—Gangtok (Sharma, 2008), multispectral satellite imagery, Google Earth imagery and field investigation
ALOS PALSAR DEM	Raster	Japan Aerospace Exploration Agency
LISS-III	Raster	National Remote Sensing Centre (NRSC)—ISRO
Geology map	Vector	Geological Survey of India
Soil map	Raster	National Bureau of Soil Survey and Land Use Planning (NBSS&LUP)
Geomorphology and lineament map	Raster and vector	Dasgupta et al. (2000), National Mission on Geomorphological and Lineament Mapping, http://bhuvan.nrsc.gov.in/gis/thematic/index.php
Road map	Raster and vector	Open Street Map and Google Earth
Rainfall data	Raster	Hazard, risk and vulnerability analysis (HRVA)—Gangtok
PGA map	Vector	Adhikari and Nath (2016)
Earthquake catalogue	Vector	Nath et al. (2017), USGS, IMD and ISC

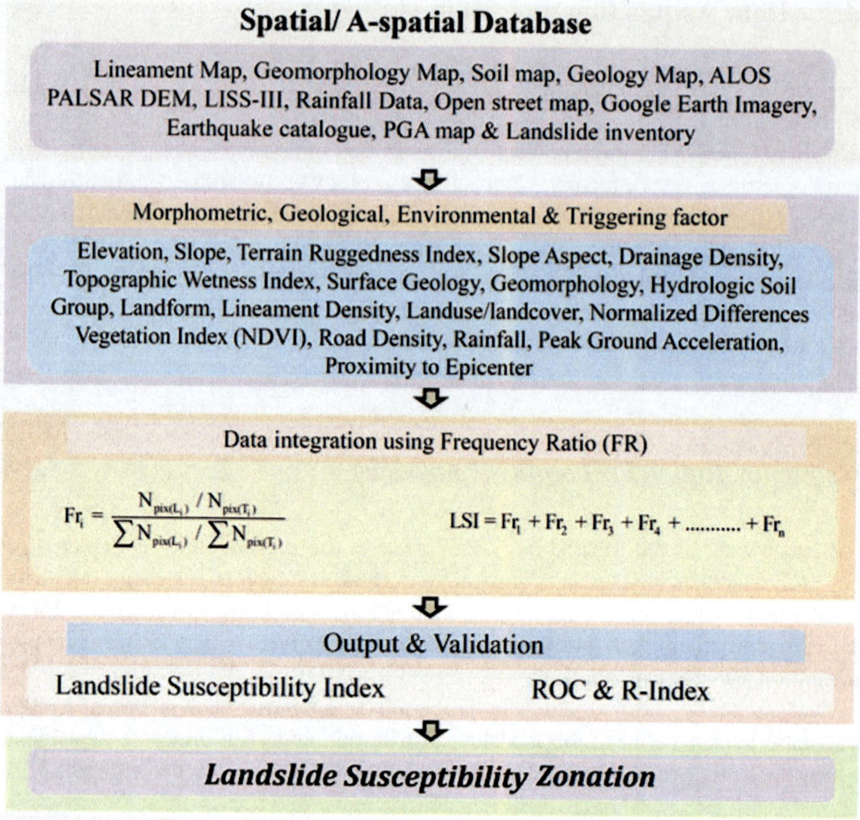

Fig. 24.2 Computational protocol of an integrated approach for landslide susceptibility zonation

landslides and the area of the class as a percentage of the entire map (Lee & Pradhan, 2007). When evaluating the probability of a landslide within a specific period and within a certain area, it is of major importance to recognize the conditions that can cause the landslide and the process that could trigger the movement (Shahabi et al., 2014; Yalcin et al., 2011). The correlation between the landslide areas and the associated factors can be allocated from the connections between the areas without past landslides and the landslide-related parameters (Shahabi et al., 2014). The frequency ratio values for each class of every causative factor have been calculated by using the following equation (Karim et al., 2011; Lee & Talib, 2005; Lee & Pradhan, 2007; Shahabi et al., 2014):

$$\mathrm{Fr_i} = \frac{\mathrm{N_{pix(L_i)}}/\mathrm{N_{pix(T_i)}}}{\sum \mathrm{N_{pix(L_i)}}/\sum \mathrm{N_{pix(T_i)}}} \quad (1)$$

where $\mathrm{N_{pix(L_i)}}$ is the number of pixels containing landslide in class (i), $\mathrm{N_{pix(T_i)}}$ is the total number of pixels having class (i) in the whole area, $\Sigma \mathrm{N_{pix(L_i)}}$ is the total number of pixels containing landslide and $\Sigma N_{pix(T_i)}$ is the total number of pixels in the whole area.

The frequency ratio of each factor type or range has been calculated from their relationship with landslide events. The frequency ratio has been calculated for the sub-criteria of parameter, and then the frequency ratio has been summed up to calculate the Landslide Susceptibility Index (LSI) (Lee & Sambath, 2006):

$$LSI = Fr_1 + Fr_2 + Fr_3 + Fr_4 + \ldots\ldots\ldots\ldots + Fr_n \quad (2)$$

where Fr is the rating of each factor type or range and n is the number of factors. According to the frequency ratio method, the ratio is that of the area where a landslide has occurred to the total area, so a value of 1 is an average value. A value greater than 1 indicates a higher correlation because the percentage of landslide is higher than the area, whereas a value less than 1 indicates a lower correlation (Shahabi et al., 2014).

Accuracy Assessment

Receiver operating characteristic (ROC) curve is a technique for visualizing, organizing and selecting classifiers based on their performance. It is defined as a plot of test sensitivity or true positive rate (TPR) as the y-coordinate versus its 1-specificity or false positive rate (FPR) as x-coordinate at various threshold settings which is an effective method for evaluating the performance of dichotomy problems (Park et al., 2004; Swets et al., 2000). It is widely used in the validation of landslide susceptibility maps and also for estimating the accuracy of GIS-based statistical models. The area under the ROC curve known as AUC is a common metric that can be used to compare different tests, and the values ranging from 0.5 to 1 will be widely employed to estimate the accuracy of the presence or absence of predictive models. An AUC close to 0.5 corresponds to a poor diagnostic test, and larger AUC is more accurate in the test. Additionally, the accuracy of the resulting LSI was evaluated using the index of relative landslide density (R-index).

24.3 Results and Discussions

Mapping of Landslide Influencing Factor

A landslide inventory is an important attribute for landslide susceptibility zonation, and also a prior knowledge is required for the miscellany of thematic layers (Guzzetti et al., 1999). However, we have used 70% of the landslide inventory as a training database for the calculation of landslide frequency of each thematic layer attribute

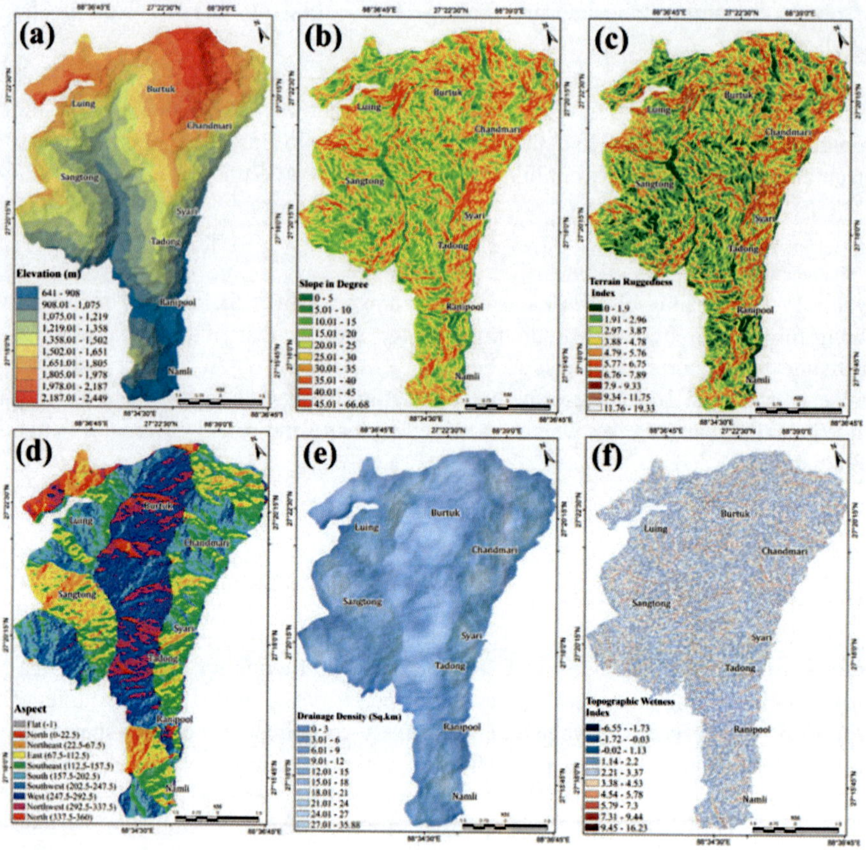

Fig. 24.3 Causative factors map used in the present study. (**a**) Elevation, (**b**) Slope, (**c**) Terrain Ruggedness Index, (**d**) Aspect, (**e**) Drainage Density and (**f**) Topographic Wetness Index

for different causative factors and the remaining 30% for the validation purpose. In the present study, based on the data arability, we have selected the landslide causative factors based on morphometric (as depicted in Fig. 24.3), geological (as shown in Fig. 24.4), environmental (as portrayed in Fig. 24.5) and triggering (as represented in Fig. 24.6) aspect as illustrated in Table 24.3.

Landslide Susceptibility Zonation Mapping Using Frequency Ratio

The results obtained through the frequency ratio method for each thematic attributes are exemplified in Table 24.4. It is observed that the morphological facet slope of 40.01–45° has the maximum weightage of frequency ratio (3.01). Therefore,

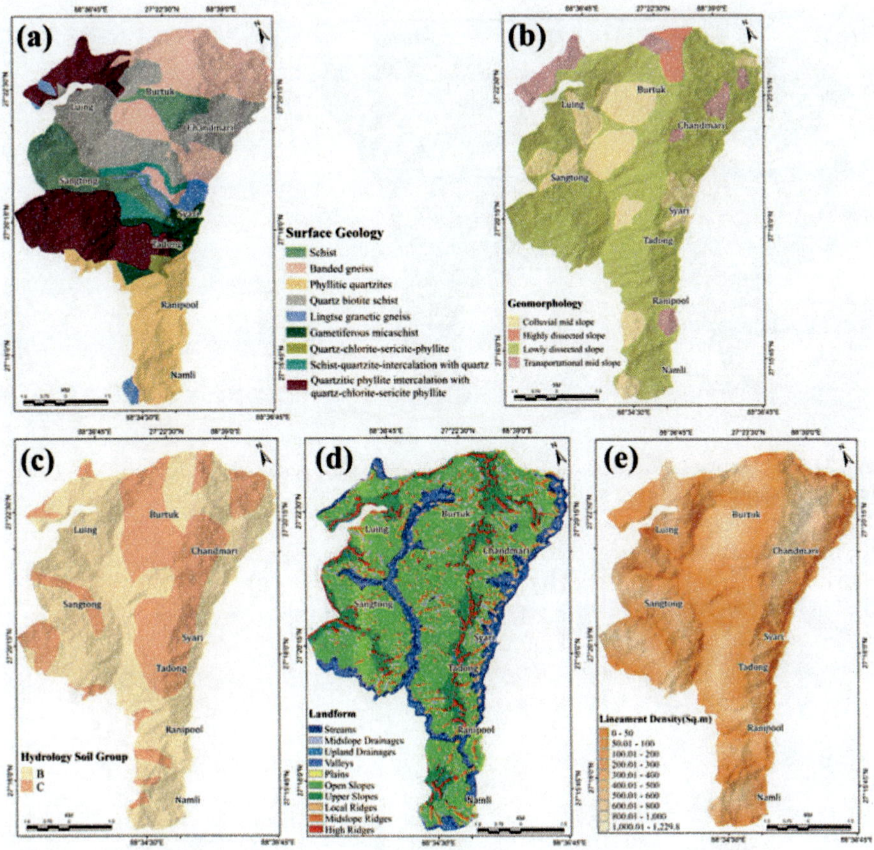

Fig. 24.4 Causative factors map used in the present study. (**a**) Surface Geology, (**b**) Geomorphology, (**c**) Hydrology Soil Group, (**d**) Landform and (**e**) Lineament Density

according to Mohammady et al. (2012), the shear stress in the soil or other unconsolidated material usually increases as the slope angle increases, whereas the gentle slopes are concomitant with low shear stress and are anticipated to have low landslide frequency. On the other hand, in terms, the steepest direction of slope, the frequency ratio is higher towards southwest (2.15) and flat in north and northeast (0.00) as most of the landslides have occurred by facing towards southwest and southeast direction. As the rivers of Gangtok continuously saturate the lower portion of the material and erode the slope, it therefore affects the stability. Thus, the evaluation in terms of drainage density highlights that the class of 12.01–15 and 18.01–21 km^2 with frequency ratio weight of (1.79) and (1.55), respectively, have high correlation for the occurrence of landslide. Although the state is the quarry of landslides during monsoon, therefore in context of rainfall it is witnessed that the frequency ratio weight is the highest (1.25) in 3400 mm of rainfall. In terms of geological aspect, it is realized that the weightage of frequency ratio is the highest

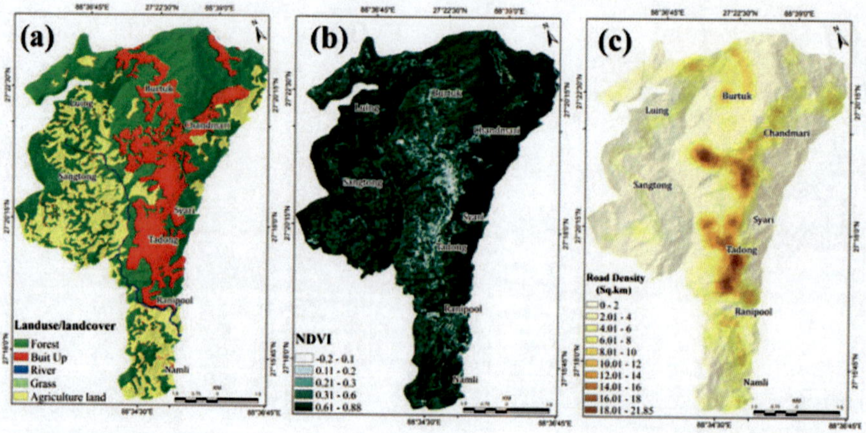

Fig. 24.5 Causative factors map used in the present study. (**a**) Landuse/landcover, (**b**) NDVI and (**c**) Road Density

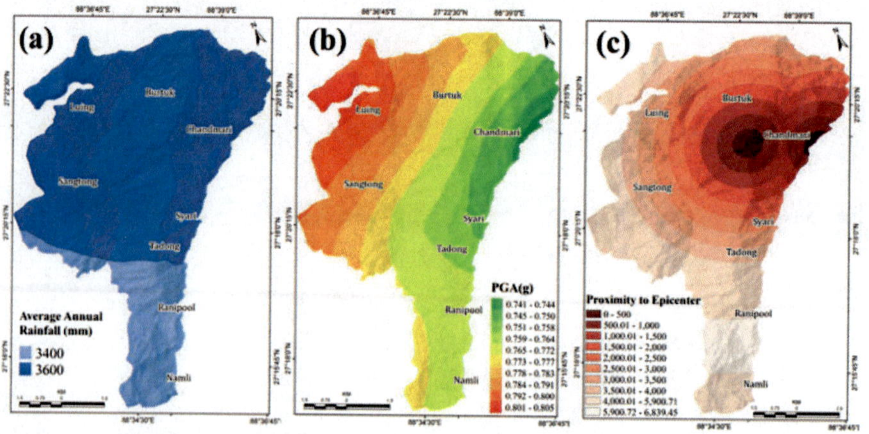

Fig. 24.6 Causative factors map used in the present study. (**a**) Average Annual Rainfall, (**b**) PGA and (**c**) Proximity to the Epicentre

(2.17) in Lingtse granitic gneiss, (1.44) in quartz-chlorite-sericite-phyllite and (1.16) in schist for the event of landslides, respectively. In case of lineament density, the class of 0–50 km^2 has a high frequency ratio weight (2.21) which shows the existence of various discontinuities and lineaments for the occurrence of landslide and the probability of slope failure. However, in the case of hydrology soil group, it is observed that frequency ratio weight is highest (1.22) in HSG-C, (0.87) in HSG-B, which indicates that low infiltration rates when thoroughly wet and consists chiefly of soils with a layer that impedes downward movement of water and soils with moderately fine to fine structure will be a brink of slope failure and mass movement. Lastly from the environmental point of view, it is observed that in context of various land use/land cover classes, the frequency ratio weight is highest (2.10) in built up,

Table 24.3 Importance of different landslide causative factors used in the present study

Causative factor	Importance of causative factor
Elevation	Elevation or altitude is one of the morphometric causative factors for landslide occurrences
Slope angle	Slope failure is a very significant issue for the landslide occurrence in the hilly terrain and is associated with the slope movement due to the gravitational forces (Catani et al., 2005)
Terrain ruggedness index (TRI)	TRI is defined as elevation variation between neighbouring cells of a digital elevation model
Slope aspect	Aspect which identifies the steepest downslope across a surface and plays an augmented role for the event of landslide
Drainage density	The propinquity of steep slope to drainage network is an additional key element for the stability analysis, and drainage network has been chosen in order to simultaneously contemplate the undercutting of hydrographic system for the role of inappropriate drainage (Shahabi et al., 2012)
Topographic wetness index	Topographic Wetness Index is an important morphometric factor within the surface run-off model (Beven & Kirkby, 1979)
Surface geology	The surface geological features have been considered as an independent variable because there might be a variance in permeability of rocks and soils which leads to a variation of lithology and structure (Ayalew & Yamagishi, 2005)
Geomorphology	Geomorphology and landform are the spatial topological interactions of landforms which involve segregating the terrain into intangible spatial objects such as chronology, composition and structure
Landform	
Hydrologic soil group	Hydrologic soil group (HSG)-B has a moderate infiltration rate when thoroughly wet and consists chiefly of moderately deep to deep, moderately well to well-drained soils with moderately fine to moderately coarse textures. HSG-C has low infiltration rates when thoroughly wet and consists chiefly of soils with a layer that impedes downward movement of water and soils with moderately fine to fine structure
Lineament density	The lineament is also an important factor for the landslide occurrence as the degree of fracturing and weathering plays a crucial role for slope failure (Bui et al., 2012)
Land use/land cover	Land use/land cover is one of the most important environment factors to govern slope stability as it controls the rate of weathering and erosion
Normalized differences vegetation index (NDVI)	NDVI can also be used as a virtuous gage for the probability of mass movement because the value of NDVI is higher in landslide-prone areas where there is a dense vegetation cover such as the areas with heavy rainfall and the soil with drenched condition (Vakhshoori & Zare, 2016)
Road density	The road segmentation is basically a significant spot of anthropogenic instability, and the numerous road construction activities such as quarrying of soil, striking of additional load, vertical segmentation of slope, dam construction and vegetation removal may lead to a slope failure which frequently

(continued)

Table 24.3 (continued)

Causative factor	Importance of causative factor
	serves as a cradle for the occurrence of landslide (Saadatkhah et al., 2014)
Rainfall	The rainfall, peak ground acceleration and proximity to epicentre are deliberated as a dynamic/triggering and eliciting factor for the mass movement
Peak ground acceleration	
Proximity to epicentre	

(0.00) in river and grass, which is due to the various anthropogenic activities and unplanned urbanization. In case of vegetation indices (NDVI), the frequency ratio weight is the highest (2.54) in the class 0.21–0.3 because the soil is in drenched condition with dense vegetation cover (Vakhshoori & Zare, 2016), and for road density, the weightage of frequency ratio is the highest, (4.80) and (1.85), between 10.01–12 and 6.01–8 km^2, respectively.

After that, the spatial distribution of Landslide Susceptibility Index (LSI) is adopted from the frequency ratio method and has been classified into five susceptible zones including low, moderate, high, very high and severe as depicted in Fig. 24.7(a). It is observed that around 7.77 and 7.90 km^2 of the total area fall under severe and very high zones, whereas 8.56, 10.00 and 20.15 km^2 of the total area fall under high, moderate and low susceptible zones, respectively, but the lower part of Greater Gangtok is covered under low susceptible zone. The susceptibility area vs. the number of landslide plot exhibits that high to severe landslide susceptibility zones is more prone to landslide, as depicted in Fig. 24.7(b). The R-index values for the severe to high susceptibility class are FR is 0.16%, 0.15% and 0.09% respectively as shown in Fig. 24.7(c). The result obtained from ROC indicates that the severe zone is more susceptible to landslide as compared to very high, high, moderate and low zones as depicted in Fig. 24.7(d). The estimated AUC of 0.751 suggests that the landslide susceptibility map obtained through the frequency ratio model is in strong agreement with the inventory and could be used for various mitigational measures.

24.4 Conclusion

The landslide susceptibility zonation is very essential to delineate the landslide-prone region in the Greater Gangtok. Based on the above studies, a landslide susceptibility zonation map has been prepared here by using the frequency ratio technique. The entire terrain has been divided into five landslide susceptibility zones such as low, moderate, high, very high and severe. The present study is an attempt towards the application of GIS for landslide susceptibility mapping based on morphometric, geological, environmental and triggering causative factors. Application of GIS involves the generation of thematic data layers and their spatial relation to determine the numerical weights of the factors in order of their influence on landslide occurrences. The zonation map is validated by determining landslide inventory data

Table 24.4 Frequency ratio for the causative factors of landslide

Factor	Attribute	No. pixels in domain	% of the pixel in the domain	No. of landslides in the class	% of landslides in the class	Frequency ratio
Slope (degree)	0–5	0	0.0000	5134	1.4714	0.00
	5.01–10	0	0.0000	13,562	3.8868	0.00
	10.01–15	156.25	3.8462	34,932	10.0112	0.38
	15.01–20	781.25	19.2308	65,879	18.8804	1.02
	20.01–25	625	15.3846	72,922	20.8989	0.74
	25.01–30	156.25	3.8462	64,285	18.4236	0.21
	30.01–35	781.25	19.2308	43,852	12.5676	1.53
	35.01–40	781.25	19.2308	25,953	7.4379	2.59
	40.01–45	468.75	11.5385	13,384	3.8357	3.01
	45.01–66.68	312.5	7.6923	9025	2.5865	2.97
Drainage density (km^2)	0–3	0	0.0000	4438	1.2719	0.00
	3.01–6	156.25	3.8462	11,505	3.2972	1.17
	6.01–9	156.25	3.8462	23,763	6.8103	0.56
	9.01–12	156.25	3.8462	50,701	14.5305	0.26
	12.01–15	1406.25	34.6154	67,446	19.3295	1.79
	15.01–18	937.5	23.0769	60,820	17.4305	1.32
	18.01–21	937.5	23.0769	52,020	14.9085	1.55
	21.01–24	156.25	3.8462	40,028	11.4717	0.34
	24.01–27	0	0.0000	23,889	6.8464	0.00
	27.01–35.88	156.25	3.8462	14,318	4.1034	0.94
Slope aspect	Flat (−1)	0	0.0000	1348	0.3863	0.00
	North (0–22.5)	0	0.0000	6806	1.9505	0.00
	Northeast (22.5–67.5)	0	0.0000	21,079	6.0411	0.00
	East (67.5–112.5)	156.25	3.8462	50,391	14.4417	0.27

(continued)

Table 24.4 (continued)

Factor	Attribute	No. pixels in domain	% of the pixel in the domain	No. of landslides in the class	% of landslides in the class	Frequency ratio
	Southeast (112.5–157.5)	937.5	23.0769	66,200	18.9724	1.22
	South (157.5–202.5)	781.25	19.2308	57,659	16.5246	1.16
	Southwest (202.5–247.5)	1250	30.7692	49,829	14.2806	2.15
	West (247.5–292.5)	625	15.3846	55,045	15.7755	0.98
	Northwest (292.5–337.5)	312.5	7.6923	33,923	9.7221	0.79
	North (337.5–360)	0	0.0000	6648	1.9053	0.00
Topographic Wetness Index	−6.55–−1.73	0	0.0000	3975	1.1392	0.00
	−1.72–−0.03	312.5	7.6923	17,004	4.8732	1.58
	−0.02–1.13	0	0.0000	27,105	7.7681	0.00
	1.14–2.2	625	15.3846	32,176	9.2214	1.67
	2.21–3.37	468.75	11.5385	30,440	8.7239	1.32
	3.38–4.53	0	0.0000	19,773	5.6668	0.00
	4.54–5.78	0	0.0000	11,759	3.3700	0.00
	5.79–7.3	0	0.0000	6439	1.8454	0.00
	7.31–9.44	156.25	3.8462	2695	0.7724	4.98
	9.45–16.23	156.25	3.8462	798	0.2287	16.82
Terrain Ruggedness Index	0–1.9	0	0.0000	33,417	9.5770	0.00
	1.91–2.96	1250	30.7692	89,713	25.7110	1.20
	2.97–3.87	312.5	7.6923	80,687	23.1243	0.33
	3.88–4.78	156.25	3.8462	60,919	17.4589	0.22
	4.79–5.76	781.25	19.2308	40,627	11.6434	1.65
	5.77–6.75	781.25	19.2308	21,508	6.1640	3.12
	6.76–7.89	468.75	11.5385	12,057	3.4554	3.34
	7.9–9.33	312.5	7.6923	6465	1.8528	4.15

	9.34–11.75	0	0.0000	2980	0.8540	0.00
	11.76–19.33	0	0.0000	555	0.1591	0.00
Lineament density (km^2)	0–50	781.25	19.2308	30,344	8.6963	2.21
	50.01–100	312.5	7.6923	31,559	9.0446	0.85
	100.01–200	781.25	19.2308	61,013	17.4858	1.10
	200.01–300	468.75	11.5385	53,846	15.4318	0.75
	300.01–400	625	15.3846	46,947	13.4546	1.14
	400.01–500	156.25	3.8462	39,813	11.4101	0.34
	500.01–600	625	15.3846	30,580	8.7640	1.76
	600.01–800	156.25	3.8462	36,824	10.5535	0.36
	800.01–1000	156.25	3.8462	13,606	3.8994	0.99
	1000.01–1229.8	0	0.0000	4509	1.2922	0.00
Hydrology Soil Group	HSG-B	2187.5	53.8462	217,083	62.2143	0.87
	HSG-C	1875	46.1538	131,845	37.7857	1.22
Epicentre proximity (m)	0–500	156.25	3.8462	6560	1.8800	2.05
	500.01–1000	156.25	3.8462	22,261	6.3798	0.60
	1000.01–1500	156.25	3.8462	31,296	8.9692	0.43
	1500.01–2000	312.5	7.6923	32,726	9.3790	0.82
	2000.01–2500	625	15.3846	36,941	10.5870	1.45
	2500.01–3000	468.75	11.5385	40,319	11.5551	1.00
	3000.01–3500	312.5	7.6923	39,056	11.1931	0.69
	3500.01–4000	312.5	7.6923	31,118	8.9182	0.86
	4000.01–5900.71	1562.5	38.4615	89,082	25.5302	1.51
	5900.72–6839.45	0	0.0000	19,622	5.6235	0.00
Rainfall (mm)	3400	937.5	23.0769	64,571	18.5055	1.25
	3600	3125	76.9231	284,357	81.4945	0.94
Landuse/landcover	Forest	1406.25	34.6154	186,860	53.5526	0.65

(continued)

Table 24.4 (continued)

Factor	Attribute	No. pixels in domain	% of the pixel in the domain	No. of landslides in the class	% of landslides in the class	Frequency ratio
	Built up	1875	46.1538	76,668	21.9724	2.10
	River	0	0.0000	2387	0.6841	0.00
	Grass	0	0.0000	2	0.0006	0.00
	Agriculture land	781.25	19.2308	82,884	23.7539	0.81
Geomorphology	Lowly dissected slope	3125.00	76.9231	251,845	72.1768	1.07
	Colluvial mid slope	625.00	15.3846	66,051	18.9297	0.81
	Transportational mid slope	312.50	7.6923	23,403	6.7071	1.15
	Highly dissected slope	0.00	0.0000	7622	2.1844	0.00
Surface geology	Schist	937.50	23.0769	69,236	19.8425	1.16
	Banded gneiss	0.00	0.0000	9467	2.7132	0.00
	Phyllitic quartzites	937.50	23.0769	71,497	20.4905	1.13
	Quartz-biotite-schist	156.25	3.8462	13,119	3.7598	1.02
	Lingtse granitic gneiss	312.50	7.6923	12,348	3.5388	2.17
	Gametiferous mica schist	0.00	0.0000	42,843	12.2785	0.00
	Quartz-chlorite-sericite-phyllite	1250.00	30.7692	74,815	21.4414	1.44
	Schist-quartzite-intercalation with quartz	0.00	0.0000	2311	0.6623	0.00
	Quartzitic phyllite intercalation with quartz-chlorite-sericite-phyllite	468.75	11.5385	53,292	15.2731	0.76
Elevation (m)	641–908	156.25	3.8462	25,185	7.2178	0.53
	908.01–1075	0.00	0.0000	36,265	10.3933	0.00
	1075.01–1219	625.00	15.3846	47,752	13.6853	1.12
	1219.01–1358	468.75	11.5385	48,575	13.9212	0.83
	1358.01–1502	781.25	19.2308	45,063	12.9147	1.49

	1502.01–1651	625.00	15.3846	42,416	12.1561	1.27
	1651.01–1805	625.00	15.3846	35,079	10.0534	1.53
	1805.01–1978	468.75	11.5385	31,669	9.0761	1.27
	1978.01–2187	156.25	3.8462	23,422	6.7126	0.57
	2187.01–2449	156.25	3.8462	13,502	3.8696	0.99
Peak ground acceleration (g)	0.741–0.744	0.00	0.0000	6813	1.9526	0.00
	0.745–0.750	0.00	0.0000	12,450	3.5681	0.00
	0.751–0.758	781.25	19.2308	39,409	11.2943	1.70
	0.759–0.764	468.75	11.5385	43,046	12.3366	0.94
	0.765–0.772	1093.75	26.9231	75,673	21.6873	1.24
	0.773–0.777	0.00	0.0000	47,435	13.5945	0.00
	0.778–0.783	468.75	11.5385	34,795	9.9720	1.16
	0.784–0.791	312.50	7.6923	31,875	9.1351	0.84
	0.792–0.800	312.50	7.6923	27,115	7.7709	0.99
	0.801–0.805	625.00	15.3846	30,317	8.6886	1.77
Normalized difference vegetation index (NDVI)	−0.2–0.1	0.00	0.0000	3045	0.8727	0.00
	0.11–0.2	156.25	3.8462	9682	2.7748	1.39
	0.21–0.3	312.50	7.6923	10,556	3.0253	2.54
	0.31–0.6	1250.00	30.7692	69,662	19.9646	1.54
	0.61–0.88	2343.75	57.6923	256,674	73.5607	0.78
Landform	Streams	156.25	3.8462	19,335	5.5413	0.69
	Midslope drainages	781.25	19.2308	24,617	7.0550	2.73
	Upland drainages	0.00	0.0000	2547	0.7300	0.00
	Valleys	312.50	7.6923	36,773	10.5389	0.73
	Plains	0.00	0.0000	795	0.2278	0.00
	Open slopes	1718.75	42.3077	179,407	51.4166	0.82
	Upper slopes	468.75	11.5385	38,490	11.0309	1.05

(continued)

Table 24.4 (continued)

Factor	Attribute	No. pixels in domain	% of the pixel in the domain	No. of landslides in the class	% of landslides in the class	Frequency ratio
	Local ridges	0.00	0.0000	3025	0.8669	0.00
	Midslope ridges	625.00	15.3846	25,774	7.3866	2.08
	High ridges	0.00	0.0000	18,856	5.4040	0.00
Road density (km^2)	0–2	1250.00	30.7692	154,617	44.3120	0.69
	2.01–4	781.25	19.2308	70,841	20.3025	0.95
	4.01–6	468.75	11.5385	46,408	13.3002	0.87
	6.01–8	625.00	15.3846	29,076	8.3330	1.85
	8.01–10	312.50	7.6923	19,773	5.6668	1.36
	10.01–12	625.00	15.3846	11,177	3.2032	4.80
	12.01–14	0.00	0.0000	6908	1.9798	0.00
	14.01–16	0.00	0.0000	5538	1.5871	0.00
	16.01–18	0.00	0.0000	3286	0.9417	0.00
	18.01–21.85	0.00	0.0000	1304	0.3737	0.00

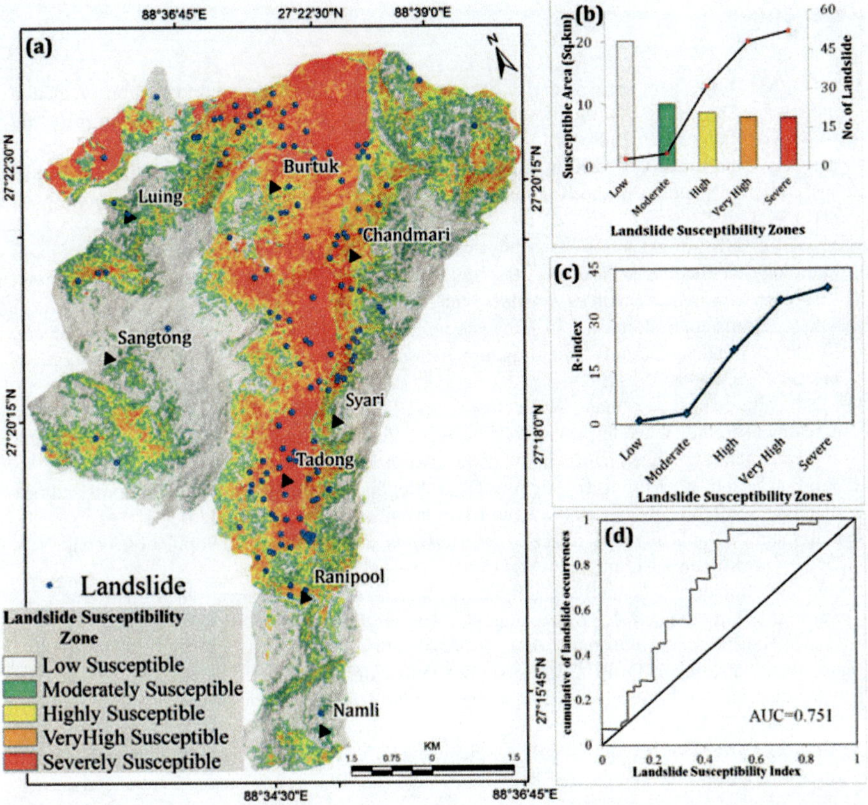

Fig. 24.7 Landslide susceptibility zonation map of the Greater Gangtok region. (**a**) The susceptibility map has been classified into five zones, viz. low, moderate, high, very high and severe. (**b**) The susceptibility area vs. the number of landslide plot. (**c**) Relative landslide density index (R-index) of the frequency ratio based susceptibility map. (**d**) ROC curve of the landslide susceptibility map

and is found to be in coherence with the ground instability conditions. The verification procedure proved that the produced landslide hazard zonation maps are substantial for land degradation management and planning of the study region. Besides this, landslide susceptibility maps can be useful for planners for selecting suitable locations for developmental activities in the region.

Acknowledgements This research work has been supported by the Ministry of Earth Sciences, Govt. of India, vide Sanction Order No.: MoES/P.O. (Seismo)/1(207)/2013. We are also thankful to Mines, Minerals and Geology Department, Govt. of Sikkim, for providing data and geological maps and also for assisting during fieldwork.

References

Adhikari, M. D., & Nath, S. K. (2016). Site-specific next generation ground motion prediction models for Darjeeling-Sikkim Himalaya using strong motion seismometry. *Journal of Indian Geophys Union, 20*(2), 151–170.

Ayalew, L., & Yamagishi, H. (2005). The application of GIS-based logistic regression for landslide susceptibility mapping in the Kakuda-Yahiko Mountains, Central Japan. *Geomorphology, 65* (1), 15–31.

Beven, K. J., & Kirkby, M. J. (1979). A physically based, variable contributing area model of basin hydrology/un modèle à base physique de zone d'appel variable de l'hydrologie du bassin versant. *Hydrological Sciences Journal, 24*(1), 43–69.

Bhasin, R., Grimstad, E., Larsen, J. O., Dhawan, A. K., Singh, R., Verma, S. K., & Venkatachalam, K. (2002). Landslide hazards and mitigation measures at Gangtok, Sikkim Himalaya. *Engineering Geology, 64*(4), 351–368.

Bui, D. T., Pradhan, B., Lofman, O., Revhaug, I., & Dick, O. B. (2012). Landslide susceptibility assessment in the HoaBinh province of Vietnam: A comparison of the Levenberg–Marquardt and Bayesian regularized neural networks. *Geomorphology, 171*, 12–29.

Catani, F., Casagli, N., Ermini, L., Righini, G., & Menduni, G. (2005). Landslide hazard and risk mapping at catchment scale in the Arno River basin. *Landslides, 2*(4), 329–342.

Chandramouli, C., & General, R. (2011). *Census of India 2011*. Provisional Population Totals. New Delhi: Government of India, pp 409–413.

Dasgupta, S., Pande, P., Ganguly, D., Iqbal, Z., Sanyal, K., Venaktraman, N. V., Dasgupta, S., Sural, B., Harendranath, L., Mazumdar, K., Sanyal, S., Roy, A., Das, L. K., Misra, P. S., & Gupta, H. (2000). *Seismotectonic atlas of India and its environs*. Geological Survey of India.

Gupta, S. K., & Shukla, D. P. (2018). Application of drone for landslide mapping, dimension estimation and its 3D reconstruction. *Journal of the Indian Society of Remote Sensing, 46*(6), 903–914.

Guzzetti, F., Carrara, A., Cardinali, M., & Reichenbach, P. (1999). Landslide hazard evaluation: a review of current techniques and their application in a multi-scale study, Central Italy. *Geomorphology, 31*(1–4), 181–216.

Karim, S., Jalileddin, S., & Ali, M. T. (2011). Zoning landslide by use of frequency ratio method (case study: Deylaman Region). *Middle-East Journal of Scientific Research, 9*(5), 578–583.

Kaur, H., Gupta, S., Parkash, S., & Thapa, R. (2018). Knowledge-driven method: A tool for landslide susceptibility zonation (LSZ). *Geology, Ecology, and Landscapes*, 1–15.

Kaur, H., Gupta, S., Parkash, S., Thapa, R., Gupta, A., & Khanal, G. C. (2019). Evaluation of landslide susceptibility in a hill city of Sikkim Himalaya with the perspective of hybrid modelling techniques. *Annals of GIS, 25*(2), 113–132.

Lee, S., & Pradhan, B. (2007). Landslide hazard mapping at Selangor, Malaysia using frequency ratio and logistic regression models. *Landslides, 4*(1), 33–41.

Lee, S., & Sambath, T. (2006). Landslide susceptibility mapping in the DamreiRomel area, Cambodia using frequency ratio and logistic regression models. *Environmental Geology, 50* (6), 847–855.

Lee, S., & Talib, J. A. (2005). Probabilistic landslide susceptibility and factor effect analysis. *Environmental Geology, 47*(7), 982–990.

Mandal, S., & Mandal, K. (2018). Modeling and mapping landslide susceptibility zones using GIS based multivariate binary logistic regression (LR) model in the Rorachu river basin of eastern Sikkim Himalaya, India. *Modeling Earth Systems and Environment, 4*(1), 69–88.

Mohammady, M., Pourghasemi, H. R., & Pradhan, B. (2012). Landslide susceptibility mapping at Golestan Province, Iran: A comparison between frequency ratio, Dempster–Shafer, and weights-of-evidence models. *Journal of Asian Earth Sciences, 61*, 221–236.

Nath, S. K., Mandal, S., Adhikari, M. D., & Maiti, S. K. (2017). A unified earthquake catalogue for South Asia covering the period 1900–2014. *Natural Hazards, 85*(3), 1787–1810.

Park, S. H., Goo, J. M., & Jo, C. H. (2004). Receiver operating characteristic (ROC) curve: Practical review for radiologists. *Korean Journal of Radiology, 5*(1), 11–18.

Saadatkhah, N., Kassim, A., & Lee, M. L. (2014). Qualitative and quantitative landslide susceptibility assessments in Hulu Kelang area, Malaysia. *Electronic Journal of Geotechnical Engineering, 19*, 545–563.

Sarkar, S., Kanungo, D. P., Patra, A. K., & Kumar, P. (2008). GIS based spatial data analysis for landslide susceptibility mapping. *Journal of Mountain Science, 5*(1), 52–62.

Shahabi, H., Ahmad, B. B., & Khezri, S. (2012). Application of satellite remote sensing for detailed landslide inventories using frequency ratio model and GIS. *International Journal of Computer Science, 9*(4), 108–117.

Shahabi, H., Khezri, S., Ahmad, B. B., & Hashim, M. (2014). Landslide susceptibility mapping at central Zab basin, Iran: A comparison between analytical hierarchy process, frequency ratio and logistic regression models. *Catena, 115*, 55–70.

Sharma, A. K. (2008). Landslide and its mitigation for disaster management using remote sensing and GIS technique-a case study of Gangtok area, East Sikkim (Doctoral dissertation, Sikkim Manipal University of Health, Medical and Technological sciences).

Sharma, A. K., Joshi, V., & Kumar, K. (2011). Landslide hazard zonation of Gangtok area, Sikkim Himalaya using remote sensing and GIS techniques. *Journal of Geomatics, 5*(2), 87–88.

Swets, J. A., Dawes, R. M., & Monahan, J. (2000). Better decisions through science. *Scientific American, 283*, 82–87.

Vakhshoori, V., & Zare, M. (2016). Landslide susceptibility mapping by comparing weight of evidence, fuzzy logic, and frequency ratio methods. *Geomatics, Natural Hazards and Risk, 7*(5), 1731–1752.

Yalcin, A., Reis, S., Aydinoglu, A. C., & Yomralioglu, T. (2011). A GIS-based comparative study of frequency ratio, analytical hierarchy process, bivariate statistics and logistics regression methods for landslide susceptibility mapping in Trabzon, NE Turkey. *Catena, 85*(3), 274–287.

Chapter 25
Glacial Lake Outburst in Uttarakhand (India): Role of Geospatial Technology for its Mitigation Strategy

Gouri Sankar Bhunia and Pravat Kumar Shit

Abstract Glacial lakes are usually found in isolated areas and higher elevations in mountainous regions. Glaciers in mountainous areas are retreating and gradually becoming lakes due to changing climate and human-induced threats, which pose a great challenge in the context of a storm surge in the case of a glacial lake outburst flood (GLOF) for downstream settlements. Owing to the insularity of the places, the large number of glacial lakes and their poor accessibility, field-based mapping and monitoring of glacial lake vulnerabilities are daunting in these areas. Glacial lakes can be physically digitised from multi-spectral or multi-polarisation satellite data such as acquired by Landsat and Sentinel-2 or Sentinel-1 from true or false colour composites (TCC/FCC), mostly accompanied by high-resolution images derived from Google Earth. After the launch of the Sentinel-1 and Sentinel-2 Copernicus satellites, data from these satellites has been progressively included as SAR can see through the clouds and has a higher spatial resolution (10 m) and revisit time (5 days with S-2A and S-2B).

Keywords Glacial lake outburst flood · Sentinel data · Google Earth · SAR data · Uttarakhand · India

25.1 Introduction

Ice from snowfall accumulates from glaciers on hilltops and the margins of valleys. In a method called ablation, this ice melts, contributing warmth to downstream water sources and rivers. The phase is ongoing and is part of the normal cycle of water on Earth. As glaciers recede, they leave a gap filled with water that becomes a glacial

G. S. Bhunia (✉)
Department of Geography, Seacom Skills University, Bolpur, West Bengal, India

P. K. Shit
PG Department of Geography, Raja N.L.Khan Womens College (Autonomous), Midnapore, West Bengal, India

P. K. Shit et al. (eds.), *Geospatial Technology for Environmental Hazards*, Advances in Geographic Information Science, https://doi.org/10.1007/978-3-030-75197-5_25

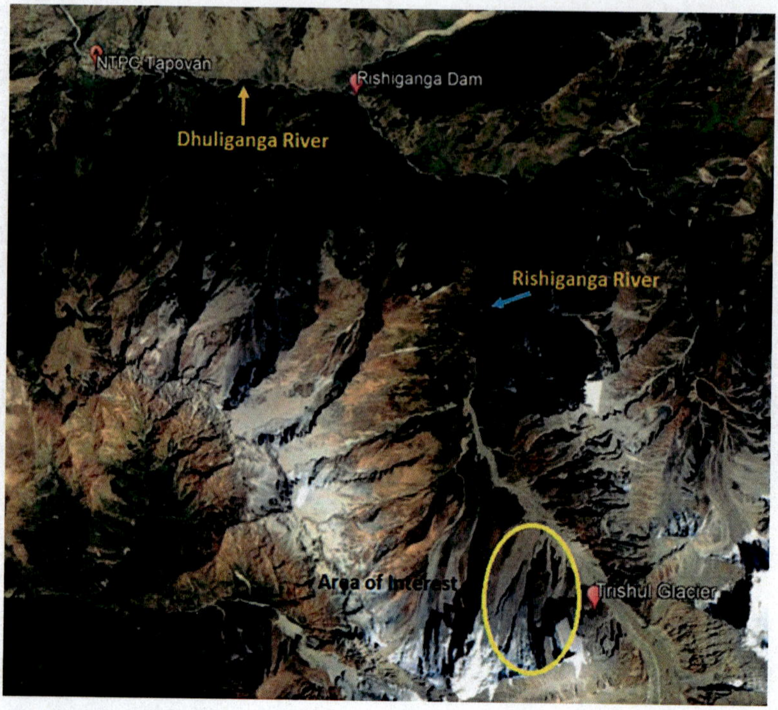

Fig. 25.1 Location map of the study area

lake. The water in glacial lakes piles up behind fragile, natural "glacial/moraine dams" consisting of ice, sand, pebbles and ice particles as glaciers retreat. The unstable architecture of the moraine dam contributes, unlike earthen dams, to the sudden collapse of the dam on top of the glacial lake, which retains vast amounts of water. A dam collapse has the capacity to spill millions of cubic metres of water in a fleeting period, inducing downstream devastating floods. A glacial lake outburst flood (GLOF) corresponds to the overflowing that takes place when abruptly the water dredged by a glacier or moraine is released.

A glacial lake outburst flood (GLOF), possibly caused by a massive slab of ice breaking from a glacier, was the flash flood of water, mud and rocks that rushed through the Rishiganga and Dhauliganga river valleys at Chamoli village (located in 30°29′16.790″N, 79°55′31.942″E) in Uttarakhand on February 9, 2021 (Fig. 25.1). At about 3700 m above sea level, floodwaters crashed into the Rishi Ganga hydro-electric dam and totally swept it down. At least 35 people working on the project are absent, according to the chief minister of Uttarakhand. The Rishi Ganga project's concrete and rubble were then taken downstream and crashed into the 520 MW Tapovan Vishnugad hydropower project under operation, inflicting severe damage. Ice and rock plunged into a glacial pool, allowing the banks to break and the steep and narrow gorges of the upper Himalayas to gush down with water. Because of climate change, the number and area of glacial lakes have risen in the mountain range (Aggarwal et al., 2016). A report undertaken in the *Science Advances* journal in 2019 cautioned that the Himalayan glaciers are deteriorating at an unprecedented pace and could occur again in the 2013 Kedarnath-like catastrophe. The visibility and awareness are that since the beginning of the twentieth century, owing to climate change, the glaciers of the Himalayas have melted nearly twice as fast (Maurer et al., 2019).

25.2 Observations of Scientific Community

In the case of the Uttarakhand glacier burst, it is suspected that the water pockets inside the Nanda Devi glacier that led to this incident may have formed. This catastrophe is also related by some scientists to climate change and global warming. High temperatures and less snowfall will contribute to an uptick in glacier melting, allowing glacial lake water to climb above levels. Based on the initial observation of glaciology and hydrology division at the Dehradun-based Wadia Institute of Himalayan Geology, satellite imaging has pointed to the probability of the flood having been exacerbated by the melting of snow and not a glacial eruption of the lake. Preliminary observation stated that satellite images have shown that on February 2, 2021, there was no snow in the valley, but on February 5 and 6, 2021, very heavy snowfall was observed. On February 7, 2021 (Fig. 25.2), this fresh snow began to melt, resulting in the regressing of the snow bank and a corresponding landslide. It gained momentum and kinetic energy as the snow bank proceeded down the valley, thus raising the amount of water and soil on the path (Roy & Sinha, 2021). Similarly, a comprehensive study performed at the Indian Institute of Science in Bengaluru by the Divecha Centre for Climate Change confirmed that the flash flood may have been triggered by the discharge of water locked in a lake underneath the northern Nanda Devi glacier. The potential of this underground lake was to hold 4.5 million cubic

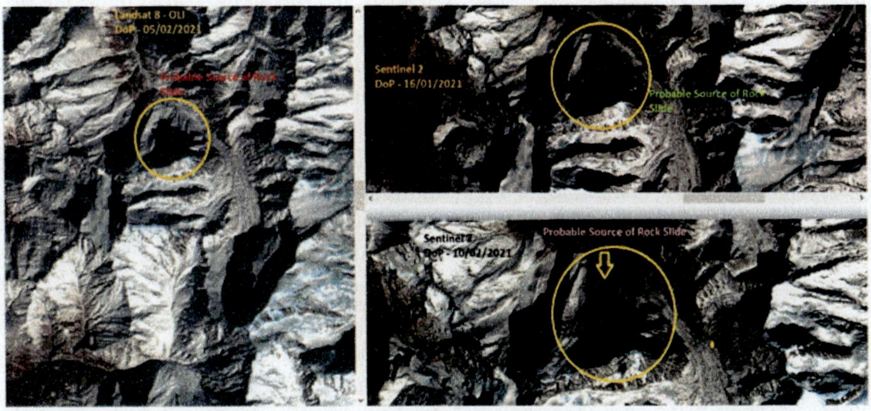

Fig. 25.2 Change analysis of Uttarakhand glacial lake outburst

Fig. 25.3 Digital elevation Model of Uttarakhand glacial burst (Source: Planet Team, 2021)

metres of water. If this water-filled depression creates enough hydrostatic pressure, the lower portion of the ablation zone (near the tip of the glacier where the snowmelt occurs) can be accelerated, potentially releasing water from the underground lake. The possible explanation for the flash flood may be this.

However, before the flooding incident, available satellite imagery did not indicate the existence of a glacial lake. Although scientists and researchers are now trying to determine what caused the flood, the leading hypothesis is that a large fragment of ice broke at an altitude above 5500 m above sea level due to a rock slide. International geologists and glaciologists analysing satellite imagery suggest the origin of the Uttarakhand flooding catastrophe in Chamoli tends to be a landslide and not, as generally assumed, a glacial eruption. Dr. Dan Shugar of the University of Calgary, who experts in glacial and geological conditions at high altitudes, carried out the first identification. In order to deduce that a landslide caused the devastating flash floods along the Alaknanda and Dhauliganga rivers, Shugar used multi-temporal satellite images from Planet Laboratories (Fig. 25.3), taken before and after the tragedy, as also shown by a path of dust visible in the satellite imagery.

Scientists warn that several of these glacial lakes that have increased in size in current history pose a possible danger to downstream environments and have suggested short-term and long-term solutions to reduce the possibility of catastrophic flooding (Fig. 25.4). In the upper reaches of the Himalayas, avalanches

Fig. 25.4 Uttarakhand glacier burst: a large piece of Nanda Devi glacier broke off on a chilly winter morning of February 2021 and fell into a river, triggering an avalanche and glacial lake outburst flood (GLOF). (Source: Photograph Rupali Pruthi on Feb 8, 2021)

are unlikely in February as the snow and ice remain frozen owing to reduced temperatures. Conversely, this year has been colder than expected, and Uttarakhand and the rest of India witnessed their maximum average temperature in 60 years in January. The warming temperatures due to climate change cause accidents more common, such as avalanches, melting ice and landslides. As with colder winters, irregular weather conditions such as increased snowfall and rainfall play a part. However, the possibility of potential catastrophic moraine breaches and flash flooding raises as the number of glacial lakes and their sizes rise.

25.3 Geospatial Technology, Glacial Lake Identification and Monitoring

Glacial lake changes are of broad significance in the sense of global warming and have become a crucial component for recognising the risks of glacial lakes (Bajracharya et al., 2007; Bolch et al., 2011). Glacial lakes can be physically digitised from multi-spectral or multi-polarisation satellite data such as acquired by Landsat (Wang et al., 2015) and Sentinel-2 or Sentinel-1 (Wangchuk et al., 2019) from true or false colour composites (TCC/FCC), mostly accompanied by high-resolution images derived from Google Earth or Bing Maps (Zhang et al., 2015). After the launch of the Sentinel-1 and Sentinel-2 Copernicus satellites, data from these satellites has been progressively included as SAR can see through the clouds and has a higher spatial resolution (10 m) and revisit time (5 days with S-2A and S-2B). Investigation on changes in glacial lakes is primarily focused on remote sensing data using techniques like multi-spectral data combination (Kargel et al., 2005), water surface index (Huggel et al., 2002), remap tables reclassification (Wu and Zhu 2008), spectral analysis (Aggarwal et al., 2016) and geospatial modelling (Rounce et al., 2017) to analyse variation in region. Satellite images that can be checked by ground survey can conveniently denote different features of glaciers and lakes, like the recognition of the snow line, glacier moraines, lake borders and area (Mal et al., 2016). The Normalised Difference Water Index (NDWI) is the most widely used semi-automated optical data technique since it increases the appearance of water sources in visible spectral imagery by using a band with comparatively high-water reflectance (e.g. the blue or green band) and one with low or no reflectance (e.g. near-infrared (NIR) or short-wave infrared (SWIR) bands) (McFeeters, 1996; Amin et al., 2020). It has been found that the use of SAR data is reliable for mapping partly frozen glacial lakes, but noise removal across glacial lakes relies on supplementary datasets like glacier outlines, digital elevation model (DEM), vegetation maps extracted from the Normalised Difference Vegetation Index (NDVI) and manual corrections (Wangchuk et al., 2019). However, some other criteria that include ground evaluations are taken into consideration when designing the GLOF model. A computer terrain model and field survey data are essential for unsteady GLOF simulation and modelling. One of the evolving specialised flood prediction and recognition methods for downstream flood zones is the Hydrologic Engineering Center's river analysis method (HEC-RAS and HEC-GeoRAS) by Matkan et al. (2009). In several fields of geoscience, for example, the machine learning algorithm (random forest) has been successfully used to detect past GLOF events in the Himalayas (Veh et al., 2018). A fully automatic system for mapping glacial lakes through alpine regions has been developed by Wangchuk and Bolch (2020), including the Python module called "GLakeMap".

25.4 Future Direction and Conclusion

Glacial lakes are usually found in isolated areas and higher elevations in mountainous regions. Glaciers in mountainous areas are retreating and gradually becoming lakes due to changing climate and human-induced threats, which pose a great challenge in the context of a storm surge in the case of a GLOF for downstream settlements. Owing to the insularity of the places, the large number of glacial lakes and their poor accessibility, field-based mapping and monitoring of glacial lake vulnerabilities are daunting in these areas. For a several glacial lakes, the demarcation of glacial lake outlines is marginally problematic (1) where the spatial disparity between the surface of the lake and the shadow is not apparent and (2) where the inter-pixel spatial interaction extends beyond a beach location. The use of SAR data to chart glacial lakes is also stable, while visual datasets depend on the elimination of misidentified pixels.

References

Aggarwal, A., Jain, S. K., Lohani, A. K., & Jain, N. (2016). Glacial lake outburst flood risk assessment using combined approaches of remote sensing, GIS and dam break modelling, geomatics. *Natural Hazards and Risk, 7*(1), 18–36. https://doi.org/10.1080/19475705.2013.862573

Amin, M., Bano, D., Hassan, S. S., et al. (2020). Mapping and monitoring of glacier lake outburst floods using geospatial modelling approach for Darkut valley, Pakistan. *Meteorological Applications, 27*, e1877.

Bajracharya, B., Shrestha, A. B., & Rajbhandari, L. (2007). Glacier lake outburst flood in the Sagarmatha region—Hazard assessment using GIS and hydrodynamic modeling. *Mountain Research and Development, 27*(4), 336–344.

Bolch, T., Peters, J., Yegorov, A., Pradhan, B., Buchroithner, M., & Blagoveshchensky, V. (2011). Identification of potentially dangerous glacial lakes in the northern Tien Shan. *Natural Hazards, 59*(3), 1691–1714. https://doi.org/10.1007/s11069-011-9860-2

Huggel, C., Kaab, A., Haeberli, W., Teysseire, P., & Paul, F. (2002). Remote sensing based assessment of hazards from glacier lake outbursts: a case study in the Swiss Alps. *The Canadian Geotechnical Journal, 39*, 316–330.

Kargel, J. S., Abrams, M. J., Bishop, M. P., Bush, A., Hamilton, G., Jiskoot, H., Kaab, A., Kieffer, H. H., Lee, E. M., Paul, F., Rau, F., Raup, B., Shroder, J. F., Soltesz, D., Stainforth, D., Steams, L., & Wessels, R. (2005). Multispectral imaging contributions to global land ice measurements from space. *Remote Sensing of Environment, 99*, 187–219.

Mal, S., Singh, R. B., & Schickhoff, U. (2016). Estimating recent glacier changes in central Himalaya, India, using remote sensing data. In *Climate change, glacier response, and vegetation dynamics in the Himalaya* (pp. 205–218). Springer.

Matkan, A., Shakiba, A., Pourali, H., & Azari, H. (2009). Flood early warning with integration of hydrologic and hydraulic models, RS and GIS (case study: Madarsoo Basin, Iran). *World Applied Sciences Journal, 6*(12), 1698–1704.

Maurer, J. M., Schaefer, J. M., Rupper, S., & Corley, A. (2019). Acceleration of ice loss across the Himalayas over the past 40 years. *Science Advances, 5*(6), eaav7266. https://doi.org/10.1126/sciadv.aav7266

McFeeters, S. K. (1996). The use of the normalized difference water index (NDWI) in the delineation of open water features. *International Journal of Remote Sensing, 17*(7), 1425–1432.

Planet Team. (2021). *Planet application program interface: In space for life on earth*. San Francisco, CA. https://www.planet.com/

Rounce, D. R., Watson, C. S., & McKinney, D. C. (2017). Identification of Hazard and risk for Glacial Lakes in the Nepal Himalaya using satellite imagery from 2000–2015. *Remote Sensing, 9*(7), 654. https://doi.org/10.3390/rs9070654

Roy, E., & Sinha, A. (2021). *Uttarakhand disaster may not be glacial lake outburst, images show snow fell off mountain, The Indian Express*. New Delhi. https://indianexpress.com/article/india/uttarakhand-flash-flood-investigation-snow-avalanche-7180414/

Veh, G., Korup, O., Roessner, S., & Walz, A. (2018). Detecting Himalayan glacial lake outburst floods from Landsat time series. *Remote Sensing of Environment, 207*, 84–89.

Wang, W., Xiang, Y., Gao, Y., Lu, A., & Yao, T. (2015). Rapid expansion of glacial lakes caused by climate and glacier retreat in the Central Himalayas. *Hydrological Processes, 29*(6), 859–874.

Wangchuk, S., & Bolch, T. (2020). Mapping of glacial lakes using Sentinel-1 and Sentinel-2 data and a random forest classifier: Strengths and challenges. *Science of Remote Sensing, 2*, 100008.

Wangchuk, S., Bolch, T., & Zawadzki, J. (2019). Towards automated mapping and monitoring of potentially dangerous glacial lakes in Bhutan Himalaya using Sentinel-1 synthetic aperture radar data. *International Journal of Remote Sensing, 40*(12), 4642–4667.

Wu, Y., & Zhu, L., (2008). The response of lake-glacier variations to climate change in Nam Co Catchment, central Tibetan Plateau, during 1970–2000. *Journal of Geographical Sciences, 18*, 177–189.

Zhang, G., Yao, T., Xie, H., Wang, W., & Yang, W. (2015). An inventory of glacial lakes in the third pole region and their changes in response to global warming. *Global and Planetary Change, 131*, 148–157.

Index

P. K. Shit et al. (eds.), *Geospatial Technology for Environmental Hazards*, Advances in Geographic Information Science, https://doi.org/10.1007/978-3-030-75197-5

B

D

E

G

S

W

Y

Z